Peptide Applications in Biomedicine, Biotechnology and Bioengineering

Related titles

Handbook of Biologically Active Peptides
(ISBN 978-0-12-385095-9)

Peptide and Protein Delivery
(ISBN 978-0-12-384935-9)

Natural-Based Polymers for Biomedical Applications
(ISBN 978-1-84569-264-3)

Woodhead Publishing Series in Biomaterials

Peptide Applications in Biomedicine, Biotechnology and Bioengineering

Edited by

Sotirios Koutsopoulos

ELSEVIER

WP
WOODHEAD
PUBLISHING
An imprint of Elsevier

Woodhead Publishing is an imprint of Elsevier
The Officers' Mess Business Centre, Royston Road, Duxford, CB22 4QH, United Kingdom
50 Hampshire Street, 5th Floor, Cambridge, MA 02139, United States
The Boulevard, Langford Lane, Kidlington, OX5 1GB, United Kingdom

Notices
Knowledge and best practice in this field are constantly changing. As new research and experience broaden our understanding, changes in research methods, professional practices, or medical treatment may become necessary.

Practitioners and researchers must always rely on their own experience and knowledge in evaluating and using any information, methods, compounds, or experiments described herein. In using such information or methods they should be mindful of their own safety and the safety of others, including parties for whom they have a professional responsibility.

To the fullest extent of the law, neither the Publisher nor the authors, contributors, or editors, assume any liability for any injury and/or damage to persons or property as a matter of products liability, negligence or otherwise, or from any use or operation of any methods, products, instructions, or ideas contained in the material herein.

British Library Cataloguing-in-Publication Data
A catalogue record for this book is available from the British Library

Library of Congress Cataloging-in-Publication Data
A catalog record for this book is available from the Library of Congress

ISBN: 978-0-08-100736-5 (print)
ISBN: 978-0-08-100742-6 (online)

Publisher: Matthew Deans
Acquisition Editor: Laura Overend
Editorial Project Manager: Natasha Welford
Production Project Manager: Omer Mukthar
Cover Designer: Victoria Pearson

Typeset by MPS Limited, Chennai, India

Contents

List of contributors

Handan Acar University of Chicago, Chicago, IL, United States

Dominic Agyei University of Otago, Dunedin, New Zealand

Kengo Akagawa University of Tokyo, Meguro, Tokyo, Japan

John C. Barrett University of Chicago, Chicago, IL, United States

Margarida Bastos University of Porto, Porto, Portugal

Corrie R. Belanger University of British Columbia, Vancouver, BC, Canada

Stephen Bleakley Teagasc Food Research Centre, Dublin, Ireland; Dublin Institute of Technology, Dublin, Ireland

Aimee L. Boyle Leiden Institute of Chemistry, Leiden, The Netherlands

Elizabete S. Cândido Catholic University of Brasília, Brasília, Brazil; Catholic University Dom Bosco, Campo Grande, Brazil

Marlon H. Cardoso University of Brasília, Brasília, Brazil; Catholic University of Brasília, Brasília, Brazil; Catholic University Dom Bosco, Campo Grande, Brazil

Barbara Claro University of Porto, Porto, Portugal

Dana M. Copolovici Aurel Vlaicu University of Arad, Arad, Romania

Karolina Corin Biomedical Engineering Research Group in the School of Electrical and Information Engineering, University of the Witwatersrand, Johannesburg, South Africa; Department of Molecular Medicine and Haematology, University of the Witwatersrand, Johannesburg, South Africa; Department of Chemistry and Biochemistry, UCLA, Los Angeles, CA, United States; Molecular Biology Institute, UCLA, Los Angeles, CA, United States

Michael K. Danquah Curtin University, Sarawak, Malaysia

Octávio L. Franco University of Brasília, Brasília, Brazil; Catholic University of Brasília, Brasília, Brazil; Catholic University Dom Bosco, Campo Grande, Brazil

Rebeca Garcia-Fandino University of Porto, Porto, Portugal

Aswathy R. Girija Future Industries Institute, University of South Australia, Mawson Lakes, SA, Australia

Paris Hamilton Nephron Pharmaceuticals Corporation, West Columbia, SC, United States

Robert E.W. Hancock University of British Columbia, Vancouver, BC, Canada

Evan F. Haney University of British Columbia, Vancouver, BC, Canada

Maria Hayes Teagasc Food Research Centre, Dublin, Ireland

Yue-Wern Huang Missouri University of Science and Technology, Rolla, MO, United States

Sotirios Koutsopoulos Massachusetts Institute of Technology, Cambridge, MA, United States

Han-Jung Lee National Dong Hwa University, Hualien, Taiwan

Shuang Li Biosensor National Special Laboratory, Key Laboratory for Biomedical Engineering of Education Ministry, Department of Biomedical Engineering, Zhejiang University, Hangzhou, China

Zhibo Li Qingdao University of Science and Technology, Qingdao, China

Qingjun Liu Biosensor National Special Laboratory, Key Laboratory for Biomedical Engineering of Education Ministry, Department of Biomedical Engineering, Zhejiang University, Hangzhou, China

Ariadna Lobo-Ruiz Department of Inorganic and Organic Chemistry, Organic Chemistry Section, University of Barcelona, Barcelona, Spain

Yanli Lu Biosensor National Special Laboratory, Key Laboratory for Biomedical Engineering of Education Ministry, Department of Biomedical Engineering, Zhejiang University, Hangzhou, China

Andreea I. Lupitu Aurel Vlaicu University of Arad, Arad, Romania

Helmut R. Maecke University Hospital Freiburg, Freiburg, Germany

Michael J. Mellas University of Chicago, Chicago, IL, United States

Reshma J. Nevagi The University of Queensland, Brisbane, QLD, Australia

Subhani M. Okarvi King Faisal Specialist Hospital and Research Centre, Riyadh, Saudi Arabia

Karen G.N. Oshiro Catholic University Dom Bosco, Campo Grande, Brazil

Sharadwata Pan Technical University of Munich, Freising, Germany

Christos Petrou University of Nicosia, Nicosia, Cyprus

Ruirui Qiao Monash University, Parkville, VIC, Australia

Samilla B. Rezende Catholic University Dom Bosco, Campo Grande, Brazil

Yiannis Sarigiannis University of Nicosia, Nicosia, Cyprus

Mariusz Skwarczynski The University of Queensland, Brisbane, QLD, Australia

Evelyn Sun University of British Columbia, Vancouver, BC, Canada

Jing Sun Qingdao University of Science and Technology, Qingdao, China

Kei-Xian Tan Curtin University, Sarawak, Malaysia

Matthew V. Tirrell University of Chicago, Chicago, IL, United States

Istvan Toth The University of Queensland, Brisbane, QLD, Australia

Judit Tulla-Puche Department of Inorganic and Organic Chemistry, Organic Chemistry Section, University of Barcelona, Barcelona, Spain

Chibuike C. Udenigwe University of Ottawa, Ottawa, ON, Canada

Kun Wang Chinese Academy of Sciences, Beijing, China

Modi Wetzler Department of Chemistry, Clemson University, Clemson, SC, United States

Jiajia Wu Biosensor National Special Laboratory, Key Laboratory for Biomedical Engineering of Education Ministry, Department of Biomedical Engineering, Zhejiang University, Hangzhou, China

Qian Zhang Biosensor National Special Laboratory, Key Laboratory for Biomedical Engineering of Education Ministry, Department of Biomedical Engineering, Zhejiang University, Hangzhou, China

Shuguang Zhang Center for Bits and Atoms, Massachusetts Institute of Technology, Cambridge, MA, United States

Jian Zhong Shanghai Ocean University, Shanghai, China

Peptide synthesis: Methods, trends, and challenges

1

Christos Petrou and Yiannis Sarigiannis
University of Nicosia, Nicosia, Cyprus

1.1 Introduction

Peptides are ubiquitous in biomedicine, drug discovery, drug delivery, cosmetics, food industry, and several other fields. They are distinguished from proteins based on their smaller size (<40 amino acids) participating in most physiological processes including signal transduction, heart rate regulation, food intake, and growth. Although more than 7000 naturally occurring peptides have been discovered, inefficient and expensive manufacturing processes have hampered the development of peptide-based therapeutics. During the last two decades peptide research has seen significant growth due to several technological improvements. Despite their limitations, i.e., proteolytic instability and poor oral bioavailability, recent advances in peptide design and synthesis (including the incorporation of nonproteinogenic modifications) have revived the field of peptide therapeutics.

With more than 60 peptide drugs already in the market, approved by the Food and Drug Administration (FDA), about 140 peptide drugs currently in clinical trials, and 500–600 in preclinical development peptide-based pharmaceuticals are becoming a growing class of therapeutic agents with high economic potential for the pharmaceutical industry [1].

This introductory part would be incomplete without citing some key events in the history of peptide synthesis. Everything began in 1901 when Emil Fischer (Nobel Prize in Chemistry 1902) and his colleagues synthesized the first dipeptide, glycylglycine, whereas in 1881 Theodor Curtius had already synthesized the first *N*-protected dipeptide, benzoylglycylglycine. Emil Fischer introduced the term "peptide bond" and "peptide" so that he is considered the father of peptide chemistry. Thirty years later, in 1932, Leonidas Zervas developed the carbobenzoxy protecting group (Cbz or Z, in honor of its inventor) and in 1953 Vincent du Vigneaud reported the synthesis in solution of the endogenous cyclic nonapeptide hormone, oxytocin. For this landmark achievement he was awarded the Nobel Prize in 1955. Ten years later, in 1963, Bruce Merrifield introduced solid-phase peptide synthesis (SPPS), and for "his development of methodology for chemical synthesis on a solid matrix" he was awarded with the Nobel Prize in Chemistry in 1984. SPPS methodology was a real revolution opening new horizons in the field of peptide and protein chemistry [2]. This simple method, which is based on the coupling of protected amino acids onto an insoluble solid support, simplifies the procedure, making the peptide synthesis compatible with automatic processes enabling large-scale

Peptide Applications in Biomedicine, Biotechnology and Bioengineering.
DOI: http://dx.doi.org/10.1016/B978-0-08-100736-5.00001-6

production. Until the early 2000s, the prices of the raw material and yields of Fmoc-based SPPS were not sufficient to support large-scale production. The main disadvantage of SPPS on an industrial scale was the requirement for expensive resins or large amounts of amino acid derivatives and coupling reagents, making the synthesis of the peptides profitless due to the high overall cost of manufacturing [3]. Nevertheless, Trimeris and Roche developed a hybrid approach (SPS and SPPS) which allowed for the production the peptide drug enfuvirtide (Fuzeon, Roche), a 36-amino-acid peptide, on a metric ton-scale in a cost-efficient way. Three small protected fragments were constructed via solid-phase peptide synthesis and then after cleavage from the solid support, purification and characterization, the three fragments were condensed in solution [4].

From a synthetic point of view, peptides can be synthesized in solid phase or in solution. However, the dominant technique is the solid-phase peptide synthesis. Small to medium-sized peptides are prepared mostly in solid phase, whereas large peptides or small proteins are synthesized by using the hybrid approach. Thus, the solid-phase approach will be covered extensively in the following paragraphs.

1.2 Solid phase peptide synthesis

Currently, Solid-phase peptide synthesis (SPPS) is the most preferable method for the chemical synthesis of peptides in the lab as well as in the industry. SPPS allows for peptide synthesis to be carried out by using automation while several posttranslational backbone modifications and insertions of non-natural amino acids are easily achieved.

The revolutionary strategy developed by Merrifield is a synthetic method occurring under heterogeneous conditions on a polymeric solid support, called resin. The resin consists of small, millimeter-scale solid beads that are treated with short organic moieties, called linkers which are necessary for the prevention of aggregation of the beads. Notably, peptide synthesis on solid support occurs from the C- to the N-terminus, unlike ribosome protein synthesis which occurs from the N- to the C-terminus. The first amino acid remains covalently coupled to the linker with its carboxylic acid terminus, so that the C-terminus is protected from possible side reactions. In this way the peptide remains covalently "immobilized" to the solid support until the sequence is complete, while liquid-phase reagents and byproducts of synthesis are flushed away. Depending on the linker, several C-terminal modifications are possible. Peptide elongation is achieved by repeated cycles of deprotection of the N-terminus protecting group and coupling of the next amino acid derivative (Fig. 1.1).

The resin is always washed between the different steps. The final crude peptide is cleaved from the resin partially or fully deprotected by using the proper cleavage solution, i.e., trifluoroacetic acid (TFA) and scavengers which trap highly reactive carbocations that are formed during the cleavage procedure and are responsible for the formation of unwanted byproducts.

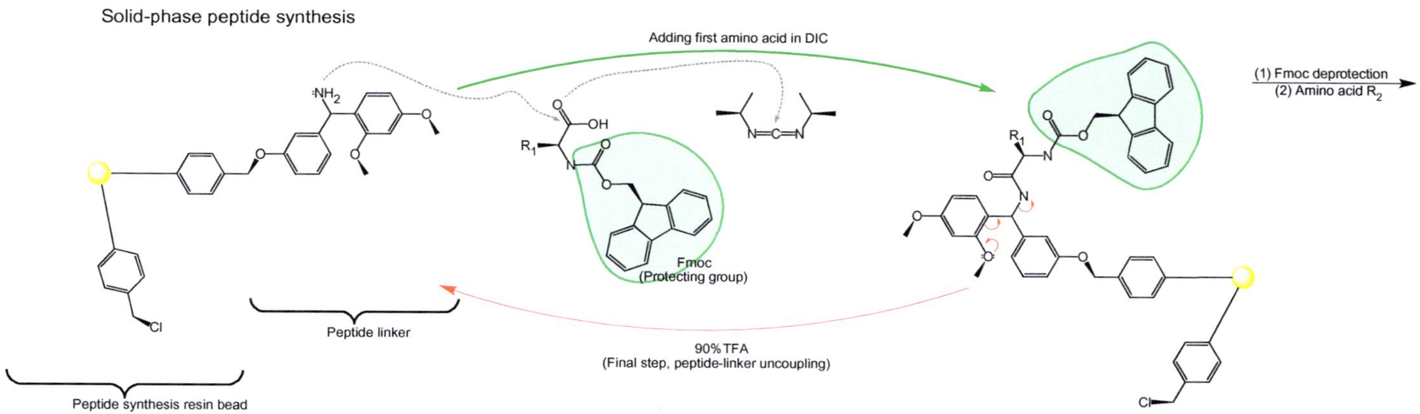

Figure 1.1 Overview of solid-phase peptide synthesis.

Protection strategies and coupling methods depend on the selected solid support. The protecting groups are necessary for the prevention of side reactions during coupling conditions. Efficient coupling of each amino acid is essential in order to achieve high yield of the final product and to reduce unwanted side reactions or deletion sequences (peptides lacking at least one of the required amino acids). N^{α}-modifications (insertion of a Fmoc- or Boc-group) act as protecting groups and are removed selectively after each successful coupling step, whereas the side-chain protecting groups provide a permanent protection against unwanted side reactions. Furthermore, special coupling reagents activate the relatively inert carboxy group by increasing its electrophilicity.

Since SPPS introduction over 40 years ago, resins, linkers between the C-terminal amino acid and polystyrene solid support, side-chain protecting groups, and new coupling reagents have been developed and refined. As a result, SPPS has become a robust technique for the efficient synthesis of peptides.

1.3 Solid supports (resins and linkers)

The proper selection of resins and linkers is key for successful peptide synthesis. The solid support should offer stability to mechanical stirring in a broad range of solvents and temperatures. In addition, the solid support should be able to swell so that the reagents can easily access the active sites. A series of resins and linkers has been developed in the past enabling a wide range of applications. Selection of protecting groups, coupling reagents, and cleavage conditions are directly associated with the selection of the polymer support and the linker. The first polymer support used by Merrifield was polystyrene, which is still in use [5]. Later, resins based on polystyrene incorporating polyethylene glycol (PEG) chains [6] and resins composed of PEG chains carrying specific crosslinkers were developed [7].

Linkers in SPPS play a pivotal role. Firstly, they provide a reversible linkage between the peptide chain and the solid support and secondly, they play a protective role against aggregation. In addition, the linker acts as a protecting group for the C-terminus carboxy group. Furthermore, the linker is used for the C-terminus modification of the peptide and determines the optimal selection of the protecting groups, coupling reagents and the cleavage conditions.

1.3.1 Resins

Based on the polymer side group, there are three types of resin: (i) polystyrene (PS)-based resins; (ii) PS-functionalized polyethylene glycol (PS-PEG) resins, and (iii) pure crosslinked PEG resins. PS resins have been widely used with great success in the synthesis mainly of short- to medium-length peptides. For medium- to long- peptides or peptides with "difficult sequences" PEG-based resins often show better performance resulting in peptides with high purity and yield.

In the case of *polystyrene resin*, the polymer is usually crosslinked with 1% of divinylbenzene (DVB). The resin is produced by suspension polymerization of styrene in the presence of divinylbenzene. This type of polymer support swells well in non polar solvents such as dichloromethane (DCM) or toluene but it is also compatible with solvents like *N,N*-dimethylformamide (DMF), dioxane, or tetrahydrofuran (THF), or *N*-methyl-2-pyrrolidone (NMP). However, this resin is not compatible with water or other polar solvents. In the case of syntheses involving highly hydrophobic amino acids, difficult peptide sequences, or peptides with a tendency to aggregate, more hydrophilic solid supports should be selected.

TentaGel (TG) [8] on the other hand behaves well in both polar and nonpolar solvents and exhibits excellent swelling properties in most solvents compatible with PEG chains. The change of the solvent from nonpolar to polar should be carried out gradually in order to preserve the optimal swelling properties of the polymer [9]. *TentaGel* is the most well studied PEG-PS resin today [10]. TG resins are prepared by grafting of PEG chains (50%–70%) to low crosslinked polystyrene by ether linkages.

The third group of polymer supports is the hydrophilic PEG-based resins, which contain no polystyrene or a very low amount of it. This group includes polymers that swell very well in water offering the advantage to study peptide–proteins interactions when the protein is up to 35–70 kDa. Main representatives of this category are poly(ethylene glycol)-poly-(*N,N*-dimethylacrylamide) copolymer (PEGA) which was developed by Meldal [11] and the crosslinked ethoxylate acrylate resin (CLEAR) which was developed by Kempe [12]. PEGA resins are very sensitive when dried and therefore they are supplied swollen in ethanol. A newer type of polymer support called ChemMatrix (CM), which was developed by Cote, combines the advantages of the two previous polymers, namely, the chemical stability of polystyrene resins and the versatility of the PEG grafted resins, making it a powerful tool for the synthesis of large peptides or difficult sequences with many hydrophobic amino acids [13]. Additionally, various polar solvents including water, DMF, THF, methanol, and acetonitrile, as well as trifluoroacetic acid (TFA) could be used with ChemMatrix because the polymer is highly polar and therefore it retains its excellent swelling properties in the presence of polar solvents [14]. Most of the above-mentioned polymer supports are marketed with a variety of different linkers, such as PAL linker, Fmoc-Rink linker, HMBA linker, etc.

1.3.2 Linkers for Fmoc-based SPPS

A linker plays a dual role during peptide synthesis. It offers protection to overcome aggregation during the elongation of the peptidic sequence while it provides a reversible linkage between the peptide chain and the solid support [15]. Linkers are easily categorized in low and high acid-labile linkers based on the conditions that are used for cleavage, usually TFA solution. Another classification is based on their linkage with the solid support. In this case the linkers are classified as integral or non-integral [16]. A third classification is based on the final C-terminal functionality of the peptide, usually peptide acid or peptide amide (Table 1.1). The classical

Table 1.1 Most common linkers for Fmoc-based SPPS

Linker name	Final C-terminal functionality	Linker structure	Cleavage conditions
Rink amide linker	Peptide amides	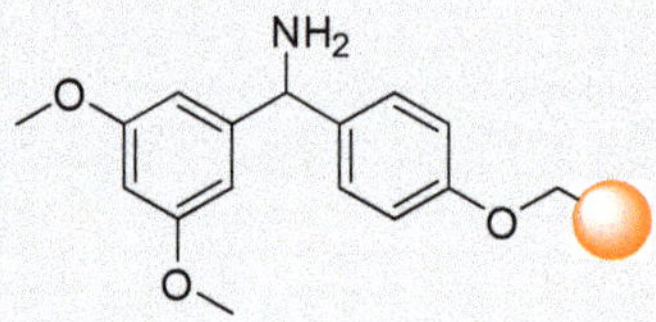	95% TFA
Sieber amide linker	Protected peptide amides	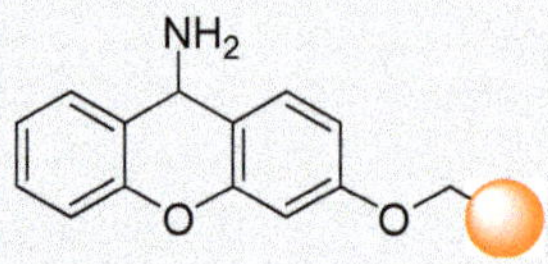	1%–3% TFA
PAL linker	Peptide amides	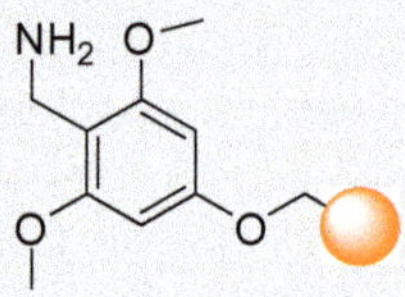	95% TFA
Wang HMPA linker	Peptide acids	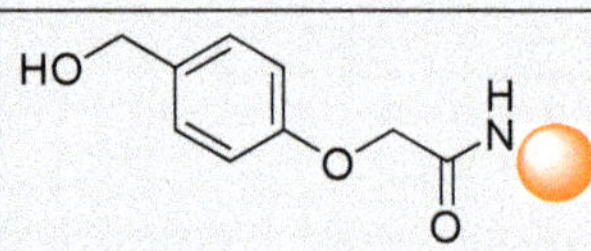	95% TFA
Wang HMBA linker	Protected peptide acids, amides, hydrazides, alcohols	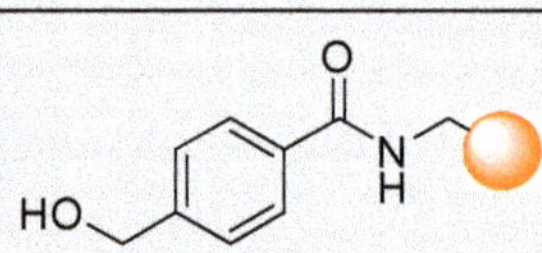	Nu
SASRIN linker	Peptide acids	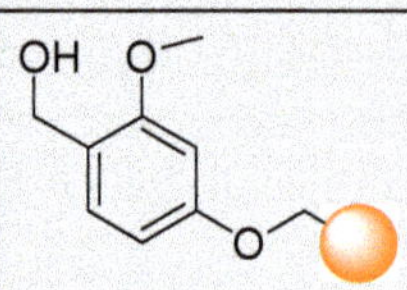	0.5%–1% TFA
2-Chlorotrityl chloride	Peptide acids	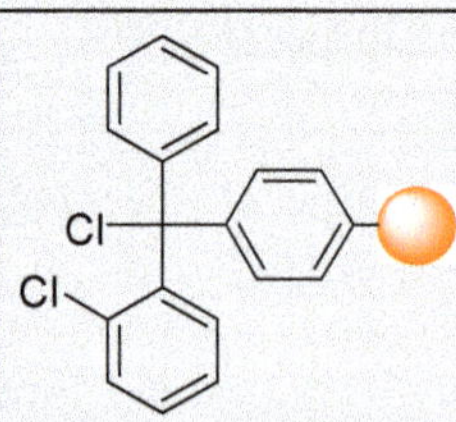	<0.5% TFA
Aryl hydrazide linker	Peptide amines or esters	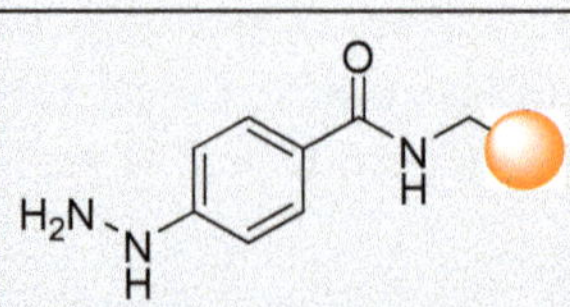	Cu(II) cat, pyridine

(*Continued*)

Table 1.1 **(Continued)**

Linker name	Final C-terminal functionality	Linker structure	Cleavage conditions
BAL linker	Peptide acids, aldehydes, thioesters	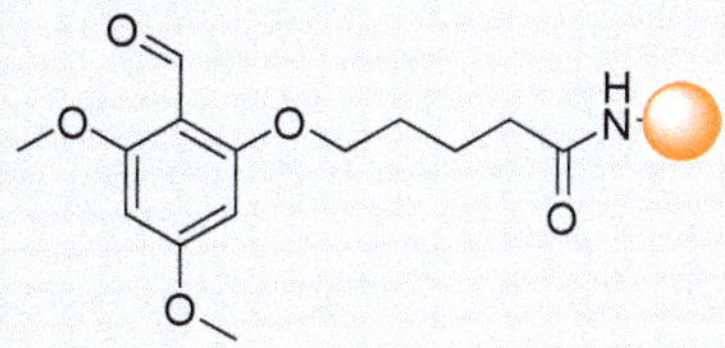	95% TFA
Safety catch linker	Peptide thioesters	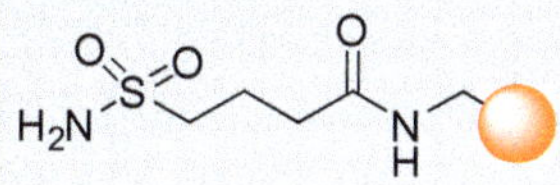	TMS-CHN2

Rink-amide linker or other aminomethyl-based linkers are tethered onto the resin with a covalent bond using standard coupling procedures. On the other hand, trityl type linkers, like the 2-chlorotrityl chloride resin (Barlos resin), are tethered to the polymer support (e.g., polystyrene) by direct synthesis [17].

The aryl hydrazide linkers are among the most useful resin-linkers for the synthesis of a wide variety of chemically modified peptides, head-to-tail cyclic peptides, as well as lipidated peptides. These linkers can be cleaved under mild oxidative conditions to provide protected peptides while they are perfectly compatible with both Boc- and Fmoc-based methodologies [18].

Also, the backbone amide linker (BAL) allows for the synthesis of C-terminal modified peptides, peptide aldehydes, or thioesters while after cleavage the C-terminal remains free for further reactions [19].

Another type of linker is the "safety catch linker" which is particularly useful for the synthesis of peptide thioesters, which are released from the resin using excess of thiol in DMF or NMP. The term "safety catch" was introduced by Kenner in 1971 [20] for peptide chemistry to describe a strategy that allows a linker to remain stable until it is activated for cleavage by chemical modification. Due to many disadvantages, such as poor loading and low reactivity in comparison to other well-known standard linkers, this linker never found a widespread application. However, nowadays it presents a useful and broadly applicable tool for the synthesis of a series of modified peptides upon treatment with different nucleophiles.

1.4 Protecting groups

1.4.1 N^{α} amino protecting groups

Based on the protection of the N^{α} amino group of an amino acid, two SPPS methods have been developed. The first method involves the use of the

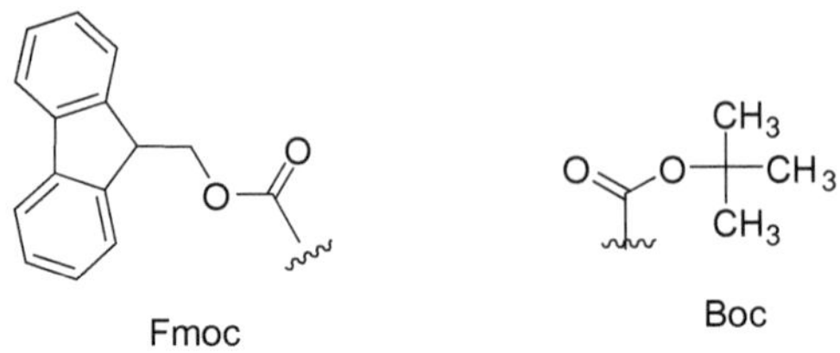

Figure 1.2 Main N^{α} amino protecting groups.

tert-butyloxycarbonyl (Boc) group which is deprotected by acid (e.g., trifluoroacetic acid) while in the second method the 9-fluorenylmethyloxycarbonyl (Fmoc) group is removed by base (e.g., piperidine) (Fig. 1.2). The method of chemical removal of these groups defines the overall strategy of the SPPS methodology.

Therefore, for the synthesis of the desired peptide after the selection of the proper resin and linker, the synthetic methodology defines the proper choice of the protecting groups of the N^a amino groups as well as the compatible side-chain protecting groups of the rest of the amino acids. Thus, two main methodologies were developed for the temporary protection and deprotection of the amino acids: the Boc/Bzl methodology, which is also called Boc-SPPS, and the Fmoc/*t*Bu methodology, which is also called Fmoc-SPPS, the latter is the most commonly used method.

In the case of the Boc/Bzl methodology, the final cleavage and deprotection requires an apparatus for corrosive and toxic hydrofluoric acid (HF). For this reason, the Fmoc/*t*Bu methodology is often preferred. The Fmoc group, introduced by Carpino, is easily removed under mild basic conditions with secondary amines, typically 20% piperidine in DMF or other solvents (DCM, THF, etc.) [21]. Under these conditions, most of the side-chain protecting groups are inert whereas they are labile to different acidic solutions, often containing TFA.

1.4.2 Side-chain protecting groups

The side-chain protecting groups for Fmoc-SPPS have been developed and studied extensively during the previous decades [22]. The main advantage of the Fmoc/*t*Bu approach is the orthogonality, a strategy allowing the deprotection of multiple protective groups at well-defined reaction conditions leaving the other protecting groups unaffected [23]. This allows for more flexibility to carry out complex strategies during synthesis. Thus, during the elongation of the sequence, diverse chemical conditions may be applied which allow for milder overall reaction conditions as well as for synthesis of partially protected peptides.

Basically, every amino acid containing a chemically reactive side-chain has to be masked temporarily with a protecting group during the peptide elongation in order to prevent side reactions and formation of byproducts. These protecting groups are orthogonal to the base-labile Fmoc-group and most of them can be removed by highly concentrated TFA solutions. Examples of such protecting groups

Table 1.2 **Common side-chain protecting groups used mainly in Fmoc-SPPS**

TFA labile		TFA stable	
Protecting groups	**Amino acids**	**Protecting groups**	**Amino acids**
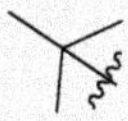 tBu	Asp, Glu: ester Ser, Thr, Tyr: ether	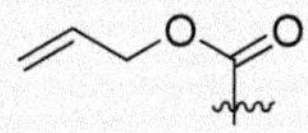 Alloc	Lys
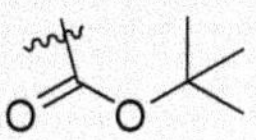 Boc	Lys, Trp	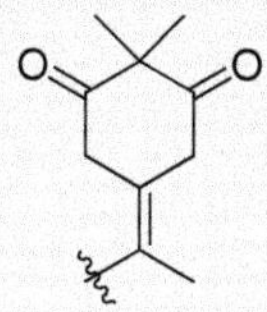 Dde	Lys
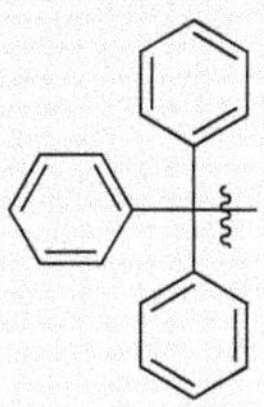 Trt	Cys, His, Asn, Gln	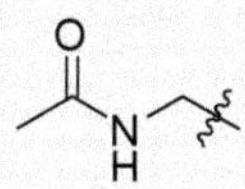 Acm	Cys
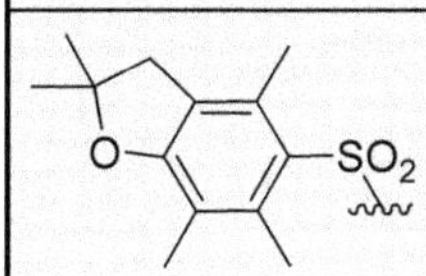 Pbf	Arg		

are the *tert*-butyl (*t*Bu) ester for Glu and Asp, the *t*Bu ether for Ser, Thr, and Tyr, the 2,2,4,6,7-pentamethyl-dihydrobenzofuran-5-sulfonyl (Pbf) for Arg, the trityl group (Trt) for Cys, Asn, Gln, and His, and the Boc group for Lys and Trp (Table 1.2). Modifications on the phenyl rings in Trt-derivatives with electron-withdrawing or electron-donating groups, increase the sensitivity to acidic solutions making their removal easier.

The 4-methoxytrityl group (Mmt) is highly acid-sensitive and can be completely removed by 1% TFA in DCM in less than 1 h, whereas only 4%–5% of the trityl group can be removed under these conditions when trityl group is used for the protection of the side-chain of cysteine [24]. The Mmt group is particularly useful when it is used in combination with Cys(S*t*Bu) in the same chain. The S*t*Bu group is removed by reduction and conversion to a Cys(5Npys) derivative by reaction with 2,2 dithiodi(5nitro) pyridine (DTNP). Treatment of the resin with 1% TFA unmasks the thiol group which then reacts spontaneously with the activated Cys(5-Npys) to give a disulfide bridge [25] (Fig. 1.3). Similar results are obtained when the 4-methyl trityl (Mtt) group is used for the protection of the side-chain of lysine [26].

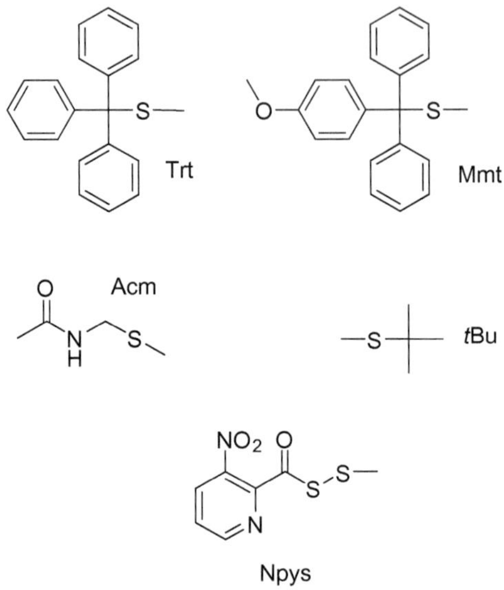

Figure 1.3 Protecting group for the thiol function of cysteine.

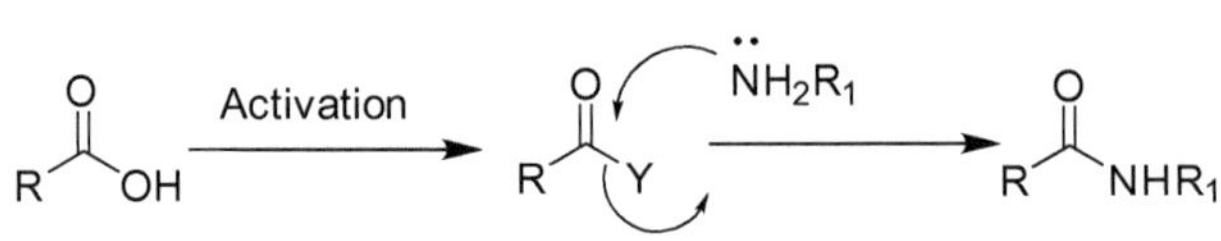

Figure 1.4 Peptide bond formation procedure.

There are some other protecting groups which are orthogonal to the cleavage conditions like the acetamidomethyl group (Acm) for Cys. The use of Acm is essential for the formation of disulfide bonds in peptides with multiple disulfide bridges. Acm is stable both in strong acidic and basic conditions and is selectively removed with thallium trifluoroacetate or iodine solution. Furthermore, the allyloxycarbonyl (Alloc) [27] and the (1-(4,4-dimethyl-2,6-dioxocyclohex-1-ylidene)-3-ethyl) (Dde) group are also stable in acidic TFA solutions and they can be used for the side protection of Lys or Ornithine (Orn) [28].

1.5 Coupling reagents

In order to form a peptide bond the carboxylic acid moiety of one amino acid should be activated with specific reagents. In this section, the most common coupling reagents currently used for forming peptide bonds are presented (Fig. 1.4).

The most common coupling agents are carbodiimide-based including dicyclohexylcarbodiimide (DCC) and diisopropylcarbodiimide (DIC) both of which contain two nitrogen atoms that are slightly basic. This attribute is useful for triggering a reaction between the acid and the carbodiimide, which results in the formation of *O*-acylisourea (Fig. 1.5). DIC is strongly preferred in automated SPPS because the resulting urea is soluble in organic solvents like DMF or DCM. In order to avoid potential side reactions (*N*-acylurea formation, epimerization, etc.) the addition of nucleophiles like 1-hydroxybenzotriazole (HOBt) or 1-hydroxy-7-azabenzotriazole (HOAt) is essential because it results in the formation of the less active benzotriazolyl ester (Fig. 1.6) [29]. The *N*-ethyl-*N*′-(3-dimethylaminopropyl) carbodiimide (EDC), which belongs to the carbodiimide family, is water-soluble and is commonly used in solution chemistry. A newer player in the field, the ethyl 2-cyano-2-(hydroxyimino) acetate (Oxyma), suppresses racemization significantly both in manual and automated synthesis [30].

After HOBt was identified as a racemization suppressant, new coupling reagents known as phosphonium or uronium salts were introduced in SPPS. Kenner and

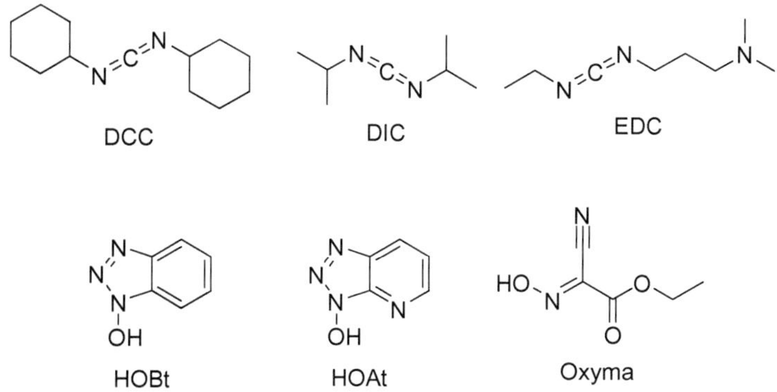

Figure 1.5 Common coupling reagents.

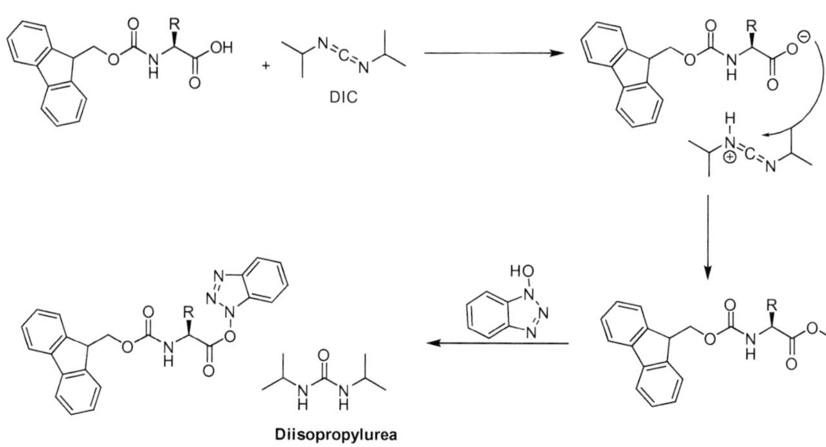

Figure 1.6 Coupling reaction by using DIC.

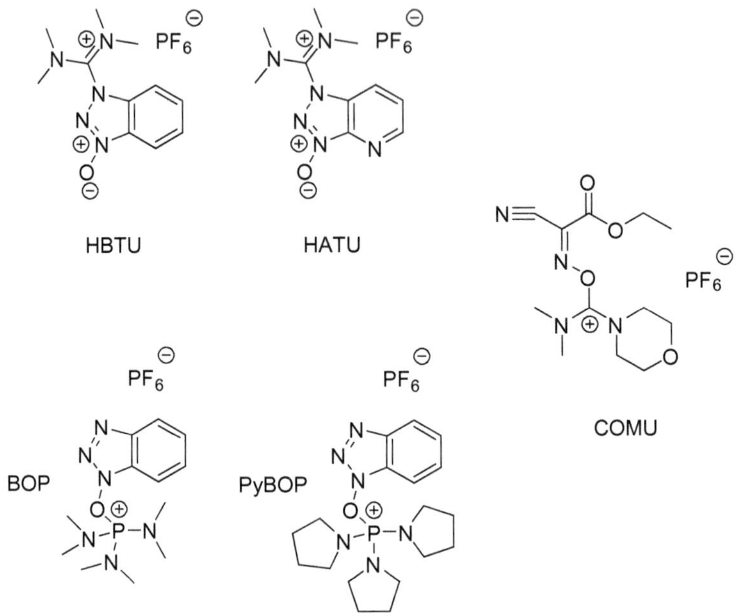

Figure 1.7 Coupling reagents of the phosphonium aminium type.

coworkers [31] developed phosphonium salts known as BOP family with main representatives 1-benzotriazolyloxy-tris-pyrrolidinophosphonium hexafluorophosphate (PyBOP) and the chloro and bromo analogs PyCloP and PyBroP, respectively. The most popular among the uronium/aminium coupling reagents are (*N*-[(1*H*-benzotriazol-1-yl)(dimethylamino)methylene]-*N*-methylmethanaminium hexafluorophosphate *N*-oxide) (HBTU) and (*N*-[(dimethylamino)-1*H*-1,2,3-triazole[4,5-b]pyridine-1-ylmethylene]-*N*-methylmethanaminium hexafluorophosphate *N*-oxide) (HATU) [32]. Initially, it was believed that HBTU and HATU are uronium salts (Fig. 1.7), however the crystal structures showed that they were aminium salts (guanidinium *N*-oxides) [33]. A novel coupling reagent introduced by Albericio and coworkers [34] is (1-(1-(cyano-2-ethoxy-2-oxoethylideneaminooxy)-dimethylamino-morpholino-methylene) methaminium hexafluorophosphate) (COMU). In a recent paper [35] the authors demonstrated that the commercial COMU and Oxyma solutions are sensitive to hydrolysis and should be stored either under inert gas or be freshly prepared. Fig. 1.8 shows the activation mechanism using HBTU.

1.6 Deprotection: Cleavage solutions

After sequence elongation the peptide has to be cleaved from the solid support and then the side-chain protecting groups have to be removed. Release of the peptide from the solid support as well as removal of the side-chain protecting groups is also

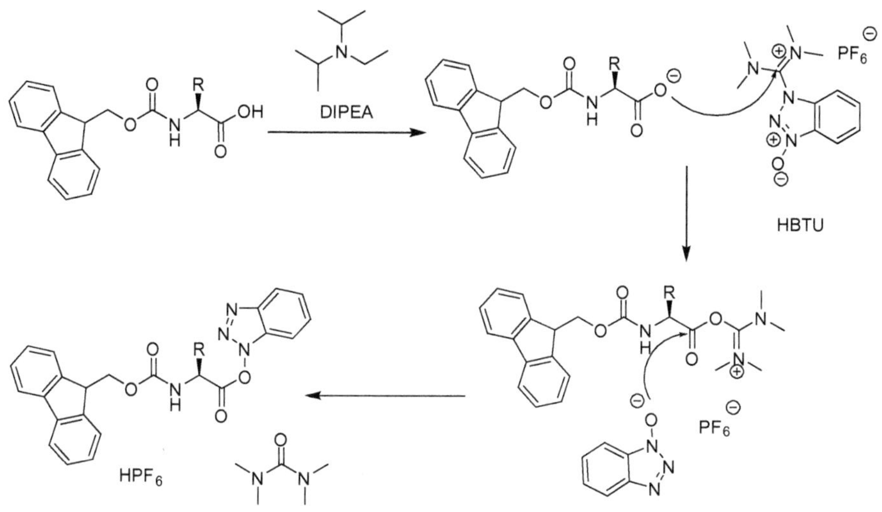

Figure 1.8 Coupling activation using HBTU.

important for increasing the yield and purity of the final crude peptide. Peptide-resin treatment with cleavage cocktail requires a series of trial experiments for the adjustment of the optimal cleavage conditions. During cleavage, several simultaneous reactions occur in the same environment. Selection of scavengers, reaction conditions, and the time required for the completion of cleavage are among the most important factors for a successful release. It is a common phenomenon that a small change in the percentage of a scavenger or in the reaction time can alter the chromatographic profile of the crude peptide. As previously noted, most cleavage strategies require different concentrations of TFA, scavengers, and reaction times. Cleavage is a process having a high-risk of side-reactions so that the use of scavengers to trap released carbocations is mandatory. The most common scavenger is water. However, other organic reagents such as triethylsilane (TES), triisopropylsilane (TIS), 1,2 ethanedithiol (EDT), thiophenol, thioanisole, and anisole work well as scavengers. An expensive but very efficient scavenger, alternative to ethanedithiol is 1,4-DL-dithiothreitol (DTT) (Fig. 1.9).

In general, the cleavage cocktail is difficult to be preselected as it depends on the type of peptide sequence, the side-chain protecting groups, and the nature of the cleavable linker [22]. However, some cleavage cocktails that are suggested in Table 1.3 have been used in the past decades with small adjustments for the peptide sequence or the resin.

The reaction time is typically 2–4 h. Longer cleavage reaction times increase the possibility of side reactions, whereas less than two hours will result in nonquantitative cleavage and side-chain deprotection. Microwave heating is used to reduce cleavage time [36], however side reactions also are accelerated. Usually, after the reaction time has been completed, the peptide is precipitated through the addition of chilled diethyl ether or methyl tert butyl ether depending on the size and the

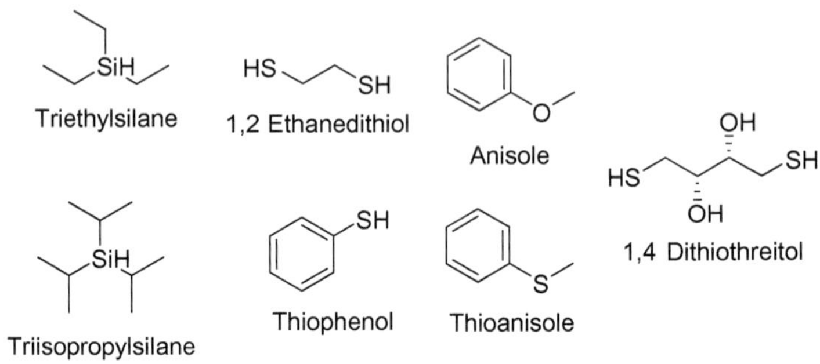

Figure 1.9 Common scavenger reagents.

Table 1.3 Cleavage cocktails: high concentrations of TFA

No	Mixture	Reaction time (h)
1	TFA/phenol/water/TIS[a] (88/5/5/2)	1–4
2	TFA/phenol/water/thioanisole/EDT (85/5/5/2.5/2.5)	1–4
3	TFA/DTT/water/TIS (90/3/5/2)	1–4
4	TFA/thioanisole/EDT/anisole (90/5/3/2)	1–4
5	TFA/water/TIS (95/2.5/2.5)	1–4
6	TFA/EDT/water/TIS (92.5/2.5/3/2)	1–4

[a]TES could be used instead of TIS.

properties of the peptide. Short peptides (4–6 residues), hydrophobic peptides, or fully protected peptides are likely to be soluble in chilled ether. Thus precipitation in chilled water and centrifugation, followed by freeze drying could be helpful for isolation of the peptide. Another way to isolate a peptide is dissolution in 10% acetic acid, extraction with chloroform or dichloromethane to remove scavengers, and freeze drying.

In peptides synthesized on Rink Amide or Wang resins using *t*Bu or Trt as the main side-chain protecting groups, TFA cocktails such as the *cocktail 5* (TFA/water/TIS [95/2.5/2.5]) for 2–3 h could be efficient for isolation of the peptide. In peptides with multiple Cys, Met, or Arg the use of *cocktails 3* or *6* is compulsory. However, optimization is always necessary.

1.7 Microwave-assisted solid-phase peptide synthesis

Microwave-assisted peptide synthesis has become one of the most useful tools for peptide chemists for the synthesis of difficult, as well as routine peptide sequences. In the early 1990s, Yu and coworkers [37] first reported the synthesis of short heptapeptides by using microwave-assisted SPPS in a kitchen microwave oven. At the conclusion of their communication letter they envisioned that "the development of a convenient method for preparation of peptide fragment will be very useful for the synthesis of big peptides." When exposed to microwave irradiation, polar molecules or ions in a liquid solution will rotate in an attempt to follow the electric or magnetic field. The polar molecules have time to align with the field but not to follow exactly the oscillating field. During the peptide synthesis there are many polar molecules and ions present that can rapidly be heated by microwave energy including polar solvents (i.e., DMF or NMP or even water), the peptide backbone which contains the highly polar peptide bond, the N-terminal amine group, bases for the coupling and deprotection reactions, polar/ionic activators, and cleavage acids. All of them are amenable to microwave heating, accelerating reaction kinetics. Furthermore, the resulting raised temperature disrupts on-resin chain aggregation due to intra- and interchain association and allows for easier access to the growing end of the peptide chain, making efficient the synthesis of "difficult" [38] and long peptidic sequences [39]. Microwave peptide synthesis may be performed on polystyrene (PS) [40], polyethylene glycol (PEG)-PS [41], PEG [42], ChemMatrix [43], and CLEAR resins. Recently, Masuda et al. reported the synthesis of a small neurosecretory protein GL of rat, a hydrophobic peptide of 80-amino-acid residues containing an intramolecular disulfide bond and amidated C-terminus [43]. The overall yield was 10%.

Due to the elevated temperatures used during microwave-assisted peptide synthesis there is an increased possibility for undesired side reactions. Racemization of the activated amino acid during the coupling step is a well-studied issue even under conventional SPPS conditions and is a common problem particularly for Cys and His residues [44]. The use of HOBt as an additive in the coupling procedure eliminates the possibility for racemization. Furthermore, studies on the risk of epimerization in peptides containing Cys and His residues during microwave-assisted SPPS suggest that His and Cys should be coupled at 50°C or maybe even at room temperature [45]. On the other hand, Hojo et al. [46] described the first aqueous microwave-assisted synthesis of histidine-containing peptides in high purity and with very low racemization. For the coupling of histidines they used as coupling reagent 4-(4,6-dimethoxy-1,3,5-triazin-2-yl)-4-methylmorpholinium chloride DMTMM, while the Fmoc amino acid derivatives were dissolved as nanoparticles in the aqueous solvent. In addition, Echalier et al. [47] noticed that elevated temperatures during microwave-assisted SPPS could promote hydrolysis of the 2-chlorotrityl ester resulting in peptide cleavage from the solid support and low overall yield of the C-terminal acid peptide. Another well-documented possible side reaction during peptide synthesis is aspartimide formation [45]. Addition of 0.1 M HOBt to the deprotection solution of piperidine reduced aspartimide formation. The use of piperazine in place of piperidine resulted in an even further decrease in the aspartimide formation. Table 1.4 summarizes briefly the advantages and disadvantages of microwave-assisted solid-phase peptide synthesis over classical solid-phase peptide synthesis.

Table 1.4 **Advantages and disadvantages of MW-SPPS**

Advantages over SPPS	**Disadvantages**
Faster reactions	Heating under elevated temperatures
Higher yields and purities	Racemization of histidine and cysteine
Efficient synthesis of difficult sequences	Safety and health hazards
Uniform and targeted heating of the sample	
Less solvents and reagents	
Greater reproducibility	
Greener solvents	

1.8 Green peptide synthesis

Solid-phase peptide synthesis is one of the most polluting and least green chemical processes [48]. DMF, NMP, or DCM are toxic, polar aprotic solvents posing environmental threats because they mix with water. Disposal of these solvents to wastewater treatment facilities results a high BOC/COD and nitrogen loading that can be problematic with the added problem of NO_x emissions [49]. They are among the chemicals of highest concern under the regulations of EU REACH (Registration, Evaluation and Authorization of Chemicals). Therefore, there is a great need to replace DMF with green alternatives [50]. Recently, Albericio and coworkers published the removal of Fmoc group by using γ-valerolactone instead of DMF during SPPS on PS and ChemMatrix resins. A different alternative for DMF during Fmoc was N-formylmorpholine which showed excellent performance only when the ChemMatrix polymeric support was used in the synthesis [51]. Previously, the same group had published the synthesis of the Aib- enkephalin pentapeptide (H-Tyr-Aib-Aib-Phe-Leu-NH_2) [52] using as an alternative solvent 2-methyl tetrahydrofuran (2-MeTHF) which can be derived from renewable resources [53]. Furthermore, they have reported the replacement of DMF with AcN and THF in the synthesis of the hindered peptides, Aib-enkephaline pentapeptide and Aib-ACP decapeptide using SPPS methodology on the ChemMatrix resin. The authors of this work showed better coupling efficiency compared with that observed when DMF was used.

Development of solid-phase peptide synthesis in water using Fmoc-protected amino acids still attracts the interest of researchers. Most of these new approaches are limited to fields of laboratory research. Hojo and coworkers [54] developed an environmentally friendly method using water-dispersible amino acid nanoparticles. The same group continued their research efforts using aqueous microwave peptide synthesis techniques proving that their methodology could easily be applied to His-containing peptides which are sensitive to racemization [46]. Furthermore, Lipshutz and coworkers reported a general method for peptide bond formation in water at ambient temperatures which may be employed for the synthesis of small peptides, mainly tri- and tetrapeptides [55].

Other methods that have been recently presented use alternative solid supports instead of common resins. Sarma and coworkers [56] demonstrated the synthesis of small dipeptides, mainly with aliphatic amino acids, at room temperature using water extract of banana (WEB)/ethylene glycol as aqueous medium with and without external base. Albericio and coworkers reported [57] that the membrane-enhanced peptide synthesis (MEPS) technique could be an alternative methodology for the synthesis of peptides on an industrial scale. As a model they used the RADA tetramer.

1.9 Future challenges: Conclusions

With increasing interest for large quantities of complex, long, chemically modified, and extra pure peptides for therapeutic uses or basic research it is imperative to search for alternative and effective ways of manufacturing peptides. Advances and improvements in synthetic peptide chemistry as well as materials and reagents used for the synthesis (resins, amino acid derivatives, coupling reagents, etc.) and purification systems have been employed to tackle current challenges in the large-scale manufacturing of peptides, with efforts focused on reducing production costs, increasing yield and purity and achieving complex modifications.

Peptides up to 15–20 amino acids with and without non-natural or modified amino acids are typically available from peptide companies at low cost. Synthesis of longer peptides or peptides carrying post-translational modifications is challenging for peptide scientists as they have to solve aggregation phenomena during the synthesis, resulting in low yields and purities and often in failure in receiving the desired peptide. Chemical or enzymatic ligation is a modern approach to synthesize longer peptides. A limitation to this methodology is the requirement for the use of Cys at the N-terminal of the first fragment. PEG resins, pseudoprolines, and addition of chaotropic salts are tools for the synthesis of longer peptides.

Although SPPS is currently, the best choice for peptide synthesis there are still challenges that need to be addressed. Peptide modifications involving the use of fatty acids and PEG moieties, especially of large sizes, cannot be automated and they are currently performed manually using traditional organic synthetic chemistry. Incorporation of N-methylated amino acids often is difficult because of steric hindrance. Acceleration of automated solid-phase peptide synthesis is another unmet challenge. Recently, it was reported a fully automated flow-based approach to solid phase peptide synthesis which reduced the total time of coupling reaction to only 40 seconds per amino acid [58]. According to the authors this could yield tens of thousands of individual 30-mer peptides per year.

This chapter summarized the progress in peptide chemistry during the last 50 years and presented innovations and developments that provided researchers in academia and pharmaceutical industry with extraordinary tools to synthesize thousand of peptides. There are more than 60 peptides in the market, around 150 peptide drugs in clinical trials, and over 500 in preclinical development. The future is bright. Peptide scientists still have a number of synthetic challenges to address including the synthesis of long peptides, bicyclic peptides, pegylated peptides, lipopeptides, glycosylated peptides, cell-penetrating peptides, and peptide–peptoid hybrids. The "Peptide Era" just started.

References

[1] K. Fosgerau, T. Hoffmann, Peptide therapeutics: current status and future directions, Drug Discov. Today 20 (1) (2015) 122–128.

[2] A.R. Mitchell, Bruce Merrifield and solid-phase peptide synthesis: a historical assessment, Biopolymers 90 (3) (2008) 175–184.

[3] S. Sachdeva, Peptides as "drugs": the journey so far, Int. J. Pept. Res. Ther. 23 (1) (2017) 49–60.

[4] T. Bruckdorfer, O. Marder, F. Albericio, From production of peptides in milligram amounts for research to multi-tons quantities for drugs of the future, Curr. Pharm. Biotechnol. 5 (1) (2004) 29–43.

[5] R.B. Merrifield, Solid phase peptide synthesis. I. The synthesis of, J. Am. Chem. Soc. 85 (14) (1963) 2149.

[6] B.G. De La Torre, A. Jakab, D. Andreu, Polyethyleneglycol-based resins as solid supports for the synthesis of difficult or long peptides, Int. J. Pept. Res. Ther. 13 (1–2) (2007) 265–270.

[7] M. Renil, et al., PEGA supports for combinatorial peptide synthesis and solid-phase enzymatic library assays, J. Pept. Sci. 4 (3) (1998) 195–210.

[8] E. Bayer, Towards the chemical synthesis of proteins, Angew. Chem. Int. Ed. Engl. 30 (2) (1991) 113–129.

[9] R. Santini, M.C. Griffith, M. Qi, A measure of solvent effects on swelling of resins for solid phase organic synthesis, Tetrahedron Lett. 39 (49) (1998) 8951–8954.

[10] S. Booth, et al., Solid-phase organic reactions III: a review of the literature Nov 96–Dec 97, Tetrahedron 54 (51) (1998) 15385–15443.

[11] F.-I. Auzanneau, et al., Synthesis and characterization of polyethylene glycol polyacrylamide copolymer (PEGA) resins containing carbohydrate ligands. Evaluation as supports for affinity chromatography, Can. J. Chem. 76 (8) (1998) 1109–1118.

[12] M. Kempe, G. Barany, CLEAR: a novel family of highly cross-linked polymeric supports for solid-phase peptide synthesis, J. Am. Chem. Soc. 118 (30) (1996) 7083–7093.

[13] F. Garcı, et al., ChemMatrix, a poly (ethylene glycol)-based support for the solid-phase synthesis of complex peptides, J. Comb. Chem. 8 (2) (2006) 213–220.

[14] Y. García-Ramos, et al., ChemMatrix® for complex peptides and combinatorial chemistry, J. Pept. Sci. 16 (12) (2010) 675–678.

[15] F. Guillier, D. Orain, M. Bradley, Linkers and cleavage strategies in solid-phase organic synthesis and combinatorial chemistry, Chem. Rev. 100 (6) (2000) 2091–2157.
[16] U. Boas, J. Brask, K.J. Jensen, Backbone amide linker in solid-phase synthesis, Chem. Rev. 109 (5) (2009) 2092–2118.
[17] K. Barlos, et al., Darstellung geschützter peptid-fragmente unter einsatz substituierter triphenylmethyl-harze, Tetrahedron Lett. 30 (30) (1989) 3943–3946.
[18] Y.H. Woo, A.R. Mitchell, J.A. Camarero, The use of aryl hydrazide linkers for the solid phase synthesis of chemically modified peptides, Int. J. Pept. Res. Ther. 13 (1–2) (2007) 181–190.
[19] U. Boas, et al., The ortho backbone amide linker (o-BAL) is an easily prepared and highly acid-labile handle for solid-phase synthesis, J. Comb. Chem. 4 (3) (2002) 223–228.
[20] P. Heidler, A. Link, N-Acyl-N-alkyl-sulfonamide anchors derived from Kenner's safety-catch linker: powerful tools in bioorganic and medicinal chemistry, Bioorg. Med. Chem. 13 (3) (2005) 585–599.
[21] L.A. Carpino, G.Y. Han, The 9-fluorenylmethoxycarbonyl amino-protecting group, J. Org. Chem. 44 (21) (1972) 3739.
[22] R. Behrendt, P. White, J. Offer, Advances in Fmoc solid-phase peptide synthesis, J. Pept. Sci. 22 (1) (2016) 4–27.
[23] V. Mäde, S. Els-Heindl, A.G. Beck-Sickinger, Automated solid-phase peptide synthesis to obtain therapeutic peptides, Beilstein J. Org. Chem. 10 (Scheme 1) (2014) 1197–1212.
[24] K. Barlos, et al., Synthesis of the very acid-sensitive Fmoc-Cys(Mmt)-OH and its application in solid-phase peptide synthesis, Int J Pept Protein Res. 47 (3) (1996) 148–153.
[25] A.K. Galande, R. Weissleder, C.H. Tung, An effective method of on-resin disulfide bond formation in peptides, J. Comb. Chem. 7 (2) (2005) 174–177.
[26] D. Li, D.L. Elbert, The kinetics of the removal of the *N*-methyltrityl (Mtt) group during the synthesis of branched peptides, J. Pept. Res. 60 (5) (2002) 300–303.
[27] K.R. Wilson, et al., Microwave-assisted cleavage of Alloc and Allyl Ester protecting groups in solid phase peptide synthesis, J. Pept. Sci. 22 (10) (2016) 622–627.
[28] J.J. Díaz-Mochón, L. Bialy, M. Bradley, Full orthogonality between Dde and Fmoc: the direct synthesis of PNA-peptide conjugates, Org. Lett. 6 (7) (2004) 1127–1129.
[29] L. Carpino, 1-Hydroxy-7-azabenzotriazole. An efficient peptide coupling additive, J. Am. Chem. Soc. 115 (13) (1993) 4397–4398.
[30] R. Subirós-Funosas, et al., Oxyma: an efficient additive for peptide synthesis to replace the benzotriazole-based HOBt and HOAt with a lower risk of explosion, Chem. Eur. J. 15 (37) (2009) 9394–9403.
[31] G. Gawen, G.W. Kenner, R. Sheppard, Acyloxyphosphonium salts as acylating agents. A new synthesis of peptides, J. Am. Chem. Soc. 91 (20) (1969) 5669–5671.
[32] V. Dourtoglou, B. Gross, V. Lambropoulou, C. Ziodrou, O-benzotriazolyl-N,N,N′,N′-tetramethyluronium hexafluorophosphate as coupling reagent for the synthesis of peptides of biological interest, Synthesis 1984 (9) (1984) 572–574.
[33] L.A. Carpino, et al., The uronium/guanidinium peptide coupling reagents: finally the true uronium salts, Angew. Chem. Int. Ed. Engl. 41 (3) (2002) 441–445.
[34] A. El-Faham, et al., COMU: a safer and more effective replacement for benzotriazole-based uronium coupling reagents, Chem. Eur. J. 15 (37) (2009) 9404–9416.
[35] R. Subirós-Funosas, et al., COMU: scope and limitations of the latest innovation in peptide acyl transfer reagents, J. Pept. Sci. 19 (7) (2013) 408–414.

[36] M. Góngora-Benítez, J. Tulla-Puche, F. Albericio, Handles for Fmoc solid-phase synthesis of protected peptides, ACS Comb. Sci. 15 (5) (2013) 217–228.
[37] H.M. Yu, S.T. Chen, K.T. Wang, Enhanced coupling efficiency in solid-phase peptide synthesis by microwave irradiation, J. Org. Chem. 57 (18) (1992) 4781–4784.
[38] J.I. Finneman, M.J. Pozzo, Novel approach for optimization of a "difficult" peptide synthesis by utilizing quantitative reaction monitoring assays, J. Pept. Sci. 18 (8) (2012) 511–518.
[39] I. Friligou, et al., Microwave-assisted solid-phase peptide synthesis of the 60-110 domain of human pleiotrophin on 2-chlorotrityl resin, Amino Acids 40 (5) (2011) 1431–1440.
[40] M. Ieronymaki, et al., Use of the 2-chlorotrityl chloride resin for microwave-assisted solid phase peptide synthesis, Biopolymers 104 (5) (2015) 506–514.
[41] X. Hemu, et al., Total synthesis of circular bacteriocins by butelase 1, J. Am. Chem. Soc. 138 (22) (2016) 6968–6971.
[42] R. Roodbeen, et al., Microwave heating in the solid-phase synthesis of N-methylated peptides: when is room temperature better?, Eur. J. Org. Chem. 2012 (36) (2012) 7106–7111.
[43] K. Masuda, et al., Microwave-assisted solid-phase peptide synthesis of neurosecretory protein GL composed of 80 amino acid residues, J. Pept. Sci. 21 (6) (2015) 454–460.
[44] S.L. Pedersen, et al., Microwave heating in solid-phase peptide synthesis, Chem. Soc. Rev. 41 (5) (2012) 1826–1844.
[45] S.A. Palasek, Z.J. Cox, J.M. Collins, Limiting racemization and aspartimide formation in microwave-enhanced Fmoc solid phase peptide synthesis, J. Pept. Sci. 13 (10) (2007) 143–148.
[46] K. Hojo, et al., Aqueous microwave-assisted solid-phase peptide synthesis using Fmoc strategy. III: racemization studies and water-based synthesis of histidine-containing peptides, Amino Acids 46 (10) (2014) 2347–2354.
[47] C. Echalier, et al., Heating and microwave assisted SPPS of C-terminal acid peptides on trityl resin: the truth behind the yield, Amino Acids 45 (6) (2013) 1395–1403.
[48] S.B. Lawrenson, R. Arav, M. North, The greening of peptide synthesis, Green Chem. 19 (2017) 1685–1691.
[49] D.J.C. Constable, C. Jimenez-Gonzalez, R.K. Henderson, Perspective on solvent use in the pharmaceutical industry, Org. Process Res. Dev. 11 (1) (2007) 133–137.
[50] D. Prat, J. Hayler, A. Wells, A survey of solvent selection guides, Green Chem. 16 (10) (2014) 4546–4551.
[51] Y.E. Jad, et al., Green solid-phase peptide synthesis (GSPPS) 3. Green solvents for Fmoc removal in peptide chemistry, Org. Process Res. Dev. 21 (3) (2017) 365–369.
[52] Y.E. Jad, et al., 2-Methyltetrahydrofuran and cyclopentyl methyl ether for green solid-phase peptide synthesis, Amino Acids 48 (2) (2016) 419–426.
[53] V. Pace, et al., 2-Methyltetrahydrofuran (2-MeTHF): a biomass-derived solvent with broad application in organic chemistry, ChemSusChem 5 (8) (2012) 1369–1379.
[54] K. Hojo, et al., Development of a method for environmentally friendly chemical peptide synthesis in water using water-dispersible amino acid nanoparticles, Chem. Cent. J. 5 (1) (2011) 49.
[55] C.M. Gabriel, et al., Amide and peptide bond formation in water at room temperature, Org. Lett. 17 (16) (2015) 3968–3971.

[56] M. Konwar, A.A. Ali, D. Sarma, A green protocol for peptide bond formation in WEB, Tetrahedron Lett. 57 (21) (2016) 2283–2285. Available from: http://dx.doi.org/10.1016/j.tetlet.2016.04.041.

[57] V. Castro, et al., Novel globular polymeric supports for membrane-enhanced peptide synthesis, Macromolecules 50 (4) (2017) 1626–1634.

[58] A.J. Mijalis, et al., A fully automated flow-based approach for accelerated peptide synthesis, Nat. Chem. Biol. 13 (5) (2017) 464–466.

Synthetic approaches of naturally and rationally designed peptides and peptidomimetics

2

Ariadna Lobo-Ruiz and Judit Tulla-Puche
Department of Inorganic and Organic Chemistry, Organic Chemistry Section, University of Barcelona, Barcelona, Spain

2.1 Natural and rationally designed peptides

In the last decade, peptides have emerged as important tools in drug discovery, as they combine the advantages of small molecules, such as stability and bioavailability, with those of proteins, such as selectivity and potency. Moreover, advances in chemical synthesis of peptides have allowed for decreased manufacturing costs, making them more attractive from an industrial point of view. In this section, we describe synthetic approaches for several bioactive peptide families, such as cyclic peptides, disulfide-rich peptides, cyclodepsipeptides, lantibiotics, and stapled peptides whereas the following section discusses strategies for achieving different families of peptidomimetics (Table 2.1).

2.1.1 Cyclic peptides

A number of natural cyclic peptides with interesting bioactivities exist in nature. With regards to their linear counterparts, cyclic peptides show higher proteolytic stability, as well as enhanced selectivity and potency [1]. A relevant natural cyclic peptide is cyclosporin A, a cyclic undecapeptide with a highly *N*-methylated backbone. Cyclosporin A was isolated from the fungal species *Tolypocladium inflatum* Gams and it is currently used as an immunosuppressant in order to prevent rejection in human transplant [2]. It is one of the few peptides that is orally absorbed, a property conferred by the high number of *N*-methyl groups present in its structure.

From a structural point of view, cyclosporin A is a head-to-tail cyclic peptide, with seven *N*Me-amino acids, a D-Ala residue, and the rare (2*S*,3*R*,4*R*,6*E*)-3-hydroxy-4-methyl-2-methylamino-6-octenoic acid (MeBmt), in addition to L-Ala and L-Val moieties. The synthesis in solution of cyclosporin A was accomplished by Wenger [3]. After several years, an efficient solid-phase approach was reported by the Rich group [4]. In their work (Scheme 2.1), the bond D-Ala-Ala was chosen for the disconnection, therefore Fmoc-Ala-OH was coupled onto a PAC-PS resin. The linear undecapeptide was assembled by regular Fmoc/*t*Bu chemistry using HATU/HOAt/DIEA as the key coupling system to enable good coupling yields over the hindered *N*Me-amino acids that are present in this peptide. Following

Peptide Applications in Biomedicine, Biotechnology and Bioengineering.
DOI: http://dx.doi.org/10.1016/B978-0-08-100736-5.00002-8

Table 2.1 Relevant bioactive peptides and peptidomimetics by type and application

Name	Type of peptide	Application	Synthetic approach	References
Cyclosporine A	Cyclic peptide	Immunosuppressant	Solid-phase	[2–5]
Didemnins A/B	Cyclodepsipeptide	Cancer	Solution	[92]
Pipecolidepsin A	Cyclodepsipeptide	Cancer	Solid-phase	[28,29]
Romidepsin	Cyclodepsipeptide	Cutaneous and peripheral T-cell lymphomas	Solution	[24–26]
Kalata B1	Cyclotide	Immunosuppressant	Solid-phase	[17,18,20]
Ziconotide	Disulfide-rich peptide	Severe chronic pain		[8,9]
Linaclotide	Disulfide-rich peptide	Chronic constipation	Solid-phase	[93]
Exenatide	Linear peptide	Diabetes mellitus type 2	Hybrid (solid-phase/ solution)	[94]
Stapled peptide MDM2/X inhibitor	Peptide	Solid tumor (phase I) & lymphoma (phase II)	Solid-phase	[46,47]
Merotocin	Peptoid–peptide hybrid	Preterm delivery (phase II)	Hybrid	[95,96]
ABT-510	Peptoid–peptide hybrid	Solid tumors (phase I), lymphoma, and melanoma (phase II).	Solid-phase	[97–100]
Pentetreotide	Peptoid–peptide hybrid	Diagonostic of Cushing syndrome (phase I & phase II).	Hybrid	[101–103]
Atazanavir (BMS-232632)	Azapeptide	Antiretroviral, anti-HIV	Solution	[104–106]
T900607	Nonpeptide-based peptidomimetic	Cancer	Solution	[107]
Emricasan	AA-peptide	Liver and kidney diseases and diabetes (phase I)	Solution	[108,109]
		Liver cirrhosis, hepatitis, hepatocellular carcinoma (phase II)		
Glasdegib	Urea-based peptidomimetic	Acute myeloid leukemia (phase II)	Solution	[110–112]

cleavage from the resin [trifluoroacetic acid (TFA)-H_2O (95:5); 3–4 h] and cyclization in solution [0.3 mM in CH_2Cl_2; propyl phosphonic anhydride $(PrPO_2)_3$ and dimethylaminopyridine (DMAP)], cyclosporin A was obtained with a yield of 10%–15%. In 2010 the Danishefsky group also described a solid-phase strategy for cyclosporin A where the same disconnection bond was chosen for macrolactamization, but in this case, isonitriles were used in the formation of amide bonds [5].

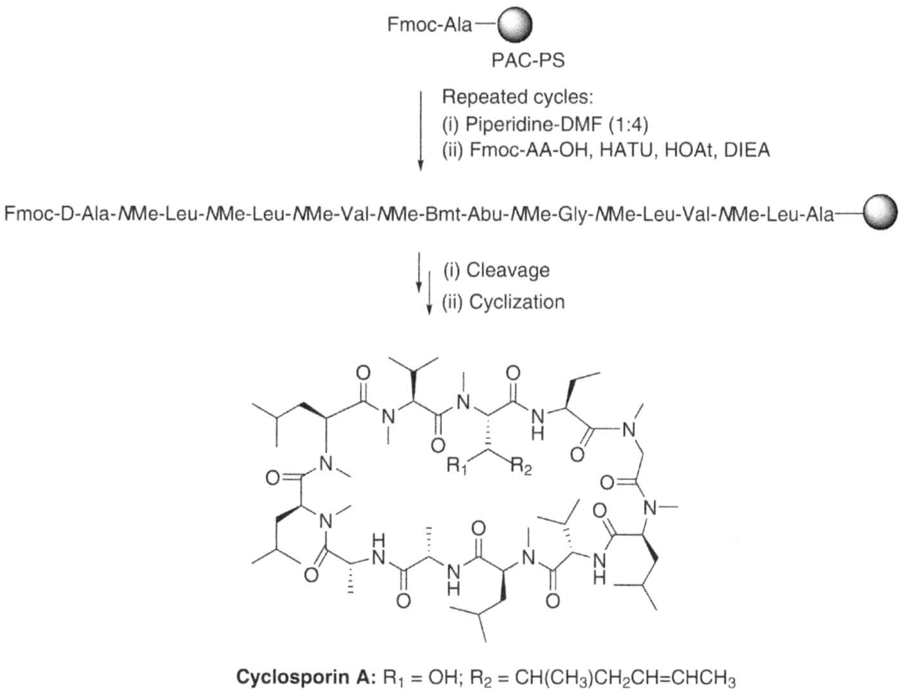

Scheme 2.1 Solid-phase synthesis of cyclosporin A.

2.1.2 Disulfide-rich peptides

Disulfide-rich peptides are privileged highly constrained peptides with many therapeutic applications [6]. They include highly knotted antimicrobial defensin peptides, knottins, plant-derived cyclotides, and venom-derived peptides.

They are formed by less than 40 residues, they lack a defined hydrophobic core, and they have unusual secondary structure elements. These architectures are stabilized mainly by two or more disulfide bonds, which act as staples that replace the hydrophobic core and assist the peptide in adopting a particular and well-defined folded structure [7]. In addition, plant-derived cyclotides and knottins contain a cystine knot, which makes them even more stable. The connectivity of disulfide bonds in these peptides and their final three-dimensional structure are highly preserved in nature, and their tasks are mainly of an offensive (e.g., prey immobilization) or defensive (e.g., against microbial predators) nature.

In the following subsections, we will discuss the synthesis of two of these highly constrained structures: conotoxins from marine cone snail venom and plant-derived cyclotides.

2.1.2.1 Conotoxins

Conotoxins, found in sea cone snails, are used by these organisms for predatory purposes: they paralyze their prey by blocking or inhibiting various targets of the nervous system. It is thought that as many as 100,000 different types of conotoxins may exist, most of them still to be discovered [8,9]. They are composed of 8–32 amino acids and they contain structural elements that are present in proteins, such as α-helixes or β-sheets. The majority of conotoxins discovered up to now contain two or three disulfide bridges, and are classified in different families according to their pharmacological indications and by disulfide bond patterns [10]. With regards to their chemical synthesis, two main strategies are employed for the correct pairing of cysteines in conotoxins and venom-derived peptides in general: random oxidation of the free cysteines or regioselective strategies involving the use of different protecting groups for each of the correct pairing of Cys. For natural peptides, random or nondirected oxidative folding of disulfide bridges is preferred, as it is thought that the correct folding is encoded in the amino acid sequence. However, for modified venom-derived peptides, the second approach is most used. Scheme 2.2A shows a

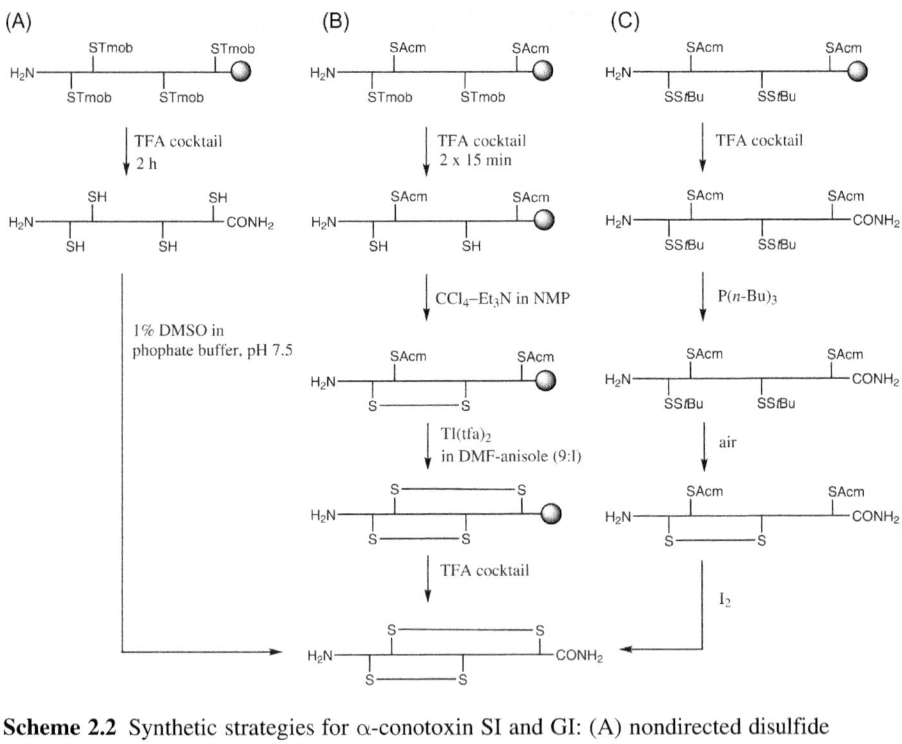

Scheme 2.2 Synthetic strategies for α-conotoxin SI and GI: (A) nondirected disulfide formation; (B) on-resin, and (C) solution regioselective disulfide formation approaches.

nondirected approach of α-conotoxin SI, in which the linear peptide was assembled by solid-phase peptide synthesis (SPPS) using a Fmoc/*t*Bu strategy and *S*-2,4,6-trimethoxybenzyl (Tmob) as the protecting group for the cysteines [6]. After acidolytic cleavage from the resin, the tetrasulfhydryl linear peptide was correctly folded under DMSO oxidation conditions. In directed regioselective approaches, different protecting groups for the cysteine pairs are used, including the *S*-acetamidomethyl (Acm) group which is the most common approach. In Scheme 2.2B, the remaining cysteine pair is protected with Tmob, which is first removed to form the first disulfide bridge on resin by treatment with CCl_4-Et_3N in *N*-methylpyrrolidone. The second disulfide is directly formed on the bis(Acm) resin-bound linear peptide by treatment with Tl $(Tfa)_2$, and finally, TFA cleavage renders the final native α-conotoxin SI [11]. A different regioselective strategy is to form the two disulfide bonds in solution after cleavage the peptide from the resin. The right combination of *S*-*t*-butylsulfenyl (SS*t*Bu) and Acm groups allow for obtaining of the correctly folded peptide (Scheme 2.2C) [12]. In regioselective directed approaches, one key factor is the correct succession of disulfide bonds. In α-conotoxin SI, for example, the oxidation of the small loop followed by the larger one is much preferred over the reversed option.

2.1.2.2 Cyclotides

Plant-derived cyclotides also have a high concentration of disulfide bonds in their small-sized framework, but additionally, cystines present a knotted arrangement (cystine knots). Moreover, a head-to-tail circularized backbone confers a highly stable framework [13]. This exceptional stability has attracted huge interest from a therapeutic point of view, as scaffolds for drug design [14]. Kalata B1, the first cyclotide to be discovered and structurally elucidated, was the active ingredient of a tea used in Africa to accelerate childbirth. From this first discovery, a large number of cyclotides have been found in plants from the Violaceae, Rubiaceae, Cucurbitaceae, Fabaceae, and Solanaceae families [15].

Cyclotides are chemically synthesized using native chemical ligation (NCL) [16] for the final head-to-tail cyclization, which requires the synthesis of the linear precursor with an *N*-terminal Cys and a *C*-terminal thioester. Under NCL conditions, folding of disulfide bonds into the native knotted arrangement also occurs [17,18]. Cyclotides can be synthesized both by Boc and Fmoc chemistry, but Boc SPPS is still preferred over the latter due to the difficulties associated over the years with obtaining the linear peptide thioester by the Fmoc strategy, although lately this limitation has been overcome [19].

Scheme 2.3 shows the synthetic strategy for obtaining Kalata B1 using Boc chemistry. Starting with Boc-Gly-phenylacetamidomethyl (Boc-Gly-PAM) resin, *S*-trityl (Trt)-β-mercaptopropionic linker is introduced after removal of the Boc group from Gly with TFA. After the Trt group is cleaved, the first amino acid is loaded by formation of a thioester bond, and elongation of the linear sequence proceeds by standard cycles of Boc SPPS. HF cleavage provides the linear thioester precursor of Kalata B1, and treatment with a 0.1 M NH_4CO_3, pH 8,2-isopropanol (1:1) solution provides the circularized and knotted native Kalata B1 in one pot.

Backbone cyclization using enzymes and inteins has also been achieved for cyclotides. These approaches are very fast and specific and they provide very clean cyclizations under mild aqueous conditions [20].

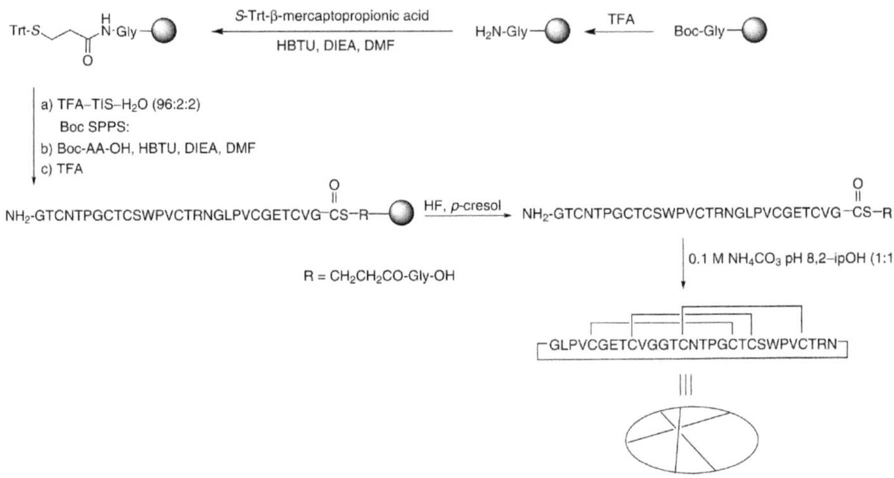

Scheme 2.3 Synthetic strategy for cyclotides by Boc SPPS, with Kalata B1 as an example.

2.1.3 Cyclodepsipeptides

Cyclodepsipeptides, cyclic peptides with at least one amide bond replaced by an ester moiety, present a broad spectrum of biological functions, which spans antitumor, anthelmintic, insecticidal, antibiotic, antifungal, immunosuppressant, antiinflammatory, anti-HIV, and antimalarial activities [21,22].

2.1.3.1 "Head-to-tail" cyclodepsipeptides

In this class of cyclodepsipeptides, the *N*-terminus is linked to the *C*-terminus, and the backbone contains at least one ester bond. A relevant peptide of this type is romidepsin (also known as FK228 or FR901228), a histone deacetylase inhibitor isolated from the culture broth of the terrestrial bacterium *Chromobacterium violaceum* [23]. In addition to their circularized backbone, romidepsin exhibits a second cycle formed by a disulfide bridge between the unusual building block, (3*S*,4*E*)-3-hydroxy-7-mercapto-4-heptenoic acid and a D-cysteine. The first synthesis of romidepsin was accomplished by Simon and coworkers [24] in solution and was improved years later by Williams et al. [25]. However, in both strategies, macrolactonization is selected to close the cycle, which significantly decreased the overall yield. It is commonly accepted that in the synthesis of cyclodepsipeptides, macrolactamization gives superior results as compared to macrolactonization. Also in 2008, Ganesan and coworkers [26] reported total synthesis of romidepsin where the amide bond between Val and the key β-hydroxy mercapto acid was chosen for macrolactamization (77% yield for this step) (Scheme 2.4). In this approach, the ester bond was constructed at a very early stage, and two fragments were first assembled from Fmoc-protected amino acids, and then coupled using PyBOP and DIEA

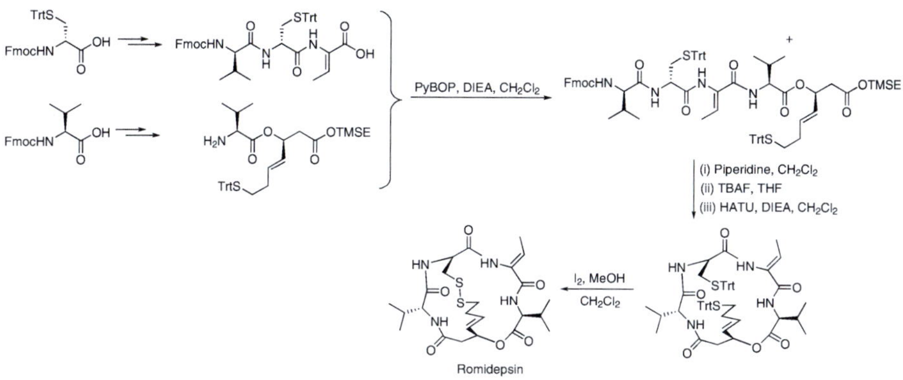

Scheme 2.4 Synthetic strategy of romidepsin.

as the coupling system. For the protection of thiols the trityl group was selected, whereas the carboxylic acid was protected with the 2-(trimethylsilyl)ethyl (TMSE) group. With the linear tetradepsipeptide in hand, the Fmoc and TMSE groups were removed and the macrolactamization was performed by treatment with HATU and DIEA. A final oxidation of the STrt moieties with I_2 delivered romidepsin in a 50% overall yield.

2.1.3.2 "Head-to-side-chain" cyclodepsipeptides

Of special complexity are the "head-to-side-chain" cyclodepsipeptides [27]. In this subfamily, the *C*-terminus is linked, by means of an ester bond, to a β-hydroxy-α-amino acid located in the middle of the chain (Fig. 2.1). The synthesis of cyclodepsipeptides of this kind requires detailed planning of the synthetic strategy: (1) careful

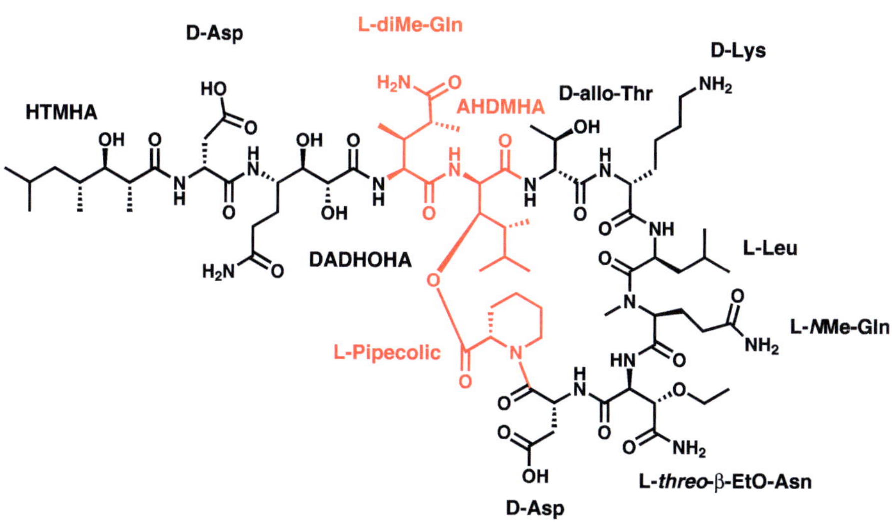

Figure 2.1 Structure of pipecolidepsin A. In red, the residues showing more complexity on the assembly.

examination and minimization of the possible risk of diketopiperazine formation due to presence of the ester bond; (2) design of a synthetic strategy for each conveniently protected nonnatural amino acid present in the cyclodepsipeptide; (3) an orthogonal protection scheme considering the lability limitations of the ester bond; and (4) a solid support that allows cleavage under mild conditions, since the stability of the nonnatural moieties under high acidic conditions is unknown [28–30].

As an example, the solid-phase synthesis of pipecolidepsin A is shown in Scheme 2.5 [28,29]. The final synthesis entailed the incorporation on the solid support of Fmoc-D-Asp(OH)-OAllyl by side chain anchoring. This allows to free the *C*-terminus once the full elongation is complete and to perform the cyclization of the solid phase. The second step was the incorporation of the following amino acids by a regular Fmoc/*t*Bu strategy and using HATU/HOAt/DIEA as coupling system: Fmoc-L-*threo*-β-EtO-Asn-OH, Fmoc-*N*Me-Gln-OH, Fmoc-L-Leu-OH, Fmoc-D-Lys(Boc)-OH, Fmoc-D-*allo*-Thr(*t*Bu)-OH, and Fmoc-AHDMHA-OH. Some of these required a previous synthesis in solution. Once the heptapeptide was in place, Fmoc-diMe-Gln-OH was coupled using *N,N′*-diisopropylcarbodiimide (DIPCDI)/1-hydroxybenzotriazole hydrate (HOBt) in order to minimize the dehydration side reaction at the side chain of

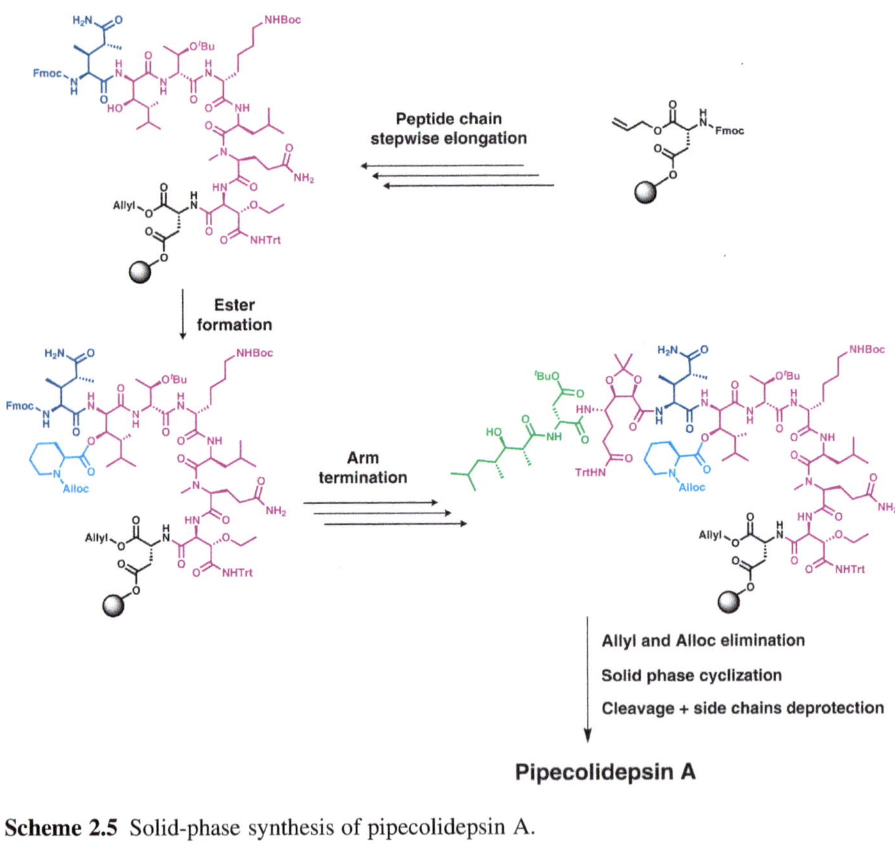

Scheme 2.5 Solid-phase synthesis of pipecolidepsin A.

Gln. Ester formation at this point required much optimization, since incorporation of pipecolic acid on the bulky AHDMHA did not proceed under standard conditions. Instead, temperature was needed, but the combination of heat (45°C) and basic conditions (DIPCDI/DMAP) caused partial cleavage of the Fmoc group of diMe-Gln, and the subsequent lactamization to produce a pyrodimethylglutamic byproduct. To minimize this side-reaction, a minimum treatment was applied (2 h 30 min), followed by a very short Fmoc cleavage single treatment (3 min), and a fast incorporation of the incoming amino acid [Fmoc-DADHOHA(acetonide,Trt)-OH] with HATU/HOAt/DIEA with previous preactivation. The last step in the elongation of the linear peptide was the coupling of the two last residues as a single moiety, using PyBOP/HOAt/DIEA for 4 h. With the linear depsipeptide in hand, the final steps involved the removal of the Allyl and (allyloxycarbonyl) Alloc groups, the cyclization on solid phase using PyBOP and the final cleavage from the solid support.

2.1.4 Lantibiotics

Lantibiotics (or lanthipeptides) are antimicrobial agents that are ribosomally synthesized and posttranslationally modified to their biologically active forms [31]. While they differ in sequences and in their spatial arrangement (Fig. 2.2), the unifying structural motif present in all lantibiotics is the unusual moiety lanthionine, which consists of two alanine residues crosslinked via a thioether linkage that connects their β-carbons (*S*-(alaninyl-3-yl)-cysteine). They usually also contain a methyl-substituted lanthionine derivative, (2*S*,3*S*,6*R*)-3-methyllanthionine, and typically contain the unsaturated amino acids 2,3-didehydroalanine (Dha) and (*Z*)-2,3-didehydrobutyrine (Dhb). Nisin is the most known and studied lantibiotic, and it is widely used as a food preservative.

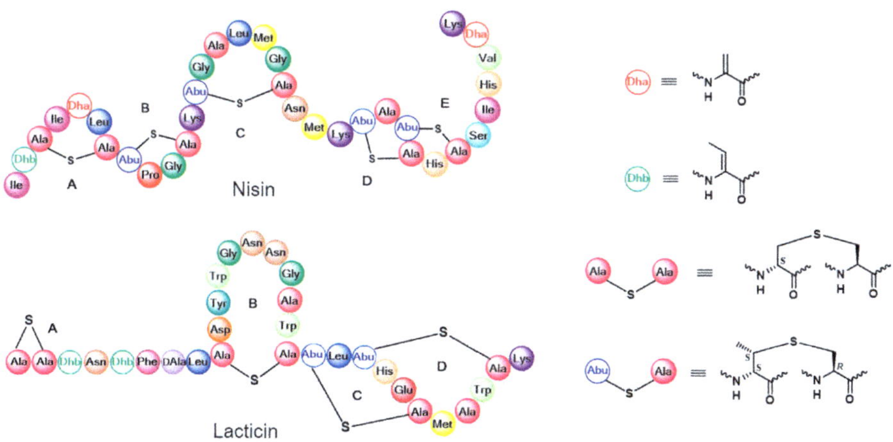

Figure 2.2 Schematic structures of nisin and lacticin 3147 A1. On the right, chemical structures of the Dha, Dhb, Lan, and MeLan moieties are shown.

Lantibiotics are produced by a large number of Gram-positive bacteria to eliminate other competing bacteria. As these modes of action are unique to lantibiotics, pathogenic bacteria have not yet evolved resistance to lantibiotics. They therefore hold considerable potential as the next generation of antibacterial agents. However, there are still limitations to overcome: lantibiotics are not ideal as therapeutic entities, as they are unstable and/or insoluble at physiological pH, prone to oxidation and proteolysis, and cannot be orally administered [32].

The chemical synthesis of nisin in solution was accomplished by Shiba and coworkers [33], after coupling five segments, four of them containing the appropriate lanthionine moieties. Thioether formation proceeded from the precursor disulfide cyclic peptide via desulfurization.

Since then, access to lantibiotics has focused on SPPS approaches [34–37]. In this sense, much effort has been devoted to achieving the Lan/Me Lan residues with the correct stereochemistry and conveniently protected, prior to the solid-phase assembly [38]. The methodologies developed for the native lantibiotics have been the starting point for the preparation of analogues in the search of compounds with better pharmacological properties [39–43].

As an example, the solid-phase synthesis of Lacticin 3147 A1 is shown in Scheme 2.6 [37]. This lantibiotic presents interlocking rings, and this implies that two levels of orthogonality are required to construct them. Therefore, orthogonal protection to Fmoc chemistry was accomplished by the Allyl and Alloc groups on the first MeLan moiety, and *p*-nitrobenzyl (pNB) and *p*-nitrobenzyloxycarbonyl (pNZ) on the second one, which are also orthogonal to the previous ones.

The synthesis started by incorporating Fmoc-Lys(Boc)-OH Wang resin with a reduced loading of 0.15 mmol/g to minimize interchain couplings. Allyl/Alloc

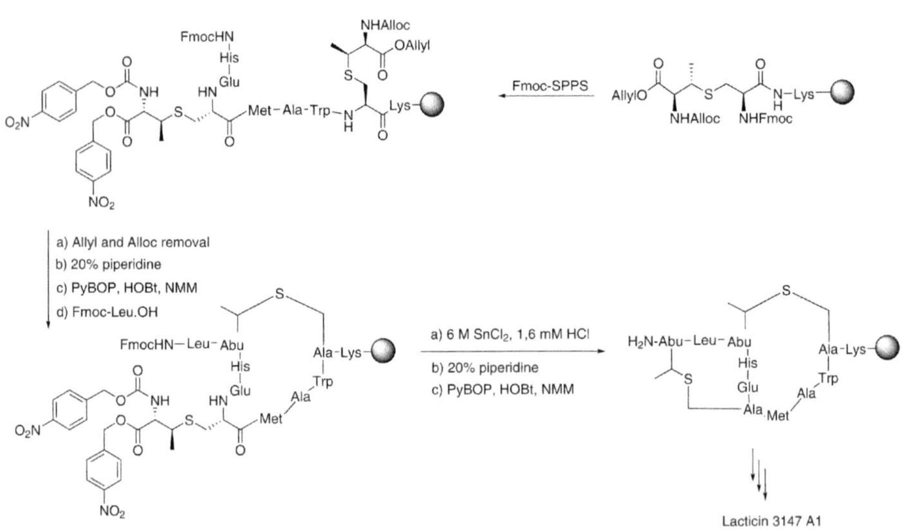

Scheme 2.6 Solid-phase synthesis of lacticin 3147 A1.

protected MeLan was then coupled, and the chain was elongated by standard Fmoc/*t*Bu until reaching the octapeptide and incorporating the pNB/pNZ protected MeLan at the appropriate position. At this point, treatment with $Pd(PPh_3)_4$ and $PhSiH_3$ cleaved the Allyl and Alloc groups and the cyclization was performed by means of PyBOP and HOBt. After introducing the following Leu residue, treatment with $SnCl_2$ removed the pNB and pNZ groups of the second MeLan, and the ring was installed as before. Subsequent SPPS and formation of the two remaining rings resulted in Lacticin 3147 A1.

2.1.5 Stapled peptides

According to Jochim and Arora [44] over 60% of the protein–protein complexes in the Protein Data Bank possess α-helical interfaces. Therefore, mimicking these α-helixes has become important as an appealing approach to target protein–protein interactions (PPIs). The limitation of short peptides, however, resides in the fact that they have little structure in solution, and therefore, rationally designed strategies have been developed to favor or to stabilize an α-helix. Of the several methodologies that have appeared over the years [45], one discovery to highlight is the development of all-hydrocarbon stapled peptides by Verdine and coworkers [46]. By incorporating α-methyl, α-alkenylglycine units separated by three or six residues, and later performing ring-closing metathesis (RCM) to link both side chains [47], these peptides are able to stabilize α-helixes and in some cases have shown enhanced potency relative to the unmodified parent polypeptide.

A careful selection of the residues that need to be replaced to incorporate the staple is required. These residues should be comprised within the α-helical binding segment of interest but should not be crucial for target recognition.

Scheme 2.7 shows the synthesis of an all-hydrocarbon stapled peptide. A Fmoc/*t*Bu strategy on a Rink amide solid support is usually used. At the appropriate

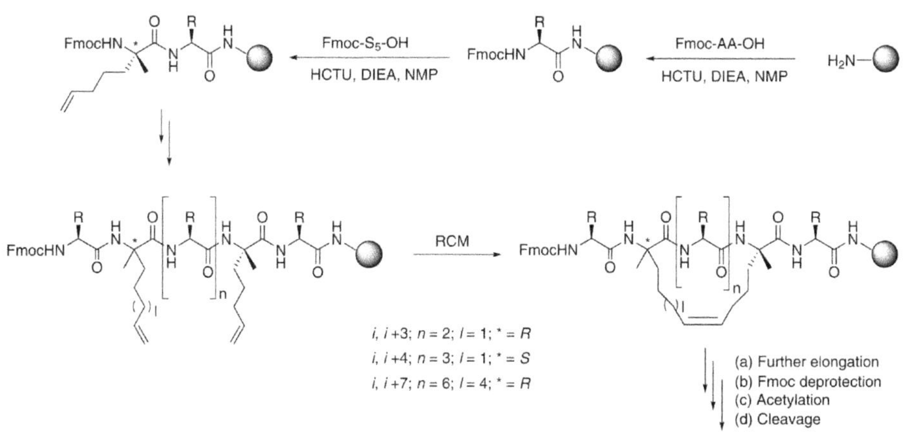

Scheme 2.7 Solid-phase synthesis of all-hydrocarbon stapled peptides.

positions, Fmoc-α-methyl, α-alkenyl moieties, previously synthesized in solution, are incorporated. After completing the elongation, the staple is formed on the solid support by RCM by a treatment with Grubbs' first-generation catalyst in 1,2-dichloroethane at room temperature (20–25°C) for 2 h. The reaction yield typically reaches ~90%, and can be driven to completion with a second treatment. The final acidic cleavage releases the stapled peptide from the resin.

Besides the all-hydrocarbon stapling technique described so far, and the classical lactamization between Lys and Glu/Asp residues [48] and disulfide bridging [49], a number of other stapling chemistries have appeared over the years [50], such as the Cu(I)-catalyzed azide-alkyne cycloaddition (CuAAc) "click reaction" [51], thioether [52] and oxime formation [53], the thiol-ene reaction [54] and Cys crosslinking [55], among others.

2.2 Peptidomimetics

In past years, many groups have put great efforts into developing peptide mimics [56–59]. Peptidomimetics arise from the modulation or chemical modification of peptides, and have the capacity to mimic their biological function with improved pharmacological properties. The design of such molecules aims at overcoming the limitations of peptides as potential drugs in the medicinal chemistry field, since peptides are prone to proteolytic degradation, present poor bioavailability, and often exhibit low selectivity [57]. Thus, the so-called peptidomimetics emerge as potential therapeutic agents. Among the most relevant types of peptidomimetics reported in the literature are: peptoids, β-peptides, γ-peptides, δ-peptides, oligoureas, azapeptides, α-aminoxy-peptides, sugar-based peptides, α/β-peptides, and polyamides [60–62]. There is extensive research in this field [56], hence the following sections focus on the description of synthetic methodologies that we consider to be most significant.

2.2.1 *Peptoids and Peptoid derivatives*

Early in the 1990s, peptoids were firstly explored by Zuckermann and coworkers [63]. Peptoids are peptidomimetics described as foldamers consisting of *N*-alkylated glycine oligomers (Fig. 2.3B). Their biological interest lies in their capability of forming a wide variety of secondary structures, including: α-helix, polyproline type I

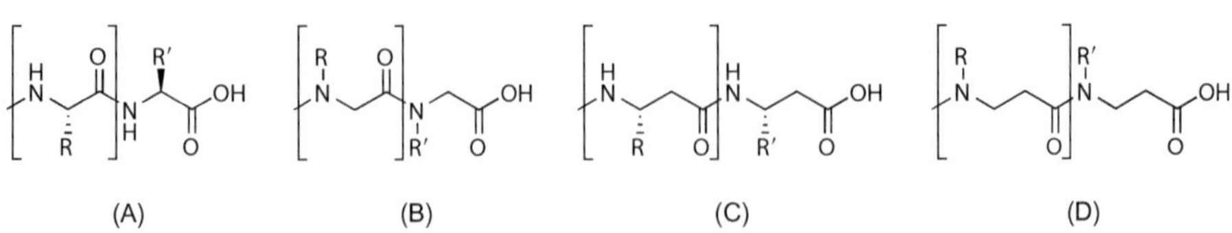

Figure 2.3 Chemical structure of (A) a peptide, (B) a peptoid, (C) a β-peptide, and (D) a β-peptoid.

helix, and type I and type III β-turns. They can also mimic more complex secondary structures [64]. Zuckermann demonstrated that single amphiphilic peptoid helices could be assembled in bundles to form a hydrophobic core. Interestingly, peptoids exhibit good cellular permeability in comparison to their analogue peptides, therefore their potential as cellular transporters has also been studied. Several peptoids have been reported as potential drug candidates due to their biological activity [61,65]. Not surprisingly, cyclic peptoids have been found to exhibit greater biological activity compared to their linear equivalents [65].

Synthetic approaches to efficiently prepare peptoid oligomers have been explored. The following criteria should be considered when facing the design of these molecules: (1) facile synthesis of lengthy oligomers; (2) easily accessible chemical libraries of a variety of monomer side chains; (3) structural and conformational stability; (4) good bioavailability and resistance against proteolytic degradation; and (5) potential biological activity [64].

Kirshenbaum and collaborators described a solid-phase approach that allows the rapid preparation of peptoid oligomer libraries [66]. They prepared peptoid oligomers by simple assembly of a haloacetic acid and a primary amine (Scheme 2.8). Bromoacetic acid is often the acid of choice, since it is an inexpensive commercially available reagent. Incorporation of the haloacetic acid onto the polymeric support, followed by the nucleophilic displacement of the amine leads to the peptoid monomer. Iterative cycles of this two-step synthesis render peptoid oligomers with the desired length. Additionally, this approach allows the preparation of peptoid oligomers containing a vast repertoire of side chains. Nonetheless, cyclic compounds can be easily prepared if the 2-chlorotrityl resin is the resin of choice by simple amidation in solution upon cleavage. It is worth mentioning, that Kirshenbaum and coworkers used the well-known CuAAC to modify peptoid oligomers [67]. They explored the conjugation of ethisterone to peptoid side chains. Functionalization of peptoid side chains with azide functionalities allowed the conjugation of ethisterone-alkyne conjugations via click chemistry. Hence, this synthetic method is presented as a standard protocol to prepare peptoid oligomer libraries.

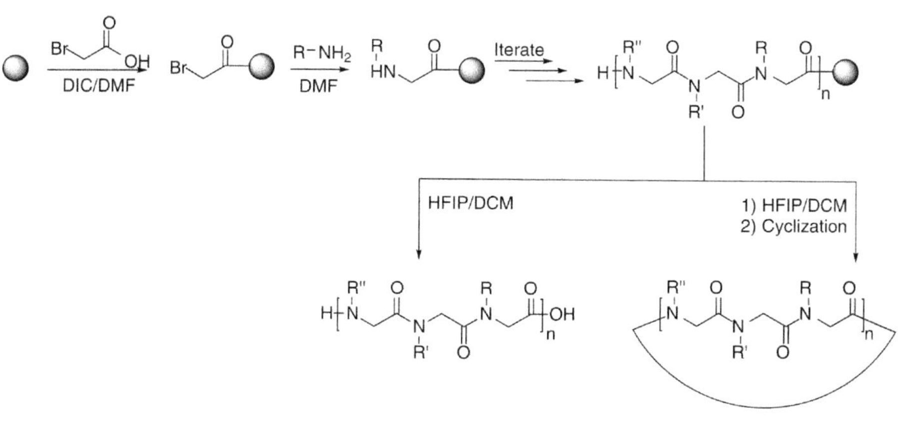

Scheme 2.8 Solid-phase approach to prepare peptoid oligomers.

β-Peptoids are an interesting class of peptoids because they combine the structural features from both β-peptides and peptoids (Fig. 2.3B–D). β-Peptides and β-peptoids incorporate an extra methylene unit at the α position within the peptide backbone which contains an *N*-alkyl modification as well [68]. They expand the existing reservoir of foldamers and have also exhibited the capability of resembling protein secondary structures. The synthetic approach to prepare β-peptoid oligomers comprises solid-phase and solution synthesis methods thus the synthesis is slightly more complex than the synthesis of their analogue peptides [69].

Kirshenbaum and coworkers described for the first time peptoid derivatives consisting of peptoid–peptide hybrids. This new type of semisynthetic peptidomimetics allows for the modulation of proteins and therefore structurally complex protein-peptoid molecules can be designed and synthesized. Peptoid–peptide hybrids can be prepared by employing chemoselective fragment condensation of a peptide with a peptoid oligomer through native amide bond formation [70]. NCL is a widely used technique to prepare large molecules. The reaction involves a *C*-terminal peptide thioester and a peptide holding an *N*-terminal cysteine residue. The ligation takes place through a spontaneous transthioesterification between the cysteine residue and the peptide thioester. Nevertheless, problems associated with the cysteine residue might be encountered, and new chemoselective ligations have been described to overcome these difficulties. Kent explored the ligation between a *C*-terminal peptide salicylaldehyde ester and an *N*-terminal serine/threonine residue. In this case, an imine intermediate product is formed, followed by an irreversible O→N acyl transfer that leads to the amide bond formation (Scheme 2.9). The described synthetic approach results in an efficient manner to prepare peptoid–peptide oligomers, as the ligation takes places with all the side chains unprotected. In addition, the resulting molecules exhibited polyproline I- or polyproline II-type helix structures, which might integrate the structural and functional attributes from both polypeptides and peptoids.

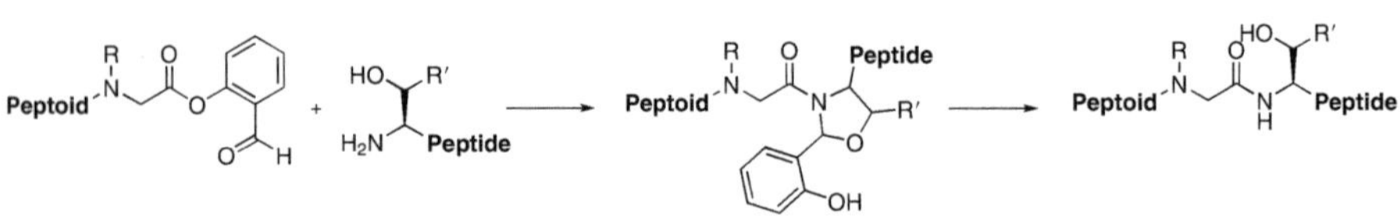

Scheme 2.9 Chemoselective fragment condensation of a peptide with a peptoid oligomer through native amide bond formation: ligation between the C-terminal peptide salicylaldehyde ester and the N-terminal serine/threonine residue.

Another example of peptoid derivatives is the so-called α-peptide/β-peptoid chimeras [71]. This new class of peptidomimetics alternates an α-amino acid unit and a β-peptoid monomer. These chimeras present the properties of peptidomimetics, and therefore they show resistance toward proteolytic degradation, possess biological activity and can resemble the secondary structure of proteins. Olsen and coworkers described the synthesis of α-peptide/β-peptoid chimeras containing lipophilic and cationic residues. A combined solid-phase/solution methodology was used. First, a versatile dimeric building block consisting of an α-peptide/β-peptoid unit was efficiently prepared in solution. The dimer was designed as follows. The *N*-terminus of the building block was protected with the Fmoc group. Hence, other dimers could be incorporated on solid-phase allowing the synthesis of oligomers of the desired length (Scheme 2.10). The side chain was protected with the 2-(trimethylsilyl)-ethoxycarbonyl (Teoc) group, making on-resin derivatization possible. The feasibility of on-resin derivatization was tested by guanidinylating and acylating the side chain. This approach turns into an efficient method to prepare this kind of peptidomimetics with a good overall yield.

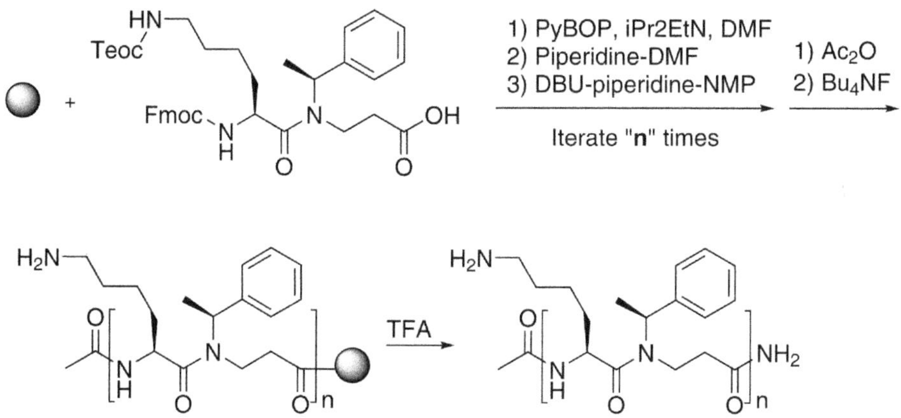

Scheme 2.10 Combined solid-phase/solution methodology to prepare α-peptide/β-peptoid chimeras.

Houghten et al. described a very useful protocol for the stepwise synthesis of a peptidomimetic library, in which all the amide functionalities within the peptide backbone were alkylated [72]. This turns into a facile and rapid approach to the preparation of a large library of peptidomimetics (over 57,500 compounds were prepared). A solid-phase approach was employed, in which each amide functionality was alkylated after every coupling step. In the literature, several methods for the alkylation of amino acids have been reported. Nonetheless, lithium *tert*-butoxide turned out to be the most efficient method to obtain the desired product. The alkylation reaction was successfully carried out with 50 different amino acids, including L- and D-amino acids. Unfortunately, alkylation was not successful on cysteine and histidine derivatives and significant side-reactions were observed.

2.2.2 AA-peptides

AA-peptides are peptidomimetics based on the chiral PNA backbone. They consist of γ-substituted-*N*-acylated-*N*-aminoethyl amino acids [62,73–75]. The repeated unit of AA-peptides is comparable to two adjacent residues of an α-peptide, in which one side chain is located on the α-carbon and the other is attached to the adjacent tertiary nitrogen as an amide moiety (Fig. 2.4). AA-peptides and α-peptides of the same length

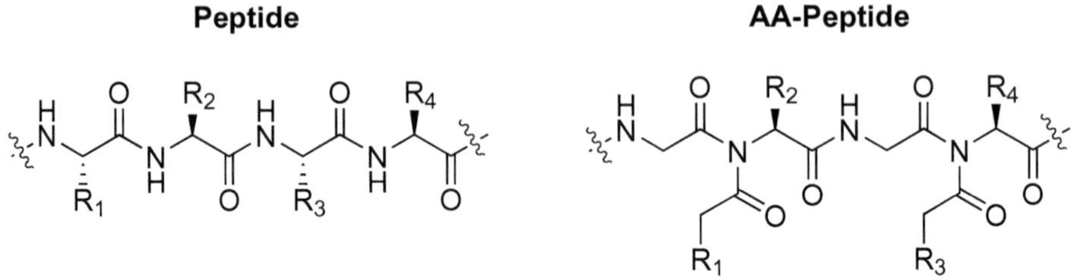

Figure 2.4 Peptide and AA-peptide structures.

project the same amount of side chains. Regarding their application in the biomedical field, AA-peptides exhibit improved pharmacological properties compared to their peptide analogues. As with most peptidomimetics, their hydrogen-bonding properties enhance their stability against proteases and they present a much more rigid structure due to the *N,N*-disubstituted amide bonds. According to the literature, some AA-peptides can cross the cell membrane [76]. Additionally, they have shown great selectivity toward the p53/MDM2 interaction [73].

Initially, the synthesis of AA-peptides was carried out on SPPS using a building block approach, in which AA-peptide monomers were prepared in solution prior to assembly on solid-phase. Although this procedure affords the desired oligomers in good yields, it is not practical, and the preparation of a library of compounds becomes tedious due to preparation of each building block in solution. Cai and coworkers described a submonomeric approach for the preparation of AA-peptides oligomers which involved Fmoc-amino aldehyde units [73]. Aldehydes are generally unstable and therefore the subunits had to be used shortly after preparation. Moreover, several steps were required for each monomer incorporation and therefore this approach was considered rather troublesome.

Recently, Cai et al. reported a synthetic methodology that combines both the building block and the submonomeric approach [77]. This new methodology circumvents the drawbacks of both methodologies. For the preparation of a library of peptidomimetics, just a few *N*-Alloc-AA-peptide building blocks need to be synthesized in solution. Unlike aldehyde subunits, *N*-alloc-AA-peptides are stable and large quantities can be prepared and stored. Incorporation of the *N*-Alloc-AA-peptide onto the polymeric support is followed by removal of the Alloc protecting group. Upon treatment with $Pd(PPh_3)_4$ and $Me_2NH \cdot BH_3$, the Alloc group is fully removed within the first 10 min. Finally, addition of a variety of carboxyl acids or acyl chlorides leads to acylation of the amine. This cycle is repeated "*n*" times until the target molecule is obtained (Scheme 2.11). A considerably large library of AA-peptides can be prepared with relatively little effort by employing this new procedure. Moreover, this combined approach has advantages over the previously described methodologies, such as reduced number of steps, shorter reaction times, and improved overall yield and purity.

Cai spent a considerable amount of time developing new AA-peptide surrogates that incorporate novel functionalities to the peptidomimetic backbone, including oligocarbamates, oligosulfonamides, and oligoureas. This new class of peptidomimetics has important applications in the biomedical field, and they are presented as promising drug candidates. These AA-peptide derivates are readily accessible, and can be prepared by employing the combined solid-phase/submonomeric approach methodology described earlier. *N*-alloc-AApeptide building blocks containing the corresponding backbone functionalities were prepared, and successfully assembled on solid-phase. This allows the preparation of large libraries of AA-peptidomimetics by simple acylation of the nitrogen moiety with a variety of commercially available sulfonyl chlorides, chloroformates, and isocyanates.

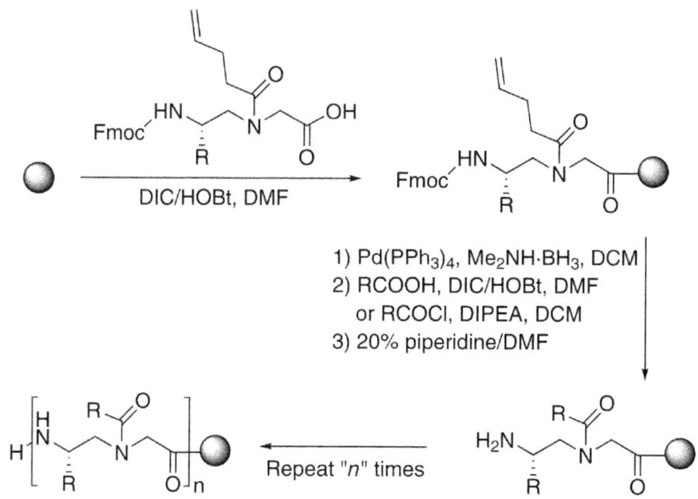

Scheme 2.11 General synthesis of AA-peptides combining a solid-phase and a submonomer methodology.

Cai et al. also explored lipidated AA-peptides, which are of great interest due to their potential as antimicrobial agents [76]. It is well known that the addition of a fatty acid to the peptide backbone enhances the antimicrobial activity of peptides toward Gram-positive and Gram-negative bacteria. Likewise, peptidomimetics holding a fatty acid chain have been reported to retain the antimicrobial activity with improved selectivity. Hence, the same combined synthetic approach was employed to obtain lipidated AA-peptides derivatives. The corresponding *N*-alloc-AA peptide subunits containing either amphiphilic or cationic residues were prepared in solution and assembled on solid-phase as previously reported. The *N*-terminus was alkylated on solid-phase with a C-16 hydrophobic tail.

2.2.3 Azapeptides

Azapeptides are a type of peptidomimetics in which the α-carbon amino acid residue is replaced by one or multiple semicarbazides (Fig. 2.5). Azapeptides exhibit metabolic stability and can induce the secondary structure of proteins, in specific

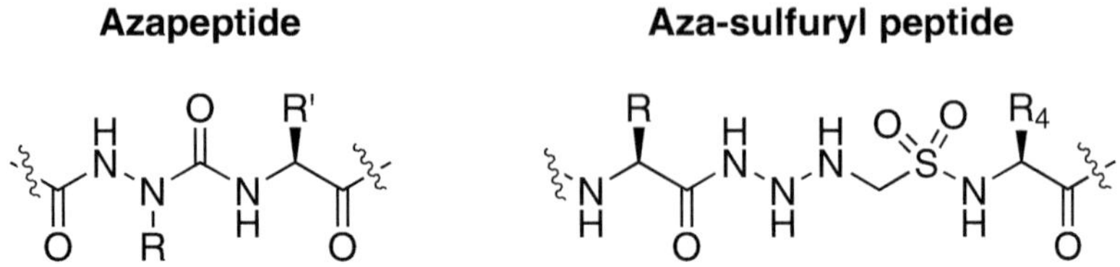

Figure 2.5 Azapeptide and aza-sulfuryl peptide structures.

they can favor the β-turn. Both linear and cyclic azapeptides are potential drug candidates and their application to the biomedical field is being explored [78,79].

Generally, a submonomer approach is carried out to prepare linear azapeptides and the diversity of the side chain is introduced at each aza-residue. Nevertheless, cyclic compounds generally exhibit greater biological activity. Lubell and coworkers explored a new approach, the so-called A3-macrocyclization, to synthesize cyclic azapeptides [78]. This novel method allows the side chain to side chain macrocyclization through the copper-catalyzed Mannich addition onto azapropargylglycine (azaPra) residues. It is an elegant and efficient manner of coupling an aldehyde, an alkyne, and an amine moiety (Scheme 2.12). Lubell pursued this multicomponent coupling to synthesize cyclic azapeptides for several reasons including: (1) aza-residues are known to favor β-turns, and therefore cyclization may be favored; (2) this approach allows the preparation of libraries of peptidomimetics; (3) the acetylene moiety and its reduction product derivatives can help to better understand the structure–activity relationship of this kind of compounds; and (4) the formed tertiary amine may improve bioavailability and biological activity.

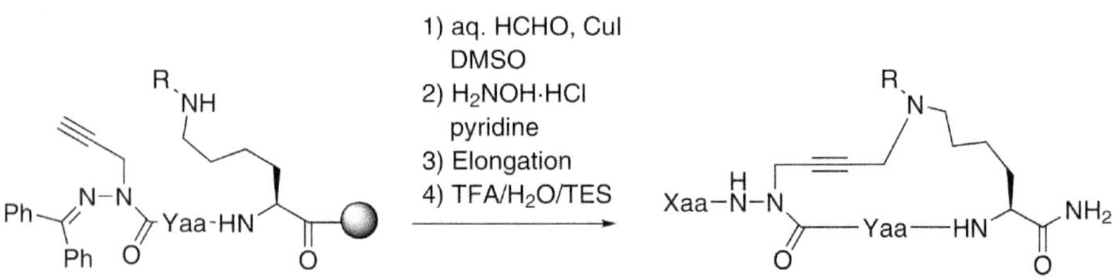

Scheme 2.12 Coupling of an aldehyde, an alkyne, and an amine moiety to prepare cyclic azapeptides.

Aza-sulfuryl peptide analogues are biologically interesting azapeptide surrogates, since they combine the properties of both azapeptides and R-sulfonamido peptides. The latter consist of the replacement of the CRH and the carbonyl group of an amino acid by a nitrogen and a sulfonyl group, respectively (Fig. 2.5) [80]. The sulfonyl group tetrahedral structure confers backbone geometry modification to aza-sulfuryl peptides, and therefore this new class of peptidomimetics emerge as potential drug candidates. Hitherto, the synthesis of large libraries of aza-sulfuryl peptidomimetics had been hampered by the lack of an efficient synthetic methodology.

Recently, Lubell reported a convergent synthesis that allows the preparation of this peptidomimetics. *N*-aminosulfamides can be synthesized by coupling of *p*-nitrophenylsulfamidates and *N*-(Boc)-aminoacid hydrazides under microwave irradiation. Next, chemoselective alkylation of the aza-sulfurylglycinyl peptides was used to add side-chain diversity and prepare other azasulfurylamino acid residues.

2.2.4 Urea-based peptidomimetics

Urea-based peptidomimetics arise as a new class of foldamers in which the α-carbon of the amino acid residues is substituted by a nitrogen moiety. Aliphatic *N,N′*-linked oligoureas are comprised within the family of foldamers, and exhibit an α-helix secondary structure [81]. Oligoureas are prepared through a fragment condensation approach that allows the synthesis of long oligourea peptidomimetic sequences (Scheme 2.13) [82]. Oligoureas can be synthesized in solution using (*S*)-succinimidyl-{2-[(tert-butoxycarbonyl)amino]-3-alkyl}carbamate units.

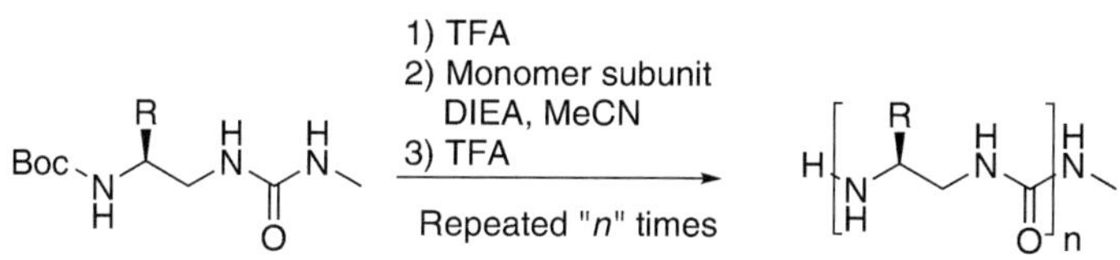

Scheme 2.13 General protocol to prepare urea-based peptidomimetics through a fragment condensation approach.

Guichard and coworkers explored to what extent the helical conformation of short aliphatic peptidomimetic oligoureas was affected by the incorporation of *N*-(pyrrolidine-2-ylmethylamine) units at certain positions within the peptide backbone of aliphatic *N,N′*-disubstituted urea oligomers. It was found that insertion of these subunits did not interfere with the helical conformation of the oligomers [83]. Additionally, Guichard et al. reported the synthesis of aliphatic oligoureas containing multiple adjacent *N*-alkylated units derived from proline [84]. A library of oligoureas containing one to seven adjacent pyrrolidine residues was prepared. Structural studies suggested that the helical conformation is compromised by the length of the repeat pyrrolidine units.

2.2.5 Nonpeptide-based peptidomimetics

Another strategy to target PPIs is the development of nonpeptide-based peptidomimetics [58,60,85,86]. Generally, the chemical structure of these peptidomimetics has no similarity to that of peptides but can mimic the most prevalent secondary structures found in nature such as α-helix, β-strands, and β-sheets. Such peptidomimetics have the same binding mode as their native protein but present improved

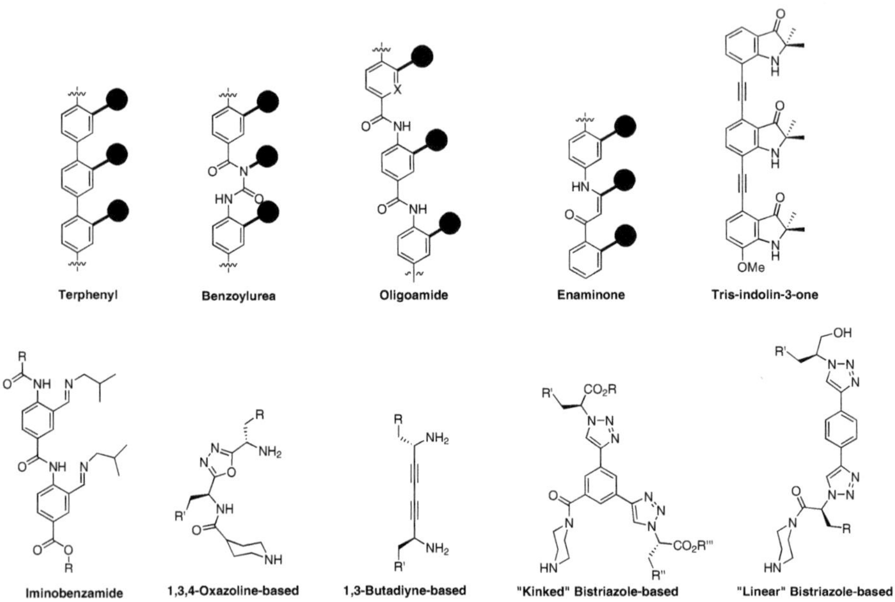

Figure 2.6 Structures of nonpeptide-based peptidomimetic scaffolds.

pharmacological properties (better bioavailability, metabolic stability, and ability to permeate cell membranes).

Several scaffolds have been developed, including: terphenyls, benzoylureas, oligobenzamides oligophenylenaminones, and imidazolidin-2-ones, among others (Fig. 2.6). In 2001, Orner introduced the terphenyl scaffold which can resemble the α-helix secondary structure and projects the important residues on a single face. The designs of many other scaffolds have been inspired by the terphenyl scaffold.

Hamilton et al. explored the benzoylurea scaffold which can be readily prepared. Moreover, the synthesis of molecular constructs of various lengths is possible. Benzoylurea-based peptidomimetics adopt a helical structure and project the important residues on a single face of the helix [86]. Following this work, Hamilton and coworkers developed a surrogate scaffold that incorporated enaminones within the backbone, namely an oligophenylenaminones-based scaffold [87]. Oligophenylenaminones exhibit α-helix mimicry. The best approach to prepare these molecular constructs was evaluated, and conjugation of an aniline to an ynone turned out to be the most efficient methodology. A series of i, $i+4$, $i+7$ peptidomimetics was constructed.

Many scaffolds that induce β-sheet structures between two attached peptides have been reported such as epindolidiones, dibenzofurans, metallopeptides, and oligoureas. However, Smith introduced for the first time a nonpeptidic scaffold consisting of oligopyrrolinone units that resembled β-sheet secondary structures [88]. In 2009 a 2,2-disubstituted-indolin-3-one scaffold was introduced by Hamilton, which links the 2,2-disubstituted-indolin-3-one groups through their

4,7-positions by an alkyne spacer [89]. The β-strand structure is promoted by the formation of hydrogen bonds between the carbonyl moiety of one indolinone and the amine group of the other. In this case, all the substituents are located on the same face of the molecular construct. A monomeric approach is employed to synthesize the scaffold.

Burgess directed his research at the development of universal peptidomimetics [89,90]. Universal peptidomimetics are described as small sets of scaffolds designed to analogue local pairs of amino acids (including noncontiguous ones) in any secondary structure. These scaffolds are very promising compounds and are considered a powerful tool for targets where exact binding conformations are unknown, including the following situations: (1) when the conformation of the key residues for the PPIs is unknown; (2) when only one of the two proteins involved in the interaction has been characterized; and (3) when cellular crystal structures are different than the ones in solution. Examples of relevant universal peptidomimetic scaffolds developed by Burgess are the 1,3,4-oxazoline-based mimic, the 1,3-butadiyne-based mimic, the "kinked" bistriazole-based mimic, and the "linear" bistriazole-based mimic [91]. These small heterocyclic structures mimic many pairs in amino acid conformation that resemble all common secondary structures. Moreover, they can be tuned with a variety of amino acid side chains to enhance their biological activity.

Acknowledgments

This work was partially supported by CICYT (CTQ2015-68677-R). J.T.-P. acknowledges a Ramon y Cajal contract (MINECO).

References

[1] J.S. Davies, The cyclization of peptides and depsipeptides, J. Pept. Sci. 9 (2003) 471–501.

[2] M. Dreyfuss, E. Harri, H. Hofmann, H. Kobel, W. Pache, H. Tscherter, Cyclosporin A and C: new metabolites from trichoderma polysporum, Eur. J. Appl. Microbiol. 3 (1976) 125–133.

[3] R.M. Wenger, Synthesis of cyclosporine. Total syntheses of 'cyclosporin A' and 'cyclosporin H', two fungal metabolites isolated from the species *Tolypocladium inflatum* GAMS, Helv. Chim. Acta 67 (1984) 502–525.

[4] Y.M. Angell, T.L. Thomas, G.R. Flentke, D.H. Rich, Solid-phase synthesis of cyclosporin peptides, J. Am. Chem. Soc. 117 (1995) 7279–7280.

[5] X. Wu, J.L. Stockdill, P. Wang, S.J. Danishefsky, Total synthesis of cyclosporine: access to *N*-methylated peptides via isonitrile coupling reactions, J. Am. Chem. Soc. 132 (2010) 4098–4100.

[6] M. Gongora-Benítez, J. Tulla-Puche, F. Albericio, Multifaceted roles of disulfide bridges. Peptides as therapeutics, Chem. Rev. 114 (2014) 901–926.

[7] S. Cheek, S.S. Krishna, N.V. Grishin, Structural classification of small, disulfide-rich protein domains, J. Mol. Biol. 359 (2006) 215–237.
[8] R. Halai, D.J. Craik, Conotoxins: natural product drug leads, Nat. Prod. Rep. 26 (2009) 526–536.
[9] K.B. Akondi, M. Muttenthaler, S. Dutertre, Q. Kaas, et al., Discovery, synthesis, and structure-activity relationships of conotoxins, Chem. Rev. 114 (2014) 5815–5847.
[10] N.L. Daly, D.J. Craik, Structural studies of conotoxins, IUBMB Life 61 (2009) 144–150.
[11] M.C. Munson, G. Barany, Synthesis of a-conotoxin SI, a bicyclic tridecapeptide amide with two disulfide bridges: illustration of novel protection schemes and oxidation strategies, J. Am. Chem. Soc. 115 (1993) 10203–10210.
[12] E. Atherton, R.C. Sheppard, P. Ward, Peptide synthesis. Part 7. Solid-phase synthesis of conotoxin G1, peptide synthesis. Part 7. Solid-phase synthesis of conotoxin G1, J. Chem. Soc. Perkin Trans. 1 (10) (1985) 2065–2073.
[13] D.J. Craik, M. Cěmažar, C.K. Wang, N.L. Daly, The cyclotide family of circular mini-proteins: nature's combinatorial peptide template, Biopolymers 84 (2006) 250–266.
[14] D.J. Craik, J.E. Swedberg, J.S. Mylne, M. Cemazar, Cyclotides as a basis for drug design,, Expert Opin. Drug Discov. 7 (2012) 179–194.
[15] J. Weidmann, D.J. Craik, Discovery, structure, function, and applications of cyclotides: circular proteins from plants, J. Exp. Bot. 67 (2016) 4801–4812.
[16] P.E. Dawson, T.W. Muir, I. Clark-Lewis, S.B. Kent, Synthesis of proteins by native chemical ligation, Science 266 (1994) 776–779.
[17] N.L. Daly, S. Love, P.F. Alewood, D.J. Craik, Chemical synthesis and folding pathways of large cyclic polypeptides: studies of the cystine knot polypeptide kalata B1, Biochemistry 38 (1999) 10606–10614.
[18] M. Akcan, D.J. Craik, Synthesis of cyclic disulfide-rich peptides, Methods Mol. Biol. 1047 (2013) 89–101.
[19] J.B. Blanco-Canosa, P.E. Dawson, An efficient Fmoc-SPPS approach for the generation of thioester peptide precursors for use in native chemical ligation, Angew. Chem. Int. Ed. 47 (2008) 6851–6855.
[20] H. Qu, B.J. Smithies, T. Durek, D.J. Craik, Synthesis and protein engineering applications of cyclotides, Aust. J. Chem. 70 (2017) 152–161.
[21] G.S.B. Andavan, R. Lemmens-Gruber, Cyclodepsipeptides from marine sponges: natural agents for drug research, Mar. Drugs 8 (2010) 810–834.
[22] S. Sivanathan, J. Scherkenbeck, Cyclodepsipeptides: a rich source of biologically active compounds for drug research, Molecules 19 (2014) 12368–12420.
[23] H. Ueda, H. Nakajima, Y. Hori, T. Fujita, M. Nishimura, T. Goto, et al., FR901228, a novel antitumor bicyclic depsipeptide produced by *Chromobacterium violaceum* No. 968. I. Taxonomy, fermentation, isolation, physico-chemical and biological properties, and antitumor activity, J. Antibiot. 47 (1994) 301–310.
[24] K.W. Li, J. Wu, W. Xing, J.A. Simon, Total synthesis of the antitumor depsipeptide FR-901,228, J. Am. Chem. Soc. 118 (1996) 7237–7238.
[25] T.J. Greshock, D.M. Johns, Y. Noguchi, R.M. Williams, Improved total synthesis of the potent HDAC inhibitor FK228 (FR-901228), Org. Lett. 10 (2008) 613–616.
[26] S. Wen, G. Packham, A. Ganesan, Macrolactamization versus macrolactonization: total synthesis of FK228, the depsipeptide histone deacetylase inhibitor, J. Org. Chem. 73 (2008) 9353–9361.
[27] M. Pelay-Gimeno, J. Tulla-Puche, F. Albericio, Head-to-side-chain" cyclodepsipeptides of marine origin, Mar. Drugs 11 (2013) 1693–1717.

[28] M. Pelay-Gimeno, Y. Garcia-Ramos, M.J. Martin, et al., The first total synthesis of the cyclodepsipeptide pipecolidepsin A, Nat. Commun. 2352 (2013). Available from: http://dx.doi.org/10.1038/ncomms3352.

[29] M. Pelay-Gimeno, F. Albericio, J. Tulla-Puche, Synthesis of complex head-to-side-chain cyclodepsipeptides, Nat. Protoc. 11 (2016) 1924–1947.

[30] M.J. Martin, R. Rodríguez-Acebes, Y. Garcia-Martin, et al., Stellatolides, a new cyclodepsipeptide family from the sponge *Ecionemia acervus*: isolation, solid-phase total synthesis, and full structural assignment of stellatolide A, J. Am. Chem. Soc. 136 (2014) 6754–6762.

[31] C. Chatterjee, M. Paul, L.L. Xie, W.A. van der Donk, Biosynthesis and mode of action of lantibiotics, Chem. Rev. 105 (2005) 633–683.

[32] J. Escano, L. Smith, Multipronged approach for engineering novel peptide analogues of existing lantibiotics, Expert Opin. Drug Discov 10 (2015) 857–870.

[33] K. Fukase, M. Kitazawa, A. Sano, K. Shimbo, et al., Total synthesis of peptide antibiotic nisin, Tetrahedr. Lett. 29 (1988) 795–798.

[34] A.B. Tabor, The challenge of the lantibiotics: synthetic approaches to thioether-bridged peptides, Org. Biomol. Chem. 9 (2011) 7606–7628, and references therein.

[35] V.R. Pattabiraman, S.M. McKinnie, J.C. Vederas, Solid-supported synthesis and biological evaluation of the lantibiotic peptide bis(desmethyl) lacticin 3147 A2, Angew. Chem. Int. Ed. 47 (2008) 9472–9475.

[36] A.C. Ross, H. Liu, V.R. Pattabiraman, J.C. Vederas, Synthesis of the lantibiotic lactocin S using peptide cyclizations on solid phase, J. Am. Chem. Soc. 132 (2009) 462–463.

[37] W. Liu, A.S.H. Chan, H. Liu, S.A. Cochrane, J.C. Vederas, Solid supported chemical syntheses of both components of the lantibiotic lacticin 3147, J. Am. Chem. Soc. 133 (2011) 14216–14219.

[38] P.J. Knerr, W.A. van der Donk, Chemical synthesis of the lantibiotic lacticin 481 reveals the importance of lanthionine stereochemistry, J. Am. Chem. Soc. 135 (2013) 7094–7097.

[39] V.R. Pattabiraman, J.L. Stymiest, D.J. Derksen, N.I. Martin, J.C. Vederas, Multiple on-resin olefin metathesis to form ring-expanded analogues of the lantibiotic peptide, lacticin 3147 A2, Org. Lett. 9 (2007) 699–702.

[40] A.C. Ross, S.M. McKinnie, J.C. Vederas, The synthesis of active and stable diaminopimelate analogues of the lantibiotic peptide lactocin S, J. Am. Chem. Soc. 134 (2012) 2008–2011.

[41] P.J. Knerr, W.A. van der Donk, Chemical synthesis and biological activity of analogues of the lantibiotic epilancin 15X, J. Am. Chem. Soc. 134 (2012) 7648–7651.

[42] A.B. Tabor, Recent advances in synthetic analogues of lantibiotics: what can we learn from these? Bioorg. Chem. 55 (2014) 39–50.

[43] T. Koopmans, T.M. Wood, P. Hart, L.H. Kleijn, A.P. Hendrickx, R.J. Willems, et al., Semisynthetic lipopeptides derived from nisin display antibacterial activity and lipid II binding on par with that of the parent compound, J. Am. Chem. Soc. 137 (2015) 9382–9389.

[44] A.L. Jochim, P.S. Arora, Assessment of helical interfaces in protein–protein interactions, Mol. Biosyst. 5 (2009) 924–926.

[45] L.K. Henchey, A.L. Jochim, et al., Contemporary strategies for the stabilization of peptides in the alpha-helical conformation, Curr. Opin. Chem. Biol. 12 (2008) 692–697.

[46] C.E. Schafmeister, J. Po, G.L. Verdine, An all-hydrocarbon cross-linking system for enhancing the helicity and metabolic stability of peptides, J. Am. Chem. Soc. 122 (2000) 5891–5892.

[47] G.L. Verdine, G.J. Hilinski, "Stapled peptides for intracellular drug targets,", Methods Enzymol. 503 (2012) 3–33, and references therein.
[48] A.M. Felix, E.P. Heimer, C.-T. Wang, T.J. Lambros, et al., Synthesis, biological activity and conformational analysis of cyclic GRF analogs, Int. J. Pept. Protein Res. 32 (1988) 441–454.
[49] D.Y. Jackson, D.S. King, J. Chmielewski, S. Singh, P.G. Schultz, General approach to the synthesis of short .alpha.-helical peptides, J. Am. Chem. Soc. 113 (1991) 9391–9392.
[50] Y.H. Lau, P. de Andrade, Y. Wu, D.R. Spring, Peptide stapling techniques based on different macrocyclisation chemistries, Chem. Soc. Rev. 44 (2015) 91–102.
[51] M. Scrima, A. Le Chevalier-Isaad, P. Rovero, A.M. Papini, M. Chorev, A.M. D'Ursi, CuI-catalyzed azide-alkyne intramolecular i-to-(i + 4) side-chain-to-side-chain cyclization promotes the formation of helix-like secondary structures, Eur. J. Org. Chem. 2010 (2010) 446–457.
[52] F.M. Brunel, P.E. Dawson, Synthesis of constrained helical peptides by thioether ligation: application to analogs of gp41, Chem. Commun. 20 (2005) 2552–2554.
[53] C.M. Haney, M.T. Loch, W.S. Horne, Promoting peptide α-helix formation with dynamic covalent oxime side-chain cross-links, Chem. Commun. 47 (2011) 10915–10917.
[54] Y. Wang, D.H.C. Chou, A Thiol-Ene coupling approach to native peptide stapling and macrocyclization, Angew. Chem. Int. Ed. 54 (2015) 10931–10934.
[55] D.P. Fairlie, A. Dantas de Araujo, Stapling peptides using cysteine crosslinking, Biopolym. (Peptide Sci.) 106 (2016) 843–853, and references therein.
[56] K. Estieu-Gionnet, G. Guichard, Stabilized helical peptides: overview of the technologies and therapeutic promises, Expert Opin. Drug Discov 6 (2011) 937–963.
[57] N.T. Ross, W.P. Katt, A.D. Hamilton, Synthetic mimetics of protein secondary structure domains, Philos Trans. A Math. Phys. Eng. Sci. 368 (2010) 989–1008.
[58] M.K.P. Jayatunga, S. Thompson, A.D. Hamilton, α-Helix mimetics: outwards and upwards, Bioorgan. Med. Chem. Lett. 24 (2014) 717–724.
[59] Y.L. Angell, K. Burgess, Peptidomimetics via copper-catalyzed azide-alkyne cycloadditions, Chem. Soc. Rev. 36 (2007) 1674–1689.
[60] O.V. Kulikov, S. Thompson, H. Xu, et al., Design and synthesis of oligoamide-based double alpha-helix mimetics, Eur. J Org. Chem. 2013 (2013) 3433–3445.
[61] B. Yoo, S.B.Y. Shin, M.L. Huang, K. Kirshenbaum, Peptoid macrocycles: making the rounds with peptidomimetic oligomers, Chem. Eur. J. 16 (2010) 5527–5537.
[62] Y. Niu, Y. Hu, R.E. Wang, et al., AApeptides as a new class of peptidomimetics to regulate protein–protein interactions, in: J. Cai, R.E. Wang (Eds.), Protein Interactions, InTech, Rijeka, 2012, pp. 155–170. ISBN: 978-953-51-0244-1.
[63] R.N. Zuckermann, J.M. Kerr, S.B.H. Kent, et al., Efficient method for the preparation of peptoids [oligo(N-substituted glycines)] by submonomer solid-phase synthesis, J. Am. Chem. Soc. 114 (1992) 10646–10647.
[64] B. Yoo, K. Kirshenbaum, Peptoid architectures: elaboration, actuation, and application, Curr. Opin. Chem. Biol. 12 (2008) 714–721.
[65] K. Andreev, M.W. Martynowycz, A. Ivankin, et al., Cyclization improves membrane permeation by antimicrobial peptoids, Langmuir 2016 (2016) 12905–12913.
[66] P.M. Levine, K. Imberg, M.J. Garabedian, K.K. Kirshenbaum, Multivalent peptidomimetic conjugates: a versatile platform for modulating androgen receptor activity, J. Am. Chem. Soc. 134 (2012) 6912–6915.
[67] H. Jang, A. Fafarman, J.M. Holub, K. Kirshenbaum, Click to fit: versatile polyvalent display on a peptidomimetic scaffold, Org. Lett. 7 (2005) 1951–1954.

[68] C.A. Olsen, Beta-peptoid "foldamers"—why the additional methylene unit? Biopolymers 96 (2011) 561–566.
[69] J.S. Laursen, J. Engel-Andreasen, C.A. Olsen, Beta-peptoid foldamers at last, Acc. Chem. Res. 48 (2015) 2696–2704.
[70] P.M. Levine, T.W. Craven, R. Bonneau, K. Kirshenbaum, Chemoselective fragment condensation between peptide and peptidomimetic oligomers, Org. Biomol. Chem. 11 (2013) 4142–4146.
[71] C.A. Olsen, G. Bonke, L. Vedel, et al., Alpha-Peptide/ beta-peptoid chimeras, Org. Lett. (2007) 1549–1552.
[72] B. Dörner, G.M. Husar, J.M. Ostresh, R.A. Houghten, The synthesis of peptidomimetic combinatorial libraries through successive amide alkylations, Bioorg. Med. Chem. 4 (1996) 709–715.
[73] Y. Hu, X. Li, S.M. Sebti, J. Chen, J. Cai, Design and synthesis of AApeptides: a new class of peptide mimics, Bioorg. Med. Chem. Lett. 21 (2011) 1469–1471.
[74] Y. Shi, P. Teng, P. Sang, F. She, L. Wei, J. Cai, γ-AApeptides: design, structure, and applications, Acc. Chem. Res. 49 (2016) 428–441.
[75] H. Wu, Q. Qiao, P. Teng, et al., New class of heterogeneous helical peptidomimetics, Org. Lett. 17 (2015) 3524–3527.
[76] Y. Hu, M.N. Amin, S. Padhee, et al., Lipidated peptidomimetics with improved antimicrobial activity, ACS Med. Chem. Lett. 3 (2012) 683–686.
[77] H. Wu, P. Teng, J. Cai, Rapid access to multiple classes of peptidomimetics from common γ-AApeptide building blocks, Eur. J. Org. Chem. 2014 (2014) 1760–1765.
[78] J. Zhang, M. Mulumba, H. Ong, W.D. Lubell, Diversity-oriented synthesis of cyclic azapeptides by A3-macrocyclization provides high-affinity CD36-modulating peptidomimetics, Angew. Chem. Int Ed. 56 (2017) 1–6.
[79] T.A. Godina, W.D. Lubell, Mimics of peptide turn backbone and side-chain geometry by a general approach for modifying azabicyclo[5.3.0]alkanone amino acids, J. Org. Chem. 76 (2011) 5846–5849.
[80] S. Turcotte, S.H. Bouayad-Gervais, W.D. Lubell, N-Aminosulfamide peptide mimic synthesis by alkylation of aza-sulfurylglycinyl peptides, Org. Lett. 14 (2012) 1318–1321.
[81] C. Hemmerlin, M. Marraud, D. Rognan, et al., Helix-forming oligoureas: temperature-dependent NMR, structure determination, and circular dichroism of a nonamer with functionalized side chains, Helv. Chim. Acta 85 (2002) 3692–3711.
[82] N. Pendem, C. Douat, P. Claudon, et al., Helix-forming propensity of aliphatic urea oligomers incorporating noncanonical residue substitution patterns, J. Am. Chem. Soc. 135 (2013) 4884–4892.
[83] N. Pendem, Y.R. Nelli, C. Douat, et al., Controlling helix formation in the gamma-peptide superfamily: heterogeneous foldamers with urea/amide and urea/carbamate backbones, Angew. Chem. Int. Ed. 52 (2013) 4147–4151.
[84] J. Fremaux, B. Kauffmann, G. Guichard, Synthesis and folding propensity of aliphatic oligoureas containing repeats of proline-type units, J. Org. Chem. 79 (2014) 5494–5502.
[85] I. Saraogi, A.D. Hamilton, P. Street, P.O. Box, N. Haven, α-Helix mimetics as inhibitors of protein–protein interactions, Biochem. Soc. Trans. 36 (2008) 1414–1417.
[86] T. Yamashita, P.C. Knipe, N. Busschaert, S. Thompson, A.D. Hamilton, A modular synthesis of conformationally preorganised extended beta-strand peptidomimetics, Chem. Eur. J. 21 (2015) 14699–14702.
[87] M.J. Adler, A.D. Hamilton, Oligophenylenaminones as scaffolds for alpha-helix mimicry, J. Org. Chem. 76 (2011) 7040–7047.

[88] A.B. Smith, M.C. Guzman, P.A. Sprengeler, T.P. Keenan, R.C. Holcomb, J.L. Wood, et al., De novo design, synthesis, and X-ray crystal structures of pyrrolinone-based b-strand peptidomimetics, J. Am. Chem. Soc 116 (1994) 9947–9962.
[89] P.N. Wyrembak, A.D. Hamilton, Alkyne-linked 2,2-disubstituted-indolin-3-one oligomers as extended Beta-strand mimetics, J. Am. Chem. Soc. 131 (2009) 4566–4567.
[90] E. Ko, J. Liu, K. Burgess, Minimalist and universal peptidomimetics, Chem. Soc. Rev. 40 (2011) 4411–4421.
[91] E. Ko, J. Liu, L.M. Perez, G. Lu, A. Schaefer, K. Burgess, Universal peptidomimetics, J. Am. Chem. Soc. 133 (2011) 462–477.
[92] J. Lee, J.N. Currano, P.J. Carroll, M.M. Joullie, Didemnins, tamandarins and related natural products, Nat. Prod. Rep. 29 (2012) 404–424.
[93] M. Góngora-Benítez, J. Tulla-Puche, M. Paradís-Bas, O. Werbitzky, M. Giraud, F. Albericio, Optimized Fmoc solid-phase synthesis of the cysteine-rich peptide linaclotide, Biopolymers 96 (2011) 69–80.
[94] D.G. Parkes, K.F. Mace, M.E. Trautmann, Discovery and development of exenatide: the first antidiabetic agent to leverage the multiple benefitsof the incretin hormone, GLP-1, Expert Opin. Drug Discov. 8 (2013) 219–244.
[95] S. Alagarsamy, R. Galyean, H. Tariga, et al., New, potent, and selective peptidic oxytocin receptor agonists, J. Med. Chem. 57 (2014) 5306–5317.
[96] M. Manning, A. Misicka, A. Olma, et al., Oxytocin and vasopressin agonists and antagonists as research tools and potential therapeutics neuroendocrinology, J. Neuroendocrinol. 24 (2012) 609–628.
[97] W.K. Nevala, C.B. Uhl, J. Allred, E.T. Creagan, A. Phase II, Study of ABT-510 (thrombospondin-1 analog) for the treatment of metastatic melanoma, Am. J. Clin. Oncol. 30 (2007) 303–309.
[98] A. Rusk, E. Mckeegan, F. Haviv, S. Majest, J. Henkin, C. Khanna, Cancer therapy: preclinical preclinical evaluation of antiangiogenic thrombospondin-1 peptide mimetics, ABT-526 and ABT-510, in companion dogs with naturally occurring cancers, Clin. Cancer Res. 12 (2006) 7444–7456.
[99] J.A. Gietema, R. Hoekstra, F.Y.F.L. De Vos, et al., A phase I study assessing the safety and pharmacokinetics of the thrombospondin-1-mimetic angiogenesis inhibitor ABT-510 with gemcitabine and cisplatin in patients with solid tumors, Ann. Oncol. 17 (2006) 1320–1327.
[100] E. Matsumoto, Y. Fujita, Y. Okada, E.I. Kauppinen, Hydrophobic benzyl amines as supports for liquid-phase C-terminal amidated peptide synthesis: application to the preparation of ABT-510, J. Pept. Sci. 21 (2015) 691–695.
[101] E. Bombardieri, V. Ambrosini, C. Aktolun, et al., In-pentetreotide scintigraphy: procedure guidelines for tumour imaging, Eur. J. Nucl. Med. Mol. Imaging 37 (2010) 1441–1448.
[102] J.R. Buscombe, M.E. Caplin, A.J.W. Hilson, Long-term efficacy of high-activity 111 in-pentetreotide therapy in patients with disseminated neuroendocrine tumors, J. Nucl. Med. 44 (2003) 1–7.
[103] C. Fang, L. Fang, D. Yanrong, et al., Preparation and experimental study of ^{111} in-pentetreotide, Chin. J. Nucl. Med. 19 (1999) 149–151.
[104] K.F. Croom, S. Dhillon, S.J. Keam, Atazanavir A review of its use in the management of HIV-1 infection, Drugs 69 (2009) 1107–1140.
[105] A.J. Busti, D. Pharm, R.G.H. Li, D. Pharm, D.M. Margolis, Atazanavir for the treatment of human immunodeficiency virus infection, Pharmacotherapy 24 (2004) 1732–1747.

[106] A. Fassler, G. Bold, H.-G. Capraro, et al., Antivirally active heterocyclic azahexane derivatives, United States Patent 5,849,911 (1998).
[107] K.A. Gelmon, K. Belanger, D. Soulieres, et al., A phase I study of T900607 given once every 3 weeks in patients with advanced refractory cancers; National Cancer Institute of Canada Clinical Trials Group (NCIC - CTG) IND 130*, Invest. New Drugs 23 (2005) 445–453.
[108] P.J. Pockros, E.R. Schiff, M.L. Shiffman, et al., Oral IDN-6556, an antiapoptotic caspase inhibitor, may lower aminotransferase activity in patients with chronic hepatitis C, Hepatology 46 (2007) 324–329.
[109] F.J. Barreyro, S. Holod, P.V. Finocchietto, A.M. Camino, J.B. Aquino, The pan-caspase inhibitor Emricasan (IDN-6556) decreases liver injury and fibrosis in a murine model of non-alcoholic steatohepatitis, Liver Int. 35 (2015) 953–966.
[110] Y. Minami, H. Minami, T. Miyamoto, et al., Phase I study of glasdegib (PF-04449913), an oral smoothened inhibitor, in Japanese patients with select hematologic malignancies, Cancer Sci. 11 (2017) 1–6.
[111] P. Chaudhry, M. Singh, T.J. Triche, M. Guzman, A.A. Merchant, GLI3 repressor determines Hedgehog pathway activation and is required for response to SMO antagonist glasdegib in AML, Blood 129 (2017) 3465–3476.
[112] J.L. Lam, A. Vaz, B. Hee, et al., Metabolism, excretion and pharmacokinetics of [C] glasdegib (PF-04449913) in healthy volunteers following oral administration Metabolism, excretion and pharmacokinetics of [14 C] glasdegib, Xenobiotica (2017) 1–13, Early Online.

Applications of de novo designed peptides

3

Aimee L. Boyle
Leiden Institute of Chemistry, Leiden, The Netherlands

3.1 Introduction

De novo designed peptides are increasingly utilized in a large number of fields, especially biomedicine and bionanotechnology [1–6]. This is for numerous reasons; arguably the most important is that there is now an increased understanding of the rules that govern peptide and protein assembly, making de novo peptide design more reliable. Related to this, large proteins have been extensively studied and functional regions identified meaning small, structurally similar peptides that mimic these regions are now common design targets. Peptides can also be programmed to self-assemble, via the formation of noncovalent interactions, into defined supramolecular assemblies that can be designed to fulfill a specific function. In addition, recent developments in solid-phase peptide synthesis now allow for 60–70 amino acid peptides to be synthesized in a matter of hours [7]. Peptides are also intrinsically biocompatible, making them suitable for a variety of biological applications, an area which has been the focus of extensive research in recent years.

In the first part of this chapter, the characteristics and design principles of numerous peptide building blocks including α-helices, β-strands, and collagens are briefly discussed. The latter part of this chapter presents, in detail, examples of how such building blocks have been designed and programmed to fulfill specific functions. All the examples presented in this chapter are summarized in Table 3.6.

3.2 Peptide building blocks

3.2.1 α-Helices, helical bundles, and coiled coils

3.2.1.1 α-Helices

The α-helix is a common element of protein secondary structure, formed when amino acids "wind up" to form a right-handed helix where the side-chains point out from the central coil (Fig. 3.1A,B). An α-helix has 3.6 residues per turn, meaning amino acid side chains that are three or four residues apart are bought together in space and so α-helices are stabilized by hydrogen bond formation between the

Peptide Applications in Biomedicine, Biotechnology and Bioengineering.
DOI: http://dx.doi.org/10.1016/B978-0-08-100736-5.00003-X

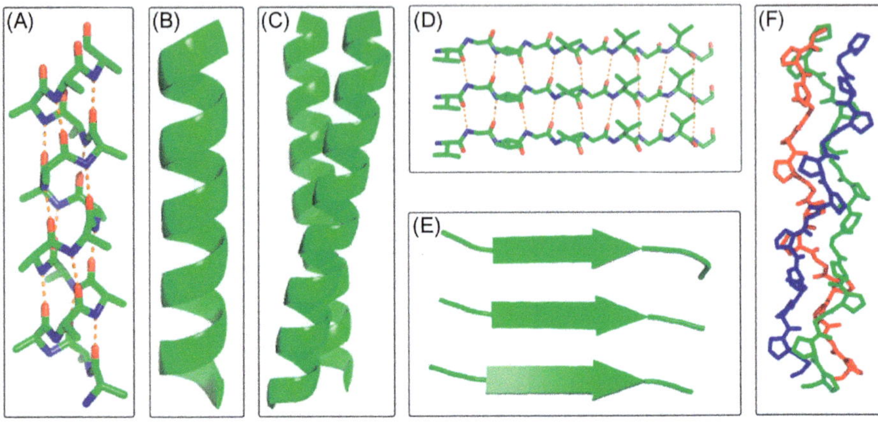

Figure 3.1 Structures of common peptide folding motifs. (A) Molecular view of an α-helix with intramolecular hydrogen bonds in orange; (B) cartoon representation of an α-helix; (C) cartoon representation of a dimeric coiled coil; (D) molecular representation of three β-strands forming a β-sheet, intermolecular hydrogen bonds are orange; (E) cartoon representation of three parallel β-strands and; (F) homotypic collagen triple helix with strands colored differently for clarity.
Source: The image in (C) is adapted from PDB file 4DZM and in (F) from 1K6F.

carbonyl oxygen of one amino acid, and the amide proton of another amino acid four residues further along the peptide chain (Fig. 3.1A).

3.2.1.2 Coiled coils

Coiled coils are formed when two or more α-helices self-assemble by winding around each other to form a left-handed supercoil (Fig. 3.1C). Although dimers, trimers, and tetramers are the most common structures, larger coiled-coils of up to seven helices can now be prescriptively designed [8].

The defining characteristic of a coiled coil is a specific packing regime known as knobs-into-holes (KIH) packing [9]; the side-chains of hydrophobic residues act as "knobs" and pack into "holes" formed by four residues from a neighboring helix. The majority of coiled coils also possess a repetitive sequence pattern, known as a heptad repeat. This comprises seven amino acids denoted *abcdefg*, where hydrophobic residues fall at the "*a*" and "*d*" positions, giving rise to a (*HPPHPPP*)n pattern [10]. Placing hydrophobic amino acids three and four residues apart means they end up on the same face of the helix, and burial of this hydrophobic face is the driving force for coiled-coil formation.

3.2.1.3 Helical bundles

Helical bundles are groups of α-helices that do not fulfill the packing and sequence criteria of coiled-coils. Additionally, the helices in a helical bundle are often not completely aligned [11,12]. This structural flexibility is important as it can allow for the formation of cavities capable of binding small molecules, for example, cytochromes are four-helix bundles that have a heme group bound. Such four-helix bundles are common design targets and many have now emulated natural proteins and demonstrated redox- and oxygen-carrying functionalities [13–15].

3.2.2 β-Strands and sheets

3.2.2.1 β-Strands

The β-strand is another common secondary structural element. For most designed β-strands, the primary structure has an alternating pattern of hydrophobic (*H*) and polar (*P*) amino acids [16,17]. This leads to an extended structure where the amino acid side-chains alternate between the two faces of the strand, extending into space at an angle of approximately 90 degrees from the face of the strand (Fig. 3.1D).

3.2.2.2 β-Sheets

β-Sheets are formed when several β-strands self-assemble, and are stabilized by interstrand hydrogen bonding, leading to the formation of extended amphipathic sheets in which hydrophobic side-chains point in one direction and polar side-chains in the other (Fig. 3.1D,E). Like coiled coils, β-sheets can have parallel, antiparallel, or mixed arrangements of the individual strands, although most natural sheets in proteins tend to be antiparallel. Apart from the repetitive sequence pattern, there are few rules governing the formation of β-strands and sheets, although it is known that amino acids with β-branched side-chains, e.g., Val (V), Thr (T), and Ile (I), are preferred [18,19]. β-Strands can also assemble into small structures known as β-hairpins, where two antiparallel strands are joined by a short disordered loop.

3.2.3 Collagens

Peptides intended to fold as α-helices and β-strands are the most common motifs to have been designed de novo, but other structures have also been exploited, especially for biomedical applications. One such structure is collagen, a protein found in connective tissue of mammals. It has a triple-helical structure, meaning three helices come together to form a collagen helix (Fig. 3.1F). The individual helices are left-handed and stabilized by intermolecular hydrogen bonds with the resulting triple helix being right-handed [20–22]. Collagens can be ether homotypic, where all three helices have the same amino acid sequence, or heterotypic, where they are different [23]. One of the reasons collagens are desirable design targets is due to their repetitive sequence pattern comprising an $(\text{Xaa-Yaa-Gly})_n$ repeat, providing an easy design handle.

3.3 De novo designed peptide structures and their applications

De novo designed peptides based on the structures outlined above fulfill applications in a variety of fields including biomedicine, biotechnology, and bioengineering—the themes of this book. Examples of peptides that have been utilized for various purposes within these fields are provided in the following sections. The design rationale for each peptide is explained, along with how this relates to its intended application. The list of applications discussed is not exhaustive but serves to present a taste of the diverse functions of de novo designed peptides and highlights well-established and newly emerging fields. Other fields in which peptides are commonly utilized are presented throughout this book.

3.3.1 Tissue engineering and regeneration

Some of the earliest examples of de novo designed peptides were created in order to fulfill tissue engineering applications; self-assembled peptides were demonstrated to be capable of supporting the attachment and proliferation of cells. A range of early and more recent designs are presented below.

3.3.1.1 Scaffolds from self-assembled β-structured components

One of the first examples came from the Zhang lab in 1993 [24]. The group developed the peptide EAK16 Ac-(A-E-A-E-A-K-A-K)$_2$-NH$_2$ based on their studies of the yeast protein Zuotin [25]. EAK16 is soluble in solution due to the high number of charged residues, and it adopts a β-structure due to the alternating *HPHPHP* pattern. The peptide self-assembles into what was described as a "macroscopic membrane," meaning a fibrous network was observed. It is proposed that fibers formed through a combination of intermolecular electrostatic interactions between Lys, (K), and Glu, (E), residues, and hydrophobic interactions between Ala, (A), residues that are found on the opposite face of the strand. It is likely that the strands are staggered, to allow for extended fibers to be formed, and this theory is supported by the fact that shorter peptides (with a lesser ability to form staggered arrangements) do not form extended structures.

A second peptide, RAD16-II, Ac-(R-A-R-A-D-A-D-A)$_2$-NH$_2$ was subsequently developed [26]. In this design, the charge pattern was reversed—positive charges are closer to the amino-terminus of the peptide, and Arg, (R), substituted for Lys, while Asp, (D), substituted for Glu. This peptide also formed fibrous networks (Fig. 3.2A), which are assumed to be arranged and stabilized in a similar fashion to the EAK16 networks. Both self-assembled scaffolds were capable of supporting the attachment and proliferation of a range of cell types [28]. The majority of subsequent work has focused on this RAD16 scaffold, with numerous modifications and additions to the peptide being explored. An early study evaluated RAD16-I Ac-(R-A-D-A)$_4$-NH$_2$ and RAD16-II. Both peptides were shown to support neuronal cell attachment and neurite outgrowth, and it was also demonstrated that these peptides

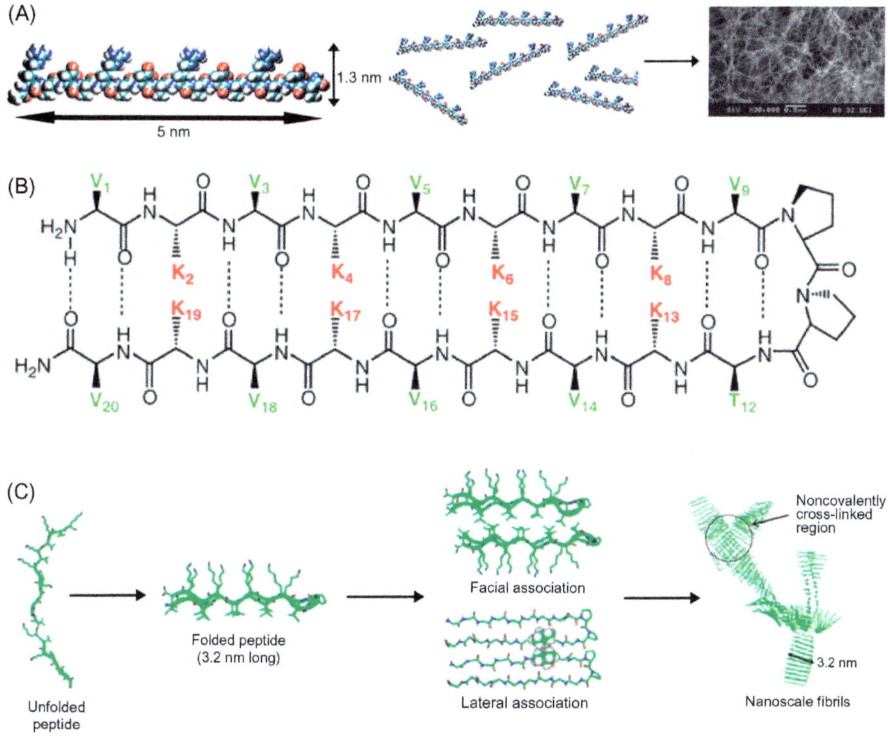

Figure 3.2 β-Structured hydrogels. (A) Representation of the RADA16-I building block; many copies of this self-assemble to form a fibrous network shown in the EM image; (B) structure of the MAX1 β-hairpin with hydrogen bonds shown as dotted lines; (C) assembly pathway of MAX1.
Source: Images are adapted and reproduced with permission from (A) R.G. Ellis-Behnke, Y.X. Liang, S.W. You, D.K.C. Tay, S.G. Zhang, K.F. So, et al., Nano neuro knitting: peptide nanofiber scaffold for brain repair and axon regeneration with functional return of vision, Proc. Natl. Acad. Sci. U.S.A. 103 (2006) 5054–5059 [27], © National Academy of Sciences; (B), (C) R. Rughani and J.P. Schenider, Molecular design of β-hairpin peptides for material construction. MRS Bulletin, 33 (2008) 530-535 [120], © Cambridge University Press.

could be injected without causing adverse effects in rats. No comparisons were made between the two variants however [29].

Further studies incorporated bioactive peptide motifs including Arg-Gly-Asp (RGD), a cell adhesion motif, and Ile-Lys-Val-Ala-Val (IKVAV), a laminin-derived motif known to promote neurite regeneration. The self-assembled matrices were then used to culture stem cells with varying degrees of success [30]. Another study showed that these peptides could be administered in vivo and promote axon regeneration and brain repair in injured animal models [27].

There are now countless examples of β-structured peptides that self-assemble to form fibers and hydrogels which are capable of promoting tissue regeneration; several reviews provide detailed information [31–35].

Fibrous hydrogels have also been designed based on β-hairpin structures, with the majority of work coming from the Schneider and Pochan groups. The first design, MAX1, is a 20-residue peptide: V-K-V-K-V-K-V-K-V-^{D}P-P-T-K-V-K-V-K-V-K-V-NH_2, the central V-^{D}P-P-T sequence is known to adopt a type-II′ β-turn conformation [36]. This turn promotes the folding of the peptide into a monomeric β-hairpin. This structure is only adopted in solution above pH 9; at this pH some of the Lys side-chains are neutral and so charge repulsion is decreased, allowing the peptide to fold intramolecularly to adopt the β-hairpin structure, which is stabilized by hydrogen bonds between backbone amide groups (Fig. 3.2B). The β-hairpins also self-assemble intermolecularly, leading to hydrogel formation (Fig. 3.2C); these higher-order assemblies are also stabilized by backbone hydrogen bonding, in addition to hydrophobic interactions between the Val-rich faces of the hairpins. If the solution is acidified and the Lys side-chains become protonated then repulsion of these charged side-chains leads to disassembly of the hydrogel and unfolding of the hairpins. Numerous variants of this original design have been produced [37–39], and the group has demonstrated the hydrogels are capable of cell growth and proliferation [40], one version also has inherent antibacterial activity and is capable of killing meticillin-resistant *Staphylococcus aureus* (MRSA) [41].

3.3.1.2 SAFs (self-assembling fibers) from α-helical building blocks

Fibrous scaffolds from helical building blocks were developed after β-structured systems. The majority of designs are based on coiled-coils as the rules for self-assembly are well-described.

One design of note is that of the self-assembling fiber (SAF) system, created by the Woolfson lab [42]. The reason this design is important is twofold: it is the first example of a heterodimeric fiber-forming system; and the way in which the peptides were designed to self-assemble, and subsequently form fibers, was unique at the time. The SAFs consist of two complementary peptides, SAF-p1 and SAF-p2 (Table 3.1) that are designed to self-assemble into a parallel, offset, coiled-coil dimer (Fig. 3.3A).

Table 3.1 The sequences of the SAF peptides, SAF-p1, SAF-p2, SAF-p2a and the hSAF peptides, $hSAF_{AAA\text{-}W}$ p1 and $hSAF_{AAA\text{-}W}$ p2

	Heptad repeat
	gabcdef gabcdef gabcdef gabcdef gabcdef gabcdef
SAF-p1	Ac-KIAALKQ KIASLKQ EIDALEY ENDALEQ-NH_2
SAF-p2	Ac-KIRALKA KNAHLKQ EIAALEQ EIAALEQ-NH_2
SAF-p2a	Ac-KIRRLKQ KNARLKQ EIAALEY EIAALEQ-NH_2
$hSAF_{AAA\text{-}w}$p1	Ac-KIAALKA KIAALKA EIAALEW ENAALEA-NH_2
$hSAF_{AAA\text{-}w}$p2	Ac-KIAALKA KNAALKA EIAALEW EIAALEA-NH_2

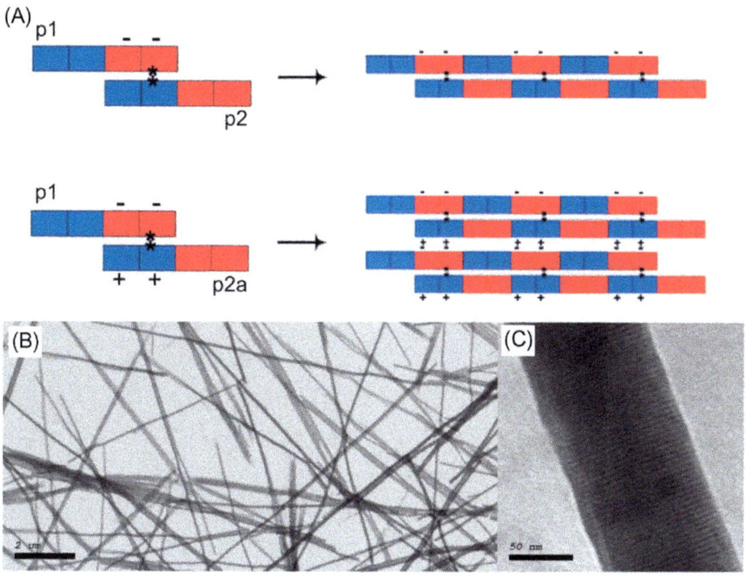

Figure 3.3 SAF peptides. (A) Assembly schematics for different SAF designs showing charge interactions allow for fiber thickening in the second-generation design; (B) representative TEM image of the fibers, with; (C) close-up of one fiber with the striations evident.
Source: Images reproduced with permission from A.M. Smith, E.F. Banwell, W.R. Edwards, M.J. Pandya, D.N. Woolfson, Engineering increased stability into self-assembled protein fibers, Adv. Funct. Mater. 16 (2006) 1022–1030 [43], © John Wiley & Sons.

This is achieved by having Ile and Leu residues at positions *a* and *d of the heptad repeat (discussed in section 3.2.1.2)*, to specify for a dimeric assembly, patterning the charges at *e* and *g* to bias a staggered assembly, and finally by including Asn (N) residues at offset *a* positions, which also specifies for a staggered assembly as the Asn residues pair to form a hydrogen bond across the hydrophobic interface.

The peptides were shown to be helical when mixed by circular dichroism (CD) spectroscopy and X-ray fiber diffraction. TEM analysis showed that long fibers were formed, although these were thicker than anticipated.

A second-generation design was produced to circumvent stability issues in the original design. This design incorporated positively charged Arg residues at *c* positions in SAF-p2 (Table 3.1), and it was assumed these would interact electrostatically with Asp residues in SAF-p1 (Fig. 3.3A). The Arg residues did indeed provide stabilizing interactions, with fibers forming at room temperature (Fig. 3.3B). TEM revealed a striated pattern within the fibers, suggestive of ordering of the individual peptides within the fiber (Fig. 3.3C) [43]. This was investigated further in a cryo-TEM study that concluded the peptides assemble along the long axis of the fibers and that the thickening is due to hexagonal packing of the coiled coils within the fiber [44].

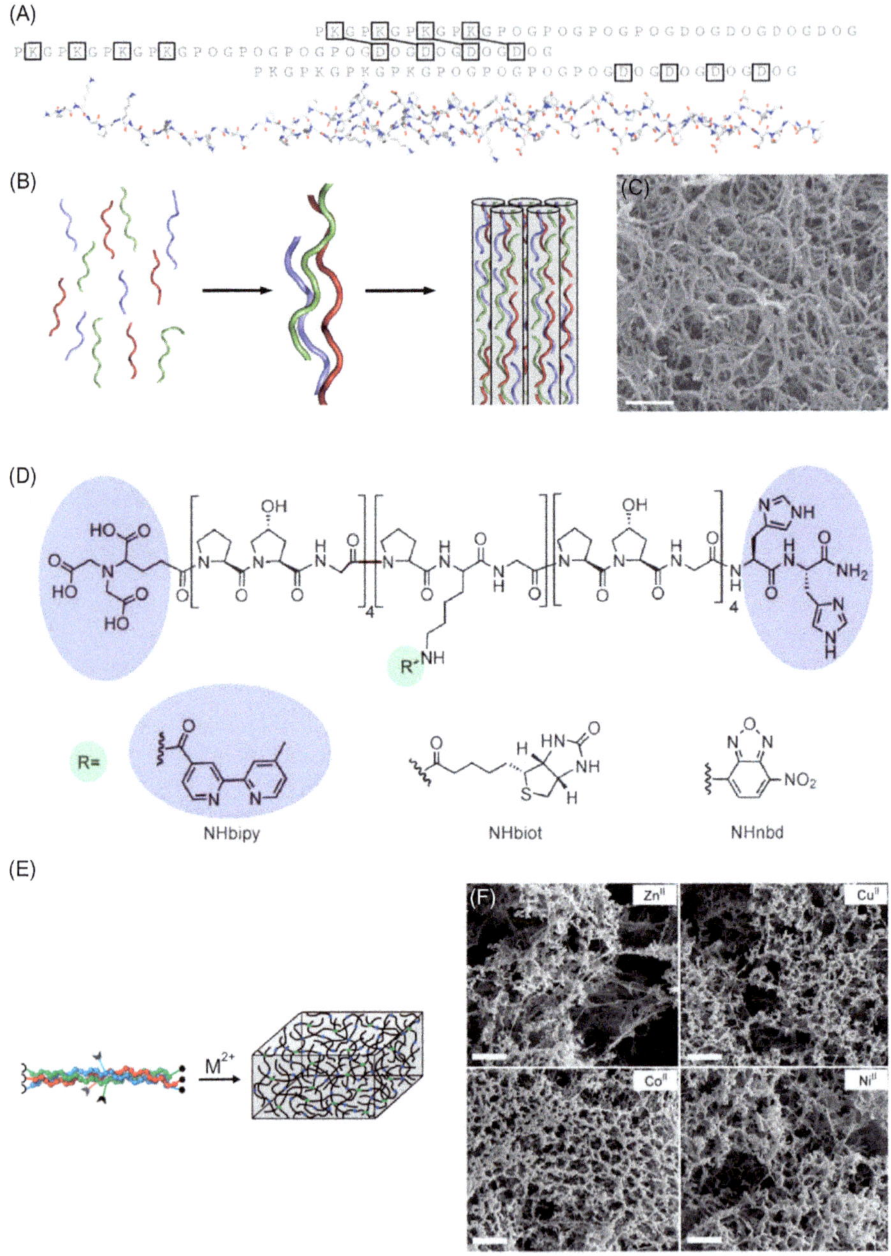

Figure 3.4 Hydrogels from collagens. (A) Assembly schematic of KOD into a collagen triple helix; (B) pathway of hydrogel assembly: short peptides; to collagen triple helices; to fibers and finally; (C) hydrogels; an SEM image of the fibrous, hydrogelating network, scale bar = 1 μm; (D) structures of the peptides used by the Chmielewski group, with metal ligands in blue; (E) schematic of the metal-binding collagen design, upon addition of divalent metals, these molecules self-assemble to form fibrous networks; (F) SEM images of the networks with different divalent metals (scale bar = 5 μ48m).

To introduce functionality to this system, the *b*, *c*, and *f* positions were mutated to Ala residues, to promote hydrophobic interactions between fibers, leading to $hSAF_{AAA-W}$ p1 and $hSAF_{AAA-W}$ p2. The resulting fibers were more flexible and thinner, and formed hydrogels which were shown to support growth and differentiation of PC12 cells [45]. The RGDS motif was later incorporated into the fibers using click chemistry and these modified hydrogels supported the growth and proliferation of PC12 cells more successfully over a 14-day period, when compared to nonmodified hydrogels [46]. These RGDS-modified gels also support the attachment, proliferation, and differentiation of embryonic neural stem cells [47].

3.3.1.3 Tissue engineering and regeneration with collagen-mimetic peptides

The Hartgerink lab has designed a collagen mimetic peptide, "KOD" which has the sequence Ac-$(P\text{-}K\text{-}G)_4(P\text{-}O\text{-}G)_4(D\text{-}O\text{-}G)_4$-$NH_2$, where "O" is the unnatural amino acid hydroxyproline (Hyp) [48]. The rationale behind this design was that the peptide would assemble to form a "sticky-ended" collagen triple helix, as the Asp and Lys would align and form stabilizing electrostatic interactions, with the sticky-ends allowing for further self-assembly (Fig. 3.4A–C). Asp and Lys were chosen as previous NMR studies revealed that Arg-Glu pairings, that have been used previously, are not optimal as the Arg residues form strong hydrogen bonds with backbone carbonyl groups from an adjacent peptide strand. It was hypothesized that Lys and Asp would provide more effective electrostatic interactions. As discussed in Section 3.2.3, the sequence of natural collagens dictates that every third position contains a Gly residue and other positions were filled with Pro (Proline, P) and Hyp to promote formation of the characteristic polyproline type II helices. The group characterized every step of formation: from the triple helix structure; to fiber formation; and to gelation at high concentrations. These gels are capable of promoting the adhesion and activation of platelets, in addition to clotting plasma and blood, indicating these collagen hydrogels could be used as a hemostat [49].

Other interesting, from a de novo design point of view, collagen mimetic scaffolds come from the Chmielewski group. They employ metal ions to trigger the assembly of collagens and form a range of fibrous and discrete structures. A scaffold based on a P-O-G repeating sequence, with an N-terminal nitriloacetic acid (NTA) moiety, a C-terminal dihistidine (His_2) motif, and a central bipyridyl (bipy)

◀ *Source*: Images adapted and reproduced with permission from (A) V.A. Kumar, N.L. Taylor, A.A. Jalan, L.K. Hwang, B.K. Wang, J.D. Hartgerink, A nanostructured synthetic collagen mimic for hemostasis, Biomacromolecules 15 (2014) 1484–1490 [49], © American Chemical Society; (B) and (C) L.E.R. O'leary, J.A. Fallas, E.L. Bakota, M.K. Kang, J.D. Hartgerink, Multi-hierarchical self-assembly of a collagen mimetic peptide from triple helix to nanofibre and hydrogel, Nat. Chem. 3 (2011) 821–828 [48], © Nature Publishing Group; and (D)–(F) M.M. Pires, D.E. Przybyla, J. Chmielewski, A metal-collagen peptide framework for three-dimensional cell culture, Angew. Chem. Int. Ed. 48 (2009) 7813–7817 [50], © John Wiley & Sons.

ligand attached to the Lys side-chain of a P-K-G repeat was created (Fig. 3.4D) [50]. These moieties were included as they are all metal-binding units. In the absence of metal, the peptides fold into discrete collagen triple helices but when various divalent metals are added, fibrous networks are formed (Fig. 3.4E). The different metal-binding ligands can also be exploited to form more complex dual-metal scaffolds (Fig. 3.4F). HeLa cells were added to these peptides and metal ions were subsequently added, triggering assembly. It was demonstrated that the HeLa cells were successfully encapsulated within the fibrous network and that they remained viable over a 5-day period.

This scaffold has been mixed with an RGD-containing scaffold and a His-tagged human epidermal growth factor (hEGF) was also incorporated, which is capable of binding to the NTA–metal complexes on the scaffolds, but gradually leaks out into the hydrogel matrix over time. They showed that the combination of all these factors was optimal to promote the adhesion and proliferation of an EGF-dependent cell line [51].

3.3.1.4 Short peptide motifs as molecular signals

In addition to peptides designed to self-assemble and form scaffolds for tissue growth and engineering a huge number of biomimetic peptides have been developed, the sequences of which are usually derived from large growth factor proteins. The entire proteins are often not used themselves due to cost, production, and stability issues, meaning peptides have become a useful alternative. Some such peptides have already been mentioned; RGD and IKVAV are two examples. There are too many different sequences to discuss here, and many are also specific to certain cell types and so are not generally applicable. Recent reviews have discussed the use of such peptides and so more information can be gained from these [52–55].

3.3.2 Drug encapsulation, release, and delivery

3.3.2.1 Drug encapsulation and release from hydrogels

Following on from the previous topic where hydrogels have been used as scaffolds for tissue engineering, they can also be employed for drug release applications.

The Schneider and Pochan groups have evolved their MAX1 hydrogel, discussed in Section 3.3.1.1 and formed MAX8. The difference between MAX1 and MAX8 is a single mutation, a Lys in position 15 was mutated to a Glu. This means the peptide has a reduced net charge, allowing it to self-assemble faster than MAX1, creating a more rigid gel with smaller mesh sizes [40]. The encapsulation and release of model proteins from a MAX8 hydrogel revealed that positively charged and neutral proteins readily diffuse out of the hydrogel due to steric and repulsive electrostatic interactions. Negatively charged peptides form favorable interactions with the hydrogel however, and so only small amounts were released [56]. The same hydrogel is also effective at encapsulating, and subsequently facilitating the sustained release of, curcumin [57], and another publication demonstrated that several growth factors, including nerve growth factor (NGF) and brain-derived neurotrophic factor (BDNF), could be encapsulated and released indicating the gel may be suitable for therapeutic purposes [58]. This was further reinforced by a recent study that showed vincristine, a chemotherapeutic, underwent a sustained release from the gel for 28 days after encapsulation, and it remained biologically active during this time [59].

The RAD-16-I hydrogels discussed in Section 3.3.1.1 have also been investigated as potential drug-delivery vehicles. Initial studies looked at the diffusion properties of different dyes within the gel, revealing that electrostatic interactions between the gel and the molecules were important for release, as was the gel concentration, indicating that release profiles could be tailored [60]. A follow-up study looked at the release of a range of proteins, and found that, in this case, diffusion was primarily dependent on the size of the protein, although gel concentration also had an effect [61].

3.3.2.2 Discrete structures for encapsulation, release, and delivery

The Chmielewski group have modified their collagen constructs and successfully formed discrete structures. To effect this, the central bipyridine ligand was removed to leave the collagen mimetic peptide, an Ac-(P-O-G)$_9$-NH$_2$ repeat, with an NTA unit at the N-terminus and a His$_2$ motif at the C-terminus (Fig. 3.5A). Upon Zn^{II} addition, micron-sized particles dubbed "microflorettes" were formed (Fig. 3.5B). The group propose that these structures are formed as a result of metal coordination by the NTA and His ligands, in a manner analogous to the NTA/His-tag system used for protein purification [64]. A follow-up publication demonstrated that these microflorettes are capable of both incorporating proteins within the particles, and displaying proteins on the surface. Release rates of these proteins are slow but the group state that this could be tuned by modifying the structures and that this is the first step towards devloping drug-delivery vehicles [62]. A different design from the same group, named Hbyp3, contains three bipyridine ligands per helix, so nine in total in a collagen triple helix. The group surmised that these bipyridine ligands could have aromatic interactions with each other and so assembly may be promoted in the absence of metals. This was indeed the case; disk-like structures were observed by microscopy. It was proposed that the edges of these disks would be populated with bipyridine ligands that are not participating in aromatic interactions and so would be capable of coordinating metal ions. Addition of metal ions to these disks produced hollow spheres [65]. These spheres have been shown to encapsulate dextran, which is released when the spheres are heated, as heat disrupts the collagen triple helical structure and so causes disassembly of the spheres. Longer collagens were found to have a higher thermal stability and so it was proposed that cargo delivery rates could be modulated by changing the peptide design [65,66].

The Aili group also use a metal-mediated approach to self-assembly. They designed a 42-amino-acid helix-turn-helix peptide, N-A-A-D-L-E-K-A-I-E-A-L-E-K-H-L-E-A-K-G-P-C-D-A-A-Q-L-E-K-Q-L-E-Q-A-F-E-A-F-E-R-A-G-CO$_2$H dubbed JR2EC.

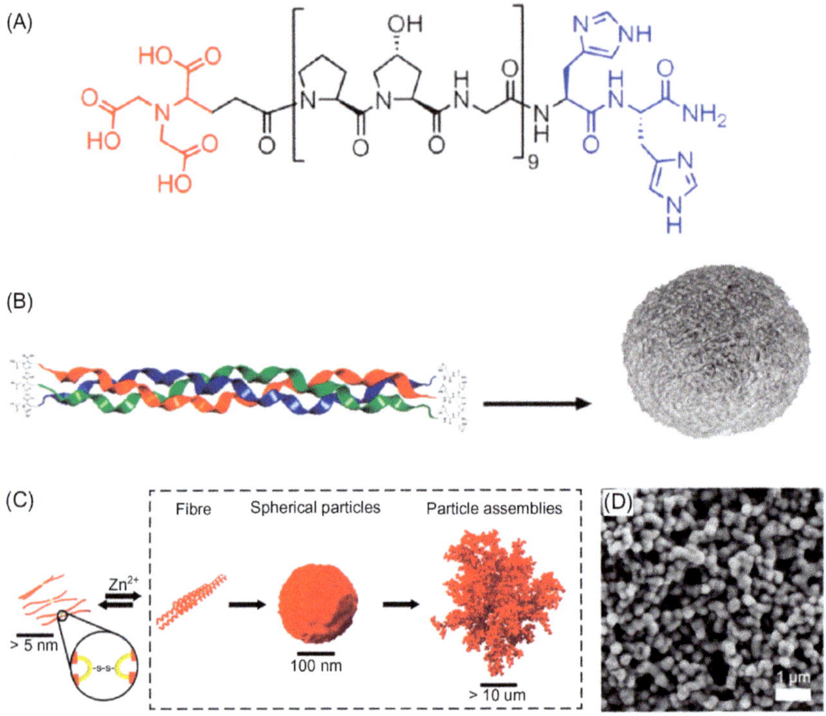

Figure 3.5 Discrete particles for drug release. (A) Structure of the collagen peptide used by the Chmielewski group; (B) schematic representation of the collagen peptide with NTA molecules at the N-terminus (left), and His tags at the C-terminus (right); these self-assemble to form "microflorettes"; (C) assembly schematic of the zinc-binding peptide, JR2EC; (D) an SEM image of the assembled particles of JR2EC in the presence of Zn^{II}.
Source: Images adapted with permission from (A) and (B) M.M. Pires, D. Ernenwein, J. Chmielewski, Selective decoration and release of His-tagged proteins from metal-assembled collagen peptide microflorettes, Biomacromolecules 12 (2011) 2429–2433 [62], © American Chemical Society; (C) and (D) C. Aronsson, R. Selegard, D. Aili, Zinc-triggered hierarchical self-assembly of fibrous helix-loop-helix peptide superstructures for controlled encapsulation and release, Macromolecules 49 (2016) 6997–7003 [63], © American Chemical Society.

The peptide was designed to fold and dimerize, leading to the formation of four-helix bundles at acidic pH, or at neutral pH in the presence of Zn^{II}. The peptide can be dimerized by oxidizing the Cys residue and at acidic pH, long fibers are seen [63]. At neutral pH in the presence of Zn^{II} however, aggregated spherical particles were observed. It was hypothesized that addition of Zn^{II} promotes the formation of short fibers and lateral interactions cause the formation of particles instead of extended fibers. Diffusion-limited aggregation then causes these individual spheres to form larger assemblies (Fig. 3.5C,D). Peptides where the Cys was replaced with a Val residue, preventing dimerization and so fiber assembly and another sequence where the L-Ala was replaced with D-Ala preventing helix formation failed to form these spherical assemblies indicating that four-helix bundle formation is critical to obtain the correct assembly. A molecule named p-FTAA, which has been shown to decrease the toxicity of prions and amyloid-β [67], and so is potentially important for the treatment of diseases such as Alzheimer's, was encapsulated within these superstructures with addition of EDTA facilitating its release. By tuning the amount of Zn^{II} used, it was shown that the release profile of p-FTAA could be altered.

3.3.2.3 Translocating and permeating peptides for drug delivery

In addition to the self-assembled peptide structures presented so far in this section, standalone peptides can also be effective at promoting drug delivery, usually by aiding the translocation of drugs across membranes.

Skin penetrating peptides (SPPs) are part of a wider family of peptides known as cell penetrating peptides (CPPs) that are discussed in detail in Chapter 13, Cell-penetrating peptides. SPPs provide a noninvasive method for delivering transdermal medicines. Such peptides are usually discovered by either phage display or by experimentally screening peptide libraries. Both techniques are time-consuming and often expensive and so alternative methods are actively sought. A recent paper detailed the computational screening of virtual peptide libraries to discover novel CPPs [68]. Virtual libraries were generated based on the sequence of SPACE, a disulfide-bonded cyclic peptide with the sequence A-C-T-G-S-T-Q-H-Q-C-G-CO_2H [69], and these libraries were initially computationally screened for interactions with keratin (a protein abundant in skin, which is thought to mediate SPP entry) and cyclosporin A (CsA), a model therapeutic used to determine the efficacy of SPPs. Several sequences were returned as promising candidates and seven were experimentally validated. All were shown to penetrate skin and enhance the dermal translocation of CsA, and three were shown to perform as well as, or better than, SPACE proving that this method of computational generation and screening of peptide libraries is a viable alternative to extensive peptide screening.

Peptides that penetrate tumors and/or enhance tumor targeting of therapeutics are also common design targets. One peptide, designed by the Ruoslahti group, has been found to be highly effective at delivering drugs to all areas of tumors. This is also a cyclic peptide and has the sequence C-R-G-D-K/R-G-P-D-C-CO_2H. The design was carefully considered: the RGD motif binds to receptors on the tumor surface and is subsequently cleaved, exposing a binding motif for a protein named neuropilin-1; this activates a tissue penetration pathway (Fig. 3.6A). When this peptide sequence was conjugated to imaging agents, their sensitivity was improved, and activity of antitumor drugs was also increased when these were conjugated to the peptide [70]. The group has subsequently designed a similar peptide with the sequence C-R-N-G-R-G-P-D-C-CO_2H, this binds to a different cell-surface receptor before activating the same tissue-penetrating pathway, and has been shown to enhance the activity of doxorubicin [72].

A different approach has been adopted by the Schneider group. They designed a peptide named SVS-1, with the sequence K-V-K-V-K-V-K-V-^{D}P-P-T-K-V-K-V-K-

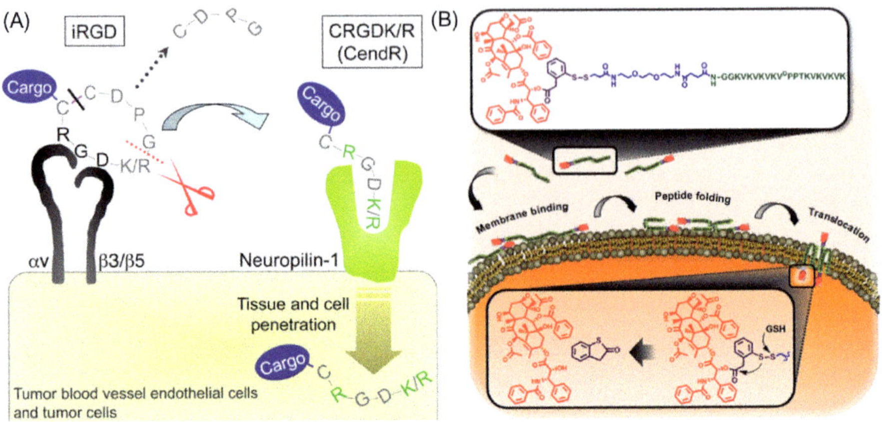

Figure 3.6 Mechanisms of entry of cell-translocating peptides. (A) Proposed mechanism of entry for the peptide designed by the Ruoslahti group; (B) entry mechanism for the peptide–drug conjugate designed by the Schneider group.
Source: Images reproduced with permission from (A) K.N. Sugahara, T. Teesalu, P.P. Karmali, V.R. Kotamraju, L. Agemy, O.M. Girard, et al., Tissue-penetrating delivery of compounds and nanoparticles into tumors, Cancer Cell. 16 (2009) 510–520 [70], © Elsevier; (B) S.H. Medina, J.P. Schneider, Cancer cell surface induced peptide folding allows intracellular translocation of drug, J. Control. Release 209 (2015) 317–326 [71], © Elsevier.

V-K-NH_2, which is similar to the MAX peptides designed by the same group. This peptide is unfolded in solution, due to electrostatic repulsion of the Lys residues, but folds at the surface of cancer cells. This folding is due to the presence of anionic lipid head groups at cancer-cell surfaces; the positively charged Lys residues interact with these and the peptide adopts a β-hairpin conformation. This folding results in the Val-rich face of the peptide being solvent-exposed, which is unfavorable, and so the peptide buries into the hydrophobic cell membrane (Fig. 3.6B). SVS-1 was originally designed as a lytic peptide [73], but it was discovered that it can cross cell membranes without lysing them if present at lower concentrations [71]. This peptide has been conjugated to paclitaxel, a poorly soluble hydrophobic cancer drug. Drug solubility was vastly improved as a result of conjugation, and it remained active.

3.3.3 Antigen display and vaccine development

Virus-like particles (VLPs) are self-assembled nanoparticles that have a regular, symmetrical structure. They have been utilized for drug delivery and targeting, imaging, and antigen display. VLPs were originally made from noninfectious viral capsid proteins, however there are numerous problems associated with this; such proteins have to be recombinantly expressed, which is time-consuming and often

Table 3.2 The sequence of the coiled-coils used in both the development of SAPNs and SVLPs

	Heptad repeat
	bcdefg abcdefg abcdefg abcdefg abcdefg abcdefg abcdefg abcd
SAPN-1	Ac-DEMLRE LQETNAA LQDVREL LRQQVKQ ITFLKCL LM-NH_2
SAPN-2	Ac-RLLCR LEELERR LEELERR LEELERR-NH_2
SVLP	GGAVSK VLHLEGE VNKIKSA LLSTNKA VVSLSNG VSVLTSK VLDLKNY IDKQ-NH_2

low-yielding, modification is not straightforward, and there are safety concerns. Producing VLPs from another material, for example peptides, may be a solution to some of these issues.

A notable example of a peptidic VLP comes from Burkhard and colleagues who developed a "self-assembling protein nanoparticle" (SAPN) from coiled-coils [74]. They postulated that it should be possible to make a repetitive nanoparticle by joining coiled coil of different oligomeric states together. Initially, a dodecahedron was designed by joining a coiled-coil known to form a homopentamer to one that forms homotrimers via a Gly-Gly linker. The sequence of both the pentamer and trimer are given in Table 3.2. They hypothesized that, according to known geometric rules, 60 copies of this construct would self-assemble to form a regular dodecahedral nanoparticle. Both Analytical Ultracentrifugation, (AUC) and TEM showed that the desired structures were formed using specific conditions (Fig. 3.7A,B). Subsequent research has investigated the formation conditions in more detail [76], and has also looked at alternative constructs that would create a similar particle [77].

The group proposed the SAPNs could be used as an antigen display or vaccine platform due to their regular, repetitive nature. Indeed the SAPNs have been used to develop prototypic vaccines for a range of diseases including SARS [78], malaria [79], and bird flu [80].

Similar structures, termed "synthetic virus-like particles" (SVLPs) have been presented by the Robinson group. These are not composed solely of coiled-coils as the peptides are fused to lipids to form amphiphilic structures. The coiled-coil sequence chosen is from a viral protein that forms trimers (Table 3.2). They added two Gly residues to the N-terminus of this peptide and then coupled, via a succinate linker, a phospholipid [75]. The rationale behind this design was that, in aqueous solvents, the coiled-coil peptide would assemble into its native trimer, and then further self-assembly would result due to the lipid chains clustering away from the aqueous environment.

The presence of regular nanoparticles was confirmed by TEM (Fig. 3.7C,D). A synthetic antigen was subsequently coupled to the C-terminal end of the peptides and experiments showed that antigen-specific antibodies were raised in rabbit models. Variants of these SVLPs have been shown to be effective as prototypic vaccines with epitope mimetics of HIV and malaria being incorporated [81–83].

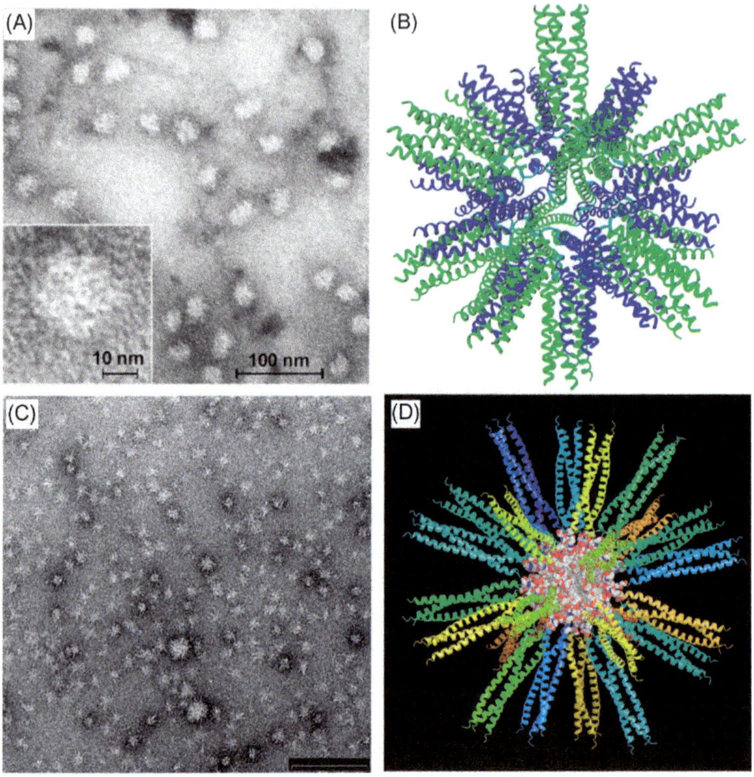

Figure 3.7 SAPNs and SVLPs. (A) TEM image of SAPNs; (B) computer model of SAPNs with pentamers in green, trimers in dark blue, and linkers in light blue; (C) TEM image of SVLPs (scale bar = 100 nm), and; (D) computer-generated image of SVLP structure with peptides extended from the lipid core.
Source: Images reproduced with permission from (A) and (B) S. Raman, G. Machaidze, A. Lustig, U. Aebi, P. Burkhard, Structure-based design of peptides that self-assemble into regular polyhedral nanoparticles, Nanomedicine 2 (2006) 95–102 [74], © Elsevier; (C) and (D) F. Boato, R.M. Thomas, A. Ghasparian, A. Freund-Renard, K. Moehle, J.A. Robinson, Synthetic virus-like particles from self-assembling coiled-coil lipopeptides and their use in antigen display to the immune system, Angew. Chem. Int. Ed. 46 (2007) 9015–9018 [75], © John Wiley & Sons.

3.3.4 Membrane protein stabilization

Relatively few structures of membrane proteins are known due to inherent solubility and stability problems and the absence of a structure makes engineering and understanding such proteins challenging. Efforts to stabilize membrane proteins have focused on engineering their sequences to enhance stability, and solubility has been improved by using detergents and surfactants, although these can affect stability. Peptide surfactants are emerging candidates for both solubilizing and stabilizing membrane proteins.

3.3.4.1 α-Helical amphipathic peptides

The first such peptide, PD1, was created in 1993 [84]. It has the sequence Ac-E-E-L-L-K-Q-A-L-Q-Q-A-Q-Q-L-L-Q-Q-A-Q-E-L-A-K-K-NH_2 and was designed to be an α-helical peptide that would pack around the highly hydrophobic regions of membrane proteins, with the aim of stabilizing and solubilizing them. The Ala residues were intended to provide a flat hydrophobic surface that membrane proteins could interact with, and the Leu residues were positioned so that they projected out from the sides of the helix and so could form hydrophobic interactions with other PD1 helices that were surrounding the same membrane protein. The hydrophobic amino acids were positioned such that they were all on one face of the helix (Fig. 3.8A). Glu and Lys were included at the helix termini in an $i-i+4$ arrangement, meaning they are capable of forming stabilizing intrahelical ionic interactions. PD1 was shown to stabilize bacteriorhodopsin and rhodopsin for 2 days, although some rhodopsin precipitate was observed. It could not stabilize another membrane protein, PhoE porin, however.

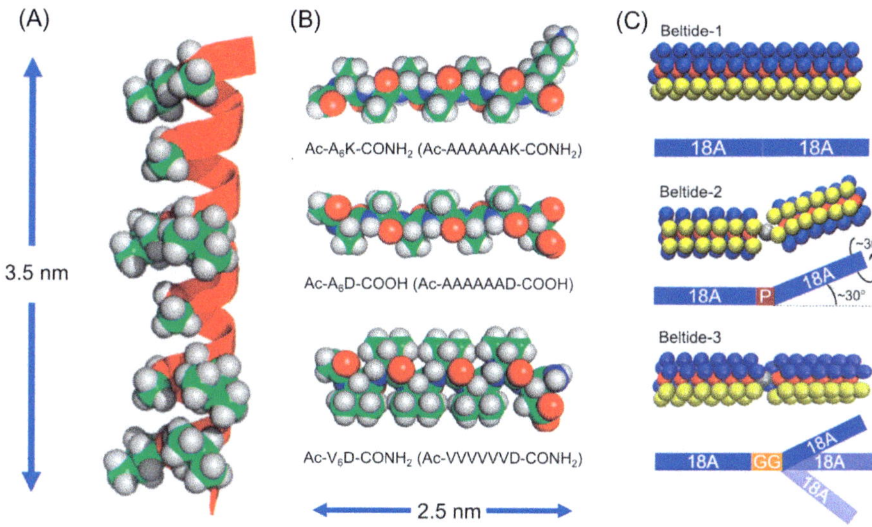

Figure 3.8 Peptides for membrane protein stabilization. (A) Structure of the helical PD1 peptide, with hydrophobic side-chains shown as spheres; (B) models of lipid-like, self-assembling peptides; (C) coarse-grained models of the beltide peptides. Color code for: (A) and (B): green = carbon, white = hydrogen, red = oxygen, blue = nitrogen; (C) blue = hydrophilic regions, yellow = hydrophobic regions, red = central portions, gray = linkers.
Source: Images adapted and reproduced with permission from (A) and (B) S. Koutsopoulos, L. Kaiser, H.M. Eriksson, and S. Zhang, Designer peptide surfactants stabilize diverse functional membrane proteins. Chem. Soc. Rev., 41 (2012) 1721-1728 [121], © The Royal Society of Chemistry; (C) A.N. Larsen, K.K. Sorensen, N.T. Johansen, A. Martel, J.J.K. Kirkensgaard, K.J. Jensen, et al., Dimeric peptides with three different linkers self-assemble with phospholipids to form peptide nanodiscs that stabilize membrane proteins, Soft. Matter. 12 (2016) 5937–5949 [85], published by The Royal Society of Chemistry.

Helical peptides have also been joined via different peptidic linkers to form structures termed "beltides" which, when combined with lipids, form peptide nanodiscs capable of stabilizing bacteriorhodopsin. The helical portion of the beltide is named 18A, and has the sequence Ac-D-W-L-K-A-F-Y-D-K-V-A-E-K-L-K-E-A-F-NH_2. This peptide has previously been shown to form nanodiscs which become unstable over time. It was hypothesized that two copies of the peptide would improve the structural stability of the resulting nanodiscs, therefore two copies of 18A were joined either: directly via a peptide bond (Beltide-1), via a Pro residue (Beltide-2), or by a Gly_2 linker (Beltide-3) (Fig. 3.8C). All three constructs formed nanodiscs when combined with DMPC lipids, however the sizes differed slightly with Beltide-2 forming the smallest nanodiscs and Beltide-1 the largest, the latter were also more stable. All three nanodiscs stabilized bacteriorhodopsin for more than 12 days [85].

3.3.4.2 Short, self-assembling peptides

Interest turned to shorter, self-assembling peptides due to the high costs involved in the production of longer, helical peptides. These short peptides were designed to resemble natural lipids, with several neutral amino acids resembling an extended, uncharged "tail" and one or two charged amino acids at one end of the peptide mimicking a lipid headgroup (Fig. 3.8B).

The first example of a short, amphiphilic peptide being employed for membrane protein stabilization used a peptide with the sequence A-A-A-A-A-A-K (A_6K). This stabilized a membrane protein named photosystem-1, in the dry state, for more than 3 weeks when used at high concentrations (up to 0.5%, corresponding to 8 mM). A second peptide V-V-V-V-V-V-D (V_6D) was not as effective at stabilization as A_6K [86]. A subsequent study looked at stabilizing the same protein in solution [87]. For this study, several different peptides were employed in an effort to determine key factors that control stabilization by these short peptides, the sequences of the peptides used are shown in Table 3.3.

It was determined that I_6K_2 was the most effective stabilizer of PS-1, followed by A_6K, V_6K_2, and V_6R_2, these peptides also enhanced the activity of the complex. This led the researchers to conclude that, for this protein complex a positively charged C-terminus was essential, and capping of the C-terminus was also necessary. They hypothesize that there must be a match between the conformational characteristics of the peptide surfactant and the protein complex for optimal stabilization.

The same group has also demonstrated that rhodopsin can be stabilized in solution using A-A-A-A-A-A-D (A_6D) [88]. Interestingly, this peptide has a negatively charged C-terminus; such peptides were not effective at stabilizing PS-1, indicating that the same peptide cannot be used for every membrane protein, they must be individually tailored and a large number of properties such as type (positive or negative), position of charge, peptide conformation, and nature of the N- and C-terminus must be taken into account.

Table 3.3 Sequences of the peptides explored for PS-1 stabilization

Peptide name	Sequence
V_6R_2	Ac-V-V-V-V-V-V-R-R-NH_2
V_6K_2	Ac-V-V-V-V-V-V-K-K-NH_2
V_6D_2	Ac-V-V-V-V-V-V-D-D-NH_2
I_6K_2	Ac-I-I-I-I-I-I-K-K-NH_2
A_6K	Ac-A-A-A-A-A-A-K-NH_2
A_6K	Ac-A-A-A-A-A-A-K-OH
A_6D	Ac-A-A-A-A-A-A-D-OH
KA_6	K-A-A-A-A-A-A-NH_2
DA_6-NH_2	D-A-A-A-A-A-NH_2

3.3.5 Imaging contrast agents

Techniques such as magnetic resonance imaging (MRI), radioisotope labeling, and fluorescence imaging are important diagnostic tools. There is an increasing need to develop probes that can be targeted to specific tissues, possess good stability, provide good contrast, and that are easily prepared and delivered. Previously, organic small molecules or antibodies have been employed as carriers for imaging probes, but peptides are of interest as they can be targeted to specific cells, are biocompatible, and can be readily engineered and modified. There are numerous reviews covering different imaging types [89–93], but some of the more important and recent developments are highlighted below.

3.3.5.1 Translocating and targeting peptide conjugates

In an effort to solve the problems of targeting efficiency and uptake, researchers have employed a variety of different peptides conjugated to imaging agents. For example, the cell-penetrating TAT peptide has been conjugated to numerous imaging agents [94], and peptides such as bombesin and the RGD motif have been used to target tumors; they both bind to receptors commonly overexpressed in various cancers [91,92].

One peptide has been developed that is now commercially available for a range of uses, including imaging. It is a cyclic peptide, called octreotide, with the sequence ^{D}P-C-F-^{D}W-K-T-C-T [95]. This peptide was developed to mimic somatostatin, a peptide hormone that has numerous roles within the body: it regulates the endocrine system; and it also interacts with G-protein-coupled receptors to affect different processes including neurotransmission. Researchers have coupled the peptide to a range of radiolabels to facilitate in vivo imaging. The peptide-radiolabel conjugate is injected, and travels through the patient's bloodstream before localizing to tumor cells that express the somatostatin receptor. It is particularly useful in locating and diagnosing pancreatic tumors.

3.3.5.2 De novo designed chelating peptides

Whilst the previous examples have a radiolabel or other imaging molecule covalently attached to the peptide, alternative peptides have been designed to noncovalently bind a class of heavy metals named lanthanides that can be used in MRI. Incorporating the metal within a targeting peptide, rather than covalently binding it may have advantages for imaging.

In one example, a de novo designed chimeric peptide was produced. The inspiration for the design comes from two sources: the first is EF-hand peptides, these are calcium-binding proteins that can bind lanthanides in place of calcium and potentially offer high relaxivity of the bound ions—an important parameter for efficient MRI imaging. The second source of inspiration came from helix-turn-helix peptides; these are small peptide motifs that often bind to DNA with high affinity, meaning the designed peptide could bind an MRI contrast agent efficiently, and be targeted to DNA within cells. The peptide was named D3W and the sequence is Ac-T-E-R-R-R-Q-Q-L-D-K-D-G-D-G-T-I-D-E-R-E-I-K-I-W-F-Q-N-K-R-A-K-I-K-NH_2, with the underlined portion being an EF-hand calcium-binding loop from the protein calmodulin. The flanking helical portions are from a DNA-binding transcription factor. Previous studies with this peptide showed it to be capable of coordinating lanthanide ions [96], as well as binding DNA [97]. In the most recent study, Gd^{III} was bound and the relaxivity of the peptide–metal complex was investigated and was found to be better than commercial agents, especially when DNA was bound to the peptide. This means that a lower concentration of the GdP3W complex can be used, compared to other MRI agents, to achieve the same contrast. A drawback, however, was that the Gd affinity for the D3W peptide was too low for biological applications meaning the affinity must be improved for this peptide to be used in vivo [98].

A second example utilizes coiled-coil peptides to bind Gd ions in their hydrophobic interior. A five-heptad design designated MB1 with the sequence Ac-G-I-A-A-I-E-Q-K-I-A-A-N-E-W-K-D-A-A-I-E-Q-K-I-A-A-I-E-Q-K-I-A-A-I-E-Q-K-G-NH_2 was created [99]. This peptide possesses Ile residues at *a* and *d* positions which favors the trimeric oligomerization state [100]. Two of these Ile residues were mutated to Asn and Asp, both of which are capable of coordinating lanthanide ions via their oxygen-rich side-chains (Fig. 3.9). This peptide is poorly folded in the absence of metal ions, but on the addition of $GdCl_3$, a well-folded α-helical structure is observed by CD spectroscopy. Further experiments designed to evaluate its properties as an MRI contrast agent showed it exhibited comparable relaxivity properties to commercially available MRI contrast agents. This design also exhibited affinities for Gd^{III} that were too low for in vivo experiments and so efforts are concentrated on improving the affinity by redesigning the coiled coil.

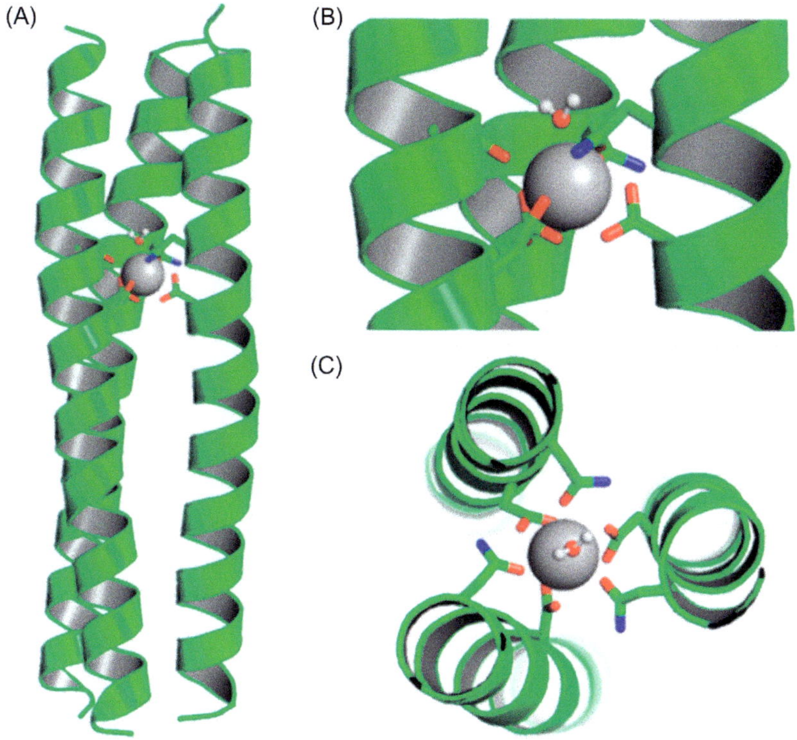

Figure 3.9 A computer-generated model of MB1 with Gd^{III} bound. (A) Side-view; (B) close-up of the metal binding site; (C) top view. Peptide chains are represented as green ribbons, Asn and Asp side-chains are shown as sticks, (oxygen atoms = red, nitrogen atoms = blue), the Gd^{III} ion is a gray sphere, a Gd-coordinated water molecule is shown as a ball-and-stick model.
Source: Image reproduced with permission from M.R. Berwick, D.J. Lewis, A.W. Jones, R.A. Parslow, T.R. Dafforn, H.J. Cooper, et al., De novo design of Ln(III) coiled coils for imaging applications. J. Am. Chem. Soc. 136 (2014) 1166–1169, © American Chemical Society.

3.3.6 3D printing—peptide bioinks

There is enormous interest in 3D printing biological constructs, from synthetic drug-screening constructs, to scaffolds for tissue engineering, and even whole organs [101–103]. To date the materials for printing, the so-called bioinks, are usually a hydrogelating material combined with cells. A variety of synthetic and natural hydrogels have been exploited and some have been modified with short peptide motifs, such as RGD, in order to promote cell adhesion.

This chapter has already examined how hydrogelating peptides can be used as cell culture scaffolds and drug-delivery vehicles, and bioprinting would allow for the more precise fabrication of such structures. The first example of a peptide

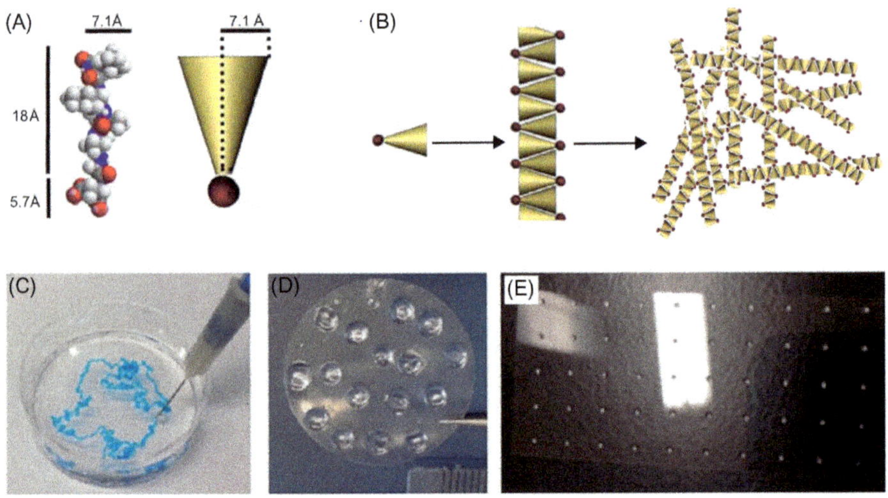

Figure 3.10 Peptide bioinks. (A) A model of a short, amphiphilic peptide employed by the Hauser group; (B) proposed mechanism of assembly from small aggregates; to fibers; to networks; (C) peptide "noodles" formed by extrusion into a salt bath; (D) and (E) arrays of printed bioink droplets.
Source: Images modified and reproduced with permission from (A) and (B) A. Mishra, Y.H. Loo, R.H. Deng, Y.J. Chuah, H.T. Hee, J.Y. Ying, et al., Ultrasmall natural peptides self-assemble to strong temperature-resistant helical fibers in scaffolds suitable for tissue engineering, Nano Today 6 (2011) 232–239 [104], © Elsevier; (C)–(E) Y.H. Loo, A. Lakshmanan, M. Ni, L.L. Toh, S. Wang, C.A.E. Hauser, Peptide bioink: self-assembling nanofibrous scaffolds for three-dimensional organotypic cultures, Nano Lett. 15 (2015) 6919–6925 [105], © American Chemical Society.

bioink was presented by the Hauser group. They have previously demonstrated that short, amphiphilic peptides comprising 2–5 hydrophobic amino acids with a polar amino acid at the C-terminus form 3D hydrogels (Fig. 3.10A,B) [104,106]. Over 100 peptides were rationally designed (although details of the design process were not provided) and evaluated for the ability to form stable gels. Two of these peptides, Ac-I-L-V-A-G-K-NH_2 and Ac-L-I-V-A-G-K-NH_2, were used for bioprinting. It was discovered that the peptide with Ile at the N-terminus gels faster, and at a lower concentration, than the peptide with Leu at the same position. Hydrogel "noodles" and droplets of these peptides were obtained by electronically dispensing the peptide solution and then mixing or washing them with PBS (Fig. 3.10C–E); the salts accelerate the gelation process, which is a matter of seconds in same cases. A wide range of potential applications have been demonstrated with these printed hydrogels; both embryonic and mesenchymal stem cells have been encapsulated into the hydrogel droplets and remain viable over several days with the mesenchymal cells aligning with, and elongating along, the hydrogel fibers. They have also prepared multicellular constructs, shown that cells remain healthy for long periods

(>3 weeks) of time, and implanted hydrogel discs into mice that remain for more than 2 months and cause no adverse effects. The authors state that these findings indicate the peptide bioinks' suitability for drug delivery, tissue engineering, and therapeutic screening applications [105].

3.3.7 Nanoparticle and nanowire synthesis

Nanoparticles and nanowires often have interesting electronic and optical properties due to a phenomenon known as quantum confinement. These properties are tuneable, and depend on the size and shape of the nanoparticles. For these reasons, such components are of interest in the field of bionanotechnology and methods to synthesize them reliably are actively being investigated. One potential method of creating such components involves using peptides as a template for their synthesis, with different types of peptides being employed for both nanowire and nanoparticle synthesis.

3.3.7.1 Peptide-directed nanoparticle synthesis

Numerous examples of peptide-mediated nanoparticle synthesis are present in literature and the subject has been widely reviewed [107–109]. An early example was inspired by the knowledge that some bacteria, when grown in the presence of silver ions, are capable of producing silver nanocrystals [110]. Instead of isolating silver-binding peptides from bacteria, as had been done previously, the authors identified putative silver-binding peptides from a phage display peptide library. To facilitate this, silver particles were incubated with the phage display library and three peptides that exhibited an affinity for silver were identified, their sequences are shown in Table 3.4.

After identification, the peptides were chemically synthesized and incubated with silver nitrate in buffer. A precipitate was observed in the solutions containing AG3 and AG4, but not AG5 although the solution was colored. Similarly, both AG3- and AG4-containing solutions gave a plasmon resonance absorption band around 440 nm, while no band was observed for AG5. TEM analysis of nanoparticles grown in the presence of AG4 revealed the particles to be 60–150 nm in diameter and a mix of hexagonal, trigonal, and spherical particles (Fig. 3.11A) [111].

It has been demonstrated that the AG3 peptide is capable of nucleating the growth of gold nanoparticles (Fig. 3.11B) [112], indeed a large amount of research

Table 3.4 Sequences of silver-binding peptides identified from a phage display library

Peptide name	Sequence
AG3	A-Y-S-S-G-A-P-P-M-P-P-F-CO_2H
AG4	N-P-S-S-L-F-R-Y-L-P-S-D-CO_2H
AG5	S-L-A-T-Q-P-P-A-T-P-P-V-CO_2H

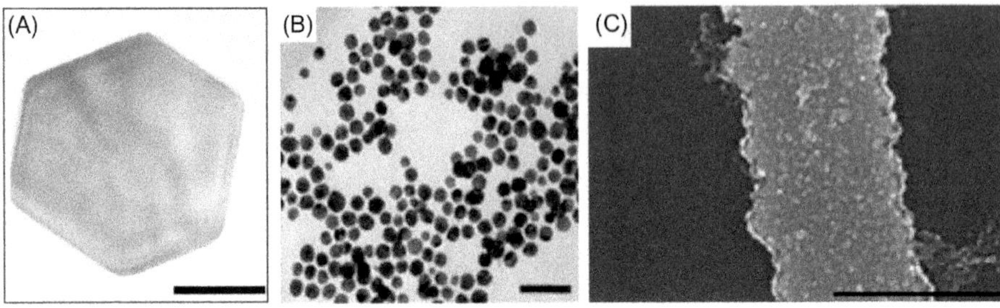

Figure 3.11 Nanoparticles formed via peptide-mediated synthesis. (A) Example of a silver nanoparticle formed using the AG4 peptide, (scale bar = 50 nm); (B) gold nanoparticles synthesized using AG3 (scale bar = 100 nm); (C) a gold-silica nanocable produced using Midas-11 (scale bar = 500 nm).
Source: Images reproduced with permission from (A) R.R. Naik, S.J. Stringer, G. Agarwal, S.E. Jones, M.O. Stone, Biomimetic synthesis and patterning of silver nanoparticles, Nat. Mater. 1 (2002) 169–172 [111], © Nature Publishing Group; (B) J.M. Slocik, M.O. Stone, R.R. Naik, Synthesis of gold nanoparticles using multifunctional peptides, Small 1 (2005) 1048–1052 [112], © Wiley; (C) J. Kim, N.V. Myung, H.G. Hur, Peptide directed synthesis of silica coated gold nanocables, Chem. Commun. 46 (2010) 4366–4368 [113], © The Royal Society of Chemistry.

into peptide-mediated nanoparticle synthesis has focused on the formation of gold nanoparticles. Many of the peptides have been identified in a similar fashion to those outlined above—by screening random peptide libraries and making mutations to improve the identified peptides. Several articles compare the better-known gold-binding peptides in an attempt to identify important sequence and structural characteristics [114,115].

3.3.7.2 Nanowire synthesis

A seminal example in this field came from Reches and Gazit in 2003 [116]. They discovered that the N-terminally Fmoc-protected dipeptide Fmoc-FF was capable of self-assembling into nanotubes. This peptide motif was originally of interest as the same researchers discovered that other, short (~4–6 amino acid) aromatic peptides formed amyloid fibrils [117], which are implicated in numerous disease states, and so they decided to investigate Fmoc-FF as it is the core recognition motif of amyloid beta (Aβ). The peptide was dissolved at high concentrations into an organic solvent, but upon dilution into aqueous medium it could be observed that the dipeptide underwent a self-assembly process. Both TEM and SEM analysis revealed nanotubes of a reasonably uniform length were formed and FT-IR analysis showed the assemblies have a β-sheet conformation. It was then demonstrated that ionic silver could be added to the nanotubes, with TEM and EDX analysis showing silver nanoparticles were formed within the tubes. This method was subsequently modified: nanotubes were added to a boiling ionic silver solution and citric acid was

Table 3.5 Peptide sequences used to synthesize different metal nanowires

Peptide	Sequence	Metal
A7	C-N-N-P-M-H-Q-N-C-CO_2H	ZnS
CP7	C-N-A-G-D-H-A-N-C-CO_2H	CoPt
J140	S-L-T-P-L-T-T-S-H-L-R-S-CO_2H	CdS
FP12	H-N-K-H-L-P-S-T-Q-P-L-A-CO_2H	FePt

subsequently added to reduce the silver and ensure its homogeneous distribution. Upon addition of proteinase K, the peptidic exterior of the tubes was proteolysed, leaving ~20 nm diameter silver nanowires, that extended for several microns, behind.

Another paper published the following year took a different approach to generating nanowires. Four peptides, specific for different metals, were identified using phage display. Two were disulfide-linked cyclic peptides and two were linear, their sequences and the metals they are specific for are given in Table 3.5. Once identified, these peptides were incorporated into the capsid of M13 bacteriophage, producing a linear template that displays peptides capable of directing nanowire synthesis. To produce ZnS and CdS nanowires, the viral peptide template was incubated with Zn- and Cd-containing salts at low temperatures. This led to the formation of small (3–5 nm) nanocrystals, which were prevented from fusing by the presence of the peptides. Heating above 350°C removed the peptide–phage template and allowed the crystals to fuse and form nanowires.

CoPt and FePt nanowires were synthesized in a similar manner; the appropriate peptide–phage template was incubated with metal salts which were reduced in situ. Again, heating removed the organic template, leaving metal nanowires behind [118].

Gold nanowires have been synthesized using a peptide with the sequence T-G-T-S-V-L-I-A-T-P-G-V-CO_2H [119]. This peptide, dubbed Midas-11, was derived from a peptide named Midas-2, which was itself also identified from a combinatorial peptide library using phage display. The difference between Midas-11 and Midas-2 is that the Gly in position 11 was originally a Tyr, (Y). This one-amino-acid substitution led to large changes in the structures observed; Midas-2 formed spherical nanoparticles at pH 7.5 in PBS, whereas Midas-11 formed a range of different gold nanostructures, depending on the pH of the solution and the concentration of $HAuCl_4$ used. Nanowires were formed using 0.5 mM $HAuCl_4$ at pH 5.7, at pH > 7, nanoparticles were formed, when the pH was dropped below 3, large hexagonal or trigonal gold platelets were formed and nanoribbons were formed at pH 5.7 when the $HAuCl_4$ concentration was increased to 30 mM. The group has since demonstrated that these nanoribbons can be insulated with a silica shell if a second, dual-functional (both gold-binding and silica-synthesizing) peptide is added once the nanoribbons have been formed (Fig. 3.11C) [113].

Table 3.6 The sequences of the de novo designed peptides discussed in this chapter, along with the design method, structure, applications, and associated references

Peptide	Design principles, structure	Applications	References
EAK16	Inspired by natural proteins	Tissue engineering	[24,28]
Ac-$(AEAEAKAK)_2$-NH_2	β-Structured, self-assembling		
RAD16-II	De novo designed, inspired by EAK16	Tissue engineering, cell culture, injury repair	[26,28]
Ac-$(RARADADA)_2$-NH_2	β-Structured, self-assembling		
RAD16-I	De novo designed, inspired by RAD16-II	Drug delivery, tissue engineering	[29,60,61]
Ac-$(RADA)_4$-NH_2	β-Structured, self-assembling		
MAX1	De novo designed	Tissue engineering, antibacterials	[36,41]
VKVKVKVKV^{D}PPTKVKVKVKV-NH_2	β-Hairpin, self-assembling, hydrogelating		
	pH responsive		
MAX8	De novo designed, related to MAX1	Drug encapsulation and release, drug delivery	[40,56–59]
VKVKVKVKV^{D}PPTKVEVKVKV-NH_2	β-Hairpin, self-assembling, hydrogelating.		
SVS-1	De novo designed, related to MAX1	Cell-penetrating, drug delivery	[71,73]
KVKVKVKV^{D}PPTKVKVKVK-NH_2	β-Hairpin		
hSAF-$_{AAAW}$ p1	De novo designed	Tissue engineering, cell culture	[42,43,45–47]
Ac-KIAALKAKIAALKAEIAALEWENAALEA-NH_2			
hSAF-$_{AAAW}$ p2	Coiled-coil peptides, self-assembling, hydrogelating		
Ac-KIAALKAKNAALKAEIAALEWEIAALEA-NH_2			

KOD	De novo designed	Tissue engineering, disease models, hemostat	[48,49]
Ac-$(PKG)_4(POG)_4(DOG)_4$-NH_2	Collagen mimetic peptide		
	Self-assembling, hydrogelating		
NHbipy	De novo designed	Tissue engineering	[50,51]
NTA-$(POG)_4$-(PK[bipy]G)-$(POG)_4$-HH-NH_2	Collagen mimetic peptide		
NTA = nitrilotriacetic acid	Self-assembling with metals		
Bipy = bipyridyl			
NCoH	De novo designed, based on NHbipy	Drug delivery	[62,64–66]
NTA-$(POG)_9$HH-NH_2	Self-assembles with metals		
NTA = nitrilotriacetic acid			
JR2EC	De novo designed	Drug encapsulation and release	[63]
NAADLEKAIEALEKHLEAKGPCDAAQLEKQLEQAFEAFERAG-CO_2H	Helix-loop-helix peptide		
	Dimerizes, assembles with Zn^{II}		
SPACE	Discovered via phage display	Transdermal drug delivery, skin penetrating	[69]
ACTGSTQHQCG-CO_2H	Cyclic peptide		
iRGD	Discovered via phage display	Drug delivery, tumor penetrating	[70]
CRGDK/RGPDC-CO_2H	Cyclic peptide		
iNGR	De novo designed, based on iRGD	Drug delivery, tumor penetrating	[72]
CRNGRGPDC-CO_2H	Cyclic peptide		

(Continued)

Table 3.6 (Continued)

Peptide	Design principles, structure	Applications	References
SAPNs	De novo designed	Antigen display, vaccine development	[74,78–80]
Ac-DEMLRELQETNAALQDVRELLRQQVKQITFLKCLLM-GG-RLLCR LEELERR LEELERR LEELERR-NH_2	Fusion of existing coiled-coil peptides Self-assembling		
SVLPs	De novo designed	Antigen display, vaccine development	[75,81–83]
GGAVSKVLHLEGEVNKIKSALLSTNKAVVSLSNGVSVLTSK VLDLKNY IDKQ-NH_2	Trimeric coiled coil Self-assembling		
PD1	De novo designed	Membrane protein stabilization	[84]
Ac-EELLKQALQQAQQLLQQAQELAKK-NH_2	α-Helical		
Beltide-1	De novo designed,	Membrane protein stabilization	[85]
Ac-DWLKAFYDKVAEKLKEAFDWLKAFYDKVAEKLKEAF-NH_2	α-Helical		
Beltide-2			
Ac-DWLKAFYDKVAEKLKEAFPDWLKAFYDKVAEKLKEAF-NH_2			
Beltide-3			
Ac-DWLKAFYDKVAEKLKEAFGGDWLKAFYDKVAEKLKEAF-NH_2			
Octreotide	Inspired by somastatin	Tumor cell imaging	[95]
^{D}PCF^{D}WKTCT-CO_2H	Cyclic peptide		
P3W	De novo designed	Gd binding, MRI imaging agent	[96–98]
Ac-TERRRQQLDKDGDGTIDEREIKIWFQNKRAKIK-NH_2	Chimeric, inspired by EF hands and helix-turn-helix peptides		

MB1	De novo designed	Gd binding, MRI imaging agent	[99]
Ac-GIAAIEQKIAANEWKDAAIEQKIAAIEQKIAAIEQKG-NH_2	Tetrameric coiled-coil		
Ultrashort Peptides	De novo designed	Bioinks, cell culture, tissue engineering	[104–106]
Ac-ILVAGK-NH_2	Self-assembling, hydrogelating		
Ac-LIVAGK-NH_2			
AG3	Discovered using phage display	Au and Ag nanoparticle formation	[111,112]
AYSSGAPPMPPFCO_2H	Inspired by bacterial peptides		
AG4			
NPSSLFRYLPSD-CO_2H			
AG5			
SLATQPPATPPV-CO_2H			
Fmoc-FF-NH_2	Inspired by amyloid peptides	Formation of gold nanowires	[116,117]
A7	Discovered by phage display	Nanowire synthesis	[118]
CNNPMHQNC-CO_2H	A7 and CP7 are cyclic peptides		
CP7			
CNAGDHANC-CO_2H			
J140			
SLTPLTTSHLR-S-CO_2H			
FP12			
HNKHLPSTQPLA-CO_2H			
Midas-11	Discovered by phage display	Formation of gold nanowires	[113,119]
TGTSVLIATPGV-CO_2H	Inspired by bacterial peptides		

3.4 Conclusions and outlook

The fact that peptides can now be created to assemble and function in a defined and programmable manner has led to a noticeable increase in the number of designs emerging. This is due to both fundamental research that has uncovered rules governing peptide folding and assembly, and to advances in peptide synthesis, particularly solid-phase peptide synthesis, meaning the production of peptides is becoming increasingly accessible.

As is evident from this chapter, the applications of such peptides, and their self-assembled structures are incredibly diverse with well-established fields such as tissue engineering and drug delivery still being actively investigated as more sophisticated structures are developed and new fields, such as employing peptides, as bioinks being established. The number of designs and potential applications can only be expected to grow as our knowledge of peptide structure and function is further advanced and understood.

3.5 Summary of de novo designed peptides

The peptides discussed in this chapter, their design method, applications and associated references are listed in Table 3.6.

References

[1] A.L. Boyle, D.N. Woolfson, De novo designed peptides for biological applications, Chem. Soc. Rev. 40 (2011) 4295–4306.
[2] C. Tamerler, M. Sarikaya, Molecular biomimetics: nanotechnology and bionanotechnology using genetically engineered peptides, Philos. Trans. R. Soc. A Math. Phys. Eng. Sci. 367 (2009) 1705–1726.
[3] R. De La Rica, H. Matsui, Applications of peptide and protein-based materials in bionanotechnology, Chem. Soc. Rev. 39 (2010) 3499–3509.
[4] X.Y. Gao, H. Matsui, Peptide-based nanotubes and their applications in bionanotechnology, Adv. Mater. 17 (2005) 2037–2050.
[5] Y. Loo, S.G. Zhang, C.A.E. Hauser, From short peptides to nanofibers to macromolecular assemblies in biomedicine, Biotechnol. Adv. 30 (2012) 593–603.
[6] R.J. Mart, R.D. Osborne, M.M. Stevens, R.V. Ulijn, Peptide-based stimuli-responsive biomaterials, Soft Matter. 2 (2006) 822–835.
[7] R. Behrendt, P. White, J. Offer, Advances in Fmoc solid-phase peptide synthesis, J. Pept. Sci. 22 (2016) 4–27.
[8] A.R. Thomson, C.W. Wood, A.J. Burton, G.J. Bartlett, R.B. Sessions, R.L. Brady, et al., Computational design of water-soluble alpha-helical barrels, Science 346 (2014) 485–488.
[9] F.H.C. Crick, The packing of alpha-helices – simple coiled-coils, Acta Crystallogr. 6 (1953) 689–697.

[10] A. Lupas, Coiled coils: new structures and new functions, Trends Biochem. Sci. 21 (1996) 375–382.
[11] W.D. Kohn, R.S. Hodges, De novo design of alpha-helical coiled coils and bundles: models for the development of protein-design principles, Trends Biotechnol. 16 (1998) 379–389.
[12] W.D. Kohn, C.T. Mant, R.S. Hodges, Alpha-helical protein assembly motifs, J. Biol. Chem. 272 (1997) 2583–2586.
[13] B.R. Gibney, Y. Isogai, F. Rabanal, K.S. Reddy, A.M. Grosset, C.C. Moser, et al., Self-assembly of heme A and heme B in a designed four-helix bundle: implications for a cytochrome c oxidase maquette, Biochemistry 39 (2000) 11041–11049.
[14] R.L. Koder, J.L.R. Anderson, L.A. Solomon, K.S. Reddy, C.C. Moser, P.L. Dutton, Design and engineering of an O-2 transport protein, Nature 458 (2009) 305–309.
[15] J.M. Shifman, C.C. Moser, W.A. Kalsbeck, D.F. Bocian, P.L. Dutton, Functionalized de novo designed proteins: mechanism of proton coupling to oxidation/reduction in heme protein maquettes, Biochemistry 37 (1998) 16815–16827.
[16] S.H. Gellman, Minimal model systems for beta sheet secondary structure in proteins, Curr. Opin. Chem. Biol. 2 (1998) 717–725.
[17] R.M. Hughes, M.L. Waters, Model systems for beta-hairpins and beta-sheets, Curr. Opin. Struct. Biol. 16 (2006) 514–524.
[18] C.K. Smith, L. Regan, Guidelines for protein design: the energetics of beta sheet side chain interactions, Science 270 (1995) 980–982.
[19] C.K. Smith, L. Regan, Construction and design of β-sheets, Acc. Chem. Res. 30 (1997) 153–161.
[20] M. Vanderrest, R. Garrone, Collagen family of proteins, FASEB J. 5 (1991) 2814–2823.
[21] K.E. Kadler, D.F. Holmes, J.A. Trotter, J.A. Chapman, Collagen fibril formation, Biochem. J. 316 (1996) 1–11.
[22] A. Rich, F.H.C. Crick, Molecular structure of collagen, J. Mol. Biol. 3 (1961) 483–506.
[23] J.A. Fallas, L.E.R. O'leary, J.D. Hartgerink, Synthetic collagen mimics: self-assembly of homotrimers, heterotrimers and higher order structures, Chem. Soc. Rev. 39 (2010) 3510–3527.
[24] S. Zhang, T. Holmes, C. Lockshin, A. Rich, Spontaneous assembly of a self-complementary oligopeptide to form a stable macroscopic membrane, Proc. Natl. Acad. Sci. U.S.A. 90 (1993) 3334–3338.
[25] S. Zhang, C. Lockshin, A. Herbert, E. Winter, A. Rich, Zuotin, a putative Z-DNA binding protein in *Saccharomyces cerevisiae*, EMBO J. 11 (1992) 3787–3796.
[26] S. Zhang, C. Lockshin, R. Cook, A. Rich, Unusually stable beta-sheet formation in an ionic self-complementary oligopeptide, Biopolymers 34 (1994) 663–672.
[27] R.G. Ellis-Behnke, Y.X. Liang, S.W. You, D.K.C. Tay, S.G. Zhang, K.F. So, et al., Nano neuro knitting: peptide nanofiber scaffold for brain repair and axon regeneration with functional return of vision, Proc. Natl. Acad. Sci. U.S.A. 103 (2006) 5054–5059.
[28] S.G. Zhang, T.C. Holmes, C.M. Dipersio, R.O. Hynes, X. Su, A. Rich, Self-complementary oligopeptide matrices support mammalian-cell attachment, Biomaterials 16 (1995) 1385–1393.
[29] T.C. Holmes, S. De Lacalle, X. Su, G.S. Liu, A. Rich, S.G. Zhang, Extensive neurite outgrowth and active synapse formation on self-assembling peptide scaffolds, Proc. Natl. Acad. Sci. U.S.A. 97 (2000) 6728–6733.

[30] F. Gelain, D. Bottai, A. Vescovi, S.G. Zhang, Designer self-assembling peptide nanofiber scaffolds for adult mouse neural stem cell 3-dimensional cultures, PLoS One. 1 (2006) e119.
[31] T. Waku, N. Tanaka, Recent advances in nanofibrous assemblies based on β-sheet-forming peptides for biomedical applications, Polym. Int. 66 (2017) 277–288.
[32] Z.Q. Yu, Z. Cai, Q.L. Chen, M.H. Liu, L. Ye, J.Y. Ren, et al., Engineering beta-sheet peptide assemblies for biomedical applications, Biomater. Sci. 4 (2016) 365–374.
[33] S.G. Zhang, Fabrication of novel biomaterials through molecular self-assembly, Nat. Biotechnol. 21 (2003) 1171–1178.
[34] P.X. Ma, Biomimetic materials for tissue engineering, Adv. Drug Deliv. Rev. 60 (2008) 184–198.
[35] Q.Y. Meng, S.L. Yao, X.M. Wang, Y.Y. Chen, RADA16: a self-assembly peptide hydrogel for the application in tissue regeneration, J. Biomater. Tissue Eng. 4 (2014) 1019–1029.
[36] J.P. Schneider, D.J. Pochan, B. Ozbas, K. Rajagopal, L. Pakstis, J. Kretsinger, Responsive hydrogels from the intramolecular folding and self-assembly of a designed peptide, J. Am. Chem. Soc. 124 (2002) 15030–15037.
[37] L.A. Haines, K. Rajagopal, B. Ozbas, D.A. Salick, D.J. Pochan, J.P. Schneider, Light-activated hydrogel formation via the triggered folding and self-assembly of a designed peptide, J. Am. Chem. Soc. 127 (2005) 17025–17029.
[38] B. Ozbas, K. Rajagopal, J.P. Schneider, D.J. Pochan, Semiflexible chain networks formed via self-assembly of beta-hairpin molecules, Phys. Rev. Lett. 93 (2004) 268106.
[39] D.J. Pochan, J.P. Schneider, J. Kretsinger, B. Ozbas, K. Rajagopal, L. Haines, Thermally reversible hydrogels via intramolecular folding and consequent self-assembly of a de novo designed peptide, J. Am. Chem. Soc. 125 (2003) 11802–11803.
[40] L. Haines-Butterick, K. Rajagopal, M. Branco, D. Salick, R. Rughani, M. Pilarz, et al., Controlling hydrogelation kinetics by peptide design for three-dimensional encapsulation and injectable delivery of cells, Proc. Natl. Acad. Sci. U.S.A. 104 (2007) 7791–7796.
[41] D.A. Salick, D.J. Pochan, J.P. Schneider, Design of an injectable beta-hairpin peptide hydrogel that kills methicillin-resistant *Staphylococcus aureus*, Adv. Mater. 21 (2009) 4120–4123.
[42] M.J. Pandya, G.M. Spooner, M. Sunde, J.R. Thorpe, A. Rodger, D.N. Woolfson, Sticky-end assembly of a designed peptide fiber provides insight into protein fibrillogenesis, Biochemistry 39 (2000) 8728–8734.
[43] A.M. Smith, E.F. Banwell, W.R. Edwards, M.J. Pandya, D.N. Woolfson, Engineering increased stability into self-assembled protein fibers, Adv. Funct. Mater. 16 (2006) 1022–1030.
[44] T.H. Sharp, M. Bruning, J. Mantell, R.B. Sessions, A.R. Thomson, N.R. Zaccai, et al., Cryo-transmission electron microscopy structure of a gigadalton peptide fiber of de novo design, Proc. Natl. Acad. Sci. U.S.A. 109 (2012) 13266–13271.
[45] E.F. Banwell, E.S. Abelardo, D.J. Adams, M.A. Birchall, A. Corrigan, A.M. Donald, et al., Rational design and application of responsive alpha-helical peptide hydrogels, Nat. Mater. 8 (2009) 596–600.
[46] N. Mehrban, E. Abelardo, A. Wasmuth, K.L. Hudson, L.M. Mullen, A.R. Thomson, et al., Assessing cellular response to functionalized alpha-helical peptide hydrogels, Adv. Healthc. Mater. 3 (2014) 1387–1391.
[47] N. Mehrban, B.F. Zhu, F. Tamagnin, F.I. Young, A. Wasmuth, K.L. Hudson, et al., Functionalized alpha-helical peptide hydrogels for neural tissue engineering, ACS Biomater. Sci. Eng. 1 (2015) 431–439.

[48] L.E.R. O'leary, J.A. Fallas, E.L. Bakota, M.K. Kang, J.D. Hartgerink, Multi-hierarchical self-assembly of a collagen mimetic peptide from triple helix to nanofibre and hydrogel, Nat. Chem. 3 (2011) 821–828.
[49] V.A. Kumar, N.L. Taylor, A.A. Jalan, L.K. Hwang, B.K. Wang, J.D. Hartgerink, A nanostructured synthetic collagen mimic for hemostasis, Biomacromolecules 15 (2014) 1484–1490.
[50] M.M. Pires, D.E. Przybyla, J. Chmielewski, A metal-collagen peptide framework for three-dimensional cell culture, Angew. Chem. Int. Ed. 48 (2009) 7813–7817.
[51] V. Hernandez-Gordillo, J. Chmielewski, Mimicking the extracellular matrix with functionalized, metal-assembled collagen peptide scaffolds, Biomaterials 35 (2014) 7363–7373.
[52] M.M. Stevens, J.H. George, Exploring and engineering the cell surface interface, Science 310 (2005) 1135–1138.
[53] K. Anselme, Osteoblast adhesion on biomaterials, Biomaterials 21 (2000) 667–681.
[54] M.P. Lutolf, J.A. Hubbell, Synthetic biomaterials as instructive extracellular microenvironments for morphogenesis in tissue engineering, Nat. Biotechnol. 23 (2005) 47–55.
[55] H. Shin, S. Jo, A.G. Mikos, Biomimetic materials for tissue engineering, Biomaterials 24 (2003) 4353–4364.
[56] M.C. Branco, D.J. Pochan, N.J. Wagner, J.P. Schneider, The effect of protein structure on their controlled release from an injectable peptide hydrogel, Biomaterials 31 (2010) 9527–9534.
[57] A. Altunbas, S.J. Lee, S.A. Rajasekaran, J.P. Schneider, D.J. Pochan, Encapsulation of curcumin in self-assembling peptide hydrogels as injectable drug delivery vehicles, Biomaterials 32 (2011) 5906–5914.
[58] S. Lindsey, J.H. Piatt, P. Worthington, C. Sonmez, S. Satheye, J.P. Schneider, et al., Beta hairpin peptide hydrogels as an injectable solid vehicle for neurotrophic growth factor delivery, Biomacromolecules 16 (2015) 2672–2683.
[59] J.E.P. Sun, B. Stewart, A. Litan, S.J. Lee, J.P. Schneider, S.A. Langhans, et al., Sustained release of active chemotherapeutics from injectable-solid beta-hairpin peptide hydrogel, Biomater. Sci. 4 (2016) 839–848.
[60] Y. Nagai, L.D. Unsworth, S. Koutsopoulos, S.G. Zhang, Slow release of molecules in self-assembling peptide nanofiber scaffold, J. Control. Release 115 (2006) 18–25.
[61] S. Koutsopoulos, L.D. Unsworth, Y. Nagai, S.G. Zhang, Controlled release of functional proteins through designer self-assembling peptide nanofiber hydrogel scaffold, Proc. Natl. Acad. Sci. U.S.A. 106 (2009) 4623–4628.
[62] M.M. Pires, D. Ernenwein, J. Chmielewski, Selective decoration and release of His-tagged proteins from metal-assembled collagen peptide microflorettes, Biomacromolecules 12 (2011) 2429–2433.
[63] C. Aronsson, R. Selegard, D. Aili, Zinc-triggered hierarchical self-assembly of fibrous helix-loop-helix peptide superstructures for controlled encapsulation and release, Macromolecules 49 (2016) 6997–7003.
[64] M.M. Pires, J. Chmielewski, Self-assembly of collagen peptides into microflorettes via metal coordination, J. Am. Chem. Soc. 131 (2009) 2706–2712.
[65] D.E. Przybyla, J. Chmielewski, Metal-triggered collagen peptide disk formation, J. Am. Chem. Soc. 132 (2010) 7866–7867.
[66] J. Gleaton, J. Chmielewski, Thermally controlled collagen peptide cages for biopolymer delivery, ACS Biomater. Sci. Eng. 1 (2015) 1002–1008.

[67] I. Margalith, C. Suter, B. Ballmer, P. Schwarz, C. Tiberi, T. Sonati, et al., Polythiophenes inhibit prion propagation by stabilizing prion protein (PrP) aggregates, J. Biol. Chem. 287 (2012) 18872–18887.

[68] S. Menegatti, M. Zakrewsky, S. Kumar, J.S. D.E. Oliveira, J.A. Muraski, S. Mitragotri, De novo design of skin-penetrating peptides for enhanced transdermal delivery of peptide drugs, Adv. Healthc. Mater. 5 (2016) 602–609.

[69] T. Hsu, S. Mitragotri, Delivery of siRNA and other macromolecules into skin and cells using a peptide enhancer, Proc. Natl. Acad. Sci. U.S.A. 108 (2011) 15816–15821.

[70] K.N. Sugahara, T. Teesalu, P.P. Karmali, V.R. Kotamraju, L. Agemy, O.M. Girard, et al., Tissue-penetrating delivery of compounds and nanoparticles into tumors, Cancer Cell. 16 (2009) 510–520.

[71] S.H. Medina, J.P. Schneider, Cancer cell surface induced peptide folding allows intracellular translocation of drug, J. Control. Release 209 (2015) 317–326.

[72] L. Alberici, L. Roth, K.N. Sugahara, L. Agemy, V.R. Kotamraju, T. Teesalu, et al., De novo design of a tumor-penetrating peptide, Cancer Res. 73 (2013) 804–812.

[73] C. Sinthuvanich, A.S. Veiga, K. Gupta, D. Gaspar, R. Blumenthal, J.P. Schneider, Anticancer beta-hairpin peptides: membrane-induced folding triggers activity, J. Am. Chem. Soc. 134 (2012) 6210–6217.

[74] S. Raman, G. Machaidze, A. Lustig, U. Aebi, P. Burkhard, Structure-based design of peptides that self-assemble into regular polyhedral nanoparticles, Nanomedicine 2 (2006) 95–102.

[75] F. Boato, R.M. Thomas, A. Ghasparian, A. Freund-Renard, K. Moehle, J.A. Robinson, Synthetic virus-like particles from self-assembling coiled-coil lipopeptides and their use in antigen display to the immune system, Angew. Chem. Int. Ed. 46 (2007) 9015–9018.

[76] Y.K. Yang, P. Ringler, S.A. Muller, P. Burkhard, Optimizing the refolding conditions of self-assembling polypeptide nanoparticles that serve as repetitive antigen display systems, J. Struct. Biol. 177 (2012) 168–176.

[77] T. Doll, R. Dey, P. Burkhard, Design and optimization of peptide nanoparticles, J. Nanobiotechnol. 13 (2015) 73.

[78] T. Pimentel, Z. Yan, S.A. Jeffers, K.V. Holmes, R.S. Hodges, P. Burkhard, Peptide nanoparticles as novel immunogens: design and analysis of a prototypic severe acute respiratory syndrome vaccine, Chem. Biol. Drug Des. 73 (2009) 53–61.

[79] S.A. Kaba, C. Brando, Q. Guo, C. Mittelholzer, S. Raman, D. Tropel, et al., A nonadjuvanted polypeptide nanoparticle vaccine confers long-lasting protection against rodent malaria, J. Immunol. 183 (2009) 7268–7277.

[80] T. Neef, S. Babapoor, C. Mittelholzer, S. Raman, M. Khan, P. Burkhard, A self-assembling polypeptide nanoparticle vaccine for avian influenza, Biophys. J. 102 (2012) 585A.

[81] S.L. Okitsu, F. Boato, M.S. Mueller, D.B. Li, D. Vogel, N. Westerfeld, et al., Antibodies elicited by a virosomally formulated *Plasmodium falciparum* serine repeat antigen-5 derived peptide detect the processed 47 kDa fragment both in sporozoites and merozoites, Peptides 28 (2007) 2051–2060.

[82] S.L. Okitsu, U. Kienzl, K. Moehle, O. Silvie, E. Peduzzi, M.S. Mueller, et al., Structure-activity-based design of a synthetic malaria peptide eliciting sporozoite inhibitory antibodies in a virosomal formulation, Chem. Biol. 14 (2007) 577–587.

[83] T. Riedel, A. Ghasparian, K. Moehle, P. Rusert, A. Trkola, J.A. Robinson, Synthetic virus-like particles and conformationally constrained peptidomimetics in vaccine design, Chembiochem 12 (2011) 2829–2836.

[84] C.E. Schafmeister, L.J.W. Miercke, R.M. Stroud, Structure at 2.5 angstrom of a designed peptide that maintains solubility of membrane-proteins, Science 262 (1993) 734–738.
[85] A.N. Larsen, K.K. Sorensen, N.T. Johansen, A. Martel, J.J.K. Kirkensgaard, K.J. Jensen, et al., Dimeric peptides with three different linkers self-assemble with phospholipids to form peptide nanodiscs that stabilize membrane proteins, Soft Matter. 12 (2016) 5937–5949.
[86] P. Kiley, X.J. Zhao, M. Vaughn, M.A. Baldo, B.D. Bruce, S.G. Zhang, Self-assembling peptide detergents stabilize isolated photosystem I on a dry surface for an extended time, PLoS Biol. 3 (2005) 1180–1186.
[87] K. Matsumoto, M. Vaughn, B.D. Bruce, S. Koutsopoulos, S.G. Zhang, Designer peptide surfactants stabilize functional photosystem-I membrane complex in aqueous solution for extended time, J. Phys. Chem. B 113 (2009) 75–83.
[88] X.J. Zhao, Y. Nagai, P.J. Reeves, P. Kiley, H.G. Khorana, S.G. Zhang, Designer short peptide surfactants stabilize G protein-coupled receptor bovine rhodopsin, Proc. Natl. Acad. Sci. U.S.A. 103 (2006) 17707–17712.
[89] K. Chen, P.S. Conti, Target-specific delivery of peptide-based probes for PET imaging, Adv. Drug Deliv. Rev. 62 (2010) 1005–1022.
[90] S.B. Fonseca, M.P. Pereira, S.O. Kelley, Recent advances in the use of cell-penetrating peptides for medical and biological applications, Adv. Drug Deliv. Rev. 61 (2009) 953–964.
[91] S. Lee, J. Xie, X.Y. Chen, Peptide-based probes for targeted molecular imaging, Biochemistry 49 (2010) 1364–1376.
[92] S. Liu, D.S. Edwards, 99mTc-Labeled small peptides as diagnostic radiopharmaceuticals, Chem. Rev. 99 (1999) 2235–2268.
[93] S.M. Okarvi, Peptide-based radiopharmaceuticals: future tools for diagnostic imaging of cancers and other diseases, Med. Res. Rev. 24 (2004) 357–397.
[94] K.E. Bullok, S.T. Gammon, S. Violini, A.M. Prantner, V.M. Villalobos, V. Sharma, et al., Permeation peptide conjugates for in vivo molecular imaging applications, Mol. Imaging 5 (2006) 1–15.
[95] E.P. Krenning, W.A.P. Breeman, P.P.M. Kooij, J.S. Lameris, W.H. Bakker, J.W. Koper, et al., Localization of endocrine-related tumors with radioiodinated analog of somatostatin, Lancet 1 (1989) 242–244.
[96] J.T. Welch, W.R. Kearney, S.J. Franklin, Lanthanide-binding helix-turn-helix peptides: solution structure of a designed metallonuclease, Proc. Natl. Acad. Sci. U.S.A. 100 (2003) 3725–3730.
[97] R.T. Kovacic, J.T. Welch, S.J. Franklin, Sequence-selective DNA cleavage by a chimeric metallopeptide, J. Am. Chem. Soc. 125 (2003) 6656–6662.
[98] P. Caravan, J.M. Greenwood, J.T. Welch, S.J. Franklin, Gadolinium-binding helix-turn-helix peptides: DNA-dependent MRI contrast agents, Chem. Commun. 20 (2003) 2574–2575.
[99] M.R. Berwick, D.J. Lewis, A.W. Jones, R.A. Parslow, T.R. Dafforn, H.J. Cooper, et al., De novo design of Ln(III) coiled coils for imaging applications, J. Am. Chem. Soc. 136 (2014) 1166–1169.
[100] P.B. Harbury, T. Zhang, P.S. Kim, T. Alber, A switch between 2-stranded, 3-stranded and 4-stranded coiled coils in GCN4 leucine-zipper mutants, Science 262 (1993) 1401–1407.
[101] B. Derby, Printing and prototyping of tissues and scaffolds, Science 338 (2012) 921–926.

[102] V. Mironov, T. Boland, T. Trusk, G. Forgacs, R.R. Markwald, Organ printing: computer-aided jet-based 3D tissue engineering, Trends Biotechnol. 21 (2003) 157–161.
[103] S.V. Murphy, A. Atala, 3D bioprinting of tissues and organs, Nat. Biotechnol. 32 (2014) 773–785.
[104] A. Mishra, Y.H. Loo, R.H. Deng, Y.J. Chuah, H.T. Hee, J.Y. Ying, et al., Ultrasmall natural peptides self-assemble to strong temperature-resistant helical fibers in scaffolds suitable for tissue engineering, Nano Today 6 (2011) 232–239.
[105] Y.H. Loo, A. Lakshmanan, M. Ni, L.L. Toh, S. Wang, C.A.E. Hauser, Peptide bioink: self-assembling nanofibrous scaffolds for three-dimensional organotypic cultures, Nano Lett. 15 (2015) 6919–6925.
[106] C.A.E. Hauser, R.S. Deng, A. Mishra, Y.H. Loo, U. Khoe, F.R. Zhuang, et al., Natural tri- to hexapeptides self-assemble in water to amyloid beta-type fiber aggregates by unexpected alpha-helical intermediate structures, Proc. Natl. Acad. Sci. U.S.A. 108 (2011) 1361–1366.
[107] C.L. Chen, N.L. Rosi, Peptide-based methods for the preparation of nanostructured inorganic materials, Angew. Chem. Int. Ed. 49 (2010) 1924–1942.
[108] M.B. Dickerson, K.H. Sandhage, R.R. Naik, Protein- and peptide-directed syntheses of inorganic materials, Chem. Rev. 108 (2008) 4935–4978.
[109] M.R. Jones, K.D. Osberg, R.J. Macfarlane, M.R. Langille, C.A. Mirkin, Templated techniques for the synthesis and assembly of plasmonic nanostructures, Chem. Rev. 111 (2011) 3736–3827.
[110] T. Klaus, R. Joerger, E. Olsson, C.G. Granqvist, Silver-based crystalline nanoparticles, microbially fabricated, Proc. Natl. Acad. Sci. U. S. A. 96 (1999) 13611–13614.
[111] R.R. Naik, S.J. Stringer, G. Agarwal, S.E. Jones, M.O. Stone, Biomimetic synthesis and patterning of silver nanoparticles, Nat. Mater. 1 (2002) 169–172.
[112] J.M. Slocik, M.O. Stone, R.R. Naik, Synthesis of gold nanoparticles using multifunctional peptides, Small 1 (2005) 1048–1052.
[113] J. Kim, N.V. Myung, H.G. Hur, Peptide directed synthesis of silica coated gold nanocables, Chem. Commun. 46 (2010) 4366–4368.
[114] Y. Li, Z.H. Tang, P.N. Prasad, M.R. Knecht, M.T. Swihart, Peptide-mediated synthesis of gold nanoparticles: effects of peptide sequence and nature of binding on physicochemical properties, Nanoscale 6 (2014) 3165–3172.
[115] N.M. Bedford, Z.E. Hughes, Z.H. Tang, Y. Li, B.D. Briggs, Y. Ren, et al., Sequence-dependent structure/function relationships of catalytic peptide-enabled gold nanoparticles generated under ambient synthetic conditions, J. Am. Chem. Soc. 138 (2016) 540–548.
[116] M. Reches, E. Gazit, Casting metal nanowires within discrete self-assembled peptide nanotubes, Science 300 (2003) 625–627.
[117] M. Reches, Y. Porat, E. Gazit, Amyloid fibril formation by pentapeptide and tetrapeptide fragments of human calcitonin, J. Biol. Chem. 277 (2002) 35475–35480.
[118] C.B. Mao, D.J. Solis, B.D. Reiss, S.T. Kottmann, R.Y. Sweeney, A. Hayhurst, et al., Virus-based toolkit for the directed synthesis of magnetic and semiconducting nanowires, Science 303 (2004) 213–217.
[119] J. Kim, Y. Rheem, B. Yoo, Y. Chong, K.N. Bozhilov, D. Kim, et al., Peptide-mediated shape- and size-tunable synthesis of gold nanostructures, Acta Biomater. 6 (2010) 2681–2689.
[120] R. Rughani, J.P. Schenider, Molecular design of β-hairpin peptides for material construction, MRS Bull. 33 (2008) 530–535.
[121] S. Koutsopoulos, L. Kaiser, H.M. Eriksson, S. Zhang, Designer peptide surfactants stabilize diverse functional membrane proteins, Chem. Soc. Rev. 41 (2012) 1721–1728.

Design and applications of cyclic peptides

4

Barbara Claro, Margarida Bastos and Rebeca Garcia-Fandino
University of Porto, Porto, Portugal

4.1 Introduction

Peptides are known for their ability to influence physiological and biochemical functions of life, as neurotransmitters, neuromodulators, during the cell–cell communication upon interaction with receptors, or during metabolism and reproduction [1]. They have the capability to bind to particular targets in vivo while not interfering so significantly with the immune system as their endogenous protein counterparts. Moreover, the high diversity of their constituent building blocks and their reduced toxicity make them promising platforms for therapeutics in different areas of science and technology. As such, they have been the subject of an increased amount of studies in the past few decades [2–7]. Despite encouraging results, one of the major barriers to the therapeutic use of linear peptides is their low stability under physiological conditions. Linear peptides are usually rather rapidly cleaved by enzymes in vivo, losing their biological activity [8,9]. To overcome these obstacles, diverse peptide modifications have been proposed [10–13]. The development of new peptide-based systems obtained through cyclization of linear peptides was recently considered one of the most promising approaches, due to a series of advantages that will be discussed below [14,15].

Constraining a peptide into a cyclic structure decreases the conformational freedom for each constituent within the ring and forces the molecule into an ordered secondary structure [16]. Although considered a simplistic approach [17,18], it was suggested that the advantage of cyclization resulted from the increase in rigidity being translated into a decrease of the entropic term of the Gibbs energy, thus providing superior binding affinities than natural binders to a receptor [19,20]. Other studies show that they present a higher resistance to endopeptidases when compared to the more flexible linear analogues [21–26]. For CPs the increase in rigidity is frequently followed by the formation of preferred side-chain conformations. Thus, the structural arrangement of the side chains is frequently the most important factor in the molecular recognition process, using the peptide backbone as a template for their orientation [27]. Cyclization was also shown to increase the propensity for β-turn formation, which is believed to improve binding affinity [28–30]. Another suggested advantage of cyclization is that the elimination of charged termini in CPs can enhance membrane permeability [31], although it has been demonstrated that a peptide does not cross the membrane better simply because it is cyclized, but because of its structural features [32]. Additionally, the lack of both amino and

Peptide Applications in Biomedicine, Biotechnology and Bioengineering.
DOI: http://dx.doi.org/10.1016/B978-0-08-100736-5.00004-1

carboxyl termini makes CPs less prone to hydrolysis by exopeptidases [33], decreasing off-target side effects [34], and thus leading to substantially enhanced metabolic stability and specificity [35–38]. Furthermore, CPs are significantly smaller compared to proteins and therefore more affordable due to lower manufacturing costs through various chemical methods [39].

4.2 Cyclic peptides in nature

Gene-encoded CPs were originally believed to be aberrations in the peptide world. The discovery of Gramicidin S in 1944 by Gause and Brazhnikova during the Second World War marked a "before and after" in the history of CPs [40]. The number of CPs disclosed has continually grown, with thousands of CPs known at present. Most arise from natural CPs and have been extensively used in pharmaceutical and agricultural applications as active ingredients obtained from natural extracts (bacteria, fungi, plants, animal venoms) [41–45]. The major advantage of Nature-derived CPs is that despite Nature never having faced the need to create a CP that encompasses all important drug-like properties (such as protection against serum enzymes), evolution has already done the work of selecting a set of bioactive peptide scaffolds that may be advantageous for drug-discovery purposes, since natural sources contain a rich diversity of CPs and CP-like structures. The structures of these naturally occurring CPs range from large peptides ($\sim$70 residues) containing mainly α-helical arrangements, to small β-hairpins (12–14 residues), as shown in Fig. 4.1 [45,46]. Most of them share a common function: the defense-related or toxic properties against microbes and/or insects, exhibiting a wide variety of unusual and potent biological activities

Gramicidin S is an example of CPs obtained from bacteria (Fig. 4.1). It is a cyclic decapeptide, extracted nonribosomally from the soil bacterium *Aneurinibacillus*, with an antiparallel β-sheet constructed by two identical pentapeptides joined head to tail, formally written as cyclo(-Val-Orn-Leu-D-Phe-Pro-)$_2$ [47,48]. Tyrocidines are antibiotics with different structures, differing only in the analogue-type substitutions of the aromatic residues—Tyrocidine A, B, and C (Fig. 4.1) [49]. They are the major constituent of tyrothricin, which also contains gramicidin, produced by the bacteria *Bacillus brevis*, found in soil [50]. Cyclosporin is an immunosuppressive agent used to prevent organ rejection in transplantation surgery. This cyclic undecapeptide with high lipophilicity is produced as a secondary metabolite by strains of fungi *Cylindrocarpum lucidum* Booth and *Tolypocladium inflatum* Gams isolated from soil samples. Seven of its eleven amide bonds are N-methylated, adopting different conformations depending on the environment in which they are located [23,51]. Some ribosomally synthesized CPs produced in bacteria include for example AS-48 [52], which is intensively used in food preservation because of its activity against *Salmonella* and *Listeria*, or lactocyclicin Q and garvicin ML, isolated from cheese and Mallard ducks, respectively, extremely stable and with antimicrobial activity [53,54]. Vancomycin [55] is a

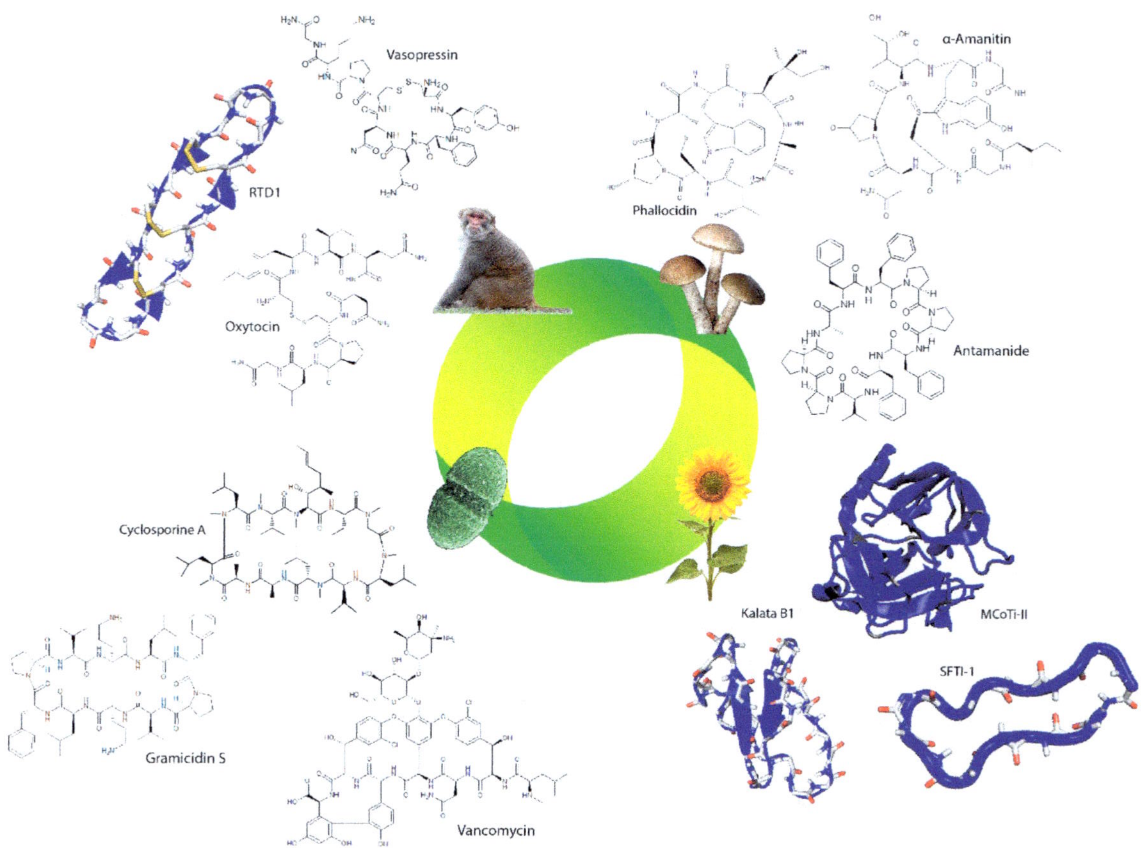

Figure 4.1 Examples of natural CPs obtained from bacteria, fungi, plants, and animals.

branched tricyclic glycosylated nonribosomal peptide produced by the Actinobacteria *Nocardia orientalis* and used as the last resort in hospitals to treat drug-resistant Gram-positive bacterial infections. Its biosynthesis occurs via different nonribosomal protein synthases (NRPSs). The molecule consists of a seven-membered peptide chain that is formed by parts of three phenylglycine systems, two chlorinated tyrosine units, aspartic acid, and N-methylleucine. Two ether bonds and a carbon–carbon bond join the various substituents on the peptide chain into three large rings. A disaccharide, composed of glucose and vancosamine, is also present but is not part of the cyclic structure.

Some CPs found in fungi are, for example, the highly toxic alpha-amanitin (Fig. 4.1) and phallocidin, targeting RNA polymerase II and F-actin, respectively [56]. They are ribosomally expressed and differ from other CPs by having an internal crosslink, which is not a disulfide bond. The mushroom metabolite antamanide [44,57], isolated from the lethal *Amanita phalloides*, is another immunosuppressant consisting on a cyclic decapeptide containing one valine residue, four proline residues, one alanine residue, and four phenylalanine residues. It has been shown that altering any of the proline residues (or the phenylalanine residues to tyrosine or L-cyclohexylalanine) reduces the immunosuppressive activity to zero [58].

The largest family of CPs found in plants is the cyclotides [59,60]. They have been found in more than 20 species from the *Violaceae*, *Rubiaceae*, *Cucurbitaceae*, *Fabaceae*, and *Solanaceae* plant families. More than 280 sequences have been reported so far (http://www.cybase.org.au/), but it has been predicted that the cyclotide class might comprise more than 50,000 members [41]. Cyclotides are 28–37 amino acids (AA) in size and have a cystine knotted arrangement of three disulfide bonds. One of the best known examples is Kalata-B1 [61], an amphipathic peptide containing 29 AA residues. Two of the largest cyclotides, MCoTI-I and MCoTI-II, are also trypsin inhibitors, and are found in the seeds of a bitter melon from Vietnam. Another type of CP in plants is sunflower trypsin inhibitors. Sunflower trypsin inhibitor-1 (SFTI-1) [62] is a 14-residue peptide with one disulfide bond that is extracted from the seeds of the common sunflower, and is a potent trypsin inhibitor.

Some primates express CPs called θ-defensins as a part of their immune system, and to date, they are the only known ribosomally synthesized CPs in mammals [63,64]. The characteristic structural motif of the θ-defensins is the cyclic cystine ladder, including a CP backbone and three parallel disulfide bonds. In contrast to the cyclic cystine knot, which characterizes the plant cyclotides, the cyclic cystine ladder has not been described as a structural motif. Some examples are RTD-1 [65], isolated from macaque leukocytes and some recent naturally occurring isoforms, isolated from baboon leukocytes [66]. Oxytocin [67] and vasopressin [68,69] are peptide hormones released by the primate posterior pituitary gland. They are nonapeptides with a single disulfide bridge, differing only by two substitutions in the amino acid sequence. Both are examples of CPs already in the market, being originally discovered as hormones found in mammals and later used as the "starting point" in drug design in the pharmaceutical industry.

4.3 Types of cyclic peptides: classification

4.3.1 Size

A number of CPs has been discovered in Nature and they vary greatly in structure and size, from the smallest CPs in fungi containing only seven residues to the largest in bacteria comprising 78 residues (Fig. 4.2) [71]. Small peptide cycles (five to six amino acids) have shown in recent studies acceptable bioavailability and good stability against enzymatic degradation, providing good skeletons for drug design [72,73]. Overall, the smaller the size of the CPs, the higher the conformational restraints and the smaller the observable flexibility [74]. Synthetic macrocycles composed by large rings (in the 600–1500 molecular weight range) are also very promising, and are very useful, for example, in protein–protein interactions [75].

Ring size is a significant factor in the success of the synthesis of a CP. Whereas six-membered rings are quite easy to synthesize, 7–15-membered rings are less accessible. In the case of small-to-medium-sized rings, the ground-state E geometry

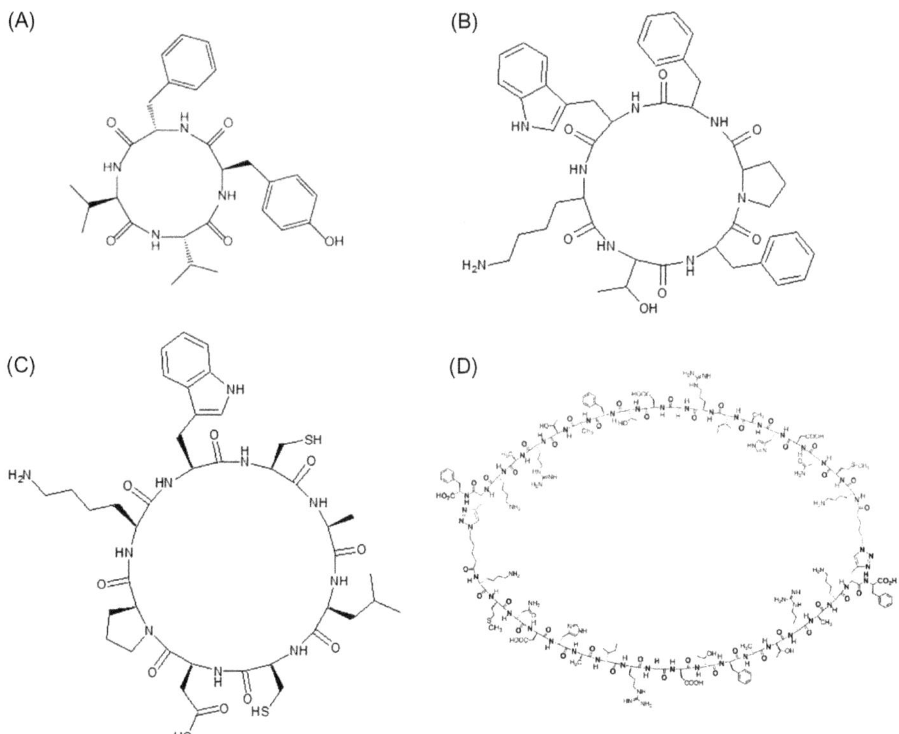

Figure 4.2 Examples of CPs with different sizes: (A) Cyclo(D-Tyr-L-Phe-D-Val-L-Val) with four residues; (B) Cyclo(Pro-Phe-Trp-Lys-Thr-Phe) with six residues; (C) Cyclo(L-Cys-L-Ala-L-Leu-L-Cys-L-Asp-L-Pro-L-Lys-L-Trp) with eight residues; and (D) "Duoboros" with 18 residues [70].

of the peptide bond is not a favorable conformation for the linear peptide to be cyclized, since the two ends that must react are too far apart from each other. This problem is not present in larger ring sizes since they can accommodate *E* peptide bonds. On the other hand, the synthesis of these large macrocycles demands special conditions to avoid intermolecular reactivity [76].

4.3.2 Number of rings

Although most of the CPs known or designed are monocyclic, Nature has also created many peptides with multicyclic motifs. Some naturally occurring bicyclic peptides are, for example, α-amanitin (Fig. 4.1) [77,78], phalloidin [58] (Fig. 4.1), bouvardin [79,80] (Fig. 4.3), moroidin [81,82] (Fig. 4.3), SFTI-1 [62], or the θ-defensins [63,64] (Fig. 4.1). Typical linkages formed in ribosomal peptides are disulfide, amide, thioether, or ester bonds. In nonribosomal peptides, additional cyclization chemistries are found, such as biaryl or biaryl ether linkages. These peptidic structures contain two macrocyclic rings with limited conformational flexibility, which can enhance their overall binding properties. Furthermore, their small size affords a number of advantages over large proteins, including access to

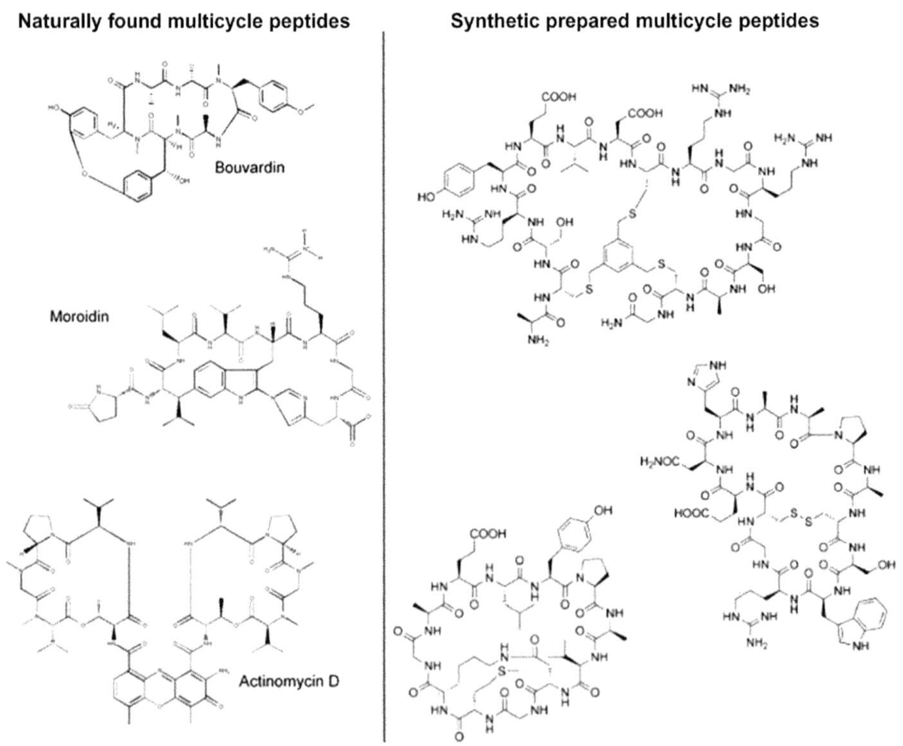

Figure 4.3 Examples of multicyclic peptides found in Nature (left) or chemically synthesized (right).

chemical synthesis, better tissue penetration, and alternative routes of administration, such as topical application. Owing to their attractive features, bicyclic peptides are attracting increased interest for biological use (Fig. 4.3) [83–88].

Examples of natural CPs containing more than two cyclic units in their structure are also known, i.e., nisin [89], mersacidin [90], actinomycin D [91], or vancomycin [55]. Many of these polycyclic peptides consist of cycles sequentially organized along the peptide backbone. In these molecules, the conformational constraints imposed by the polycyclic backbone enable them to adopt a defined structure in solution, which is crucial for their biological activity and advantageous toward enhancing their cell penetration properties and stability against proteolysis. A recent review has summarized the properties of existing polycyclic peptide drugs relating them to their interesting molecular architectures [92].

4.3.3 Physical properties

Based on their structures and biological function, CPs have been classified into different families (Table 4.1) [93]:

1. Highly charged (cationic or anionic) CPs with membrane-disrupting (antimicrobial peptides (AMPs)) and membrane-penetrating properties (CPPs). Cyclic AMPs have emerged as good antimicrobial candidates due to their robust secondary structure and high activity [94–98]. One of the most representative examples is CPs composed of amino acids of opposite alternating chirality (L and D), in which all side chains are exposed to the outer surface of the structure leaving an empty internal hole, which favors the formation of a nanotube by self-assembly of the different units at the membrane site (see Section 4.7) [99,100].
2. Nonpolar CPs enclosing lipophilic modifications to the amide backbone. These CPs exhibit a wide range of biological activities, many of them associated with penetrating properties into eukaryotic cells by passive diffusion. Nonpolar CPs are characterized by a predominance of aliphatic residues, the presence of D-amino acids, and N-methylation of backbone amides. The CPs in this category, whose crystal structures have been solved, often contain one or more β-turns that are templates for various patterns of intramolecular hydrogen bonds. Since they do not interact with membranes electrostatically, it has been speculated that the penetration mechanism is similar to most small drug molecules. Some of the most representative examples of the nonpolar CPs are the already-mentioned cyclosporin A [101] or the K^+ ionophore valinomycin. Other examples include argyrin B, cyclomarin or CSA (homodetic "all-amide" peptides) destruxin B, aureobasidin A, HUN-7293/CT08 or guangomide A (depsipeptides), lissoclinamide 7, patellamide D, ascidiacyclamide or trunkamide (pattelamide-like CPs, derived from ribosomally synthesized sequences of alternating lipophilic and Ser, Thr, or Cys residues), phakellistatins (proline-rich peptides), coibamide A, didemnin B, YM254890, selenamide, luzopeptin, callynormine A (heterodetic CPs) [93]. A detailed review has summarized a large number of examples of CPs of this family [93].
3. Mixed-polarity CPs, with amphiphilic features but with action not limited to microbial targets. Their mechanism of action does not depend on membrane-disrupting or membrane-penetrating ability. They exhibit activity in mammalian cells that suggests intracellular targets, as they are supposed to enter the cell through a passive membrane diffusion or other mechanisms not well known yet. Some representative examples of this group are

Table 4.1 Classification of CPs according to their physical properties

Classification of CPs		Structural features	Representative cyclic peptides	Biological activitiy
Charge	Cationic	Cationic residues present (e.g., Arg or Lys)	Gramicidin S, polymyxin B	AMP/CPP
	Anionic	Anionic residues present (e.g., Asp or Glu)	Gramicidin A	AMP/CPP
Mixed polarity		May contain a lipid tail	Largamide, pseudodesmin A, microcin, papuamides, thiostrepton	Activity in mammalian cells but not limited to microbial targets
Nonpolar		Predominance of aliphatic residues (e.g., Gly, Ala, Val, Leu or Ile), presence of D-amino acids, or N-methylation of backbone amides	Aureobasidin A, argyrin B, aureobasidin, guangomide A, cyclomarin, CSA, Destruxin B	Penetrating properties into eukaryotic cells by passive diffusion or other mechanisms not well known
Cyclotides and cysteine-knot proteins		Multiple disulfide bonds and multiloop conserved structures	Kalata B1, Bowman–Birk trypsin inhibitor	Membrane permeation of the cysteine knot

kahalilide F [102] and also other CPs with a lipid tail appended, such as largamide, pseudodesmin A, the papuamides, thiostrepton or microcin [103].

4. Cyclotides and cysteine-knot proteins, are small (2–8 kDa) proteins with unusual topologies and notable oral activity. They are backbone-cyclized microproteins that contain multiple disulfide bonds and a conserved multiloop structure [61,104,105]. Some representative examples of CPs from this family are the already mentioned Kalata B1 [61] or the Bowman–Birk inhibitor. The mechanism of membrane permeation of this type of CPs is not yet well established. A recent review has summarized the available chemical or biological methods for production of these interesting CPs [106].

4.3.4 Type of cyclization

The ring structure of a CP can be formed by linking one end of the peptide to the other by an amide bond, or other chemically stable bonds such as, e.g., lactone, ether, or thioether disulfide [107]. Based on the type of bond that comprises the ring, CPs can be divided into three main categories: (1) homodetic (containing only peptide bonds); (2) heterodetic (diverse functional groups are also employed to connect the amino acids); and (3) complex, consisting of a mixture of homodetic and heterodetic linkages (Fig. 4.4). Linear peptides can be cyclized not only by joining the N-termini connects with the C-termini of the terminal peptides (head-to-tail bond), but also by linking amine and carboxylic functions in amino acid side chains, giving side-chain-to-head or side-chain-to-tail connections. It has been demonstrated that the position of cyclization within the sequence will constrain particular motifs and alternative geometries and will affect the binding affinity of CPs [108,109]. Tapeinou et al. have recently summarized the cyclization strategies according to the following classification [19,27,110]:

- N-to-C or head-to-tail cyclization (homodetic)—this is the simplest and most common cyclization, where the peptide bond is formed between the N-terminus amino group and the C-terminus carboxyl group [111,112]. It has been inferred that the ring size is the most important factor for the head-to-tail macrocyclization. For peptides containing more than seven amino acids the ring closure is not problematic, although for smaller peptides the ring connection should be chosen carefully to avoid problems. Furthermore, the incorporation of turn-inducing structural elements embedded along the linear precursor can also result in a more efficient macrocyclization [13,113].
- Cyclotides—the peptide bond is formed head-to-tail and the ring is usually strengthened by three disulfide bonds [59,114].
- Side chain-to-one of the termini cyclization (head-to-side chain or side chain-to-tail (heterodetic))—the bond is formed between the N- or C-terminus and the side chain functional group of amino acids [26].
- Side-chain-to-side-chain or backbone cyclization—the bond is formed between two side chains of amino acids (homodetic and heterodetic) [115–118]. Through this methodology, a bridge is created from a backbone amide nitrogen or a Cα carbon atom to any other position, introducing restrains into the linear peptide and leaving the peptide chain termini free [119]. In order to be effective, the side-chain-to-side-chain macrocyclization can involve a condensation reaction between side chains under high dilution to minimize unwanted intermolecular processes such as oligo- and polymerizations [120]. This cyclization is advantageous for peptide hormones that are often found as N-terminal primary amide, for example [121].
- Disulfide—the disulfide bond is formed between two thiol groups (cysteine) [122]. Disulfide cyclizations are easy to generate but require that two cysteines be included in the sequence, and the disulfide bond may be subject to reduction and reoxidation into unintended configurations.
- Thioether—the thioether bond is formed between the side chain thiol group of a cysteine and the α-carbon atom of an amino acid [123].

4.3.5 Type of building blocks

To improve the potency of a CP, additional diversity has been obtained through the incorporation of noncanonical amino acids (e.g., D-amino acids, ornithine, hydroxyproline, N-methylated residues, or other non-natural amino acids) [13,124–129] or

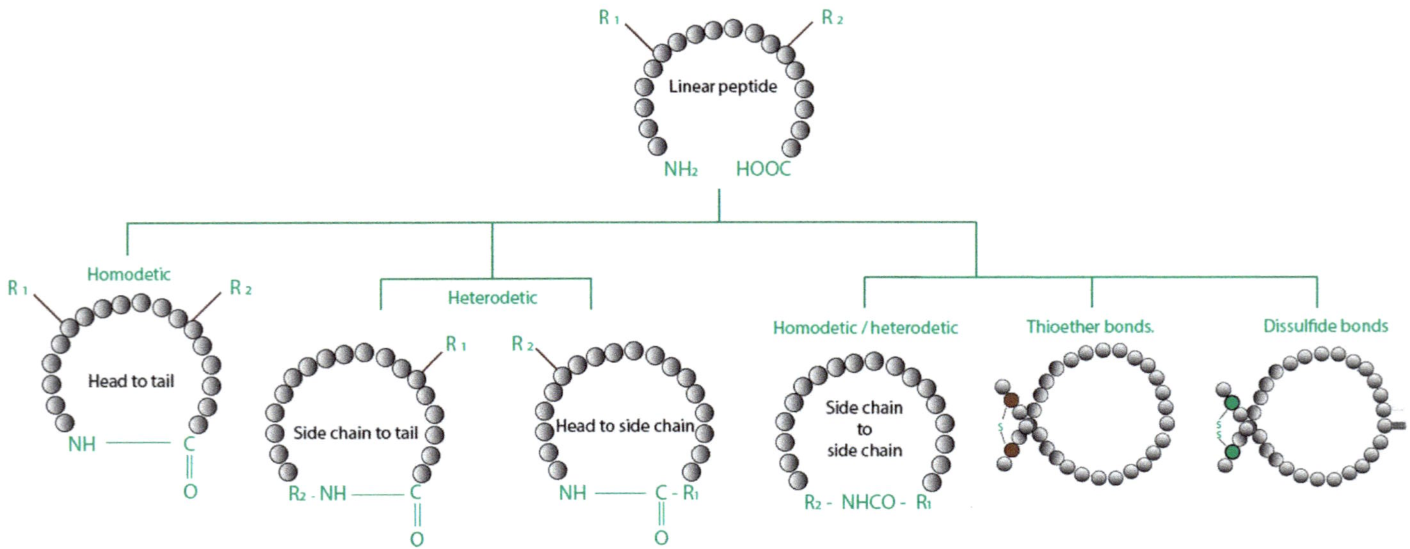

Figure 4.4 Principal categories of cyclization according to the type of bonds that comprise the ring of CP.

by means of peptide structural modifications. The introduction of small groups or the replacement of the canonical amino acids by noncanonical ones influences the entire CP backbone, unlike for the linear peptides where it only affects the insertion area [130]. Amino acid side-chain alterations also change the behavior of the CPs. The most commonly reported side-chain alteration in CPs is cysteine bridging, which is easily formed by oxidation of two thiol functions of the amino acid side chains. Disulfide bonds play an important role in constraining peptide conformations, although the resulting cycles usually retain considerable flexibility and are easily cleaved under physiological conditions [76,131].

Several reviews have been published about CPs containing noncanonical amino acids, mainly related to the application of peptide nanotubes by self-assembly (self-assembled cyclic peptide nanotubes (SCPNs), see Section 4.7) [132,133]. SCPNs have been assembled from CPs with α-amino acids (α-CPs) comprising 4 [134], 6 [135–137], 8 [138–141], 10 [142,143], or 12 [144] residues, resulting in internal diameters ranging from 2 up to 13 Å. The chemistry of SCPNs can also be altered by using chiral β^3-amino acid residues (Fig. 4.5) [145,146]. The inner surfaces of D, L-α- and β-PNTs are hydrophilic, and it is not possible to modify their pore properties by introducing functional groups because all the amino acid side chains have to point outwards and pseudoequatorially. However, this shortcoming is overcome if cyclic γ-amino acids are used, as has been proven by Granja's group through the design, synthesis, and characterization of a new class of CP composed of alternating α-amino acids and *cis*-3-aminocyclohexanecarboxylic acid (γ-Ach) [147]. The use

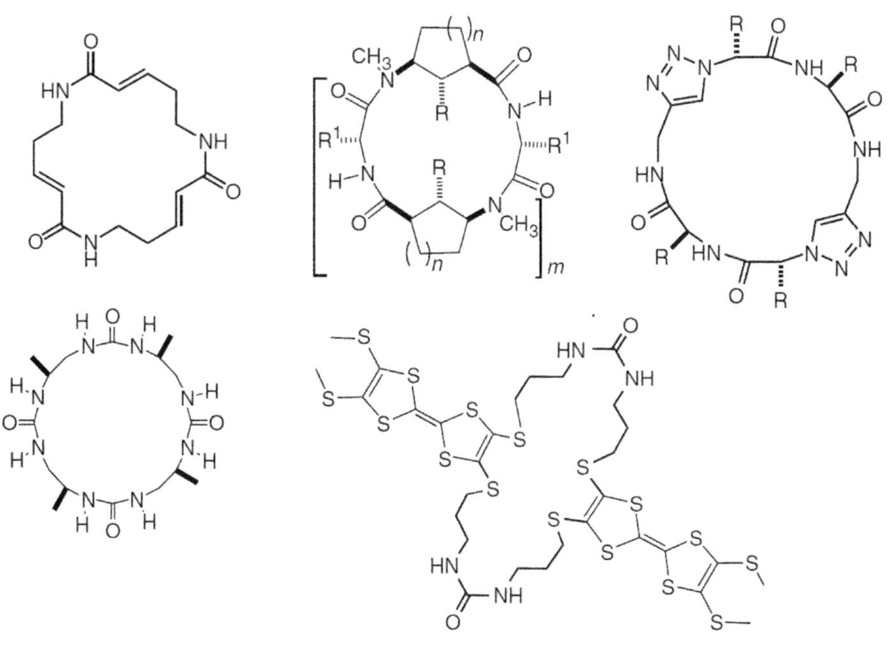

Figure 4.5 Examples of CPs containing noncanonical amino acids.

of a γ-amino acid with a hydroxyl group at C2 has also allowed the incorporation of different groups that modify the internal properties of the CPs [148,149]. Another synthetic route to modify the lumen chemistry of CPs was obtained by means of the insertion into the peptide sequence of an unnatural amino acid containing a methyl functional group pointing toward the inside of the pore (Fig. 4.5) [150]. Other examples of modifications of the original Ghadiri CPs include the use of δ-amino acids [151], ε-amino acids [152,153], CPs derived from dialkylmalonic acid and containing disulfide bridges between the two terminal cysteines to close the ring [154], or pseudopeptidic macrocyclics [155] (Fig. 4.5). The introduction of moieties that are not amino acids, e.g., to generate peptidomimetics, can enhance diversity [156], specificity, and functionality, but it may be accompanied by an increase in toxicity compared to that associated with CPs of natural amino acids. On the other hand, N-methylation patterns that facilitate the formation of transannular (intramolecular) hydrogen bonds are related with good passive permeability and oral bioavailability [157]. Recent reviews address N-methylation in peptides from synthetic and conformational perspectives [73,158,159].

4.3.6 Type of secondary structure

CPs are now known to reproduce peptide backbones constrained to strand, sheet, helix, turn, or loop conformations (Fig. 4.6). A recent review summarizes the most significative examples over the last 20 years of CPs that mimic this type of structure, and identifies some additional restraints incorporated into natural cyclic

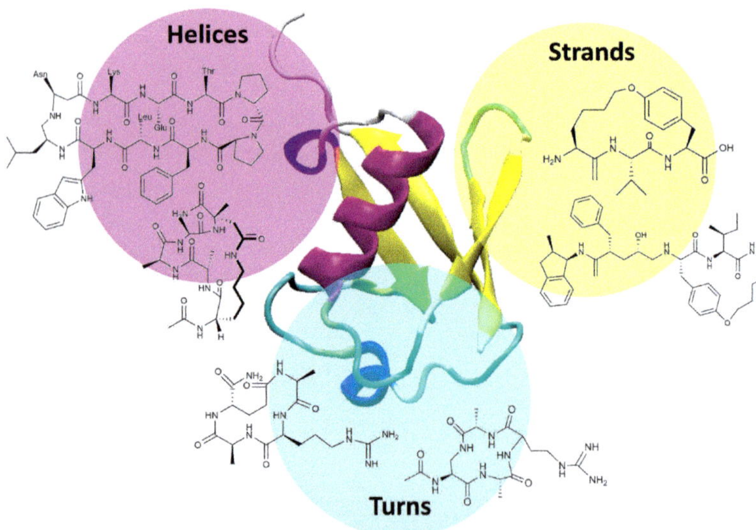

Figure 4.6 Examples of short synthetic peptide sequences that can be induced to fold into protein-like bioactive conformations (strands, helices, turns) by cyclization [160].

peptides and synthetic macrocyclic peptidomimetics that refine peptide structure and confer desired biological properties [160].

4.4 Methods for the design of cyclic peptides

One of the main sources of inspiration for designing CPs is Nature itself. Cybase [161] is an online database of known natural and non-natural head-to-tail cyclic proteins, i.e., proteins that possess a cyclic backbone in which the N- and C-termini have been joined with a conventional amide bond. This database includes several cyclotide entries (plant-made defense proteins (such as the kalata proteins), primate rhesus θ-defensins (RTDs) with increased biological activities, such as anti-HIV activity), which are used as lead molecules for drug design [114].

There are several approaches to the design of CPs depending on the application for which they are aimed. Presently there are two main approaches in drug design. The first consists of intentional modifications of known cyclic analogues that may or may not have affinity against a target. Rational design requires the knowledge of the role of each residue so that cyclization occurs at the amino acid positions least important for the retention of the activity of the linear peptide. Some examples have included mutated or grafted natural cyclotides [162,163] or modified cyclic hormones to obtain enhanced compounds towards different targets [164,165] (i.e., melanocortin [166,167], gonadotropin releasing hormone [168–172], or somatostatin [173]). In many cases cyclization of linear peptides was used to obtain enhanced pharmacological agents, many of them targeting protein–protein interactions (PPIs) [168–170,172,174–181]. The second approach is used when little is known about the target (e.g., receptor) or the ligand (the CP to be designed). This second approach uses combinatorial syntheses to identify lead compounds and usually involves: (1) screening methods to identify biological function; (2) synthesis of libraries based on linear peptides; and (3) mutagenesis studies driven by trial and error [182,183]. Several groups have reviewed the topic of library generation [184–187], with *phage display* (see below) or *combinatorial methods* being some of the most common methods to create screening libraries. The two approaches are not mutually exclusive. Lead compounds identified by combinatorial synthesis can be optimized through rational approaches to determine the best analogue. Likewise, peptides designed through rational approaches can be optimized by iterative replacement of each amino acid (Fig. 4.7) [188]. The combination of both approaches has led to successful examples of CPs with pharmacological activity [189].

During the design of CP many variables need to be optimized, experimentally generated, tested, and analyzed, making the overall process tedious and time consuming. Furthermore, the ambitious challenge of understanding the mechanisms of action of CPs would require atomic resolution, which is still unattainable by most experimental techniques known today. Hence, computational approaches to the design of new CPs can provide a better understanding of their sequence–structure relationships and would greatly benefit CP development attempts [190,191].

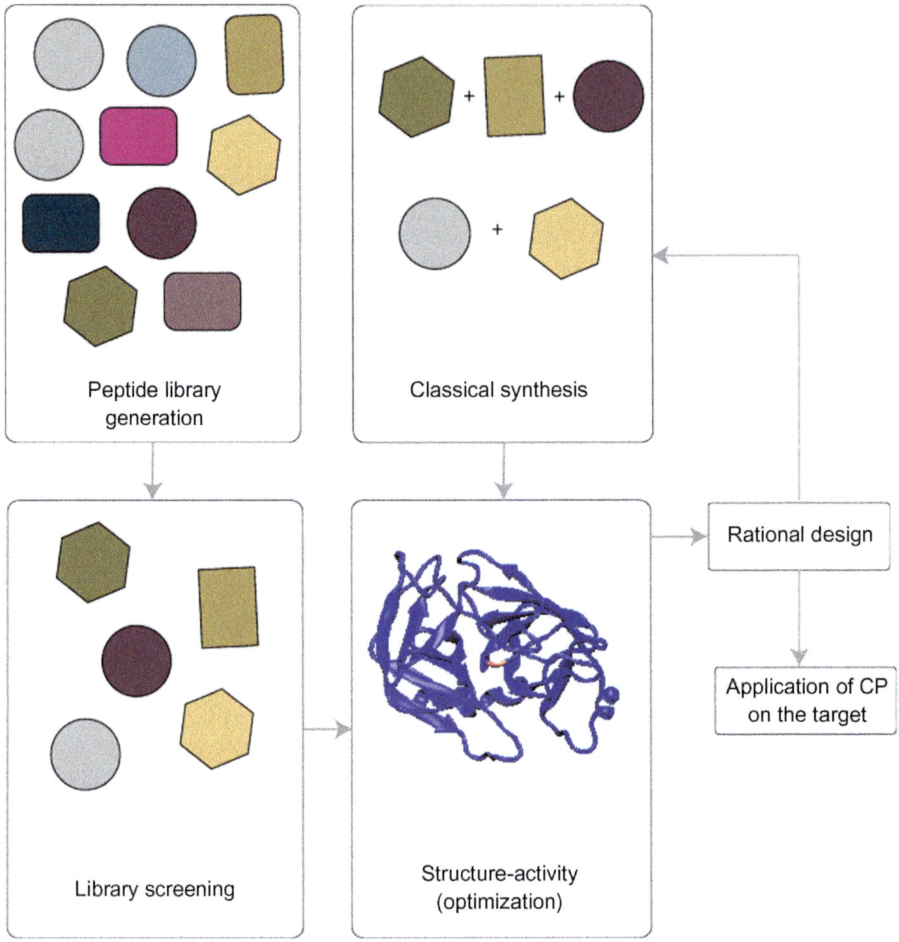

Figure 4.7 Main approaches to identify lead compounds using combinatorial synthesis and/or rational design.

Several recent reviews have summarized computational methods to generate CP libraries for virtual screening, as well as current efforts to accurately predict the conformations adopted by CPs [192–195]. Although several software tools have been developed (i.e., CycloP [196], PLOP [197,198], Peplook [199,200], PEPstrMOD [201], PEP-FOLD [202,203], I-TASSER [204]) and constitute valuable complementary tools, the reliability of these prediction algorithms is highly related to sampling CP conformational space (using algorithms based on distance geometry [205–207], stochastic search [208], or systematic search [209,210], among others) and to accurately describing CP energetics. Furthermore, it is important to remember that most biological systems are not static. To understand how structure and function are related one ought to visualize molecules in action. Complete understanding of the processes at the molecular level requires the ability to watch molecules as they react, interact, or transform in real time by structurally characterizing the often short-lived transient species and intermediates. Indeed, obtaining high-resolution (temporal and spatial) information about dynamic processes is extremely challenging. In this regard, molecular modeling techniques, mainly molecular dynamics (MD) simulations, may provide useful working hypotheses of the mechanism of action in different environments.

Over the last few years, computational chemistry has played an important role in drug design with validation in wet laboratories. Docking methods employed in computational drug design have been coupled with MD simulations to predict the occurrence of complexes [192,211]. MD simulations have also been employed to investigate the mechanism of cyclic AMPs at the membrane level (see some examples in Fig. 4.8) [212–215,217,218]. Nevertheless, the invaluable insights provided by computer modeling in this area are still in their early days. For example, the absence of reliable systematic studies focused on the comparison of different computational methodologies, or on AMP interactions with different types of lipid bilayers, just to name a few cases, is striking. There are also very few attempts that included more than one membrane-active molecule [219], and there is a lack of quantitative information at comparable time-scales as those employed in wet-labs due to the difficulty in simulating events that span this relevant range. To address the difficulties introduced by the ring strain in the free energy barriers between

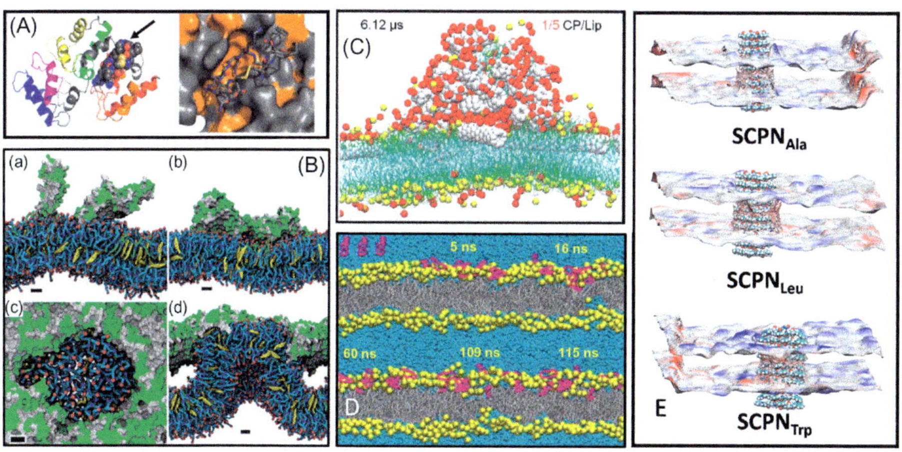

Figure 4.8 Some examples of Docking and MD simulations applied to the study of CPs on membranes. (A) Complex between the calcium-binding protein calbindin-D28K and the cyclic peptide (CKFSIKNRQC) shown as spheres, and detail of the binding site, representing in orange the hydrophobic patches [212]. (B) CG-MD simulation of the progressive membrane association and disruption activity of kB1 [213]. (C) CG-MD simulations of a CP [RRKWLWLW] aggregate interacting with a POPE-POPG membrane [214]. (D) MD simulation of the BPC194 CP interacting with a lipid bilayer [215]. (E) Local membrane thickness for three SCPNs composed of α,γ-CPs [216].

conformations, a variety of enhanced sampling methods, such as REMD [220–224] or META and BE-META simulations [15,225,226] have been applied to help accelerate the sampling process [227–231]. It is clear that for a full understanding of functional dynamics a combination of these methods with structural and spectroscopic techniques is required, to allow the generation of dynamic structural models consistent with all available experimental data. Algorithm improvements, parallelization and specialized hardware have been tried extensively to systematically explore events in the microsecond scale, and MD simulation times and size are now beginning to extend into ranges where direct comparison and experimental testing (EPR, NMR, OCD, AFM, and SAXS) [232,233] are starting to become feasible. Advances in coarse-grain (CG) methodology, which considers appropriately chosen groups of atoms as single interacting centers, have allowed access to unprecedented time-scales in MD simulations, extending them towards biologically relevant time-scales (see, e.g., Fig. 4.8B,C) [213,214,234]. Furthermore, while CG models offer researchers means of accelerating the discovery process, further acceleration is possible through the use of advanced computer hardware, such as graphical processing units (GPUs) [235]. By intelligently combining these techniques it might be possible to address the main limitations of computational techniques applied to CPs, such as structure generation, sampling, and also the accuracy of applying to CPs the force-fields commonly employed for linear peptides and proteins, which remains to be determined [236,2,37].

4.5 Approaches to the preparation of CPs

Most of the CPs produced by Nature involve ribosomal synthesis, comprising mRNA translation to peptide chains of L-amino acids, often posttranslationally modified and cyclized by enzymes [238,239]. However, Nature also widely uses nonribosomal synthesis, making use of enzymes to catalyze the assembly of non-proteinogenic amino acids and derivatives followed by cyclization [240,241]. Encouraged by the existence of natural CPs with biological activity, synthetic chemists and biologists have invested great effort in the development of tools to obtain biologically active CP compounds. These efforts have included both genetic and synthetic methods [23].

The genetic methods (Fig. 4.9) include phage display, intein-based CPs, and mRNA display.

In *phage display technology* [242] (Fig. 4.9A) a gene encoding a protein of interest is inserted into a phage coat protein gene, inducing the phage to display the protein (or peptide) on its surface while containing the gene for the protein (or the peptide) on its inside. Usually, peptides are displayed on the N-terminus, middle, or C-terminus of coat proteins, and are not meant to be cyclic. However, cyclization can be obtained by disulfide bridge formation. One of the main disadvantages of this method is that phage display is limited to natural, ribosomal amino acids. An important point in this methodology is that phage display has been used to generate

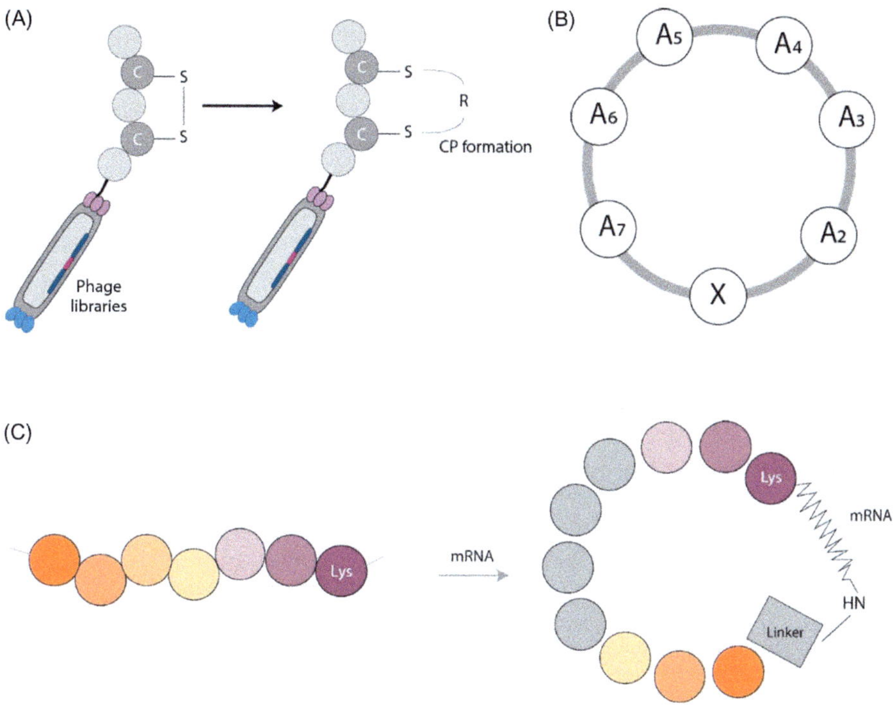

Figure 4.9 Genetic methods to produce CPs: (A) phage display, (B) intein-based CPs, and (C) mRNA display.

screening libraries [243,244]. Furthermore, databases and computational tools have played an important role in phage display study [245,246].

Inteins refer to protein sequences that are spliced out during maturation [247,248]. SICLOPPS (split-intein circuit ligation of peptides and proteins) is a valuable tool to develop CP libraries using a *trans*-intein DnaE protein [249]. The CP formed from SICLOPPS takes a general structure cyclo ($XA_1A_2A_3$...An), where X is either Cys or Ser (Fig. 4.9B). The main disadvantages of these methods are the requirement of Cys (or Ser) residue in the sequences, the limitation of the choice of the amino acids to ribosomal amino acids and that CPs can only be prepared in vivo.

The method of *mRNA display* (Fig. 4.9C) is an in vitro method of displaying peptides/proteins coupled to the encoding mRNA that has been recently modified to prepare CPs [250,251]. Some of its limitation are poor efficiency, failure of the cyclization reaction when there are multiple Lys residues in the sequence, and undesired dimerization between two mRNA–peptide hybrid molecules mediated by crosslinker.

The synthetic method includes chemical synthesis strategies in *solution* and *solid-phase* [19]. Several groups have summarized the various synthetic methods used to generate CPs [76,107,131,252]. The *solution-phase peptide synthesis* method has been widely used to synthesize individual peptides. Popular synthetic strategies are based on orthogonally protected precursors at specific functional groups, which are selectively deprotected during peptide synthesis [120,253]. However, the long synthesis time and the laboriousness of the purification, usually reducing overall yields, has led to almost complete substitution of solution-phase by solid-phase peptide synthesis in most research laboratories [254]. The main advantage of solid-supported peptide synthesis and macrocyclizations is that simple washing and filtration are often enough for reasonable purification [255]. To cyclize peptides on a solid support, the linear precursor is most commonly anchored to the support through the side chain of a trifunctional amino acid. A protecting-group strategy of at least three dimensions of orthogonality is required to construct the linear peptide, deprotect the N- and C-termini, cyclize in a head-to-tail, and finally cleave the product from the solid support [76,253,256]. Innovative straightforward protocols involving solid-phase peptide synthesis, followed by in-solution fragment coupling, have been introduced to improve the yield and facilitate the synthesis of large polypeptides [76,257–260].

Solid-phase peptide synthesis in combination with intramolecular native chemical ligation [261] is one of the most common methods of producing synthetic CPs (Fig. 4.10). It allows partial automation and the incorporation of non-native or posttranslationally modified amino acid equivalents that would be difficult to incorporate in peptides produced recombinantly.

The synthetic methods involve individual synthesis (individual compounds are synthesized separately), parallel synthesis (the synthetic intermediates are split during the synthesis and used in later steps), as well as split-and-pool synthesis (where the starting material is split in *n* portions, reacted with *n* building blocks, and recombined in one flask for the second step, repeating this procedure many times. It allows the preparation of peptide libraries in large scales). Recent development of a CP library based on split-and-pool synthesis allows on-bead screening, in-solution screening, and microarray screening of CPs for biological activity [23].

4.6 Limitations of cyclic peptides

Although CPs often show improvements in binding affinity, specificity, or stability when compared to their linear analogues, due to the imposition of constrained geometries, it is important to note that cyclization does not necessarily lead to improvements in all, or sometimes even in any, of these properties [262]. Certain linear peptides can contain sequences that can support rigid structures recognized by targeted receptors without the need for cyclization [263]. There are also cases where peptide cyclization worsens the biological activity of the linear analogues [264] and even linear analogues of native CPs displaying a similar or superior binding affinity to targeted receptors as compared with their cyclic counterparts [265]. Furthermore, peptide cyclization was shown to sometimes decrease the degree of cell membrane penetration. It has been suggested that the lower cell penetration of

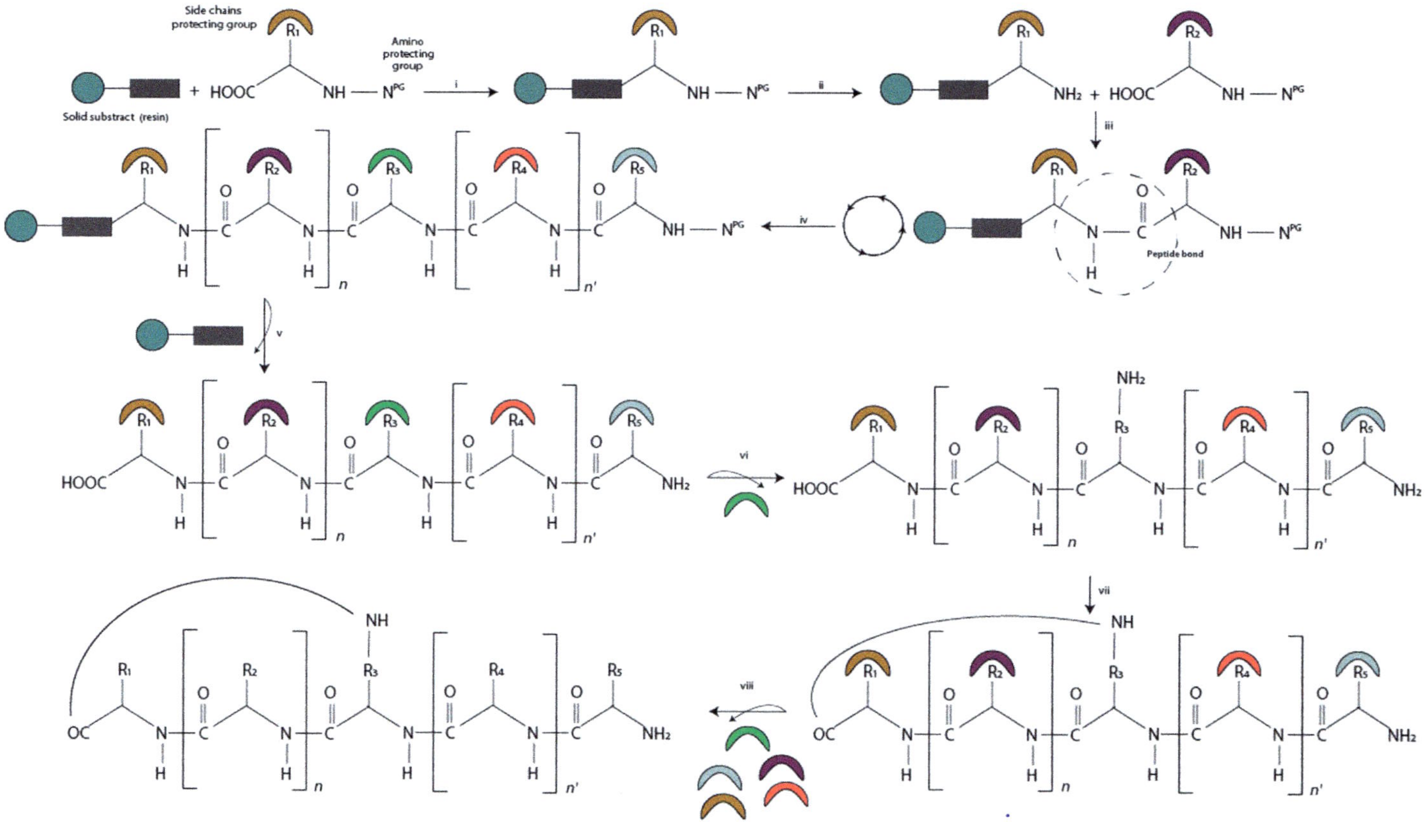

Figure 4.10 Visual representation of the solid-state synthesis of a side chain-to-tail CP. (i) The first amino acid is loaded to the solid support via the carboxyl group, followed by (ii) the removal of the NPG–amino protected group. Afterwards, (iii) the amino group with the activated carboxyl group, of the second amino acid, is coupled and (iv) so as the following amino acids. (v) A cleavage of the resins occurs and also (vi) a selective removal of the side chain amino protected groups. A (vii) cyclization between the side chain amino group and the C-terminal carboxyl group takes place and (viii) a final deprotection of the side chain of the protected groups finishes the cyclization.

particular CPs could be due to a reduction in the enthalpy of interaction with the membrane, contributing to a less favorable internalization [266].

Another important limitation is the fact that scaling-up the production of CPs from a research laboratory to an industrial level is not as accessible as in noncyclized peptides. CPs generally demand a ligation strategy that must be integrated within the synthesis procedure, complicating the chemistry and thus the scaling-up process. Furthermore, the absence of charges on the N- and C-termini reduces the effectiveness of proteomic sequencing techniques, such as mass spectrometry and Edman degradation, making quality control and characterization efforts more complicated than in linear peptides.

4.7 Applications of cyclic peptides

CPs have been widely exploited over recent years, enabling their use in different applications, such as therapeutics, diagnostics, vaccines, and agriculture, among others.

CPs have played a rather important role in *pharmaceutical research* as biomedically useful agents or as lead compounds in drug development [6,23,267]. In the past peptides have been considered poor drug molecules due to various reasons, with probably the most important drawbacks being their quick metabolization/degradation and the fact that many peptides do not usually cross the cell membrane as some small molecules do. However, there has been a rebirth of interest in peptides and CPs in recent years. This is partly because both pharmaceutical companies and academic researchers have realized that these molecules afford quick access to the underexploited level intermediate between traditional small organic molecules and the higher molecular weight space of proteins and antibodies [93,268]. CPs are considered good drug leads and useful therapeutics and biochemical tools, due to the structural features mentioned in the previous sections. Structural rigidity combined with different peptide sequences can provide binding toward target molecules. They can also be more selective when compared to small molecules, while being smaller than protein molecules, such as antibodies and growth factors, and they can more closely reproduce specific interactions involved in PPIs [23]. The constant progress on synthesis and purification strategies, together with a significant decrease in amino acid prices have promoted sustained growth in the use of CPs as therapeutically relevant targets in the drug market [92]. Some representative examples used in the current pharmaceutical market to treat different diseases include *cyclosporine A* and cyclic citrullinated peptides for rheumatoid arthritis [269,270], *azathioprine* for ulcerative colitis [271], *somatostatin* analogues for advanced metastatic colorectal cancer, diabetic retinopathy or pancreatic and gastrointestinal cancers [272,273], *alcitonin* used in osteoporosis and hypercalcemia treatments [274], *ziconotide* indicated for severe chronic pain [275], *eptifibatide* for myocardial infarction [276], and analogues of myelin basic protein (MBP) and myelin oligodendrocyte glycoprotein (MOG) which have shown promising results for multiple sclerosis [277,278]. *Cyclotides* also exhibit different therapeutic activities such as antimicrobial, [279] anti-HIV [280], and anticancer [281], but they are not in use as drugs because of their cytotoxicity [104].

Nearly all known natural CPs display high *antibacterial activity*. Tyrothricin [49], the first commercialized antibiotic and still used in the clinic today (Tyrosur) or vancomycin [55] are examples of CPs with antimicrobial activity already used in therapeutics. AMPs share amphiphilicity and cationic structural properties with CPPs. Thus, CPPs can present potential application as AMPs with dual actions as both antibiotics and possible molecular transporter properties [282]. However, many of the known CPs are also hemolytic and thus lack the selectivity required for use as a human antibiotic. Efforts to develop CPs as antibiotics in vivo are directed towards the development of analogues that possess greater selectivity for bacterial cells over erythrocytes, and represent an emerging field with very promising initial results.

In recent years it was proposed to use CPs as *agricultural fungicides*, in the control of plant diseases [283]. Antimicrobial CPs that are active against plant pathogens have a wide spectrum of activity (e.g., *Alternaria*, *Botrytis*, *Cochliobolus*, *Geotrichum*, *Penicillium*, *Sclerotinia*, *Fusarium* species, and *Magnaporthe grisea*), probably due to similarity with the fungal cell wall, where antimicrobial CPs are active. However, there are several concerns that need to be solved prior to extended application of CPs in agriculture, such as the selective toxicity for pathogenic fungi over their hosts and the synergistic effects with other fungicides.

The cyclization of peptides has a very promising potential in *drug delivery*. CPP-mediated intracellular nano-drug delivery systems (nano-DDS) can improve delivery of water-insoluble drugs, negatively charged molecules (i.e., DNA, siRNA, phosphopeptides) and proteins [284]. Furthermore, cyclic CPPs containing specific amino acids have been shown to have different geometries, different affinities, and different mechanisms for similar targets as compared to linear CPPs. Thus, CPs can be designed to be used as nuclear delivery vehicles of anticancer compounds targeting DNA. The functionalization of CPs with tumor-targeting moieties can be used as a strategy for selective cancer cell targeting and to enhance nuclear targeting of anticancer drugs. Cyclic CPPs can be also covalently conjugated to active drug cargos to generate prodrugs (chemically modified analogues of an active metabolite) that can improve the pharmacokinetic and pharmacodynamic (PK/PD) properties of the drug. Therefore, an increase in the application of a new generation of CPs for the delivery of a wide range of drugs and biomolecules is to be expected. A representative example is, e.g., a novel drug-delivery system comprised of alternating tryptophan and arginine in a cyclic octapeptide [WR]4 recently presented by Parang and coworkers [285]. They showed that the optimal balance between electrostatic and hydrophobic interactions of the CPs (drug carrier) and phosphopeptides (transporters which give on/off signals to many enzymes through interactions with protein kinases), led to the formation of circular vesicle-like nanostructures (25–60 nm in diameter) with improved intracellular phosphopeptide delivery. Compared to its linear counterpart this delivery system had higher enzymatic stability, improved cell permeability, bypassed endosomal uptake and allowed the nuclear targeting and cellular delivery of impermeable phosphopeptides [286]. Cyclic [W(RW)4] was conjugated with Dox through an appropriate linker to afford cyclic [W(RW)4]–Dox, a cyclic prodrug ([W(RW)4] – Dox) that can be used to enhance the cellular retention of the drug in the treatment of ovarian cancer [287]. Very recently, it has been found that a novel class of pH-sensitive CPs containing tryptophan and glutamic acid residues have potential applications for targeting tumors and translocation of polar cargo molecules across the cellular membrane. These peptides might have application not only in targeting of acidic diseased tissue, but might also find very interesting applications in cosmetics and skin care products [288].

CPs have also been used as surfactants, showing promising ability for stabilization of silver nanoparticles [283]. These studies established a new class of surfactant-like cyclic peptide that self-assembled into nanostructures and could have potential application for the stabilization of silver nanoparticles and protein biomolecules. Since then, CPs have been used in nano-DDS-containing metal nanoparticles (metal nano-DDSs), which have emerged as effective tools for the treatment of various diseases. The formation of [WR]5-capped gold nanoparticles was shown to enhance the intracellular delivery of model anti-HIV drugs, phosphopeptides, and siRNA [289,290], exhibiting higher molecular transporting potency when compared with the linear counterparts. It has been recently demonstrated that a noncell-penetrating cyclic peptide can be converted to an efficient intracellular drug transporter through capping of the gold nanoparticles [289]. In an aqueous solution of chloroaurate this cyclic peptide formed a gold-capped CP delivery system. It was loaded with hydrophobic drugs (including doxorubicin, lamivudine, emtricitabine, and stavudine) in an equal molar ratio and showed improved cellular uptake and retention when used as a molecular transporter [291]. The generated cyclic peptide-capped selenium nanoparticles (CP–SeNPs) using a CP composed of five tryptophan, four arginine, and one cysteine [W5R4C], can potentially be used as nanosized delivery tools for negatively charged biomolecules and anticancer drugs [292].

Other applications behind CPs have been multifunctional nanoparticles for targeted cancer imaging, cancer therapy, and to identify peptide-targeting ligands. The application of CPs in the *imaging of angiogenesis* (formation of new blood vessels) is another promising approach. Cyclic arginine–glycine–asparatic acid, known as RGD-peptide, is responsible for binding integrin $\alpha_v\beta_3$. Integrins are heterodimeric molecular marker membrane receptors comprised of an α- and a β-subunit that mediate interactions between cells and the extracellular matrix [293]. This CP plays a key role in tumor angiogenesis and metastasis, whereas its expression is significantly upregulated in invasive tumor cells of certain cancers. Cyclic RGD-peptide has also been used for tumor cell targeting due to its high selectivity and affinity for different integrin subtypes [294–296]. The ability to visualize noninvasively and yet to quantify $\alpha_v\beta_3$ integrin expression level in vivo is a property that allows the understanding of the relationship between integrin level and tumor growth and spread. A series of cyclic RGD-peptides were labeled with different radionuclides for positron emission tomography and single-photon emission computed tomography imaging showing tumor targeting efficacy and in vivo pharmacokinetic profiles [297–300].

In addition, cyclo-RGD compound can be used to deliver nanoparticles filled with anticancer drug to the area where the tumor is growing [301].

Many studies have investigated the self-assembly process of CPs. Modification and functionalization of the surface of CP nanostructures can be achieved through reactions with the various side chains of the amino acids present on the CP [302]. By carefully designing the repeating CP unit and optimizing the self-assembly conditions, CP nanostructures can be tailored to meet the needs of specific applications (Fig. 4.11). However, few applications have been developed so far using these self-assembled nanostructures. This is due to the present lack of understanding of the mechanisms of self-assembly and the influence of various parameters on the self-assembling process [303]. It has been recently demonstrated that it is possible to control the supramolecular peptide architecture and the formation of nanotubes or nanoparticles using three different methods—phase equilibrium, pH-driven, and pH-sensitive methods [304]. The dimensions of the self-assembled nanostructures were found to be strongly influenced by the CP concentration, side chain modification, pH value, reaction time, stirring intensity, and sonication time.

SCPNs (Fig. 4.11) are structures built up by stacking CPs, reported for the first time by Ghadiri and coworkers, based on cyclic polypeptides with an even number of alternating D- and L-amino acids [99,305]. They are stabilized by hydrogen bonds—the amino acid side chains are directed outwards while the carbonyl and the amino groups of the peptide backbone are directed perpendicularly to the ring. In this conformation, antiparallel β-sheet-type interactions can be established between CPs in such a way that they stack on top of each other. These structures have attracted special attention due to the ease of modification of their structural and functional properties and because they can adopt a particular orientation by

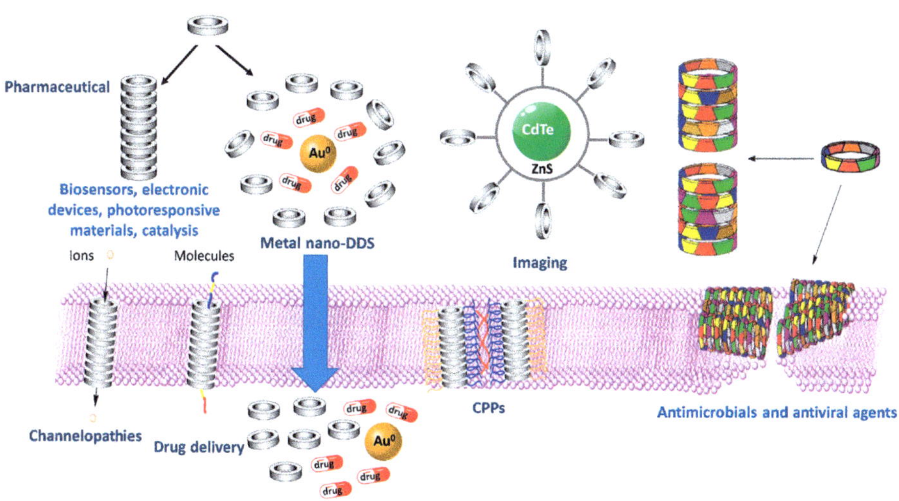

Figure 4.11 Schematic representation of some of the most important applications of self-assembled CPs into SCPNs or nanoparticles.

appropriate choice of the amino acid side chain functionalities and the ring size of the peptide subunit employed. In addition, due to the *Cn* symmetry of the backbone skeleton, nonsymmetrical CPs can form an infinite number of different structures, due to the interstrand rotation between two consecutive CPs, to form nonequivalent interactions for each β-sheet. Appropriate unit design and optimization of conditions for self-assembly allows the nanotube properties to be tailored for specific applications [306]. These nanostructures can be used in artificial photosystems, biosensors, antimicrobials and antiviral agents, electronic devices, photoresponsive materials, selective transmembrane transport channels, catalysis, and drug delivery [135,302,307,308]. Regarding their use in drug delivery, they can be used as artificial ion channels and pores [138]. For example, SCPNs composed of (W-D-L)4-Q-D-L and applied to the delivery of the antitumor drug 5-fluorouracil (5-FU) rapidly reached a high level of penetration into tumor cells where the drug effect was strengthened as the dosage of the cyclic peptide increased. This increase was associated with the CP improving the transport of the drug into the target cell [309]. Results from MD simulations showed that the 5-FU molecule was transported by hopping through different potential energy minima distributed along the subunits of the nanotube [309].

While almost all the SCPNs that have been developed so far have hydrophilic inner surfaces and can only permeate polar molecules, it is possible to modulate the inner properties of particular peptide nanotubes. As was discussed in Section 4.3, α,γ-SCPN [147,310,311] are composed of CPs alternating D-α-amino acids and (1*S*,3*S*)-3-aminocycloalkanecarboxylic acids. γ-Amino acids impose each ring in antiparallel orientation between each two subunits and the side chains of the amino acids are on the outside of the nanotube. This structure generates a partly hydrophobic cavity, which can be modulated by simple chemical modification of the β-carbon of the cyclic γ-amino acids, allowing the tuning of the transport properties of the nanotube. The group of Prof. JR Granja has recently developed a synthetic strategy that allows preparation of γ-amino acids modified at C2 position [148,149], α,γ-SCPN, with promising applications ranging from ion separation to antimicrobial therapy [133,312,313]. Other approximations to create peptide nanotubes with tunable interiors are, for example, the addition of 3-amino-2-methyl benzoic acid in the D,L-alternating primary sequence of the cyclic peptide [150]. The modulation of the external properties of SCPNs allows the design of CPs that self-assemble into lipid membranes, changing their permeability and resistance properties [314]. In these studies it was shown that depending on their sequences the formed SCPNs can be oriented either perpendicularly to the lipid membrane (hydrophobic CPs) [135,141] or in parallel (amphipathic CPs) (Fig. 4.11) [315]. The consequences of this orientation determine the effect caused in the lipid membrane. Hydrophobic CPs form hydrophobic transmembrane channels that can efficiently transport ions and small polar molecules. On the other hand, amphipathic CPs have disruptive properties on the membrane, conferring them a high antimicrobial activity. It was shown that cationic amphiphilic CPs selectively interact with bacterial membranes of Gram-positive bacteria and provided good protection against mice infected with MRSA (meticillin-resistant *Staphylococcus aureus*) [316]. The structural studies suggest a *carpet-type mechanism* that decreases the membrane stability, dissipating the membrane potential and leading eventually to cell death. One key feature that makes these CPs particularly promising antimicrobial peptides is that in principle the CPs by themselves should not be toxic. It is the membrane interaction that triggers the nanotube formation, having a supramolecular structure in the active form.

The use of SCPNs as *biosensors* has also been described. Studies based on ion channels modified with cyclic D,L-α-octapeptides have shown the ability to detect various polyanionic organic molecules, including the second messenger inositol tris-phosphate. These nanotubes inserted into self-assembled organosulfur monolayers supported on gold films have also shown the capability of diffusion-limited size-selective ion sensing [315]. Overall, these new results are very important to the development of a new class of biosensors and optical devices.

4.8 Concluding remarks

CPs have several structural features that improve the hydrophobicity of native linear counterparts providing conformational stability and enhanced affinity to a certain target. Cyclization of linear peptides also increases the resistance to cleavage by proteolytic enzymes and membrane permeability, leading to superior bioavailability.

Several natural CPs have recently emerged as templates for drug design due to their resistance to chemical or enzymatic hydrolysis and high selectivity to receptors. Apart from several naturally occurring cyclic peptides already in use, biologically active cyclic peptides were developed in recent years with genetic and synthetic approaches. Although not all cyclization strategies and constrained geometries enhance these properties to the same extent, CPs may be obtained with particular characteristics that allow them to be used in various important applications, such as therapeutics, agriculture, or diagnostics.

The development of practical protocols that mimic the power of Nature's strategies remains paramount for the advancement of novel CPs. From increased knowledge of the action of CPs in Nature, it might be possible to optimize ways they could be used in different applications. However, in most cases, the three-dimensional structures of naturally occurring CPs are still unknown, and efforts to determine their structures can provide many important new lessons for synthetic chemists. Furthermore, without the ability to understand dynamic structural changes, the design and creation of new CPs will rely on empirical rule-based approaches as well as on a fair degree of serendipity. More importantly, developing a greater understanding of the relationship between the structure of CPs and their activity and combining this knowledge with the current advances in rational drug design, peptide synthesis, and structure determination will result in the development of novel and more potent CPs. This will contribute to improved treatment of human diseases, such as cancer, infection, neurodegeneration and autoimmunity, as well as to other important applications related to nanotechnology.

References

[1] N. Sewald, H.-D. Jakubke, Introduction and background, Peptides: Chemistry and Biology, Wiley-VCH Verlag GmbH & Co. KGaA, 2009, pp. 1–4.

[2] B. Zhong, C. Zhang, S. Guo, C. Zhang, Rational design of cyclic peptides to disrupt TGF-B/SMAD7 signaling in heterotopic ossification, J. Mol. Graph. Model. 72 (2017) 25–31.

[3] R.P. Hicks, Antibacterial and anticancer activity of a series of novel peptides incorporating cyclic tetra-substituted Cα amino acids, Bioorg. Med. Chem. 24 (18) (2016) 4056–4065.

[4] M. Bastos, G. Bai, P. Gomes, D. Andreu, E. Goormaghtigh, M. Prieto, Energetics and partition of two cecropin-melittin hybrid peptides to model membranes of different composition, Biophys. J. 94 (6) (2008) 2128–2141.

[5] M. Bastos, T. Silva, V. Teixeira, K. Nazmi, J.G. Bolscher, S.S. Funari, et al., Lactoferrin-derived antimicrobial peptide induces a micellar cubic phase in a model membrane system, Biophys. J. 101 (3) (2011) L20–L22.

[6] K. Fosgerau, T. Hoffmann, Peptide therapeutics: current status and future directions, Drug Discov. Today 20 (1) (2015) 122–128.

[7] R.E.W. Hancock, H.G. Sahl, Antimicrobial and host-defense peptides as new anti-infective therapeutic strategies, Nat. Biotechnol. 24 (12) (2006) 1551–1557.

[8] A.A. Zompra, A.S. Galanis, O. Werbitzky, F. Albericio, Manufacturing peptides as active pharmaceutical ingredients, Future Med. Chem. 1 (2) (2009) 361–377.

[9] L. Otvos, Peptide-based drug design: here and now, in: L. Otvos (Ed.), Peptide-Based Drug Design, Humana Press, Totowa, NJ, 2008, pp. 1–8.

[10] D. Goodwin, P. Simerska, I. Toth, Peptides as therapeutics with enhanced bioactivity, Curr. Med. Chem. 19 (26) (2012) 4451–4461.

[11] V.M. Ahrens, R. Frank, S. Boehnke, C.L. Schütz, G. Hampel, D.S. Iffland, et al., Receptor-mediated uptake of boron-rich neuropeptide Y analogues for boron neutron capture therapy, ChemMedChem 10 (1) (2015) 164–172.

[12] J.M. Mason, Design and development of peptides and peptide mimetics as antagonists for therapeutic intervention, Future Med. Chem. 2 (12) (2010) 1813–1822.

[13] G. Luca, M. Rossella De, C. Lucia, Chemical modifications designed to improve peptide stability: incorporation of non-natural amino acids, pseudo-peptide bonds, and cyclization, Curr. Pharm. Des. 16 (28) (2010) 3185–3203.

[14] M. Brindisi, S. Maramai, S. Brogi, E. Fanigliulo, S. Butini, E. Guarino, et al., Development of novel cyclic peptides as pro-apoptotic agents, Eur. J. Med. Chem. 117 (2016) 301–320.

[15] H. Yu, Y.-S. Lin, Toward structure prediction of cyclic peptides, Phys. Chem. Chem. Phys. 17 (6) (2015) 4210–4219.

[16] M. Guharoy, P. Chakrabarti, Secondary structure based analysis and classification of biological interfaces: identification of binding motifs in protein–protein interactions, Bioinformatics 23 (15) (2007) 1909–1918.

[17] J.E. DeLorbe, J.H. Clements, B.B. Whiddon, S.F. Martin, Thermodynamic and structural effects of macrocyclic constraints in protein–ligand interactions, ACS Med. Chem. Lett. 1 (8) (2010) 448–452.

[18] J. DeLorbe, J. Clements, M. Teresk, A. Benfield, H. Plake, L. Millspaugh, et al., Thermodynamic and structural effects of conformational constraints in protein–ligand interactions. Entropic paradoxy associated with ligand preorganization, J. Am. Chem. Soc. 131 (46) (2009) 16758–16770.

[19] A. Tapeinou, M.-T. Matsoukas, C. Simal, T. Tselios, Review cyclic peptides on a merry-go-round; towards drug design, Pept. Sci. 104 (5) (2015) 453–461.
[20] X. Du, Y. Li, Y.-L. Xia, S.-M. Ai, J. Liang, P. Sang, et al., Insights into protein–ligand interactions: mechanisms, models, and methods, Int. J. Mol. Sci. 17 (2) (2016) 144.
[21] P. Edman, Chemistry of amino acids and peptides, Annu. Rev. Biochem. 28 (1959) 69–96.
[22] D.A. Horton, G.T. Bourne, M.L. Smythe, Exploring privileged structures: the combinatorial synthesis of cyclic peptides, Mol. Divers. 5 (4) (2000) 289–304.
[23] S.H. Joo, Cyclic peptides as therapeutic agents and biochemical tools, Biomol. Ther. 20 (1) (2012) 19–26.
[24] F.J. Duffy, M. Devocelle, D.C. Shields, Computational approaches to developing short cyclic peptide modulators of protein–protein interactions, in: P. Zhou, J. Huang (Eds.), Computational Peptidology, Springer New York, New York, 2015, pp. 241–271.
[25] T.S. Young, D.D. Young, I. Ahmad, J.M. Louis, S.J. Benkovic, P.G. Schultz, Evolution of cyclic peptide protease inhibitors, Proc. Nat. Acad. Sci. U.S.A. 108 (27) (2011) 11052–11056.
[26] K. Shibata, T. Suzawa, S. Soga, T. Mizukami, K. Yamada, N. Hanai, et al., Improvement of biological activity and proteolytic stability of peptides by coupling with a cyclic peptide, Bioorg. Med. Chem. Lett. 13 (15) (2003) 2583–2586.
[27] K. Jensen, Peptide and Protein Design for Biopharmaceutical Applications, Wiley, 2009.
[28] M.S. Lee, B. Gardner, M. Kahn, H. Nakanishi, The three-dimensional solution structure of a constrained peptidomimetic in water and in chloroform observation of solvent induced hydrophobic cluster, FEBS Lett. 359 (2–3) (1995) 113–118.
[29] K. Uma, R. Kishore, P. Balaram, Stereochemical constraints in peptide design: analysis of the influence of a disulfide bridge and an α-aminoisobutyryl residue on the conformation of a hexapeptide, Biopolymers 33 (6) (1993) 865–871.
[30] D. Lau, L. Guo, R. Liu, J. Marik, K. Lam, Peptide ligands targeting integrin α3β1 in non-small cell lung cancer, Lung Cancer 52 (3) (2006) 291–297.
[31] T. Rezai, B. Yu, G.L. Millhauser, M.P. Jacobson, R.S. Lokey, Testing the conformational hypothesis of passive membrane permeability using synthetic cyclic peptide diastereomers, J. Am. Chem. Soc. 128 (8) (2006) 2510–2511.
[32] Y.-U. Kwon, T. Kodadek, Quantitative comparison of the relative cell permeability of cyclic and linear peptides, Chem. Biol. 14 (6) (2007) 671–677.
[33] G. Hussack, T. Hirama, W. Ding, R. MacKenzie, J. Tanha, Engineered single-domain antibodies with high protease resistance and thermal stability, PLoS One 6 (11) (2011) e28218.
[34] M. Lindgren, M. Hällbrink, A. Prochiantz, Ü. Langel, Cell-penetrating peptides, Trends Pharmacol. Sci. 21 (3) (2000) 99–103.
[35] E. Cini, G. Bifulco, G. Menchi, M. Rodriquez, M. Taddei, Synthesis of enantiopure 7-substituted azepane-2-carboxylic acids as templates for conformationally constrained peptidomimetics, Eur. J. Org. Chem. 2012 (11) (2012) 2133–2141.
[36] G. Grigoryan, A.W. Reinke, A.E. Keating, Design of protein-interaction specificity gives selective bZIP-binding peptides, Nature 458 (7240) (2009) 859–864.
[37] A. Grauer, B. König, A. Peptidomimetics, Versatile route to biologically active compounds, Eur. J. Org. Chem. 2009 (30) (2009) 5099–5111.
[38] D.J. Mandell, T. Kortemme, Computer-aided design of functional protein interactions, Nat. Chem. Biol. 5 (11) (2009) 797–807.
[39] N. Terrett, Drugs in middle space, MedChemComm 4 (3) (2013) 474–475.

[40] G.F. Gause, M.G. Brazhnikova, Gramicidin S and its use in the treatment of infected wounds, Nature 154 (3918) (1944) 703–803.
[41] K. Thell, R. Hellinger, E. Sahin, P. Michenthaler, M. Gold-Binder, T. Haider, et al., Oral activity of a nature-derived cyclic peptide for the treatment of multiple sclerosis, Proc. Nat. Acad. Sci. U.S.A. 113 (15) (2016) 3960–3965.
[42] H.-S. Lee, H.J. Shin, K.H. Jang, T.S. Kim, K.-B. Oh, J. Shin, Cyclic peptides of the nocardamine class from a marine-derived bacterium of the genus *Streptomyces*, J. Nat. Prod. 68 (4) (2005) 623–625.
[43] V.A. Burns, B.G. Bobay, A. Basso, J. Cavanagh, C. Melander, Targeting RNA with cysteine-constrained peptides, Bioorg. Med. Chem. Lett. 18 (2) (2008) 565–567.
[44] D.J. Craik, Seamless proteins tie up their loose ends, Science 311 (5767) (2006) 1563.
[45] N. Daly, D. Wilson, D. Craik, Cyclic peptides from plants and their promise in drug design, Aust. Biochem. 43 (2) (2012) 7–9.
[46] D. Craik, A. Poth, M. Colgrave, M. Akcan, B. Oku, A. Chan, et al., Discovery and applications of naturally occurring cyclic peptides, Planta Med. 77 (12) (2011) 1231–2231.
[47] T. Mogi, K. Kita, Gramicidin S and polymyxins: the revival of cationic cyclic peptide antibiotics, Cell. Mol. Life Sci. 66 (23) (2009) 3821.
[48] E. Conti, T. Stachelhaus, M.A. Marahiel, P. Brick, Structural basis for the activation of phenylalanine in the non-ribosomal biosynthesis of gramicidin S, EMBO J. 16 (14) (1997) 4174.
[49] H.D. Mootz, M.A. Marahiel, The tyrocidine biosynthesis operon of *Bacillus brevis*: complete nucleotide sequence and biochemical characterization of functional internal adenylation domains, J. Bacteriol. 179 (21) (1997) 6843–6850.
[50] F. Lipmann, R. Roskoski, W. Gevers, H. Kleinkauf, Tyrocidine biosynthesis by three complementary fractions from *Bacillus brevis* (ATCC 8185), Biochemistry 9 (25) (1970) 4839–4845.
[51] S.A. Survase, L.D. Kagliwal, U.S. Annapure, R.S. Singhal, Cyclosporin A: a review on fermentative production, downstream processing and pharmacological applications, Biotechnol. Adv. 29 (4) (2011) 418–435.
[52] P. Martinez Viedma, H. Abriouel, N. Ben Omar, R. Lucas Lopez, A. Galvez, Inhibition of spoilage and toxigenic *Bacillus* species in dough from wheat flour by the cyclic peptide enterocin AS-48, Food Control. 22 (5) (2011) 756–761.
[53] J. Borrero, D.A. Brede, M. Skaugen, D.B. Diep, C. Herranz, I.F. Nes, et al., Characterization of Garvicin ML, a novel circular bacteriocin produced by *Lactococcus garvieae* DCC43, isolated from Mallard ducks (*Anas platyrhynchos*), Appl. Environ. Microbiol. 77 (1) (2011) 369–373.
[54] N. Sawa, T. Zendo, J. Kiyofuji, K. Fujita, K. Himeno, J. Nakayama, et al., Identification and characterization of lactocyclicin Q, a novel cyclic bacteriocin produced by *Lactococcus* sp strain QU 12, Appl. Environ. Microbiol. 75 (6) (2009) 1552–1558.
[55] A.V.R. Rao, M.K. Gurjar, K.L. Reddy, A.S. Rao, Studies directed toward the synthesis of vancomycin and related cyclic-peptides, Chem. Rev. 95 (6) (1995) 2135–2167.
[56] H.E. Hallen, H. Luo, J.S. Scott-Craig, J.D. Walton, Gene family encoding the major toxins of lethal Amanita mushrooms, Proc. Nat. Acad. Sci. U.S.A. 104 (48) (2007) 19097–19101.
[57] A.T. Bockus, C.M. McEwen, R.S. Lokey, Form and function in cyclic peptide natural products: a pharmacokinetic perspective, Curr. Top. Med. Chem. 13 (7) (2013) 821–836.

[58] T. Wieland, H. Faulstich, Amatoxins, phallotoxins, phallolysin, and antamanide – biologically-active components of poisonous amanita mushrooms, CRC Crit. Rev. Biochem. 5 (3) (1978) 185–260.

[59] D.J. Craik, N.L. Daly, T. Bond, C. Waine, Plant cyclotides: a unique family of cyclic and knotted proteins that defines the cyclic cystine knot structural motif, J. Mol. Biol. 294 (5) (1999) 1327–1336.

[60] T. Mahatmanto, Review seed biopharmaceutical cyclic peptides: from discovery to applications, Biopolymers 104 (6) (2015) 804–814.

[61] O. Saether, D.J. Craik, I.D. Campbell, K. Sletten, J. Juul, D.G. Norman, Elucidation of the primary and three-dimensional structure of the uterotonic polypeptide Kalata B1, Biochemistry 34 (13) (1995) 4147–4158.

[62] M.L.J. Korsinczky, H.J. Schirra, D.J. Craik, Sunflower trypsin inhibitor-1, Curr. Protein Pept. Sci. 5 (5) (2004) 351–364.

[63] N. Venkataraman, A.L. Cole, P. Ruchala, A.J. Waring, R.I. Lehrer, O. Stuchlik, et al., Reawakening retrocyclins: ancestral human defensins active against HIV-1, PLoS Biol. 7 (4) (2009) 720–729.

[64] A.C. Conibear, K.J. Rosengren, P.J. Harvey, D.J. Craik, Structural characterization of the cyclic cystine ladder motif of theta-defensins, Biochemistry 51 (48) (2012) 9718–9726.

[65] Y.Q. Tang, J. Yuan, G. Osapay, K. Osapay, D. Tran, C.J. Miller, et al., A cyclic antimicrobial peptide produced in primate leukocytes by the ligation of two truncated alpha-defensins, Science 286 (5439) (1999) 498–502.

[66] A.E. Garcia, G. Osapay, P.A. Tran, J. Yuan, M.E. Selsted, Isolation, synthesis, and antimicrobial activities of naturally occurring theta-defensin isoforms from baboon leukocytes, Infect. Immun. 76 (12) (2008) 5883–5891.

[67] D.W. Urry, M. Ohnishi, R. Walter, Secondary structure of the cyclic moiety of the peptide hormone oxytocin and its deamino analog, Proc. Nat. Acad. Sci. 66 (1) (1970) 111–116.

[68] S. Granier, S. Terrillon, R. Pascal, H. Déméné, M. Bouvier, G. Guillon, et al., A cyclic peptide mimicking the third intracellular loop of the V2 vasopressin receptor inhibits signaling through its interaction with receptor dimer and G protein, J. Biol. Chem. 279 (49) (2004) 50904–50914.

[69] M. Manning, J.P. Przybylski, A. Olma, W.A. Klis, M. Kruszynski, N.C. Wo, et al., No requirement of cyclic conformation of antagonists in binding to vasopressin receptors, Nature 329 (6142) (1987) 839–840.

[70] S. Punna, J. Kuzelka, Q. Wang, M.G. Finn, Head-to-tail peptide cyclodimerization by copper-catalyzed azide-alkyne cycloaddition, Angew. Chem. Int. Ed. 44 (15) (2005) 2215–2220.

[71] D.J. Craik, N.L. Daly, I. Saska, M. Trabi, K.J. Rosengren, Structures of naturally occurring circular proteins from bacteria, J. Bacteriol. 185 (14) (2003) 4011–4021.

[72] A. Bryan, L. Joseph, J.A. Bennett, H.I. Jacobson, T.T. Andersen, Design and synthesis of biologically active peptides: a "tail" of amino acids can modulate activity of synthetic cyclic peptides, Peptides 32 (12) (2011) 2504–2510.

[73] J.E. Bock, J. Gavenonis, J.A. Kritzer, Getting in shape: controlling peptide bioactivity and bioavailability using conformational constraints, ACS Chem. Biol. 8 (3) (2013) 488–499.

[74] C. Ramakrishnan, P.K.C. Paul, K. Ramnarayan, Cyclic peptides — small and big and their conformational aspects, J. Biosci. 8 (1) (1985) 239–251.

[75] E.M. Driggers, S.P. Hale, J. Lee, N.K. Terrett, The exploration of macrocycles for drug discovery—an underexploited structural class, Nat. Rev. Drug Discov. 7 (7) (2008) 608–624.
[76] C.J. White, A.K. Yudin, Contemporary strategies for peptide macrocyclization, Nat. Chem. 3 (7) (2011) 509–524.
[77] D.A. Bushnell, P. Cramer, R.D. Kornberg, Structural basis of transcription: alpha-amanitin-RNA polymerase II cocrystal at 2.8 A resolution, Proc. Nat. Acad. Sci. U.S.A. 99 (3) (2002) 1218–1222.
[78] F. Enjalbert, S. Rapior, J. Nouguier-Soule, S. Guillon, N. Amouroux, Treatment of amatoxin poisoning: 20-year retrospective analysis, J. Toxicol. Clin. Toxicol. 40 (6) (2002) 715–757.
[79] S.D. Jolad, J.J. Hoffmann, S.J. Torrance, R.M. Wiedhopf, J.R. Cole, S.K. Arora, et al., Bouvardin and deoxybouvardin, anti-tumor cyclic hexapeptides from bouvardia-ternifolia (*Rubiaceae*), J. Am. Chem. Soc. 99 (24) (1977) 8040–8045.
[80] M. Zalacain, E. Zaera, D. Vazquez, A. Jimenez, The mode of action of the anti-tumor drug bouvardin, an inhibitor of protein-synthesis in eukaryotic cells, FEBS Lett. 148 (1) (1982) 95–97.
[81] T.W.C. Leung, D.H. Williams, J.C.J. Barna, S. Foti, P.B. Oelrichs, Structural studies on the peptide moroidin from laportea moroides, Tetrahedron 42 (12) (1986) 3333–3348.
[82] S.D. Kahn, P.M. Booth, J.P. Waltho, D.H. Williams, Computer-assisted structure determination – structure of the peptide moroidin from laportea-moroides, J. Org. Chem. 54 (8) (1989) 1901–1904.
[83] C. Heinis, T. Rutherford, S. Freund, G. Winter, Phage-encoded combinatorial chemical libraries based on bicyclic peptides, Nat. Chem. Biol. 5 (7) (2009) 502–507.
[84] A. Angelini, L. Cendron, S. Chen, J. Touati, G. Winter, G. Zanotti, et al., Bicyclic peptide inhibitor reveals large contact interface with a protease target, ACS Chem. Biol. 7 (5) (2012) 817–821.
[85] S. Chen, J. Morales-Sanfrutos, A. Angelini, B. Cutting, C. Heinis, Structurally diverse cyclisation linkers impose different backbone conformations in bicyclic peptides, ChemBioChem 13 (7) (2012) 1032–1038.
[86] W. Lian, P. Upadhyaya, C.A. Rhodes, Y. Liu, D. Pei, Screening bicyclic peptide libraries for protein–protein interaction inhibitors: discovery of a tumor necrosis factor-alpha antagonist, J. Am. Chem. Soc. 135 (32) (2013) 11990–11995.
[87] J.S. Quartararo, P. Wu, J.A. Kritzer, Peptide bicycles that inhibit the Grb2 SH2 domain, ChemBioChem 13 (10) (2012) 1490–1496.
[88] R. Roodbeen, B. Paaske, L. Jiang, J.K. Jensen, A. Christensen, J.T. Nielsen, et al., Bicyclic peptide inhibitor of urokinase-type plasminogen activator: mode of action, ChemBioChem 14 (16) (2013) 2179–2188.
[89] J. Lubelski, R. Rink, R. Khusainov, G.N. Moll, O.P. Kuipers, Biosynthesis, immunity, regulation, mode of action and engineering of the model lantibiotic nisin, Cell. Mol. Life Sci. 65 (3) (2008) 455–476.
[90] C. Chatterjee, M. Paul, L.L. Xie, W.A. van der Donk, Biosynthesis and mode of action of lantibiotics, Chem. Rev. 105 (2) (2005) 633–683.
[91] H. Lackner, 3-Dimensional structure of actinomycins, Angew. Chem. Int. Ed. Engl. 14 (6) (1975) 375–386.
[92] V. Baeriswyl, C. Heinis, Polycyclic peptide therapeutics, ChemMedChem 8 (3) (2013) 377–384.

[93] T.B. Andrew, M.M. Cayla, R.S. Lokey, Form and function in cyclic peptide natural products: a pharmacokinetic perspective, Curr. Top. Med. Chem. 13 (7) (2013) 821–836.
[94] A. Stavrakoudis, I.G. Tsoulos, Z.O. Shenkarev, T.V. Ovchinnikova, Molecular dynamics simulation of antimicrobial peptide arenicin-2: β-hairpin stabilization by noncovalent interactions, Pept. Sci. 92 (3) (2009) 143–155.
[95] R.E.W. Hancock, M.G. Scott, The role of antimicrobial peptides in animal defenses, Proc. Nat. Acad. Sci. U.S.A. 97 (16) (2000) 8856–8861.
[96] K. Matsuzaki, Control of cell selectivity of antimicrobial peptides, Biochim. Biophys. Acta 1788 (8) (2009) 1687–1692.
[97] M.R. Yeaman, N.Y. Yount, Mechanisms of antimicrobial peptide action and resistance, Pharmacol. Rev. 55 (1) (2003) 27.
[98] M. Zasloff, Antimicrobial peptides of multicellular organisms, Nature 415 (6870) (2002) 389–395.
[99] M.R. Ghadiri, J.R. Granja, R.A. Milligan, D.E. McRee, N. Khazanovich, Self-assembling organic nanotubes based on a cyclic peptide architecture, Nature 366 (6453) (1993) 324–327.
[100] K. Rosenthal-Aizman, G. Svensson, A. Undén, Self-assembling peptide nanotubes from enantiomeric pairs of cyclic peptides with alternating D and L amino acid residues, J. Am. Chem. Soc. 126 (11) (2004) 3372–3373.
[101] P.F. Augustijns, T.P. Bradshaw, L.S.L. Gan, R.W. Hendren, D.R. Thakker, Evidence for a polarized efflux system in Caco-2 cells capable of modulating cyclosporine A transport, Biochem. Biophys. Res. Commun. 197 (2) (1993) 360–365.
[102] J. Gao, M.T. Hamann, Chemistry and biology of kahalalides, Chem. Rev. 111 (5) (2011) 3208–3235.
[103] D.W. Lee, B.S. Kim, Antimicrobial cyclic peptides for plant disease control, Plant Pathol. J. 31 (1) (2015) 1–11.
[104] N.L. Daly, K.J. Rosengren, D.J. Craik, Discovery, structure and biological activities of cyclotides, Adv. Drug Deliv. Rev. 61 (11) (2009) 918–930.
[105] L. Skjeldal, L. Gran, K. Sletten, B.F. Volkman, Refined structure and metal binding site of the Kalata B1 peptide, Arch. Biochem. Biophys. 399 (2) (2002) 142–148.
[106] Y.L. Li, T. Bi, J.A. Camarero, Chemical and biological production of cyclotides, Adv. Bot. Res. 76 (2015) 271–303.
[107] J.S. Davies, The cyclization of peptides and depsipeptides, J. Pept. Sci. 9 (8) (2003) 471–501.
[108] A. Kumar, G. Ye, Y. Wang, X. Lin, G. Sun, K. Parang, Synthesis and structure–activity relationships of linear and conformationally constrained peptide analogues of CIYKYY as Src tyrosine kinase inhibitors, J. Med. Chem. 49 (11) (2006) 3395–3401.
[109] N. Assa-Munt, X. Jia, P. Laakkonen, E. Ruoslahti, Solution structures and integrin binding activities of an RGD peptide with two isomers, Biochemistry 40 (8) (2001) 2373–2378.
[110] C. Gilon, C. Mang, E. Lohof, A. Friedler, H. Kessler, Synthesis of cyclic peptides, in: M. Goodman, C. Toniolo, L. Moroder, A. Felix (Eds.), Synthesis of Peptides and Peptidomimetics, G. Thieme, 2002.
[111] Y.-U. Kwon, T. Kodadek, Encoded combinatorial libraries for the construction of cyclic peptoid microarrays, Chem. Commun. (Camb). (44) (2008) 5704–5706.
[112] C.M. Deber, V. Madison, E.R. Blout, Why cyclic peptides? Complementary approaches to conformations, Acc. Chem. Res. 9 (3) (1976) 106–113.

[113] S. Jiang, Z. Li, K. Ding, P.P. Roller, Recent progress of synthetic studies to peptide and peptidomimetic cyclization, Curr. Org. Chem. 12 (17) (2008) 1502–1542.
[114] D.J. Craik, J.E. Swedberg, J.S. Mylne, M. Cemazar, Cyclotides as a basis for drug design, Expert Opin. Drug Discov. 7 (3) (2012) 179–194.
[115] K.S. Harris, T. Durek, Q. Kaas, A.G. Poth, E.K. Gilding, B.F. Conlan, et al., Efficient backbone cyclization of linear peptides by a recombinant asparaginyl endopeptidase, Nat. Commun. 6 (2015) 10199.
[116] A. Piserchio, G.D. Salinas, T. Li, J. Marshall, M.R. Spaller, D.F. Mierke, Targeting specific PDZ domains of PSD-95: structural basis for enhanced affinity and enzymatic stability of a cyclic peptide, Chem. Biol. 11 (4) (2004) 469–473.
[117] P. Grieco, P.M. Gitu, V.J. Hruby, Preparation of "side-chain-to-side-chain" cyclic peptides by Allyl and Alloc strategy: potential for library synthesis, J. Pept. Res. 57 (3) (2001) 250–256.
[118] S.R. Tala, S.M. Schnell, C. Haskell-Luevano, Microwave-assisted solid-phase synthesis of side-chain to side-chain lactam-bridge cyclic peptides, Bioorg. Med. Chem. Lett. 25 (24) (2015) 5708–5711.
[119] C. Gilon, D. Halle, M. Chorev, Z. Selincer, G. Byk, Backbone cyclization: a new method for conferring conformational constraint on peptides, Biopolymers 31 (6) (1991) 745–750.
[120] Lundquist, J.C. Pelletier, A new tri-orthogonal strategy for peptide cyclization, Org. Lett. 4 (19) (2002) 3219–3221.
[121] C.G. Ullman, L. Frigotto, R.N. Cooley, In vitro methods for peptide display and their applications, Brief. Funct. Genomics 10 (3) (2011) 125–134.
[122] M. Hahn, D. Winkler, K. Welfle, R. Misselwitz, H. Welfle, H. Wessner, et al., Cross-reactive binding of cyclic peptides to an anti-TGFα antibody Fab fragment: an X-ray structural and thermodynamic analysis, J. Mol. Biol. 314 (2) (2001) 293–309.
[123] M. Akamatsu, P.P. Roller, L. Chen, Z.-Y. Zhang, B. Ye, T.R. Burke Jr, Potent inhibition of protein-tyrosine phosphatase by phosphotyrosine-mimic containing cyclic peptides, Bioorg. Med. Chem. 5 (1) (1997) 157–163.
[124] G. Luca, C. Giuliana, T. Alessandra, S. Federico, M. Rossella De, C. Gianpaolo, Cyclopeptide analogs for generating new molecular and 3D diversity, Comb. Chem. High Throughput Screen 12 (10) (2009) 929–939.
[125] A. Fernández-Tejada, F. Corzana, J.H. Busto, A. Avenoza, J.M. Peregrina, Conformational effects of the non-natural α-methylserine on small peptides and glycopeptides, J. Org. Chem. 74 (24) (2009) 9305–9313.
[126] A. Fernández-Tejada, F. Corzana, J.H. Busto, G. Jiménez-Osés, J.M. Peregrina, A. Avenoza, Non-natural amino acids as modulating agents of the conformational space of model glycopeptides, Chemistry 14 (23) (2008) 7042–7058.
[127] J. Schwochert, R. Turner, M. Thang, R.F. Berkeley, A.R. Ponkey, K.M. Rodriguez, et al., Peptide to peptoid substitutions increase cell permeability in cyclic hexapeptides, Org. Lett. 17 (12) (2015) 2928–2931.
[128] S. Chen, D. Gfeller, S.A. Buth, O. Michielin, P.G. Leiman, C. Heinis, Improving binding affinity and stability of peptide ligands by substituting glycines with D-amino acids, ChemBioChem 14 (11) (2013) 1316–1322.
[129] A.C. Rand, S.S.F. Leung, H. Eng, C.J. Rotter, R. Sharma, A.S. Kalgutkar, et al., Optimizing PK properties of cyclic peptides: the effect of side chain substitutions on permeability and clearance, MedChemComm 3 (10) (2012) 1282–1289.

[130] O. Demmer, A.O. Frank, H. Kessler, Design of cyclic peptides, in: K. Jensen (Ed.), Peptide and Protein Design for Biopharmaceutical Applications, John Wiley & Sons, Ltd., 2009.

[131] J.N. Lambert, J.P. Mitchell, K.D. Roberts, The synthesis of cyclic peptides, J. Chem. Soc. Perkin Trans. 1 (5) (2001) 471–484.

[132] R. Garcia-Fandino, M. Amorin, J.R. Granja, Synthesis of supramolecular nanotubes, Supramolecular Chemistry: From Molecules to Nanomaterials, John Wiley & Sons, Ltd, 2012.

[133] J. Montenegro, M.R. Ghadiri, J.R. Granja, Ion channel models based on self-assembling cyclic peptide nanotubes, Acc. Chem. Res. 46 (12) (2013) 2955–2965.

[134] T. Suga, S. Osada, H. Kodama, Formation of ion-selective channel using cyclic tetrapeptides, Bioorg. Med. Chem. 20 (1) (2012) 42–46.

[135] S. Fernandez-Lopez, H.-S. Kim, E.C. Choi, M. Delgado, J.R. Granja, A. Khasanov, et al., Antibacterial agents based on the cyclic D,L-[alpha]-peptide architecture, Nature 412 (6845) (2001) 452–455.

[136] J. Taira, S. Osada, R. Hayashi, T. Ueda, M. Jelokhani-Niaraki, H. Aoyagi, et al., *Trans*-bilayer ion conduction by proline containing cyclic hexapeptides and effects of amino acid substitutions on ion conducting properties, Bull. Chem. Soc. Jpn. 83 (6) (2010) 683–688.

[137] J.T. Fletcher, J.A. Finlay, M.E. Callow, J.A. Callow, M.R. Ghadiri, A combinatorial approach to the discovery of biocidal six-residue cyclic D,L-α-peptides against bacteria, MRSA and *E. coli* and the aiofouling algae *Ulva linza* and *Navicula perminuta*, Chemistry 13 (14) (2007) 4008–4013.

[138] M.R. Ghadiri, J.R. Granja, L.K. Buehler, Artificial transmembrane ion channels from self-assembling peptide nanotubes, Nature 369 (6478) (1994) 301–304.

[139] J. Sanchez-Quesada, M.R. Ghadiri, H. Bayley, O. Braha, Cyclic peptides as molecular adapters for a pore-forming protein, J. Am. Chem. Soc. 122 (48) (2000) 11757–11766.

[140] H.S. Kim, J.D. Hartgerink, M.R. Ghadiri, Oriented self-assembly of cyclic peptide nanotubes in lipid membranes, J. Am. Chem. Soc. 120 (18) (1998) 4417–4424.

[141] L. Motiei, S. Rahimipour, D.A. Thayer, C.-H. Wong, M.R. Ghadiri, Antibacterial cyclic D,L-[small alpha]-glycopeptides, Chem. Commun. (25) (2009) 3693–3695.

[142] J.R. Granja, M.R. Ghadiri, Channel-mediated transport of glucose across lipid bilayers, J. Am. Chem. Soc. 116 (23) (1994) 10785–10786.

[143] J. Sánchez-Quesada, H. Sun Kim, M.R. Ghadiri, A synthetic pore-mediated transmembrane transport of glutamic acid, Angew. Chem. Int. Ed. 40 (13) (2001) 2503–2506.

[144] N. Khazanovich, J.R. Granja, D.E. McRee, R.A. Milligan, M.R. Ghadiri, Nanoscale tubular ensembles with specified internal diameters. Design of a self-assembled nanotube with a 13-[angstrom] pore, J. Am. Chem. Soc. 116 (1994) 6011–6012.

[145] D. Seebach, E. Dubost, R.I. Mathad, B. Jaun, M. Limbach, M. Löweneck, et al., New open-chain and cyclic tetrapeptides, consisting of α-, β2-, and β3-amino-acid residues, as somatostatin mimics – a survey, Helv. Chim. Acta 91 (9) (2008) 1736–1786.

[146] K. Gademann, M. Ernst, D. Hoyer, D. Seebach, Synthesis and biological evaluation of a cyclo-β-tetrapeptide as a somatostatin analogue, Angew. Chem. Int. Ed. 38 (9) (1999) 1223–1226.

[147] M. Amorín, L. Castedo, J.R. Granja, New cyclic peptide assemblies with hydrophobic cavities: the structural and thermodynamic basis of a new class of peptide tanotubes, J. Am. Chem. Soc. 125 (10) (2003) 2844–2845.

[148] C. Reiriz, M. Amorin, R. Garcia-Fandino, L. Castedo, J.R. Granja, Alpha-gamma-cyclic peptide ensembles with a hydroxylated cavity, Org. Biomol. Chem. 7 (21) (2009) 4358–4361.
[149] N. Rodriguez-Vazquez, R. Garcia-Fandino, M. Amorin, J.R. Granja, Self-assembling alpha, gamma-cyclic peptides that generate cavities with tunable properties, Chem. Sci. 7 (1) (2016) 183–187.
[150] R. Hourani, C. Zhang, R. van der Weegen, L. Ruiz, C. Li, S. Keten, et al., Processable cyclic peptide nanotubes with tunable interiors, J. Am. Chem. Soc. 133 (39) (2011) 15296–15299.
[151] S. Leclair, P. Baillargeon, R. Skouta, D. Gauthier, Y. Zhao, Y.L. Dory, Micrometer-sized hexagonal tubes self-assembled by a cyclic peptide in a liquid crystal, Angew. Chem. Int. Ed. 43 (3) (2004) 349–353.
[152] W.S. Horne, C.D. Stout, M.R. Ghadiri, A heterocyclic peptide nanotube, J. Am. Chem. Soc. 125 (31) (2003) 9372–9376.
[153] J.H. van Maarseveen, W.S. Horne, M.R. Ghadiri, Efficient route to C2 symmetric heterocyclic backbone modified cyclic peptides, Org. Lett. 7 (20) (2005) 4503–4506.
[154] S. Guha, A. Banerjee, Self-assembled robust dipeptide nanotubes and fabrication of dipeptide-capped gold nanoparticles on the surface of these nanotubes, Adv. Funct. Mater. 19 (12) (2009) 1949–1961.
[155] I. Alfonso, M. Bru, M.I. Burguete, E. García-Verdugo, S.V. Luis, Structural diversity in the self-assembly of pseudopeptidic macrocycles, Chemistry 16 (4) (2010) 1246–1255.
[156] C. Qin, X. Bu, X. Wu, Z. Guo, A chemical approach to generate molecular diversity based on the scaffold of cyclic decapeptide antibiotic tyrocidine A, J. Comb. Chem. 5 (4) (2003) 353–355.
[157] T.R. White, C.M. Renzelman, A.C. Rand, T. Rezai, C.M. McEwen, V.M. Gelev, et al., On-resin N-methylation of cyclic peptides for discovery of orally bioavailable scaffolds, Nat. Chem. Biol. 7 (11) (2011) 810–817.
[158] J. Chatterjee, F. Rechenmacher, H. Kessler, N-methylation of peptides and proteins: an important element for modulating biological functions, Angew. Chem. Int. Ed. 52 (1) (2013) 254–269.
[159] E. Biron, J. Chatterjee, O. Ovadia, D. Langenegger, J. Brueggen, D. Hoyer, et al., Improving oral bioavailability of peptides by multiple N-methylation: somatostatin analogues, Angew. Chem. Int. Ed. 47 (14) (2008) 2595–2599.
[160] T.A. Hill, N.E. Shepherd, F. Diness, D.P. Fairlie, Constraining cyclic peptides to mimic protein structure motifs, Angew. Chem. Int. Ed. 53 (48) (2014) 13020–13041.
[161] C.K.L. Wang, Q. Kaas, L. Chiche, D.J. Craik, CyBase: a database of cyclic protein sequences and structures, with applications in protein discovery and engineering, Nucleic Acids Res. 36 (Database issue) (2008) D206–D210.
[162] A.G. Poth, L.Y. Chan, D.J. Craik, Cyclotides as grafting frameworks for protein engineering and drug design applications, Pept. Sci. 100 (5) (2013) 480–491.
[163] L.Y. Chan, S. Gunasekera, S.T. Henriques, N.F. Worth, S.-J. Le, R.J. Clark, et al., Engineering pro-angiogenic peptides using stable, disulfide-rich cyclic scaffolds, Blood 118 (25) (2011) 6709.
[164] E. Emanuele, M. Arra, S. Pesenti, Vasopressin and oxytocin as neurohormonal mediators of MDMA (ecstasy) sociosexual behavioural effects, Med. Hypotheses 67 (5) (2006) 1250–1251.

[165] M. Manning, A. Misicka, A. Olma, K. Bankowski, S. Stoev, B. Chini, et al., Oxytocin and vasopressin agonists and antagonists as research tools and potential therapeutics, J. Neuroendocrinol. 24 (4) (2012) 609–628.

[166] R. Eliasen, N.L. Daly, B.S. Wulff, T.L. Andresen, K.W. Conde-Frieboes, D.J. Craik, Design, synthesis, structural and functional characterization of novel melanocortin agonists based on the cyclotide Kalata B1, J. Biol. Chem. 287 (48) (2012) 40493–40501.

[167] A.V. Mayorov, M. Cai, E.S. Palmer, Z. Liu, J.P. Cain, J. Vagner, et al., Solid-phase peptide head-to-side chain cyclodimerization: discovery of C(2)-symmetric cyclic lactam hybrid α-melanocyte-stimulating hormone (MSH)/agouti-signaling protein (ASIP) analogues with potent activities at the human melanocortin receptors, Peptides 31 (10) (2010) 1894–1905.

[168] D. Laimou, T. Katsila, J. Matsoukas, A. Schally, K. Gkountelias, G. Liapakis, et al., Rationally designed cyclic analogues of luteinizing hormone-releasing hormone: enhanced enzymatic stability and biological properties, Eur. J. Med. Chem. 58 (2012) 237–247.

[169] S.C. Koerber, J. Rizo, R.S. Struthers, J.E. Rivier, Consensus bioactive conformation of cyclic GnRH antagonists defined by NMR and molecular modeling, J. Med. Chem. 43 (5) (2000) 819–828.

[170] J.E. Rivier, R.S. Struthers, J. Porter, S.L. Lahrichi, G. Jiang, L.A. Cervini, et al., Design of potent dicyclic (4–10/5–8) gonadotropin releasing hormone (GnRH) antagonists, J. Med. Chem. 43 (5) (2000) 784–796.

[171] J.E. Rivier, J. Porter, L.A. Cervini, S.L. Lahrichi, D.A. Kirby, R.S. Struthers, et al., Design of monocyclic (1–3) and dicyclic (1–3/4–10) gonadotropin releasing hormone (GnRH) antagonists, J. Med. Chem. 43 (5) (2000) 797–806.

[172] R.J. Bienstock, J. Rizo, S.C. Koerber, J.E. Rivier, A.T. Hagler, L.M. Gierasch, Conformational analysis of a highly potent dicyclic gonadotropin-releasing hormone antagonist by nuclear magnetic resonance and molecular dynamics, J. Med. Chem. 36 (22) (1993) 3265–3273.

[173] D. D'Addona, A. Carotenuto, E. Novellino, V. Piccand, J.C. Reubi, A. Di Cianni, et al., Novel sst5-selective somatostatin dicarba-analogues: synthesis and conformation–affinity relationships, J. Med. Chem. 51 (3) (2008) 512–520.

[174] A. Kling, P. Lukat, D.V. Almeida, A. Bauer, E. Fontaine, S. Sordello, et al., Targeting DnaN for tuberculosis therapy using novel griselimycins, Science 348 (6239) (2015) 1106.

[175] O. Demmer, A.O. Frank, F. Hagn, M. Schottelius, L. Marinelli, S. Cosconati, et al., A conformationally frozen peptoid boosts CXCR4 affinity and anti-HIV activity, Angew. Chem. Int. Ed. 51 (32) (2012) 8110–8113.

[176] E. Miranda, I.K. Nordgren, A.L. Male, C.E. Lawrence, F. Hoakwie, F. Cuda, et al., A cyclic peptide inhibitor of HIF-1 heterodimerization that inhibits hypoxia signaling in cancer cells, J. Am. Chem. Soc. 135 (28) (2013) 10418–10425.

[177] R.P. Morse, J.L.E. Willett, P.M. Johnson, M. Zheng, A. Credali, A. Iniguez, et al., Diversification of β-augmentation interactions between CDI toxin/immunity proteins, J. Mol. Biol. 427 (23) (2015) 3766–3784.

[178] J.E. Rivier, G. Jiang, R.S. Struthers, S.C. Koerber, J. Porter, L.A. Cervini, et al., Design of potent dicyclic (1–5/4–10) gonadotropin releasing hormone (GnRH) antagonists, J. Med. Chem. 43 (5) (2000) 807–818.

[179] B. Zhao, X. Wei, W. Li, R.S. Udan, Q. Yang, J. Kim, et al., Inactivation of YAP oncoprotein by the Hippo pathway is involved in cell contact inhibition and tissue growth control, Genes Dev. 21 (21) (2007) 2747–2761.

[180] Z. Zhang, Z. Lin, Z. Zhou, H.C. Shen, S.F. Yan, A.V. Mayweg, et al., Structure-based design and synthesis of potent cyclic peptides inhibiting the YAP–TEAD protein–protein interaction, ACS Med. Chem. Lett. 5 (9) (2014) 993–998.
[181] S. Bonetto, L. Spadola, A.G. Buchanan, L. Jermutus, J. Lund, Identification of cyclic peptides able to mimic the functional epitope of IgG1-Fc for human Fc gamma RI, FASEB J. 23 (2) (2009) 575–585.
[182] S.E. Blondelle, K. Lohner, Combinatorial libraries: a tool to design antimicrobial and antifungal peptide analogues having lyric specificities for structure-activity relationship studies, Biopolymers 55 (1) (2000) 74–87.
[183] S.E. Blondelle, C. Pinilla, C.S. Boggiano, Synthetic combinatorial libraries as an alternative strategy for the development of novel treatments for infectious diseases, Comb. Chem. 369 (2003) 322–344.
[184] M.A. Gallop, R.W. Barrett, W.J. Dower, S.P.A. Fodor, E.M. Gordon, Applications of combinatorial technologies to drug discovery. 1. Background and peptide combinatorial libraries, J. Med. Chem. 37 (9) (1994) 1233–1251.
[185] C.P. Scott, E. Abel-Santos, A.D. Jones, S.J. Benkovic, Structural requirements for the biosynthesis of backbone cyclic peptide libraries, Chem. Biol. 8 (8) (2001) 801–815.
[186] O.H. Aina, J. Marik, R. Liu, D.H. Lau, K.S. Lam, Identification of novel targeting peptides for human ovarian cancer cells using "one-bead one-compound" combinatorial libraries, Mol. Cancer Ther. 4 (5) (2005) 806.
[187] O.H. Aina, R. Liu, J.L. Sutcliffe, J. Marik, C.-X. Pan, K.S. Lam, From combinatorial chemistry to cancer-targeting peptides, Mol. Pharm. 4 (5) (2007) 631–651.
[188] C. Qin, X. Bu, X. Zhong, N.L. Joyce Ng, Z. Guo, Optimization of antibacterial cyclic decapeptides, J. Comb. Chem. 6 (3) (2004) 398–406.
[189] M.D. Galsky, N.J. Vogelzang, P. Conkling, E. Raddad, J. Polzer, S. Roberson, et al., A phase I trial of LY2510924, a CXCR4 peptide antagonist, in patients with advanced cancer, Clin. Cancer Res. 20 (13) (2014) 3581.
[190] K. Barreto, C.R. Geyer, Screening combinatorial libraries of cyclic peptides using the yeast two-hybrid assay, in: W. Xiao (Ed.), Yeast Protocols, Springer New York, New York, 2014, pp. 273–309.
[191] K. Grabowska, A.K. Puszko, P.F.J. Lipiński, A.K. Laskowska, B. Wileńska, E. Witkowska, et al., Design, synthesis and in vitro biological evaluation of a small cyclic peptide as inhibitor of vascular endothelial growth factor binding to neuropilin-1, Bioorg. Med. Chem. Lett. 26 (12) (2016) 2843–2846.
[192] S.M. McHugh, J.R. Rogers, S.A. Solomon, H. Yu, Y.-S. Lin, Computational methods to design cyclic peptides, Curr. Opin. Chem. Biol. 34 (2016) 95–102.
[193] S.E. Allen, N.V. Dokholyan, A.A. Bowers, Dynamic docking of conformationally constrained macrocycles: methods and applications, ACS Chem. Biol. 11 (1) (2016) 10–24.
[194] G. Sliwoski, S. Kothiwale, J. Meiler, E.W. Lowe, Computational methods in drug discovery, Pharmacol. Rev. 66 (1) (2014) 334–395.
[195] S. Kalyaanamoorthy, Y.-P.P. Chen, Structure-based drug design to augment hit discovery, Drug Discov. Today 16 (17–18) (2011) 831–839.
[196] F.J. Duffy, M. Verniere, M. Devocelle, E. Bernard, D.C. Shields, A.J. Chubb, CycloPs: generating virtual libraries of cyclized and constrained peptides including nonnatural amino acids, J. Chem. Inf. Model. 51 (4) (2011) 829–836.
[197] D.J. Mandell, E.A. Coutsias, T. Kortemme, Sub-angstrom accuracy in protein loop reconstruction by robotics-inspired conformational sampling, Nat. Methods 6 (8) (2009) 551–552.

[198] M.P. Jacobson, D.L. Pincus, C.S. Rapp, T.J.F. Day, B. Honig, D.E. Shaw, et al., A hierarchical approach to all-atom protein loop prediction, Proteins 55 (2) (2004) 351–367.

[199] J. Beaufays, L. Lins, A. Thomas, R. Brasseur, In silico predictions of 3D structures of linear and cyclic peptides with natural and non-proteinogenic residues, J. Pept. Sci. 18 (1) (2012) 17–24.

[200] A. Thomas, S. Deshayes, M. Decaffmeyer, M.H. Van Eyck, B. Charloteaux, R. Brasseur, Prediction of peptide structure: how far are we? Proteins 65 (4) (2006) 889–897.

[201] S. Singh, H. Singh, A. Tuknait, K. Chaudhary, B. Singh, S. Kumaran, et al., PEPstrMOD: structure prediction of peptides containing natural, non-natural and modified residues, Biol. Direct 10 (1) (2015) 73.

[202] P. Thevenet, Y. Shen, J. Maupetit, F. Guyon, P. Derreumaux, P. Tuffery, PEP-FOLD: an updated de novo structure prediction server for both linear and disulfide bonded cyclic peptides, Nucleic Acids Res. 40 (2012) W288–W293.

[203] Y. Shen, J. Maupetit, P. Derreumaux, P. Tufféry, Improved PEP-FOLD approach for peptide and miniprotein structure prediction, J. Chem. Theory Comput. 10 (10) (2014) 4745–4758.

[204] Y. Zhang, I-TASSER server for protein 3D structure prediction, BMC Bioinformatics 9 (2008) 40.

[205] S. Izrailev, F. Zhu, D.K. Agrafiotis, A distance geometry heuristic for expanding the range of geometries sampled during conformational search, J. Comput. Chem. 27 (16) (2006) 1962–1969.

[206] F. Zhu, D.K. Agrafiotis, Self-organizing superimposition algorithm for conformational sampling, J. Comput. Chem. 28 (7) (2007) 1234–1239.

[207] P. Bonnet, D.K. Agrafiotis, F. Zhu, E. Martin, Conformational analysis of macrocycles: finding what common search methods miss, J. Chem. Inf. Model. 49 (10) (2009) 2242–2259.

[208] A. Rayan, H. Senderowitz, A. Goldblum, Exploring the conformational space of cyclic peptides by a stochastic search method, J. Mol. Graph. Model. 22 (5) (2004) 319–333.

[209] J. Li, T. Ehlers, J. Sutter, S. Varma-O'Brien, J. Kirchmair, CAESAR: a new conformer generation algorithm based on recursive buildup and local rotational symmetry consideration, J. Chem. Inf. Model. 47 (5) (2007) 1923–1932.

[210] P.C.D. Hawkins, A.G. Skillman, G.L. Warren, B.A. Ellingson, M.T. Stahl, Conformer generation with OMEGA: algorithm and validation using high quality structures from the Protein Databank and Cambridge Structural Database, J. Chem. Inf. Model. 50 (4) (2010) 572–584.

[211] U.S.F. Tambunan, A.A. Parikesit, Y. Dephinto, F.R.P. Sipahutar, Computational design of drug candidates for influenza A virus subtype H1N1 by inhibiting the viral neuraminidase-1 enzyme, Acta Pharm. 64 (2) (2014) 157–172.

[212] B.G. Bobay, L.R. Butler, J. Cavana, Computational design of cyclic peptide inhibitors of the anti-apoptotic protein Calbindin-D28K, Biochem. Pharmacol. 3 (4) (2014) 142.

[213] W. Nawae, S. Hannongbua, M. Ruengjitchatchawalya, Defining the membrane disruption mechanism of kalata B1 via coarse-grained molecular dynamics simulations, Sci. Rep. 4 (2014) 3933.

[214] A. Khalfa, M. Tarek, On the antibacterial action of cyclic peptides: insights from coarse-grained MD simulations, J. Phys. Chem. B. 114 (8) (2010) 2676–2684.

[215] A.D. Cirac, G. Moiset, J.T. Mika, A. Kocer, P. Salvador, B. Poolman, et al., The molecular basis for antimicrobial activity of pore-forming cyclic peptides, Biophys. J. 100 (10) (2011) 2422–2431.

[216] R. Garcia-Fandino, A. Pineiro, J.L. Trick, M.S.P. Sansom, Lipid bilayer membrane perturbation by embedded nanopores: a simulation study, ACS Nano 10 (3) (2016) 3693–3701.

[217] C.D. Fjell, J.A. Hiss, R.E.W. Hancock, G. Schneider, Designing antimicrobial peptides: form follows function, Nat. Rev. Drug Discov. 11 (1) (2012) 37–51.

[218] G. Maccari, M. Di Luca, R. Nifosi, F. Cardarelli, G. Signore, C. Boccardi, et al., Antimicrobial peptides design by evolutionary multiobjective optimization, PLoS Comput. Biol. 9 (9) (2013) e1003212.

[219] C.F. Lopez, S.O. Nielsen, G. Srinivas, W.F. Degrado, M.L. Klein, Probing membrane insertion activity of antimicrobial polymers via coarse-grain molecular dynamics, J. Chem. Theory Comput. 2 (3) (2006) 649–655.

[220] V.A. Voelz, K.A. Dill, I. Chorny, Peptoid conformational free energy landscapes from implicit-solvent molecular simulations in AMBER, Biopolymers 96 (5) (2011) 639–650.

[221] J.M. Damas, L.C.S. Filipe, S.R.R. Campos, D. Lousa, B.L. Victor, A.M. Baptista, et al., Predicting the thermodynamics and kinetics of helix formation in a cyclic peptide model, J. Chem. Theory Comput. 9 (11) (2013) 5148–5157.

[222] A.M. Razavi, W.M. Wuest, V.A. Voelz, Computational screening and selection of cyclic peptide hairpin mimetics by molecular simulation and kinetic network models, J. Chem. Inf. Model. 54 (5) (2014) 1425–1432.

[223] E. Yedvabny, P.S. Nerenberg, C. So, T. Head-Gordon, Disordered structural ensembles of vasopressin and oxytocin and their mutants, J. Phys. Chem. B. 119 (3) (2015) 896–905.

[224] A.E. Wakefield, W.M. Wuest, V.A. Voelz, Molecular simulation of conformational pre-organization in cyclic RGD peptides, J. Chem. Inf. Model. 55 (4) (2015) 806–813.

[225] A. Spitaleri, M. Ghitti, S. Mari, L. Alberici, C. Traversari, G.-P. Rizzardi, et al., Use of metadynamics in the design of isoDGR-based $\alpha_v\beta_3$ antagonists to fine-tune the conformational ensemble, Angew. Chem. Int. Ed. 50 (8) (2011) 1832–1836.

[226] C. Paissoni, M. Ghitti, L. Belvisi, A. Spitaleri, G. Musco, Metadynamics simulations rationalise the conformational effects induced by N-methylation of RGD cyclic hexapeptides, Chemistry 21 (40) (2015) 14165–14170.

[227] Y. Sugita, Y. Okamoto, Replica-exchange molecular dynamics method for protein folding, Chem. Phys. Lett. 314 (1–2) (1999) 141–151.

[228] D.J. Earl, M.W. Deem, Parallel tempering: theory, applications, and new perspectives, Phys. Chem. Chem. Phys. 7 (23) (2005) 3910–3916.

[229] A. Laio, M. Parrinello, Escaping free-energy minima, Proc. Natl. Acad. Sci. U.S.A. 99 (20) (2002) 12562–12566.

[230] S. Piana, A. Laio, A bias-exchange approach to protein folding, J. Phys. Chem. B. 111 (17) (2007) 4553–4559.

[231] P. Labute, LowModeMD—Implicit low-mode velocity filtering applied to conformational search of macrocycles and protein loops, J. Chem. Inf. Model. 50 (5) (2010) 792–800.

[232] J. Burck, S. Roth, P. Wadhwani, S. Afonin, N. Kanithasen, E. Strandberg, et al., Conformation and membrane orientation of amphiphilic helical peptides by oriented circular dichroism, Biophys. J. 95 (8) (2008) 3872–3881.

[233] A.S. Ulrich, Solid state 19F NMR methods for studying biomembranes, Prog. Nucl. Magn. Reson. Spectrosc. 46 (1) (2005) 1–21.
[234] V. Tozzini, Coarse-grained models for proteins, Curr. Opin. Struct. Biol. 15 (2) (2005) 144–150.
[235] J.D. Owens, M. Houston, D. Luebke, S. Green, J.E. Stone, J.C. Phillips, GPU computing, Proc. IEEE 96 (5) (2008) 879–899.
[236] Y. Goldtzvik, M. Goldstein, R. Benny Gerber, On the crystallographic accuracy of structure prediction by implicit water models: tests for cyclic peptides, Chem. Phys. 415 (2013) 168–172.
[237] H. Geng, F. Jiang, Y.-D. Wu, Accurate structure prediction and conformational snalysis of cyclic peptides with residue-specific force fields, J. Phys. Chem. Lett. 7 (10) (2016) 1805–1810.
[238] W.Y. Xu, L.L. Li, L.C. Du, N.H. Tan, Various mechanisms in cyclopeptide production from precursors synthesized independently of non-ribosomal peptide synthetases, Acta Biochim. Biophys. Sin. (Shanghai) 43 (10) (2011) 757–762.
[239] P.G. Arnison, M.J. Bibb, G. Bierbaum, A.A. Bowers, T.S. Bugni, G. Bulaj, et al., Ribosomally synthesized and post-translationally modified peptide natural products: overview and recommendations for a universal nomenclature, Nat. Prod. Rep. 30 (1) (2013) 108–160.
[240] S.A. Sieber, M.A. Marahiel, Molecular mechanisms underlying nonribosomal peptide synthesis: approaches to new antibiotics, Chem. Rev. 105 (2) (2005) 715–738.
[241] M.A. Fischbach, C.T. Walsh, Assembly-line enzymology for polyketide and nonribosomal peptide antibiotics: logic, machinery, and mechanisms, Chem. Rev. 106 (8) (2006) 3468–3496.
[242] G.P. Smith, Filamentous fusion phage: novel expression vectors that display cloned antigens on the virion surface, Science 228 (4705) (1985) 1315–1317.
[243] J.R. Hsiao, Y. Chang, Y.L. Chen, S.H. Hsieh, K.F. Hsu, C.F. Wang, et al., Cyclic alphavbeta6-targeting peptide selected from biopanning with clinical potential for head and neck squamous cell carcinoma, Head Neck 32 (2) (2010) 160–172.
[244] K.E. Duncan, B.R. Dempsey, L.E. Killip, J. Adams, M.L. Bailey, G.A. Lajoie, et al., Discovery and characterization of a nonphosphorylated cyclic peptide inhibitor of the peptidylprolyl isomerase, Pin1, J. Med. Chem. 54 (11) (2011) 3854–3865.
[245] J. Huang, B.B. Ru, P. Dai, Bioinformatics resources and tools for phage display, Molecules 16 (1) (2011) 694–709.
[246] J. Huang, B.B. Ru, P. Zhu, F.L. Nie, J. Yang, X.Y. Wang, et al., MimoDB 2.0: a mimotope database and beyond, Nucleic Acids Res. 40 (D1) (2012) D271–D277.
[247] F.B. Perler, E.O. Davis, G.E. Dean, F.S. Gimble, W.E. Jack, N. Neff, et al., Protein splicing elements: inteins and exteins—a definition of terms and recommended nomenclature, Nucleic Acids Res. 22 (7) (1994) 1125–1127.
[248] F.B. Perler, Protein splicing mechanisms and applications, IUBMB Life 57 (7) (2005) 469–476.
[249] C.P. Scott, E. Abel-Santos, M. Wall, D.C. Wahnon, S.J. Benkovic, Production of cyclic peptides and proteins in vivo, Proc. Nat. Acad. Sci. U. S. A. 96 (24) (1999) 13638–13643.
[250] S.W. Millward, T.T. Takahashi, R.W. Roberts, A general route for post-translational cyclization of mRNA display libraries, J. Am. Chem. Soc. 127 (41) (2005) 14142–14143.
[251] S.W. Millward, S. Fiacco, R.J. Austin, R.W. Roberts, Design of cyclic peptides that bind protein surfaces with antibody-like affinity, ACS Chem. Biol. 2 (9) (2007) 625–634.

[252] K.D. Kopple, Synthesis of cyclic peptides, J. Pharm. Sci. 61 (9) (1972) 1345–1356.
[253] S.A. Kates, N.A. Solé, C.R. Johnson, D. Hudson, G. Barany, F. Albericio, A novel, convenient, three-dimensional orthogonal strategy for solid-phase synthesis of cyclic peptides, Tetrahedron Lett. 34 (10) (1993) 1549–1552.
[254] R.B. Merrifield, Solid phase peptide synthesis. I. The synthesis of a tetrapeptide, J. Am. Chem. Soc. 85 (14) (1963) 2149–2154.
[255] G.T. Bourne, S.W. Golding, R.P. McGeary, W.D.F. Meutermans, A. Jones, G.R. Marshall, et al., The development and application of a novel safety-catch linker for BOC-based assembly of libraries of cyclic peptides,1, J. Org. Chem. 66 (23) (2001) 7706–7713.
[256] O. Demmer, I. Dijkgraaf, M. Schottelius, H.J. Wester, H. Kessler, Introduction of functional groups into peptides via N-alkylation, Org. Lett. 10 (10) (2008) 2015–2018.
[257] G. Koopmanschap, E. Ruijter, R.V.A. Orru, Isocyanide-based multicomponent reactions towards cyclic constrained peptidomimetics, Beilstein J. Org. Chem. 10 (2014) 544–598.
[258] J.M. Smith, J.R. Frost, R. Fasan, Emerging strategies to access peptide macrocycles from genetically encoded polypeptides, J. Org. Chem. 78 (8) (2013) 3525–3531.
[259] V.R. Pattabiraman, J.W. Bode, Rethinking amide bond synthesis, Nature 480 (7378) (2011) 471–479.
[260] J.M. Smith, F. Vitali, S.A. Archer, R. Fasan, Modular assembly of macrocyclic organo–peptide hybrids using synthetic and genetically encoded precursors, Angew. Chem. Int. Ed. 50 (22) (2011) 5075–5080.
[261] R.J. Clark, J. Jensen, S.T. Nevin, B.P. Callaghan, D.J. Adams, D.J. Craik, The engineering of an orally active conotoxin for the treatment of neuropathic pain, Angew. Chem. Int. Ed. 49 (37) (2010) 6545–6548.
[262] Á. Roxin, G. Zheng, Flexible or fixed: a comparative review of linear and cyclic cancer-targeting peptides, Future Med. Chem. 4 (12) (2012) 1601–1618.
[263] G. Colombo, F. Curnis, G.M. De Mori, A. Gasparri, C. Longoni, A. Sacchi, et al., Structure-activity relationships of linear and cyclic peptides containing the NGR tumor-homing motif, J. Biol. Chem. 277 (49) (2002) 47891–47897.
[264] G.J. Mizejewski, M. Muehlemann, M. Dauphinee, Update of alpha fetoprotein growth-inhibitory peptides as biotherapeutic agents for tumor growth and metastasis, Chemotherapy 52 (2) (2006) 83–90.
[265] W.R. Baumbach, T.A. Carrick, M.H. Pausch, B. Bingham, D. Carmignac, I.C. Robinson, et al., A linear hexapeptide somatostatin antagonist blocks somatostatin activity in vitro and influences growth hormone release in rats, Mol. Pharmacol. 54 (5) (1998) 864–873.
[266] J. Zhong, Y. Chau, Antitumor activity of a membrane lytic peptide cyclized with a linker sensitive to membrane type 1-matrix metalloproteinase, Mol. Cancer Ther. 7 (9) (2008) 2933–2940.
[267] M. Katsara, T. Tselios, S. Deraos, G. Deraos, M.T. Matsoukas, E. Lazoura, et al., Round and round we go: cyclic peptides in disease, Curr. Med. Chem. 13 (19) (2006) 2221–2232.
[268] C.A. Lipinski, F. Lombardo, B.W. Dominy, P.J. Feeney, Experimental and computational approaches to estimate solubility and permeability in drug discovery and development settings, Adv. Drug Deliv. Rev. 64 (2012) 4–17.
[269] A.R. Hamel, F. Hubler, A. Carrupt, R.M. Wenger, M. Mutter, Cyclosporin A prodrugs: design, synthesis and biophysical properties, J. Pept. Res. 63 (2) (2004) 147–154.

[270] T.W.J. Huizinga, C.I. Amos, A.H.M. van der Helm-van Mil, W. Chen, F.A. van Gaalen, D. Jawaheer, et al., Refining the complex rheumatoid arthritis phenotype based on specificity of the HLA-DRB1 shared epitope for antibodies to citrullinated proteins, Arthritis. Rheum. 52 (11) (2005) 3433–3438.

[271] E.F. Stange, S.P.L. Travis, S. Vermeire, W. Reinisch, K. Geboes, A. Barakauskiene, et al., European evidence-based consensus on the diagnosis and management of ulcerative colitis: definitions and diagnosis, J. Crohns Colitis 2 (1) (2008) 1–23.

[272] R. Colucci, C. Blandizzi, N. Ghisu, T. Florio, M. Del Tacca, Somatostatin inhibits colon cancer cell growth through cyclooxygenase-2 downregulation, Br. J. Pharmacol. 155 (2) (2008) 198–209.

[273] M.D. Culler, K. Oberg, R. Arnold, E.P. Krenning, I. Sevilla, J.A. Diaz, Somatostatin analogs for the treatment of neuroendocrine tumors, Cancer Metastasis Rev. 30 (1) (2011) 9–17.

[274] C.H. Chesnut, M. Azria, S. Silverman, M. Engelhardt, M. Olson, L. Mindeholm, Salmon calcitonin: a review of current and future therapeutic indications, Osteoporos. Int. 19 (4) (2008) 479–491.

[275] A. Schmidtko, J. Lotsch, R. Freynhagen, G. Geisslinger, Ziconotide for treatment of severe chronic pain, Lancet 375 (9725) (2010) 1569–1577.

[276] K. Ley, J. Rivera-Nieves, W.J. Sandborn, S. Shattil, Integrin-based therapeutics: biological basis, clinical use and new drugs, Nat. Rev. Drug Discov. 15 (3) (2016) 173–183.

[277] T. Tselios, V. Apostolopoulos, I. Daliani, S. Deraos, S. Grdadolnik, T. Mavromoustakos, et al., Antagonistic effects of human cyclic MBP87-99 altered peptide ligands in experimental allergic encephalomyelitis and human T-cell proliferation, J. Med. Chem. 45 (2) (2002) 275–283.

[278] T. Tselios, M. Aggelidakis, A. Tapeinou, V. Tseveleki, I. Kanistras, D. Gatos, et al., Rational design and synthesis of altered peptide ligands based on uman myelin oligodendrocyte glycoprotein 35-55 epitope: inhibition of chronic experimental autoimmune encephalomyelitis in mice, Molecules 19 (11) (2014) 17968–17984.

[279] J.P. Tam, Y.A. Lu, J.L. Yang, K.W. Chiu, An unusual structural motif of antimicrobial peptides containing end-to-end macrocycle and cystine-knot disulfides, Proc. Nat. Acad. Sci. U.S.A. 96 (16) (1999) 8913–8918.

[280] C.K.L. Wang, M.L. Colgrave, K.R. Gustafson, D.C. Ireland, U. Goransson, D.J. Craik, Anti-HIV cyclotides from the Chinese medicinal herb *Viola yedoensis*, J. Nat. Prod. 71 (1) (2008) 47–52.

[281] S.L. Gerlach, R. Rathinakumar, G. Chakravarty, U. Goransson, W.C. Wimley, S.P. Darwin, et al., Anticancer and chemosensitizing abilities of cycloviolacin O2 from *Viola odorata* and psyle cyclotides from *Psychotria leptothyrsa*, Biopolymers 94 (5) (2010) 617–625.

[282] D. Oh, J.D. Sun, A.N. Shirazi, K.L. LaPlante, D.C. Rowley, K. Parang, Antibacterial activities of amphiphilic cyclic cell-penetrating peptides against multidrug-resistant pathogens, Mol. Pharm. 11 (10) (2014) 3528–3536.

[283] D. Mandal, R.K. Tiwari, A.N. Shirazi, D. Oh, G.F. Ye, A. Banerjee, et al., Self-assembled surfactant cyclic peptide nanostructures as stabilizing agents, Soft Matter 9 (39) (2013) 9465–9475.

[284] D. Mandal, A.N. Shirazi, K. Parang, Cell-penetrating homochiral cyclic peptides as nuclear-targeting molecular transporters, Angew. Chem. Int. Ed. 50 (41) (2011) 9633–9637.

[285] A.N. Shirazi, R.K. Tiwari, D. Oh, A. Banerjee, A. Yadav, K. Parang, Efficient delivery of cell impermeable phosphopeptides by a cyclic peptide amphiphile containing tryptophan and arginine, Mol. Pharm. 10 (5) (2013) 2008–2020.

[286] A.N. Shirazi, D. Oh, R.K. Tiwari, B. Sullivan, A. Gupta, G.D. Bothun, et al., Peptide amphiphile containing arginine and fatty acyl chains as molecular transporters, Mol. Pharm. 10 (12) (2013) 4717–4727.
[287] A.N. Shirazi, R. Tiwari, B.S. Chhikara, D. Mandal, K. Parang, Design and biological evaluation of cell-penetrating peptide-doxorubicin conjugates as prodrugs, Mol. Pharm. 10 (2) (2013) 488–499.
[288] D. Weerakkody, A. Moshnikova, N.S. El-Sayed, R.C. Adochite, G. Slaybaugh, J. Golijanin, et al., Novel pH-sensitive cyclic peptides, Sci. Rep. 6 (2016) 31322.
[289] A.N. Shirazi, D. Mandal, R.K. Tiwari, L.R. Guo, W. Lu, K. Parang, Cyclic peptide-capped gold nanoparticles as drug delivery systems, Mol. Pharm. 10 (2) (2013) 500–511.
[290] A.N. Shirazi, K.L. Paquin, N.G. Howlett, D. Mandal, K. Parang, Cyclic peptide-capped gold nanoparticles for enhanced siRNA delivery, Molecules 19 (9) (2014) 13319–13331.
[291] A.N. Shirazi, R.K. Tiwari, D. Oh, B. Sullivan, K. McCaffrey, D. Mandal, et al., Surface decorated gold nanoparticles by linear and cyclic peptides as molecular transporters, Mol. Pharm. 10 (8) (2013) 3137–3151.
[292] A.N. Shirazi, R.K. Tiwari, D. Oh, B. Sullivan, A. Kumar, Y.A. Beni, et al., Cyclic peptide-selenium nanoparticles as drug transporters, Mol. Pharm. 11 (10) (2014) 3631–3641.
[293] Y. Takada, X. Ye, S. Simon, The integrins, Genome Biol. 8 (5) (2007) 215.
[294] J.D. Hood, D.A. Cheresh, Role of integrins in cell invasion and migration, Nat. Rev. Cancer 2 (2) (2002) 91–100.
[295] H. Jin, J. Varner, Integrins: roles in cancer development and as treatment targets, Br. J. Cancer 90 (3) (2004) 561–565.
[296] X. Chen, P.S. Conti, R.A. Moats, In vivo near-infrared fluorescence imaging of integrin $\alpha_v\beta_3$ in brain tumor xenografts, Cancer Res. 64 (21) (2004) 8009.
[297] X. Chen, R. Park, M. Tohme, A.H. Shahinian, J.R. Bading, P.S. Conti, MicroPET and autoradiographic imaging of breast cancer αv-integrin expression using 18F- and 64Cu-labeled RGD peptide, Bioconjug. Chem. 15 (1) (2004) 41–49.
[298] X. Chen, R. Park, A.H. Shahinian, M. Tohme, V. Khankaldyyan, M.H. Bozorgzadeh, et al., 18F-labeled RGD peptide: initial evaluation for imaging brain tumor angiogenesis, Nucl. Med. Biol. 31 (2) (2004) 179–189.
[299] R. Haubner, B. Kuhnast, C. Mang, W.A. Weber, H. Kessler, H.-J. Wester, et al., [18F]Galacto-RGD: synthesis, radiolabeling, metabolic stability, and radiation dose estimates, Bioconjug. Chem. 15 (1) (2004) 61–69.
[300] M.L. Janssen, W.J. Oyen, I. Dijkgraaf, L.F. Massuger, C. Frielink, D.S. Edwards, et al., Tumor targeting with radiolabeled αvβ3 integrin binding peptides in a nude mouse model, Cancer Res. 62 (21) (2002) 6146.
[301] E.A. Murphy, B.K. Majeti, L.A. Barnes, M. Makale, S.M. Weis, K. Lutu-Fuga, et al., Nanoparticle-mediated drug delivery to tumor vasculature suppresses metastasis, Proc. Nat. Acad. Sci. U.S.A. 105 (27) (2008) 9343–9348.
[302] R.J. Brea, C. Reiriz, J.R. Granja, Towards functional bionanomaterials based on self-assembling cyclic peptide nanotubes, Chem. Soc. Rev. 39 (5) (2010) 1448–1456.
[303] P. Kumaraswamy, R. Lakshmanan, S. Sethuraman, U.M. Krishnan, Self-assembly of peptides: influence of substrate, pH and medium on the formation of supramolecular assemblies, Soft Matter 7 (6) (2011) 2744–2754.

[304] L.M. Sun, Z. Fan, Y.Z. Wang, Y.J. Huang, M. Schmidt, M.J. Zhang, Tunable synthesis of self-assembled cyclic peptide nanotubes and nanoparticles, Soft Matter 11 (19) (2015) 3822–3832.
[305] J.D. Hartgerink, J.R. Granja, R.A. Milligan, M.R. Ghadiri, Self-assembling peptide nanotubes, J. Am. Chem. Soc. 118 (1) (1996) 43–50.
[306] J. Cheng, J. Zhu, B. Liu, Z. Liao, Z. Lai, Structure of a self-assembled single nanotube of cyclo[(-D-Ala-L-Ala)4-], Mol. Simul. 35 (8) (2009) 625–630.
[307] A. Montero, P. Gastaminza, M. Law, G.F. Cheng, F.V. Chisari, M.R. Ghadiri, Self-assembling peptide nanotubes with antiviral activity against hepatitis C virus, Chem. Biol. 18 (11) (2011) 1453–1462.
[308] C.R. Martin, P. Kohli, The emerging field of nanotube biotechnology, Nat. Rev. Drug Discov. 2 (1) (2003) 29–37.
[309] H.F. Liu, J.A. Chen, Q. Shen, W. Fu, W. Wu, Molecular insights on the cyclic peptide nanotube-mediated transportation of antitumor drug 5-fluorouracil, Mol. Pharm. 7 (6) (2010) 1985–1994.
[310] R.J. Brea, L. Castedo, J.R. Granja, M.A. Herranz, L. Sanchez, N. Martin, et al., Electron transfer in Me-blocked heterodimeric alpha,gamma-peptide nanotubular donor-acceptor hybrids, Proc. Nat. Acad. Sci. U.S.A. 104 (13) (2007) 5291–5294.
[311] R.J. Brea, L. Castedo, J.R. Granja, Large-diameter self-assembled dimers of alpha, gamma-cyclic peptides, with the nanotubular solid-state structure of cyclo- (L-Leu-D-N-Me-gamma-Acp)(4-) 4CHCl(2)COOH, Chem. Commun. (31) (2007) 3267–3269.
[312] M. Calvelo, S. Vazquez, R. Garcia-Fandino, Molecular dynamics simulations for designing biomimetic pores based on internally functionalized self-assembling alpha, gamma-peptide nanotubes, Phys. Chem. Chem. Phys. 17 (43) (2015) 28586–28601.
[313] R. Garcia-Fandino, M. Amorin, L. Castedo, J.R. Granja, Transmembrane ion transport by self-assembling alpha,gamma-peptide nanotubes, Chem. Sci. 3 (11) (2012) 3280–3285.
[314] D. Seebach, J.L. Matthews, A. Meden, T. Wessels, C. Baerlocher, L.B. McCusker, Cyclo-beta-peptides: structure and tubular stacking of cyclic tetramers of 3-aminobutanoic acid as determined from powder diffraction data, Helv. Chim. Acta 80 (1) (1997) 173–182.
[315] J. Sánchez-Quesada, M.P. Isler, M.R. Ghadiri, Modulating ion channel properties of transmembrane peptide nanotubes through heteromeric supramolecular assemblies, J. Am. Chem. Soc. 124 (34) (2002) 10004–10005.
[316] N. Malanovic, K. Lohner, Antimicrobial peptides targeting Gram-positive bacteria, Pharmaceuticals (Basel) 9 (3) (2016) 59.

Further reading

P. De Santis, S. Morosetti, R. Rizzo, Conformational analysis of regular enantiomeric sequences, Macromolecules 7 (1) (1974) 52–58.

Peptides containing D-amino acids and retro-inverso peptides: General applications and special focus on antimicrobial peptides

5

Marlon H. Cardoso[1,2,3], *Elizabete S. Cândido*[2,3], *Karen G.N. Oshiro*[3], *Samilla B. Rezende*[3] *and Octávio L. Franco*[1,2,3]
[1]University of Brasília, Brasília, Brazil, [2]Catholic University of Brasília, Brasília, Brazil, [3]Catholic University Dom Bosco, Campo Grande, Brazil

5.1 Introduction and overview

According to the central dogma established by classical naturalists and biologists, proteins/peptides were expected to be exclusively constituted by 20 building blocks, including 19 L-amino acids (L-AAs) and glycine, which is not chiral. At that time, special cases were reported for D-amino acids (D-AAs) containing proteins/peptides such as the presence of D-alanine in octopine derivatives from octopus [1]. Over the years, however, increasing reports were published revealing the natural occurrence of D-AAs in several organisms [2,3]. Since there is no codon for D-AAs synthesis, a gap regarding the biogenesis of D-AAs containing proteins/peptides encouraged further studies, which have led to diverse hypotheses. Among them, we can cite the incorporation of D-AAs during the translation of L-proteins/peptides, the isomerization of L-AAs within polypeptide chains by nonenzymatic pathways, as well as by the action of specific isomerases [4]. From a functional point of view, these post-translational and chemical modifications in D-AAs containing proteins/peptides confer on them an unusual stereochemistry mostly related to enhanced resistance to enzyme-catalyzed breakdown, which is indicated as a desirable property when it comes to pharmacological investments.

Structurally, apart from the conserved physicochemical properties between L- and D-enantiomers, each enantiomer rotates plane-polarized light in opposite directions. In other words, it means that in α-helical peptides, for example, the left-handed helix of the parent L-peptide could be presented as a right-handed helix in the D-peptide analogues [5]. Interestingly, experimental structural analyses have shown that D-AAs may act as α-helical and β-sheet breakers in D-AAs containing peptides, contributing to a completely opposite spatial orientation when compared to their parent L-peptides, thus favoring proteolytic stability and, consequently, bioavailability in vivo [6]. On the other hand, there are suggestions indicating that D-AAs stabilize turns, for example, in the $(i + 1)$ or $(i + 2)$ positions of types II' and II β-turns, respectively [7]. In addition, more recently, Makwana and Mahalakshmi

Peptide Applications in Biomedicine, Biotechnology and Bioengineering.
DOI: http://dx.doi.org/10.1016/B978-0-08-100736-5.00005-3

[8] demonstrated that the N-terminal switch from L- to D-AA aliphatic residues in β-hairpin peptides causes an increase in their structural stability. All these findings clearly indicate some controversies regarding the influence of D-AAs within peptides' structure, which might explain the multiple actions and mechanisms of these molecules, also encouraging deeper studies regarding the design of bioactive D-AA peptides.

However, D-AAs require time-consuming and expensive synthesis, which may discourage their application in clinical trials. To overcome this obstacle, in 1979, Goodman and Chorev [9] proposed the synthesis of retro-inverso (RI) peptides, also known as all-D-retro peptides or retro-enantio peptides, which are composed of D-AAs assembled in the reverse order to that of their parent L-sequence (Fig. 5.1) [9]. This fusion could guarantee structural stability, spatial orientation, side-chain topology, and consequently, the peptide's bioactivity, but would retain its resistance to the enzymatic degradation characteristic of the D-peptides [10–13].

In this scenario, this chapter will describe, in a general way, the design and engineering of partial D-AAs (diastereomeric), all-D-AAs, and RI peptides, with focus on antimicrobial peptides (AMPs) to shed some light on their pros and cons in terms of synthesis, bioactivities, and industrial application as a new class of promising medicines.

5.2 Designing D-AAs containing peptides

Rational design has emerged as a powerful tool in addressing peptides' structure–function relationships. In this context, several methods have been applied for this purpose, mostly computational-based methods including evolutionary and genetic algorithms [14], linguistic models [15], random sequence generation [16], and amino acid replacements [16]. Amino acid replacement design approaches take into account folding stability, as well as energy-based probabilities in order to generate novel synthetic peptides that are compatible with referenced peptides [17]. This strategy is the one most commonly applied to the generation of diastereomeric and all-D-AAs peptides. Hong and coworkers, for example, based on a membrane-active AMP named KLSK (KKVVFKVKFKK), investigated the cleavage sites of this peptide in the presence of serum. By doing that, the authors could generate one enantiomeric and six diastereomeric analogues presenting D-lysine substitutions for improved serine proteases digestion resistance. Interestingly, apart from the increased stability in serum already expected due to such modifications, the authors reported that diastereomeric analogues with substitutions at the N- and C-terminal regions had little effect on their α-helical content, while the D-enantiomer adopted a left-handed α-helix, thus preserving the antimicrobial properties of the parent L-peptide [18]. In contrast, substituting the central L-lysine at position 6 led to complete loss α-helical contents and antimicrobial activity [18]. In this work, the idea of using cleavage sites for amino acid chiral changes illustrates well the concept behind the rational design.

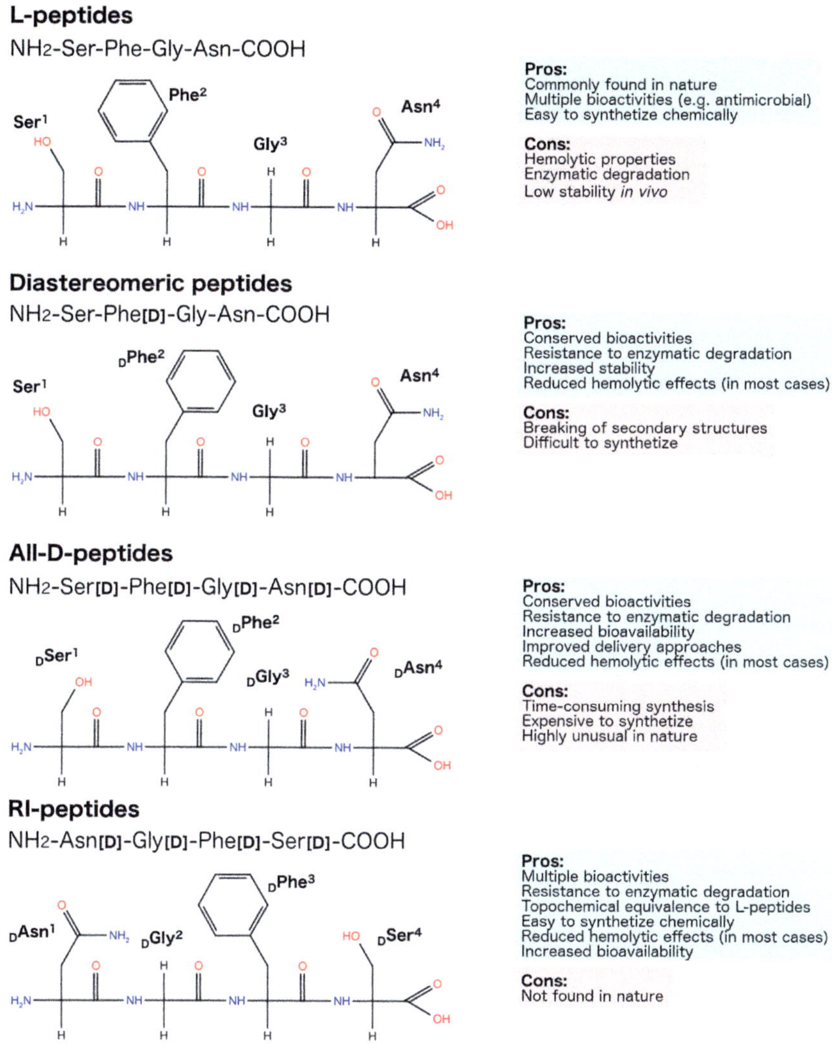

Figure 5.1 Representation of the planar structure and chirality of hypothetic L-, diastereomeric, all-D- and RI-peptides, also highlighting their pros and cons.

Apart from the field of antimicrobials, another example of rational design involving application of D-AAs can be seen in the works that have been performed by Daggett's research group [19,20]. In recent years, this group has explored design strategies to generate and characterize hairpin peptides able to inhibit aggregation of the β-amyloid peptide Aβ42, which could be related to human amyloid diseases that might lead to neurodegeneration and organ dysfunction [21]. Based on that, a novel 23-residue hairpin peptide, named AP90, was designed presenting alternating L- and D-AAs. The arrangement distribution of the amino acids within AP90's sequence favored conformation propensities to the adoption of an α-sheet structure with a designed turn [22]. α-Sheet is considered a novel secondary structure that is defined by the alternation of α_L and α_R structural organizations, favoring a parallel alignment between all carbonyl groups, as well as amide protons on opposite faces of the peptide strand. The authors suggest that this unusual structural signature might explain the potential amyloid inhibition of AP90 [22].

All these modifications based on the use of D-AAs have shown several advantages. However, it is important to consider the structural/functional relationship that makes such molecules so attractive. In this field, experimental studies have tried to join computational-based designing methods and molecular modeling studies to fill this gap related to the structural arrangement of D-AA peptides. In this context, recently, Singh and coworkers came up with a new computational tool, named PEPstrMOD. This software offers unique advantages in the prediction of atomic coordinates of peptides constituted by natural, non-natural modified and D-AA residues, as well as terminal and posttranslational modifications. Altogether, these tools may help researchers to have preliminary insights regarding the structure adopted by the designed D-AA peptides, leading to a better understanding for further therapy applications [23].

On the other hand, there are works based on the generation of D-AAs containing peptides that lack the "rational" thought expected in the design of bioactive molecules. One example is the synthesis of several diastereomeric peptides based on a template sequence by substituting every L-AA by its D-enantiomer one by one, in a trial and error manner. It is important to highlight that this approach is more likely a screening for improved molecules or novel activities, which is highly desirable in many fields, but it is also a blind and unspecific methodology. In the field of AMPs, the main focus of this chapter, blind chiral modifications that do not take into account the peptide's charge, hydrophobicity, hydrophobic moment, selectivity, and structural arrangement could lead to a library of ineffective molecules. In the following topics, readers will find selected works on diastereomeric, all-D-AAs, and retro-inverso general peptides and AMPs that are template-based designed, with a description of their activities, toxicities, structures, and applications as promising molecules in the pharmacology industry.

5.3 General applications of D-AAs containing peptides

D-AAs containing peptides may display several functions, acting as anticancer [24] and antiinflammatory agents [25], inhibitors of amyloidogenesis [22], and delivery agents [26], among others (Table 5.1). Veine and coworkers, for example, reported that chiral changes in histidine and cysteine residues in the peptide PHSCN (Ac-P_DHS_DCN-NH_2) improved by up to 150,000-fold the potential of this diastereomeric peptide to inhibit α5β1 integrin in human prostate cancer invasion (Table 5.1) [33]. Moreover, it is known that one of the biggest challenges in working with anticancer agents regards their selectivity for cancerous cells. In this

Table 5.1 Multifunctional properties of D-AAs containing peptides

Peptides	Application	References
AP90	Inhibition of β-amyloid peptide Aβ42 aggregation	[22]
$_D$Leu^9	Antibacterial/antifungal	[6]
GS14$_D$K^4	Antibacterial	[28]
[D]-HC-1(R) and [D]-HC-3(R)	Antibacterial	[29]
Diastereomeric granulysin-derived peptide	Antibacterial	[10]
all-D-magainin	Antibacterial	[30]
D-LAK	Tuberculosis	[60]
D-LL-37	Antibiofilm	[54]
DJK-5 and DJK6	Antibiofilm	[53]
D-magainin-NH_2 and D-melittin	Antibacterial	[31]
RI-catestatin	Antihypertensive properties	[71]
all-D-prosaptide	Central nervous system neurodegeneration	[13]
RI-JK3	Antibiofilm	[53]
RI-BMAP28	Antibacterial/antileishmanial	[32]
RI-VPI-variants	Immunogenic peptide	[75]
RI-peptide corresponding to the influenza virus hemagglutinin B-cell epitope	Immune responses	[77]
D-PHSCN	Prostate cancer	[33]
Enzyme-instructed self-assembly (EISA) of small D-peptides	Selective killing of cancer cells	[24]
D-AAs host defense peptides	Apoptosis in human leukemia cells	[34]
9Pbw3	Antiinflammatory	[25]
Delmitide	Antiinflammatory	[35]
dfTAT	Delivery agent	[26]

context, Zhou and coworkers recently described the enzyme-instructed self-assembly (EISA) of small D-peptides as a multiple-step process for selectively killing cancer cells. In this study, the authors designed two D-tetrapeptides containing up to two phosphotyrosine residues, as well as a capping naphthyl group at the

N-terminal [24]. Thus, by enzymatic dephosphorylating reactions the peptides were self-assembled to form nanofibers in water. Further, these formulations were evaluated toward several cancer cell lines and one normal cell line. As a result, it was observed that unphosphorylated D-tetrapeptides had no effects against the cancerous cell. On the other hand, mono- and diphosphorylated D-tetrapeptides selectively killed cancer cells, but not the normal cells, with more potent activities observed for the monophosphorylated D-tetrapeptides (Table 5.1) [24]. Interestingly, reports have also shown that D-AA host defense peptides are able to induce apoptosis in human leukemia cells [34].

In addition, Lee and coworkers have reported the advantages in using an all-D-enantiomer peptide of a synthetic peptide, named 9Pbw3 (RLWLAIWRR-NH_2), designed based on an insect defensin from the larvae of the beetle *Protaetia brevitarsis*. The enantiomer of 9Pbw3 (9Pbw3-D) presented lower cytotoxic activities toward RAW 264.7 cell lines, also showing a stronger inhibition of nitric oxide (NO), as well as inflammation-induced cytokine production in RAW 264.7 cell lines stimulated by lipopolysaccharide (LPS), compared to its parent L-peptide (Table 5.1) [25]. Another example of antiinflammatory peptides is the semisynthetic D-AA decapeptide derived from HLA class I B2702, named Delmitide (RDP58) (Table 5.1) [35]. Delmitide has completed phase II clinical trials whose application was focused on inflammatory bowel disease. This peptide is able to inhibit the synthesis of proinflammatory cytokines, including TNF-α, IFN-γ, IL-12, and IL2, by preventing the formation of the MyD88/IRAK/TRAF6 protein complex, thus inhibiting IkB, JNK, and p38 map kinase pathways [35].

D-AA peptides have also been engineered to act as cell-penetrating peptides (CCPs), which are considered important molecules for the delivery of enzymes, transcription factors, peptides, and cell-impermeable small molecules within target cells with high efficiencies [36]. In this context, Najjar and coworkers reported that by inverting the chirality of a model CPP, named dfTAT, it was possible to generate a CPP resistant to proteases, but with conserved overall modes of action, including cell permeation through an endocytic-uptake mechanism, endosomal escape, and accession to the cytosol (Table 5.1) [26]. Moreover, D-dfTAT was retained in the cells' cytosol for days, while its parent L-CPP could only resist for a few hours under the same conditions. Such findings indicate the powerful source in D-AA peptide engineering for the development of delivery agents with high stability and half-life for future applications.

5.4 AMPs as promising bioinspired molecules

The increase in diseases caused by pathogenic microorganisms has become a public concern, mainly due to the multidrug-resistance phenomenon emerging from the indiscriminate use of certain medicines worldwide [37]. One of the biggest challenges in medical science and research is to find favorable alternatives to restrain infections caused by multidrug-resistant bacterial strains such as *Escherichia coli*,

Staphylococcus aureus, and *Pseudomonas aeruginosa*. In addition, pathogenic bacteria can also exist in their biofilm state, and are found in abundance attached to biotic or abiotic surfaces, including medical devices and human tissues [37]. Apart from bacterial diseases, infections caused by fungi (*Candida* spp. [38] and *Aspergillus* spp. [39]), protozoa (*Trypanosoma cruzi* [40] and *Leishmania major* [41], and viruses (H1N1 influenza [42]) also play an important role in mortality rates.

Due to their cationic and amphipathic properties, AMPs have gained great attention as multifunctional molecules against pathogenic microorganisms, including free-floating bacterial cells and biofilm, fungi [43], protozoa [40], and viruses [42,44]. Regarding their modes of action, AMPs can act, for example, in several biosynthetic pathways, differently from those mechanisms observed for conventional drugs. They are also able to overcome physical microbial barriers and act on specific inner targets such as bacterial ribosomes and fungal nucleus, as well as performing membrane-related activities [45]. Some studies regarding the rational design of AMPs have helped to identify optimal AMP physicochemical properties, including appropriate hydrophobicity, charge and amphipathic structural arrangement, which are all directly related with their antimicrobial activities [46]. Template-based designs taking into account charge and amphipathicity based on well-known AMPs, for example, have been applied to improve their activities [17].

5.5 AMPs partially constituted of D-AAs (diastereomeric AMPs)

It is known that thousands of peptides occur naturally in organisms, acting in the processes of cell differentiation, hormonal and immunological regulation, and defense response [11]. These peptides may present two chiral forms, known as L- and D-enantiomers (Fig. 5.1) [47]. The first report of natural D-AA biosynthesis dates from 1927, by Morizawa [1] and, albeit an unusual finding at that time [1], a series of subsequent works started revealing that plenty of organisms could generate diastereomeric proteins such as the gramicidins [48], gratisins [49], and tyrocidines [50], as well as diastereomeric peptides from the bacterium *Bacillus brevis* [4,51]. In animals, the first diastereomeric peptide reported was an opiate-like peptide, named dermorphin ($TA_{[D]}$FGTPS-NH2), isolated in 1981 by Montecucchi and coworkers [52] from the skin methanol extracts of *Phyllomedusa sauvagei*. Since then, many studies regarding naturally occurring, as well as synthetic, diastereomeric bioactive peptides have been in progress, including the prospection and design of diastereomeric AMPs.

It is known that AMPs present several properties that make them promising alternatives to no-longer-effective antibiotics. However, AMPs have limitations including toxicity toward human cells, low bioavailability and stability, as well as enzymatic degradation. In this scenario, works have shown that parent L-AMPs can be used as models for amino acid substitutions (Fig. 5.1), generating diastereomeric

Table 5.2 Antibacterial activities comparison between AMPs containing D-AAs and its parent L-AMPs against Gram-negative and -positive strains

	MIC (μM and μg/mL[a])										
	References	*Acinetobacter junii*	*Bacillus megaterium*	*Escherichia coli*	*Enterococcus faecalis*	*Pseudomonas aeruginosa*	*Staphylococcus aureus*	*Staphylococcus epidermidis*	*Serratia marcescens*	*Streptococcus pyogenes*	*Salmonella typhimurium*
Ascaphin-8	[27]	–	–	6	–	–	3	–	–	–	–
Ascaphin ($_{D}Lys^{4}$)		–	–	6	–	–	3	–	–	–	-
Ascaphin ($_{D}Lys^{8}$)		–	–	6	–	–	3	–	–	–	–
Ascaphin ($_{D}Lys^{10}$)		–	–	25	–	–	>100	–	–	–	–
Ascaphin ($_{D}Lys^{14}$)		–	–	12	–	–	>100	–	–	–	–
Ascaphin ($_{D}Lys^{18}$)		–	–	6	–	–	100	–	–	–	–
Temporin L	[6]	3	1.5	12	–	–	3	–	–	6	–
Temporin L ($_{D}Leu^{9}$)		3	1.5	12	–	–	6	–	–	6	–
L-GS14	[28]	–	–	>200[a]	1.5[a]	>200[a]	>200[a]	>200[a]	–	–	>200[a]
$GS14_{D}K^{4}$		–	–	3.1[a]	1.5[a]	3.1[a]	2[a]	2[a]	–	–	4[a]
Phd-1 (L-AAs)	[29]	–	–	15	–	nd	20	–	–	–	–
[D]-HC-1(R)		–	–	5	–	5	5	–	–	–	–
Phd-3 (L-AAs)		–	–	17	–	nd	17	–	–	–	–

[D]-HC-3(R)		–	–	2.5	–	5	2.5	–	–	–	–
granulysin (32–42)	[10]	–	–	20[a]	40[a]	–	40[a]	–	40[a]	10[a]	40[a]
D-granulysin		–	–	>20[a]	>40[a]	–	40[a]	–	>40[a]	15[a]	20[a]
all-L-magainin	[30]	–	–	12.5[a]	–	6.25[a]	25[a]	–	–	–	–
all-D-magainin		–	–	12.5[a]	–	3.12[a]	50[a]	–	–	–	–
L-magainin-NH_2	[31]	–	–	4	–	30	300	–	–	–	–
D-magainin-NH_2		–	–	1	–	30	100	–	–	–	–
L-melittin		–	–	0.8	–	3	0.2	–	–	–	–
D-melittin		–	–	1	–	2	0.1	–	–	–	–
BMAP28	[32]	5[a]	–	–	–	9.5	50	–	10[a]	–	6[a]
RI-BMAP28		10[a]	–	–	–	9.5	43	–	12[a]	–	18.5[a]

MIC, minimum inhibitory concentration; nd: not determined; – denotes conditions that were not tested.
[a]MIC values expressed in μg/mL.

Table 5.3 Antifungal and antiprotozoa activities comparison between AMPs containing D-AAs and its parent L-AMPs

		MIC (μM and μg/mL[a])			
	References	***Candida albicans***	***Schizosaccharomyces pombe***	***Saccharomyces cerevisiae***	***Leishmania major***
Ascaphin-8	[27]	12	–	–	–
Ascaphin ($_D$Lys^4)		6	–	–	–
Ascaphin ($_D$Lys^8)		6	–	–	–
Ascaphin ($_D$Lys^{10})		50	–	–	–
Ascaphin ($_D$Lys^{14})		25	–	–	–
Ascaphin ($_D$Lys^{18})		12	–	–	–
Temporin L	[6]	3	6	6	–
Temporin L ($_D$Leu^9)		3	6	6	–
L-GS14	[28]	150[a]	–	–	–
GS14$_D$K^4		6.2[a]	–	–	–
Phd-1 (L-AAs)	[29]	18	–	–	–
[D]-HC-1(R)		5	–	–	–
Phd-3 (L-AAs)		16	–	–	–
[D]-HC-3(R)		2.5	–	–	–
BMAP28	[32]	–	–	–	2[b]
RI-BMAP28		–	–	–	2[c]

MIC, minimum inhibitory concentration; – denotes conditions that were not tested.
[a]MIC values expressed in μg/mL.
[b]Concentration (μM) to reduce *L. major* cells viability in ~55%.
[c]Concentration (μM) to reduce *L. major* cells viability in ~80%.

AMPs with improved therapeutic index (TIs). Based on the work developed by Conlon and colleagues single substitutions of Asp^4, Gly^8, Ala^{10}, Val^{14}, and Leu^{18} by D-lysine residues in the frog skin AMP ascaphin-8 could generate a set of AMPs with retained minimum inhibitory concentrations (MICs) against *E. coli*, *S. aureus*, and *Candida albicans* (Tables 5.2 and 5.3) [27]. However, a 10-fold decrease in cytotoxicity was observed toward human erythrocytes compared to L-ascaphin-8 [27]. Similarly, Grieco and coworkers observed that replacements of Lys^7, Phe^8, Leu^9, Arg^{11}, Ile^{12}, and Leu^{13} by their respective D-enantiomers in the frog skin AMP temporin L almost abolished the hemolytic effect of their parent L-peptide at 48 μM, with more than a 75% decrease, except for the $_D$Phe^8 analogue (94% hemolysis) [6]. Otherwise, in this study, these modifications appeared also to interfere in the original antimicrobial properties of temporin L, with fourfold lower activities in some cases. The most active analogue was $_D$Leu^9, which maintained the same MIC values of its precursor against three yeast strains, as well as one and four Gram-positive and -negative bacterial strains (Tables 5.2, 5.3 and 5.1), respectively, but only with 9% hemolysis at 48 μM [6]. The authors also used circular dichroism (CD) and nuclear magnetic resonance (NMR) to show that D-AAs act as helix breakers and, depending on the breaking position, this could lead to one activity or another, which might explain the wide range of hemolytic effects of the temporin L diastereomeric analogues [6].

Such structural changes due to the insertion of D-AAs within diastereomeric peptides leading to different patterns of hemolytic/antimicrobial activities have gained great attention. The aim is to better understand how systematic physicochemical alterations, such as specific chirality changes on preexisting AMPs would enhance or diminish their TI. As an example, we can cite the structural and functional investigation of diastereomeric analogue peptides derived from the synthetic cyclic gramicidin 14 (GS14) (sequence cyclo(VKLKV$_D$YPLKVKL$_D$YP)) [28] which were designed based on the naturally occurring gramicidin S (GS) (sequence cyclo(V-Orn-L$_D$FP)$_2$) first reported in 1942 by Georgyi Frantsevitch Gause [55]. In that work Kondejewski and coworkers generated 14 analogues based on the highly amphipathic β-sheet GS14 by a series of single changes from L- to D-AAs along the GS14 sequence, thus evaluating large amphipathicity, activities, and structural changes [28]. CD and NMR studies in hydrophilic conditions revealed that all analogues, even those presenting essentially identical spectra, were highly disordered structurally, indicating a disruption in the β-sheet conformation characteristic of their parent L-GS14. The same disruption was observed under hydrophobic conditions, but here with considerably different spectra, indicating several backbone conformations depending on the position of the enantiomeric substitutions [28]. This variety in the 14 analogues' structural conformations also seemed to influence their hydrophobicity and, consequently, amphipathicity. Interestingly, the authors could conclude that the higher the amphipathicity, the higher the affinity of the D-analogues for lipopolysaccharides (LPS) from bacterial surfaces, and the higher their hemolytic properties toward human erythrocytes [28]. Promising findings were observed in the antimicrobial assays, and the diastereomeric peptide GS14$_D$K^4 was the most effective, revealing 6500-fold increased TI toward *P. aeruginosa* and *E. coli* strains, as well as up to 10,000-fold increased TI against Gram-positive bacteria compared to its parent L-GS14 (Tables 5.1 and 5.2) [28]. Similar results were also reported by this research group some years later [56], once more suggesting the crucial role of D-AA positions within diastereomeric peptide sequences for optimal hydrophobicity, amphipathicity, and nonpolar faces for microbial specificity and high TI, also highlighting the relevance of β-sheet structure disruptions for the antimicrobial activities displayed by the gramicidin-derived D-peptides [28,56].

Apart from the cytotoxic and hemolytic issues regarding the usage of AMPs in human treatments, studies have also shown that these molecules, when in the presence of physiological salt concentrations, may have their antibacterial and antifungal properties attenuated, as is the case of the well-known human β-defensin (HBD) host defense peptides [57,58]. Interestingly, studies have reported that peptides derived from the C-terminal regions of HBD-1 and -3 presenting L-arginines and L-isoleucines replaced by their D-enantiomers ([D]-HC-1(R) and [D]-HC-3(R)) were shown to be from three- to eightfold more active against *E. coli*, *S. aureus*, and *C. albicans* compared to their parent L-peptides (Table 5.2, 5.3 and 5.1) [29]. More importantly, in this same work, Olli and colleagues also observed that both analogues retained more than 80% of their antibacterial properties toward Gram-negative strains in the presence of 150 mM of NaCl [29]. Here, substitutions from L- to D-AAs in the analogue peptides also appeared to affect their structure greatly, both presenting CD spectra characteristic of flexible random coil conformations in all solvents tested, differentially from the β-sheet and α-helical findings for their parent L-peptide [29]. Interestingly, however, such structural divergences seemed not to interfere in their modes of action toward both bacteria and yeast, where all peptides were found to translocate the microbial surfaces in order to perform intracellular killing mechanisms, whose specific targets are yet unknown [29].

Another obstacle to be overcome, regarding the pharmacological interest of AMPs, is their susceptibility to enzymatic degradation, which is not unusual in oral and/or intravenous administrations. Thus, not only have diastereomeric AMPs been designed to improve their TI, but studies have also explored the D-enantiomeric strategy aiming to generate promising AMPs resistant to proteolytic events, thus improving their stability and bioavailability in vivo. Taking this challenge into account, Hamamoto and colleagues performed several L- to D-AA substitutions on an 11-amino-acid residue arginine-rich peptide (RVCRTGRSRWR) corresponding to a loop region of the human granulysin (residues 32–42) [10]. One diastereomeric peptide with six chiral changes ($R_DV_DCR_DTG_DR_DSR_DWR$) was synthesized and evaluated, and it was revealed to retain and even to improve its antimicrobial potential against *E. coli*, *Salmonella typhimurium*, *Serratia marcescens*, *S. aureus*, *Streptococcus pyogenes*, and *Enterococcus faecalis*, when compared to L-granulysin (Tables 5.2, 5.3 and 5.1). Moreover, when treated with 50 μg/mL of trypsin and further evaluated against *E. coli* and *S. aureus* strains, only diastereomeric granulysin preserved its original activities almost entirely, while L-granulysin had its antimicrobial potential completely abolished, thus confirming the proteolytic resistance conferred by D-AA substitutions [10]. In addition, the authors also incubated both L- and diastereomeric peptides in the presence of 15% fetal calf serum containing several proteases. As for the trypsin susceptibility assay, granulysin containing D-AAs retained much of its antibacterial potential toward *E. coli* and *S. aureus*, while L-granulysin drastically decreased its functions [10]. Apart from this, several other studies have shown how effective D-AA AMPs are even under proteolytic conditions; most of them are entirely constituted of D-AAs (all-D-AMPs) and are addressed in the topic below [18,30].

5.6 All-D-amino acid AMPs

Studies have shown that D-AA AMP analogues have similar activities in relation to their parent L-peptides found in nature [59]. However, one of the limitations in the use of L-natural peptides is due to their susceptibility to proteolytic degradation during drug administration and in the infection sites [53]. Thus, the development of synthetic peptides with changes in their chirality became attractive over the years. Bessale and coworkers obtained findings for an all-D-magainin analogue, which revealed similar and even lower MIC values against free-floating strains of *E. coli*, *P. aeruginosa*, and *S. aureus* when compared to its all-L-magainin parental peptide (Tables 5.2 and 5.1) [30]. In addition, it was demonstrated that all-D-magainin lost its hemolytic effect toward human erythrocytes, as well as being resistant to proteolysis [30].

Taking into account this parallel between resistance to proteases and AMP effectiveness [6], all-D-AMPs have also been studied as potential antituberculosis (anti-TB) agents [60]. Lan and coauthors tested six all-D-AMPs (D-LAK family) in vitro against resistant clinical strains of *Mycobacterium tuberculosis*, as well as ex vivo using THP-1 cells (macrophage model) [60]. As a result, the authors observed inhibition in the growth of susceptible and resistant *M. tuberculosis* in vitro at concentrations ranging from 50 to 100 μM, also suggesting a detergent-like mechanism of action for these all-D-AMPs (Table 5.1). Based on that, a synergistic evaluation was proposed by associating these detergent-like D-AMPs with the intracellular anti-TB drug isoniazid (INH), leading to increased activities against resistant *M. tuberculosis*, allied to low cytotoxicity toward macrophage-like THP-1 cells [60]. In addition, a year later, the same research group reported that the most promising all-D-AMPs that resulted from their previous study (D-LAK120-HP13 and D-LAK120-A) could be successfully associated with spray formulations (mannitol as a pulverized agent) for rapid pulmonary administration, thus being considered a powerful new drug against tuberculosis [61].

Although some studies have shown an improvement and/or similarity regarding the results in the inhibition of free-floating bacterial cells and biofilm formation, it is worth noting the importance of chirality in relation to antimicrobial compounds' resistance to degradation, as well as its advantages. In this context, Dean and coauthors showed that LL-37 peptides in their L- and D-forms were equivalent in their inhibition of *P. aeruginosa* biofilm formation, as well as decreasing the number of preformed biofilms (Table 5.4) [54]. However, it was observed that D-LL-37 promoted increased bacterial cell motility, also revealing improved resistance to proteases by in vivo *Galleria mellonella* biological assays [54]. Years later, de la Fuente-Núñez and coauthors [53] performed a large screening for antibiofilm action of all-D-peptides based on the L-amino acid host defense peptide IDR-1018 [53]. The functions of IDR-1018 had been widely investigated previously, revealing a significant reduction in the LPS-induced TNF-α response [62], modulation, expression, and markers of neutrophil adhesion and activation [63] activity against biofilms [64] in synergistic association for the inhibition and eradication of bacterial biofilms [65]. Thus, based on the high biological potential of IDR-1018, but considering its susceptibility to proteolytic degradation, a series of all-D-amino acid peptides was designed, named from DJK1 to DJK6, all presenting better antibiofilm properties at 10 μg/mL when compared to their parent L-peptides. In some cases, as for DJK-5 and DJK6, the results were highly promising; with inhibition percentages >98% against both *P. aeruginosa* and *Klebsiella pneumoniae* strains (Tables 5.4

and 5.1). The same pattern was also observed in the biofilm eradication assays, thus revealing the effectiveness of these engineered D-peptides against preformed Gram-negative biofilms [53].

Some studies suggest that the biological pathways of genes and enzymes involved in the synthesis and metabolism of D-amino acids may be promising targets in the discovery of effective antimicrobial agents [66]. From a functional point of view, the use of D-AA-containing molecules is attractive due to their resistance to enzymatic degradation [7], making their application possible in medical devices and in the development of inhalants and oral medicines. However, although studies have shown a clear advantage in the stability, bioavailability, and bioactivity regarding D-enantiomer AMPs, less promising findings can also be found in chirality alteration. Wade and coauthors showed examples of all-D-AMPs that maintained their bioactive properties against Gram-positive and -negative bacteria, whereas both D-magainin-NH_2 and D-melittin were, at lower concentrations, more effective ony against *E. coli*, *S. aureus*, and *P. aeruginosa* (Tables 5.2 and 5.1). Also, being more resistant to enzymatic degradation (trypsin) compared to their L-enantiomers [31]. However, apart from these promising results, a high hemolytic activity could be observed for D-melittin [31]. Furthermore, other obstacles must also be taken into consideration regarding all-D-AMPs, including the high costs of synthesis [67,68], the difficulty of synthesizing large D-AMPs on a major scale while maintaining their actions [69], and the scarcity of experimental studies related to the structure–function relationship of these molecules [70].

5.7 Retro-inverso AMPs

As described above, both diastereomeric and all-D-AMPs have shown promising pharmacological potential compared to their L-enantiomers, as well as some conventional medicines. However, such classes of AMPs require time-consuming and expensive synthesis, which may discourage their application in further advanced clinical trials. As an alternative, in 1979 Goodman and Chorev proposed a novel, effective approach where D-AA peptides would be synthesized as the reverse sequence of their L-AAs peptides, also known as retro-inverso (RI) peptides [9]. Since then, studies have reported that RI peptides present high topochemical equivalence to L-peptides, suggesting similar bioactivities allied to proteolytic resistance [67]. Interestingly, some studies have also shown that binding specificity to molecular targets and bioavailability in vivo of RI peptides are equal to or even higher than those of their precursors (Fig. 5.1). As an example, Biswas and coworkers developed a novel RI peptide, named RI-catestatin, which clearly enhanced the antihypertensive properties of its L-catestatin in vivo (Table 5.1) [71]. Moreover, Taylor and colleagues also reported the improved stability, biological activity, and blood–brain barrier of a series of RI prosaptide peptides compared to an all-D-prosaptide, indicating the importance of the RI strategy in the development of new therapeutic candidates for central nervous system neurodegeneration (Table 5.1) [13].

More recently, the RI peptides have gained increasing attention in the field of antimicrobials. One of the major works regarding this application was developed by de la Fuente-Núñez and coauthors, where a large screening for L-AMPs, all-D-AMPs, and RI-AMPs with antibiofilm activities was performed based on the host defense peptide IDR-1018 [53]. In this study, nine RI-AMPs were designed and evaluated for their potential to inhibit *P. aeruginosa* and *K. pneumoniae* biofilm formation at 10 μg/mL (Tables 5.4 and 5.1). Among these, five RI-AMPs were able to inhibit more than 70% of biofilm formation; of these, RI-JK3 $(RIVRVRIARLQV)_D$ was the most active peptide, with inhibition percentages higher than 99% against both multidrug-resistant strains (Table 5.4) [53]. Interestingly, the authors also observed that the RI peptide of IDR-1018 (RI-1018—$RRWIRVAVILRV)_D$, besides being capable of inhibiting *P. aeruginosa* biofilm formation in 95% at 10 μg/mL, was also able to protect two nonvertebrate models from lethal infections by this bacterial strain, with survival rates >80% and >50% for *C. elegans* and *G. mellonella* in the first 24 h of treatment, respectively [53]. Kindrachuk and colleagues also performed antibacterial studies with a RI-AMP, named RI-BMAP28, based on the bovine myeloid antimicrobial peptide 28 (BMAP28) [32]. RI-BMAP28 was evaluated against eight Gram-negative and four Gram-positive bacterial strains presenting MIC values (from 6.5 to >50 μg/mL) very similar to those from the original BMAP28 (Tables 5.2 and 5.1), but revealing high stability toward trypsin, proteinase K, and proteinase XXV. Moreover, the hemolytic and cytotoxic activities observed for BMAP28 were completely eliminated in the RI-BMAP28 analogue, clearly indicating its higher TI compared to its parental L-AMP [32]. Furthermore, structurally, both BMAP28 and RI-BMAP28 adopted random coil and α-helical conformation (right-handed helix for BMAP28 and left-handed helix for RI-BMAP28) in hydrophilic and membrane mimicking environments, respectively [32].

In addition to the multidrug-resistant free-floating bacterial cells and biofilm, protozoa parasites, such as *Leishmania major*, are also considered a global public health problem, mainly in Asia, Africa, the Americas, and Europe, where leishmaniasis, for example, affects up to 20 million people [41,72]. In this situation, Lynn and coworkers reported that the above-cited RI-BMAP28 peptide derived from the cathelicidin BMAP28 could also act as a promising antileishmanial RI peptide [41]. In the bioassays the authors reported that RI-BMAP28, at 2 μM, could reduce *L. major* promastigote cell viability by 80%, which can also be expressed as a 1.5-fold higher activity compared to the original BMAP28 (Tables 5.3 and 5.1) [41]. Moreover, significant membrane integrity disruption, followed by loss of cytosolic contents, vacuolar swelling and, finally, cell death could be observed in *L. major* cells treated with RI-BMAP28 [41].

Taking into account the high similarity between the side chains of RI peptides and their parent L-peptides, studies have also suggested a conserved level of antigenic crossreactivity [73], which raised the hypothesis of the use of RI peptides as synthetic vaccines [74]. Muller and colleagues explored this hypothesis by synthesizing an RI-variant corresponding to the immunodominant loop (region 141–159) of the VP1 protein expressed on the capsid of foot-and-mouth disease virus (FMDV) [75]. As a result, the authors reported that the RI-variant of this protein

Table 5.4 Antibiofilm activities comparison between AMPs containing D-AAs and its parent L-AMPs

MBIC (μg/mL), percentages of inhibition			
	References	***Klebsiella pneumoniae***	***Pseudomonas aeruginosa***
Inhibition percentages at 10 μg/mL			
IDR-1018	[53]	99	99
RI-1018		5	95
IDR-1012		1	41
RI-1012		0	95
IDR-1002		0	7
RI-1002		73	72
L-JK1		1	6
D-JK1		87	85
RI-JK1		–	0
L-JK2		–	43
D-JK2		–	87
RI-JK2		91	91
L-JK3		–	0
D-JK3		–	45
RI-JK3		99	100
L-JK4		0	12.7
D-JK4		99	99
RI-JK4		71	99.8
L-JK5		–	0
D-JK5		99.8	99.7
RI-JK5		–	0
L-JK6		0	69
D-JK6		98	98.4
RI-JK6		92	74
Inhibition percentages at 1 μg/mL			
L-LL-37	[54]	–	50
D-LL-37		–	50

MBIC, minimum biofilm inhibitory concentration; RI, retro-inverso peptides; – denotes conditions that were not tested.

segment could not only crossreact equally compared with its parent L-peptide, but also increase and extend the duration of the IgG response induced by the RI-variant. Moreover, it was observed that concentrations of antipeptide antisera were considerably higher, clearly indicating that RI strategies would be applied in the enhancement of immunogenic peptides [75]. Years later, the same research group [76] also reported that a single inoculation of the VP1 protein RI-variant could induce improved levels of neutralizing antibodies when compared to its corresponding L-peptide. In addition, VP1 RI-variants, not being susceptible to proteolytic events, were also able to protect guinea pigs substantially against FMDV [76]. Interestingly, this approach for the development of peptide-based vaccines was also shown to be useful in the case of influenza virus immunization. In this work, Ben-Yedidia and coworkers synthesized an RI-peptide corresponding to the influenza virus hemagglutinin B-cell epitope and demonstrated that it could induce immune responses in mice at both mucosal and systemic levels after intranasal immunization [77]. Furthermore, it was observed that the half-life of the RI-peptide in the presence of lung homogenate proteases was 700-fold higher compared to its parent L-peptide [77]. Put together, all these findings demonstrate the high pharmacological potential of RI-peptides not only as antibacterial molecules, but also as antibiofilm and antiprotozoa molecules, as well as providing novel alternatives for synthetic peptide-based vaccine development (Table 5.4).

5.8 Future trends and biomedical applications

Bioinspired peptides are considered as the most promising alternative for drug design nowadays. However, it is essential to search for efficient and precise strategies to deliver these bioactive molecules into the host system and to maintain their bioavailability. This could occur since these molecules are naturally susceptible to the action of several enzymes, especially the proteases—mainly when the administration is oral or parenteral or by dietary therapy [78]. Although engineered bioactive peptides are excellent alternatives for drug development, only a few of them have advanced to clinical trial tests, due to problems with biostability and time taken for delivery. One way of making these molecules available to the drug development market is the use of D-AA isoforms [5,78]. In nature, the D-AAs are quite rare and considered miraculous by some authors, who have been stimulating the use and deep study of this differential arrangement of molecules for clinical and biomedical applications [67]. As occurs with the usual bioactive peptides, which have become attractive and relatively low-cost molecules for the development of new drugs when compared to protein therapies, the use of D-AA peptides presents great advantages, such as biostability, reduction of complexity, increased bacterial selectivity, antiinflammatory activity, and improvement of electrostatic interactions [5,79]. Moreover, recent reports have explored the usage of D-AAs in self-assembled hydrogel biomaterials in order to search for higher stability against enzymatic degradation. This feature, in particular, could be sufficient motivation to incorporate just a few D-AAs into the molecules. In contrast, the typical approach consists of using the full D-enantiomer of an established self-assembling motif, i.e., substituting all L-amino acids with their D-counterparts [80].

The enhanced biostability of D-AA peptides compared with their L- counterparts seems to be a remarkable option for drug-delivery systems in vivo. The substitution of exoprotease-sensitive residues for noncanonical residues appears to be a wise choice, providing one strategy to improve the stability of bioactive peptides toward protease actions [5,79]. Among the alternatives for the acquisition of more stable peptides is the use of all-D-peptide isomers, involving the retro-inverso sequences and also the processes of acetylation and amidation of the N- and C-terminal of the peptides to singly reduce susceptibility to amino and carboxy-exopeptidases [78]. Techniques such as high-pressure liquid chromatography (HPLC) and mass spectrometry (MS) analysis can be used to determine the plasma peptide half-lives and enzymatic site cleavage, respectively. These approaches could provide an important advance in the design of the bioavailable peptides [78]. It was also demonstrated that the use of MALDI-ToF/ToF-mass spectrometry is a suitable tool for identifying the presence of D-AA residues within natural peptides at position 2. Specifically, this tool has been successfully applied to the analysis of all peptides, where the presence of epimeric forms has been proposed [81].

From a medical perspective, it is well known that the incorporation of D-glutamic acid and D-alanine into the bacterial cell wall could protect these pathogens from exoprotease action. Moreover, it was reported that D-serine and D-aspartic acid control neurotransmission in the brain, where these peptides are naturally present. The great advantage of insertion of D-AAs into diastereomeric peptides is their ability to create motifs with biological activities that promote cell adhesion and can control stem cell differentiation. It is important to make clear that studies about chirality effects at the supramolecular level are still seeking to elucidate their characteristics and biologic relationships. This field of research is still in its infancy, albeit evolving very quickly [80].

Studies performed with D-enantiomeric forms of peptides against clinical bacterial strains showed important results, demonstrating that D-peptides exhibited the same self-association ability as their L- counterparts. They thus produce similar biological activities. The fact that the hemolytic/antimicrobial activities of D-peptides against red blood and microbial cells, respectively, were quantitatively equivalent to their L-enantiomers confirms that no chirality selectivity by membrane or other stereoselective interactions at the cytoplasm occur during these processes [5]. Therefore, within a clinical context, we can use as an example the study of clinical strains of *P. aeruginosa*, which are associated mainly with infections in cystic fibrosis, cancer, and burn injury patients; these are infections whose control remains a huge problem in the nosocomial environment. In an effort to overcome this situation, promising results have been observed using D-AA peptides. Their association with nanostructures to produce drug carriers is another potential application. Thus, the importance of the biostability of these molecules in vivo is made very clear, for example, in the treatment of life-threatening diseases, including cancers [67].

Due to the mismatch of their chirality, D-AA peptides are considered poor substrates for natural enzymes such as proteases, which can be seen as an advantage, as reported by Li and coworkers, who explored the interactions between alkaline phosphatases and their substrates, using L-AA peptides and their D- counterparts [82]. All-D-peptides have been tested for their immunomodulatory activity, associated with antimicrobial and anticancer activities, as reported for the modified sculentin-1a. The authors of that study showed that the replacement of two amino acids was enough to promote the reduction of mammalian cell toxicity, increasing effectiveness against the biofilm forms of *P. aeruginosa*, and still maintaining high activity against the free-floating form of this pathogen. It was also found that this modification enhanced the peptide's stability and improved its immunomodulatory potential by promoting the migration of lung epithelial cells, apparently boosting its competence to repair the integrity of damaged lung tissue [83].

Furthermore, in the attempt to combat viruses such as influenza, Gupta and colleagues successfully used D-AAs in the elaboration of high-affinity protein or bacterial ligands, also called synbodies. The main objective was the development of a virus–ligand system, due to the great stability against proteases conferred by the D-AAs within the peptide structure. This suggests that the approach with virus-binding synbodies could be conjugated with transport proteins or other moieties that might increase the therapeutic profile of the developing virus–ligand system [42].

It has also been reported that the use of peptides containing D-AAs can present the capacity of self-assembly for amyloidosis therapy. Researchers believe that small peptides with similar sequences can bind to the Aβ peptide, avoiding the formation of amyloid aggregates [80]. Furthermore, the use of hydrogels for drug delivery is widely used, encouraging its association with D-AA applications. The use of systems of gelling peptides bearing D-AAs designed for drug delivery is already known, as described by Basu and coworkers [84] for a cancer soft drug biomaterial delivery vehicle. In addition, the selective formation of supramolecular nanofibers and hydrogels in pericellular spaces has been shown to inhibit cancer cells [85]. Moreover, Stephan and Irvine [86] described the development of a smart supramolecular biomaterial to instruct cell fates by varying the cohesive forces within nanofibers of supramolecular materials. Following this same idea, Wang and coworkers [87] have discovered that self-assembly of peptides containing D-AAs can be improved in cell membranes, confirming the possibility of using such D-AA-based peptides to engineer a cell surface with well-designed peptides or proteins.

However, some obstacles still have to be overcome in the use D-AA peptides, since a large number of them present high molecular weight and length or require some kind of ligand, such as a lipid or liposomes, to deliver them to the target [67]. Also pro-drug-based approaches, including cell-penetrating peptide conjugates, and chemical modifications via site-specific polyethylene glycol [PEG]ylation, lipidation and glycosylation have been used in peptide drug elaboration. Nonetheless, unfavorable consequences have been reported for these excipients, including reduced bioactivity, aggregation, or reduced solubility/stability of formulated peptides and toxicity-related problems, among the main concerns [78].

In summary, even taking into account a number of challenges to be studied in the future, here we describe the huge advantages of bioactive peptides containing D-AAs (with focus on AMPs) as promising antibacterial, antibiofilm, antiprotozoa, antifungal, anticancer, and antiinflammatory molecules, and even as new candidates for vaccine and hydrogel formulations, clearly indicating their potential pharmacological properties that may be useful in a near future.

References

[1] K. Morizawa, The extractive substances in *Octopus octopodia*, Acta Sch. Med. Univ. Imp. Kyoto 9 (1927) 285–298.

[2] N. Fujii, K. Satoh, K. Harada, Y. Ishibashi, Simultaneous stereoinversion and isomerization at specific aspartic acid residues in alpha A-crystallin from human lens, J. Biochem. 116 (1994) 663–669.

[3] K. Richter, R. Egger, G. Kreil, D-alanine in the frog skin peptide dermorphin is derived from L-alanine in the precursor, Science 238 (1987) 200–202.

[4] C. Ollivaux, D. Soyez, J.Y. Toullec, Biogenesis of D-amino acid containing peptides/proteins: where, when and how? J. Pept. Sci. 20 (2014) 595–612.

[5] Y. Chen, A.I. Vasil, L. Rehaume, C.T. Mant, J.L. Burns, M.L. Vasil, et al., Comparison of biophysical and biologic properties of alpha-helical enantiomeric antimicrobial peptides, Chem. Biol. Drug Des. 67 (2006) 162–173.

[6] P. Grieco, A. Carotenuto, L. Auriemma, M.R. Saviello, P. Campiglia, I.M. Gomez-Monterrey, et al., The effect of D-amino acid substitution on the selectivity of temporin L towards target cells: identification of a potent anti-*Candida* peptide, Biochim. Biophys. Acta 1828 (2013) 652–660.

[7] J.B. Mitchell, J. Smith, D-amino acid residues in peptides and proteins, Proteins: Struct. Funct. Bioinf. 50 (2003) 563–571.

[8] K.M. Makwana, R. Mahalakshmi, NMR analysis of tuning cross-strand Phe/Tyr/Trp-Trp interactions in designed beta-hairpin peptides: Terminal switch from L to D amino acid as a strategy for beta-hairpin capping, J. Phys. Chem. B. 119 (2015) 5376–5385.

[9] M. Goodman, M. Chorev, On the concept of linear modified retro-peptide structures, Acc. Chem. Res. 12 (1979) 7.

[10] K. Hamamoto, Y. Kida, Y. Zhang, T. Shimizu, K. Kuwano, Antimicrobial activity and stability to proteolysis of small linear cationic peptides with D-amino acid substitutions, Microbiol. Immunol. 46 (2002) 741–749.

[11] M. Liu, X. Li, Z. Xie, C. Xie, C. Zhan, X. Hu, et al., D-peptides as recognition molecules and therapeutic agents, Chem. Rec. 16 (2016) 1772–1786.

[12] M. Rahnamaeian, Antimicrobial peptides: modes of mechanism, modulation of defense responses, Plant Signal Behav. 6 (2011) 1325–1332.

[13] E.M. Taylor, D.A. Otero, W.A. Banks, J.S. O'brien, Retro-inverso prosaptide peptides retain bioactivity, are stable In vivo, and are blood-brain barrier permeable, J. Pharmacol. Exp. Ther. 295 (2000) 190–194.

[14] G. Schneider, W. Schrodl, G. Wallukat, J. Muller, E. Nissen, W. Ronspeck, et al., Peptide design by artificial neural networks and computer-based evolutionary search, Proc. Natl. Acad. Sci. U.S.A. 95 (1998) 12179–12184.

[15] C. Loose, K. Jensen, I. Rigoutsos, G. Stephanopoulos, A linguistic model for the rational design of antimicrobial peptides, Nature 443 (2006) 867–869.

[16] E.E. Oren, C. Tamerler, D. Sahin, M. Hnilova, U.O. Seker, M. Sarikaya, et al., A novel knowledge-based approach to design inorganic-binding peptides, Bioinformatics 23 (2007) 2816–2822.

[17] C.D. Fjell, J.A. Hiss, R.E. Hancock, G. Schneider, Designing antimicrobial peptides: form follows function, Nat. Rev. Drug Discov. 11 (2012) 37–51.
[18] S.Y. Hong, J.E. Oh, K.H. Lee, Effect of D-amino acid substitution on the stability, the secondary structure, and the activity of membrane-active peptide, Biochem. Pharmacol. 58 (1999) 1775–1780.
[19] R.S. Armen, B.M. Bernard, R. Day, D.O. Alonso, V. Daggett, Characterization of a possible amyloidogenic precursor in glutamine-repeat neurodegenerative diseases, Proc. Natl. Acad. Sci. U.S.A. 102 (2005) 13433–13438.
[20] R.S. Armen, M.L. Demarco, D.O. Alonso, V. Daggett, Pauling and Corey's alpha-pleated sheet structure may define the prefibrillar amyloidogenic intermediate in amyloid disease, Proc. Natl. Acad. Sci. U.S.A. 101 (2004) 11622–11627.
[21] F. Chiti, C.M. Dobson, Amyloid formation by globular proteins under native conditions, Nat. Chem. Biol. 5 (2009) 15–22.
[22] J. Kellock, G. Hopping, B. Caughey, V. Daggett, Peptides composed of alternating L- and D-amino acids inhibit amyloidogenesis in three distinct amyloid systems independent of sequence, J. Mol. Biol. 428 (2016) 2317–2328.
[23] S. Singh, H. Singh, A. Tuknait, K. Chaudhary, B. Singh, S. Kumaran, et al., PEPstrMOD: structure prediction of peptides containing natural, non-natural and modified residues, Biol. Direct 10 (2015) 73.
[24] J. Zhou, X. Du, N. Yamagata, B. Xu, Enzyme-instructed self-assembly of small D-peptides as a multiple-step process for selectively killing cancer cells, J. Am. Chem. Soc. 138 (2016) 3813–3823.
[25] E. Lee, J.K. Kim, S. Shin, K.W. Jeong, J. Lee, D.G. Lee, et al., Enantiomeric 9-mer peptide analogs of protaetiamycine with bacterial cell selectivities and anti-inflammatory activities, J. Pept. Sci. 17 (2011) 675–682.
[26] K. Najjar, A. Erazo-Oliveras, D.J. Brock, T.Y. Wang, J.P. Pellois, An L- to D-amino acid conversion in an endosomolytic analog of the cell-penetrating peptide TAT influences proteolytic stability, endocytic uptake, and endosomal escape, J. Biol. Chem. 292 (2017) 847–861.
[27] J.M. Conlon, S. Galadari, H. Raza, E. Condamine, Design of potent, non-toxic antimicrobial agents based upon the naturally occurring frog skin peptides, ascaphin-8 and peptide XT-7, Chem. Biol. Drug Des. 72 (2008) 58–64.
[28] L.H. Kondejewski, M. Jelokhani-Niaraki, S.W. Farmer, B. Lix, C.M. Kay, B.D. Sykes, et al., Dissociation of antimicrobial and hemolytic activities in cyclic peptide diastereomers by systematic alterations in amphipathicity, J. Biol. Chem. 274 (1999) 13181–13192.
[29] S. Olli, N. Rangaraj, R. Nagaraj, Effect of selectively introducing arginine and D-amino acids on the antimicrobial activity and salt sensitivity in analogs of human beta-defensins, PLoS One 8 (2013) e77031.
[30] R. Bessalle, A. Kapitkovsky, A. Gorea, I. Shalit, M. Fridkin, All-D-magainin: chirality, antimicrobial activity and proteolytic resistance, FEBS Lett. 274 (1990) 151–155.
[31] D. Wade, A. Boman, B. Wåhlin, C. Drain, D. Andreu, H.G. Boman, et al., All-D amino acid-containing channel-forming antibiotic peptides, Proc. Natl. Acad. Sci. U.S.A. 87 (1990) 4761–4765.
[32] J. Kindrachuk, E. Scruten, S. Attah-Poku, K. Bell, A. Potter, L.A. Babiuk, et al., Stability, toxicity, and biological activity of host defense peptide BMAP28 and its inversed and retro-inversed isomers, Biopolymers 96 (2011) 14–24.

[33] D.M. Veine, H. Yao, D.R. Stafford, K.S. Fay, D.L. Livant, A D-amino acid containing peptide as a potent, noncovalent inhibitor of alpha5beta1 integrin in human prostate cancer invasion and lung colonization, Clin. Exp. Metastasis 31 (2014) 379–393.

[34] S.J. Kim, I.W. Kim, Y.N. Kwon, E.Y. Yun, J.S. Hwang, Synthetic Coprisin analog peptide, D-CopA3 has antimicrobial activity and pro-apoptotic effects in human leukemia cells, J. Microbiol. Biotechnol. 22 (2012) 264–269.

[35] S. Travis, L.M. Yap, C. Hawkey, B. Warren, M. Lazarov, T. Fong, et al., RDP58 is a novel and potentially effective oral therapy for ulcerative colitis, Inflamm. Bowel. Dis. 11 (2005) 713–719.

[36] C. Bechara, S. Sagan, Cell-penetrating peptides: 20 years later, where do we stand? FEBS Lett. 587 (2013) 1693–1702.

[37] C. de la Fuente-Núñez, F. Reffuveille, L. Fernandez, R.E. Hancock, Bacterial biofilm development as a multicellular adaptation: antibiotic resistance and new therapeutic strategies, Curr. Opin. Microbiol. 16 (2013) 580–589.

[38] K.Y. Lum, S.T. Tay, C.F. Le, V.S. Lee, N.H. Sabri, R.D. Velayuthan, et al., Activity of novel synthetic peptides against *Candida albicans*, Sci. Rep. 5 (2015) 9657.

[39] M.E. Falagas, F. Ntziora, G.I. Betsi, G. Samonis, Caspofungin for the treatment of fungal infections: a systematic review of randomized controlled trials, Int. J. Antimicrob. Agents 29 (2007) 136–143.

[40] J.R.S. Leite, L.P. Silva, M.I.S. Rodrigues, M.V. Prates, G.D. Brand, B.M. Lacava, et al., Phylloseptins: a novel class of anti-bacterial and anti-protozoan peptides from the *Phyllomedusa* genus, Peptides 26 (2005) 565–573.

[41] M.A. Lynn, J. Kindrachuk, A.K. Marr, H. Jenssen, N. Pante, M.R. Elliott, et al., Effect of BMAP-28 antimicrobial peptides on *Leishmania major* promastigote and amastigote growth: role of leishmanolysin in parasite survival, PLoS Negl. Trop. Dis. 5 (2011) e1141.

[42] N. Gupta, J. Lainson, V. Domenyuk, Z.G. Zhao, S.A. Johnston, C.W. Diehnelt, Whole-virus screening to develop synbodies for the influenza virus, Bioconjug. Chem. 27 (2016) 2505–2512.

[43] A.J. De Lucca, T.J. Walsh, Antifungal peptides: novel therapeutic compounds against emerging pathogens, Antimicrob. Agents Chemother. 43 (1999) 1–11.

[44] N. Sun, S.A. Funke, D. Willbold, Mirror image phage display—Generating stable therapeutically and diagnostically active peptides with biotechnological means, J. Biotechnol. 161 (2012) 121–125.

[45] E. de Souza Cândido, M.H. e Silva Cardoso, D.A. Sousa, J.C. Viana, N.G. de Oliveira-Júnior, V. Miranda, et al., The use of versatile plant antimicrobial peptides in agribusiness and human health, Peptides 55 (2014) 65–78.

[46] M.H. Cardoso, S.M. Ribeiro, D.O. Nolasco, C. de la Fuente-Núñez, M.R. Felicio, S. Gonçalves, A polyalanine peptide derived from polar fish with anti-infectious activities, Sci. Rep. 6 (2016) 21385.

[47] G.A. Petsko, On the other hand, Science 256 (1992) 1403–1404.

[48] S.B. Hladky, D.A. Haydon, Ion transfer across lipid membranes in the presence of gramicidin A. I. Studies of the unit conductance channel, Biochim. Biophys. Acta 274 (1972) 294–312.

[49] M. Tamaki, Y. Imazeki, A. Shirane, K. Fujinuma, M. Shindo, M. Kimura, et al., Novel gratisin derivatives with high antimicrobial activity and low hemolytic activity, Bioorg. Med. Chem. Lett. 21 (2011) 440–443.

[50] P.J. Loll, E.C. Upton, V. Nahoum, N.J. Economou, S. Cocklin, The high resolution structure of tyrocidine A reveals an amphipathic dimer, Biochim. Biophys. Acta 1838 (2014) 1199–1207.
[51] H. Li, N. Anuwongcharoen, A.A. Malik, V. Prachayasittikul, J.E. Wikberg, C. Nantasenamat, Roles of D-amino acids on the bioactivity of host defense peptides, Int. J. Mol. Sci. (2016) 17.
[52] P.C. Montecucchi, R. De Castiglione, S. Piani, L. Gozzini, V. Erspamer, Amino acid composition and sequence of dermorphin, a novel opiate-like peptide from the skin of *Phyllomedusa sauvagei*, Int. J. Pept. Protein Res. 17 (1981) 275–283.
[53] C. de la Fuente-Núñez, F. Reffuveille, S.C. Mansour, S.L. Reckseidler-Zentene, D. Hernández, G. Brackman, et al., D-enantiomeric peptides that eradicate wild-type and multidrug-resistant biofilms and protect against lethal *Pseudomonas aeruginosa* infections, Chem. Biol. 22 (2015) 196–205.
[54] S. Dean, B. Bishop, M.H. Van Regenmortel, M. Hoek, Susceptibility of *Pseudomonas aeruginosa* biofilm to alpha-helical peptides: D-enantiomer of LL-37, Front. Microbiol. (2011) 2.
[55] Y.M. Gall, M.B. Konashev, The discovery of Gramicidin S: the intellectual transformation of G.F. Gause from biologist to researcher of antibiotics and on its meaning for the fate of Russian genetics, Hist. Philos. Life Sci. 23 (2001) 137–150.
[56] D.L. Lee, J.P. Powers, K. Pflegerl, M.L. Vasil, R.E. Hancock, R.S. Hodges, Effects of single D-amino acid substitutions on disruption of beta-sheet structure and hydrophobicity in cyclic 14-residue antimicrobial peptide analogs related to gramicidin S, J. Pept. Res. 63 (2004) 69–84.
[57] M.J. Goldman, G.M. Anderson, E.D. Stolzenberg, U.P. Kari, M. Zasloff, J.M. Wilson, Human beta-defensin-1 is a salt-sensitive antibiotic in lung that is inactivated in cystic fibrosis, Cell 88 (1997) 553–560.
[58] T. Tomita, S. Hitomi, T. Nagase, H. Matsui, T. Matsuse, S. Kimura, et al., Effect of ions on antibacterial activity of human beta defensin 2, Microbiol. Immunol. 44 (2000) 749–754.
[59] D.J. Triggle, Stereoselectivity of drug action, Drug Discov. Today 2 (1997) 138–147.
[60] Y. Lan, J.T. Lam, G.K. Siu, W.C. Yam, A.J. Mason, J.K. Lam, Cationic amphipathic D-enantiomeric antimicrobial peptides with in vitro and ex vivo activity against drug-resistant *Mycobacterium tuberculosis*, Tuberculosis 94 (2014) 678–689.
[61] P.C.L. Kwok, A. Grabarek, M.Y. Chow, Y. Lan, J.C. Li, L. Casettari, et al., Inhalable spray-dried formulation of D-LAK antimicrobial peptides targeting tuberculosis, Int. J. Pharm. 491 (2015) 367–374.
[62] M. Wieczorek, H. Jenssen, J. Kindrachuk, W.R. Scott, M. Elliott, K. Hilpert, et al., Structural studies of a peptide with immune modulating and direct antimicrobial activity, Chem. Biol. 17 (2010) 970–980.
[63] F. Niyonsaba, L. Madera, N. Afacan, K. Okumura, H. Ogawa, R.E. Hancock, The innate defense regulator peptides IDR-HH2, IDR-1002, and IDR-1018 modulate human neutrophil functions, J. Leukoc. Biol. 94 (2013) 159–170.
[64] C. de la Fuente-Núñez, F. Reffuveille, E.F. Haney, S.K. Straus, R.E. Hancock, Broad-spectrum anti-biofilm peptide that targets a cellular stress response, PLoS Pathog. 10 (2014) e1004152.
[65] F. Reffuveille, C. de la Fuente-Núñez, S. Mansour, R.E. Hancock, A broad-spectrum antibiofilm peptide enhances antibiotic action against bacterial biofilms, Antimicrob. Agents Chemother. 58 (2014) 5363–5371.
[66] M. Friedman, Origin, microbiology, nutrition, and pharmacology of D-amino acids, Chem. Biodivers. 7 (2010) 1491–1530.

[67] Z. Feng, B. Xu, Inspiration from the mirror: D-amino acid containing peptides in biomedical approaches, Biomol. Concepts 7 (2016) 179–187.

[68] R. Mahalakshmi, P. Balaram, The use of D-amino acids in peptide design. D-amino acids: a new frontier in amino acid and protein research – practical methods and protocols, Cell Biol. (2007) 415–430.

[69] M.T. Weinstock, M.T. Jacobsen, M.S. Kay, Synthesis and folding of a mirror-image enzyme reveals ambidextrous chaperone activity, Proc. Natl. Acad. Sci. U.S.A. 111 (2014) 11679–11684.

[70] D.E. Mortenson, J.D. Steinkruger, D.F. Kreitler, D.V. Perroni, G.P. Sorenson, L. Huang, et al., High-resolution structures of a heterochiral coiled coil, Proc. Natl. Acad. Sci. U. S. A. 112 (2015) 13144–13149.

[71] N. Biswas, J. Gayen, M. Mahata, Y. Su, S.K. Mahata, D.T. O'connor, Novel peptide isomer strategy for stable inhibition of catecholamine release: application to hypertension, Hypertension 60 (2012) 1552–1559.

[72] M.J. Mcconville, E. Handman, The molecular basis of *Leishmania* pathogenesis, Int. J. Parasitol. 37 (2007) 1047–1051.

[73] G. Guichard, N. Benkirane, G. Zeder-Lutz, M.H. Van Regenmortel, J.P. Briand, S. Muller, Antigenic mimicry of natural L-peptides with retro-inverso-peptidomimetics, Proc. Natl. Acad. Sci. U. S. A. 91 (1994) 9765–9769.

[74] S. Muller, N. Benkirane, G. Guichard, M.H. Van Regenmortel, F. Brown, The potential of retro-inverso peptides as synthetic vaccines, Expert. Opin. Investig. Drugs 7 (1998) 1429–1438.

[75] S. Muller, G. Guichard, N. Benkirane, F. Brown, M.H. Van Regenmortel, J.P. Briand, Enhanced immunogenicity and cross-reactivity of retro-inverso peptidomimetics of the major antigenic site of foot-and-mouth disease virus, Pept. Res. 8 (1995) 138–144.

[76] J.P. Briand, N. Benkirane, G. Guichard, J.F. Newman, M.H. Van Regenmortel, F. Brown, et al., A retro-inverso peptide corresponding to the GH loop of foot-and-mouth disease virus elicits high levels of long-lasting protective neutralizing antibodies, Proc. Natl. Acad. Sci. U. S. A. 94 (1997) 12545–12550.

[77] T. Ben-Yedidia, A.S. Beignon, C.D. Partidos, S. Muller, R. Arnon, A retro-inverso peptide analogue of influenza virus hemagglutinin B-cell epitope 91-108 induces a strong mucosal and systemic immune response and confers protection in mice after intranasal immunization, Mol. Immunol. 39 (2002) 323–331.

[78] M. Bagheri, R.E. Hancock, High-performance liquid chromatography and mass spectrometry-based design of proteolytically stable antimicrobial peptides, Methods. Mol. Biol. 1548 (2017) 61–71.

[79] B. Jacob, G. Rajasekaran, E.Y. Kim, I.S. Park, J.K. Bang, S.Y. Shin, The stereochemical effect of SMAP-29 and SMAP-18 on bacterial selectivity, membrane interaction and anti-inflammatory activity, Amino Acids 48 (2016) 1241–1251.

[80] M. Melchionna, K.E. Styan, S. Marchesan, The unexpected advantages of using D-amino acids for peptide self- assembly into nanostructured hydrogels for medicine, Curr. Top. Med. Chem. 16 (2016) 2009–2018.

[81] J. Koehbach, C.W. Gruber, C. Becker, D.P. Kreil, A. Jilek, MALDI TOF/TOF-based approach for the identification of D-amino acids in biologically active peptides and proteins, J. Proteome. Res. 15 (2016) 1487–1496.

[82] J. Li, Y. Gao, Y. Kuang, J. Shi, X. Du, J. Zhou, et al., Dephosphorylation of D-peptide derivatives to form biofunctional, supramolecular nanofibers/hydrogels and their potential applications for intracellular imaging and intratumoral chemotherapy, J. Am. Chem. Soc. 135 (2013) 9907–9914.

[83] A. Di Grazia, F. Cappiello, H. Cohen, B. Casciaro, V. Luca, A. Pini, et al., D-Amino acids incorporation in the frog skin-derived peptide esculentin-1a(1-21)NH2 is beneficial for its multiple functions, Amino Acids 47 (2015) 2505–2519.
[84] K. Basu, A. Baral, S. Basak, A. Dehsorkhi, J. Nanda, D. Bhunia, et al., Peptide based hydrogels for cancer drug release: modulation of stiffness, drug release and proteolytic stability of hydrogels by incorporating D-amino acid residue(s), Chem. Commun. (Camb) 52 (2016) 5045–5048.
[85] X.D. Xu, B.B. Lin, J. Feng, Y. Wang, S.X. Cheng, X.Z. Zhang, et al., Biological glucose metabolism regulated peptide self-assembly as a simple visual biosensor for glucose detection, Macromol. Rapid Commun. 33 (2012) 426–431.
[86] M.T. Stephan, D.J. Irvine, Enhancing cell therapies from the outside in: Cell surface engineering using synthetic nanomaterials, Nano Today 6 (2011) 309–325.
[87] H. Wang, Y. Wang, A. Han, Y. Cai, N. Xiao, L. Wang, et al., Cellular membrane enrichment of self-assembling D-peptides for cell surface engineering, ACS Appl. Mater. Interfaces 6 (2014) 9815–9821.

Peptide nutraceuticals

6

Aswathy R. Girija
Future Industries Institute, University of South Australia, Mawson Lakes, SA, Australia

6.1 Introduction

Nutraceuticals is the name coined for a class of healthy foods by Stephen DeFelice in 1989, and is a combination of the words "nutrition" and "pharmaceutical." A technical definition of nutraceutical from a *Science* policy forum states: "Nutraceuticals are those diet supplements that deliver a concentrated form of a biologically active component of food which is present in a non-food matrix to enhance health" [1]. A distinction between functional food and nutraceuticals is that the former offers a health benefit by providing one or more active ingredients within the food matrix in addition to nutrition and the latter are those isolated or concentrated ingredients/components (e.g., bioactive components) that offer both medical benefits by preventing diseases and for the treatment of diseases. Nutraceuticals include dietary fiber, polyunsaturated fatty acids (PUFA, fish oil), proteins, peptides, amino acids, keto-acids, minerals, antioxidant vitamins, etc. Research is being focused on the innovation through interdisciplinary technologies including biotechnology, metabolic engineering, and nanotechnology to form a different platform for the isolation of novel nutraceutical components, large-scale production, and smart delivery of nutraceuticals. Currently nutraceuticals range from isolated or concentrated nutrients, dietary supplements, genetically engineered "designer" foods, processed or supplemented product food materials to nanotechnology-driven nanomaterials as smart nanocarriers for enhanced bioavailability of the same. In this chapter we focus on a class of healthy foods—nutraceuticals—with special emphasis on peptide nutraceuticals.

Bioactive peptides derived from food consist of amino acid chains that stay indolent within the parent protein. Most of the bioactive peptides are short amino acid residues of 2–20 amino acids and some peptides are large and may be up to 40 amino acids [2]. Bioactive peptide nutraceuticals are formed mainly from protein precursors by digestive enzymes (by gastrointestinal digestion), during the processing of food (such as fermentation, ripening, cooking), storage, or by in vitro hydrolysis by several proteolytic enzymes (Fig. 6.1) [3–6].

Several microorganisms produce proteases that are released and end up in the proteolysis and generation of bioactive peptides. However, when compared with microorganisms, the proteolytic enzymes can be supplanted for proteolysis owing to the rapid and safe production of bioactive peptides. Owing to the lack of residual toxic end product, enzymatic hydrolysis is highly preferred in the food industry. The proteinases used for protein hydrolysis include pepsin, trypsin, chymotrypsin, neutrase, alcalase, papain, etc. The parameter for the efficient enzymatic hydrolysis

Peptide Applications in Biomedicine, Biotechnology and Bioengineering.
DOI: http://dx.doi.org/10.1016/B978-0-08-100736-5.00006-5

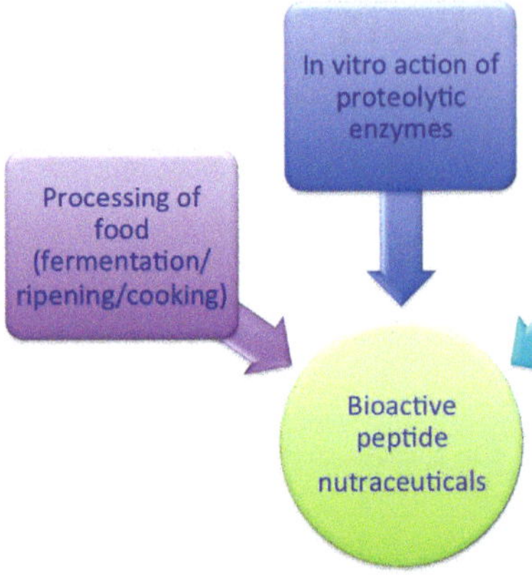

Figure 6.1 Generation of peptide nutraceuticals.

is yet another area of vast research for desired products with maximum yield [7]. Peptide nutraceuticals can be classified into several types, based on their origin, activity, etc. Depending on their origin, peptide nutraceuticals can be broadly categorized into plant, animal, and microbial based. Depending on their bioactivity, peptide nutraceuticals can be categorized into several categories including antioxidant, antihypertensive, anticancer/antiproliferative, antiinflammatory, antimicrobial, antithrombotic, and immunomodulatory peptides, etc. (Fig. 6.2). A brief description of some of the major categories with an emphasis on their activities is given in this chapter.

6.2 Antioxidant peptides

Antioxidants are those components that protect the body against reactive oxygen species (ROS). ROS are generated in the body during oxidative metabolism and inflammation and exert oxidative damage to membrane lipids, proteins, and DNA. ROS thus initiate and/or aid in the progression of several chronic diseases such as cardiovascular, neurodegenerative, inflammatory, diabetes, and cancer, and are also associated with aging [8]. Research in the field of natural peptide antioxidants can explore the potential for wellbeing. Recently, the huge demand for natural antioxidants over synthetic antioxidants has gained attention for the consumption of natural food. The health risks associated with synthetic antioxidants also account for the shift from synthetic to natural antioxidants. Biologically active peptides with antioxidant properties are obtained from plant and animal proteins. The antioxidant peptides act as direct scavengers for free radicals, inhibitors of lipid peroxidation, and/or metal ion chelating agents that catalyse the formation of free radicals [9–11]. The antioxidant property of nutraceutical peptides is highly associated with hydrophobic amino acids and histidine residues.

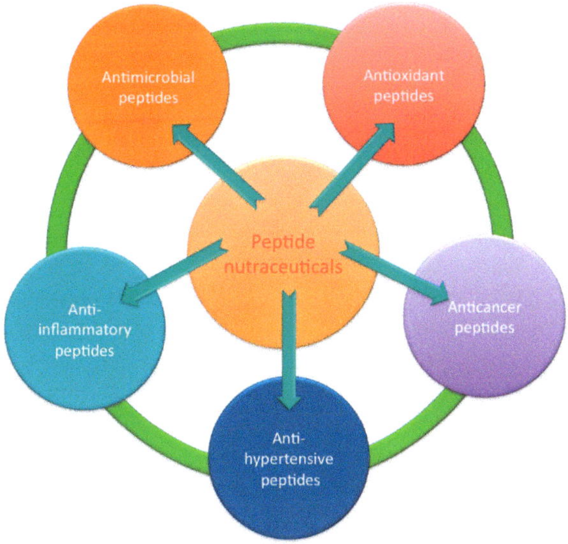

Figure 6.2 Role of peptide nutraceuticals.

Several peptide nutraceuticals have been investigated under in vitro and in vivo conditions for their antioxidant potential and functional properties. Milk peptides derived from αs-casein exhibited free radical scavenging and inhibited both enzymatic and nonenzymatic lipid peroxidation. Whey protein that was processed under low temperature retained high levels of glutamylcysteine that stimulated the synthesis of glutathione, a potent antioxidant [12]. Several studies have demonstrated that protein hydrolysates from marine proteins (from fish, oyster, etc.) are antioxidant peptide sources. Antioxidant peptides soluble in trichloroacetic acid from *Crassostrea talienwhanensis* subtilism digest was studied by Wang et al. [13]. Amino acid sequences of the peptides Pro-Val-Met-Gly-Ala and Glu-His-Gly-Val of 518 and 440 Da, respectively, were identified and exhibited high antioxidant actions based on their hydroxyl and 1,1-diphenyl-2-picrylhydrazyl (DPPH) radical scavenging activities [13]. Peptide from *Otolithes ruber* muscle protein hydrolysate was purified and antioxidant activity was assessed both by in vitro and in vivo studies. The antioxidant potential was contributed by the peptide with amino acid sequence, Lys-Thr-Phe-Cys-Gly-Arg-His (861.6 Da). The peptide quenched DPPH and hydroxyl radicals (84.5 ± 1.2% and 62.4 ± 2.9%), and inhibited DNA damage and lipid peroxidation under in vitro studies [14]. In a recent study, three antioxidant peptides with high antioxidant activities were isolated from *Pseudosciaena crocea* muscle using pepsin and alcalase. The peptide sequences of three antioxidant peptides were Tyr-Leu-Met-Ser-Arg (PC-1), Val-Leu-Tyr-Glu-Glu (PC-2), and Met-Ile-Leu-Met-Arg (PC-3), with molecular weights of 651.77, 668.82, and 662.92 Da, respectively. The antioxidant activities of purified peptides were evaluated by several methods and it was observed that PC-1 presented the maximum scavenging activities on DPPH

(EC_{50} = 1.35 mg/mL), superoxide (EC_{50} = 0.450 mg/mL), and 2,2′-azino-*bis*(3-ethylbenzthiazoline-6-sulfonic acid) (ABTS) (EC_{50} = 0.312 mg/mL) radicals. Among the three peptides, PC-2 exhibited the highest hydroxyl radical scavenging activity (EC_{50} = 0.353 mg/mL). The researchers suggest that the beneficial properties of isolated peptides might be due to their reduced size, and hydrophobic and aromatic amino acids in the sequences [15].

Recently, the antioxidant properties of vegetable proteins and peptides have gained great interest and beenextensively explored both in in vitro and in vivo studies [16]. Hydrolysates and peptides from enzymatic hydrolysis from plant sources offer several beneficial properties including improved digestibility, enhanced sensory qualities, reduced allergic components; superior biological activities including antioxidant potential. Hydrolysates with antioxidant activity produced from rice bran protein were hydrolyzed with proteases M, N, P, S, and pepsin [17]. Protease M hydrolysates displayed a high degree of hydrolysis with low antioxidant activity and hydrolysates with a degree of hydrolysis less than 10% had superior antioxidant activity than hydrolysates above 20%. This result signifies that the antioxidant activity of the hydrolysates depends on amino acid sequences of peptides and protease specificities [17]. The defatted rice endosperm protein (REP) was digested effectively with neutrase for the production of antioxidant peptides. The identified antioxidant peptides were Phe-Arg-Asp-Glu-His-Lys-Lys and Lys-His-Asp-Arg-Gly-Asp-Glu-Phe with 959.5 Da and 1002.5 Da, respectively. Neutrase hydrolysate from rice endosperm protein (NHREP) exhibited superior DPPH and hydroxyl radical scavenging activities and the inhibition percentage of auto-oxidation of NHREP was 80.09%. Chemically synthesized peptide inhibited lipid peroxidation significantly and exhibited improved cellular viability of *t*-butylhydroperoxide (*t*-BHP)-induced cells (human embryonic lung fibroblasts, MRC-5-74.38% and mouse macrophage RAW 264.7−78.39 at 80 μg/mL) [18]. Antioxidant peptide in hemp seed protein hydrolysate (HPH) was produced through simulated gastrointestinal tract digestion of hemp seed proteins. Antioxidant peptides Trp-Val-Tyr-Tyr and Pro-Ser-Leu-Pro-Ala with DPPH scavenging (67% and 58%) and metal chelation activity (94% and 96%, respectively) were produced [19].

It was demonstrated that peptides from a microbial fermentation possess several functional properties, including antioxidant activity. Antioxidant peptides obtained from lactic acid bacteria are well established. Microbes act on plant origin substrates or food materials and produce peptides with functional properties. Antioxidant peptides isolated from a traditional Chinese salt-fermented soybean food (douchi) has been investigated in in vitro and in vivo conditions [20]. The in vitro IC_{50} values for DPPH and ABTS radical scavenging activities and the chelating ability of ferrous ions with Douchi extracts were 0.658, 0.204, and 206 mg/mL, respectively. Antioxidant enzymatic activities of extracts in cholesterol-fed rats showed that Douchi extracts had excellent antioxidant activities which could affect the activities of antioxidant enzymes and lipid peroxidation, and mitigate the degree of lipidosis of hepatocytes with improvement of lipid metabolism in the liver. The first report on the biological activity of semifermented dry cacao protein fractions was by Preza et al. [21]. The albumin fraction presented free radical scavenging capacity and antitumor activity. Antioxidant peptides from cacao beans exhibited antitumor activity against murine lymphoma in mice with an oral dose of 25 mg/kg per day for a 15-day treatment. Antioxidant peptide release by microbial fermentation of rice protein, grape must, and soybean by *Bacillus pumilus*, *Oenococcus oenis*, and *Lactobacillus plantarum*, respectively, has been reported [22–24].

6.3 Antihypertensive peptides

Hypertension is often diagnosed when systolic and diastolic blood pressures are $\geq$ 140 and 90 mmHg, respectively, and the blood pressure in the arteries is raised. Systemic hypertension involves the renin–angiotensin system (RAS), which is a major endocrine system for controlling blood pressure. During metabolic disorders, when the blood flow is reduced, secreted renin converts angiotensinogen into angiotensin I. Another RAS enzyme, angiotensin converting enzyme (ACE), subsequently removes histidyl-leucine from angiotensin I and converts to physiologically active octapeptide angiotensin II that binds to vascular wall receptors leading to contraction of blood vessels and resulting in high blood pressure. ACE-inhibitory activity from naturally occurring peptides by the proteolytic degradation products of proteins from animal (e.g., dairy, fish, and meat), plant (e.g., rice, wheat, maize, soy and garlic), or microbial (e.g., lactic acid bacteria, yeast) origins have been studied extensively. Most of the research into ACE-inhibitory peptides is from dairy products which are either produced by in vitro enzymatic digestion of milk proteins or by the chemical synthesis of peptide analogues. Bioactivity of the native form of milk proteins is almost incomplete, however its activity is established by proteolytic digestion to release and activate bioactive peptides from the parent protein. Owing to their physicochemical versatility, milk peptides are considered as significant components in food or pharmaceutical applications. The ACE-inhibitors of milk proteins are contributed from the fragments of casein (casokinins) or whey proteins (lactokinins) [25].

Apart from milk peptides, several animal sources of ACE peptides have been identified. Thermolysin hydrolysates of porcine skeletal muscle proteins exhibited antihypertensive activities. Two ACE inhibitory peptides, Met-Asn-Pro-Pro-Lys and Ile-Thr-Thr-Asn-Pro, exhibited a blood pressure-reducing effect in spontaneous hypertensive rats (SHR) when administered in single oral doses [26]. In another study, chicken breast meat extract treated with fungal protease and gastric proteases displayed an ACE-inhibitory effect. Three ACE-inhibitory peptides with a common sequence of Gly-X-X-Gly-X-X-Gly-X-X have been identified and the peptide Gly-Phe-Hyp-Gly-Thr-Hyp-Gly-Leu-Hyp-Gly-Phe showed enhanced ACE-inhibitory activity [27]. A high level of ACE-inhibitory activity from hydrolysates of chicken leg bones by alcalase enzyme was reported by Cheng et al. [28]. Arihara et al. reported two ACE-inhibitory pentapeptides from porcine myosin by thermolysin digestion. The sequences of inhibitory peptides were termed myopentapeptides A and B, where Met-Asn-Pro-Pro-Lys corresponded to positions 79–83 on the myosin heavy chain and Ile-Thr-Thr-Asn-Pro corresponded to positions 306–310 on the myosin heavy chain [29]. The peptides exhibited IC_{50} at concentration of 945.5 and 549.0 μM, respectively. Another ACE-inhibitory peptide from the pepsin hydrolysate of crude porcine troponin was described by Katayama et al. [30]. The amino acid sequence of Lys-Arg-Glu-Lys-Tyr-Asp-Ile corresponded to amino acid positions 198–204 of troponin T.

Marine fish-derived bioactive peptides have also been studied extensively as antihypertensive components. Hydrolysis of skin gelatin of *Okamejei kenojei* with alcalase and protease yielded two peptides with antihypertensive and antioxidant activities [31]. The identified peptides Met-Val-Gly-Ser-Ala-Pro-Gly-Val-Leu (829-Da) and Leu-Gly-Pro-Leu-Gly-His-Gln (720-Da) exhibited ACE-inhibitory activity with IC_{50} values of 3.09 μM and 4.22 μM, respectively, and could be used as antihypertensive agents in cardiovascular diseases. ACE-inhibitory peptides have been isolated from hydrolysates of several fishery sources such as salmon, sardine, yellowfin sole, and dried bonito [32]. Salmon hydrolysate exhibited high inhibitory activity against angiotensin I-converting enzyme ($IC_{50} = 27.9$ protein μg/mL) under in vitro studies. The researchers isolated six active dipeptides containing Trp: Trp-Ala, Val-Trp, Trp-Met, Met-Trp, Ile-Trp, and Leu-Trp. A reduction of systolic blood pressure was observed when the hydrolysate was orally administered to spontaneously hypertensive rats [32]. Peptides with ACE activity were isolated and identified by alcalase hydrolysis of protein concentrate obtained from industrial manufacturing effluent of cuttlefish [33]. Four fractions (A–D) with significant ACE-inhibitory activity were isolated. The IC_{50} values of the fractions were in the 1.92–8.83 μg/mL range, however more studies are required to identify which peptides were responsible for effective antihypertensive activity of identified fractions.

Many peptides derived from plant proteins exhibit antihypertensive activity. Li et al. demonstrated the in vitro ACE-inhibitory activity of peptide Thr-Gln-Val-Tyr derived from rice protein hydrolyzed by protease alcalase [34]. The antihypertensive effect of rice protein hydrolysate was examined in spontaneously hypertensive rats and alcalase-generated hydrolysate exhibited enhanced in vitro ACE-inhibitory activity with an IC_{50} value of 0.14 mg/mL. A reduction in systolic blood pressure in spontaneously hypertensive rats was observed following a single oral administration of the hydrolysate. Two ACE-inhibitory tripeptides (Val-Asn-Pro and Val-Trp-Pro) from rice protein hydrolysate were isolated and their antihypertensive effects in vivo were evaluated [35]. Both peptides exhibited competitive ACE inhibition and were stable against ACE and gastrointestinal proteases, pepsin and chymotrypsin. Single oral administration of the tripeptides Val-Asn-Pro and Val-Trp-Pro in spontaneously hypertensive rats lowered the systolic blood pressure and the antihypertensive effects of tripeptides lasted for 8 h. A novel cholecystokinin CCK-dependent vasorelaxing peptide, Ile-His-Arg-Phe, reduced blood pressure and food intake in rodents as reported by Kontani et al. [36]. It was reported that the dipeptides, Arg-Phe and Ile-His-Arg-Phe, were generated in the chymotrypsin digest of the partial structure of rice glutelin. Ile-His-Arg-Phe, corresponding to rice glutelin (155–158), exhibited vasorelaxing activity in the mesenteric artery of spontaneous hypertensive rats and orally administered IHRF reduced systolic blood pressure in SHRs. The antihypertensive activity of Ile-His-Arg-Phe was more effective and sustained for a longer period than Arg-Phe. It enhanced the intracellular Ca^{2+} flux and release of CCK in the enteroendocrine cell STC-1.

Antihypertensive peptides derived from wheat have been extensively studied. Wheat germ hydrolysate prepared with an alcaline protease resulted in an ACE-inhibitory tripeptide Ile-Val-Tyr [37], and wheat gliadin hydrolysate with acid protease yielded Ile-Ala-Pro that exhibited ACE-inhibitory activity [38]. Nogata et al. investigated the production of ACE-inhibitory peptides by autolysis reactions from wheat milling by-products with aspartic proteases. From the hydrolysate of the by-product fraction, six peptides Leu-Gln-Pro, Ile-Gln-Pro, Leu-Arg-Pro, Val-Pro, Ile-Tyr, and Thr-Phe with ACE-inhibitory activity were isolated [39]. Antihypertensive tripeptides such as Leu-Ala-Tyr, Leu-Gln-Pro, Leu-Arg-Pro, and Leu-Ser-Pro were isolated from maize endosperm protein (α-zein) hydrolysate prepared with thermolysin [40]. The hypotensive activity of Leu-Gln-Pro was demonstrated in spontaneously hypertensive rats in which the blood pressure was lowered by 15 mmHg after a 30 mg/kg intravenous injection. In another study, alcalase-mediated hydrolysis of corn gluten meal (CGM) after the removal of starch yielded an ACE-inhibitory dipeptide (Ala-Tyr) with IC_{50} of 14.2 μM. The activity of the peptide was retained even after oral administration to spontaneously hypertensive rats. A reduction of systolic blood pressure of 9.5 mmHg was observed 2 h after oral administration of dipeptide at doses of 50 mg/kg [41].

Although most of the studies were performed with enzymatic hydrolysates, there are reports on the generation of ACE-inhibitory peptides through microbial fermentation. Rizzello et al. investigated the synthesis of ACE-inhibitory peptides during sourdough fermentation of white wheat, wholemeal wheat, and rye flours and reported that the highest ACE-inhibitory activity was observed by fermenting flour under semiliquid conditions by using wholemeal wheat flour [42,43]. Fourteen peptides (IC_{50} 0.19–0.54 mg/mL) were identified from wholemeal wheat sourdough fermentation and all the identified peptides retained the antihypertensive epitope Val-Ala-Pro. Fermentation and hydrolysis of pea seeds protein by *L. plantarum* 299v were reported by Jakubczyk et al. [44]. Fermentation improved the ACE-inhibitory activity of peptides from pea proteins. The purified and active peptide fraction Lys-Glu-Asp-Asp-Glu-Glu-Glu-Glu-Gln-Gly-Glu-Glu-Glu obtained from pea proteins hydrolysates exhibited an IC_{50} of 64.04 g/mL. Soymilk and its fermented product have several beneficial biological properties and are a great source of bioactive peptides. Antihypertensive peptides are released from soybean products including soymilk, soy-yogurt, and soy sauce through fermentation with lactic acid bacteria (*Lactobacillus casei*, *Lactobacillus acidophilus*, *Lactobacillus bulgaricus*, *Streptococcus thermophilus*) and also with *Bacillus* species such as *Bacillus natto*, *Bacillus subtilis*, *Bifidobacterium longum*, etc., as starters [24].

6.4 Anticancer/antiproliferative peptides

Cancer is one of the major health issues affecting the lives of millions of people throughout the world. Cancer is the condition recognized by the uncontrolled division of cells and the spread of abnormal cells, which may be due to several external environmental factors (chemicals, radiation, infectious organisms, etc.) or internal factors (mutations, hormones, and distorted immunity) [45]. The survival rate of cancer patients is highly dependent on the stage at which the cancer is diagnosed. In general, the survival rate is around 90% with early cancer diagnosis, whereas it steadily drops to 8% during late diagnosis and associated tumor metastases [46,47]. Although several natural and synthetic chemotherapeutic agents are now available to treat cancer, the need for natural food products or components during digestion can inhibit the prognosis of cancer by enhancing apoptosis, thus inhibiting cancer metastases. The anticancer potential of dietary proteins and peptides generated by fermentation, enzymatic hydrolysis, or gastrointestinal digestion, has been shown to prevent different stages of cancer from initiation, promotion, and progression. Peptides from animal (fish, meat, milk), plant (plant protease inhibitors, plant lectins, lunasin), and microbial origins have been reported to mediate in the apoptosis, angiogenesis, and metastasis of tumors.

Lactoferrin (Lf), an iron-binding milk protein, is stable against degradation by proteolytic enzymes. Owing to its unique receptor-mediated mechanism, it passes through the gastrointestinal tract and affects gene transcription [48–50]. Although the real mechanism of action of Lf is not clear, it was proposed that Lf acts as an antiproliferative agent through its iron-binding capability, thus damaging the nucleic acid structure. Lf has also been reported to induce apoptosis and inhibit angiogenesis [51–53]. The hydrolysis of milk protein, β-casein, yielded β-casomorphins (β-CMs) that functioned as exogenous opioid-like peptides. The primary amino acid sequence of β-CM is NH_2-Tyr-Pro-Phe-Pro-Gly-Pro-Ile-Pro-Asn-Ser-Leu-COOH that is placed at positions 60–70 in bovine β-casein. Owing to the high content of prolin, their level is abundant in stomach and is rather resistant to proteolysis [54]. α-Casomorphins are separated from peptic hydrolysates of α-casein fractions. α- and β-casomorphins are produced by the effect of different proteases released from tumor cells [55,56]. The peptides released from casein execute antiproliferative activities by opioid receptors and apoptosis [57]. Casomorphin peptides generated from both α- and β-caseins reduced the proliferation of prostate carcinoma cell lines.

Casein phosphopeptides (CPPs) are bioactive peptides with several phosphorylated sites with the capability to bind and solubilize calcium and are released during the digestion of casein [58]. It was demonstrated that in HT-29 cells and in a primary human colon cancer cell line (AZ-97), the activation of voltage-activated L-type calcium channels that mediated the calcium influx was associated with apoptosis and suppression of the growth of colon cancer cells [59]. Perego and coworkers reported that CPPs offered protection to differentiated intestinal cells from toxicity induced by calcium overload and inhibited apoptosis and in undifferentiated cancerous cells they induced apoptosis [60]. The studies confirm the benefits of CPPs as effective nutraceuticals and as modulators of cellular processes associated with the cancer phenotype. Antitumor effects of bovine lactoferrin by inhibiting angiogenesis and reducing endothelial cell proliferation have been reported [61,62]. In another study, bovine lactoferrin suppressed the proliferation of breast cancer cells in a dose-dependent manner by inducing apoptosis [63].

It has been reported that several peptides separated from meat and marine organisms, specifically fish, exhibited cytotoxicity activity, thus inhibiting cancer cell proliferation [2,64,65]. The cytotoxic effect of four peptides from beef sarcoplasmic proteins in different cell lines (breast adenocarcinoma [MCF-7], stomach adenocarcinoma [AGS], and lung carcinoma [A549] cells) was reported by Jang et al. [66]. In their study, the peptide Gly-Phe-His-Ile exhibited highest cytotoxic effect on MCF-7 cells and also reduced the cellular viability of AGS cells. The peptide Gly-Leu-Ser-Asp-Gly-Glu-Trp-Gln inhibited proliferation of AGS cells, whereas none of the peptides exerted cytotoxic effect on A549 cells. Su et al. extracted an anticancer bioactive peptide (ACBP) isolated from goat spleens with immunization by human gastric cancer extracts [67]. The bioactive peptide exhibited cytotoxicity in the human gastric cancer line BGC-823 in a dose-dependent fashion. The peptide suppressed the growth of gastric tumor in an in vivo xenograft model and suggested that the ACBP peptide could serve as a potential anticancer peptide inducing apoptosis and antiproliferation. Yu et al. extracted anticancer bioactive peptide-3 (ACBP-3) from goat liver and the antitumor potential in vitro and in vivo on gastric cancer stem cells was reported [68]. The ratio of CD44(+) cells was decreased and resulted in the reduction of spheroid colonies, which eventually blocked their clone-forming capability. In in vivo studies when the cells were treated with ACBP-3 or in combination with cisplatin, tumor growth was considerably suppressed. Hsu and coworkers reported the antiproliferative potential in human breast cancer cell line MCF-7 by the hydrolysate of tuna dark muscle byproduct by papain (PA) and Protease XXIII (PR) [69]. The highest antiproliferative activities were presented by the peptide fractions with molecular weight ranging between 400 and 1400 Da. The antiproliferative peptides from papain hydrolysate and Protease XXIII were Leu-Pro-His-Val-Leu-Thr-Pro-Glu-Ala-Gly-Ala-Thr (1206 Da) and Pro-Thr-Ala-Glu-Val-Tyr-Met-Val-Thr (1124 Da) and exhibited dose-dependent inhibition of the MCF-7 cells with IC_{50} of 8.1 and 8.8 μM, respectively. In a study, researchers developed a special kind of TFD (Thomsen-Friedenreich disaccharide), called TFD100 from polar fish, Pacific cod. TFD100 binds to protein galectin-3, which is expressed in excess in prostate cancer cells, and suppresses the interaction with TFD antigen present on the cell surface. It blocked cancer cells from attaching to the vessel walls and suppressed the death of T cells, eventually boosting the immune response [70]. In mice models, TFD100 at nanomolar range inhibited angiogenesis, tumor–endothelial cell interactions, and metastasis of prostate cancer cells. Bukowski reported a shark cartilage extract, AE-941, displayed inhibition of matrix protease, VEGF binding to receptor, initiation of endothelial cell apoptosis, and instigation of production of angiostatin [71]. In vitro studies on peptides isolated from shark cartilage exhibited antiangiogenic property by the suppression of vascular endothelial growth factor (VEGF)-induced migration and tubulogenesis of human umbilical vein endothelial cells (HUVEC) and inhibited outgrowth of microvessels from rat aorta [72,73].

Among bioactive plant peptides, lunasin is widely studied for its anticancer activities. Lunasin is an Arg-Gly-Asp (RGD) peptide comprised of 43 amino acid residues, isolated from soybean. It includes distinctive amino acid sequences with cell adhesion motif of Arg-Gly-Asp and a carboxylic acid tail of nine aspartic acid residues. The amino acid sequence 23–31 of lunasin targets histones H3 and H4, the Arg-Gly-Asp motif accounts for the cellular internalization, and the polyaspartic acid tail is responsible for binding of lunasin to core histones, thus disturbing cell division and leading to cell death. Galvez and de Lumen demonstrated a plausible role of lunasin in cancer prevention by the transfection of lunasin plasmid into cancer cells (mouse hepatoma cancer cells, mouse fibroblast cells, and human cancer cells) and reported the inhibition of cell division [74]. Synthetic lunasin also presented adherence to chromatin, and inhibited cell proliferation. The exogenous use of synthetic lunasin stopped the transformation of mouse fibroblasts to cancerous cells by internalization of lunasin via the Arg-Gly-Asp motif [75]. In the study, lunasin also prevented tumorigenesis in a SENCAR mouse skin cancer model. In another study, lunasin arrested the G_2/M phase of the cell cycle in a dose-dependent manner in L1210 leukemia cells by inducing apoptosis in L1210 leukemia cells through the activation of caspase-3 [76]. Lunasin induced apoptosis in HT-29 human colon cancer cells by activating apoptosis through the mitochondrial pathway and resulted in G_2 arrest of the cell cycle [77]. The study demonstrated that lunasin can modify the Bax/Bcl-2 ratio, where the expression of proapoptotic Bax was increased and the expression of antiapoptotic Bcl-2 proteins was decreased, thus activating apoptosis.

Anticancer peptides have been isolated from enzymatic hydrolysates of plant proteins, especially from rice, soy, and marine algae. An antiproliferative peptide, Glu-Gln-Arg-Pro-Arg, has been isolated from rice bran proteins digested with alcalase [78]. The peptide demonstrated 84% inhibition in colon cancer cells (Caco-2, HCT-116), 80% in breast cancer cells (MCF-7, MDA-MB-231), and 84% in liver cancer cells (HepG-2) at a concentration range of 600–700 g/mL. Another anticancer peptide, X-Met-Leu-Pro-Ser-Tyr-Ser-Pro-Tyr, was isolated from the defatted soy protein hydrolyzed with thermoase. In vitro cytotoxicity in P388D1, a mouse monocyte macrophage cell line, exhibited very high cytotoxicity by disturbing the G_2/M phase [79]. Peptides derived from marine algae have also demonstrated anticancer potential. The polypeptide (Cys-Pro-Ala-Pro) from *Chlorella pyrenoidosa* protein hydrolysate presented antiproliferative activity on human liver cancer HepG2 cells by CPAP-induced apoptosis and necrosis [80]. The peptide Val-Glu-Cys-Tyr-Gly-Pro-Asn-Arg-Pro-Gln-Phe from *Chlorella vulgaris* protein waste demonstrated dose- dependent cytotoxicity and stimulated post-G_1 cell cycle arrest in gastric cancer AGS cells. Under in vitro studies, the peptide was not toxic to normal lung fibroblast WI-38 cells [81].

Although lunasin have been confirmed as an antiproliferative peptide, the high cost associated with its production limits its use in chemotherapy and in nutritional treatments [82]. As an alternative to the chemical synthesis, available natural sources of lunasin from various vegetable sources by microbial fermentation have been reported. Fermentation by lactic acid bacteria from cereal and nonconventional (pseudo cereals and legumes) flours to release lunasin was reported by Rizzello et al. [83]. The concentration of lunasin enhanced up to several times during fermentation when compared to untreated flours. *Lactobacillus curvatus* SAL33 and

Lactobacillus brevis AM7, with nonspecific aminopeptidase activity, released the highest concentrations of lunasin from wholemeal wheat, soybean, barley, amaranth, and rye flours. In another study, flours from *Pisum vulgaris*, *Cicer arietinum*, *Lathyrus sativus*, *Lens culinaris*, and *Pisum sativum* species were subjected to fermentation with *L. plantarum* C48 and *L. brevis* AM7 strains [84]. The researchers reported nine different lunasin-like polypeptides similar to lunasin were identified and exhibited antiproliferation of human adenocarcinoma Caco-2 cells.

6.5 Antiinflammatory peptides

Inflammation is the part of the body's complex biological response to remove harmful stimuli characterized by a protective response involving augmented endothelial permeability, seepage of protein-rich exudates, and intrusion of immune cells into extravascular tissues to initiate the repair mechanism. The vascular endothelium plays a major role in the extravasation of leukocytes, which is a characteristic of inflammation. Thus, a hallmark of inflammation comprises the instigation of proinflammatory signaling cascades, upregulation of leukocyte adhesion molecules, infiltration of leukocytes into tissue, and enhanced levels of cytokines and chemokines in the circulation. Inflammation is crucial for imparting resistance against microbial infections and for wound healing. However, uncontrolled inflammation is one of the major causes of chronic disorders including asthma, cardiovascular disease, obesity, diabetes, inflammatory bowel disease, cancer, osteoporosis, and neurological diseases like Parkinson's [85,86]. Food-derived bioactive peptides mediate antiinflammation by blocking or lowering the expression of inflammatory biomarkers and/or by regulating transcription factors.

Milk-derived bioactive peptides from casein and whey proteins are widely studied for their antiinflammatory properties. The role of the tripeptide, Val-Pro-Pro, derived from the bacterial fermentation of casein, has demonstrated attenuation of leukocyte–endothelial interactions by inhibition of the proinflammatory c-Jun N-terminal kinase pathway [87]. The tripeptide (Val-Pro-Pro and Ile-Pro-Pro) offered protection against atherosclerosis in apolipoprotein E (ApoE) knockout mice through combined action of modulation of inflammatory and hypertensive pathways [88]. Casein hydrolysates, by enzymatic digestion, presented antiinflammatory properties on activated macrophages [89]. Whey protein hydrolysates demonstrated inhibition of inflammatory responses in respiratory and intestinal epithelial cells [90,91]. Shimizu et al. demonstrated the inhibition of dermatitis in NC/Nga mice by whey protein hydrolysates [92]. Casein hydrolysates from microbial fermentation demonstrated positive signs in the treatment of adjuvant arthritis in rats and chemically induced colitis in mice by chronic and acute inflammatory responses [93,94]. Bioactive peptide lactoferricin from lactoferrin exhibited antiinflammatory effects on human cartilage and synovial cells [95,96]. Tripeptides (Ile-Arg-Trp and Ile-Gln-Trp) from ovotransferrin in egg white exhibited downregulation of the expression of cytokine-induced inflammatory protein in vascular endothelium partially by modulating the NF-κB pathway [97,98].

In one study, an antiinflammatory tripeptide (Pro-Ala-Tyr) from salmon pectoral fin byproduct protein hydrolysate was isolated by pepsin hydrolysis. The tripeptide treatment in lipopolysaccharide induced nitric oxide (NO) and prostaglandin E2 (PGE2) in RAW264.7 cells inhibited the NO by 63.80% and PGE2 by 45.33%. The peptide mediated the downregulation of protein expression of inducible nitric oxide synthase and cyclooxygenase-2 responsible for the generation of NO and PGE2. Also, the tripeptide reduced the level of proinflammatory cytokines, including tumor necrosis factor-α, and interleukin-6 and -1β [99]. Antiinflammatory peptides were isolated from enzymatic hydrolysates of tuna cooking juice and their secretory and cellular responses in murine peritoneal macrophages RAW264.7 was demonstrated by Cheng et al. [100]. Alcalase hydrolysate (AH) presented the best inhibitory effect and two peptide fractions, Pro-Arg-Arg-Thr-Arg-Met-Met-Asn-Gly-Gly-Arg (1543.8 Da) and Met-Gly-Pro-Ala-Met-Met-Arg-Thr-Met-Pro-Gly (1211.5 Da), demonstrated the highest antiinflammatory activity.

Plant protein hydrolysates including soybean and corn enzymatic hydrolysates have shown antiinflammatory effects both in in vitro and in vivo studies. However, only a few peptides have been identified with antiinflammatory activities [101−104]. Soybean tripeptide Val-Pro-Tyr, a novel PepT1 substrate, inhibited proinflammatory mediators in vitro in Caco-2 intestinal epithelial and THP-1 macrophages, and diminished colitis in mice by mediating the downregulation of proinflammatory cytokines in the colon, suggesting Val-Pro-Tyr can be an ideal candiate for the treatment of inflammatory bowel disease (IBD). Antiinflammatory peptide pyroGlu-Leu in the enzymatic hydrolysate of wheat gluten was orally administered to study the improvement of colitis in mice. Administration of peptide stabilized the population of *Bacteroidetes* and *Firmicutes* in the colon. The peptide offered protection against dextran sulfate-induced colitis in mice and chemically induced hepatitis in rats [105,106].

Lunasin from defatted soybean flour presented its in vitro antiinflammatory activity using RAW 264.7 macrophages as reported by Dia et al. [82]. Treatment of lipopolysaccharide (LPS)-induced RAW 264.7 macrophages with lunasin reduced the level of nitric oxide (NO), prostaglandin E2 (PGE_2), expression of inducible nitric oxide synthase (iNOS), and cyclooxygenase-2 (COX-2), thus inhibiting COX-2/PGE_2 and iNOS/NO pathways. Hernandez-Ledesma and coworkers demonstrated that lunasin inhibited the production of TNF-α and IL-6 [107]. Lunasin presented antiinflammatory effects in lipopolysaccharide (LPS)-induced human THP-1 macrophages by inhibiting the Akt-mediated NF-κB pathway [108]. Lunasin interacted with αVβ3 integrin that is linked with activation of inflammatory pathways such as Akt/NF-κB.

Spirulina is an exceptional source of protein and an ideal candidate to be hydrolyzed into bioactive peptides. In a study by Vo et al., two peptides, Leu-Asp-Ala-Val-Asn-Arg (P1) and Met-Met-Leu-Asp-Phe (P2), from enzymatic hydrolysate of

Spirulina maxima were found to be effectual against early atherosclerotic responses by inhibiting mast-cell degranulation by reducing the release of histamine and boosting the level of intracellular Ca^{2+} [109]. The mechanism in the downregulation of P1 and P2 on high-affinity IgE receptor-mediated allergic reaction was investigated. P1 interfered with calcium- and microtubule-dependent signaling pathways, while P2 was involved in the inhibition of phospholipase Cγ activation and reactive oxygen species formation. Also, the peptides resulted in the depression of nuclear factor-κB translocation and generation of IL-4.

A novel peptide, thalassospiramide G (**1**), along with thalassospiramides A and D (**2**–**3**), was isolated from a large culture of marine unicellular bacteria *Thalassospira* sp. Thalassospiramides A and D (**2**–**3**) suppressed lipopolysaccharide-induced nitric oxide generation mouse macrophage RAW 264.7 cells, with IC_{50} values of 16.4 and 4.8 μM, respectively [110].

6.6 Antimicrobial peptides

Antimicrobial peptides identified from natural sources have been widely studied and are used as antimicrobials including antibacterial, antifungal, and antiviral agents. The efficiency and the mode of action of antimicrobial peptides may differ based on their structural features, selectivity, and sensitivity on target microorganisms. Most antimicrobial peptides comprise of less than 50 amino acids with hydrophobic amino acids (~50%) and often in amphipathic 3D confirmation [111]. It was reported that animal-derived antimicrobial peptides display inhibitory effects against a broader spectrum of microorganisms than those isolated from microorganisms [111]. In terms of sensitivity, antimicrobial peptides from microorganisms exhibit superior efficiency even at low concentrations in the nanomolar range [112].

Antimicrobial peptides derived from milk are widely studied. There are several reports on the control of microbial infections and inhibition of pathogens (both Gram-positive and Gram-negative). Lactoferricin derived from milk protein is the most widely studied antimicrobial peptide. Lactoferrampin, derived from lactoferrin, presented antimicrobial activity against *Streptococcus mutans*, *Escherichia coli*, *B. subtilis*, and *Pseudomonas aeruginosa* [113]. Several casein-derived peptides also offered antimicrobial properties. Caseicidin peptide released from casein demonstrated antimicrobial activity against several pathogens including *Staphylococcus* spp., *Sarcina* spp., *Diplococcus pneumoniae*, *B. subtilis*, and *Streptococcus pyogenes*. Nonimmunogenic antibacterial peptides from chymosin digest of αs1-casein B, named "isracidin," demonstrated inhibition against pathogens such as *Staphylococcus aureus*, *S. pyogenes*, and *Listeria monocytogenes* [114]. In addition to bacterial infections, isracidin and lactoferricin B inhibited *Candida albicans*. In another study by Zucht et al., a cationic fragment of casein, casocidin-I, inhibited the growth of *E. coli* and *Staphylococcus carnosus* [115].

Jang et al. demonstrated antimicrobial activity of peptides derived from a bovine meat source. Antibacterial activities of the peptides Gly-Phe-His-Ile, Asp-Phe-His-Ile-Asn-Gly, Phe-His-Gly, and Gly-Leu-Ser-Asp-Gly-Glu-Trp-Gln against Gram-positive (*Bacillus cereus*, *L. monocytogenes*, and *S. aureus*) and Gram-negative (*Salmonella typhimurium*, *E. coli*, and *P. aeruginosa*) bacteria were reported in their study [66]. Of these four peptides, Gly-Leu-Ser-Asp-Gly-Glu-Trp-Gln inhibited both Gram-positive and Gram-negative pathogens. Antimicrobial activity of peptide NKLP27 NK-lysin, isolated from tongue sole (*Cynoglossus semilaevis*), was demonstrated against both Gram-negative and Gram-positive bacteria. The antibacterial activity of NKLP27 was dependent on the C-terminal five residues and the peptide damaged the cell membrane, infiltrated into the cytoplasm, and degraded the nucleic acid. In vivo studies also demonstrated a reduction in infection and the peptide upregulated the expression of the immune genes [116]. An antimicrobial peptide with 55 amino acid residues and molecular mass of 6621.55 Da from hemolymph of *Mytilus coruscus* displayed antimicrobial activity against fungi and Gram-positive bacteria. The mRNA expression of precursor mytichitin-1 was upregulated in the gonad after bacterial infection, indicating the involvement of a host immune response against bacterial infection [117].

The bacterium *Bacillus amyloliquefaciens* isolated from a mangrove system secreted bioactive cyclic lipopeptide of a heptapeptide and a 15-carbon 3-hydroxy fatty acid. The cyclic lipopeptide inhibited the growth of yeast strains including *C. albicans*, *Candida tropicalis*, *Metschnikowia bicuspidata*, *Saccharomyces cerevisiae*, and *Yarrowia lipolytica* [118]. In plants there are several proteins and peptides that function as a defense system. Several antifungal proteins have been isolated from leguminous plants. These include chitinases, glucanases, thionins, defensins, vicilins, allergen-like proteins, cyclophilin-like proteins, etc. Hydrolysate of pea proteins by protease from *Aspergillus oryzae* exhibited antifungal activity [119]. The antifungal activity was contributed to by three proteins (pea defensins 1 and 2, and a nonspecific lipid transfer protein (nsLTP)) and a blend of peptides released during hydrolysis. The pea flour hydrolysate inhibited the growth of molds and extended the shelf-life of bread [119].

Lactic acid bacteria act as excellent biopreservative organisms owing to their capability for the production and release of antimicrobial molecules. Although the exact mechanism for antifungal activity is not clear, it is considered that peptides and organic acids generated during the fermentation of sour dough by lactic acid bacteria mediate an antifungal effect. Fermentation of wheat germ with two autochthonous lactic acid bacteria (*L. plantarum* LB1 and *Lactobacillus rossiae* LB5) exhibited an antifungal property [120]. The four peptides released by lactic acid bacteria fermentation in wheat-germ mediated inhibition of fungi including *Penicillium*, *Aspergillus*, *Monilia*, *Mucor*, *Endomyces*, *Cladosporium*, *Fusarium*, and *Rhizopus* that spoil baked food and extended the shelf-life of breads [120–122].

The peptide nutraceuticals with different functions described in the chapter are summarized in Table 6.1.

Table 6.1 Summary of peptide nutraceuticals presented in this chapter

Application	Peptide	Source	References
Antioxidant	High levels of glutamylcysteine	Whey	[12]
	Pro-Val-Met-Gly-Ala and Glu-His-Gly-Val	*Crassostrea talienwhanensis* subtilism digest	[13]
	Lys-Thr-Phe-Cys-Gly-Arg-His	*Otolithes ruber* muscle protein hydrolysate	[14]
	Tyr-Leu-Met-Ser-Arg (PC-1), Val-Leu-Tyr-Glu-Glu (PC-2), and Met-Ile-Leu-Met-Arg (PC-3)	*Pseudosciaena crocea* muscle using pepsin and alcalase	[15]
	Phe-Arg-Asp-Glu-His-Lys-Lys and Lys-His-Asp-Arg-Gly-Asp-Glu-Phe	Defatted rice endosperm protein (REP) digested with neutrase	[18]
	Trp-Val-Tyr-Tyr and Pro-Ser-Leu-Pro-Ala	Hemp seed protein hydrolysate (HPH)	[19]
	Other antioxidant peptides	Douchi extracts, semifermented dry cacao protein fractions, fermentation of rice protein by *Bacillus pumilus*, fermentation of grape must by *Oenococcus oenis*, fermentation of soyabean by *Lactobacillus plantarum*	[20–24]
Antihypertensive peptides	Casein (casokinins)	Milk proteins	[25]
	Lactokinins	Whey proteins	[25]
	Met-Asn-Pro-Pro-Lys and Ile-Thr-Thr-Asn-Pro	Porcine skeletal muscle proteins	[26]
	Gly-Phe-Hyp-Gly-Thr-Hyp-Gly-Leu-Hyp-Gly-Phe	Chicken breast meat extract	[27]
	Met-Asn-Pro-Pro-Lys and Ile-Thr-Thr-Asn-Pro	Porcine myosin	[29]
	Lys-Arg-Glu-Lys-Tyr-Asp-Ile	Porcine troponin	[30]
	Met-Val-Gly-Ser-Ala-Pro-Gly-Val-Leu and Leu-Gly-Pro-Leu-Gly-His-Gln	*Okamejei kenojei*	[31]

(*Continued*)

Table 6.1 **(Continued)**

Application	Peptide	Source	References
	Six dipeptide: Trp-Ala, Val-Trp, Trp-Met, Met-Trp, Ile-Trp, and Leu-Trp	Salmon hydrolysate	[32]
	Thr-Gln-Val-Tyr	Rice protein	[34]
	Val-Asn-Pro and Val-Trp-Pro	Rice protein	[35]
	Arg-Phe and Ile-His-Arg-Phe	Rice glutelin	[36]
	Ile-Val-Tyr	Wheat germ hydrolysate	[37]
	Ile-Ala-Pro	Wheat gliadin hydrolysate	[38]
	Six peptides: Leu-Gln-Pro, Ile-Gln-Pro, Leu-Arg-Pro, Val-Pro, Ile-Tyr, and Thr-Phe	Wheat milling byproducts	[39]
	Leu-Ala-Tyr, Leu-Gln-Pro, Leu-Arg-Pro and Leu-Ser-Pro	Maize endosperm protein (α-zein)	[40]
	Ala-Tyr	Corn gluten meal	[41]
	Lys-Glu-Asp-Asp-Glu-Glu-Glu-Glu-Gln-Gly-Glu-Glu-Glu	Fermentation and hydrolysis of pea seeds protein by *L. plantarum* 299v	[44]
Anticancer/ Antiproliferative peptides	β-Casomorphins (β-CMs), α-casomorphins	Milk protein	[54–56]
	Casein phosphopeptides	Milk protein	[58]
	Gly-Phe-His-Ile, Gly-Leu-Ser-Asp-Gly-Glu-Trp-Gln	Beef sarcoplasmic proteins	[66]
	Anticancer bioactive peptide (ACBP)	Goat spleen	[67]
	Anticancer bioactive peptide-3 (ACBP-3)	Goat liver	[68]
	Leu-Pro-His-Val-Leu-Thr-Pro-Glu-Ala-Gly-Ala-Thr and Pro-Thr-Ala-Glu-Val-Tyr-Met-Val-Thr	Tuna dark muscle	[69]

(*Continued*)

Table 6.1 **(Continued)**

Application	Peptide	Source	References
	TFD100	Pacific cod	[70]
	AE-941	Shark cartilage extract	[71]
	Lunasin	Soybean	[74–77]
	Glu-Gln-Arg-Pro-Arg	Rice bran proteins	[78]
	X-Met-Leu-Pro-Ser-Tyr-Ser-Pro-Tyr	Defatted soy protein	[79]
	Cys-Pro-Ala-Pro	*Chlorella pyrenoidosa*	[80]
	Val-Glu-Cys-Tyr-Gly-Pro-Asn-Arg-Pro-Gln-Phe	*Chlorella vulgaris*	[81]
Antiinflammatory peptides	Val-Pro-Pro	Bacterial fermentation of casein	[87]
	Ile-Arg-Trp and Ile-Gln-Trp	Egg white	[97,98]
	Pro-Ala-Tyr	Salmon pectoral fin	[99]
	Pro-Arg-Arg-Thr-Arg-Met-Met-Asn-Gly-Gly-Arg and Met-Gly-Pro-Ala-Met-Met-Arg-Thr-Met-Pro-Gly	Tuna cooking juice	[100]
	Val-Pro-Tyr	Soybean	[105,106]
	Lunasin	Defatted soybean flour	[82,107,108]
	Leu-Asp-Ala-Val-Asn-Arg (P1) and Met-Met-Leu-Asp-Phe (P2)	*Spirulina maxima*	[109]
	Thalassospiramide G (**1**), A and D (**2**–**3**)	*Thalassospira* sp.	[110]
Antimicrobial peptides	Lactoferrampin	Milk protein	[113]
	Isracidin	Casein	[114]
	Casocidin-I	Casein	[115]
	Gly-Phe-His-Ile, Asp-Phe-His-Ile-Asn-Gly, Phe-His-Gly and Gly-Leu-Ser-Asp-Gly-Glu-Trp-Gln	Bovine meat	[66]
	NKLP27 NK-lysin	*Cynoglossus semilaevis*	[116]
	Cyclic lipopeptide	*Bacillus amyloliquefaciens*	[118]

6.7 Future prospectives

Bioactive peptides are ideal nutraceutical candidates that have garnered considerable attention owing to their diverse health-promoting properties and natural abundance. Though the efficacy of bioactive peptides is less compared with specific drugs, bioactive peptides do not create deleterious and undesired side effects. Although the production of bioactive peptide nutraceuticals was focused entirely on milk and animal protein sources, the possibility of isolation of the same from plant and microbial sources with advances in biotechnology has opened new avenues for the exploration of the same. Apart from the advances in biotechnology, the recent developments in proteomics have enabled the detection and quantification of peptides, permitting sensitive and highly specific identification of bioactive peptide sequences. The emerging technologies in food science for the production of nutraceuticals without compromising safety and consumer acceptance are some challenges still to be addressed. Some antioxidant peptides are permitted for fortification of food products as additives in some countries. However, the antioxidant peptides derived from animal sources are in their infancy and have a long way to go. There is a need for animal experiments and human clinical trials for testing their bioavailability and safe application. Also, the structure–function association of peptides is critical and has to be elucidated and research on, focusing more on the structural details of antioxidant peptides from the different protein sources. For peptide nutraceuticals with an antihypertensive role, research effort to concentrate pure peptide fractions that function as modulators of the renin–angiotensin system in hypertension is essential. Also, the necessity to generate peptide databases on identified bioactive peptide sequences with demonstrated blood pressure-reducing effects in humans or animal models facilitate the modeling of peptide structure–function relationships, thus enabling the location of active peptide sequences in food proteins. The bioactive peptides play a vital role as functional foods that could treat or improve the impact of tumors. The isolation and identification of anticancer peptides is anticipated to create a huge impact in the tumor therapeutic markets. However, high-level molecular research is essential to elucidate the plausible mechanisms of the anticancer activities of bioactive peptides. Further investigations and clinical trials are crucial to improve and develop cost-effective safe bioactive peptides.

References

[1] S.H. Zeisel, Regulation of "nutraceuticals", Science 285 (5435) (1999) 1853–1855.

[2] F. Shahidi, Y. Zhong, Bioactive peptides, J. AOAC Int. 91 (4) (2008) 914–931.

[3] H. Korhonen, A. Pihlanto, Bioactive peptides from food proteins, in: Y.H. Hui (Ed.), Handbook of Food Products Manufacturing: Health, Meat, Milk, Poultry, Seafood, and Vegetables, John Wiley & Sons, Inc., Hoboken, NJ, 2007, pp. 5–38.

[4] M. Gobbetti, F. Minervini, C.G. Rizzello, Bioactive peptides in dairy products, in: Y.H. Hui (Ed.), Handbook of Food Products Manufacturing: Health, Meat, Milk, Poultry, Seafood, and Vegetables, John Wiley & Sons, Inc., Hoboken, NJ, 2007, pp. 489–517.

[5] J. Carrasco-Castilla, A.J. Hernandez-Alvarez, C. Jimenez-Martınez, et al., Use of proteomics and peptidomics methods in food bioactive peptide science and engineering, Food Eng. Rev. 4 (4) (2012) 224–243.
[6] R. Hartmann, H. Meisel, Food-derived peptides with biological activity: from research to food applications, Curr. Opin. Biotechnol. 18 (2) (2007) 163–169.
[7] C.G. Rizzello, D. Tagliazucchi, E. Babini, et al., Bioactive peptides from vegetable food matrices: research trends and novel biotechnologies for synthesis and recovery, J. Funct. Foods 27 (2016) 549–569.
[8] M. Valko, M. Izakovic, M. Mazur, et al., Role of oxygen radicals in DNA damage and cancer incidence, Mol. Cell. Biochem. 266 (1–2) (2004) 37–56.
[9] S. Ranathunga, N. Rajapakse, S.K. Kim, Purification and characterization of antioxidative peptide derived from muscle of conger eel (*Conger myriaster*), Eur. Food Res. Technol. 222 (3) (2006) 310–315.
[10] S. Zhou, E.A. Decker, Ability of amino acids, dipeptides, polyamines, and sulfhydryls to quench hexanal, a saturated aldehydic lipid oxidation product, J. Agric. Food Chem. 47 (5) (1999) 1932–1936.
[11] H.M. Chen, K. Muramoto, F. Yamauchi, Structural analysis of antioxidative peptides from soybean beta-conglycinin, J. Agric. Food Chem. 43 (3) (1995) 574–578.
[12] G. Bounous, P. Gold, The biological activity of undenatured dietary whey proteins: role of glutathione, Clin. Invest. Med. 14 (4) (1991) 296–309.
[13] Q. Wang, W. Li, Y. He, et al., Novel antioxidative peptides from the protein hydrolysate of oysters (*Crassostrea talienwhanensis*), Food Chem. 145 (2014) 991–996.
[14] R.A. Nazeer, N.S. Kumar, R. Jai Ganesh, In vitro and in vivo studies on the antioxidant activity of fish peptide isolated from the croaker (*Otolithes ruber*) muscle protein hydrolysate, Peptides 35 (2) (2012) 261–268.
[15] C.F. Chi, F.Y. Hu, B. Wang, et al., Purification and characterization of three antioxidant peptides from protein hydrolyzate of croceine croaker (*Pseudosciaena crocea*) muscle, Food Chem. 168 (2015) 662–667.
[16] M.C. García, P. Puchalska, C. Esteve, et al., Vegetable foods: a cheap source of proteins and peptides with anti-hypertensive, antioxidant, and other less occurrence bioactivities, Talanta 106 (2013) 328–349.
[17] A.P. Adebiyi, A.O. Adebiyi, J. Yamashita, et al., Purification and characterization of antioxidative peptides derived from rice bran protein hydrolysates, Eur. Food Res. Technol. 228 (2009) 553–563.
[18] J. Zhang, H. Zhang, L. Wang, et al., Isolation and identification of antioxidative peptide from rice endosperm protein enzymatic hydrolysate by consecutive chromatography and MALDI-TOF/TOF MS/MS, Food Chem. 119 (1) (2010) 226–234.
[19] A.T. Girgih, R. He, S.A. Malomo, et al., Structural and functional characterization of hemp seed (*Cannabis sativa* L.) protein-derived antioxidant and anti-hypertensive peptides, J. Funct. Foods 6 (2014) 384–394.
[20] D. Wang, L. Wang, F. Zhu, et al., In vitro and in vivo studies on the antioxidant activities of the aqueous extracts of Douchi (a traditional Chinese salt-fermented soybean food), Food Chem. 107 (4) (2008) 1421–1428.
[21] A.M. Preza, M.E. Jaramillo, A.M. Puebla, et al., Antitumor activity against murine lymphoma L5178Y model of proteins from cacao (*Theobroma cacao* L.) seeds in relation with in vitro antioxidant activity, BMC Complement. Altern. Med. 10 (2010) 61.
[22] L. Dei Più, A. Tassoni, D.I. Serrazanetti, et al., Exploitation of starch industry by-product to produce bioactive peptides from rice hydrolyzed proteins, Food Chem. 155 (2014) 199–206.

[23] G.R. Apud, M.J. Vaquero, G. Rollan, et al., Increase in antioxidant and anti-hypertensive peptides from Argentinean wines by *Oenococcus oeni*, Int. J. Food Microbiol. 163 (2–3) (2013) 166–170.

[24] B.P. Singh, S. Vij, S. Hati, Functional significance of bioactive peptides derived from soybean, Peptides 54 (2014) 171–179.

[25] Y.W. Park, M.S. Nam, Bioactive peptides in milk and dairy products: a review., Korean J. Food Sci. Anim. Resour. 35 (6) (2015) 831–840.

[26] Y. Nakashima, K. Arihara, A. Sasaki, et al., Anti-hypertensive activities of peptides derived from porcine skeletal muscle myosin in spontaneously hypertensive rats, J. Food Sci. 67 (1) (2002) 434–437.

[27] A. Saiga, T. Okumura, T. Makihara, et al., Angiotensin I-converting enzymes inhibitory peptides in a hydrolyzed chicken breast muscle extract, J. Agric. Food Chem. 51 (6) (2003) 1741–1745.

[28] F.Y. Cheng, Y.T. Liu, T.C. Wan, et al., The development of angiotensin I-converting enzyme inhibitor derived from chicken bone protein, Anim. Sci. J. 79 (1) (2008) 122–128.

[29] K. Arihara, Y. Nakashima, T. Mukai, et al., Peptide inhibitors for angiotensin I-converting enzyme from enzymatic hydrolysates of porcine skeletal muscle proteins, Meat Sci. 57 (3) (2001) 319–324.

[30] K. Katayama, H.E. Anggraeni, T. Mori, et al., Porcine skeletal muscle troponin is a good source of peptides with angiotensin-I converting enzyme inhibitory activity and anti hypertensive effects in spontaneously hypertensive rats, J. Agric. Food Chem. 56 (2) (2008) 355–360.

[31] D.H. Ngo, B. Ryu, S.K. Kim, Active peptides from skate (*Okamejei kenojei*) skin gelatin diminish angio tensin-I converting enzyme activity and intracellular free radical-mediated oxidation, Food Chem. 143 (2014) 246–255.

[32] S. Ono, M. Hosokawa, K. Miyashita, et al., Isolation of peptides with angiotensin I-converting enzyme inhibitory effect derived from hydrolyzate of upstream chum salmon muscle, J. Food Sci. 68 (5) (2003) 1611–1614.

[33] I.R. Amado, J.A. Vazquez, P. Gonzalez, et al., Identification of the major ACE-inhibitory peptides produced by enzymatic hydrolysis of a protein concentrate from cuttlefish wastewater, Mar. Drugs 12 (3) (2014) 1390–1405.

[34] G.H. Li, M.R. Qu, J.Z. Wan, et al., Anti-hypertensive effect of rice protein hydrolysate with in vitro angiotensin I-converting enzyme inhibitory activity in spontaneously hypertensive rats, Asia Pac. J. Clin. Nutr. 16 (Suppl 1) (2007) 275–280.

[35] J. Chen, S. Liu, R. Ye, et al., Angiotensin-I converting enzyme (ACE) inhibitory tripeptides from rice protein hydrolysate: purification and characterization, J. Funct. Foods 5 (4) (2013) 1684–1692.

[36] N. Kontani, R. Omae, T. Kagebayashi, et al., Characterization of Ile-His-Arg-Phe, a novel rice-derived vasorelaxing peptide with hypotensive and anorexigenic activities, Mol. Nutr. Food Res. 58 (2) (2014) 359–364.

[37] T. Matsui, C.H. Li, Y. Osajima, Preparation and characterization of novel bioactive peptides responsible for angiotensin I-converting enzyme inhibition from wheat germ, J. Pept. Sci. 5 (7) (1999) 289–297.

[38] H. Motoi, T. Kodama, Isolation and characterization of angiotensin I-converting enzyme inhibitory peptides from wheat gliadin hydrolysate, Mol. Nutr. Food Res. 47 (5) (2003) 354–358.

[39] Y. Nogata, T. Nagamine, M. Yanaka, et al., Angiotensin I-converting inhibitory peptides produced by autolysis reaction from wheat bran, J. Agric. Food Chem. 57 (15) (2009) 6618–6622.

[40] S. Miyoshi, H. Ishikawa, T. Kaneko, et al., Structures and activity of angiotensin converting enzyme inhibitors in an α-zein hydrolysate, Agric. Biol. Chem. 55 (5) (1991) 1313–1318.

[41] Y. Yang, G. Tao, P. Liu, et al., Peptide with angiotensin I-converting enzyme inhibitory activity from hydrolyzed corn gluten meal, J. Agric. Food Chem. 55 (19) (2007) 7891–7895.

[42] C.G. Rizzello, M. De Angelis, R. Di Cagno, et al., Highly efficient gluten degradation by lactobacilli and fungal proteases during food processing: new perspectives for celiac disease, Appl. Environ. Microbiol. 73 (14) (2007) 4499–4507.

[43] C.G. Rizzello, A. Cassone, R. Di Cagno, et al., Synthesis of angiotensin I-converting enzyme (ACE)-inhibitory peptides and γ-aminobutyric acid (GABA) during sourdough fermentation by selected lactic acid bacteria, J. Agric. Food Chem. 56 (16) (2008) 6936–6943.

[44] A. Jakubczyk, M. Karas, B. Baraniak, et al., The impact of fermentation and in vitro digestion on formation angiotensin converting enzyme (ACE) inhibitory peptides from pea proteins, Food Chem. 141 (4) (2013) 3774–3780.

[45] T. Tanaka, Chemoprevention of human cancer: biology and therapy, Crit. Rev. Oncol. Hematol. 25 (3) (1997) 139–174.

[46] J.A. Meyerhardt, R.J. Mayer, Systemic therapy for colorectal cancer, N. Engl. J. Med. 352 (5) (2005) 476–487.

[47] J.B. O'Connell, M.A. Maggard, C.Y. Ko, Colon cancer survival rates with the new American Joint Committee on Cancer sixth edition staging, J. Natl. Cancer Inst. 96 (19) (2004) 1420–1425.

[48] K. Ashida, H. Sazaki, Y.A. Suzuki, et al., Cellular internalization of lactoferrin in intestinal epithelial cells, Biometals 17 (3) (2004) 311–315.

[49] Y.A. Suzuki, B. Lopez, B. Lonnerdal, Mammalian lactoferrin receptors: structure and function, Cell. Mol. Life Sci. 62 (22) (2005) 2560–2575.

[50] K.A. Son, J. Park, C.K. Chung, et al., Human lactoferrin activates transcription of IL-1 beta gene in mammalian cells, Biochem. Biophys. Res. Commun. 290 (1) (2002) 236–241.

[51] S.A. Gonzalez-Chavez, S. Arevalo-Gallegos, Q. Rascon-Cruz, Lactoferrin: structure, function and applications, Int. J. Antimicrob. Agents 33 (2009) 301–306.

[52] B. Lonnerdal, S. Iyer, Lactoferrin—molecular structure and biological function, Annu. Rev. Nutr. 15 (1995) 93–110.

[53] L. Rodirgues, J. Teixera, S. Fernando, et al., Lactoferin and cancer disease prevention, Crit. Rev. Food Sci. Nutr. 49 (3) (2009) 203–217.

[54] Z. Sun, Z. Zhang, X. Wang, et al., Relation of β-casomorphin to apnea in sudden infant death syndrome, Peptides 24 (6) (2003) 937–943.

[55] A.K. Tandon, G.M. Clark, G.C. Chamness, et al., Cathepsin D and prognosis in breast cancer, N. Engl. J. Med. 322 (5) (1990) 297–302.

[56] F. Jänicke, M. Schmitt, L. Pache, et al., Urokinase (uPA) and its inhibitor PAI-1 are strong and independent prognostic factors in node-negative breast cancer, Breast Cancer Res. Treat. 24 (3) (1992) 195–208.

[57] E. Sienkiewicz-Szlapka, B. Jarmolowska, S. Krawczuk, et al., Transport of bovine milk-derived opioid peptides across a Caco-2 monolayer, Int. Dairy J. 19 (4) (2009) 252–257.

[58] R. Berrocal, S. Chanton, M.A. Juillerat, et al., Tryptic phosphopeptides from whole casein. II. Physicochemical properties related to the solubilization of calcium, J. Dairy Res. 56 (3) (1989) 335–341.

[59] A. Zawadzki, Q. Liu, Y. Wang, et al., Verapamil inhibits L-type calcium channel mediated apoptosis in human colon cancer cells, Dis. Colon Rectum 51 (11) (2008) 1696–1702.

[60] S. Perego, S. Cosentino, A. Fiorilli, et al., Casein phosphopeptides modulate proliferation and apoptosis in HT-29 cell line through their interaction with voltage-operated L-type calcium channels, J. Nutr. Biochem. 23 (7) (2012) 808–816.

[61] D.W. Hoskin, A. Ramamoorthy, Studies on anticancer activities of antimicrobial peptides, Biochim. Biophys. Acta 1778 (2) (2008) 357–375.

[62] M. Shimamura, Y. Yamamoto, H. Ashino, et al., Bovine lactoferrin inhibits tumor-induced angiogenesis, Int. J. Cancer 111 (1) (2004) 111–116.

[63] Y. Zhang, C.F. Lima, L.R. Rodrigues, In vitro evaluation of bovine lactoferrin potential as an anticancer agent, Int. Dairy J. 40 (2015) 6–15.

[64] L. Najafian, A.S. Babji, A review of fish-derived antioxidant and antimicrobial peptides: their production, assessment and applications, Peptides 33 (1) (2012) 178–185.

[65] J.T. Ryan, R.P. Ross, D. Bolton, et al., Bioactive peptides from muscle sources: meat and fish, Nutrients 3 (9) (2011) 765–791.

[66] A. Jang, C. Jo, K.S. Kang, et al., Antimicrobial and human cancer cell cytotoxic effect of synthetic angiotensin-converting enzyme (ACE) inhibitory peptides, Food Chem. 107 (1) (2008) 327–336.

[67] X. Su, C. Dong, J. Zhang, et al., Combination therapy of anti-cancer bioactive peptide with Cisplatin decreases chemotherapy dosing and toxicity to improve the quality of life in xenograft nude mice bearing human gastric cancer, Cell Biosci. 4 (1) (2014) 7.

[68] L. Yu, L. Yang, W. An, et al., Anticancer bioactive peptide-3 inhibits human gastric cancer growth by suppressing gastric cancer stem cells, J. Cell. Biochem. 115 (4) (2014) 697–711.

[69] K.C. Hsu, E.C.Y. Li-Chan, C.L. Jao, Antiproliferative activity of peptides prepared from enzymatic hydrolysates of tuna dark muscle on human breast cancer cell line MCF-7, Food Chem. 126 (2) (2011) 617–622.

[70] P. Guha, E. Kaptan, G. Bandyopadhyaya, et al., Cod glycopeptide with picomolar affinity to galectin-3 suppresses T-cell apoptosis and prostate cancer metastasis, Proc. Natl. Acad. Sci. U.S.A. 110 (13) (2013) 5052–5057.

[71] R.M. Bukowski, AE-941, a multifunctional antiangiogenic compound: trials in renal cell carcinoma, Expert Opin. Investig. Drugs 12 (8) (2003) 1403–1411.

[72] L. Zheng, P. Ling, Z. Wang, et al., A novel polypeptide from shark cartilage with potent anti-angiogenic activity, Cancer Biol. Ther. 6 (5) (2007) 775–780.

[73] A. Rabbani-Chadegani, S. Abdossamadi, A. Bargahi, et al., Identification of low-molecular weight protein (SCP1) from shark cartilage with anti-angiogenesis activity and sequence similarity to parvalbumin, J. Pharm. Biomed. Anal. 46 (3) (2008) 563–567.

[74] A.F. Galvez, B.O. de Lumen, A soybean cDNA encoding a chromatin-binding peptide inhibits mitosis of mammalian cells, Nat. Biotechnol. 17 (5) (1999) 495–500.

[75] A.F. Galvez, N. Chen, J. Macasieb, et al., Chemopreventive property of a soybean peptide (lunasin) that binds to deacetylated histones and inhibits acetylation, Cancer Res. 61 (20) (2001) 7473–7478.

[76] E.G. de Mejia, W.Y. Wang, V.P. Dia, Lunasin, with an arginine–glycine–aspartic acid motif, causes apoptosis to L1210 leukemia cells by activation of caspase-3, Mol. Nutr. Food Res. 54 (3) (2010) 406–414.

[77] V.P. Dia, E.G. de Mejia, Lunasin promotes apoptosis in human colon cancer cells by mitochondrial pathway activation and induction of nuclear clusterin expression, Cancer Lett. 295 (1) (2010) 44–53.

[78] A. Kannan, N.S. Hettiarachchy, J.O. Lay, et al., Human cancer cell proliferation inhibition by a pentapeptide isolated and characterized from rice bran, Peptides 31 (9) (2010) 1629−1634.
[79] S.E. Kim, H.H. Kim, J.Y. Kim, et al., Anticancer activity of hydrophobic peptides from soy proteins, Biofactors (Oxford, England) 12 (1−4) (2000) 151−155.
[80] X. Wang, X. Zhang, Separation, antitumor activities, and encapsulation of polypeptide from *Chlorella pyrenoidosa*, Biotechnol. Prog. 29 (3) (2013) 681−687.
[81] I.C. Sheih, T.J. Fang, T.K. Wu, et al., Anticancer and antioxidant activities of the peptide fraction from algae protein in waste, J. Agric. Food Chem. 58 (2) (2010) 1202−1207.
[82] V.P. Dia, W. Wanga, V.L. Oh, et al., Isolation, purification and characterisation of lunasin from defatted soybean flour and in vitro evaluation of its anti-inflammatory activity, Food Chem. 114 (1) (2009) 108−115.
[83] C.G. Rizzello, L. Nionelli, R. Coda, et al., Synthesis of the cancer preventive peptide lunasin by lactic acid bacteria during sourdough fermentation, Nutr. Cancer 64 (1) (2012) 111−120.
[84] C.G. Rizzello, B. Hernández-Ledesma, S. Fernández-Tomé, et al., Italian legumes: effect of sourdough fermentation on lunasin-like polypeptides, Microb. Cell Fact. 14 (2015) 168.
[85] K. Majumder, Y. Mine, J. Wu, The potential of food protein derived anti-inflammatory peptides against various chronic inflammatory diseases, J. Sci. Food Agric. 96 (7) (2016) 2303−2311.
[86] S. Chakrabarti, F. Jahandideh, J. Wu, Food-derived bioactive peptides on inflammation and oxidative stress, BioMed Res. Int. 2014 (2014) 11. Available from: http://dx.doi.org/10.1155/2014/608979.
[87] K. Aihara, H. Ishii, M. Yoshida, Casein-derived tripeptide, Val-Pro-Pro (VPP), modulates monocyte adhesion to vascular endothelium, J. Atheroscler. Thromb. 16 (5) (2009) 594−603.
[88] T. Nakamura, T. Hirota, K. Mizushima, et al., Milk-derived peptides, Val-Pro-Pro and Ile-Pro-Pro,atten uate atherosclerosis development in apolipoprotein e-deficient mice: a preliminary study, J. Med. Food 16 (5) (2013) 396−403.
[89] D.S. Nielsen, P.K. Theil, L.B. Larsen, et al., Effect of milk hydrolysates on inflammation markers and drug-induced transcriptional alterations in cell-based models, J. Anim. Sci. 90 (Suppl 4) (2012) 403−405.
[90] M.M. Iskandar, N. Dauletbaev, S. Kubow, et al., Whey protein hydrolysates decrease IL-8 secretion in lipopolysaccharide (LPS)-stimulated respiratory epithelial cells by affecting LPS binding to Toll-like receptor 4, Br. J. Nutr. 110 (1) (2013) 58−68.
[91] A.F. Piccolomini, M.M. Iskandar, L.C. Lands, et al., High hydrostatic pressure pretreatment of whey proteins enhances whey protein hydrolysate inhibition of oxidative stress and IL-8 secretion in intestinal epithelial cells, Food Nutr. Res. (2012) 56. Available from: http://dx.doi.org/10.3402/fnr.v56i0.17549.
[92] N. Shimizu, K. Dairiki, S. Ogawa, et al., Dietary whey protein hydrolysate suppresses development of atopic dermatitis-like skin lesions induced by mite antigen in NC/Nga mice, Allergol. Int. 55 (2) (2006) 185−189.
[93] M. Pescuma, M.B. Espeche Turbay, F. Mozzi, et al., Diversity in proteinase specificity of thermophilic lactobacilli as revealed by hydrolysis of dairy and vegetable proteins, Appl. Microbiol. Biotechnol. 97 (17) (2013) 7831−7844.
[94] M. Hatori, K. Ohki, S.I. Hirano, et al., Effects of a casein hydrolysate prepared from *Aspergillus oryzae* protease on adjuvant arthritis in rats, Biosci. Biotechnol. Biochem. 72 (8) (2008) 1983−1991.

[95] J.S. Kim, M.B. Ellman, D. Yan, et al., Lactoferricin mediates anti-inflammatory and anti-catabolic effects via inhibition of IL-1 and LPS activity in the intervertebral disc, J. Cell. Physiol. 228 (9) (2013) 1884–1896.
[96] D. Yan, D. Chen, J. Shen, et al., Bovine lactoferricin is anti-inflammatory and anti-catabolic in human articular cartilage and synovium, J. Cell. Physiol. 228 (2) (2013) 447–456.
[97] W. Huang, S. Chakrabarti, K. Majumder, et al., Egg-derived peptide IRW inhibits TNF-α-induced inflammatory response and oxidative stress in endothelial cells, J. Agric. Food Chem. 58 (20) (2010) 10840–10846.
[98] K. Majumder, S. Chakrabarti, S.T. Davidge, et al., Structure and activity study of egg protein ovotrans ferrin derived peptides (IRW and IQW) on endothelial inflammatory response and oxidative stress, J. Agric. Food Chem. 61 (9) (2013) 2120–2129.
[99] C.B. Ahn, Y.S. Cho, J.Y. Je, Purification and anti-inflammatory action of tripeptide from salmon pectoral fin byproduct protein hydrolysate, Food Chem. 168 (2015) 151–156.
[100] M.L. Cheng, H.C. Wang, K.C. Hsu, et al., Anti-inflammatory peptides from enzymatic hydrolysates of tuna cooking juice, Food Agric. Immunol. 26 (6) (2015) 770–781.
[101] V.P. Dia, N.A. Bringe, E.G. de Mejia, Peptides in pepsin–pancreatin hydrolysates from commercially available soy products that inhibit lipopolysaccharide-induced inflammation in macrophages, Food Chem. 152 (2014) 423–431.
[102] M.M. Shahi, M.R. Rashidi, S. Mahboob, et al., Protective effect of soy protein on collagen-induced arthritis in rat, Rheumatol. Int. 32 (8) (2012) 2407–2414.
[103] M.G. Vernaza, V.P. Dia, E. Gonzalez de Mejia, et al., Antioxidant and antiinflammatory properties of germinated and hydrolysed Brazilian soybean flours, Food Chem. 134 (4) (2012) 2217–2225.
[104] J.S. Hwang, H.J. Yoo, H.J. Songa, et al., Inflammation-related signalling pathways implicating TGFβ are revealed in the expression profiling of MCF7 cell treated with fermented soybean, Chungkookjang, Nutr. Cancer 63 (4) (2011) 645–652.
[105] S. Wada, K. Sato, R. Ohta, et al., Ingestion of low dose pyroglutamyl leucine improves dextran sulfate sodium-induced colitis and intestinal microbiota in mice, J. Agric. Food Chem. 61 (37) (2013) 8807–8813.
[106] M. Sato, T. Hosokawa, T. Yamaguchi, et al., Angiotensin I-converting enzyme inhibitory peptides derived fromWakame (*Undaria pinnatifida*) and their anti-hypertensive effect in spontaneously hypertensive rats, J. Agric. Food Chem. 50 (21) (2002) 6245–6252.
[107] B. Hernandez-Ledesma, C.C. Hsieh, B.O. de Lumen, Antioxidant and anti-inflammatory properties of cancer preventive peptide lunasin in RAW 264.7 macrophages, Biochem. Biophys. Res. Commun. 390 (3) (2009) 803–808.
[108] A. Cam, E.G. de Mejia, RGD-peptide lunasin inhibits Akt-mediated NF-κB activation in human macrophages through interaction with the $\alpha V\beta 3$ integrin, Mol. Nutr. Food Res. 56 (10) (2012) 1569–1581.
[109] T.S. Vo, D.H. Ngo, K.H. Kang, et al., The role of peptides derived from *Spirulina maxima* in downregulation of fFcεRI-mediated allergic responses, Mol. Nutr. Food Res. 58 (2014) 2226–2234.
[110] S. Um, Y. Pyee, E.H. Kim, et al., Thalassospiramide G, a new gamma-amino-acid-bearing peptide from the marine bacterium *Thalassospira* sp, Mar. Drugs 11 (3) (2013) 611–622.
[111] T. Rydlo, J. Miltz, A. Mor, Eukaryotic antimicrobial peptides: promises and premises in food safety, J. Food Sci. 71 (9) (2006) R125–R135.

[112] J.I. Nagao, S.M. Asaduzzaman, Y. Aso, et al., Lantibiotics: insight and foresight for new paradigm, J. Biosci. Bioeng. 102 (3) (2006) 139–149.

[113] M.I. Van der Kraan, K. Nazmi, A. Teeken, et al., Lactoferrampin, an antimicrobial peptide of bovine lactoferrin, exerts its candidacidal activity by a cluster of positively charged residues at the C-terminus in combination with a helix-facilitating N-terminal part, Biol. Chem. 386 (2) (2005) 137–142.

[114] E. Lahov, W. Regelson, Antibacterial and immunostimulating casein-derived substances from milk: caseicidin, isracidin peptides, Food Chem. Toxicol. 34 (1) (1996) 131–145.

[115] H.D. Zucht, M. Raida, K. Adermann, et al., Casocidin-I: a casein αs2-derived peptide exhibits antibacterial activity, FEBS Lett. 372 (2–3) (1995) 185–188.

[116] M. Zhang, M.F. Li, L. Sun, NKLP27: a teleost NK-lysin peptide that modulates immune response, induces degradation of bacterial DNA, and inhibits bacterial and viral infection, PLoS One 9 (9) (2014) e106543.

[117] C.L. Qin, W. Huang, S.Q. Zhou, et al., Characterization of a novel antimicrobial peptide with chiting-biding domain from *Mytilus coruscus*, Fish Shellfish Immunol. 41 (2) (2014) 362–370.

[118] B. Song, Y.J. Rong, M.X. Zhao, et al., Antifungal activity of the lipopeptides produced by *Bacillus amyloliquefaciens* anti-CA against *Candida albicans* isolated from clinic, Appl. Microbiol. Biotechnol. 97 (16) (2013) 7141–7150.

[119] C.G. Rizzello, A. Lavecchia, V. Gramaglia, et al., Long-term fungal inhibitory activity of *Pisum sativum* flour hydrolysate during storage of wheat flour bread, Appl. Environ. Microbiol. 81 (12) (2015) 4195–4206.

[120] C.G. Rizzello, A. Cassone, R. Coda, et al., Antifungal activity of sourdough fermented wheat germ used as an ingredient for bread making, Food Chem. 127 (3) (2011) 952–959.

[121] R. Coda, A. Cassone, C.G. Rizzello, et al., Antifungal activity of *Wickerhamomyces anomalus* and *Lactobacillus plantarum* during sourdough fermentation: identification of novel compounds and longterm effect during storage of wheat bread, Appl. Environ. Microbiol. 77 (10) (2011) 3484–3492.

[122] R. Coda, C.G. Rizzello, F. Nigro, et al., Long-term fungi inhibitory activity of water-soluble extract from *Phaseolus vulgaris* cv Pinto and sourdough lactic acid bacteria during bread storage, Appl. Environ. Microbiol. 74 (23) (2008) 7391–7398.

Peptoid applications in biomedicine and nanotechnology

7

Jing Sun and Zhibo Li
Qingdao University of Science and Technology, Qingdao, China

7.1 Introduction

Recent advances in synthetic chemistry offer a great opportunity to generate biomimetic structures and functions from synthetic molecules. Peptoids, also referred to as *N*-substituted glycines, are peptidomimetic polymers that offer many advantages for application in biomedicine and nanoscience [1–10]. The peptoid backbone is identical to that of a polyglycine, one of the polypeptides, but the side chain is covalently attached to the amide nitrogen. This difference removes the hydrogen bonding donor in the backbone and the main chain chirality [11,12]. Unlike polypeptides, the side chain identity and monomer sequence of the pepotids, instead of hydrogen bonding, dominates the properties of peptoids, which allows for simplicity of engineering and design, good solubility in many common solvents, and accessible thermal processability [13]. These advantages make peptoids potential candidates for biomedical applications with excellent biocompatibility and potent biological activities [14,15].

Two fundamentally different synthetic approaches are generally used to prepare peptoids: solid-phase submonomer synthesis and polymerization in solution. Zuckermann et al. developed a two-step submonomer synthetic method that excludes main chain protecting groups based on solid-phase peptide synthesis (SPPS) [16]. The peptoids are readily obtained for chains of shorter lengths [17–20]. The precise control over sequence and near absolute monodispersity enables high tunability on the functional properties of the peptoids [11,21,22]. As the side chain group is introduced from a primary amine via an S_N2 displacement reaction, a tremendous range of chemical diversity of peptoids is accessible because hundreds of primary amines are readily available. With this approach, peptoids of up to 50 monomers can be easily synthesized [7,23]. In contrast, polymerization techniques can achieve high molecular weight of polypeptoids, but the control over the sequence is limited [24]. Therefore the combination of solid-phase submonomer synthesis with polymerization techniques offers excellent opportunities to tailor peptoid design for various applications in biomedicine and nanomaterial science. In this chapter we will discuss recent developments in synthesis and function of peptoids, focusing on applications in biomedicine and nanotechology.

Peptide Applications in Biomedicine, Biotechnology and Bioengineering.
DOI: http://dx.doi.org/10.1016/B978-0-08-100736-5.00007-7

7.2 Synthesis and functional properties of peptoids

7.2.1 Solid-phase synthesis of sequence-specific peptoids

Based on the well-established Merrifield method of SPPS, Zuckermann et al. developed a two-step method to synthesize peptoids using *N*-protected *N*-substituted glycine monomers without main chain protecting groups [1,2]. The two-step monomer addition cycle involves an acylation step with a haloacetic acid (bromoacetic acid is preferred), followed by a displacement reaction with a primary amine, which incorporates the functional side chains (Fig. 7.1) [16,25,26]. In the S_N2 displacement step, the functional groups are incorporated into the side chains by reactive primary amines. Therefore, hundreds of commercial available primary amines enable a wide range of chemical and sequence diversity in peptoids. The coupling efficiency is generally >99% per monomer addition cycle, which allows for up to 50 monomers to be synthesized with good yields. Chemical coupling has also been used to obtain long chains of peptoids [27]. The displacement reaction is not air or moisture sensitive and no heating is typically required. Additionally, the two-step submonomer cycle involves only the simple pipetting of stable reagent solutions at room temperature. Therefore, the entire procedure can be accomplished either manually or automatically, and can be readily adapted to almost any commercial peptide synthesizer [3,18,28].

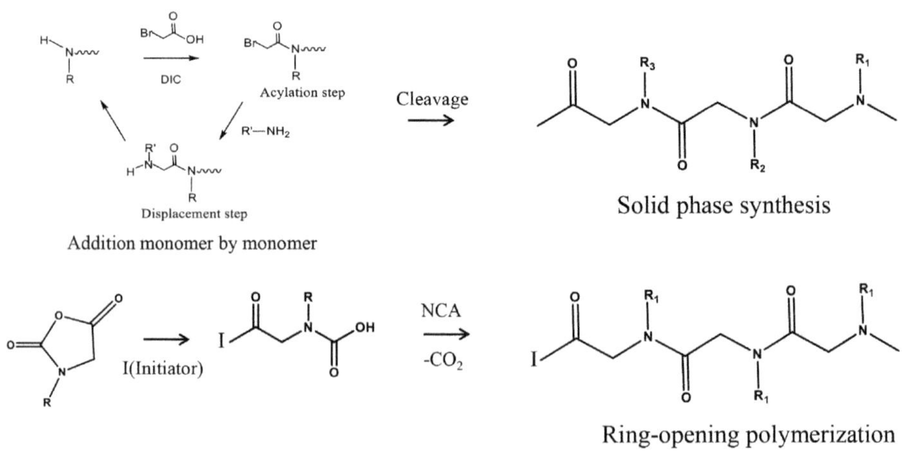

Figure 7.1 The solid-phase submonomer method allows for the synthesis of sequence-specific peptoids from cheap and readily available materials (amines and haloacetic acid submonomers). A repeating two-step of monomer addition cycle of acylation followed by nucleophilic S_N2 displacement is performed at room temperature, and can be performed manually or by automated synthesizers. The living ring-opening polymerization of *N*-substituted *N*-carboxy anhydride monomers with a primary amine as an initiator is used for the synthesis of polypeptoids. DIC = *N*,*N'*-diisopropylcarbodiimide.

7.2.2 Solution polymerization of polypeptoids

In addition to a solid-phase synthetic approach, the ring opening polymerization (ROP) has been used to synthesize polypeptoids from *N*-substituted carboxyanhydride (NCA) monomers (Fig. 7.1) [29]. Poly(*N*-methyl glycine), also referred to as polysarcosine, was first synthesized due to its synthetic accessibility [30–32]. It is reported that polysarcosine has excellent biocompatibility and water solubility, which has been used for the application of drug delivery [33–37]. In recent years, polypeptoids with a variety of side chains have been reported.

Zhang's group [24,38–40] has approached *N*-heterocyclic carbine (NHC)-mediated polymerization of *N*-substituted NCA monomers (Fig. 7.2). The molecular weight and polymerization efficiency can be well controlled in solvents with low dielectric constants. It is further observed that cyclic polymer chains can be obtained due to the presence of a cyclic topology of zwitterionic intermediates. The cyclic structure is confirmed by atomic force microscopy (AFM) with modification of poly(ethylene glycol) [41]. The primary amines were also used to initiate a series of polypeptoid polymerizations with alkyl side chains by Luxenhofer et al. [42,43].

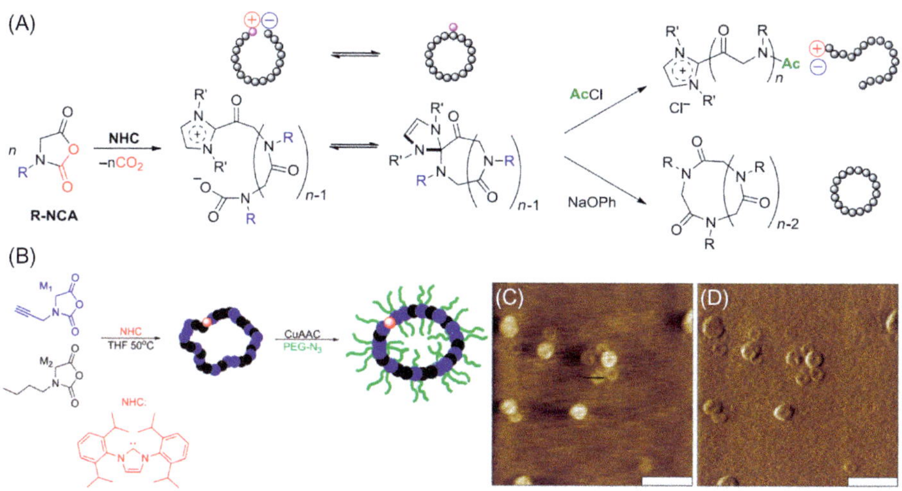

Figure 7.2 (A) The linear and cyclic polypeptoids from *N*-heterocyclic carbene-mediated polymerization of *N*-substituted *N*-carboxyanhydrides with different end-capping reagents. R represents the side chain. (B) Scheme of a large cycle of brush-like polypeptoids from click chemistry with PEG; (C) and (D) representative AFM topographic and amplitude images of the cyclic structure with scale bar of 500 nm.
Source: (A) Reproduced with permission from J. Sun, R.N. Zuckermann, Peptoid polymers: a highly designable bioinspired material, ACS Nano 7 (2013) 4715–4732. Copyright (2013) ACS. (B)–(D) Reproduced with permission from S.H. Lahasky, W.K. Serem, L. Guo, J.C. Garno, D. Zhang, Synthesis and characterization of cyclic brush-like polymers by N-heterocyclic carbene-mediated zwitterionic polymerization of N-propargyl N-carboxyanhydride and the grafting-to approach, Macromolecules 44 (2011) 9063–9074 [41]. Copyright (2011) ACS.

The polymerization kinetics shows that it occurs in a controlled manner, by which the polymer molecular weights are well controlled and obey a Poisson distribution. A series of block copolypeptoids with different aliphatic side chains were further prepared by successive ROP of *N*-substituted NCAs, and their thermal properties were systematically studied [44]. Schlaad and coworkers [45,46] synthesized well-defined glycopeptoids by ROP of *N*-allylglycine NCA under homo- or heterophase conditions and subsequent postmodification. Instead of *N*-substituted NCAs, *N*-substituted glycine *N*-thiocarboxyanhydrides (*N*-substituted NTAs) were also used to synthesize homopolypeptoids and copolypeptoids by Ling et al. Polymerizations of *N*-substituted NTAs required higher temperatures due to lower reactivity of *N*-substituted NTAs [47]. In addition, polypeptoids have also been synthesized by metal-catalyzed alternating copolymerization with imines and carbon monoxide (CO) as monomers [48,49].

7.2.3 Secondary structure in sequence-specific peptoids

The peptoid backbone is generally flexible due to lack of main chain chirality and hydrogen bonding donor [21,50]. It was reported that the persistence length of (*R*)-*N*-(1-phenylethyl)glycine-containing peptoids is determined to be 0.5–1 nm, which is similar to that of polystyrene (1 nm) [51]. Unlike the peptide that prefers *trans* amide bond, the conformational heterogeneity of the peptoid backbone is dependent on the *cis/trans* isomerization of the tertiary backbone amides [52]. It has been reported that the chemical structure of side chains can effectively control the *cis* versus *trans* geometry of the backbone amide bonds (Fig. 7.3) [54,58–61]. An oligo (*S*)-*N*-(1-phenylyethyl)glycine (Nspe) helix containing bulky α-chiral side chains is entirely composed of *cis*-amide bonds with a periodicity of three residues per turn, determined by both theoretical [53,62] and experimental [52] studies. In addition to the chiral aromatic side chain, a short peptoid with an α-chiral aliphatic side chain ((*R*)-*N*-(1-cyclohexylethyl)glycine), can also form a left-handed helix with all *cis*-amide bonds [63]. It is demonstrated that all these peptoid helices are stabilized by steric interactions of side chains [64]. Blackwell et al. demonstrated that an $n \rightarrow \pi^*$ interaction between the carbonyl oxygen of the backbone and the π^* orbital of an adjacent aromatic side chain can result in a stabilized *cis* amide conformation [59]. Kirshenbaum and coworkers obtained 90% *trans*-amide conformer in the peptoid backbone by incorporating *N*-aryl side chains [61]. A variety of polar and reactive functional groups have been introduced into chiral, aromatic amine submonomers, which were used to create covalent helix dimers and to gain increased water solubility. Peptoid loops [65] and turns [66,67] have also been discovered to establish new structural scaffolds. An unusual "threaded loop" structure was reported with a peptoid nonamer that contains bulky, α-chiral side chains [57]. Blackwell's group demonstrated a unique acyclic peptoid reverse-turn conformation with all *trans*-amide backbone by incorporating *N*-aryl side chains [54]. The precise control over the backbone conformation and side chain functionality can be used as the basis to design peptoids for applications in biomedicine and nanotechnology.

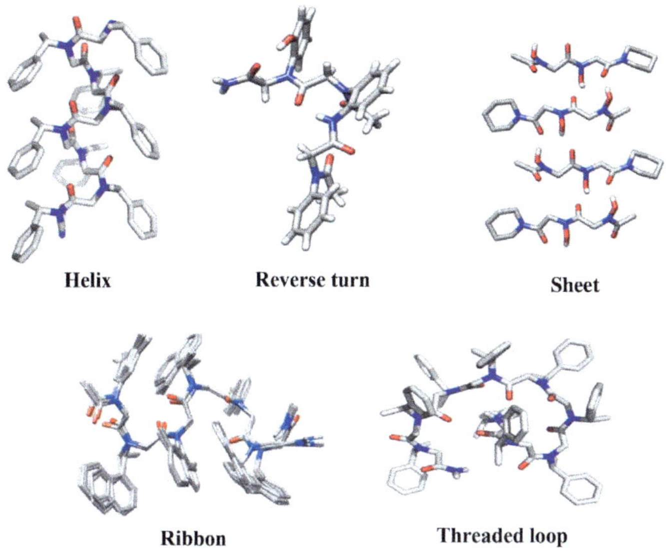

Figure 7.3 A variety of peptoid secondary structures: a peptoid helix from a homopolymer of *N*-(*S*)-(1-phenylethyl)glycine (Nspe) [53]; the *N*-aryl peptoid reverse turn [54]; *N*-hydroxyamide sheet-like structures [55]; alternating *N*-aryl/*N*-1-naphthylethyl peptoid ribbon structure [56]; and the peptoid threaded loop structure of $(Nspe)_9$ [57].
Source: Reproduced with permission from J. Sun, R.N. Zuckermann, Peptoid polymers: a highly designable bioinspired material, ACS Nano 7 (2013) 4715–4732. Copyright (2013) ACS.

7.2.4 Physical properties of peptoids

Crystallization behavior of peptoids

The sequence specificity and monodispersity of the peptoid make it an excellent platform to study the structure–property relationships of the materials for many applications. Peptoids are shown to be stable at very high temperatures (300°C) and are able to crystallize. Rosales et al. investigated the influence of monomer, composition, and distribution on the crystallization behavior of peptoids in the solid state [11]. They demonstrated that the crystallinity can be readily controlled through the insertion of defects at precise locations along the peptoid backbone. The melting point decreases with increasing defect content, consistent with Flory's theory of crystallization. Sun and coworkers explored the impact of side chain structure of peptoids on crystallization behavior with a series of peptoid diblock copolymers poly(*N*-decylglycine)-*block*-poly(*N*-2-(2-(2-methoxyethoxy) ethoxy)ethylglycine). They demonstrated that all of these block copolymers can self-assemble into lamellar microphases driven by crystallization of poly(*N*-decylglycine). Crystallization of the normally amorphous poly(*N*-2-(2-(2-methoxyethoxy)ethoxy)ethylglycine) chain is induced by the preorganization of the crystalline poly(*N*-decylglycine) chain [68,69]. Zhang's group synthesized a series of polypeptoids with aliphatic side chains by ring-opening polymerization. An interesting side chain and main chain coupled crystallization behavior was observed in the polypeptoids with longer aliphatic side chains. They further studied the influence of the cyclic and linear architecture of both main chains and side chains on the crystallization properties [70]. Schlaad et al. obtained rose-bud type morphology from a crystalline polypeptoid solution [46].

Thermal-responsive properties of polypeptoids

The polypeptoid with various aliphatic side chains was demonstrated to show lower critical solution temperature (LCST) behavior. Zhang's group reported the random cyclic copolypeptoid cyclo-[poly(*N*-ethylglycine)-*r*-poly(*N*-butylglycine)] with cloud point temperatures (CPs) in the range of 20–60°C [71]. They compared its CP with that of a random linear co-polypeptoid analogue. Both polypeptoids show a comparable CP with a difference of ca. 5°C. They further demonstrated that the cloud point transition window is much broader than that of the benchmark thermal-responsive polymer PNIPAM. Copolymers of methyl and butyl monomers were also found to show LCST behavior [47]. In addition, the peptoid homopolymer poly(*N*-C3 glycine)s (C3 includes c *n*-propyl, allyl, and isopropyl groups) show CPs within the range from 15 to 60°C. The CPs were observed to increase in the order of *n*-propyl, allyl, and isopropyl groups substituted polypeptoids [46].

7.3 Applications in biomedicine and nanotechology

7.3.1 Combinatorial libraries for peptoid ligand screening for therapeutics

Combinatorial library synthesis offers an efficient way to achieve great molecular diversity for the discovery of biomolecular ligands. The peptoids are an ideal class of molecules for combinatorial approaches as large libraries of peptoids can be synthesized easily from readily available primary amines. With the development of advanced screen methodology, highly specific protein-binding molecules can be inexpensively and quickly screened from hundreds of thousands of compounds for therapeutics.

In 1988, researchers at Chiron Corp, one of the original biotechnology companies and now part of Novartis AG, developed small-molecule drugs by new drug-discovery technologies. The solid-phase submonomer synthesis methods facilitated the library synthesis. A variety of library synthesis approaches have been developed, such as SPOT peptoid arrays [72], equimolar mixture screening [73,74], OBOC (one-bead-one-compound) screening [75,76], spotted microarrays [77], and positional scanning libraries [78,79]. Zuckermann and coworkers synthesized large libraries of chemically diverse short peptoid oligomers and screened them for biological activities. They discovered several potent peptoid trimer ligands for G-protein-coupled receptors [74] and the urokinase receptor [80], which was one of the first investigations of high-affinity ligands from a diverse combinatorial library

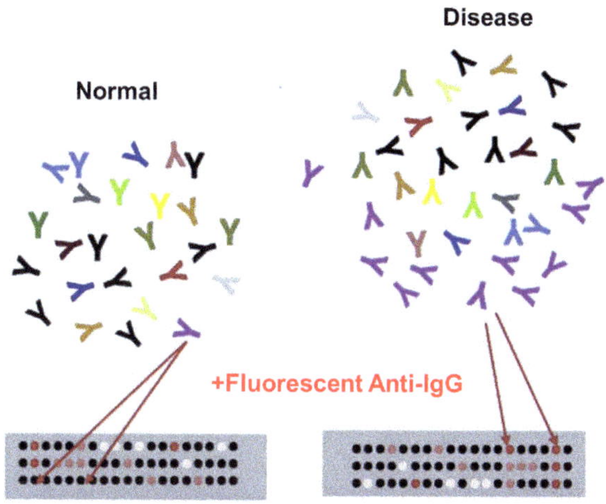

Figure 7.4 Schematic representation of the strategy employed to identify synthetic molecules that capture antibody biomarkers [77].
Source: Reproduced with permission from M.M. Reddy, R. Wilson, J. Wilson, S. Connell, A. Gocke, L. Hynan, et al., Identification of candidate IgG biomarkers for Alzheimer's disease via combinatorial library screening, Cell 144 (2011) 132–142. Copyright (2010) Cell.

of synthetic compounds for pharmaceutically relevant receptors. More research groups have been inspired to work on the discovery of potential therapeutics from peptoid libraries [81–85]. Kodadek et al. developed a general approach to identify diagnostically useful antibodies in the absence of antigen identification. They further discovered the potentially useful diagnostic biomarkers in humans via identification of candidate IgG biomarkers for Alzheimer's disease (Fig. 7.4) [77]. Peptoid–fluorophore conjugates have also been prepared as reporters for intracellular delivery [86] or the folding into secondary structures [87,88]. All these examples demonstrate that peptoid libraries are a powerful, inexpensive, and convenient source of diverse ligands for a variety of biomedical applications.

7.3.2 Peptoids for antifouling coating applications

Peptoids that can prevent the adhesion of proteins, bacteria, and marine microbes have received increasing interest for antifouling applications. An early report on poly(*N*-(2-methoxyethyl) glycine) oligomers showed that it can enhance antifouling behaviors given the excellent water solubility, nonionic feature, and potent hydrogen bonding. The Messersmith group first prepared the peptoids for the application of antifouling materials. They focused on a class of poly(*N*-(2-methoxyethyl) glycine) oligomers modified with short five monomers. These peptoids can mimic mussel adhesion peptides that contains L-3,4-dihydroxyphenylalanine and basic primary amines. The obtained peptoids exhibit excellent protein resistance, and its ability to resist fouling can be maintained for several months under frequent challenge of fresh serum and cells [89]. They further explored more antifouling systems such as the peptoids with hydroxyl-containing monomers and the heteropolymer with alternating charged monomers [90]. Van Zoelen et al. designed peptoid–polystyrene diblock copolymer thin films. The amount and sequence of fluorinated monomers were finely tuned for antifouling coating applications [91]. They demonstrated that the presence of three fluorinated groups at the end of the peptoid with a chain length of 45 allowed for maximal peptoid surface display and rapid surface reconstruction. As the number of such groups increases to five, the speed of reconstruction was found to be slower by an order of magnitude.

7.3.3 Peptoids for antimicrobial application

The discovery of antibiotics has had a dramatic positive impact on human health. However, the emergence of "superbugs" exhibits resistance to many antibiotics. In the past few decades, an emerging class of antimicrobial peptides (AMPs) has been studied as potential antibiotic compounds. AMPs are short peptides with an amphipathic structure that allows for permeabilization of bacterial cytoplasmic membranes. The AMPs are less likely to be affected by bacterial resistance, which offers great advantages for antimicrobial application. However, AMPs generally have a short half-life in vivo due to the poor pharmacokinetics, which motivates the effort to discover peptoid mimics of AMPs as antibiotics.

Peptoids show great potential to be used as pharmaceuticals due to their improved stability and highly tunable side-chain chemistry. Winter and coworkers discovered peptoid trimers with amine groups in the side chain, which inhibit bacterial growth by combinatorial library screening. The MIC (minimal inhibitory concentration) of the representative peptoid was reported to be in the range of 5–40 mM. Increasing hydrophobicity of the peptoids can cause the system to be hemolytic. Barron's group successfully mimicked a cationic, amphiphilic helical AMP and prepared a library of ampeptoids (antimicrobial peptoid oligomers) with helical structures. The obtained novel compounds exhibit low-micromolar MIC value and low mammalian cytotoxicity. They are demonstrated to be a promising class of antimicrobial therapeutics with broad-spectrum activity against six clinically relevant BSL2 pathogens [92] (Fig. 7.5).

Macrocyclic AMPs can easily bind to a biological target due to the ordered conformation, which may offer enhanced antimicrobial activity. The Kirschenbaum group has synthesized a large number of potent antimicrobial peptoids with cationic/hydrophobic faces on opposing faces of the planar macrocycle [93]. The antimicrobial property of cyclic peptoids was further compared to that of their linear precursors. They demonstrated that the cyclization improves antibacterial properties.

7.3.4 Peptoids as antifungal agents

In comparison to bacterial infections, it is normally more difficult to treat fungal infections due to the limited number and great toxicity of antifungal drugs. Barron and Larabell et al. used soft X-ray tomography to study the cellular response of *Candida albicans* to antifungal peptoids [94]. They demonstrated that the peptoid

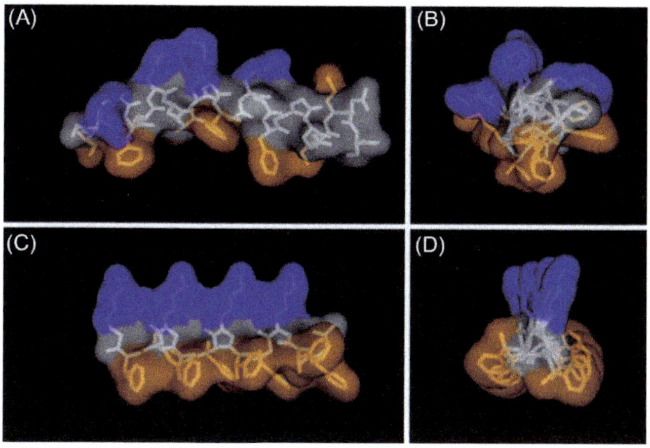

Figure 7.5 (A, B) NMR structure of magainin-2 in DPPC micelles (Protein Data Bank (PDB) ID Code 2MAG) (A, parallel, B, perpendicular to its helical axis). (C, D) Similar views of a model structure of ampetoid from a published NMR structure of a peptoid helix.

treatment can suppress the formation of the pathogenic hyphal phenotype that resulted in dramatic changes in cell and organelle morphology. Ramírez and coworkers synthesized a library of peptoids containing four glycine residues and acylated with a fatty acid by two consecutive Ugi reactions [95]. The peptoids showed remarkable in vitro activity against two fungal plant pathogens, *Fusarium virguliforme* and *Fusarium lateritium* with MIC values ranging from 7 to 38 μM.

More recently, Bicker's group obtained the antifungal peptoid against *Cryptococcus neoformans* using a high-throughput assay from combinatorial libraries. The identified peptoid has excellent stability, minimal cytotoxicity in mammalian cells, and a low hemolytic activity to human erythrocytes, which exhibit comparable potency to existing clinical antifungal agents [96].

7.3.5 Peptoids as anticancer drugs

Albericio et al. synthesized a library of small peptoid monomers and dipeptoids on a BAL resin [97]. By structure design, the dipeptoids exclude the formation of DKP, a problematic side reaction typically observed in peptoid synthesis. These peptoids were assayed for antiproliferative activity against representative human neoplastic cell lines. It was demonstrated that the C-terminal *N*-alkyl amide peptoids are cytotoxic and antimitotic and they are the smallest peptoids reported to show such activity. A dual function of antioxidant conjugate agent with a chelator was also reported for anticancer applications [98]. Gonzalgo et al. reported that a novel, synthetically derived peptoid molecule exhibited significant growth inhibition to human bladder cancer cells [99]. This may be caused by disruption of the cancer cell membrane, and tumor cell metabolism may also be reduced in vivo.

7.3.6 Peptoid-modulated crystal growth for biominerals and antifreeze agents

The substantial structural diversity of peptoids and their feasible preparation offer a unique opportunity to mimic the key role of biomineralization in proteins [100,101]. The carboxylic acid-rich amphiphilic peptoids were used to mineralize CO_2 in the presence of calcium ion to form calcite (Fig. 7.6) [101]. It was demonstrated that the peptoids bearing a balance of charge and hydrophobicity offer both a high degree of control over calcite growth morphology and an unprecedented acceleration of crystal growth rates at very low concentrations. In addition, the peptoids were also used to tune the morphology and growth rate of ice crystals [100]. Dual-action antifreeze effects of ice growth inhibition and reduced melting temperature were observed, which were prove to inhibit the formation of ice crystals and the growth of ice crystals once formed.

7.3.7 Peptoid architectures for biomimetic materials research

In addition to their linear structure, peptoids with various architectures have also been explored. Kodadek et al. reported a new approach named "one bead two compounds" to synthesize combinatorial libraries of cyclic peptoids [86]. Kirshenbaum and coworkers [66] have prepared head-to-tail macrocyclic peptoids by solution cyclization of the secondary amine to the free carboxyl group. Cyclic 5- to 20mers have been reported with high yields and excellent cyclization efficiencies over diverse sequences and chain lengths. X-ray crystallography was used to elucidate that macrocyclic peptoids adopt rigid and compact folded conformations. The macrocyclic conformation also results in an enhancement of binding affinities and specificities [15]. A crystalline tube-like array formed by macrocyclic peptoids permits reversible sequestration of cocrystallized water molecules through a single-crystal-to-single-crystal transformation [102]. Cyclic peptoids can also form stable complexes with metal ions in the solid state to form 1D metal organic framework structures (Fig. 7.7A) [106]. "Hairy" architectures of peptoids, also known as polymer brush-like structures, have been synthesized by chemical functionalization of peptoid scaffolds [107]. The "click chemistry" strategy has been employed as an efficient approach to incorporate bioactive ligands in the application of biosensors [108].

Peptoids with specific sequences have been used as scaffolds to generate protein-like function. Kang et al. employed a type of peptoid with helical structure as a scaffold to construct a precisely defined cofacial array of porphyrins for many potential applications (Fig. 7.7C) [109]. Kirshenbaum et al. synthesized a series of peptoids with excellent catalytic efficiency by incorporating TEMPO (2,2,6,6-tetramethylpiperidine-1-oxyl), a well-known catalyst for oxidative transformations to various sites on the backbone of helical peptoids (Fig. 7.7D) [110]. They demonstrated that the enantioselectivity of the catalytic peptoids is dominated by the chirality of the helical backbone, the position of the catalytic center along the backbone, and the degree of conformational ordering of the peptoid scaffold.

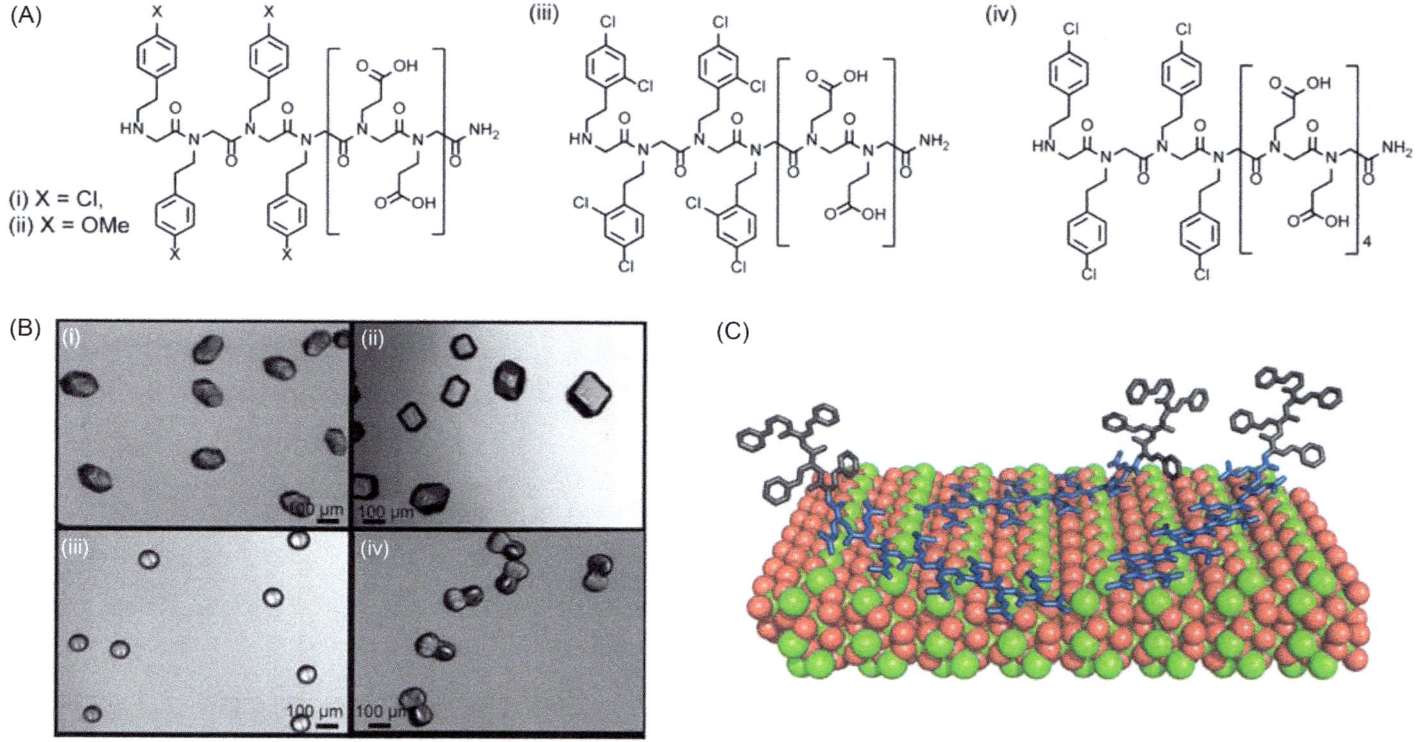

Figure 7.6 The nucleation and growth of calcite has been modulated by anionic and aromatic peptoids. (A) The chemical structures. (B) The morphology of $CaCO_3$ crystals grown from aqueous solutions are shown. (C) A model for the interaction between a peptoid structure and the calcite is shown, where the hydrophobic peptoid regions are out of the plane and the hydrophilic regions are in plane.
Source: (A) Reproduced with permission from A.S. Knight, E.Y. Zhou, M.B. Francis, R.N. Zuckermann, Sequence programmable peptoid polymers for diverse materials applications, Adv. Mater. 27 (2015) 5665–5691. Copyright (2015) Wiley. (B) Reproduced with permission *from A.S. Knight, E.Y. Zhou, M.B. Francis, R.N. Zuckermann, Sequence programmable peptoid polymers for diverse materials applications, Adv. Mater. 27 (2015) 5665–5691.* Copyright (2011) ACS. (C) Reproduced with permission *from A.S. Knight, E.Y. Zhou, M.B. Francis, R.N. Zuckermann, Sequence programmable peptoid polymers for diverse materials applications, Adv. Mater. 27 (2015) 5665–5691*. Copyright (2014) Nature Publishing Group.

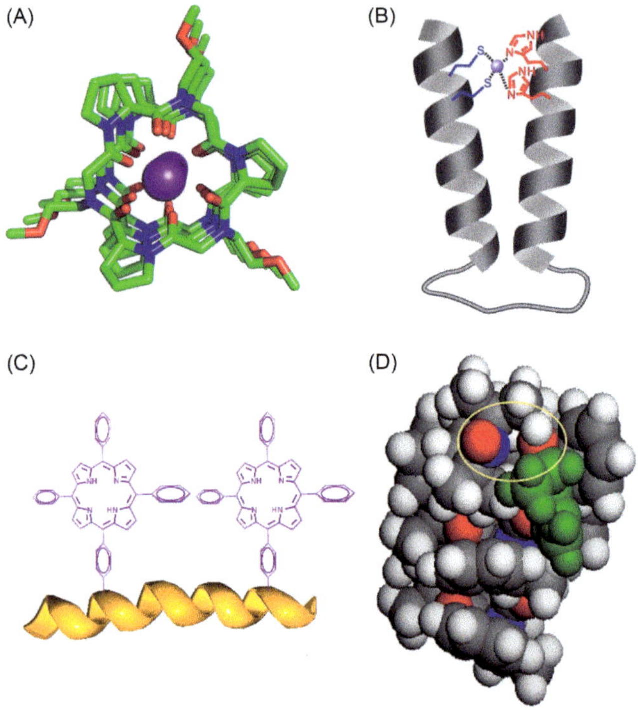

Figure 7.7 Structured peptoids modified with functional moieties. (A) 1D metal-organic framework structure derived from a proline-containing cyclic hexapeptoid and sodium atoms [97]. (B) Peptoid two-helix bundles functionalized with thiols and imidazoles have been identified that selectively bind zinc ion with nanomolar affinity [103]. (C) Porphyrin groups were conjugated to a peptoid helix to study electron transfer [104]. (D) Energy-minimized structure of a TEMPO-functionalized catalytic peptoid with the docked substrate 1-phenylethanol [105].
Source: (A) and (D) Reproduced with permission from J. Sun, R.N. Zuckermann, Peptoid polymers: a highly designable bioinspired material, ACS Nano 7 (2013) 4715–4732. Copyright (2013) ACS. (B) Reproduced with permission from H.K. Murnen, A.M. Rosales, A.V. Dobrynin, R.N. Zuckermann, R.A. Segalman, Persistence length of polyelectrolytes with precisely located charges, Soft Matter 9 (2013) 90–98. Copyright (2008) ACS. (C) Reproduced with permission from K.A. Dill, Dominant forces in protein folding, Biochemistry 29 (1990) 7133–7155. Copyright (2013) ACS.

Globule formation is considered to be the key to the protein-folding problem. Hydrophobic collapse is a predominant driving force of folding into compact globular structures [104,105]. Great efforts have been focused on exploring the impact of the sequence patterning of hydrophobic and polar residues on folding [111,112]. The multihelical bundles were screened by combinatorial libraries of amphiphilic peptoid helices with threefold periodicity and subsequently covalently linked together to form single-chain sequences [113]. The functional groups have also been incorporated into the precise positions in the two-helix bundle. A high-affinity selective zinc-binding site that mimics the natural zinc finger domains in DNA-binding proteins was created (Fig. 7.7B) [114]. This is one of the first tertiary folds prepared from a non-natural single chain polymer for specific molecular recognition.

Khokhlov and Khalatur predicted that polymers with blocky distributions of monomers form more stable globules than random analogues where the monomers are evenly distributed along the chain [115]. An "HP model" was used to describe simplified interactions in peptides, where H represents hydrophobic and P represents polar. In order to explore the effect of hydrophobic sequence patterning on globule formation in the absence of other factors (e.g., chirality, intrachain hydrogen bonding), Zuckermann's group designed protein-like peptoids that mimic the HP model and systematically investigated the coil-to-globule transitions (Fig. 7.8A) [27]. Two HP peptoid sequences with chain lengths of 100 were synthesized. One of them was "protein-like" with a longer blocky sequence of both hydrophobic and polar monomers. The other was a "repeating" sequence containing the sequence of both monomers as short as possible, which distributed more evenly. They show that the protein-like sequence collapsed into a more compact globule with smaller gyration radius in aqueous solution than that with the repeating sequence. The coil-to-globule

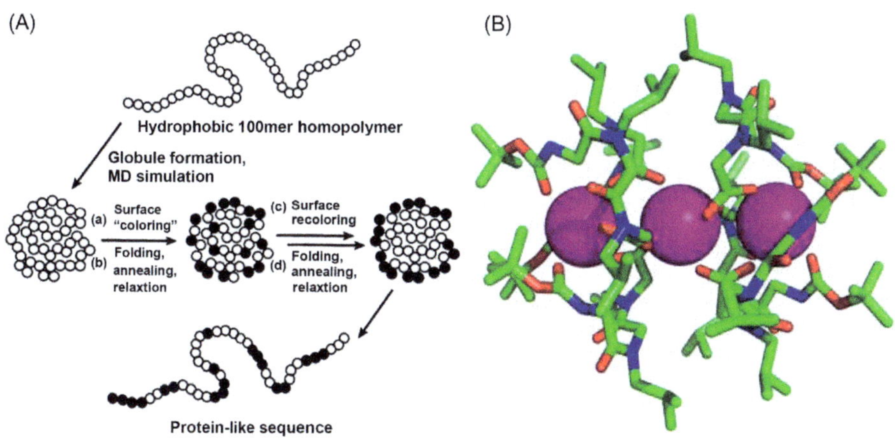

Figure 7.8 (A) Protein-like peptoid sequences were designed using all-atom molecular dynamics simulation [27]. White circles represent the hydrophobic monomer with a methyl side chain; black circles represent the polar monomer with a carboxylate side chain.
(B) Collagen-mimetic peptoid-containing trimer motif can form a crystalline complex with calcium ions [116].
Source: (A) Reproduced with permission from H.K. Murnen, A.R. Khokhlov, P.G. Khalatur, R.A. Segalman, R.N. Zuckermann, Impact of hydrophobic sequence patterning on the coil-to-globule transition of protein-like polymers, Macromolecules 45 (2012) 5229–5236. Copyright (2012) ACS. (b) Reproduced with permission from J. Sun, R.N. Zuckermann, Peptoid polymers: a highly designable bioinspired material, ACS Nano 7 (2013) 4715–4732. Copyright (2013) ACS.

transition window of the protein-like sequence was more cooperative, narrow with a significantly higher unfolding free energy than that of the peptoid with repeating sequence. Nile Red fluorescence was further included in the system, showing an increased emission peak from the peptoid with protein-like sequence. This suggests that the protein-like chain with longer blocks of H and P monomers can fold into a tighter globule.

The impact of electrostatic interactions on conformations of charged polymers is also of great interest for applications ranging from ion-exchange membranes to biopolymers such as DNA and proteins. Interactions between ionized groups play a key role in polyelectrolyte conformations. Zuckermann and Segalman et al. studied the effect of charge distribution on polyelectrolyte conformation in aqueous solutions with a series of peptoids bearing different distributions of ionizable groups along the molecular chain [103]. Small-angle neutron scattering was used to investigate the persistence length. At low ionic strengths, the peptoids with high-density charge segments show a larger persistence length due to the increasing electrostatic repulsion between ionized groups. At higher ionic strengths, all the peptoids show a decrease in persistence length, which is in agreement with previous theoretical and experimental results. The peptoids offer a platform to understand the fundamental behavior of hydrophobic and ionic sequence patterning proteins with non-natural polymers, which provides insight into the design of a new generation of protein-mimetic materials.

7.3.8 Peptoids for delivery of nucleic acids

Control over sequence patterning between hydrophobic and cationic side chains is critical for application of intracellular delivery vehicles. Many synthetic peptoids are able to condense plasmid DNA into small particles and protect it from nuclease degradation. In particular, the most efficient peptoid was a 36-mer that contained 12 cationic aminoethyl side chains with a repeating triplet sequence (cationic–hydrophobic–hydrophobic) [7]. This peptoid condensed plasmid DNA into uniform spherical nanoparticles of 50–100 nm in diameter and mediated the transfection of a number of cell lines with good efficiency. A small library of lipid–peptoid conjugates was subsequently synthesized [117]. The complex with DNA also showed a spherical shape with a diameter of ~100 nm. Interestingly, the most active compound contained the same repeating triplet motif (cationic–aromatic–aromatic). The compound exhibited much higher transfection efficiencies, which can transfect primary cells and deliver siRNA [118]. However, the structure–activity relationships in both systems are not clear and further examination is required [119]. A peptoid dendrimer was reported by microwave-assisted solid-phase methods [120]. The third-generation compound displayed minimal cytotoxicity and was demonstrated to be an efficient mediator of DNA transfection. These versatile peptoid architectures with various functional groups offer great advantages for molecular recognition, scaffolds, and supramolecular assemblies [86,116,121–126].

7.3.9 Peptoids as collagen mimetics

Collagen is one of the most abundant fibrous proteins in the human body, and is responsible for providing the scaffolding matrix. The collagen mimetic materials have received great attention in the applications of drug delivery, biomedical devices, and tissue engineering. Goodman and coworkers first synthesized collagen peptidomimetics by incorporating peptoid residue *N*-isobutylglycine (Nleu) into a series of collagen mimetics composed of Gly-Pro-Nleu, Gly-Nleu-Pro, and Gly-Nleu-Nleu trimer repeat units [127,128]. These collagen peptidomimetics exhibited potent biological activity and enhanced stability against enzymatic degradation. Additionally, the achiral peptoids were used to simplify the synthetic strategies by reducing racemization problems. Interestingly, a single peptoid-containing trimer repeat unit, boc-Gly-Nleu-Nleu, was shown to be capable of forming a crystalline complex with calcium ions (Fig. 7.8B) [129]. They further demonstrated that the contributions of steric and hydrophobic effects are the important factors for the triple helix formation. The discovery of collagen mimetics and metal-binding ability offers new opportunities in the design of novel collagen mimetic complexes.

7.3.10 Nanostructure based on self-assembly of peptoids

Nanostructures based on self-assembly of peptoids in bulk

Microphase-separated block copolymers have a wide array of potential applications ranging from nanoscale materials to energy storage. The sequence specificity of peptoids allows for tenability of other physical properties, such as the interaction strength between two blocks of a diblock copolymer. Rosales et al. synthesized sequence-defined diblock copolymers containing polystyrene and peptoids with *N*-2-methoxyethyl side chains, ranging from 18 to 48 monomers in length [130]. The block copolymers can be readily self-assembled into hexagonally packed and lamellar morphologies, in good agreement with the classical block copolymer phase diagram. *N*-(2-phenylethyl)glycine (Npe) residues were further introduced to increase the compatibility with the polystyrene block. As the strength of segregation decreased, there was a decrease in the order–disorder transition temperature and an increase in the domain spacing. Sun et al. presented a systematic study of the relationship between chemical structure and morphology in the solid state using peptoid diblock copolymers poly(*N*-2-(2-(2-methoxyethoxy)ethoxy)ethylglycine)-block-poly(*N*-(2-ethylhexyl)glycine) (pNte-*b*-pNeh) with PDIs of ≤ 1.00013 [131]. The main chain length was held fixed at 36 with the volume fraction of the Nte block (ϕ_{Nte}) varying from 0.11 to 0.65. All the ordered phases show a lamellar morphology, even in the case of $\phi_{Nte} = 0.11$, which is in qualitative disagreement with all known theories of microphase separation in block copolymers. The peptoid block copolymers provide a tunable platform for studies on the effect of composition and sequence design on self-assembly of block copolymers.

Nanostructures based on the solution self-assembly of peptoids

Zuckermann's group recently obtained a family of peptoid sequences that self-assemble into robust and precisely ordered nanosheets in aqueous media by screening from different amphiphilic sequence patterns [132]. A twofold periodic sequence pattern which alternates between an aromatic hydrophobic monomer (*N*-(2-phenylthyl) glycine (Npe)) and one of two oppositely charged ionic polar monomers (*N*-(2-aminoethyl)glycine (Nae) and *N*-(2-carboxyethyl)glycine (Nce)) is the key to generating sheet-forming sequences (Fig. 7.9). This results in a highly ordered, planar bilayer structure with hundreds of microns in length and width of only 3 nm thick. The nanosheets were then functionalized with a streptavidin-binding peptide sequence cyclo-[CHPQFC]- at the *N*-terminus. Recent research has discovered that a single polymer strand combining all three monomers (aromatic positive and negative) arranged in the same twofold sequence pattern can also form nanosheets, which is different from former designs [133]. The stability of the block charge pattern design was

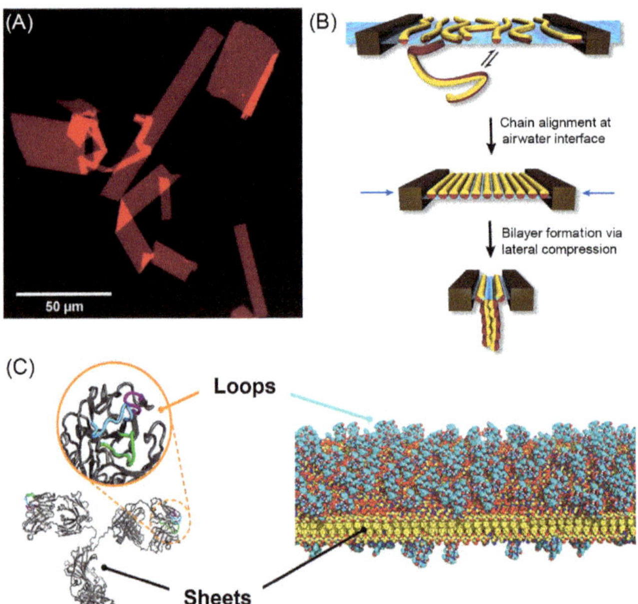

Figure 7.9 (A) Two-dimensional crystalline nanosheet bilayers observed by fluorescence microscopy (stained with Nile red) [126]. (B) Peptoid nanosheets are formed by a unique assembly mechanism involving the formation of an ordered monolayer at the air–water interface, followed by a buckling event induced by lateral compression of the monolayer to produce a stable bilayer [128]. (C) The nanosheets were used as scaffords to display loops. *Source*: (A) Reproduced with permission from J. Sun, R.N. Zuckermann, Peptoid polymers: a highly designable bioinspired material, ACS Nano 7 (2013) 4715–4732. Copyright (2013) ACS. (B) Reproduced with permission from G. Melacini, Y. Feng, M. Goodman, Collagen-based structures containing the peptoid residue N-isobutylglycine (Nleu). 6. Conformational analysis of Gly-Pro-Nleu sequences by 1H NMR, CD, and molecular modeling, J. Am. Chem. Soc. 118 (1996) 10725–10732. Copyright (2011) ACS. (C) Reproduced with permission from E.A. Jefferson, P. Gantzel, E. Benedetti, M. Goodman, A multinuclear Ca^{2+} complex of a linear N-protected glycyl-dipeptoid derivative, J. Am. Chem. Soc. 119 (1997) 3187–3188. Copyright (2015) ACS.

consistent with expectations based on theoretical considerations of the molecules' electrostatic interactions. The mechanism of the formation of the nanosheet was further investigated [134]. The key intermediate in nanosheet formation is that a small but important fraction of the dissolved compound spontaneously adsorbs to the air–water interface. Distinct stages of the compression/expansion cycle were revealed by the monitoring of the surface pressure isotherm in a Langmuir trough. The peptoid monolayer reaches a solid phase upon compression, which brings two adjacent hydrophobic moieties together into a bilayer covered by a zwitterionic polar surface these peptoid nanosheets, with only two molecules thick and chemically defined interior and exterior, are one of the largest two-dimensional organic crystals known.

The peptoid offer a highly engineerable 2D nanomaterial platform that can potentially serve many applications including biosensing, membrane-based separations, drug delivery, and templates to grow inorganic materials. The conformationally constrained peptide and peptoid were displayed on the surface of free-floating nanosheets as a variety of functional loop sequences that serve as substrates for enzymes (Fig. 7.9) [135]. Chen et al. reported the self-assembly of 12-mer peptoids with alternating cationic and aromatic monomers into networks of hexagonally patterned nanoribbons on mica surfaces. They demonstrated that the peptoids assembled into nanoparticles by subsequently transforming into hexagonally patterned nanoribbons on mica surfaces [136].

Zuckermann and coworkers further synthesized an amphiphilic peptoid diblock copolymer [*N*-(2-phenylthyl)glycine]$_{15}$-*b*-[*N*-(2-carboxyethyl)glycine]$_{15}$ [137]. The peptoid can dissolve in aqueous solution and self-assemble into nanosheets, which further forms the superhelical structures with a diameter of 624 ± 69 nm and lengths ranging from 2 to 20 μm. Increasing the chain length of each block results in a similar superhelical structure with an increase in the lamellar spacing. Helix formation only occurred in a pH range where the carboxylates were partially charged. Interestingly, the superhelices are remarkably homochiral despite the achiral nature of all peptoids. To understand the superhelix internal structure and the role of charge location and density in the self-assembly, various chemical modifications in the form of closely related analogues were made. A nonionic side chain and a hydrogen bond donor/acceptor side chain were used as hydrophilic blocks. Neither of them can self-assemble into well-ordered structures, demonstrating that ionic interactions are exclusively important for helix self-assembly.

Sun et al. recently prepared purified, monodisperse amphiphilic peptoid diblock copolymers, which can behave like molecular tile units that self-assemble into hollow, crystalline nanotubes in aqueous media (Fig. 7.10) [138]. The driving force for self-assembly of peptoid nanotubes excludes hydrophobic core, chirality, hydrogen bonding, electrostatic, and $\pi-\pi$ interactions. Instead, the nanotubes are held together primarily by van der Waals interactions of side chains. These results are indicative of the remarkable structure-directing influence of polymer self-assembly, and further suggested that flexible, low-molecular-weight polymers can serve as molecular tile units that assemble into precision supramolecular architectures.

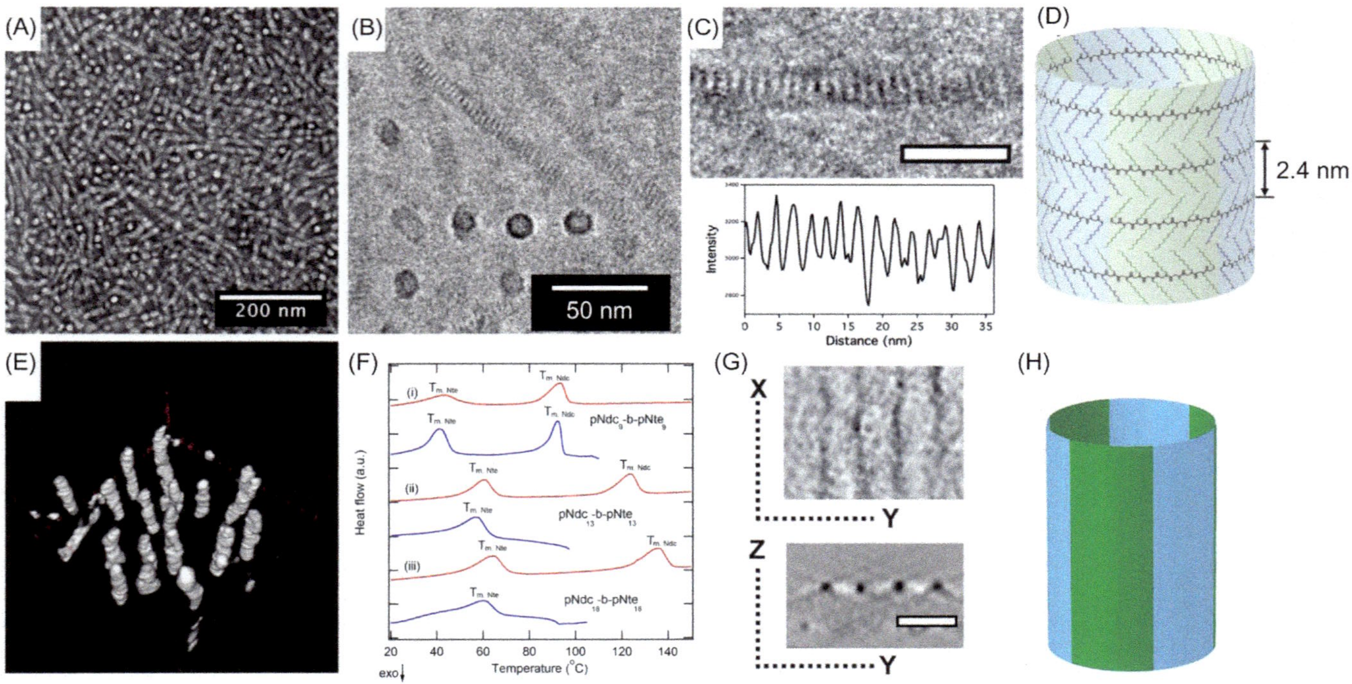

Figure 7.10 (A) Uranyl acetate negative-stained energy-filtered TEM micrograph of dried nanotubes. (B) Cryo-EM 2D projection image of nanotubes. (C) Cryo-EM close-up of a single nanotube shows regular latitudinal stripes (scale bar, 20 nm). A line profile along the longitudinal direction of the nanotube indicates a periodicity of 2.4 ± 0.2 nm. (D) A model of the proposed mechanism for the formation of a nanotube. (E) Three-dimensional visualized image of reconstructed nanotubes in a vitreous specimen. (F) Thermal analysis of nanotube solutions. (G) Slices in tomogram show both the top view and cross-section of ruthenium tetroxide-stained nanotubes, and indicate longitudinal striping. (H) Cartoon showing the origin of the longitudinal striping [132].

Source: Reproduced with permission from K.T. Nam, S.A. Shelby, P.H. Choi, A.B. Marciel, R. Chen, L. Tan, et al., Free-floating ultrathin two-dimensional crystals from sequence-specific peptoid polymers, Nat. Mater. 9 (2010) 454–460. Copyright (2016) National Academy of Sciences.

Zhang's group investigated the self-assembly properties of cyclic copolypeptoids in an aqueous solution [71]. The cylindrical micelles were obtained with the amphiphilic diblock cyclic copolymer cyclo-[poly(*N*-methylglycine)-*b*-poly(*N*-decylglycine)] in methanol. It was demonstrated that the transition from spherical to cylindrical micelle is due to the crystallized core of the micelle.

7.3.11 Polyethylene glycol mimetic peptoids for biopharmaceuticals

PEGylated peptide/protein-based biopharmaceuticals have been extensively investigated for a variety of applications, due to their enhanced proteolytic stability and therapeutic efficacy, and reduced rates of renal clearance. However, the polydispersity of PEG results in heterogeneous preparations, which requires a complicated purification processing. Barron's group have synthesized poly[*N*-(2-methoxyethyl) glycine] oligomers for an excellent PEG-mimetic material due to its similar chemical structure to PEG and homogeneous distribution [139]. They further incorporated an *N*-2-methoxyethyl glycine (Nmeg) oligomer with different chain lengths covalently onto a short therapeutic peptide that is neither soluble nor stable to serum [140]. Comparable therapeutic efficacy relative to the parent peptide was observed with one Nmeg monomer attached, but with a prolonged half-life. Instead, longer Nmeg oligomer chains were shown to interfere with peptide binding to the target protein.

In recent years, poly(ethyleneoxide) (PEO)-based materials have received a great deal of interest for use as battery solid-polymer electrolyte applications, due to their high ionic conductivity. Sun et al. synthesized a series of well-defined PEO-mimetic peptoids with varying main chain lengths and number of ethylene oxide (EO) units in the side chains by a submonomer solid-phase method. They investigated the conductive ability of peptoids as electrolytes with small amounts of lithium salts dissolved in the solid state [22]. All peptoids are amorphous, with decreased glass transition temperature (T_g) as the side-chain EO unit length increases. The optimum ionic conductivity of 2.6×10^{-4} S/cm was achieved by poly(*N*-2-(2-(2-methoxyethoxy)ethoxy)ethylglycine)-Li salt complex at 100°C, which is two orders of magnitude higher than previously reported for comb-like PEO mimetic polypeptides. This is due to the absence of hydrogen bond donors along the peptoid main chain that results in a flexible backbone, allowing for the increased segment motion of the polymer chains. They demonstrated that the conductivity of the system is strongly dependent on the glass transition temperature, and not the other factors such as complexation with salt or the length of the pendant $(EO)_n$ chains. Recently, Zhang's group also synthesized a series of polypeptoids containing oligomeric ethylene glycol side chains with a controlled molecular weight by ROP of the corresponding *N*-carboxyanhydrides bearing OEG side chains using primary amine initiators [141]. The polypeptoids show no appreciable adsorption of lysozyme and minimal cytotoxicity toward HEp2 cells.

7.3.12 Simulation of peptoid architectures

The computational tool is becoming one of the most useful methods to predict folding and assembly of peptoids. Molecular dynamics simulation was used to predict the stable helical fold induced by chiral side chains, from which a noctamer of (*S*)-*N*-(1-phenylethyl)glycine can adopt a polyproline type I helical conformation [53], and that *N*-aryl peptoids may form polyproline type II helices [61]. Both are in agreement with subsequent experimental findings. A compete landscape of peptoid backbone energies provided a detailed understanding of the local backbone dihedral angle preferences [142]. Molecular dynamics force fields from peptides have been used to predict the structures of several short peptoid oligomers. A combination of replica exchange molecular dynamics simulation (REMD) and quantum mechanical refinement was able to predict the structure of the nonamer to within 0.2 Å. It also predicted the structure of three short peptoid sequences, the pattern of *cis/trans* backbone amides, and the structure of a cyclic peptoid nonamer to an accuracy of 1.0 Å rmsd-backbone [143]. This work is therefore considered as a tangible milestone for computational peptoid design. Zuckerman and Whitelam used molecular dynamics simulations to determine the atomic-resolution structure of the recently discovered peptoid nanosheet [144]. It was shown that nanosheets are structurally and dynamically heterogeneous, and potentially porous to water and ions. Moreover, their formation is enabled by the peptoids' adoption of a secondary structure that is not seen in Nature, which is a zigzag pattern called a $\sum$ ("sigma")-strand.

Peptoids	Applications	References
CHIR 2279, CHIR 2283, CHIR 2276, CHIR 4531, CHIR 4534, CHIR4537	Peptoid ligand screening for therapeutics	[74]
Nlys-Nlys-Nmba-Nmba-Nleu-Nlys-Nser-Nleu, Nlys-Nlys-Nall-Nall-Nall-Npip-Nser-Nall, etc.		[77]
CHIR5585		[80]
A hybrid peptide–peptoid backbone Y-E-V-P-P-P-V-P-P-R-R-R (prolines are replaced by various *N*-substituted glycines)		[81,82]
JM79 homodimers, JM79 homotrimers, JM79 heterodimers, JM81 heterodimers		[83]
The dimerized peptoids dB2c, dB1, dB2, dB3, and dB4		[84]
NMOP1, NMOP2, NMOP3, NMOP4, NMOP5, NMOP6, NMOP7, NMOP8, NMOP9, NMOP10		[85]

Peptoids	Applications	References
β-Ala-Cys-Glu(Biotin)-cyclo(Glu-Nmea-Npip-Nall-Nphe-Nall-Nffa-Nmea), β-Ala-Cys-Glu(Biotin)-cyclo(Glu-Nall-Nleu-Npip-Nphe-Nleu-Nleu-Nmea), etc.		[86]
Aromatic fluorescent/spirolactam/guanidinylated fluorescent/aromatic guanidinylated spirolactam peptoid and linear analogues		[87]
Nae-Nspe-$(Nsce)_2$-Nspe-$(Nae)_2$-Nspe-$(Nsce)_2$-Nspe-$(Nae)_2$-Nspe-Nsce, Ndmn-Nspe-$(Nsce)_2$-Nspe-$(Nae)_2$-Nspe-$(Nsce)_2$-Nspe-$(Nae)_2$-Nspe-Nsce, etc.		[88]
A 20-mer *N*-methoxyethyl glycine peptoid and *L*-3,4-dihydroxyphenylalanine	Antifouling surface coatings	[89]
Peptide consisting of alternating residues of 3,4-dihydroxyphenylalanine, lysine, and *N*-methoxyethyl glycine peptoid or peptoid containing hydroxyl side chains		[90]
PSA-*b*-(EGx-NFy) or PSA-*b*-(EGx-NFy-EGx)		[91]
H-(NLys-Nspe-Nspe)$_4$-NH_2, H-(NLys-Nssb-Nspe-NLys-Nssb-Nsna)$_2$-NH_2, etc.	Antimicrobials for biomedical applications	[92]
Short amphiphilic linear and cyclic peptoid oligomers composed of 6–10 residues and cyclic decapeptide gramicidin S		[93]
Phenotypes of peptoid	Antifungal agents	[94]
N-substituted tetraglycines acylated with a pentadecane residue		[95]
NPhe-NPhe-NapNPhe-NFur and NTri-NLys-NFur		[96]
N-(3,3-Diphenylpropyl)glycyl-*N*-(4-methoxyphenethyl)glycinamide, *N*-(4-Methoxyphenethyl)glycyl-*N*-(3,3-diphenylpropyl)glycinamide, etc.	Peptoids as anticancer drugs	[98]
Ethylenediaminetetraacetic acid bis-(5-aminosalicylic acid) methyl ester		[99]
Ac(Nser)3, Ac(Sar)3, Ac(Nme)3, Ac(Ser)3	Peptoid modulated of crystal growth for biominerals and antifreeze agents	[100]
N-[2-(X-phenylethyl)]glycine, *N*-(2-carboxyethyl)glycine (Nce)		[101]

Peptoids	**Applications**	**References**
$(Nae)_n$, $(NchmNchmNae)_{12}$	Delivery of nucleic acids	[7]
Cyclic semipeptoids		[124]
Neutral and cationic, linear and cyclic peptoids (*N*-alkylglycine peptidomimetics)		[125]
Peptoid dendrimers using *N*-Fmoc–*N*-(6-*N'*-Fmoc-aminohexyl)-glycine		[109]
Cyclic hexameric *N*-methoxyethyl glycine oligomer	Biomimetic materials research	[106]
Ethisterone—peptidomimetic conjugates		[108]
Porphyrin–peptoid conjugates		[109]
$Ntempo(Nspe)_6$, $NspeNtempo(Nspe)_5$, etc.		[110]
A peptoid (*N*-substituted glycine polymer) two-helix bundle with zinc-binding		[113,114]
Poly(*N*-(2-carboxyethyl)glycine)-*block*-poly(*N*-methylglycine)		[27]
Gly-Pro-Nleu, Gly-Nleu-Pro, and Gly-Nleu-Nleu	Collagen mimetics	[127,128]
$Ca^{2+}{}_3(Boc\text{-}Gly\text{-}Nleu\text{-}Nleu\text{-}O^-)_6$		[129]
Polystyrene block with peptoids containing *N*-2-methoxyethyl side chains	Nanostructures based on self-assembly in bulk	[130]
Poly(*N*-2-(2-(2-methoxyethoxy)ethoxy) ethylglycine)-*block*-poly(*N*-(2-ethylhexyl)glycine)		[131]
$(Nae\text{-}Npe)_n$, $(Nce\text{-}Npe)_n$, $(Nae\text{-}Npe\text{-}Npe)_n$, etc.	Nanostructures based on the solution self-assembly	[132]
$(Nae\text{-}Npe)_9\text{-}(Nce\text{-}Npe)_9$		[133]
$(Nae\text{-}Npe\text{-}Nce\text{-}Npe)_9$		
$(Nae\text{-}Npe)_{18}$, $(Nce\text{-}Npe)_{18}$		[134]
$(Nae\text{-}Npe)_x\text{-}Nme_y\text{-}(Nce\text{-}Npe)_z$		[136]
$(Nae\text{-}Npe)_4$-Nme-βAla-Lys-Thr-Gln-Ala-Ser-βAla-$(NME)_2$-$(Nce\text{-}Npe)_4$		
$Npe_m\text{-}Npe_n$, $Npe_m\text{-}Nce_n$		[137]

Peptoids	Applications	References
Poly(*N*-decylglycine)-block-poly(*N*-2-(2-(2-methoxyethoxy)ethoxy)ethyl)glycine		[138]
Cyclic and linear copolymer poly[(*N*-ethyl glycine)-*r*-(*N*-butyl glycine)]		[71]
Poly[*N*-(2-methoxyethyl)glycine] oligomers, poly[*N*-(2-methoxyethyl) glycine] oligomers with short therapeutic peptide	Application of biopharmaceuticals	[139,140]
Peptoids with ethylene oxide in the side chains		[141]

References

[1] R.N. Zuckermann, Peptoid origins, Pept. Sci. 96 (2011) 545–555.

[2] R.J. Simon, R.S. Kania, R.N. Zuckermann, V.D. Huebner, D.A. Jewell, S. Banville, et al., Peptoids: a modular approach to drug discovery, Proc. Natl. Acad. Sci. U.S.A. 89 (1992) 9367–9371.

[3] R.N. Zuckermann, The chemical synthesis of peptidomimetic libraries, Curr. Opin. Struc. Biol. 3 (1993) 580–584.

[4] R.N. Zuckermann, T. Kodadek, Peptoids as potential therapeutics, Curr. Opin. Mol. Ther. 11 (2009) 299–307.

[5] K. Kirshenbaum, R.N. Zuckermann, K.A. Dill, Designing polymers that mimic biomolecules, Curr. Opin. Struc. Biol. 9 (1999) 530–535.

[6] A.E. Barron, R.N. Zuckerman, Bioinspired polymeric materials: in-between proteins and plastics, Curr. Opin. Chem. Biol. 3 (1999) 681–687.

[7] J.E. Murphy, T. Uno, J.D. Hamer, F.E. Cohen, V. Dwarki, R.N. Zuckermann, A combinatorial approach to the discovery of efficient cationic peptoid reagents for gene delivery, Proc. Natl. Acad. Sci. U.S.A. 95 (1998) 1517–1522.

[8] B.C. Lee, M.D. Connolly, R.N. Zuckermann, Bio-inspired polymers for nanoscience research., Proc. NSTI Nanotech. 2 (2007) 28–31.

[9] D. Zhang, S.H. Lahasky, L. Guo, C.U. Lee, M. Lavan, Polypeptoid materials: current status and future perspectives, Macromolecules 45 (2012) 5833–5841.

[10] R. Luxenhofer, C. Fetsch, A. Grossmann, Polypeptoids: a perfect match for molecular definition and macromolecular engineering? J. Polym. Sci., Part A: Polym. Chem. 51 (2013) 2731–2752.

[11] A.M. Rosales, H.K. Murnen, R.N. Zuckermann, R.A. Segalman, Control of crystallization and melting behavior in sequence specific polypeptoids, Macromolecules 43 (2010) 5627–5636.

[12] J. Sun, R.N. Zuckermann, Peptoid polymers: a highly designable bioinspired material, ACS Nano 7 (2013) 4715–4732.

[13] A.S. Knight, E.Y. Zhou, M.B. Francis, R.N. Zuckermann, Sequence programmable peptoid polymers for diverse materials applications, Adv. Mater. 27 (2015) 5665–5691.

[14] J. Seo, B.C. Lee, R.N. Zuckermann, Peptoids: synthesis, characterization, and nanostructures, in: P. Ducheyne, K.E. Healy, D.W. Hutmacher, D.W. Grainger, C.J. Kirkpatrick (Eds.), Comprehensive Biomaterials, vol. 2, Elsevier, Amsterdam, 2011, pp. 53–76.

[15] S.A. Fowler, H.E. Blackwell, Structure–function relationships in peptoids: recent advances toward deciphering the structural requirements for biological function, Org. Biomol. Chem. 7 (2009) 1508–1524.

[16] R.N. Zuckermann, J.M. Kerr, S.B.H. Kent, W.H. Moos, Efficient method for the preparation of peptoids [oligo (N-substituted) glycines] by submonomer solid-phase synthesis, J. Am. Chem. Soc. 114 (1992) 10646–10647.

[17] R.N. Zuckermann, D.A. Goff, Synthesis of (N-substituted) glycine polymers of defined length and sequence, Polym. Prepr. 35 (1994) 975–976.

[18] G.M. Figliozzi, R. Goldsmith, S.C. Ng, S.C. Banville, R.N. Zuckermann, Synthesis of N-substituted glycine peptoid libraries, Methods Enzymol. 267 (1996) 437–447.

[19] L.S. Richter, D.C. Spellmeyer, E.J. Martin, G.M. Figliozzi, R.N. Zuckermann, Automated synthesis of nonnatural oligomer libraries: the peptoid concept, Comb. Pept. Nonpept. Libr. (1996) 387–404.

[20] A.S. Culf, R.J. Ouellette, Solid-phase synthesis of N-substituted glycine oligomers (alpha-peptoids) and derivatives, Molecules 15 (2010) 5282–5335.

[21] A.M. Rosales, H.K. Murnen, S.R. Kline, R.N. Zuckermann, R.A. Segalman, Determination of the persistence length of helical and non-helical polypeptoids in solution, Soft Matter 8 (2012) 3673–3680.

[22] J. Sun, G.M. Stone, N.P. Balsara, R.N. Zuckermann, Structure–conductivity relationship for peptoid-based PEO-mimetic polymer electrolytes, Macromolecules 45 (2012) 5151–5156.

[23] B.C. Lee, K.A. Dill, R.N. Zuckermann, Synthesis of long non-natural sequence-specific heteropolymers, Polym. Prepr. 46 (2005) 174–175.

[24] L. Guo, D. Zhang, Cyclic poly (alpha-peptoid)s and their block copolymers from N-heterocyclic carbene-mediated ring-opening polymerizations of N-substituted N-carboxylanhydrides, J. Am. Chem. Soc. 131 (2009) 18072–18074.

[25] R.N. Zuckermann, G.M. Figliozzi, S.C. Banville, J.M. Kerr, M.A. Siani, E.J. Martin, et al., Automated tools for the production of non-natural molecular diversity, in: R. Epton (Ed.), Innovations and Perspectives in Solid-Phase Synthesis, Mayflower Worldwide Ltd, Oxford, 1994, pp. 397–402.

[26] T. Uno, E. Beausoleil, R.A. Goldsmith, B.H. Levine, R.N. Zuckermann, New submonomers for poly N-substituted glycines (peptoids), Tetrahedron Lett. 40 (1999) 1475–1478.

[27] H.K. Murnen, A.R. Khokhlov, P.G. Khalatur, R.A. Segalman, R.N. Zuckermann, Impact of hydrophobic sequence patterning on the coil-to-globule transition of protein-like polymers, Macromolecules 45 (2012) 5229–5236.

[28] H. Tran, S.L. Gael, M.D. Connolly, R.N. Zuckermann, Solid-phase submonomer synthesis of peptoid polymers and their self-assembly into highly-ordered nanosheets, J. Vis. Exp. 57 (2011) e3373.

[29] H.R. Kricheldorf, C. von Lossow, G. Schwarz, Primary amine-initiated polymerizations of alanine-NCA and sarcosine-NCA, Macromol. Chem. Phys. 205 (2004) 918–924.

[30] S.G. Waley, J. Watson, The kinetics of the polymerization of sarcosine carbonic anhydride, Proc. Roy. Soc. Lond. A 199 (1949) 499–517.

[31] S.G. Waley, J. Watson, The kinetics of the polymerization of carbonic anhydrides, J. Am. Chem. Soc. 70 (1948) 2299–2300.

[32] M. Sisido, Y. Imanishi, T. Higashimura, Molecular weight distribution of polysarcosine obtained by NCA polymerization, Makromol. Chem. 178 (1977) 3107–3114.
[33] D.G.H. Ballard, C.H. Bamford, Stereochemical aspects of the reactions between alpha-N-carboxy-amino-acid anhydrides and primary and secondary bases, J. Chem. Soc. (1958) 355–360.
[34] M. Stahmann, Polyaminoacids, Polypeptides, and Proteins, University of Wisconsin Press, Madison, WI, 1962.
[35] T. Kidchob, S. Kimura, Y. Imanishi, Preparation, structure and release profile of polypeptide microcapsules, J. Control. Release 40 (1996) 285–291.
[36] H. Tanisaka, S. Kizaka-Kondoh, A. Makino, S. Tanaka, M. Hiraoka, S. Kimura, Near-infrared fluorescent labeled peptosome for application to cancer imaging, Bioconjug. Chem. 19 (2007) 109–117.
[37] A. Makino, S. Kizaka-Kondoh, R. Yamahara, I. Hara, T. Kanzaki, E. Ozeki, et al., Near-infrared fluorescence tumor imaging using nanocarrier composed of poly (L-lactic acid)-block-poly (sarcosine) amphiphilic polydepsipeptide, Biomaterials 30 (2009) 5156–5160.
[38] L. Guo, J. Li, Z. Brown, K. Ghale, D. Zhang, Synthesis and characterization of cyclic and linear helical poly (a-peptoid) s by N-heterocyclic carbene-mediated ring-opening polymerizations of N-substituted N-carboxyanhydrides, Pept. Sci. 96 (2011) 596–603.
[39] L. Guo, S.H. Lahasky, K. Ghale, D. Zhang, N-heterocyclic carbene-mediated zwitterionic polymerization of N-substituted N-carboxyanhydrides towards poly(a-peptoid)s: kinetic, mechanism and architectural control, J. Am. Chem. Soc. 134 (2012) 9163–9171.
[40] C.U. Lee, T.P. Smart, L. Guo, T.H. Epps Iii, D. Zhang, Synthesis and characterization of amphiphilic cyclic diblock copolypeptoids from N-heterocyclic carbene-mediated zwitterionic polymerization of N-substituted N-carboxyanhydride, Macromolecules 44 (2011) 9063–9074.
[41] S.H. Lahasky, W.K. Serem, L. Guo, J.C. Garno, D. Zhang, Synthesis and characterization of cyclic brush-like polymers by N-heterocyclic carbene-mediated zwitterionic polymerization of N-propargyl N-carboxyanhydride and the grafting-to approach, Macromolecules 44 (2011) 9063–9074.
[42] C. Fetsch, A. Grossmann, L. Holz, J.F. Nawroth, R. Luxenhofer, Polypeptoids from N-substituted glycine N-carboxyanhydrides: hydrophilic, hydrophobic, and amphiphilic polymers with Poisson distribution, Macromolecules 44 (2011) 6746–6758.
[43] C. Fetsch, R. Luxenhofer, Highly defined multiblock copolypeptoids: pushing the limits of living nucleophilic ring-opening polymerization, Macromol. Rapid Commun. 19 (2012) 1708–1713.
[44] C. Fetsch, R. Luxenhofer, Thermal properties of aliphatic polypeptoids, Polymers 5 (2013) 112–127.
[45] J.W. Robinson, H. Schlaad, A versatile polypeptoid platform based on N-allyl glycine, Chem. Commun. 48 (2012) 7835–7837.
[46] J.W. Robinson, C. Secker, S. Weidner, H. Schlaad, Thermoresponsive poly(N-C3 glycine)s, Macromolecules 46 (2013) 580–587.
[47] X. Tao, Y. Deng, Z. Shen, J. Ling, Controlled polymerization of N-substituted glycine N-thiocarboxyanhydrides initiated by rare earth borohydrides toward hydrophilic and hydrophobic polypeptoids, Macromolecules 47 (2016) 6173–6180.
[48] H. Sun, J. Zhang, Q. Liu, L. Yu, J. Zhao, Metal-catalyzed copolymerization of imines and CO: a non-amino acid route to polypeptides, Angew. Chem. Int. Ed. Engl. 119 (2007) 6180–6184.

[49] J. Chai, G. Liu, K. Chaicharoen, C. Wesdemiotis, L. Jia, Cobalt-catalyzed carbonylative polymerization of azetidines, Macromolecules 23 (2008) 8980–8985.
[50] Q. Sui, D. Borchardt, D.L. Rabenstein, Kinetics and equilibria of cis/trans isomerization of backbone amide bonds in peptoids, J. Am. Chem. Soc. 129 (2007) 12042–12048.
[51] A. Brulet, F. Boue, J.P. Cotton, About the experimental determination of the persistence length of wormlike chains of polystyrene, J. Phys. II 6 (1996) 885–891.
[52] P. Armand, K. Kirshenbaum, R.A. Goldsmith, S. Farr-Jones, A.E. Barron, K.T.V. Truong, et al., NMR determination of the major solution conformation of a peptoid pentamer with chiral side chains, Proc. Natl. Acad. Sci. U.S.A. 95 (1998) 4309–4314.
[53] P. Armand, K. Kirshenbaum, A. Falicov, R.L. Dunbrack, K.A. Dill, R.N. Zuckermann, et al., Chiral N-substituted glycines can form stable helical conformations, Fold. Des. 2 (1997) 369–375.
[54] J.R. Stringer, J.A. Crapster, I.A. Guzei, H.E. Blackwell, Construction of peptoids with all *trans*-amide backbones and peptoid reverse turns via the tactical incorporation of N-aryl side chains capable of hydrogen bonding, J. Org. Chem. 75 (2010) 6068–6078.
[55] J.A. Crapster, J.R. Stringer, I.A. Guzei, H.E. Blackwell, Design and conformational analysis of peptoids containing N-hydroxy amides reveals a unique sheet-like secondary structure, Pept. Sci. 96 (2011) 604–616.
[56] J.A. Crapster, I.A. Guzei, H.E. Blackwell, A peptoid ribbon secondary structure, Angew. Chem. Int. Ed. Engl. 125 (2013) 5183–5188.
[57] K. Huang, C.W. Wu, T.J. Sanborn, A. James, K. Kirshenbaum, R.N. Zuckermann, et al., A threaded loop conformation adopted by a family of peptoid nonamers, J. Am. Chem. Soc. 128 (2006) 1733–1738.
[58] J.R. Stringer, J.A. Crapster, I.A. Guzei, H.E. Blackwell, Extraordinarily robust polyproline type I peptoid helices generated via the incorporation of -chiral aromatic N-1-naphthylethyl side chains, J. Am. Chem. Soc. 133 (2011) 15559–15567.
[59] B.C. Gorske, J.R. Stringer, B.L. Bastian, S.A. Fowler, H.E. Blackwell, New strategies for the design of folded peptoids revealed by a survey of noncovalent interactions in model systems, J. Am. Chem. Soc. 131 (2009) 16555–16567.
[60] P.A. Jordan, B. Paul, G.L. Butterfoss, P.D. Renfrew, R. Bonneau, K. Kirshenbaum, Oligo (N-alkoxy glycines): trans substantiating peptoid conformations, Pept. Sci. 96 (2011) 617–626.
[61] N.H. Shah, G.L. Butterfoss, K. Nguyen, B. Yoo, R. Bonneau, D.L. Rabenstein, et al., Oligo (N-aryl glycines): a new twist on structured peptoids, J. Am. Chem. Soc. 130 (2008) 16622–16632.
[62] S.H. Park, I. Szleifer, Structural and dynamical characteristics of peptoid oligomers with achiral aliphatic side chains studied by molecular dynamics simulation, J. Phys. Chem. B 115 (2011) 10967–10975.
[63] C.W. Wu, K. Kirshenbaum, T.J. Sanborn, A. James, K. Huang, K.A. Dill, et al., Structural and spectroscopic studies of peptoid oligomers with alpha-chiral aliphatic side chains, J. Am. Chem. Soc. 125 (2003) 13525–13530.
[64] T.J. Sanborn, C.W. Wu, R.N. Zuckermann, A.E. Barron, Extreme stability of helices formed by water-soluble poly-N-substituted glycines (polypeptoids) with alpha-chiral side chains, Biopolymers 63 (2002) 12–20.
[65] S.A. Fowler, R. Luechapanichkul, H.E. Blackwell, Synthesis and characterization of nitroaromatic peptoids: fine tuning peptoid secondary structure through monomer position and functionality, J. Org. Chem. 74 (2009) 1440–1449.

[66] S.B.Y. Shin, B. Yoo, L.J. Todaro, K. Kirshenbaum, Cyclic peptoids, J. Am. Chem. Soc. 129 (2007) 3218–3225.
[67] J.K. Pokorski, L.M.M. Jenkins, H. Feng, S.R. Durell, Y. Bai, D.H. Appella, Introduction of a triazole amino acid into a peptoid oligomer induces turn formation in aqueous solution, Org. Lett. 9 (2007) 2381–2383.
[68] J. Sun, X. Liao, A.M. Minor, N.P. Balsara, R.N. Zuckermann, Morphology–conductivity relationship in crystalline and amorphous sequence-defined peptoid block copolymer electrolytes, J. Am. Chem. Soc. 136 (2014) 14990–14997.
[69] J. Sun, A.A. Teran, X. Liao, N.P. Balsara, R.N. Zuckermann, Crystallization in sequence-defined peptoid diblock copolymers induced by microphase separation, J. Am. Chem. Soc. 136 (2014) 2070–2077.
[70] C.U. Lee, A. Li, K. Ghale, D. Zhang, Crystallization and melting behaviors of cyclic and linear polypeptoids with alkyl side chains, Macromolecules 46 (2013) 8213–8223.
[71] S.H. Lahasky, X. Hu, D. Zhang, Thermoresponsive poly(alpha-peptoid)s: tuning the cloud point temperatures by composition and architecture, ACS Macro Lett. 1 (2012) 580–584.
[72] N. Heine, T. Ast, J. Schneider-Mergener, U. Reineke, L. Germeroth, H. Wenschuh, Synthesis and screening of peptoid arrays on cellulose membranes, Tetrahedron 59 (2003) 9919–9930.
[73] T.S. Burkoth, E. Beausoleil, S. Kaur, D. Tang, F.E. Cohen, R.N. Zuckermann, Toward the synthesis of artificial proteins: the siscovery of an amphiphilic helical peptoid assembly, Chem. Biol. 9 (2002) 647–654.
[74] R.N. Zuckermann, E.J. Martin, D.C. Spellmeyer, G.B. Stauber, K.R. Shoemaker, J.M. Kerr, et al., Discovery of nanomolar ligands for 7-transmembrane G-protein-coupled receptors from a diverse N-(substituted) glycine peptoid library, J. Med. Chem. 37 (1994) 2678–2685.
[75] P.G. Alluri, M.M. Reddy, K. Bachhawat-Sikder, H.J. Olivos, T. Kodadek, Isolation of protein ligands from large peptoid libraries, J. Am. Chem. Soc. 125 (2003) 13995–14004.
[76] M.M. Reddy, K. Bachhawat-Sikder, T. Kodadek, Transformation of low-affinity lead compounds into high-affinity protein capture agents, Chem. Biol. 11 (2004) 1127–1137.
[77] M.M. Reddy, R. Wilson, J. Wilson, S. Connell, A. Gocke, L. Hynan, et al., Identification of candidate IgG biomarkers for Alzheimer's disease via combinatorial library screening, Cell 144 (2011) 132–142.
[78] I. Masip, N. Cortes, M.J. Abad, M. Guardiola, E. Perez-Paya, J. Ferragut, et al., Design and synthesis of an optimized positional scanning library of peptoids: identification of novel multidrug resistance reversal agents, Bioorg. Med. Chem. 13 (2005) 1923–1929.
[79] M. Humet, T. Carbonell, I. Masip, F. Senchez-Baeza, P. Mora, E. Canten, et al., A positional scanning combinatorial library of peptoids as a source of biological active molecules: identification of antimicrobials, J. Comb. Chem. 5 (2003) 597–605.
[80] T.M. Ross, R.N. Zuckermann, C. Reinhard, W.H. Frey, Intranasal administration delivers peptoids to the rat central nervous system, Neurosci. Lett. 439 (2008) 30–33.
[81] J.T. Nguyen, C.W. Turck, F.E. Cohen, R.N. Zuckermann, W.A. Lim, Exploiting the basis of proline recognition by SH3 and WW domains: design of N-substituted inhibitors, Science 282 (1998) 2088–2092.
[82] J.T. Nguyen, M. Porter, M. Amoui, W.T. Miller, R.N. Zuckermann, W.A. Lim, Improving SH3 domain ligand selectivity using a non-natural scaffold, Chem. Biol. 7 (2000) 463–473.

[83] J.C. Hooks, J.P. Matharage, D.G. Udugamasooriya, Development of homomultimers and heteromultimers of lung cancer-specific peptoids, Pept. Sci. 96 (2011) 567–577.
[84] M.T. Dohm, S.L. Seurynck-Servoss, J. Seo, R.N. Zuckermann, A.E. Barron, Close mimicry of lung surfactant protein B by "clicked" dimers of helical, cationic peptoids, Pept. Sci. 92 (2009) 538–553.
[85] B.L. Raveendra, H. Wu, R. Baccala, M.M. Reddy, J. Schilke, J.L. Bennett, et al., Discovery of peptoid ligands for anti-aquaporin 4 antibodies, Chem. Biol. 20 (2013) 351–359.
[86] Y.U. Kwon, T. Kodadek, Encoded combinatorial libraries for the construction of cyclic peptoid microarrays, Chem. Commun. 44 (2008) 5704–5706.
[87] E. Birtalan, B. Rudat, D.K. Kamel, D. Fritz, S.B.L. Vollrath, U. Schepers, et al., Investigating rhodamine B-labeled peptoids: scopes and limitations of its applications, Pept. Sci. 96 (2011) 694–701.
[88] A.A. Fuller, F.J. Seidl, P.A. Bruno, M.A. Plescia, K.S. Palla, Use of the environmentally sensitive fluorophore 4-N,N-dimethylamino-1,8-naphthalimide to study peptoid helix structures, Pept. Sci. 96 (2011) 627–638.
[89] A.R. Statz, R.J. Meagher, A.E. Barron, P.B. Messersmith, New peptidomimetic polymers for antifouling surfaces, J. Am. Chem. Soc. 127 (2005) 7972–7973.
[90] A.R. Statz, A.E. Barron, P.B. Messersmith, Protein, cell and bacterial fouling resistance of polypeptoid-modified surfaces: effect of side-chain chemistry, Soft Matter 4 (2008) 131–139.
[91] W. van Zoelen, R.N. Zuckermann, R.A. Segalman, Tunable surface properties from sequence-specific polypeptoid-polystyrene block copolymer thin films, Macromolecules 45 (2012) 7072–7082.
[92] N.P. Chongsiriwatana, J.A. Patch, A.M. Czyzewski, M.T. Dohm, A. Ivankin, D. Gidalevitz, et al., Peptoids that mimic the structure, function, and mechanism of helical antimicrobial peptides, Proc. Natl. Acad. Sci. U.S.A. 105 (2008) 2794–2799.
[93] M.L. Huang, S.B.Y. Shin, M.A. Benson, V.J. Torres, K. Kirshenbaum, A comparison of linear and cyclic peptoid oligomers as potent antimicrobial agents, Chemmedchem 7 (2012) 114–122.
[94] M. Uchida, G. Mcdermott, M. Wetzler, G.M. Le, M. Myllys, C. Knoechel, et al., Soft X-ray tomography of phenotypic switching and the cellular response to antifungal peptoids in *Candida albicans*, Proc. Natl. Acad. Sci. U.S.A. 106 (2009) 19375–19380.
[95] M.D. Galetti, A.M. Cirigliano, G.M. Cabrera, J.A. Ramírez, Multicomponent synthesis of acylated short peptoids with antifungal activity against plant pathogens, Mol. Divers. 16 (2012) 113–119.
[96] A.E. Corson, S.A. Armstrong, M.E. Wright, E.E. Mcclelland, K.L. Bicker, Discovery and characterization of a peptoid with antifungal activity against *Cryptococcus neoformans*, ACS Med. Chem. Lett. 7 (2016) 1139–1144.
[97] C. Mas-Moruno, L.J. Cruz, P. Mora, A. Francesch, A. Messeguer, E. Pérezpayá, et al., Smallest peptoids with antiproliferative activity on human neoplastic cells, J. Med. Chem. 50 (2007) 2443–2449.
[98] E. Leman, W. Huang, H. Contreras-Trujillo, J.P. Volksmer, S. Willingham, A. Barron, et al., 1072 Inhibition of bladder cancer cell growth by treatment with synthetically derived anti-cancer peptoids, J. Urol. 187 (2012) e435–e436.
[99] M.A. Bailey, M.J. Ingram, D.P. Naughton, A novel anti-oxidant and anti-cancer strategy: a peptoid anti-inflammatory drug conjugate with SOD mimic activity, Biochem. Biophys. Res. Commun. 317 (2004) 1155–1158.

[100] M.L. Huang, D. Ehre, Q. Jiang, C. Hu, K. Kirshenbaum, M.D. Ward, Biomimetic peptoid oligomers as dual-action antifreeze agents, Proc. Natl. Acad. Sci. U. S. A. 109 (2012) 19922–19927.
[101] C.L. Chen, J. Qi, R.N. Zuckermann, J.J. DeYoreo, Engineered biomimetic polymers as tunable agents for controlling $CaCO_3$ mineralization, J. Am. Chem. Soc. 133 (2011) 5214–5217.
[102] S.B.L. Vollrath, C. Hu, S. Brase, K. Kirshenbaum, Peptoid nanotubes: an oligomer macrocycle that reversibly sequesters water via single-crystal-to-single-crystal transformations, Chem. Commun. 49 (2013) 2317–2319.
[103] H.K. Murnen, A.M. Rosales, A.V. Dobrynin, R.N. Zuckermann, R.A. Segalman, Persistence length of polyelectrolytes with precisely located charges, Soft Matter 9 (2013) 90–98.
[104] K.A. Dill, Dominant forces in protein folding, Biochemistry 29 (1990) 7133–7155.
[105] M. Arai, K. Kuwajima, Role of the molten globule state in protein folding, Adv. Protein Chem. 53 (2000) 209–282.
[106] I. Izzo, G. Ianniello, C. De Cola, B. Nardone, L. Erra, G. Vaughan, et al., Structural effects of proline substitution and metal binding on hexameric cyclic peptoids, Org. Lett. 15 (2013) 598–601.
[107] J.M. Holub, H. Jang, K. Kirshenbaum, Clickity-click: highly functionalized peptoid oligomers generated by sequential conjugation reactions on solid-phase support, Org. Biomol. Chem. 4 (2006) 1497–1502.
[108] J.M. Holub, M.J. Garabedian, K. Kirshenbaum, Peptoids on steroids: precise multivalent estradiol-peptidomimetic conjugates generated via azide-alkyne [3 + 2] cycloaddition reactions, QSAR Comb. Sci. 26 (2007) 1175–1180.
[109] B. Kang, S. Chung, Y.D. Ahn, J. Lee, J. Seo, Porphyrin-peptoid conjugates: face-to-face display of porphyrins on peptoid helices, Org. Lett. 15 (2013) 1670–1673.
[110] G. Maayan, M.D. Ward, K. Kirshenbaum, Folded biomimetic oligomers for enantioselective catalysis, Proc. Natl. Acad. Sci. U.S.A. 106 (2009) 13679–13684.
[111] D. Baker, A surprising simplicity to protein folding, Nature 405 (2000) 39–42.
[112] J. Venkatraman, S.C. Shankaramma, P. Balaram, Design of folded peptides, Chem. Rev. 101 (2001) 3131–3152.
[113] B.C. Lee, R.N. Zuckermann, K.A. Dill, Folding a nonbiological polymer into a compact multihelical structure, J. Am. Chem. Soc. 127 (2005) 10999–11009.
[114] B.C. Lee, T.K. Chu, K.A. Dill, R.N. Zuckermann, Biomimetic nanostructures: creating a high-affinity zinc-binding site in a folded nonbiological polymer, J. Am. Chem. Soc. 130 (2008) 8847–8855.
[115] A.R. Khokhlov, P.G. Khalatur, Conformation-dependent sequence design (engineering) of AB copolymers, Phys. Rev. Lett. 82 (1999) 3456–3459.
[116] T. Hjelmgaard, S. Faure, C. Caumes, E. De Santis, A.A. Edwards, C. Taillefumier, Convenient solution-phase synthesis and conformational studies of novel linear and cyclic a,b-alternating peptoids, Org. Lett. 11 (2009) 4100–4103.
[117] C.Y. Huang, T. Uno, J.E. Murphy, S. Lee, J.D. Hamer, J.A. Escobedo, et al., Lipitoids—Novel cationic lipids for cellular delivery of plasmid DNA in vitro, Chem. Biol. 5 (1998) 345–354.
[118] Y. Utku, E. Dehan, O. Ouerfelli, F. Piano, R.N. Zuckermann, M. Pagano, et al., A peptidomimetic siRNA transfection reagent for highly effective gene silencing, Mol. BioSyst. 2 (2006) 312–317.

[119] B.A. Lobo, J.A. Vetro, D.M. Suich, R.N. Zuckermann, C.R. Middaugh, Structure/function analysis of peptoid/lipitoid: DNA complexes, J. Pharm. Sci. 92 (2003) 1905–1918.

[120] J.J. Diaz-Mochon, M.A. Fara, R.M. Sanchez-Martin, M. Bradley, Peptoid dendrimers—microwave-assisted solid-phase synthesis and transfection agent evaluation, Tetrahedron Lett. 49 (2008) 923–926.

[121] B. Yoo, S.B.Y. Shin, M.L. Huang, K. Kirshenbaum, Peptoid macrocycles: making the rounds with peptidomimetic oligomers, Chem. Eur. J. 16 (2010) 5528–5537.

[122] D.G. Rivera, L.A. Wessjohann, Supramolecular compounds from multiple Ugi multicomponent macrocyclizations: peptoid-based cryptands, cages, and cryptophanes, J. Am. Chem. Soc. 128 (2006) 7122–7123.

[123] S. Cho, J. Choi, A. Kim, Y. Lee, Y.U. Kwon, Efficient solid-phase synthesis of a series of cyclic and linear peptoid-dexamethasone conjugates for the cell permeability studies, J. Comb. Chem. 12 (2010) 321–326.

[124] E. Nnanabu, K. Burgess, Cyclic semipeptoids: peptoid-organic hybrid macrocycles, Org. Lett. 8 (2006) 1259–1262.

[125] D. Comegna, M. Benincasa, R. Gennaro, I. Izzo, F. De Riccardis, Design, synthesis and antimicrobial properties of non-hemolytic cationic[alpha]-cyclopeptoids, Bioorg. Med. Chem. 18 (2010) 2010–2018.

[126] R.H. Mattern, T.A. Tran, M. Goodman, Conformational analyses of cyclic hexapeptide analogs of somatostatin containing arylalkyl peptoid and naphthylalanine residues, J. Pept. Sci. 5 (1999) 161–175.

[127] M. Goodman, M. Bhumralkar, E.A. Jefferson, J. Kwak, E. Locardi, Collagen mimetics, Pept. Sci. 47 (1998) 127–142.

[128] G. Melacini, Y. Feng, M. Goodman, Collagen-based structures containing the peptoid residue N-isobutylglycine (Nleu). 6. Conformational analysis of Gly-Pro-Nleu sequences by 1H NMR, CD, and molecular modeling, J. Am. Chem. Soc. 118 (1996) 10725–10732.

[129] E.A. Jefferson, P. Gantzel, E. Benedetti, M. Goodman, A multinuclear Ca^{2+} complex of a linear N-protected glycyl-dipeptoid derivative, J. Am. Chem. Soc. 119 (1997) 3187–3188.

[130] A.M. Rosales, B.L. McCulloch, R.N. Zuckermann, R.A. Segalman, Tunable phase behavior of polystyrene-polypeptoid block copolymers, Macromolecules 45 (2012) 6027–6035.

[131] J. Sun, A.A. Teran, X. Liao, N.P. Balsara, R.N. Zuckermann, Nanoscale phase separation in sequence-defined peptoid diblock copolymers, J. Am. Chem. Soc. 135 (2013) 14119–14124.

[132] K.T. Nam, S.A. Shelby, P.H. Choi, A.B. Marciel, R. Chen, L. Tan, et al., Free-floating ultrathin two-dimensional crystals from sequence-specific peptoid polymers, Nat. Mater. 9 (2010) 454–460.

[133] R. Kudirka, H. Tran, B. Sanii, K.T. Nam, P.H. Choi, N. Venkateswaran, et al., Folding of a single-chain, information-rich polypeptoid sequence into a highly ordered nanosheet, Pept. Sci. 96 (2011) 586–595.

[134] B. Sanii, R. Kudirka, A. Cho, N. Venkateswaran, G.K. Olivier, A.M. Olson, et al., Shaken, not stirred: collapsing a peptoid monolayer to produce free-floating, stable nanosheets, J. Am. Chem. Soc. 133 (2011) 20808–20815.

[135] G.K. Olivier, A. Cho, B. Sanii, M.D. Connolly, H. Tran, R.N. Zuckermann, Antibody-mimetic peptoid nanosheets for molecular recognition, ACS Nano 7 (2015) 9276–9286.

[136] C.L. Chen, R.N. Zuckermann, J.J. Deyoreo, Surface-directed assembly of sequence-defined synthetic polymers into networks of hexagonally patterned nanoribbons with controlled functionalities, ACS Nano 10 (2016) 5314–5320.
[137] H.K. Murnen, A.M. Rosales, J.N. Jaworski, R.A. Segalman, R.N. Zuckermann, Hierarchical self-assembly of a biomimetic diblock copolypeptoid into homochiral superhelices, J. Am. Chem. Soc. 132 (2010) 16112–16119.
[138] J. Sun, X. Jiang, R. Lund, K.H. Downing, N.P. Balsara, R.N. Zuckermann, Self-assembly of crystalline nanotubes from monodisperse amphiphilic diblock copolypeptoid tiles, Proc. Natl. Acad. Sci. U.S.A. 113 (2016) 3954–3959.
[139] A.R. Statz, J.P. Park, N.P. Chongsiriwatana, A.E. Barron, P.B. Messersmith, Surface-immobilised antimicrobial peptoids, Biofouling 24 (2008) 439–448.
[140] R.D. Haynes, R.J. Meagher, A.E. Barron, A chemically synthesized peptoid-based drag-tag enhances free-solution DNA sequencing by capillary electrophoresis, Pept. Sci. 96 (2011) 702–707.
[141] S. Xuan, S. Gupta, X. Li, M. Bleuel, G.J. Schneider, D. Zhang, Synthesis and characterization of well-defined PEGylated polypeptoids as protein-resistant polymers, Biomacromolecules 18 (2017) 951–964.
[142] G.L. Butterfoss, P.D. Renfrew, B. Kuhlman, K. Kirshenbaum, R. Bonneau, A preliminary survey of the peptoid folding landscape, J. Am. Chem. Soc. 131 (2009) 16798–16807.
[143] G.L. Butterfoss, B. Yoo, J.N. Jaworski, I. Chorny, K.A. Dill, R.N. Zuckermann, et al., De novo structure prediction and experimental characterization of folded peptoid oligomers, Proc. Natl. Acad. Sci. U. S. A. 109 (2012) 14320–14325.
[144] R.V. Mannige, T.K. Haxton, C. Proulx, E.J. Robertson, A. Battigelli, G.L. Butterfoss, et al., Peptoid nanosheets exhibit a new secondary-structure motif, Nature 526 (2015) 415–420.

Peptides as therapeutics

8

Modi Wetzler[1] *and Paris Hamilton*[2]
[1]Department of Chemistry, Clemson University, Clemson, SC, United States,
[2]Nephron Pharmaceuticals Corporation, West Columbia, SC, United States

8.1 The role of natural peptides in the body

Over 7000 natural peptides have been identified thus far in the human body—exhibiting a wide array of biological activities. The major role of peptides within biological systems is as signaling molecules; typically binding cell-surface receptors (often to G-protein-coupled receptors) and triggering an intracellular cascade of reactions. Some of the broad classifications of these peptides include hormones, metabolic peptides, and antimicrobial peptides.

Early research on small-molecule and protein hormones, such as testosterone and growth hormone, respectively, sparked knowledge of the endocrine system that led to identification of peptide hormones decades later. The pituitary gland within the brain was correctly thought to be highly important to hormone production. A major role of the pituitary gland is to regulate the amount of hormones within the body: ensuring that not too much (or too little) is in circulation at a time. Unlike small-molecule and protein hormones, however, most peptides are rapidly cleaved in the blood and were often degraded before a blood sample could be analyzed. By the 1980s, however, the paramount role of peptide hormones in human physiology was being recognized [1].

Peptide hormones (Fig. 8.1), such as glucagon and growth hormone-releasing hormone, are secreted primarily from various organs within the neuroendocrine system, but can also be secreted by the heart, gastrointestinal tract, and various other organs. The peptide hormones are released only in response to some extracellular stimuli; for a hormone such as insulin, the stimulus is a high blood-glucose level. Upon their release into the plasma, these peptide hormones allow for the maintenance of homeostasis and the regulation of countless cellular functions. In contrast to steroid-type hormones that are more lipid soluble and act on intracellular targets, peptide hormones typically act on cell surfaces.

In the late 19th century and very early 20th century, diabetes was being recognized as a devastating killer of the population with no treatment options in sight. Around 1900 it was suggested that diabetes resulted from a lack of a pancreatic secretion/hormone. Dr. Frederick Banting, a physician, and his lab assistant Charles Best, a medical student, surgically removed the pancreas of stray dogs—later from calves at a slaughterhouse—to extract the hormone from pancreatic islets without destroying the peptide. This work led them to the isolation and discovery of insulin in 1921. In 1922, Leonard Thompson—on the brink of death—was recorded as the

Peptide Applications in Biomedicine, Biotechnology and Bioengineering.
DOI: http://dx.doi.org/10.1016/B978-0-08-100736-5.00008-9

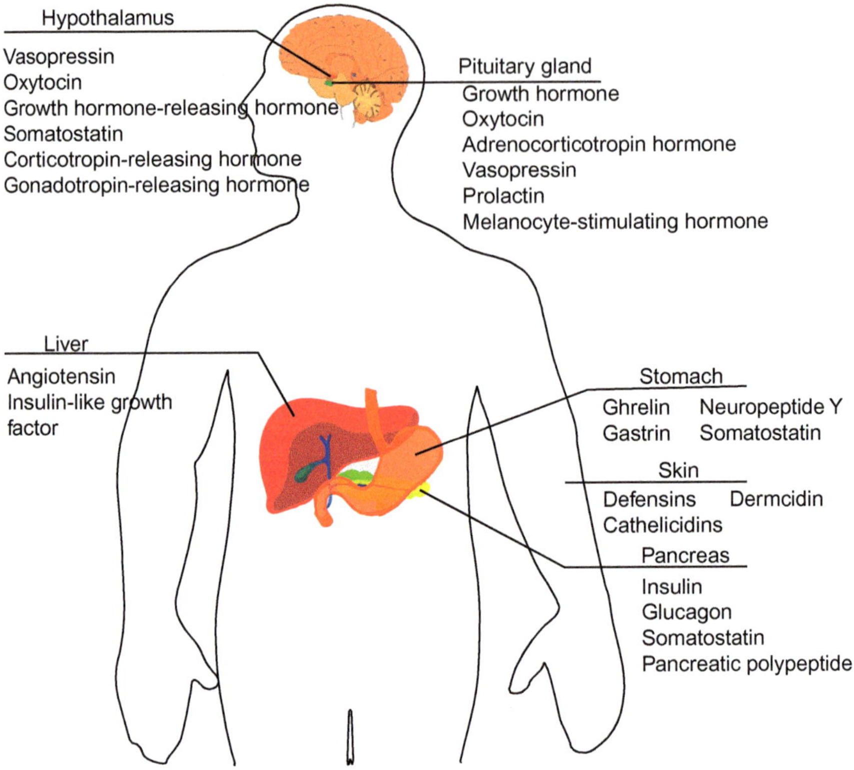

Figure 8.1 Select examples of natural peptide hormones within the body.

first individual to receive an insulin injection for the successful treatment of type I diabetes. Banting went on to receive the 1923 Nobel Prize in Physiology or Medicine for his discovery. The turnaround time between isolation of insulin and the first patient treatment was incredibly fast and is no longer representative of the time-to-market found in today's drug discovery pipeline. Treatment with the peptide insulin is estimated to have saved over 200 million lives already, and merely represents the tip of the iceberg when it comes to just diabetes.

There is a variety of additional key players for diabetes, such as incretin hormones like glucagon-like peptide-1 (GLP-1) and glucose-dependent insulinotropic peptide (GIP), that stimulate a decrease in blood glucose in response to the biological environment. GLP-1 and GIP balance out the action of eating and glucagon itself, which increases blood glucose levels. Additionally, the human body has peptide hormones like ghrelin and neuropeptide Y that increase appetite. Ghrelin is considered "the hunger hormone" because it makes us feel hungry, while GLP-1 has demonstrated the ability to decrease gastric emptying and make us feel full. Many of these peptides have known agonists, antagonists, and inverse agonists, demonstrating the rich potential both as scientific tools and therapeutic candidates.

Peptide hormones, like atrial and brain natriuretic peptides, have major roles in cardiovascular health as well as metabolism regulation [2]. Additionally, a 32-amino-acid peptide, nesiritide, is used for the treatment of congestive heart failure. Similarly, a peptide such as vasopressin, or antidiuretic hormone, has shown versatile physiological effects with not only its cardiovascular role of increasing vasoconstriction, but also in the regulation of water retention via acting on the kidney. Vasopressin and its counterpart, oxytocin, also help to regulate homeostasis between aggression and happiness, respectively, and are critical in mental development and its pathologies, such as autism spectrum disorders.

Antimicrobial peptides (AMPs), increasingly called host defense peptides (HDPs) due to their immunomodulatory effects, are an important component of our human immune system [3]. Typically, AMP/HDPs are cationic and exist primarily in two major classes: defensins (cysteine-rich; ß-sheet) and cathelicidins (noncysteine-rich; α-helix) [4]. These natural peptides have evolved in vivo to offer broad-spectrum protection against microbial infection, traditionally thought to occur through membrane permeabilization [5,6]. With antimicrobial agents, there is always the possibility of bacteria developing resistance; AMPs are no exception [7], but by not binding a specific bacterial target they require more complex adaptation than a single mutation to confer resistance.

Nature has optimized peptides to exhibit a high selectivity for their native receptors, which leads to less adverse side reactions. With over 600 proteases existing in the human body, most peptides are readily degraded. Therefore, peptides exhibit low accumulation in various tissues within biological systems and their degradation pathways are well understood [8]. Consistent with their natural role in maintaining homeostasis, however, many of the disadvantages of peptides arise from their advantages. Their selectivity comes from their large surface area and molecular weight—shortcomings for oral bioavailability. Furthermore, while the proteolytic degradation reduces accumulation leading to potential toxicity, it greatly reduces the serum half-lives. These challenges require different approaches to solve than those traditionally employed for small molecules in the pharmaceutical industry, meaning there is a wealth of opportunities to be mined. In many ways, beyond their high affinity and selectivity for native receptors, peptides are within a "goldilocks zone" between small-molecule and protein therapeutics (Fig. 8.2).

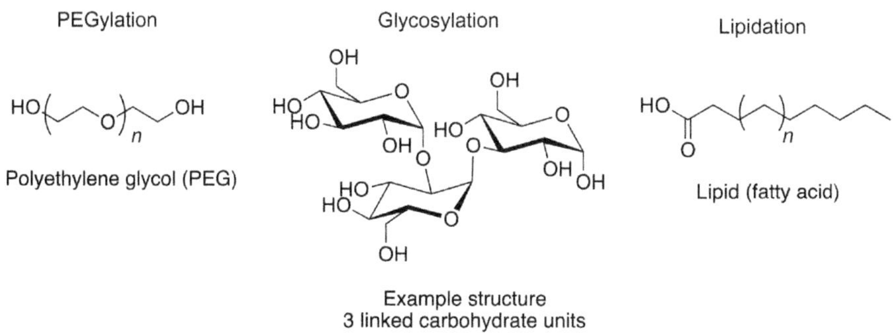

Figure 8.2 General structures for common half-life extension strategies: PEGylation, glycosylation, and lipidation.

8.2 Where peptide therapeutics fit in the spectrum between small molecules and proteins

Standard conventions identify peptides as amino acids linked through amide bonds between 2 and 100 residues in length before being considered a protein. However, the pharmaceutical industry blurs the lines even further by classifying peptides under ~1300 Daltons (Da) as small molecules; gonadorelin (Factrel) is one such peptide therapeutic that falls in that category. Alternatively, protein therapeutics are typically classified as having molecular weights over 10,000 Da. Peptides usually violate Lipinski's rule of five which predicts oral bioavailability [9]. While the rule of five was derived for oral availability via passive diffusion across the intestinal membranes, peptides are usually orally not bioavailable due to proteolytic digestion in the gut. Despite this handicap they are definitely a highly sought-after drug class for targeted therapies with minimal side effects.

The current global peptide therapeutic market is ~$19 billion (US) as of 2015 and is expected to grow to over $23 billion (US) by 2020 [10]. To date, 63 peptide drugs have been US Food and Drug Administration (FDA) approved, and with over 700 clinical trials currently studying peptide therapeutics, the field will likely grow in the number of approved drugs for years to come. Nevertheless, amongst large pharmaceutical companies, peptide molecules are still the minority of drugs being developed.

Peptides of varying length have found usefulness within diverse therapeutic indications from osteoporosis (teriparatide) to cancer (degarelix) and even type II diabetes (exenatide, lixisenatide, abliglutide, dulaglutide, and liraglutide, the first noninsulin peptide blockbuster drug), and obesity [11]. Because of high specificity to their native receptors due to their high surface area, peptides exhibit minimal off-target effects that would traditionally produce side effects in patients. Given the increasing concerns about toxicity (off-target effects) and drug–drug interactions in the pharmaceutical industry, coupled with the higher relative costs of producing protein therapeutics that also inherently mitigate these concerns, one would expect peptide therapeutics to grow for the foreseeable future.

Short natural peptides like oxytocin and vancomycin are approved for usage in various indications, such as labor induction and the treatment of infection, respectively. Some peptide drugs are synthetic (e.g., oxytocin), while others, due to complex modifications, are still isolated from fermentation (e.g., vancomycin).

It has been known for nearly two decades that there are peptides able to trigger cellular apoptosis through caspase activation [12,13]. Research in the area of apoptosis is associated with a growing number of human diseases such as cancer, neurodegenerative disease, and autoimmune disorders. Short peptides such as leuprolide have been approved for the treatment of prostate cancer, but in general anticancer peptides are a nascent field of research.

In fields such as Alzheimer's disease research, amyloid-β peptides are known to play an important but debated role. Disruption of plaque formation or disruption of amyloid precursor protein fragmentation continue to be a challenge [14,15], with notable phase III clinical trials in 2016 of aß plaque removal agents, such as Eli

Lilly's solanezumab, that successfully cleared plaques but did not improve mental function [16]. There are, however, some FDA-approved small-molecule drugs for the treatment of symptoms, such as rivastigmine (Exelon) and memantine (Namenda) [17].

Natural peptides have some very noteworthy benefits when it comes to their usage as therapeutics. Amongst these benefits, they have a potentially improved safety margin, in addition to potentially higher and more efficacious dosing. Therapeutics are rated in accordance with their safety margin—a measure of the difference between the minimum toxic dosage and minimum therapeutic dosage of a drug [18]. In opposition to the high specificity demonstrated by peptides (as discussed above), small molecules generally lack specificity within a biological system—binding to numerous proteins in vivo and leading to off-target effects. However, small molecules are orally bioavailable, have low manufacturing costs, and rest on a century of synthetic and biochemical expertise dating back to Paul Ehrlich. Both small molecules and peptides have high potency towards their targets, but peptides and proteins have exquisite specificity for their native receptors. As drug–drug interactions in aging populations become a dominant concern in the pharmaceutical industry, the high efficacy, safety, and predictable metabolism of peptides is expected to become increasingly appealing.

A majority of the biologically relevant protein–protein interactions occur within the cell as opposed to outside of the cell (or on the cell surface). However, current strategies involving proteins and antibodies are not equipped to handle these intracellular interactions. In the late 1980s, it was discovered that certain proteins could cross the cellular membrane and, subsequently, that only a portion of the protein was necessary for crossing the membrane [19–21]. It is known that cell permeability sharply decreases at higher molecular weights [22,23]. Therefore, large molecules, such as proteins, have encountered difficulties when it comes to crossing the cellular membrane, where potential drug targets may reside. Therefore, researchers are looking into the possibility of targeted delivery through conjugation of small molecules and biologics with cell-penetrating peptides (CPPs). For this strategy to work, the cytostolic delivery efficiency is a metric that needs to be improved. If a CPP has low delivery efficiency, then large dosages must be used to get enough of the drug into the cell. Large dosages, however, tend to lead to increased toxicity. Some of the earliest examples showed poor cytostolic delivery efficiencies (e.g., 1%–5%) [21]. The stabilization of peptides that can cross into intracellular space has been explored using a variety of strategies to optimize peptides for disrupting protein–protein interactions [24].

Therefore, even though peptides may offer advantages where their protein and small-molecule counterparts have limitations, the characteristic shortcomings of natural peptides need to be overcome before they can be useful as therapeutic agents.

8.3 The limitations of peptides as therapeutics

Peptides have some notable shortcomings when it comes to their potential usage as therapeutics [25]. Economically, peptides generally have higher costs associated with their production compared to small molecules. For peptides synthesized through biological methods, the manufacturing process is more complex, and thus more expensive, than a chemical synthesis production [26]. For example, the production cost of a 500 Da molecular weight small molecule is 10-fold lower than that of a 5000 Da peptide [27]. However, for smaller peptides, the cost should be more on par with that of small molecules. The solid-phase synthesis of batch amounts of peptides is faster and cheaper than the recombinant methods used for some longer sequences, especially when economy of scale is taken into consideration, yielding the ability to reduce production costs to <$1 (USD) per gram per amino acid residue [27–29]. Furthermore, peptides obtained from Solid Phase Peptide Synthesis (SPPS) have better purity profiles than those from recombinant methods [30].

Most peptides tend to exhibit poor metabolic stability, membrane permeability, and oral bioavailability. A notable exception to these downfalls is the peptide therapeutic enfuvirtide (Fuzeon), used in the treatment of HIV-1 infection, which exhibits an 84% bioavailability and an impressive 3.8-h biological half-life. In contrast, the half-life of many native peptides is on the order of minutes in blood. The poor stability of peptides to native proteases and peptidases leads to inactivation as well as rapid clearance. For example, approximately a dozen FDA-approved small-molecule diabetic medications are inhibitors of the enzyme dipeptidyl peptidase 4 which inactivates GLP1, showing the important role of just preserving the natural reserves of only one peptide in vivo.

The poor stability of most peptides typically requires constant or frequent dosing for beneficial pharmaceutical effects. The practical effectiveness of constant/frequent dosing relies heavily on patient compliance—the ability of the patient to follow the advice of a medical professional. Realistically, the frequent dosings necessary with most peptide drugs often lead to low patient compliance, which is already a substantial problem in the medical field. Peptide therapeutics are usually injected with only a few orally available peptides to date, such as cyclosporin (Neoral) and desmopressin (Minirini). The route of administration is known to have a significant effect on the pharmacokinetic and pharmacodynamic profile of peptides [31]. An increase in the availability of orally administered peptide therapies is a highly sought-after goal amongst researchers in the field because it would lead to improved patient compliance. However, due to the charge and polarity of peptides, they tend to exhibit low permeability across gut membranes. Companies, such as Protagonist Therapeutics, are aiming to develop oral peptide therapeutics for gastrointestinal disorders and diseases—they currently have one peptide entering phase IIb trials and another oral candidate entering phase I.

Small-molecule therapeutics are typically favored when a binding pocket is targeted for (de)activation deep within a protein, whereas biologics are better for the targeting and disruption of protein–protein interactions, a highly underexploited area of drug design [32,33]. Peptides are preferred when traditional small-molecule approaches fail to produce a successful clinical candidate for a particular therapeutic indication. For reasons such as those mentioned, peptides are ripe to fill a void left by a mounting interest in targets that are not effectively drugged by small molecules or antibodies.

A major concern in the peptide therapeutic area is the possibility of peptide immunogenicity—the ability of peptides to provoke unwanted immune responses. There are debatable views on assessing a product's immunogenicity, leading to nonuniform assessment and descriptions; however, recent efforts have attempted to standardize the type of data collection necessary to report on immunogenicity [34]. If there remains a perception of immunogenicity amongst pharma/biotech investors, then the full potential of peptide therapeutics will not be realized. There is, however, potential for epitopes to be chemically attached to peptides for usage as vaccines [35].

The growth of cardiovascular and metabolic diseases is likely to induce a rise in peptide therapeutic research due to high overlap with natural peptide hormone involvement in the regulation of these biological systems. Additionally, peptide therapeutics are increasingly being studied for their ability to treat gastrointestinal diseases and disorders, such as the 2012 FDA approval of linaclotide (Linzess), a 14-amino-acid peptide guanylate cyclase 2C agonist [36,37]. Various recent advances to overcome the shortcoming of peptide therapeutics have renewed interest in the usage of peptides as potential pharmaceutical agents.

8.4 Efforts to overcome peptide limitations

The challenges to using peptides as therapeutics can be overcome through various chemical modification strategies such as (but not limited to) half-life extension (HLE) strategies [38]. Some common peptide HLE strategies include cyclization, PEGylation (Fig. 8.2), *O*-glycosylation (Fig. 8.2), and backbone modifications (e.g., D-amino acids, *N*-methylation, Cα-methylation, etc.). The HLE strategies are extremely important, in the case of peptides versus proteins especially, because metabolic degradation is known to increase with a decrease in size [39]. The discussion below focuses on chemical modifications at the expense of other important and successful HLE strategies such as formulation development as well as delivery (e.g., by injecting peptide drugs into muscle with slow diffusion into blood instead of directly into blood).

Direct chemical modification (postsynthetic) strategies happen later in the peptide therapeutic optimization process, which can often lead to a reduction in potency for lead compounds—sending researchers back to the proverbial square one. Selective glycosylation—the addition of saccharide units—is one such strategy that has been demonstrated on interferon β-1a to improve various pharmacologically relevant parameters such as half-life, solubility, and stability [40]. This strategy capitalizes on a method which is thought to be a major route of protein elimination from circulation, as well as the importance of sugars in various cellular processes.

Another strategy that employs the conjugation of a known polymeric unit is selective PEGylation; referring to the attachment of polyethylene glycol (PEG) units to a target molecule. These PEG units typically exist as large polymer chains and have found considerable usage within the pharmaceutical industry, as laxatives, and within cosmetics due to their water-soluble, hydro- and lipophilic, generally nontoxic nature. Lysine residues are by far amongst the most PEGylated site on peptides, however, this can typically reduce efficacy, especially if the original lysine was part of a salt-bridge, binding pocket, or active site, and was converted to an uncharged amide as is usually the case. As an alternative, the site-specific, long-chain PEGylation of glutamine has been achieved successfully on a range of peptides (and proteins) by several groups [41–44]. Recent research has identified that a non-neglible percentage of healthy individuals likely have some antibodies to PEG, which could diminish the effectiveness of any PEGylated therapeutics used on these individuals [45–47].

For the pharmaceutical industry to minimize financial risks associated with failed drug candidates late in the development pipeline it is crucial to move certain modification strategies and testing to earlier in the drug development process. The incorporation of unnatural amino acids directly into SPPS procedures is a more facile and straightforward modification strategy than post-SPPS derivatization. This presynthetic modification method has allowed researchers to introduce various functionalities and modifications such as D-amino acids, *N*-methylated backbones, Cα-methylated amino acids, lipidation (Fig. 8.2), alkene side-chains for peptide stapling, and more recently short PEG-chains.

Peptide stapling is used to stabilize secondary structure and improve protease resistance, by incorporating alkene side chains in two locations and using an alkene metathesis catalyst to covalently link them together. The approach has been widely investigated in the literature, but there are no clinically successful candidates with this modification yet. It has also spawned a new type of thinking about covalent crosslinking of peptides. For example, recent studies have introduced an *N*-arylation method capable of macrocyclizing unprotected peptides, leading to improved proteolytic stability and improved binding affinity [48–51]. Strategies like *N*-methylation/alkylation and D-amino acid incorporation are often used in combination with other strategies in order to achieve the maximum beneficial effects, such as in the widely used cyclosporin A and recently approved etelcalcetide (Parsabiv, Fig. 8.3). Usually, however, swapping out L-amino acids for their D-counterparts causes local kinks in structure which then affect native activity. To contrast this inherent disadvantage, D-amino acids can be readily incorporated using standard SPPS protocols so these modifications can be screened early in the hit-to-lead optimization process.

Lipid conjugation is one practical approach that has circumvented some of the delivery issues associated with unfavorable peptide pharmacokinetics. The rationale for this approach is that the lipid chain increases the association with albumin, which gives the therapeutic agent increased protection from blood-circulating peptidases and proteases. This strategy has led to two notable lipid–peptide conjugates currently on the market; recombinant insulin detemir (Levemir) and liraglutide (Victoza), a blockbuster peptide drug. Only the latter is formed through direct chemical modification with a C16 fatty acid chain after expression of recombinant DNA in *Saccharomyces cerevisiae*. The C14 fatty acid chain of insulin detemir is installed

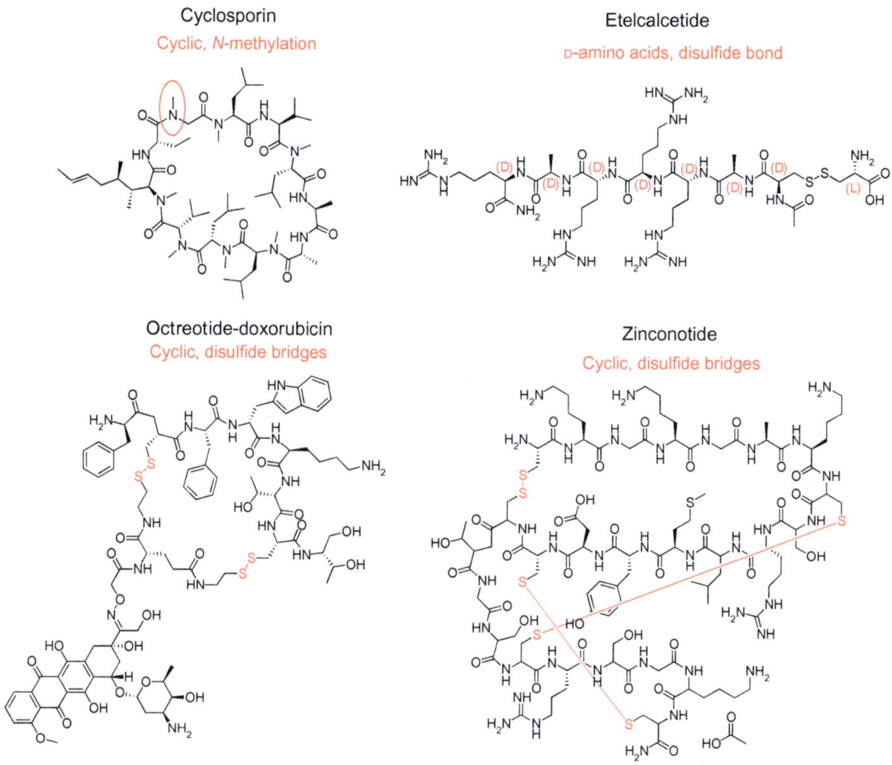

Figure 8.3 Select examples of strategies to overcome limitations of peptide therapeutics.

at a lysine residue during its complete biosynthesis within yeast cells (all insulins are at least largely biosynthesized due to their complex disulfide pattern) [52].

Cyclic peptides have proven to be a great strategy for HLE of short peptide sequences such as octreotide (Sandostatin) and lanreotide (Somatuline), both of which are long-acting analogues of somatostatin. As is the case with linaclotide, the incorporation of multiple disulfide bonds has demonstrated the ability to improve metabolic resistance. Research into this area of multiple disulfide bonds can be traced back to the work of David J. Craik and coworkers for their late 20th century work on plant peptides known as cyclotides, especially kalata B1 [53,54]. Despite overwhelming skepticism in the pharmaceutical industry at the time, their group was able to demonstrate that peptides with this structural motif were highly stable to extreme conditions [55]. Spawning from these early discoveries, pharmaceutical researchers worldwide have gained increasing interest in trying to modify these (and other) cyclic peptide structures in order to overcome stability limitations. Alanine scans and lysine scans have been investigated to identify potential sites important for bioactivity as well as for optimization [56,57]. The chemical synthesis of these cysteine knot motifs has been investigated with current SPPS technology and other strategies [58–66], but promising biological synthesis approaches have also been developed [67–70]. Cyclotides have a diverse array of biological activities from antimicrobial to antitumor, which could lend inspiration to current and future research within the broader peptide therapeutics community [71].

Ziconotide (Prialt, Fig. 8.3), a conotoxin derived from a cone snail (*Conus magnus*), is similar to cyclotides such as kalata B1 in that it has multiple disulfide bridges. This particular cyclic peptide has a biological half-life of several hours compared to the typical minutes of uncyclized peptides. Cyclosporin (Fig. 8.3) is another cyclic peptide (albeit with no disulfide bridges) that is approved for multiple indications, however its backbone is also N-methylated. Research has also suggested that ester bonds in the backbone of cyclic peptides, such as jasplankinolide, are stable to protease resistance, and possibly to the action of lipases/esterases, which are also present in serum [72]. Stereochemistry of the side chains on peptide macrocycles is also known to affect physiochemical properties like cellular permeability.

More recent work has combined the strategies of cyclization and CPPs to improve proteolytic stability as well as cytostolic delivery efficiency [73–75]. These cyclic CPPs can deliver small-molecule cargos as well as proteins and short peptides [73,74]. Historically, individual groups have employed one of the strategies for enhancing peptide stability within their comfort zones, but combining multiple techniques has been rare. In the future, the opposite is more likely to be true by using every tool in the toolbox to really push peptide therapeutics forward.

Peptide–drug conjugates are a method for introducing multiple functionalities within the same molecule. For instance, one molecule may be highly specific for a desired receptor, whereas another molecule is toxic to a cell so bridging the two—through a linker molecule—yields a molecule with potentially synergistic effects. These conjugates have found usefulness as targeted delivery systems [76] for research areas such as targeted tumor therapy [48]. Octreotide–doxorubicin (Fig. 8.3) is one such conjugate for tumor-targeted delivery [77].

A future research area for peptide therapeutics could likely be with bicyclic and polycyclic peptides—a currently underdeveloped class that displays potential in the form of very high affinity, relatively small size (1.5–2 kDa), and easy tunability [78–80]. Multiple cyclizations can be achieved with the same chemical approach such as disulfide bonds (evident in the multiple disulfide bonds present in the conotoxins discussed above, as well as cysteine knot proteins and other families), or by incorporating multiple chemically orthogonal approaches such as stapling, disulfides, and click chemistry. Bicyclic peptides can be cyclized using disulfide bridges through traditional solid-phase peptide synthesis [81]. The bicyclic peptides, as a platform, have a lot of potential for multiple indications although most research at the moment is been directed towards tumor-targeting [82–84]. Bicyclic peptides have even displayed the ability to interrupt protein–protein interactions [85]. For example, Preclinical testing of BT1718, a lead bicycle drug conjugate by Bicycle Therapeutics, has shown the ability to effectively transport cytotoxic payloads into tumors and is expected to enter clinical trials in 2017 [86].

8.5 Conclusion

In summary, the usage of peptides as therapeutics is a market full of growth potential. Historically, peptides have flown under the radar in terms of therapeutic applications, but as technology has improved, peptide limitations are becoming more manageable. Owing to their great receptor specificity and binding affinity, peptide therapeutics will continue to gain growing importance within the pharmaceutical industry; especially as human life expectancy increases and individuals are taking progressively more medications for chronic illnesses. There is no one best peptide modification strategy that will overcome the issues associated with using peptides as therapeutics, but a mixture of strategies could lead to dramatically improved clinical outcomes for drug developers. Yet, for the peptide therapeutic market to be more effectively mined: (1) new scaffolds need to be elucidated from natural sources; (2) the threat of peptide immunogenicity needs to be addressed; (3) chemical modification strategies need to be expanded and combined at earlier stages of the drug development process; and (4) the cost of large-scale peptide manufacturing will need to decrease as technology improves. Since many of these approaches and solutions have independently seen rapid improvements in the past 10–20 years, the future for peptide therapeutics looks bright.

References

[1] H.D. Niall, The evolution of peptide hormones, Ann. Rev. Physiol. 44 (1982) 615–624.
[2] T. Suzuki, T. Yamazaki, Y. Yazaki, The role of natriuretic peptides in the cardiovascular system, Cardiovasc. Res. 51 (2001) 489–494.
[3] J. Wiesner, A. Vilcinskas, Antimicrobial peptides: the ancient arm of the human immune system, Virulence 1 (2010) 440–464.
[4] E. Guani-Guerra, T. Santos-Mendoza, S.O. Lugo-Reyes, L.M. Teran, Antimicrobial pepties: general overview and clinical implications in human health and disease, Clin. Immunol. 135 (2010) 1–11.
[5] R.L. Gallo, M. Murakami, T. Ohtake, M. Zaiou, Biology and clinical relevance of naturally occurring antimicrobial peptides, J. Allergy Clin. Immunol. 110 (2002) 823–831.
[6] G. Diamond, N. Beckloff, A. Weinberg, K.O. Kisich, The roles of antimicrobial peptides in innate host defense, Curr. Pharm. Des. 15 (2009) 2377–2392.
[7] S. Gruenheid, H. Le Moual, Resistance to antimicrobial peptides in gram-negative bacteria, FEMS Microbiol. Lett. 330 (2012) 81–89.
[8] C. Lopez-Otin, L.M. Matrisian, Emerging roles of proteases in tumour suppression, Nat. Rev. Cancer 7 (2007) 800–808.
[9] C.A. Lipinski, F. Lombardo, B.W. Dominy, P.J. Feeney, Experimental and computational approaches to estimate solubility and permeability in drug discovery and development settings, Adv. Drug Deliv. Rev. 23 (1997) 3–25.
[10] Transparency Market Research, Peptide therapeutics market—increasing demand for peptide therapeutics in cancer and diabetes treatment to boost sales; global industry analysis, size, share, growth, trends and forecast 2020: TMR. GlobeNewswire, Internet,

12 September 2016. <https://globenewswire.com/news-release/2016/09/12/871199/0/en/Peptide-Therapeutics-Market-Increasing-Demand-for-Peptide-Therapeutics-in-Cancer-and-Diabetes-Treatment-to-Boost-Sales-Global-Industry-Analysis-Size-Share-Growth-Trends-and-Forecas.html>.

[11] J. Thundimadathil, Cancer treatment using peptides: current therapies and future prospects, J. Amino Acids 2012 (2012) 967347.

[12] C.D. Buckley, D. Pilling, N.V. Henriquez, G. Parsonage, K. Threlfall, D. Scheel-Toellner, et al., RGD peptides induce apoptosis by direct caspase-3 activation, Nature 397 (1999) 534–539.

[13] A. Philchenkov, Caspases: potential targets for regulating cell death, J. Cell. Mol. Med. 8 (2004) 432–444.

[14] S.A. Jacobson, M.N. Sabbagh, Investigational drugs for the treatment of AD: what can we learn from negative trials? Alzheimers Res. Ther. 3 (2011) 14.

[15] K. Mullane, M. Williams, Alzheimer's therapeutics: continued clinical failures question the validity of the amyloid hypothesis—but what lies beyond? Biochem. Pharmacol. 85 (2013) 289–305.

[16] E.R. Siemers, K.L. Sundell, C. Carlson, M. Case, G. Sethuraman, H. Liu-Seifert, et al., Phase 3 solanezumab trials: secondary outcomes in mild Alzheimer's disease patients, Alzheimers Dement. 12 (2016) 110–120.

[17] C. Hyde, J. Peters, M. Bond, G. Rogers, M. Hoyle, R. Anderson, et al., Evolution of the evidence on the effectiveness and cost-effectiveness of acetylcholinesterase inhibitors and memantine for Alzheimer's disease: systematic review and economic model, Age Ageing 42 (2013) 14–20.

[18] K. Fosgerau, T. Hoffmann, Peptide therapeutics: current status and future directions, Drug Discov. Today 20 (2015) 122–128.

[19] M. Green, P.M. Loewenstein, Autonomous functional domains of chemically synthesized human immunodeficiency virus tat trans-activator protein, Cell 55 (1988) 1179–1188.

[20] A.D. Frankel, C.O. Pabo, Cellular uptake of the tat protein from human immunodeficiency virus, Cell 55 (1988) 1189–1193.

[21] J.R. La Rochelle, G.B. Cobb, A. Steinauer, E. Rhoades, A. Schepartz, Fluorescence correlation spectroscopy reveals highly efficient cytosolic delivery of certain penta-arg proteins and stapled peptides, J. Am. Chem. Soc. 137 (2015) 2536–2541.

[22] E.A. Villar, D. Beglov, S. Chennamadhavuni, J.A.J. Porco, D. Kozakov, S. Vajda, et al., How proteins bind macrocycles, Nat. Chem. Biol. 10 (2014) 723–731.

[23] P. Matsson, J. Kihlberg, How big is too big for cell permeability? J. Med. Chem. 60 (2017) 1662–1664.

[24] N. Tsomaia, Peptide therapeutics: targeting the undruggable space, Eur. J. Med. Chem. 94 (2015) 459–470.

[25] L. Otvos, J.D. Wade, Current challenges in peptide-based drug discovery, Front. Chem. 2 (2014) 1–4.

[26] R.P. Rozek, Economic aspects of small and large molecule pharmaceutical technologies, Adv. Econ. Bus. 1 (2013) 258–269.

[27] B.L. Bray, Large-scale manufacture of peptide therapeutics by chemical synthesis, Rev. Drug Discov. 2 (2003) 587–593.

[28] L. Otvos, Peptide-based drug research and development: relative costs, comparative value, Pharm. Outsourcing 15 (2014) 16–20.

[29] R. Lax, The future of peptide development in the pharmaceutical industry, Pharma. Int. Pept. Rev. (2010) 10–15.

[30] F. Guzman, S. Barberis, A. Illanes, Peptide synthesis: chemical or enzymatic, J. Biotechnol. 10 (2007) 279–314.
[31] M. Grant, A. Leone-Bay, Peptide therapeutics: it's all in the delivery, Ther. Deliv. 3 (2012) 981–996.
[32] A. Loregian, G. Palù, Disruption of protein–protein interactions: towards new targets for chemotherapy, J. Cell. Physiol. 204 (2005) 750–762.
[33] P. Wójcik, Peptide-based inhibitors of protein–protein interactions, Bioorg. Med. Chem. Lett. 26 (2016) 707–713.
[34] G. Shankar, S. Arkin, L. Cocea, V. Devanarayan, S. Kirshner, A. Kromminga, et al., Assessment and reporting of the clinical immunogenicity of therapeutic proteins and peptides—harmonized terminology and tactical recommendations, AAPS J. 16 (2014) 658–673.
[35] N.L. Dudek, P. Perlmutter, M.- Aguilar, N.P. Croft, A.W. Purcell, Epitope discovery and their use in peptide based vaccines, Curr. Pharm. Des. 16 (2010) 3149–3157.
[36] B.L. Love, A. Johnson, L.S. Smith, Linaclotide: a novel agent for chronic constipation and irritable bowel syndrome, Am. J. Health Syst. Pharm. 71 (2014) 1081–1091.
[37] S.W. Yu, S.S. Rao, Advances in the management of constipation-predominant irritable bowel syndrome: the role of linaclotide, Therap. Adv. Gastroenterol. 7 (2014) 193–205.
[38] S.B. Gunnoo, A. Madder, Bioconjugation – using selective chemistry to enhance the properties of proteins and peptides as therapeutics and carriers, Org. Biomol. Chem. 14 (2016) 8002–8013.
[39] D. Meijer, K. Ziegler, Biological Barriers to Protein Delivery, Plenium Press, New York, 1993.
[40] K. Song, I.S. Yoon, N.A. Kim, D.H. Kim, J. Lee, H.J. Lee, et al., Glycoengineering of interferon-β 1a improves its biophysical and pharmacokinetic properties, PLoS One 9 (2014), e96967/1–e96967/14.
[41] S. Fabio, S. Silvia, S. Rodolfo, T. Giancarlo, O. Gaetano, S. Riccardo, et al., Enzymatic mono-pegylation of glucagon-like peptide 1 towards long lasting treatment of type 2 diabetes, Results Pharma Sci. 2 (2012) 58–65.
[42] H. Sato, Enzymatic procedure for site-specific PEGylation of proteins, Adv. Drug Deliv. Rev. 54 (2002) 487–504.
[43] A. Fontana, B. Spolaore, A. Mero, F.M. Veronese, Site-specific modification and PEGylation of pharmaceutical proteins mediated by transglutaminase, Adv. Drug Deliv. Rev. 60 (2008) 13–28.
[44] A. Mero, B. Spolaore, F.M. Veronese, A. Fontana, Transglutaminase-mediated PEGylation of proteins: direct identification of the sites of protein modification by mass spectrometry using a novel monodisperse PEG, Bioconjug. Chem. 20 (2009) 384–389.
[45] Q. Yang, S.K. Lai, Anti-PEG immunity: emergence, characteristics, and unaddressed questions, Wiley Interdiscip. Rev. Nanomed. Nanobiotechnol. 7 (2015) 655–677.
[46] C. Lubich, P. Allacher, M. De la Rosa, A. Bauer, T. Prenninger, F.M. Horling, et al., The mystery of antibodies against polyethylene glycol (PEG) – what do we know? Pharm. Res. 33 (2016) 2239–2249.
[47] J.L. Price, E.T. Powers, J.W. Kelly, N-PEGylation of a reverse turn is stabilizing in multiple sequence contexts, unlike N-GlcNAcylation, ACS Chem. Biol. 6 (2011) 1188–1192.
[48] D. Böhme, A.G. Beck-Sickinger, Drug delivery and release systems for targeted tumor therapy, J. Pept. Sci. 21 (2015) 186–200.

[49] G. Lautrette, F. Toute, G.G. Lee, P. Dai, B.L. Pentelute, Nitrogen arylation for macrocyclization of unprotected peptides, J. Am. Chem. Soc. 138 (2016) 8340–8343.
[50] M. Nahrwold, C. Weib, T. Bogner, F. Mertink, J. Conradi, B. Sammet, et al., Conjugates of modified cryptophycins and RGD-peptides enter target cells by endocytosis, J. Med. Chem. 56 (2013) 1853–1864.
[51] D. Zwanziger, U.K. Irfan, I. Neundorf, S. Sieger, L. Lehmann, M. Friebe, et al., Novel chemically modified analogues of neuropeptide Y for tumor targeting, Bioconjug. Chem. 19 (2008) 1430–1438.
[52] A.T. Bockus, K.W. Lexa, C.R. Pye, A.S. Kalgutkar, J.W. Gardner, K.C.R. Hund, et al., Probing the physicochemical boundaries of cell permeability and oral bioavailability in lipophilic macrocycles inspired by natural products, J. Med. Chem. 58 (2015) 4581–4589.
[53] O. Saether, D.J. Craik, I.D. Campbell, K. Sletten, J. Juul, D.G. Norman, Elucidation of the primary and three-dimensional structure of the uterotonic polypeptide kalata B1, Biochemistry 34 (1995) 4147–4158.
[54] D.J. Craik, N.L. Daly, T. Bond, C. Waine, Plant cyclotides: a unique family of cyclic and knotted proteins that defines the cyclic cysteine knot structural motif, J. Mol. Biol. 294 (1999) 1327–1336.
[55] M.L. Colgrave, D.J. Craik, Thermal, chemical, and enzymatic stability of the cyclotide kalata B1: the importance of the cyclic cysteine knot, Biochemistry 43 (2004) 5965–5975.
[56] S.M. Simonsen, L. Sando, K.J. Rosengren, C.K. Wang, M.L. Colgrave, N.L. Daly, et al., Alanine scanning mutagenesis of the prototypic cyclotide reveals a cluster of residues essential for bioactivity, J. Biol. Chem. 283 (2008) 9805–9813.
[57] Y.H. Huang, M.L. Colgrave, R.J. Clark, A.C. Kotze, D.J. Craik, Lysine-scanning mutagenesis reveals a previously unidentified amendable face of the cyclotide kalata B1 for the optimization of nematocidal activity, J. Biol. Chem. 285 (2010) 10797–10805.
[58] J.P. Tam, Y.A. Lu, Synthesis of large cyclic cystine-knot peptide by orthogonal coupling strategy using unprotected peptide precursors, Tetrahedron Lett. 38 (1997) 5599–5602.
[59] J.P. Tam, Y.A. Lu, A biomimetic strategy in the synthesis and fragmentation of cyclic protein, Protein Sci. 7 (1998) 1583–1592.
[60] N.L. Daly, S. Love, P.F. Alewood, D.J. Craik, Chemical synthesis and folding pathways of large cyclic polypeptides: studies of the cystine knot polypeptide kalata B1, Biochemistry 38 (1999) 10606–10614.
[61] J.P. Tam, Y.A. Lu, Q. Yu, Thia zip reaction for synthesis of large cyclic peptides: mechanisms and applications, J. Am. Chem. Soc. 121 (1999) 4316–4324.
[62] P. Thongyoo, E.W. Tate, R.J. Leatherbarrow, Total synthesis of the macrocyclic cysteine knot microprotein MCoTI-II, Chem. Commun. (Camb.) (2006) 2848–2850.
[63] O. Avrutina, H.U. Schmoldt, D. Gabrijelcic-Geiger, A. Wentzel, H. Frauendorf, C.P. Sommerhoff, et al., Head-to-tail cyclized cystine-knot peptides by a combined recombinant and chemical route of synthesis, ChemBioChem 9 (2008) 33–37.
[64] M. Cemazar, D.J. Craik, Microwave-assisted boc-solid phase peptide synthesis of cyclic cysteine-rich peptides, J. Pept. Sci. 14 (2008) 683–689.
[65] J.S. Zheng, H.N. Chang, J. Shi, L. Liu, Chemical synthesis of a cyclotide via intramolecular cyclization of peptide O-esters, Sci. China Chem. 55 (2012) 64–69.

[66] J.S. Zheng, S. Tang, Y. Guo, H.N. Chang, L. Liu, Synthesis of cyclic peptides and cyclic proteins via ligation of peptide hydrazides, ChemBioChem 13 (2012) 542–546.
[67] R.H. Kimura, A.T. Tran, J.A. Camarero, Biosynthesis of the cyclotide kalata B1 by using protein splicing, Angew. Chem. Int. Ed. 118 (2006) 987–990.
[68] J.A. Camarero, R.H. Kimura, Y.H. Woo, J. Cantor, E. Steenblock, A cell based approach for the biosynthesis/screening of cyclic peptide libraries against bacterial toxins, Chim. Oggi. 25 (2007) 20–23.
[69] J.A. Camarero, R.H. Kimura, Y.H. Woo, A. Shekhtman, J. Cantor, Biosynthesis of a fully functional cyclotide inside living bacterial cells, ChemBioChem 8 (2007) 1363–1366.
[70] T.L. Aboye, J.A. Camarero, Biological synthesis of circular polypeptides, J. Biol. Chem. 287 (2012) 27026–27032.
[71] D.J. Craik, Joseph rudinger memorial lecture: discovery and applications of cyclotides, J. Pept. Sci. 19 (2013) 393–407.
[72] A.T. Bockus, C.M. McEwen, R.S. Lokey, Form and function in cyclic peptide natural products: a pharmacokinetic perspective, Curr. Top. Med. Chem. 13 (2013) 821–836.
[73] Z. Qian, T. Liu, Y.Y. Liu, R. Briesewitz, A.M. Barrios, S.M. Jhiang, et al., Efficient delivery of cyclic peptides into mammalian cells with short sequence motifs, ACS Chem. Biol. 8 (2013) 423–431.
[74] Z. Qian, J.R. LaRochelle, B. Jiang, W. Lian, R.L. Hard, N.G. Selner, et al., Early endosomal escape of a cyclic cell-penetrating peptide allows effective cytosolic cargo delivery, Biochemistry 53 (2014) 4034–4046.
[75] Z. Qian, A. Martyna, R.L. Hard, J. Wang, G. Appiah-Kubi, C. Coss, et al., Discovery and mechanism of highly efficient cyclic cell-penetrating peptides, Biochemistry 55 (2016) 2601–2612.
[76] Y. Wang, A.G. Cheetham, G. Angacian, H. Su, L. Xie, H. Cui, Peptide-drug conjugates as effective prodrug strategies for targeted delivery, Adv. Drug Deliv. Rev. 110–111 (2017) 112–126.
[77] M. Lelle, S. Kaloyanova, C. Freidel, M. Theodoropoulou, M. Musheev, C. Niehrs, et al., Octreotide-mediated tumor-targeted drug delivery via a cleavable doxorubicin–peptide conjugate, Mol. Pharm. 12 (2015) 4290–4300.
[78] C. Heinis, T. Rutherford, S. Freund, G. Winter, Phage-encoded combinatorial chemical libraries based on bicyclic peptides, Nat. Chem. Biol. 5 (2009) 502–507.
[79] V. Baeriswyl, H. Rapley, L. Pollaro, C. Stace, D. Teufel, E. Walker, et al., Bicyclic peptides with optimized ring size inhibit human plasma kallikrein and its orthologues while sparing paralogous proteases, ChemMedChem 7 (2012) 1173–1176.
[80] S. Chen, D. Bertoldo, A. Angelini, F. Pojer, C. Heinis, Peptide ligands stabilized by small molecules, Angew. Chem. Int. Ed. 53 (2014) 1602–1606.
[81] R.M.J. Liskamp, Peptides: bicycling into cells, Nat. Chem. 6 (2014) 855–857.
[82] T.B. Trinh, P. Upadhyaya, Z. Qian, D. Pei, Discovery of a direct ras inhibitor by screening a combinatorial library of cell-permeable bicyclic peptides, ACS Comb. Sci. 18 (2016) 75–85.
[83] Z. Qian, C.A. Rhodes, L.C. McCroskey, J. Wen, G. Appiah-Kubi, D.J. Wang, et al., Enhancing the cell permeability and metabolic stability of peptidyl drugs by reversible bicyclization, Angew. Chem. Int. Ed. 55 (2016) 1–6.

[84] L. Pollaro, S. Raghunathan, J. Morales-Sanfrutos, A. Angelini, S. Kontos, C. Heinis, Bicyclic peptide conjugated to an albumin-binding tag diffuse efficiently into solid tumors, Mol. Cancer Ther. 14 (2015) 151–161.

[85] W. Lian, P. Upadhyaya, C.A. Rhodes, Y. Liu, D. Pei, Protein–protein interaction inhibitors: discovery of a tumor necrosis factor-α antagonist, J. Am. Chem. Soc. 135 (2013) 11990–11995.

[86] B. Adams, CRUK research arm starts first-in-class bicycle therapeutics trial in cancer. FierceBiotech, 2016. <http://www.fiercebiotech.com/biotech/cruk-research-arm-starts-first-class-bicycle-therapeutics-trial-cancer>.

Peptides for biopharmaceutical applications

9

Dominic Agyei[1], Kei-Xian Tan[2], Sharadwata Pan[3], Chibuike C. Udenigwe[4] and Michael K. Danquah[2]
[1]University of Otago, Dunedin, New Zealand, [2]Curtin University, Sarawak, Malaysia, [3]Technical University of Munich, Freising, Germany, [4]University of Ottawa, Ottawa, ON, Canada

9.1 Introduction

The art and practices of the apothecary have undergone several changes since medieval times. In modern time, different classes of drugs have been developed and approved for use in the treatment of ill health. Among these, biologically active peptides represent a diverse class of therapeutic agents, which in recent times has drawn the attention of pharmaceutical manufacturers for a number of reasons. First, there is a wide diversity of peptides for use in treating a plethora of diseases. In living host systems, bioactive peptides are able to trigger a wide range of biological activities such as antioxidant, antihypertensive, antimicrobial, cytomodulatory, antidiabetic, opioid agonist, anxiolytic, immunomodulatory, anticancer, and antiinflammatory activities, among others [1−4]. In fact, target disease areas for therapeutic proteins and peptides approved by the United States Food and Drug Administration (FDA) range from disorders (metabolic, immunological, hormonal, hematological, genetic, osteological, cardiac) to cancers and infectious diseases [5]. Moreover, certain bioactive peptides (so-called multifunctional peptides) are able to trigger more than one physiological response in living systems. Examples of multifunctional peptides includes hydrolysates of lactoferricin, a 25-sequence pepsin hydrolysate of lactoferrin, with demonstrated anticancer, antitumor, antiparasitical, and antimicrobial properties [6,7]; and lunacin, a soy-derived 43-sequence peptide that has shown antiinflammatory, antioxidant, and antihypertensive properties [8,9]. Multifunctional peptides are excellent candidates for drug development because most disease conditions develop through complex pathological mechanisms that present a multitude of symptoms, and these peptides can be used to simultaneously target such multisymptom diseases [10,11].

Second, there has been a rise in regulatory approval of peptide drugs over the past few decades [12]. Third, there has been growth in technological breakthroughs for the bioprocessing of peptides. Several production techniques have been used for the production of peptide drugs (Fig. 9.1). Some of these production strategies (e.g., enzymatic hydrolysis of proteins and microbial fermentation of proteins) are relatively low cost and therefore can address the issue of relatively high synthesis costs [2]. Also, combinatorial production strategies such as chemoenzymatic methods can

Peptide Applications in Biomedicine, Biotechnology and Bioengineering.
DOI: http://dx.doi.org/10.1016/B978-0-08-100736-5.00009-0

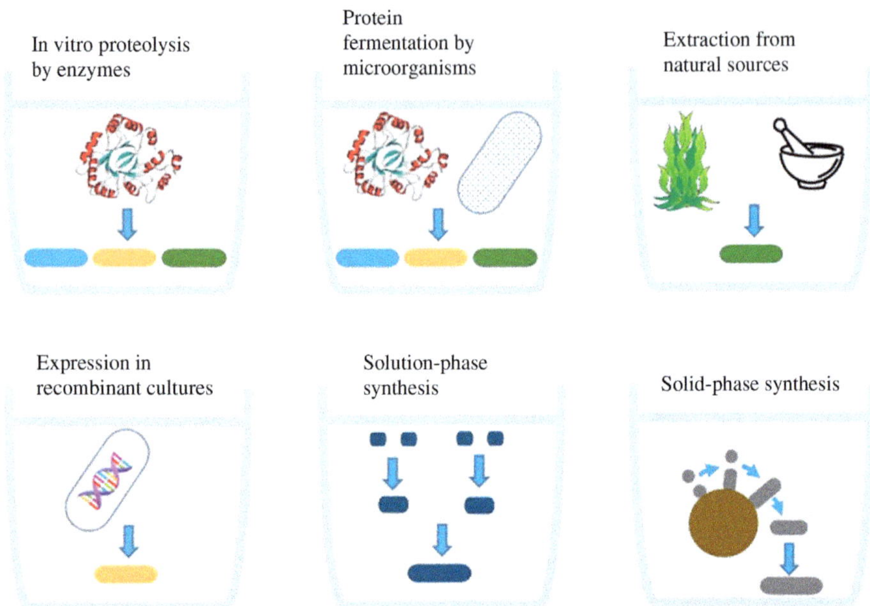

Figure 9.1 Strategies for the production of biologically active peptides.
Source: Adapted from M.K. Danquah, D. Agyei, Pharmaceutical applications of bioactive peptides, OA Biotechnol. 1 (2012) 5 [13]; T. Uhlig, T. Kyprianou, F.G. Martinelli, C.A. Oppici, D. Heiligers, D. Hills, et al., The emergence of peptides in the pharmaceutical business: from exploration to exploitation, EuPA Open Proteom. 4 (2014) 58–69 [14].

lead to the production of peptides with high solubility and improved absorption, distribution, metabolism, and excretion (ADME) profiles [3]. Advances in bioinformatics and in silico tools for simulating the binding mechanisms and metabolic profile of peptides, while studying and relating the structure of peptides to their activities [15,16] have also opened an avenue that stimulates the development and promotion of peptides as biopharmaceuticals. With these drivers, peptides are anticipated to enjoy an increased share of use in the treatment of health conditions, as well as performance on the therapeutics market. This chapter gives an overview of the drivers and restrains in the application of peptides as biopharmaceutical agents (schematized in Fig. 9.2). State-of-the-art advances made in the modification, formulation, and delivery of peptide drugs are also discussed.

9.2 Incentives for the use of peptides as biopharmaceutical products

9.2.1 Structural properties of peptides

Peptides have a modular structure made up of numerous amino acids. In general, there are 20 types of natural amino acids that are joined together via peptide bonds to

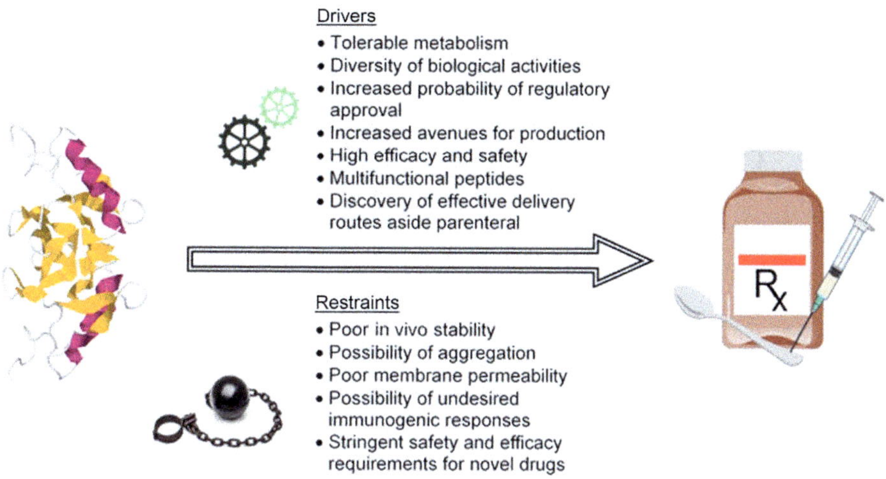

Figure 9.2 The drivers and restrains to the exploitation of peptides as therapeutic products. *Source*: Adapted from D. Agyei, I. Ahmed, Z. Akram, H.M.N. Iqbal, M.K. Danquah, Protein and peptide biopharmaceuticals: an overview, Protein Pept. Lett. 24 (2017) 94–101; K. Fosgerau, T. Hoffmann, Peptide therapeutics: current status and future directions, Drug Discov. Today 20 (2015) 122–128.

form peptides or proteins [17]. Peptides are smaller than proteins and have between 2 and 40 amino acid residues and not more than 5000 Da molecular weight, whilst proteins contain at least 40 amino acid residues with larger molecular sizes [18]. Peptides possess diversified structures due to their amino acid sequence, posttranslational modifications (PTMs), and resultant spatial conformation [14,19]. Moreover, they often fold specifically into unique 3D conformations. The biological functions of peptides and proteins are directly related to their conformational structures and, thus, conformational changes can significantly affect their functions. The structure of peptides can be classified into four different levels, namely primary, secondary, tertiary, and quaternary structures. The primary structure is the linear peptide sequence containing amino acid residues joined from the N-terminus to the C-terminus by peptide bonds. The secondary structure of peptides refers to the formation of intramolecular hydrogen bonds which results in three types of motifs including α-helices, β-sheets, and coils. The tertiary structure is formed when peptide sequences fold into 3D conformation via the interactions between side chains and the surrounding environment in addition to disulfide bonds. Hydrophilic side chains point-out in a polar environment and point-in in a nonpolar environment, whereas hydrophobic side chains point-in in a polar environment and vice versa. Lastly, quaternary structure is formed when a number of polypeptide chains assemble together via intermolecular noncovalent bonding, such as hydrogen bonding and van der Waal forces [17,20].

The 3D structure of proteins and peptides confers on them a strong structure–activity correlation. The primary organization of peptides (i.e., the type, amount, and sequence of amino acids) determines the biological properties of peptides [4,7], whereas the peptide binding behavior is influenced by the secondary rather than the tertiary structure [21]. The presence of some amino acids also determines the type of biological properties of peptides. For example, peptides with antihypertensive properties (through inhibition of angiotensin converting enzyme, ACE) often have proline and valine in their sequences [7,22]. Glycine also plays a crucial role in some immunomodulatory peptides [23], whereas the presence of an imidazole ring of histidine produces peptides with antioxidant properties [24]. Also, most opioid peptides have an N-terminal tyrosine residue [25]. The knowledge that the presence of specific amino acids influences the biological properties of peptides is vital for the rational design of peptides with desired properties, and could also assist with the development of bioinformatics tools and models to study and predict peptide structure–activity relationships.

9.2.2 Potency and selectivity

Peptides offer great potential to enhance biomedical research, pharmaceutical applications, and medical treatments due to their unique characteristics. Natural peptides are highly selective signaling ligands that target specific cell surface receptors to induce intracellular effects. For instance, cell-penetrating peptides (CPPs) possess high capability in delivering various therapeutic molecules into mammalian cells due to the insertion of penetratin or transcription trans-activating sequence [3,19]. Also, many peptides have been reported to possess their natural bioactivity after being isolated from their living hosts [14]. Hence, a tremendous number of antimicrobial peptides have been identified for medical applications. The development of therapeutic peptides is expanding, with more than 600 peptides in the preclinical stages and 140 peptides undergoing clinical trials, whereas more than 100 therapeutic peptides are available in the markets of Europe, the USA, and Japan [3,14,17,26]. Peptides possess unique characteristics for therapeutic applications such as acting as targeting ligands, antimicrobial agents, and vaccines. The tunable and modifiable properties of peptides allow them to be functionalized easily via combinatorial chemistry for better cell targeting, gene expression, and drug delivery [14]. Peptides are highly recommended as drug molecules with their benefits including high target specificity, strong binding affinity, low immunogenicity, broad range of cognate targets, lower potential of drug–drug interaction, and high stability, tolerability, and safety, whilst only a small concentration of peptide is needed to give effective therapeutic outcomes [3,27,28]. This aspect clearly demonstrates the competitiveness of peptides over traditional small molecules. In comparison to protein-based drugs, the production cost of peptide drugs is lower as there is less production complexity, and standard synthetic protocols are used [3].

9.2.3 Tolerable ADME-Tox profile

Naturally occurring peptides have challenging properties of low ADME which results in short plasma half-lives, rapid renal filtration, metabolic instability, poor solubility, and low permeability [29]. However, various in vitro, in vivo, and in silico methods have been developed to overcome the ADME limitations of peptides in order to optimize their potential as therapeutic peptides. In general, peptides consist of amino acids, which are naturally occurring and with tolerable metabolisms. They are nontoxic compounds that do not result in severe human immune responses and are therefore suitable for use as therapeutic drugs. In addition, their side effects are often associated with their concentration or local responses at the site of injection [12].

9.2.4 Low bioaccumulation

Peptides are ideal candidates for therapeutic applications as their circulation half-life is short. As a result, they do not accumulate in the mammalian system due to their low biodeposition in cells and tissues. Consequently, systemic cytotoxicity can be significantly minimized [13]. Also, peptides are biodegradable since they are molecules that are readily degraded in the mammalian body with the presence of approximately 600 types of proteases [30]. Hence, they can be easily degraded into simple amino acids via protease-mediated peptide catabolism, and subsequently be recycled for body metabolism including protein synthesis.

9.2.5 Increased probability of regulatory approval

Peptides are not prioritized in the traditional pharmaceutical development due to several limitations including their accelerated enzymatic degradation in the human body and difficulties in their administration to patients. Nonetheless, 21st-century pharmaceutical research has switched to revisit the use of peptides as therapeutics due to technological advancements such as recombinant technology and genetic engineering, which circumvent some of these limitations, thereby facilitating the approval of novel peptides with therapeutic potentials [12]. Consequently, peptide therapeutics have a 20% chance of receiving regulatory approval, a probability that is 50% higher than that for the approval of small molecules [12]. The probability of regulatory approval is calculated based on the individual approvals during a fixed timeframe and comparison of that with the amount of peptide therapeutics in the phase of clinical trials 12 years previously [31]. Generally, it takes 10–12 years for a peptide candidate to undergo the whole process starting from clinical trials to its final approval or failure stage [32].

In regulatory affairs, peptide drugs can be classified as biological entities or chemically synthesized medicinal products, depending on their natural properties and synthesis [14]. However, the FDA recognizes peptides as conventional drugs instead of biological products based on their structures and compositions [18]. There is an increasing interest in the use of peptide pharmaceuticals for medical applications, with more than 60 peptide drugs approved and more than 100 peptide therapeutics in the clinical trial phase [12–14]. This notably indicates the increased probability of regulatory approval of peptide-based therapeutics to advance the pharmaceutical industry.

9.3 Challenges to the use of peptides as biopharmaceutics

9.3.1 Poor in vivo stability

One of the most significant impediments in realization of the full commercial potential of any standard peptide biotherapeutic formulation is the issue of in vitro and in vivo stability of the peptide therapeutics [33–36]. However, the current discussion will be limited to in vivo stability because of its significance in the delivery and metabolism of peptide drugs. Typically, peptide therapeutics possess short a in vivo half-life, ranging from a few minutes to a few hours [37]. In fact, the successful use of a therapeutic formulation depends on its long-term chemical and physical stability and storage [33]. Principal factors that affect in vivo stability of peptides include pH, temperature, interaction with other compounds, and contamination from additives [38,39]. In addition, variations in the formulation as a direct and indirect consequence of protein structures, induced aggregation, and protein misfolding have also been reported [40,41]. Particularly, low stability in the gastrointestinal tract is often a great concern in addition to aggregation [40]. Other factors include protein concentration and shear forces generated during processes such as vortexing [42]. Poor in vivo stability has also been attributed to the enzyme-mediated modification that leads to short half-lives and to dominant protein structural conformation which can stimulate aggregation [36]. Furthermore, the low pH of the gastric fluid and presence of protein-digesting enzymes have also been reported to affect the stability and effective oral delivery of these molecules [43]. The proteases along with fast renal excretion are also responsible for the brief half-life of peptide therapeutics [35,37]. It has been indicated that a thorough understanding of biomodifications for purely natural amino acid constituting biotherapeutics helps to decipher the underlying causes of poor in vivo stability [34].

9.3.2 Poor oral bioavailability

In addition to the stability issues, reduced bioavailability is a major concern that downgrades effective use of protein or peptide therapeutics [33,35,44–47]. The nonparenteral methods of peptide therapeutic delivery are known to be nonbeneficial, and are associated with poor bioavailability [33]. In many cases, proteolytic enzymes are responsible for this observation. For instance, the delivery of therapeutic molecules through either inhalation or nasal routes is affected by the cleaving actions of the pulmonary or nasal mucosal cavity enzymes, and an ensemble of factors including diverse excretion principles, smaller adsorption area, and modest infiltration play crucial roles [48,49].

In fact, the most common causes for low oral bioavailability have been highlighted as degradation of the desired properties of the therapeutic molecules even before they enter into the system; poor penetration rates and poor absorption in the GI tract due to inability to cross the epithelial layers; and the presence of efflux pumps and proteolytic actions of the metabolic enzymes [44,45,47]. These, along with a diminutive circulatory half-life, are the dominant reasons for choosing parenteral means for drug delivery [50]. Although the introduction of peptide or protein inhibitors or novel functional groups has proved to be successful, safety concerns of this approach do persist [45]. Also, inappropriate absorption may be a consequence of charges on the functional groups, high molecular weights, and reduced solubility in lipids [46,51]. Furthermore, the microstructural features that determine the protein conformation must not be overlooked. Factors such as noncovalent coupling with blood components, opsonization, detachment of subunit proteins, and obliteration of alterable lateral-groups are known to reduce bioavailability [27,44,52].

9.3.3 Low membrane permeability

Several delivery routes and strategies are available for peptide therapeutic administration, including mucosal, nasal, oral, and pulmonary delivery. For an in-depth recent discussion, the reader is referred to the review by Thwala et al. [53]. In addition, several policies have been adopted to improve robustness and implementation of the desired therapeutic molecules through these routes: either using chemical means, such as PEGylation and hyperglycosylation, and/or by employing colloidal transporters such as micro- and nanoparticles, liposomes, micelles, and carbon nanotubes [35]. In spite of these considerable and diverse delivery mechanisms, poor membrane penetration or permeability creates havoc and diminishes the performance capability of the desired molecules [33,35,53,54]. Although opinions differ, one of the principal causes of the poor membrane permeability has been reported to be the inconveniencies associated with countering the proteolytic action of the enzymes at the target sites, and/or overcoming the material blockades in the form of endothelial and epithelial cell layers [33]. The high molecular weight of the peptide drugs poses a major challenge, is associated with much lower penetration rates across skin, cell, and/or mucosal membranes, and this ultimately affects the delivery to specific intracellular target sites [54]. In this context, it is appropriate to consider the challenges encountered in a specific execution pathway. The mucosal membrane and the enzymatic blockade are the main impediments to mucosal delivery, resulting in protein hydrolysis, low bioavailability, and aggregation [55,56]. Particularly, the water-attracting properties and negative charges associated with the mucus membrane result in low dissemination and consequentially poor absorption [57]. Transport of drug molecules across the nasal route is currently limited to less than 1000 Da, above which low permeability is a persistent observation and has been correlated to inherent proteolytic enzyme activities and high mucociliary excretion rates [53]. Although pulmonary delivery of therapeutics is considered a better option than nasal and/or mucosal ones, the action of protein-degrading enzymes in the lungs, rapid excretion from the macrophages in the lung alveoli and air–blood barricade are the major challenges to increased membrane permeability [53,58]. Lastly, the oral route is most preferred due to the ensemble of advantages it manifests: comfortable and uncomplicated, enhanced bioavailability, and expediency and attainability of larger surface area for drug absorption [53]. The main challenge for this pathway is to discover ways that would maintain drug viability (minimize drug degradation as a consequence of proteolytic enzymatic actions) and at the same time favor accessibility of higher surface area for drug absorption [59].

9.3.4 Poor solubility

Currently, the mandate in manufacturing peptide biotherapeutics is to develop concentrated (at least about 10% w/w) peptide solutions for subcutaneous delivery [60,61]. Subcutaneous delivery is considered effective from a therapeutic perspective [53,54]. However, developing any therapeutic formulation with solubility at this concentration level poses significant challenges [62]. In particular, two hindrances stand out: first, the increase in the solution viscosity (>20 centipoise) is the most difficult problem ultimately leading to peptide aggregation and affecting downstream processing; and second, the shear hitch at the syringe wall, limiting the injecting capability of the drug. Poor peptide solubility, especially for larger peptides, is the most crucial factor (the other being peptide stability) that affects the final fate of the molecules and may be attributed to the extent of aggregation and improper characterization of peptide–peptide interactions [62,63].

9.3.5 Possibilities of undesired immunogenic responses

Even if all other challenges related to the formulation and successful delivery of the peptide therapeutic to the desired target site have been countered, the possibility of triggering an undesired host immune response exists [33,64,65]. The response may be minimal, severe, or, in some cases, life-threatening [66]. Although several reasons could be cited, crucial factors that stimulate immunogenicity can be categorized into either product- and/or process-related, patient-related, and pathway-related [64]. The product- and process-related factors may be associated with the intrinsic product characteristics or may be induced during the synthesis procedures [67,68]. In addition, the genetic constitution and food conventions of the concerned patient [69,70], and the route of delivery [66,71] may also be contributing factors. However, undoubtedly, the involvement of PTM is the most significant factor that may be correlated with immunogenicity [64,65]. Protein structure influences the PTM, which has a profound effect on the subsequent immunogenicity [33]. The generation of antibodies against the administered drug contributes to immunogenicity and may even counteract the efficacy of the peptide drug [72–74]. A large number of PTMs, such as amidation, acetylation, glycosylation, carboxylation, phosphorylation, and disulfide-bond formation, affect peptide immunogenicity. For a recent collation of useful information, refer to the recent review by Kuriakose et al. [64]. Broadly, two principles of immunogenicity manifestation have been identified for biotherapeutics: T-cell contingent or autonomous [75]. While the T-cell-governed immune response is usually observed in the case of exogenous peptides and is enduring [76], the T-cell-liberated response is swifter and is associated with molecules from microbial sources [77]. The primary and secondary correlation between PTM and subsequent immune response has been exhaustively investigated. It has been broadly proposed that the immune response may be generated either because of the modifications in the tertiary peptide structure, or due to the altered part of the peptide molecule itself [78]. A summary of the incentives and challenges to the use of peptides as pharmaceuticals is captured in Table 9.1.

Table 9.1 Peptides as biopharmaceuticals: merits and demerits

Advantages	Safety, potency and high selectivity to targets	
	Tolerable ADME-Tox profile with low bioaccumulation	
	Metabolic and structural similarity to body signal transducers such as hormones and cytokines	
	Increased probability of regulatory approval for protein and peptide drugs	
	Unlimited number of peptides with diverse biological functions	Potential for treating a large range of different pathological diseases
		Opportunity for discovery of new potent peptides via in silico and peptidomic techniques
		Exploitation of multifunctional proteins and peptides
	Relatively low cost, novel production techniques that overcome synthetic challenges	Improved synthetic chemistry for the efficient synthesis of difficult sequences
		Moderate production costs, i.e., in vitro via the action of microbial proteases and food proteins
		Advances in peptide conjugation and modification techniques to overcome challenges of stability, solubility, specificity, and membrane permeability
		Advances in bioinformatic techniques and peptidomics to assist targeted production of peptides with desired bioactivities
Disadvantages	Poor in vivo stability and high susceptibility to hydrolysis	Low bioavailability
		Short half-life
		Fast elimination from body systems
		Difficulty to administer orally
	Someo peptides are poorly soluble	Challenges in delivery to targets
		Low membrane permeability
		Possibility of aggregation
	Possibilities of undesired immunogenic responses	

Source: Adapted from D. Agyei, I. Ahmed, Z. Akram, H.M.N. Iqbal, M.K. Danquah, Protein and peptide biopharmaceuticals: an overview, Protein Pept. Lett. 24 (2017) 94–101; D. Agyei, M.K. Danquah, Industrial-scale manufacturing of pharmaceutical-grade bioactive peptides, Biotechnol. Adv. 29 (2011) 272–277; K. Fosgerau, T. Hoffmann, Peptide therapeutics: current status and future directions, Drug Discov. Today 20 (2015) 122–128; R.E.W. Hancock, H.-G. Sahl, Antimicrobial and host-defense peptides as new anti-infective therapeutic strategies, Nat. Biotechnol. 24 (2006) 1551–1557; R. Lax, The future of peptide development in the pharmaceutical industry [Online]. PolyPeptide Group. Available from: <http://www.polypeptide.com/assets/002/5188.pdf>, 2010 (accessed).

9.4 State-of-the-art techniques for overcoming the aforementioned challenges

9.4.1 Peptide engineering via amino acid substitution

Naturally occurring peptides are often susceptible to protease-mediated degradation, which results in short circulation half-lives. Hence, amino acid substitution is one of the strategies used to avoid degradation by identifying the molecular cleavage sites in the peptide sequence followed by amino acid substitution at the location [3]. This includes D-amino acid substitution, olefenic substitution, and dehydro amino acid substitution in order to chemically modify peptide drugs, resulting in better stability and longer plasma half-lives [17]. As a case in point, the amino acids proline and hydroxyproline are known to have better resistance to degradation by proteases, thus improving their half-life en route in the body [79]. Also, D-amino acids are utilized to substitute natural L-amino acids in the peptide sequences to minimize substrate recognition by proteases. N-acetylation and C-amidation are often included in conjunction with the amino acid substitution in order to improve resistance towards exopeptidase- and endopeptidase-mediated proteolysis.

9.4.2 Peptide conjugation

Polymers such as poly(lactic-co-glycolic) acid (PLGA), polylactic acid, polyethylene glycol (PEG), and polyvinylpyrrolidone (PVP) or fatty acids are often employed to conjugate or encapsulate peptide drugs to formulate as liposomes, nano/microparticles, and micelles with a larger molecular weight [80,81]. This method leads to enhanced half-lives, pharmaco-distribution, and bioavailability of peptide drugs in the biological system [30]. PEGylation of peptides can effectively prevent immunogenicity, retain biological functionality, and minimize globular filtration and enzymatic clearance of peptides from the blood circulation [17]. An ideal carrier of peptide drugs is capable of carrying an adequate dosage of peptides without affecting their bioactivity whilst protecting formulated peptides from protease-mediated degradation. Also, polymeric formulations exhibit tunable properties allowing them to have tailored features. For instance, the cationic chitosan is able to improve muco-adhesiveness of peptide drug formulations by interacting with the anionic sialic acid groups of mucin [82]. The biodegradability and bioavailability of biopolymers assure the biosafety of peptide–polymeric formulations. PolyXen is made up of natural polymer polysialic acid and is used for improving the stability and active life of peptide biopharmaceuticals as well as protecting peptide drugs from the human immune system. ImuXen is another formulation made up of liposomes for delivering peptides by providing strong protection from degradation and efficient delivery [17].

Conjugation of peptides with lipids has been shown to yield lipopeptide conjugates that have unique structural and biological properties. Conjugation offers the creation of novel therapeutic biomolecules that combine the biological properties of peptides and lipids whilst yielding products with improved potency and selectivity.

Lipidation of peptides produces amphiphilic peptide conjugates with enhanced bioavailability; increased resistance of the peptide moiety against endopeptidase degradation; enhanced plasma shelf-life; and increased transport across cell membranes leading to higher cellular penetration [83]. The subject is of growing commercial interest and there is a booming market for lipopeptides. In fact, a number of products are already on the market or under development by pharmaceutical companies, exploiting the potential of lipid–peptide conjugates as drug leads [84]. For instance, Daptomycin is the first FDA-approved 10-membered cyclic lipopeptide antibiotic used to treat blood and skin diseases caused by Gram-positive bacteria [85]. Another example is the amphipathic cyclic Surfactin made up of lipoheptapeptide. Surfactin expresses great biological activities such as antimicrobial, antiviral, and antitumor activities due to its superior surface-interface and membrane-active characteristics [86]. Lipo-Nter is a recombinant lipopeptide that can trigger a high antitumor immunity due to its lipid structure and tryptic N-terminal fragment [87].

9.4.3 Hydrocarbon stapling

The susceptibility of peptides towards proteolytic cleavage can also be addressed by modifying the secondary structures via a folding process. This method includes hydrocarbon stapling of peptide sequences; lactam bridges; cyclization; N-methylation; enhanced intramolecular hydrogen bonding; increased hydrophobic faces; and insertion of probe-tail structure [3,29]. For instance, the permeability of peptide biopharmaceuticals can be improved via N-methylation that minimizes hydrogen bonding; intramolecular hydrogen bonding that prevents intermolecular hydrogen bonding; and cyclization that enhances rigidity by constraining the conformation and flexibility of peptides [88,89]. Hydrocarbon stapling aims to link two amino acids in order to improve the stability, solubility, potency, and helicity of the peptides.

9.4.4 Novel formulation and alternate delivery strategies

In order to promote promising peptides into pharmaceutical developments and advances, a variety of novel formulations have been innovated to overcome the physicochemical challenges faced by peptide drugs. It has been reported that chemical incorporation of salts (sodium citrate, potassium phosphate), sugars (glucose, maltose, sucrose), and agents (heparin) into the peptide formulations can significantly improve the in vivo bioavailability, solubility, and stability of peptides [14]. In addition, the transportation of peptides across cellular membranes can be improved with the insertion of both cationic and anionic surfactants (sodium dodecyl sulfate, cetrimide). The addition of protease inhibitors (e.g., bacitracin, sodium glycocholate) can protect peptides from degradation [14]. There are several alternate delivery approaches including the mucosal track via pulmonary or sublingual delivery; oral administration route via gastrointestinal tract penetration enhancers or protease inhibitors; and transdermal route via patches [30,90]. The pulmonary route is highly beneficial with a better absorption, higher bioavailability, vascularization, and more permeable surface in addition to bypassing the first-pass metabolism. Consequently, peptide drug molecules can be distributed uniformly via a more sustainable drug-release pattern with lowered dosing frequency [91,92].

9.4.5 Strategies for targeted delivery of peptide biopharmaceuticals

A smart biopolymer-based targeted delivery system is a promising strategy for delivery of various kinds of bioactive molecules including peptides. In addition, peptide biopharmaceuticals are widely investigated for targeted drug delivery as targeting agents or active ingredients of drugs. Peptides are highly potential targeting ligands and, therefore, they can be used as "add-ons" to other therapeutic agents to navigate the active ingredient towards targeted cells, and to transverse cellular barriers. A targeted delivery system allows specific delivery of active ingredients for a site-targeted action via prolonged drug release and results in minimized dosage of drugs in the body for better patient compliance. For instance, a peptide agonist of receptor neurotensin 1 (NT1) has been conjugated with radioactive ligand to treat pancreatic tumor cells. The peptide agonist directs the delivery of the radioligand to the NT1-expressing cancerous cells, resulting in a lower systemic cytotoxicity and higher dosing of therapeutic molecules at the targeted site [93]. Goserelin peptide is a synthetic analogue and agonist of luteinizing hormone-releasing hormone (LHRH), and has been used to target and treat prostate cancer by suppressing the production of sex hormones [17]. Leuprolide (also called leuprorelin) is another therapeutic peptide that is an analogue of gonadotropin-release hormone (GnRH) and works by decreasing gonadotropin and therefore suppressing the expression of hormones such as follicle-stimulating hormone and luteinizing hormone. Leuprolide is therefore used for the treatment of diseases and conditions such as endometriosis, prostate cancer, breast cancer, and uterine fibroids [94]. Kim et al. [95] reported a novel anticancer peptide that acts as both targeting ligand and anticancer drug. It is a pepsin hydrolysate extract from *Mytilus coruscus* shellfish that triggers cell death in several tumor cells including lung, prostate, and breast carcinoma cells without affecting normal cells. In a research study by Bidwell et al. [96], elastin-like polypeptide (ELP) polymer was used to conjugate with a peptide targeting specific renal cells to develop a drug carrier. The results indicate a higher binding affinity and targeting capability towards desired kidney cells with the presence of targeting peptides as compared to nontargeting drug carriers [96]. In addition, liposomes carrying a peptide that acts as targeting ligand demonstrate the targeting power of several peptides for selective drug delivery. For example, PEG liposomes encapsulating nucleic acids and conjugated with a targeting ligand made up of antagonist G peptide has resulted in an enhanced internalization in small-cell lung cancer cells as compared to other nontargeted formulations [97]. In a recent study by King et al. [98], tumor-homing peptides (CGKRK and iRGD sequences) conjugated liposomes selectively targeted specific epitopes onto the placental surface and delivered the payload to the targeted placental cells, resulting in enhanced placental growth

Table 9.2 Therapeutic applications of some peptides described in this chapter

Type of peptide	Bioactivities and application	References
Cell-penetrating peptides (short peptide sequences (10–16 residues long) rich in arginine and lysine; e.g., Tyr-Gly-Arg-Lys-Lys-Arg-Arg-Gln-Arg-Arg-Arg from TAT viral proteins)	Highly efficient for intracellular delivery, capable of crossing cellular membranes to enhance cellular internalization of active molecular ingredients	[99,100]
Daptomycin (a cyclic lipopeptide containing 13-amino acid sequences linked to a 10-carbon lipophilic tail)	Antibacterial properties, for use as an antibiotic for treating skin and blood diseases triggered by Gram-positive bacteria	[85]
Surfactin (7 amino acids made up of L-Glu-L-Leu-D-Leu-L-Val-L-Asp-D-Leu-L-Leu, in addition to a 13–15 carbon-long β-fatty acid chain)	Antitumor, antiviral, and antimicrobial properties, useful for therapeutics and food safety applications	[86]
Lipo-Nter (*N*-acyl-*S*-diacylglyceryl cysteine lipid structure with different unsaturated fatty acids chain in the R2 group)	Antitumor properties, useful for the development of the next-generation vaccines that enhance antitumor immunity	[87]
Goserelin (Glu-His-Trp-Ser-Tyr-D-Ser(*t*Bu)-Leu-Arg-Pro-AzGly1-NH_2)	Gonadotropin-releasing hormone agonist, for targeting and curing prostate tumor cells	[17]
Leuprolide (Pyr-His-Trp-Ser-Tyr-D-Leu-Arg-Pro-NHEt)	Gonadotropin-releasing hormone agonist, for suppressing and treating endometriosis, prostate cancer, and other hormone-responsive cancers	[94]
Pepsin hydrolysates (Ala-Phe-Asn-Ile-His-Asn-Arg-Asn-Leu-Leu)	Anticancer peptide, capable of targeting and treating various kinds of cancer cells	[95]

without affecting the normal development of the fetus. The therapeutic applications of the peptides described in this chapter are summarized in Table 9.2.

9.4.6 Strategies for peptide-assisted transdermal drug delivery

From the perspective of pervading the outermost skin layer, the stratum corneum, a wide array of peptide transport tactics have been reported [35,44,54]. For a recent collation of relevant work about the numerous available delivery strategies for the pulmonary, buccal, transdermal, and sublingual delivery of peptide therapeutics, the reader is referred to the recent review in Ref. [101] and references therein. The authors of this review noted the superiority of the chemical formulation-dependent strategies for the fact that these are cost-effective and do not necessitate elaborate device requirements. Furthermore, the chemical means seem better adaptable to treat usual and small skin patches. This approach does not appear to be very efficient when the size of the desired molecules encompasses the acceptable range, i.e., in the case of macromolecular drug penetration. Hence, future research in this direction should incorporate the ideas of generating novel and innovative macromolecular drug-delivery schemes. For peptide-assisted drug delivery across the pulmonary route, the complementary use of devices such as nebulizers and inhalers (either dry powder or compressed meter dose) may be beneficial provided that the device does not interfere with patient compliance and does not cause pain [101]. The use of microneedles has particularly shown promise for alleviating this problem to a large extent, thanks to its macromolecular transfer mechanism without backups or devices [102,103]. Innovative delivery approaches will always be a significant factor in driving the efficacy and market for the emerging peptide therapeutics.

9.5 Future outlook

As at 2015, approximately 60 FDA-approved peptide drugs had reached full market potential with maximum commercialization [3]. Furthermore, 130–140 peptide molecules were in the production pipeline [3,26]. Although considered to be more or less a stout pipeline, in comparison to the antibody therapeutics market, further work needs to be done in the context of the discovery of more novel peptides and clinical testing. Despite the fact that several peptide therapeutics have already entered clinical trials, the goal of reaching an optimized production-requirement stage is not near. Nevertheless, the peptide research and innovation domain stresses on ongoing peptide development with "out of the box" pathogen targeting and encountering mechanisms and novel combinations. For instance, the concept of peptide–drug conjugates (PDCs) and peptides with adaptive characteristics have recently attracted considerable attention [26,104,105]. Particularly, efforts are underway to develop peptides with enhanced chemical functionality with a more elaborate target-range. The design and production of innovative PDCs will be boosted by antibodies and cytotoxin targeting in addition to the efforts underway to develop novel antibody–drug conjugates (ADCs) [105]. It has been predicted that, in the future, clinical studies involving PDCs will be at the forefront of peptide therapeutic research [26]. Past studies have focused on the ability to target intracellular processes, although with some limitations. Furthermore, the stability against protease actions, rendering robustness to PDCs, has been studied [106]. It is worth noting that these isolated research outcomes need to be convincingly established, since the exact mechanism of cellular penetration by the desired therapeutic entities is still unclear. The current mandate is to return to the basics and investigate the fundamental biological principles of active and passive drug diffusion in order to search for smart peptide drug therapies for designing the next generation of bioactive peptides with better characteristics against the target [26].

It is no surprise that the current peptide therapeutic market, although nascent but on the rise, has generated undivided appreciation from the global drug sector [3,26,105,107]. In addition, the extraordinary growth probability is considered to be the major line of anticipation to meet the critical demand for peptide therapeutics in treatment of perennial causes of death such as cancers [93]. Although several reasons may be attributed, one key factor is the relatively safe but at the same time effective mechanism of action of peptides [3]. Future advances will include effective development and modifications of the chemical and physical characteristics of peptides, further enhancements of the already strong features, addressing the limitations of peptide drugs, and focus on developing peptides and PDCs with a broad range of action [3]. However, it is noteworthy that synthesis and subsequent scale up of novel multifunctional peptides might pose a risk because of the pharmacological implications associated with more than one recipient receptor as well as the hitherto unrecognized host immune response that might be triggered as a consequence of the in vitro to in vivo transformation [108]. Particularly, two new areas demand further investigation: searching peptide application pathways other than the traditional parenertal pathway and building conjugates of CPPs, which can facilitate the transport of small molecules or antibodies through cell membranes [3]. In addition to the development and marketing of peptide therapeutics, the delivery approaches applicable for these molecules to their target sites should constitute an integral consideration for future research and development.

References

[1] D. Agyei, I. Ahmed, Z. Akram, H.M.N. Iqbal, M.K. Danquah, Protein and peptide biopharmaceuticals: an overview, Protein Pept. Lett. 24 (2017) 94–101.

[2] D. Agyei, M.K. Danquah, Industrial-scale manufacturing of pharmaceutical-grade bioactive peptides, Biotechnol. Adv. 29 (2011) 272–277.

[3] K. Fosgerau, T. Hoffmann, Peptide therapeutics: current status and future directions, Drug Discov. Today 20 (2015) 122–128.

[4] R.E.W. Hancock, H.-G. Sahl, Antimicrobial and host-defense peptides as new anti-infective therapeutic strategies, Nat. Biotechnol. 24 (2006) 1551–1557.

[5] THPdb, THPdb: a database of FDA approved therapeutic peptides and proteins [Online]. Available from: <http://crdd.osdd.net/raghava/thpdb/index.html>, 2016 (accessed 06.12.16).

[6] L.T. Eliassen, G. Berge, B. Sveinbjornsson, J.S. Svendsen, L.H. Vorland, O. Rekdal, Evidence for a direct antitumor mechanism of action of bovine lactoferricin, Anticancer Res. 22 (2002) 2703–2710.

[7] H. Korhonen, A. Pihlanto, Bioactive peptides: production and functionality, Int. Dairy J. 16 (2006) 945–960.

[8] B. Hernandez-Ledesma, C.C. Hsieh, B.O. De Lumen, Antioxidant and anti-inflammatory properties of cancer preventive peptide lunasin in RAW 264.7 macrophages, Biochem. Biophys. Res. Commun. 390 (2009) 803–808.
[9] L.E. Seber, B.W. Barnett, E.J. Mcconnell, S.D. Hume, J. Cai, K. Boles, et al., Scalable purification and characterization of the anticancer lunasin peptide from soybean, PLoS One 7 (2012) e35409.
[10] H. Li, R.E. Aluko, Identification and inhibitory properties of multifunctional peptides from pea protein hydrolysate, J. Agric. Food Chem. 58 (2010) 11471–11476.
[11] S. Sistla, Structure–activity relationships of αs-casein peptides with multifunctional biological activities, Mol. Cell. Biochem. 384 (2013) 29–38.
[12] R. Lax, C. Meenan, Challenges for therapeutic peptides part 1: on the inside, looking out, Innov. Pharm. Pharmaceut. Technol. 42 (2012) 54–56.
[13] M.K. Danquah, D. Agyei, Pharmaceutical applications of bioactive peptides, OA Biotechnol. 1 (2012) 5.
[14] T. Uhlig, T. Kyprianou, F.G. Martinelli, C.A. Oppici, D. Heiligers, D. Hills, et al., The emergence of peptides in the pharmaceutical business: from exploration to exploitation, EuPA Open Proteom. 4 (2014) 58–69.
[15] C.C. Udenigwe, Bioinformatics approaches, prospects and challenges of food bioactive peptide research, Trends Food Sci. Technol. 36 (2014) 137–143.
[16] P. Vlieghe, V. Lisowski, J. Martinez, M. Khrestchatisky, Synthetic therapeutic peptides: science and market, Drug Discov. Today 15 (2010) 40–56.
[17] M.P. Ratnaparkhi, S.P. Chaudhari, V.A. Pandya, Peptides and proteins in pharmaceuticals, Int. J. Curr. Pharmaceut. Res. 3 (2011) 1–9.
[18] FDA, Biosimilars: questions and answers regarding implementation of the biologics price competition and innovation act of 2009, guidance for industry [Online]. Available from: <https://www.fda.gov/downloads/drugs/guidances/ucm444661.pdf>, 2015 (accessed 05.06.17).
[19] C. Recio, F. Maione, A.J. Iqbal, N. Mascolo, V. De Feo, The potential therapeutic application of peptides and peptidomimetics in cardiovascular disease, Front. Pharmacol. 7 (2016) 526.
[20] J.M. Berg, J.L. Tymoczko, L. Stryer, Section 3.2 Primary structure: amino acids are linked by peptide bonds to form polypeptide chains, Biochemistry (2002), fifth ed.
[21] H. Kaur, A. Garg, G.P.S. Raghava, PEPstr: a *de novo* method for tertiary structure prediction of small bioactive peptides, Proteins Pept. Lett. 14 (2007) 626–631.
[22] N.-Y. Lee, J.-T. Cheng, T. Enomoto, I. Nakamura, Essential of proline and valine residues in the peptide derived from lactoferrin for angiotensin converting enzyme inhibition, J. Chin. Chem. Soc. 53 (2006) 515–518.
[23] A. Pihlanto-Leppälä, Milk proteins: bioactive peptides, in: R. Hubert (Ed.), Encyclopedia of Dairy Sciences, Elsevier, Oxford, 2002.
[24] J.-Y. Je, P.-J. Park, S.-K. Kim, Antioxidant activity of a peptide isolated from Alaska pollack (*Theragra chalcogramma*) frame protein hydrolysate, Food Res. Int. 38 (2005) 45–50.
[25] Y.W. Park, M.S. Nam, Bioactive peptides in milk and dairy products: a review, Korean J. Food Sci. Anim. Resour. 35 (2015) 831–840.
[26] A.A. Kaspar, J.M. Reichert, Future directions for peptide therapeutics development, Drug Discov. Today 18 (2013) 807–817.
[27] D.J. Craik, D.P. Fairlie, S. Liras, D. Price, The future of peptide-based drugs, Chem. Biol. Drug Des. 81 (2013) 136–147.
[28] L.C. Sun, Peptide-based drug development, Mod. Chem. Appl. 1 (2013) 1–2.

[29] L. Di, Strategic approaches to optimizing peptide ADME properties, AAPS J. 17 (2015) 134–143.
[30] Z. Antosova, M. Mackova, V. Kral, T. Macek, Therapeutic application of peptides and proteins: parenteral forever? Trends Biotechnol. 27 (2009) 628–635.
[31] R. Lax, The future of peptide development in the pharmaceutical industry [Online]. PolyPeptide Group. Available from: <http://www.polypeptide.com/assets/002/5188.pdf>, 2010 (accessed).
[32] M. Angell, The Truth About the Drug Companies: How They Deceive us and What to do About It., Random House, New York, 2004.
[33] S. Frokjaer, D.E. Otzen, Protein drug stability: a formulation challenge, Nat. Rev. Drug Discov. 4 (2005) 298–306.
[34] M.P. Hall, Biotransformation and in vivo stability of protein biotherapeutics: impact on candidate selection and pharmacokinetic profiling, Drug Metab. Dispos. 42 (2014) 1873–1880.
[35] A. Patel, K. Cholkar, A.K. Mitra, Recent developments in protein and peptide parenteral delivery approaches, Ther. Deliv. 5 (2014) 337–365.
[36] R. Vaishya, V. Khurana, S. Patel, A.K. Mitra, Long-term delivery of protein therapeutics, Exp. Opin. Drug Deliv. 12 (2015) 415–440.
[37] A.H. Zalta, C.P. Sweeney, A.K. Zalta, A.H. Kaufman, Intracameral tissue plasminogen activator use in a large series of eyes with valved glaucoma drainage implants, Arch. Ophthalmol. 120 (2002) 1487–1493.
[38] W. Wang, Instability, stabilization, and formulation of liquid protein pharmaceuticals, Int. J. Pharm. 185 (1999) 129–188.
[39] R. Krishnamurthy, M.C. Manning, The stability factor: importance in formulation development, Curr. Pharm. Biotechnol. 3 (2002) 361–371.
[40] J.L. Cleland, M.F. Powell, S.J. Shire, The development of stable protein formulations: a close look at protein aggregation, deamidation, and oxidation, Crit. Rev. Ther. Drug Carrier Syst. 10 (1993) 307–377.
[41] S. Hermeling, D.J. Crommelin, H. Schellekens, W. Jiskoot, Structure–immunogenicity relationships of therapeutic proteins, Pharm. Res. 21 (2004) 897–903.
[42] W. Wang, Protein aggregation and its inhibition in biopharmaceutics, Int. J. Pharm. 289 (2005) 1–30.
[43] H.A. Schiffter, The delivery of drugs–peptides and proteins, in: M. Moo-Young (Ed.), Comprehensive Biotechnology, Elsevier BV, Amsterdam, 2011.
[44] B.J. Bruno, G.D. Miller, C.S. Lim, Basics and recent advances in peptide and protein drug delivery, Ther. Deliv. 4 (2013) 1443–1467.
[45] S. Gupta, A. Jain, M. Chakraborty, J.K. Sahni, J. Ali, S. Dang, Oral delivery of therapeutic proteins and peptides: a review on recent developments, Drug Deliv. 20 (2013) 237–246.
[46] A. Muheem, F. Shakeel, M.A. Jahangir, M. Anwar, N. Mallick, G.K. Jain, et al., A review on the strategies for oral delivery of proteins and peptides and their clinical perspectives, Saudi Pharm. J. 24 (2016) 413–428.
[47] M. Philippart, J. Schmidt, B. Bittner, Oral delivery of therapeutic proteins and peptides: an overview of current technologies and recommendations for bridging from approved intravenous or subcutaneous administration to novel oral regimens, Drug Res. (Stuttg) 66 (2016) 113–120.
[48] F. Depreter, G. Pilcer, K. Amighi, Inhaled proteins: challenges and perspectives, Int. J. Pharm. 447 (2013) 251–280.

[49] J.J. Lochhead, R.G. Thorne, Intranasal delivery of biologics to the central nervous system, Adv. Drug Deliv. Rev. 64 (2012) 614–628.

[50] J.M. Rabanel, V. Aoun, I. Elkin, M. Mokhtar, P. Hildgen, Drug-loaded nanocarriers: passive targeting and crossing of biological barriers, Curr. Med. Chem. 19 (2012) 3070–3102.

[51] B.J. Aungst, H. Saitoh, D.L. Burcham, S.-M. Huang, S.A. Mousa, M.A. Hussain, Enhancement of the intestinal absorption of peptides and nonpeptides, J. Control. Release 41 (1996) 19–31.

[52] V. Torchilin, Intracellular delivery of protein and peptide therapeutics, Drug Discov. Today Technol. 5 (2008) e95–e103.

[53] L.N. Thwala, V. Preat, N.S. Csaba, Emerging delivery platforms for mucosal administration of biopharmaceuticals: a critical update on nasal, pulmonary and oral routes, Exp. Opin. Drug Deliv. 14 (2017) 23–36.

[54] S. Mitragotri, P.A. Burke, R. Langer, Overcoming the challenges in administering biopharmaceuticals: formulation and delivery strategies, Nat. Rev. Drug Discov. 13 (2014) 655–672.

[55] V. Lassmann-Vague, D. Raccah, Alternatives routes of insulin delivery, Diabetes Metab. 32 (2006) 513–522.

[56] H. Sohi, A. Ahuja, F.J. Ahmad, R.K. Khar, Critical evaluation of permeation enhancers for oral mucosal drug delivery, Drug Dev. Ind. Pharm. 36 (2010) 254–282.

[57] K. Maisel, L. Ensign, M. Reddy, R. Cone, J. Hanes, Effect of surface chemistry on nanoparticle interaction with gastrointestinal mucus and distribution in the gastrointestinal tract following oral and rectal administration in the mouse, J. Control. Release 197 (2015) 48–57.

[58] J.S. Patton, Mechanisms of macromolecule absorption by the lungs, Adv. Drug Deliv. Rev. 19 (1996) 3–36.

[59] A. Sosnik, J. Das Neves, B. Sarmento, Mucoadhesive polymers in the design of nano-drug delivery systems for administration by non-parenteral routes: a review, Prog. Polym. Sci. 39 (2014) 2030–2075.

[60] J. Griesser, G. Hetényi, M. Moser, F. Demarne, V. Jannin, A. Bernkop-Schnürch, Hydrophobic ion pairing: key to highly payloaded self-emulsifying peptide drug delivery systems, Int. J. Pharm. 520 (2017) 267–274.

[61] C.L. Stevenson, S.A. Tao, S.J. Prestrelski, J.B. Eckenhoff, J.C. Wright, J.J. Leonard JR, Aqueous formulations of peptides, 2000. Google Patents.

[62] D.H. Johnson, W.W. Wilson, L.J. Delucas, Protein solubilization: a novel approach, J. Chromatogr. B Analyt. Technol. Biomed. Life Sci. 971 (2014) 99–106.

[63] C.M. O'driscoll, B.T. Griffin, Biopharmaceutical challenges associated with drugs with low aqueous solubility—the potential impact of lipid-based formulations, Adv. Drug Deliv. Rev. 60 (2008) 617–624.

[64] A. Kuriakose, N. Chirmule, P. Nair, Immunogenicity of biotherapeutics: causes and association with posttranslational modifications, J. Immunol. Res. 2016 (2016) 1298473.

[65] Y. Lu, L.A. Khawli, S. Purushothama, F.P. Theil, M.A. Partridge, Recent advances in assessing immunogenicity of therapeutic proteins: impact on biotherapeutic development, J. Immunol. Res. 2016 (2016) 8141269.

[66] S.K. Singh, Impact of product-related factors on immunogenicity of biotherapeutics, J. Pharm. Sci. 100 (2011) 354–387.

[67] P. Eggleton, R. Haigh, P.G. Winyard, Consequence of neo-antigenicity of the 'altered self', Rheumatology (Oxford) 47 (2008) 567–571.

[68] F.A. Harding, M.M. Stickler, J. Razo, R.B. Dubridge, The immunogenicity of humanized and fully human antibodies: residual immunogenicity resides in the CDR regions, MAbs 2 (2010) 256–265.
[69] C.L. Wagner, A. Schantz, E. Barnathan, A. Olson, M.A. Mascelli, J. Ford, et al., Consequences of immunogenicity to the therapeutic monoclonal antibodies ReoPro and Remicade, Dev. Biol. (Basel) 112 (2003) 37–53.
[70] G. Antonelli, F. Dianzani, Development of antibodies to interferon beta in patients: technical and biological aspects, Eur. Cytokine Netw. 10 (1999) 413–422.
[71] R. Singh, S. Singh, J.W. Lillard JR., Past, present, and future technologies for oral delivery of therapeutic proteins, J. Pharm. Sci. 97 (2008) 2497–2523.
[72] J. Holm, M. Gajhede, M. Ferreras, A. Henriksen, H. Ipsen, J.N. Larsen, et al., Allergy vaccine engineering: epitope modulation of recombinant Bet v 1 reduces IgE binding but retains protein folding pattern for induction of protective blocking-antibody responses, J. Immunol. 173 (2004) 5258–5267.
[73] C. Ross, K.M. Clemmesen, M. Svenson, P.S. Sorensen, N. Koch-Henriksen, G.L. Skovgaard, et al., Immunogenicity of interferon-beta in multiple sclerosis patients: influence of preparation, dosage, dose frequency, and route of administration. Danish multiple sclerosis study group, Ann. Neurol. 48 (2000) 706–712.
[74] P.S. Sorensen, C. Ross, K.M. Clemmesen, K. Bendtzen, J.L. Frederiksen, K. Jensen, et al., Clinical importance of neutralising antibodies against interferon beta in patients with relapsing-remitting multiple sclerosis, Lancet 362 (2003) 1184–1191.
[75] V. Jawa, L.P. Cousens, M. Awwad, E. Wakshull, H. Kropshofer, A.S. De Groot, T-cell dependent immunogenicity of protein therapeutics: preclinical assessment and mitigation, Clin. Immunol. 149 (2013) 534–555.
[76] Q. Vos, A. Lees, Z.Q. Wu, C.M. Snapper, J.J. Mond, B-cell activation by T-cell-independent type 2 antigens as an integral part of the humoral immune response to pathogenic microorganisms, Immunol. Rev. 176 (2000) 154–170.
[77] I. Mukovozov, T. Sabljic, G. Hortelano, F.A. Ofosu, Factors that contribute to the immmunogenicity of therapeutic recombinant human proteins, Thromb. Haemost. 99 (2008) 874–882.
[78] M.M. Van Beers, M. Bardor, Minimizing immunogenicity of biopharmaceuticals by controlling critical quality attributes of proteins, Biotechnol. J. 7 (2012) 1473–1484.
[79] D. Stepniak, L. Spaenij-Dekking, C. Mitea, M. Moester, A. De Ru, R. Baak-Pablo, et al., Highly efficient gluten degradation with a newly identified prolyl endoprotease: implications for celiac disease, Am. J. Physiol. Gastrointest. Liver Physiol. 291 (2006) G621–G629.
[80] W.B. Liechty, D.R. Kryscio, B.V. Slaughter, N.A. Peppas, Polymers for drug delivery systems, Ann. Rev. Chem. Biomol. Eng. 1 (2010) 149–173.
[81] A. Sundarram, T.P.K. Murthy, Alpha-amylase production and applications: a review, J. Appl. Environ. Microbiol. 2 (2014) 166–175.
[82] M.C. Chen, K. Sonaje, K.J. Chen, H.W. Sung, A review of the prospects for polymeric nanoparticle platforms in oral insulin delivery, Biomaterials 32 (2011) 9826–9838.
[83] B.P. Ward, N.L. Ottaway, D. Perez-Tilve, D. Ma, V.M. Gelfanov, M.H. Tschöp, et al., Peptide lipidation stabilizes structure to enhance biological function, Mol. Metab. 2 (2013) 468–479.
[84] L. Zhang, G. Bulaj, Converting peptides into drug leads by lipidation, Curr. Med. Chem. 19 (2012) 1602–1618.

[85] R.H. Baltz, V. Miao, S.K. Wrigley, Natural products to drugs: daptomycin and related lipopeptide antibiotics, Nat. Prod. Rep. 22 (2005) 717–741.
[86] K.R. Meena, S.S. Kanwar, Lipopeptides as the antifungal and antibacterial agents: Applications in food safety and therapeutics, BioMed Res. Int. 2015 (2015) 9.
[87] Y.-C. Song, H.-H. Liu, I.-H. Chen, H.-W. Chen, P. Chong, C.-H. Leng, et al., A purified recombinant lipopeptide as adjuvant for cancer immunotherapy, BioMed Res. Int. 2014 (2014) 10.
[88] J.G. Beck, J. Chatterjee, B. Laufer, M.U. Kiran, A.O. Frank, S. Neubauer, et al., Intestinal permeability of cyclic peptides: common key backbone motifs identified, J. Am. Chem. Soc. 134 (2012) 12125–12133.
[89] S.B. Rafi, B.R. Hearn, P. Vedantham, M.P. Jacobson, A.R. Renslo, Predicting and improving the membrane permeability of peptidic small molecules, J. Med. Chem. 55 (2012) 3163–3169.
[90] P. Vlieghe, V. Lisowski, J. Martinez, M. Khrestchatisky, Synthetic therapeutic peptides: science and market, Drug Discov. Today 15 (2010) 40–56.
[91] F. Andrade, M. Videira, D. Ferreira, B. Sarmento, Nanocarriers for pulmonary administration of peptides and therapeutic proteins, Nanomedicine (Lond) 6 (2011) 123–141.
[92] O. Kammona, C. Kiparissides, Recent advances in nanocarrier-based mucosal delivery of biomolecules, J. Control. Release 161 (2012) 781–794.
[93] S.M. Okarvi, Peptide-based radiopharmaceuticals and cytotoxic conjugates: potential tools against cancer, Cancer Treat. Rev. 34 (2008) 13–26.
[94] P. Periti, T. Mazzei, E. Mini, Clinical pharmacokinetics of depot leuprorelin, Clin. Pharmacokinet. 41 (2002) 485–504.
[95] E.K. Kim, H.J. Joung, Y.S. Kim, J.W. Hwang, C.B. Ahn, Y.J. Jeon, et al., Purification of a novel anticancer peptide from enzymatic hydrolysate of *Mytilus coruscus*, J. Microbiol. Biotechnol. 22 (2012) 1381–1387.
[96] G. Bidwell, C. Reese, Q. Shao, A. Chade, A kidney-targeted protein biopolymer drug delivery system, FASEB J. (2015) 29.
[97] A.O. Santos, L.C.G. Da Silva, L.M. Bimbo, M.C.P. De Lima, S. Simões, J.N. Moreira, Design of peptide-targeted liposomes containing nucleic acids, Biochim. Biophys. Acta (BBA) - Biomembranes 1798 (2010) 433–441.
[98] A. King, C. Ndifon, S. Lui, K. Widdows, V.R. Kotamraju, L. Agemy, et al., Tumor-homing peptides as tools for targeted delivery of payloads to the placenta, Sci. Adv. (2016) 2.
[99] S.R. Schwarze, K.A. Hruska, S.F. Dowdy, Protein transduction: unrestricted delivery into all cells? Trends Cell Biol. 10 (2000) 290–295.
[100] S. Trabulo, A.L. Cardoso, M. Mano, M.C.P. De Lima, Cell-penetrating peptides—Mechanisms of cellular uptake and generation of delivery systems, Pharmaceuticals 3 (2010) 961.
[101] J.O. Morales, K.R. Fathe, A. Brunaugh, S. Ferrati, S. Li, M. Montenegro-Nicolini, et al., Challenges and future prospects for the delivery of biologics: oral, mucosal, pulmonary, and transdermal routes, AAPS J. 19 (2017) 652–668.
[102] S. Hirobe, H. Azukizawa, T. Hanafusa, K. Matsuo, Y.S. Quan, F. Kamiyama, et al., Clinical study and stability assessment of a novel transcutaneous influenza vaccination using a dissolving microneedle patch, Biomaterials 57 (2015) 50–58.
[103] J. Arya, M.R. Prausnitz, Microneedle patches for vaccination in developing countries, J. Control. Release 240 (2016) 135–141.

[104] J. Engel, G. Emons, J. Pinski, A.V. Schally, AEZS-108: a targeted cytotoxic analog of LHRH for the treatment of cancers positive for LHRH receptors, Exp. Opin. Investig. Drugs 21 (2012) 891–899.
[105] J.M. Reichert, Bispecific antibodies and ADCs: once and future kings? MAbs 3 (2011) 329–330.
[106] T.K. Sawyer, AILERON therapeutics, Chem. Biol. Drug Des. 73 (2009) 3–6.
[107] C. Sheridan, Proof of concept for next-generation nanoparticle drugs in humans, Nat. Biotechnol. 30 (2012) 471–473.
[108] L. Zhou, L.M. Bohn, Functional selectivity of GPCR signaling in animals, Curr. Opin. Cell Biol. 27 (2014) 102–108.

Host defense (antimicrobial) peptides

10

Evelyn Sun, Corrie R. Belanger*, Evan F. Haney and Robert E.W. Hancock*
University of British Columbia, Vancouver, BC, Canada

10.1 Overview of host defense peptides

The increasing threat of antibiotic resistance and emergence of multidrug-resistant bacteria in hospital- and community-acquired infections is a growing medical concern. In 2014, the World Health Organization released a global report on antimicrobial resistance emphasizing the increasing threat posed by resistant bacterial, parasitic, viral, and fungal pathogens and suggested that a postantibiotic era may be on the horizon [1]. Subsequently, in 2016 the United Nations recognized the threat posed by antimicrobial resistance to human health, development, and global stability, and committed to foster innovative ways to address this global threat [2]. One promising antiinfective approach is the use of antimicrobial peptides (AMPs). These are short polypeptides found in all species of complex life including plants, insects, crustaceans, and animals (including humans), and are integral components of their innate immune systems [3,4]. Originally appreciated for their direct antimicrobial activity against planktonic bacteria [5], natural AMPs have also been shown to have potent immunomodulatory functions both in vitro and in vivo [5]. Therefore, we prefer to use the term host defense peptide (HDP) to describe these molecules to better reflect the broad range of biological activities that they mediate.

Individual HDPs can exhibit a wide range of activities that are uniquely determined, but often overlapping within a single molecule. These activities encompass various functions including direct antimicrobial activity towards bacteria, viruses, and fungi, antibiofilm activity as well as a variety of immunomodulatory functions. Here we summarize the different types of activities that have been observed for natural and synthetic HDPs, and highlight current and future applications of these multifaceted molecules with a particular emphasis on their potential use as novel antiinfective agents.

* These authors contributed equally to this work.

Peptide Applications in Biomedicine, Biotechnology and Bioengineering.
DOI: http://dx.doi.org/10.1016/B978-0-08-100736-5.00010-7

10.2 General features of HDPs

Natural HDPs are typically 12–50 amino acid residues long, possess a net positive charge (due to an abundance of Arg and Lys residues), and have a substantial proportion of hydrophobic residues (usually >50%) which allows them to fold into amphipathic conformations [6]. The primary and secondary structures of both synthetic and natural HDPs vary widely. Consequently, the relationships between structure of HDPs and their functions are exceptionally complex and of considerable current interest in peptide research, prompting extensive structure–activity relationship studies. Generally, these secondary structures, often induced upon contact with cell membranes, have been used to classify peptide families, though some peptides have multiple structural features and can be associated with more than one category. The majority of HDPs have been classified into four major structural groups defined as α-helical (e.g., magainin, cecropins, and certain cathelicidins), β-stranded (e.g., α- and β-defensins and polyphemusins), extended (e.g., indolicin), and looped structures (e.g., thanatin) [7,8].

The need for new antimicrobial and antiinfective, immunomodulating agents and the increased throughput of new peptide design methods [9] is encouraging the development of HDPs to treat recalcitrant and resistant infections. The different functional categories of peptides and their uses are discussed here.

10.3 Host defense peptides as immunomodulators

Many HDPs were originally found to demonstrate direct antimicrobial activity in dilute medium in vitro, and this direct activity against pathogens was long considered their primary biological role in vivo. However, at the high salt and glycosaminoglycan concentrations normally encountered in vivo, the typical concentrations of natural peptides, such as the human cathelicidin LL-37 that is released from epithelial cells and neutrophils at sites of infection (2–5 μg/mL), have no direct antimicrobial effects [10]. (NB: certain peptides may be present at very high concentrations, e.g., defensins in the intestinal crypts and in the vicinity of degranulating neutrophils, and thus may have meaningful antimicrobial activities.) Importantly, many of these peptides exhibit immunomodulatory activities under in vivo-like conditions (e.g., tissue culture medium) at concentrations that are physiologically meaningful and much lower than those necessary for direct antimicrobial activity [11]. For example, LL-37 can selectively modulate inflammatory responses in macrophages, lung epithelial cells, peripheral blood mononuclear cells (PBMCs), and whole blood leukocytes by dampening Toll-like receptor (TLR) responses, modulating mitogen-activated protein kinase (MAPK) pathways, and tumor necrosis factor (TNF) and interleukin (IL) responses [12,13]. Indeed, many natural HDPs exhibit potent immunomodulatory properties and we consider this to be the primary role of these molecules in vivo [3,4]. In keeping with this, virtually all known immunomodulatory activities have been demonstrated to operate in vivo [4].

Within normal tissues and fluids, HDPs are thought to mainly function as signaling mediators involved in innate immune defenses and their interactions with various immune cells and host molecules allow them to be multifunctional under diverse circumstances, including playing important roles in various diseases and inflammatory conditions [3]. Synthetic peptide mimics termed innate defense regulator (IDR) peptides have also been demonstrated with overlapping activities, although it is becoming increasingly clear that not all peptides have identical activities. Importantly, a peptide named IDR-1 was shown to have absolutely no antimicrobial activity even in dilute medium, but protected against infection in animal models by upregulating immune cell recruitment while dampening proinflammatory responses [14].

Generally, HDPs interact with and act on many different target cell types including monocytes, macrophages, dendritic cells (DCs), epithelial cells, neutrophils, and keratinocytes (Fig. 10.1) [4]. Such peptides can freely translocate across the plasma membrane of cells and/or interact with membrane receptors such as G-protein-coupled receptors, purinorceptor 7 (P2X7), IL receptor CXCR2, or TLRs [14,15]. After membrane translocation, many HDPs bind to intracellular receptors, such as GAPDH or SQSTM1, and stimulate a variety of signal transduction pathways important in the innate immune response including the p38, Erk1/2, and JNK MAP-kinases, NFκB, PI3-kinase, two Src family kinases, TRIF-IRF, TREM pathways, as well as autophagy [12,16]. This signaling can lead to downstream activation of various transcription factors such as NFκB, Creb, IRF4, AP-1, AP-2, Are, E2F1, SP1, Gre, Elk, PPARγ, STAT3, etc., and result in the potential dysregulation of more than 900 host genes directly, and even more indirectly [3,12]. Thus, the action of HDPs, like innate immunity itself, is very complex. It is therefore somewhat expected that HDPs can influence a diverse range of innate immune responses such as selectively modulating innate immunity, dampening bacterially induced proinflammatory cytokines, enhancing the production and release of various chemokines and cytokines, recruiting innate and adaptive immune cells, promoting wound healing, suppressing or increasing apoptosis, influencing angiogenesis, causing mast cell degranulation, and promoting and causing polarization of downstream adaptive immune responses. Some of the known mechanisms of immunomodulatory activity of HDPs are summarized in Fig. 10.1.

10.3.1 Effects of HDPs on inflammatory responses

Inflammation is a natural local response to injury allowing the body to kill foreign microbes and then heal by eliminating the cause of cell damage as well as clearing damaged cells and necrotic tissue [17]. However, excessive inflammation is pathological and dysregulation of inflammation is a feature of virtually every human disease [4]. Inflammation is usually triggered by an event, such as infection or local tissue damage, which stimulates the production of various factors that chemoattract immune cells, promote the loosening of blood vessel walls to ease the passage of immune cells from the blood (diapedesis/extravasation), promote activation of these immune cells, and promote nonopsonic phagocytosis and fibrin

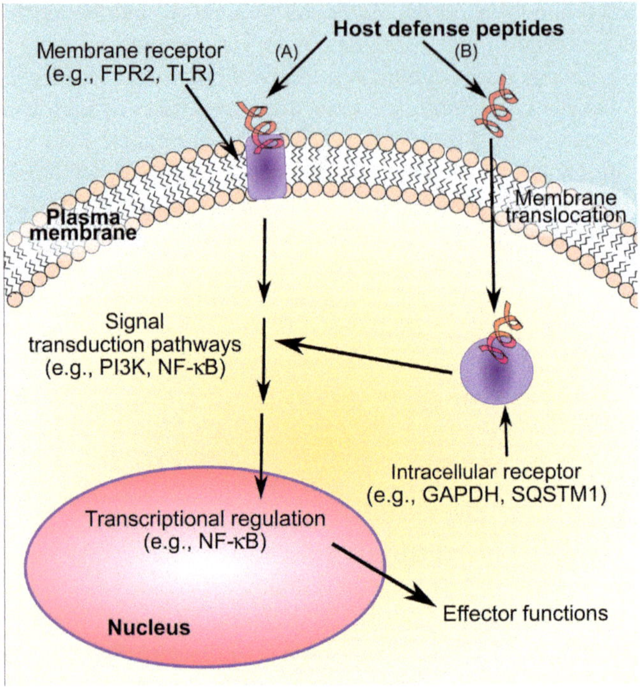

Figure 10.1 Host defense peptides interact with neutrophils or macrophages by (A) triggering membrane receptors such as G-protein-coupled receptor *N*-formyl peptide receptor 2 (FPR2) or Toll-like receptors (TLR) or (B) by spontaneously translocating across the membrane. Intracellularly localized HDPs can target receptors such as glyceraldehyde 3-phosphate dehydrogenase (GAPDH) and sequestosome 1 (SQSTM1) or signal transduction pathways such as phosphoinositide 3-kinase pathway (PI3K) and nuclear factor-κB pathway (NF-κB), to ultimately alter transcriptional regulation patterns and effect a multitude of pathways in the cell involved in various immunomodulatory functions.

clot formation in an attempt to locally contain any infectious agent. Immune cells involved in inflammatory responses initially include resident tissue macrophages, dendritic cells (DCs), and recruited neutrophils, and subsequently monocytes and lymphocytes. Activation of diverse signaling pathways within these cells leads to the transcription of early response genes encoding for numerous proteins including chemokines, cytokines, acute-phase proteins, cell adhesion molecules, costimulatory molecules, negative feedback proteins, and, of course, HDPs [12].

Natural HDPs, such as LL-37, are neither pro- nor antiinflammatory but rather selectively modulate inflammatory mechanisms. Thus HDPs can enhance inflammation by enhancing the production of chemokines, influencing diapedesis, polarizing macrophage and DC differentiation, and promoting enhanced phagocytosis, or act directly as chemokines attracting neutrophils and other immune cells [4]. At the same time, HDPs exhibit antiinflammatory effects by dampening proinflammatory cytokine responses through multiple mechanisms, such as inducing antiinflammatory cytokines, MAPK, and PI3-kinase signaling pathway responses, and blocking LPS binding to receptor proteins and LPS-binding protein [12,18].

In attempts to mimic the modulation of innate immunity by natural HDPs, many synthetic peptides have been developed which retain these key biological functions. OPR-145, for example, is a derivative of LL-37 that was designed to maintain the antimicrobial and antiinflammatory activity of LL-37 while increasing the stability against proteolytic degradation. This peptide has been through phase I/II clinical trials for the treatment of otitis media and was found to decrease the production of IL-8 in whole blood samples stimulated by UV-killed *Staphylococcus aureus*, an activity that was proposed to be linked to the peptide's ability to bind bacterial cell wall components and block activation of receptors on macrophages [19].

Another class of synthetic HDPs with inflammatory effects is the IDR peptides. IDR-1, -2, -1018, and -HH2 are derivatives of linear bactenecin, Bac2a, and all have demonstrated in vivo immunomodulatory activity in a variety of animal models of infection and inflammation [14–16,20]. IDR-1 enhanced the production of chemokines involved in clearing *S. aureus*, vancomycin-resistant *Enterococcus* and *Salmonella* infections and suppressed the production of proinflammatory cytokines in mice, while having no direct antibacterial activity [14]. IDR-1002, -1018, and -HH2 were further refined from the same template peptide, and all have been found to have increased antiinfective, antiinflammatory, and wound-healing activity compared to IDR-1 [14–16,20,21]. Indeed, collectively they have shown activity in animal models versus Gram-negative and Gram-positive infections, cerebral malaria, and LPS-hypoxia ischemia-mediated brain injury (a preterm birth model). The role of HDPs/IDRs in pro- and antiinflammatory functions can involve many different pathways and processes, and researchers are actively studying them as antiinfective treatments [22–24]. However, it is worth pointing out that the influence of HDPs on the inflammatory response has also spurred interested in developing these agents as treatments for various inflammatory diseases such as arthritis, chronic obstructive pulmonary disease (COPD), and asthma [25,26].

10.3.2 HDPs can exhibit direct chemoattractant activity

Although many of the anti- or proinflammatory activities of natural and synthetic peptides are linked to their indirect recruitment of immune cells, certain natural HDPs can also directly chemoattract immune cells as observed for LL-37, cathelin-related antimicrobial peptide (CRAMP), and defensins [27–30]. Furthermore, some researchers propose that the structural similarities between chemokines and antimicrobial peptides may help elucidate the structure–function relationships that allows HDPs to have chemoattractant properties. Both are amphipathic cations, and in many cases have shared mechanisms in humans [31]. Defensins, in particular, share various characteristics with chemokines, including size, structure, disulfide bonds, interferon-inducing properties, and overall cationic charge [32]. Human β-defensins 2 and 3, and mouse β-defensin 4 are able to chemoattract keratinocytes; and α-defensins are chemotactic for human monocytes and mast cells [28,33]. Indeed, structural studies of human β-defensin 2 demonstrate that it shares a CCR6 receptor with chemokine CCL-20 and can induce chemotaxis of dendritic cells and monocytes by targeting the CCR6 and CCR2 receptors, respectively [32].

10.3.3 HDPs promote wound healing and angiogenesis

Wound healing involves multiple steps from inflammation to regeneration [34] and is another complex process that can be enhanced by HDPs. Initially, after an injury to the skin, vascular permeability is increased and platelet and fibrin aggregation occurs. This is quickly followed by the release of several growth factors from platelets that attract neutrophils to the wound and induce inflammation. Eventually, macrophages replace neutrophils as the primary inflammatory cells and debris is removed from the wound followed by reorganization of extracellular matrices [34]. HDPs are often found around oral and cutaneous wounds, and they have been shown to play various roles in promoting wound healing including reducing the bacterial burden, increasing neutrophil and macrophage recruitment, interacting with growth factors, inducing chemotaxis of epithelial cells, promoting production of metalloproteases that restructure the extracellular matrix, and promoting angiogenesis [4,18,35]. Some of the specific processes mediated by HDPs are highlighted below.

During the initial stages of wound repair, inflammation, and proliferation, the production of natural HDPs can be induced in wounded keratinocytes by growth factors [24,36]. The presence of LL-37 at wound sites has been shown to induce migration and proliferation of fibroblasts, human microvascular endothelial cells, and human umbilical vein endothelial cells [36]. Defensins, and particularly human β-defensin 2, promote keratinocyte migration and proliferation through the phosphorylation of the epidermal growth factor (EGF) receptor and activation of STAT1 and STAT3, which are important mediators of immunity, proliferation, apoptosis, and differentiation [33,37].

In the next stages of wound healing, angiogenesis, vascularization, and reepithelialization occur, all of which can be significantly impacted by HDPs [18,24]. For example, HB-107 is a fragment of the α-helical moth peptide, cecropin B, that has no direct antimicrobial activity at physiological concentrations, but can increase leukocyte infiltration and simulate IL-8 secretion from endothelial cells, leading to improved wound repair [23]. Recently, another synthetic peptide, angiogenic peptide 30 (AG-30) and its derivatives, AG-30/5C and SR-0379, were also studied for their angiogenic, wound healing, and antimicrobial functions [38,39]. Experiments in vitro showed that SR-0379 promoted vessel formation, granulated tissue formation, and proliferation of endothelial cells and fibroblasts in rats. SR-0379 was also found to be antiinfective against *Escherichia coli*, *S. aureus*, and *Pseudomonas aeruginosa* in vitro and offered protection against *S. aureus* in rat infection models [39].

The wound-healing properties of natural and synthetic HDPs are proving beneficial in a clinical context as well. For instance, LL-37 treatment enhanced wound healing of hard-to-treat venous leg ulcers compared to placebos in phase I/II randomized clinical trials [22]. Also, the synthetic peptide IDR-1018 significantly improved healing in mouse and pig wounds and had enhanced activity compared to

LL-37 or HB-107. This peptide was found to induce superior wound closure in these models and was less toxic against human keratinocytes and primary human fibroblasts, while having no effect on bacterial colonization [24]. Taken together, it is evident that natural and synthetic HDPs have high potential as wound-healing agents and their inclusion in wound dressings and ointments to prevent wound-associated infections and enhance wound closure are areas of active therapeutic development [40].

10.3.4 The roles of HDPs in autophagy, apoptosis, and oxidative stress

Autophagy is a natural process used by cells to recycle dysfunctional cellular components and preserve cellular energy. This process involves the sequestration of cellular components into vesicles called autophagosomes that fuse with lysosomes to hydrolyze and recycle cytosolic materials, and it relies on signaling pathways and highly conserved *Atg* genes [41]. It is often activated by infections of intracellular organisms such as *Mycobacterium tuberculosis*, and therefore can be considered a part of the innate defense mechanism [6]. Alternatively, cells can undergo apoptosis, which is a process of programmed cell death wherein caspases cleave hundreds of target proteins, cells shrink, bleb, and degrade DNA, RNA, and other cellular material [41]. These two modes of cellular death are believed to be highly interconnected and certain HDPs are known to influence these natural cellular degradation pathways.

Peptide IDR-1018 was demonstrated to rescue the dysfunctional autophagy associated with cystic fibrosis cells and consequently attenuate the typical hyperinflammatory responses exhibited by these cells. In particular, treatment of CF epithelial cells with IDR-1018 abolished the accumulation of LC3 (indicating stalled autophagy) induced by the bacterial inflammatory mediator flagellin. Conversely, LL-37 can induce autophagy in neutrophils through nucleotide scavenging receptor P2X7 and G-protein-coupled receptors, and can promote autophagy by activating Beclin-1 and Atg5 in a vitamin D3-dependent manner [42,43].

Cellular apoptosis can be promoted by LL-37 [44] through the activation of caspases 3 and 9 in the airway epithelium infected with *P. aeruginosa* to promote pathogen clearance [42]. The opposite has also been found in keratinocytes where LL-37 suppresses caspase-3 activity, potentially through upregulation of cyclooxygenase-2 (COX-2) expression, ultimately leading to the upregulation of inhibitor of apoptosis-2 (IAP-2) [45]. Similarly, LL-37 inhibits neutrophil cell death increasing their longevity, while suppressing inflammatory activity [46], although other data indicate the induction of apoptosis in neutrophils [47].

Evidently, the influence of natural HDPs on autophagy and apoptosis is complex and has only recently become appreciated. However, the potential to promote cell death through the use of autophagy or apoptosis inducing synthetic HDPs, could potentially be used to treat infection by promoting clearance of infected (or cancerous) cells, or by triggering connected immunomodulatory signaling pathways.

10.3.5 Modulation of the adaptive immune response by HDPs

Adaptive immunity is an acquired immune response utilizing specialized cells that respond to antigen exposure by rearrangement of specific genes encoding recognition elements and enhancing specific antibody production and cell-mediated immunity. Cells required in cellular immunity include antigen-presenting cells, specialized lymphocytes including helper T-cells (Th1 and Th2) and B-cells involved in antigen-specific recognition and response to eliminate pathogens while retaining immunological memory to prevent reinfection [48]. While their primary role in vivo appears to be modulation of innate immunity, HDPs have been shown to play multiple roles in adaptive immune responses such as promoting adjuvant responses to enhance adaptive immunity by directing immune functions towards Th1, Th2, or mixed Th1/Th2 responses, which depend on the class of helper T-cells activated and aiding in monocyte uptake of antigens [18].

Research has demonstrated that cells overexpressing defensins promote a strong Th1 response and induce cytotoxic T-cells, NK activity, and IL-12 and IFN-γ production in mice, while increasing their protection against leukemia and tumor cells [49]. Alternatively, defensins have been found to promote Th2-type responses by indirectly inducing IL-5 and IL-10 secretion in mice intranasally treated with ovalbumin and human defensins [50]. A synthetic peptide, $KLKL_5KLK$, was also found to induce a Th2-specific activity in response to coinjected antigen ovalbumin in vaccinated mice [51]. However, this same peptide used as an adjuvant in mice vaccinated against *M. tuberculosis* also showed improved and prolonged Th1 responses with increased IFN-γ-producing cells and antigen-specific IgG compared to vaccine without adjuvant [52].

Due to their mixed influences on adaptive immunity, the use of HDPs as adjuvants in vaccine formulations in order to produce balanced cytokine and antibody profiles from both Th1- and Th2-type cells is of increasing interest. For example, the natural peptides, melittin, mouse CRAMP, LL-37, and defensins have all been found to produce mixed Th1 and Th2 responses in adaptive immunity [53–56]. To utilize this effect, Kovacs-Nolan et al. formulated vaccine adjuvant combinations of CpG oligodeoxynucleotides (ODN), polyphosphazine, and indolicidin which showed improved antigen-specific humoral responses and extended cell immune responses in vaccinated cattle [57]. More recently, vaccine formulations containing synthetic IDR peptides IDR-HH2, -HH18, or -1002 along with combinations of CpG ODN and polyphosphazenes were also found to improve adaptive immune responses in mice stimulated with detoxified pertussis toxoid (PTd). Specifically, adjuvant combinations of CpG with peptides exhibited earlier IgG2a responses in both neonatal and adult mice than any of the adjuvants alone and resulted in substantially enhanced, and more protective, Th1/Th2 responses than those to PTd without adjuvant or with traditional adjuvant formulations [58]. As a result of this research multiple IDR peptides are now being developed in adjuvant formulations for cattle vaccines. These examples support the idea that HDPs may play additional roles in mediating the adaptive immune response, expanding their role beyond key components of the innate immune system.

10.4 Direct antimicrobial activities of HDPs

Evidently, as previously mentioned, HDPs were originally studied as antimicrobial peptides (AMPs), and many of them exhibit notable direct antimicrobial activity at high concentrations in vitro and can reduce bacterial burden from pathogenic infections in vivo [59] (although the latter activities may stem in part or whole from immune modulation as discussed earlier). Mechanistically, membrane disruption is probably the most studied mechanism of direct AMP activity. However, over the years, new mechanisms of action have been proposed, such as targeting intracellular processes including synthesis of DNA, RNA, and proteins as well as inhibiting cell wall biosynthesis, cell surface structures, and cell division machinery [60–63]. Modern perspectives view HDPs as highly versatile and multifunctional with the potential to exhibit activity against multiple microbial targets.

10.4.1 Bacterial cell membrane disruption by HDPs

Important differences between microbial and eukaryotic cell surfaces allow AMPs to be selective even in complex environments. Microbial cell surfaces, for instance, are more negatively charged due to the abundance of anionic lipids, whereas eukaryotic surfaces are rich in zwitterionic lipids resulting in an overall reduced negative charge [64]. Therefore, cationic HDPs can distinguish between prokaryotic and eukaryotic cell surfaces through preferential electrostatic interactions [8]. Other features of bacteria such as their high transmembrane electrical potential gradient (oriented internal negative) and lack of cholesterol also favor membrane interaction and/or translocation. After interacting with a target cell surface, HDPs can either directly kill a cell through a lytic mechanism or by entering the cell where they interact with various intracellular targets and inhibit key cellular processes. In our experience virtually all HDPs can disrupt membranes at high enough concentrations but many act at concentrations that do not completely disrupt membranes.

Numerous models have been proposed to describe how peptides disrupt the cell membrane and a brief summary of the most commonly described membrane interaction models is provided here [7,8]. The barrel stave model (Fig. 10.2A) posits that a peptide-lined pore forms in which the hydrophobic face of the peptides directly interacts with the acyl chains in the phospholipid core while the hydrophilic side of the peptides faces towards a water-filled pore. It should be mentioned that the barrel stave model is a highly discussed mechanism of membrane destabilization but is not well supported by the data [65]. The toroidal pore model (Fig. 10.2B) is analogous to the barrel-stave model in that peptides align themselves perpendicular to the membrane with their hydrophobic regions interacting with phospholipid heads and hydrophilic regions facing the pore [7]. However, in this case, the inner and outer leaflets of the phospholipid bilayer curve towards each other and mix, creating a peptide–lipid-lined aqueous pore that allows for leakage of cellular components. Examples of peptides that have been proposed to induce such pores include: magainins, protegrins, and melittin [66–68], although

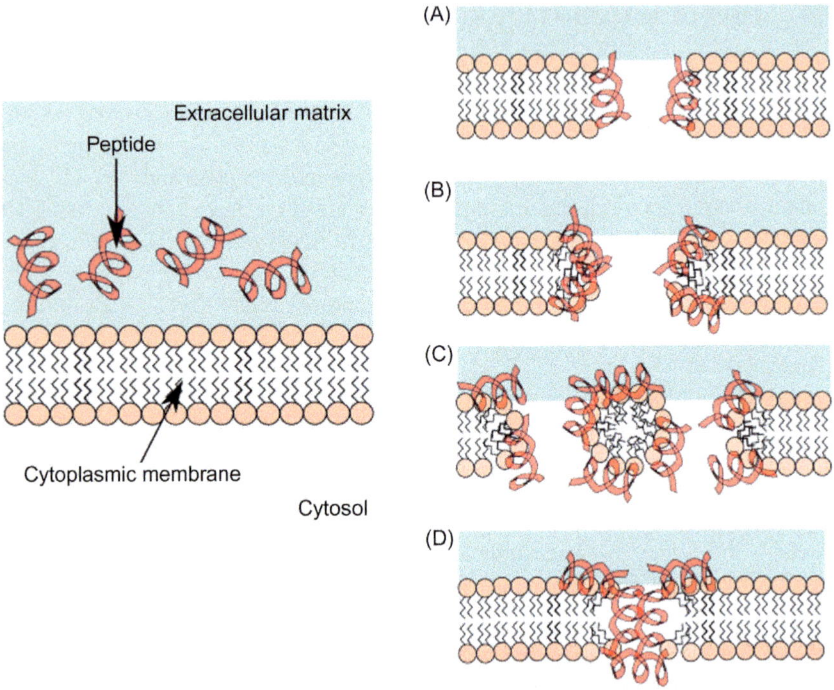

Figure 10.2 Models of peptide-induced membrane disruption or interfacial activity. (A) Barrel-stave model. Peptide forms a pore by inserting perpendicular into the membrane. (B) Toroidal pore model. Peptides form a pore where the hydrophilic regions face the interior of the pore and the hydrophobic regions remain in contact with the phospholipid headgroups. (C) Carpet model. Peptides dissolve the membrane into micelle-like structures in a detergent-like manner. (D) Aggregate model. Peptides aggregate at the membrane surface and reorient to interact with phospholipid acyl chains. This can lead to the formation of informal pores as the peptide aggregates enter the bacterial cell.

molecular dynamic simulations suggest that the actual pore structure is more disordered [69] than the oligomeric form proposed in many cartoon representations. The carpet model is used to describe the activity of HDPs that disrupt the membrane in a detergent-like manner by disintegrating it into micelles resulting in leakage and cell death (Fig. 10.2C) [70]; although this model is largely applicable only at high peptide concentrations [7,8]. Finally, our group proposed the aggregate model (Fig. 10.2D) wherein peptide oligomers aggregate on the surface of a bilayer in a concentration- and voltage-dependent manner [59]. These aggregates interact with the acyl chains of the phospholipid bilayer and can induce the formation of informal and transient aqueous channels that would allow for leakage of cellular component. Importantly, this model can be applied to peptides of all lengths and not only those that are long enough to span the width of the membrane. It also offers a mechanism to describe how some peptides can translocate into cells to reach their intracellular targets [59].

Although models of interfacial activity for peptides such as these have been extensively described in the literature, they are limited to experimentally testable situations in artificial model membrane systems (e.g., vesicles), which poorly reflect natural systems [71]. However, it is generally accepted that membrane interactions with HDPs are a crucial component of their overall mechanism of action and studying these interactions remains an active area of research in the field. In the following sections, we will outline some of the nonmembrane targets of HDPs and examine how they contribute to the overall antibacterial activities of this class of peptides.

10.4.2 Inhibition of cell wall formation by HDPs

The cell wall of Gram-positive bacteria is characterized by a thick mesh of peptidoglycan (PG) that has a major role in cell shape and osmotic stability and is synthesized by enzymes that are targeted by β-lactam antibiotics [72]. PG synthesis involves a cytoplasmic membrane-associated intermediate termed lipid II [73]. Certain HDPs interfere with PG biosynthesis, e.g., by binding of mammalian defensins, fungal defensin plectasin [74], and highly modified bacterially derived lantibiotics like nisin to lipid II [61]. For example, the amide backbone of the N-terminal ring of nisin forms hydrogen bonds with the pyrophosphate moiety of lipid II [61]. This leads to the inhibition of PG synthesis but also triggers the assembly of nisin in the membrane triggering pore formation and the efflux of ions such as K^+ and PO_4^{3-} as well as ATP [75,76]. In addition, by nisin binding to lipid III and lipid IV it interferes with teichoic and lipoteichoic acid biosynthesis [77].

Lantibiotics have been highly studied for their potential uses in food preservation and, indeed, nisin has been commonly used as a preserving agent for many years [78]. Lantibiotics have also garnered interest as potential pharmaceuticals to treat pathogenic infections. Examples include Nai-107 produced by Sentinella Phamaceuticals or MU1140 and OG253 produced by Oragenics [79]. Nai-107 has been shown to be effective against meticillin-resistant *S. aureus* (MRSA) and other multidrug-resistant bacteria and is currently being tested to treat Gram-negative bacterial infections via intravenous administration [80]. MU1140 works well against highly resistant *S. aureus* and *Streptococcus pneumoniae*, and is currently being tested to treat infections by these particular bacterial species [81]. OG253 is a promising lantibiotic that has shown preclinical efficacy against *Clostridium difficile* infections and enteritis in animal models [82].

Synthetic peptides composed of natural amino acids have also been suggested to target the cell envelope of Gram-positive bacteria. A proteomics approach was recently used to examine the shift in protein expression of *Bacillus subtilis* cells treated with a synthetic hexapeptide (RWRWRW-NH_2, called MP196) revealing upregulation of a number of proteins that were representative of cell-envelope stress and energy limitation [83]. Interestingly, the bactericidal activity of MP196 was not related with the formation of pores or ion leakage as found for several other HDPs. Instead, treatment of cells with MP196 caused displacement of cytochrome C from the outer leaflet of the membrane, thereby disrupting the electron transport chain and limiting energy production within cells. Additionally, MP196 treatment caused the peripheral membrane protein MurG, the enzyme in *B. subtilis* responsible for converting intermediates lipid I to lipid II, to delocalize from the membrane resulting in decreased glucosamine attachment and reduced cell-wall integrity [83]. Thus certain HDPs have the ability to directly inhibit cell wall biosynthesis either by directly interacting with cell-wall precursors or interfering with enzymes and energy sources required to assemble cell wall components. It is worth noting, however, that peptides often have complex mechanisms of action with multiple targets [84].

10.4.3 Antimicrobial HDPs targeting intracellular processes

As mentioned above, most cationic amphipathic peptides will disrupt membranes at sufficiently high concentrations and many "mechanistic" studies do not always examine concentrations around the minimal inhibitory concentrations. In contrast, many HDPs have been shown to enter bacterial cells without excessive membrane disruption, at their minimal effective doses, and to target intracellular processes such as DNA or RNA synthesis, protein translation, protein folding, or various enzymatic reactions [85]. For example, Bac7 is a bovine proline-rich cathelicidin peptide that enters bacteria by interacting with an ABC transporter, SbmA, on the surface of the cell followed by an intracellular interaction with DnaK which is a chaperone for ATPase [86,87]. This, in turn, results in the accumulation of misfolded proteins and reduction of cellular viability [87]. Other peptides that target intracellular processes include the hexapeptide WRWYCR, which targets DNA repair mechanisms by binding to Holliday junctions [88]; buforin II, which diffuses through the plasma membrane to bind RNA and DNA [89]; and polyphemusin, which enters *E. coli* cells without disrupting the membrane [90].

Other peptides both perturb the cytoplasmic membrane and target intracellular processes depending on the concentrations applied. Indolicidin, for example, can cause both membrane depolarization and inhibition of DNA synthesis [91,92], whereas other peptides, including the cathelicidins LL-37 and apidaecin, can inhibit DNA and/or protein synthesis without disrupting the membrane [93]. Additionally, the enzymatic targets of some HDPs can have additional effects benefitting the host aside from directly killing the invading pathogen. For instance, histatin 5, a natural HDP found in saliva, prevents tissue destruction in the mouth by inhibiting the activity of proteases produced by oral bacteria [94].

10.5 Methods of bacterial resistance to HDPs

Although resistance mechanisms to conventional antibiotics are often more readily attained, bacteria are able to develop resistance HDPs as well. There are two major types of resistance mechanisms against peptides: inducible and constitutive (Fig. 10.3). Inducible resistance includes substitutions and modifications like the arabinosaminylation and acylation of membrane molecules, overproduction of

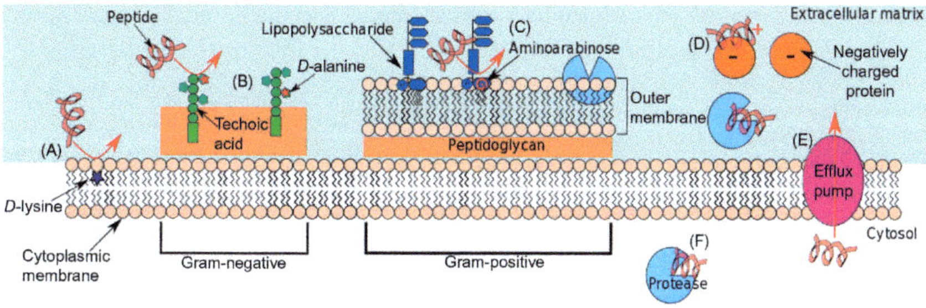

Figure 10.3 Mechanisms of resistance against HDPs. (A) Substituting D-lysine on phospholipid acyl chains. (B) Modifying teichoic acid with D-alanine in Gram-positive bacteria. (C) Modification of LPS by adding aminoarabinose (or glucosamine of phosphoethanolamine) to lipid A of the LPS at the outer membrane of Gram-negative bacteria. (D) Inducing the secretion of negatively charged molecules by host cells into the environment to bind to cationic HDPs. (E) Secreting out HDPs via efflux. (F) Degrading HDPs using intracellular proteases.

proteolytic enzymes or efflux pumps, and modification of intracellular targets. Constitutive resistance includes electrostatic shielding, reductions in membrane potential, and biofilm formation [7].

10.5.1 Bacterial surface remodeling to inhibit binding

Since the initial step in HDP binding to a bacterial cell involves an electrostatic attraction, one way that bacteria have evolved to circumvent this is to remodel surface molecules to reduce their overall negative charge. Teichoic acid, found in the PG layer of Gram-positive bacteria, is normally polyanionic because of its many phosphate groups, but several Gram-positive bacteria can modify teichoic acid by incorporating D-alanine via ester bonds, partially neutralizing surface charge and reducing peptide binding (Fig. 10.2B) [95]. Additionally, the *S. aureus* protein, MprF, a peptide resistance factor, mediates the addition of L-lysine to the phospholipid phosphatidylglycerol components (Fig. 10.1A) in order to reduce the membrane's overall negative charge and affinity for cationic HDPs [96]. Under specific adaptive conditions (e.g., divalent cation deficiency, exposure to HDPs, etc.), Gram-negative bacteria can modify the membrane proximal lipid A moiety of surface lipopolysaccharide molecules by adding L-arabinosamine, glucosamine, or phosphoethanolamine groups to one or both phosphates of the lipid A, effectively reducing the negative charge from the phosphate moiety and preventing peptide uptake via the self-promoted uptake route (Fig. 10.2C) [97–99]. Gram-negatives can also alter the acylation of lipid A through palmitoyl transferase PagP, to confer further resistance [98]. While these mechanisms are adaptive (i.e., they occur due to specific growth conditions and revert when the conditions revert), they can become fixed by mutations that result in constitutive alteration of surface molecules, and the same mechanisms that make cells resistant to the polycationic lipopeptide antibiotic polymyxin generally lead to resistance to HDPs. For example, chronic infections with *P. aeruginosa* can produce hexa-acylated lipid A molecules that contain palmitate and arabinosamine that make it resistant to certain HDPs [97].

10.5.2 Active efflux and degradation of HDPs

Two commonly used mechanisms by which bacteria resist antibiotics are active expulsion of compounds using efflux pumps and the use of specific enzymes to degrade or inactivate them. These mechanisms are not common for HDPs but have been described for some of these molecules. Examples of HDP efflux mechanisms include (Fig. 10.3E) the MtrCDE efflux system in *Nesseria gonorrhoeae* [100] and the QacA efflux pump in *S. aureus* [101]. In addition proteases and peptidases produced by bacteria can often degrade and inactivate natural HDPs (Fig. 10.3F) [70,102]. For example, the human cathelicidin LL-37 can be degraded by a number of different bacterial proteases including gelatinase from *Enterococcus faecalis*, metalloprotease from *Proteus mirabilis*, and elastase from *P. aeruginosa* [103]. *Salmonella enterica* uses the virulence factor PgtE, an outer-membrane protease, to cleave and inactivate C18G and other synthetic α-helical AMPs [104]. Proteases can be present extracellularly [103], embedded in the outer membrane [104], or intracellular in the case of SapA-mediated degradation of cationic HDPs in *Haemophilus influenzae* [105] (Fig. 10.3F). It is also true that host proteases and peptidases can cause degradation of both synthetic and natural HDPs.

10.5.3 Additional HDP resistance mechanisms

Some bacteria can also have mechanisms to sequester peptides and inhibit their antibacterial activity (Fig. 10.3D). For example, *S. aureus* produces the exoprotein, staphylokinase, which activates host plasminogen to plasmin, a negatively charged enzyme which in turn directly binds to host α-defensins and prevents them from interacting directly with *S. aureus* cells [106]. Conversely, *P. aeruginosa* and *E. faecalis* secrete proteases that degrade proteoglycans triggering the release of anionic glycosaminoglycans, such as negatively charged dermatan sulfate, which bind to and inactivate α-defensins [103]. Bacteria can also interfere with the biosynthesis of HDPs by the host to decrease the effective concentration of HDP present in the vicinity of the pathogen. For example, *Shigella dysenteriae* downregulates the expression of LL-37 and β-defensin 1 during early infection in vitro [107] and *P. aeruginosa* can induce the accumulation of cysteine proteases secreted by macrophages in the airway to degrade β-defensin 2 and β-defensin 3 [108].

From a pharmacological perspective peptide resistance is not very common but these resistance mechanisms show they are possible and create a potential limitation for the future clinical use of peptides. Nevertheless, the discovery of these resistance mechanisms provides insight into HDP mechanisms and this might potentiate the development of novel HDPs. For instance, novel peptide designs can avoid the induction of adaptive resistance mechanisms in bacteria. Therefore, it is crucial to consider existing resistance mechanisms when designing HDPs as novel antimicrobial therapies.

10.6 Antibiofilm activities of HDPs

The term antimicrobial peptides refers to their activity against free-swimming (planktonic) bacteria. However, bacteria in nature often adopt a distinct growth lifestyle as biofilms. Biofilms are multicellular communities of bacteria encased in extracellular polymeric matrices composed of polysaccharides, proteins, and DNA, which allows them to adhere to various surfaces [109,110]. The physiological switch from planktonic to biofilm growth is accompanied by large adaptive changes in gene expression, including genes involved in stress response mechanisms and antibiotic resistance genes, and has been proposed to be the prominent reason that biofilms exhibit such high (10- to 1000-fold) adaptive resistance to antibiotics [109]. Clinically, it has been estimated that up to 65% of all human infections are caused by bacteria growing within biofilms [109] and their intrinsic resistance makes them highly recalcitrant to conventional antiinfective therapies. As a result, biofilms are often associated with various chronic infections such as pneumonia in cystic fibrosis (CF) patients [111], otitis media [112], chronic ulcers and wounds, as well as infections derived from biofilms growing on the surfaces of implanted medical devices and catheters [113]. Due to the tremendous clinical relevance and a paucity of antimicrobial compounds capable of treating biofilm-associated infections, identifying novel antibiofilm agents will be of great interest in the future.

In contrast to conventional antibiotics, many natural and synthetic HDPs are able to prevent biofilm formation as well as possess the ability to kill and eradicate preformed biofilms [114–116]. This phenomenon was first described for human LL-37, which inhibited *P. aeruginosa* biofilms at subinhibitory concentrations in vitro [117]. Subsequently it was found that antibiofilm agents were independently determined compared to antimicrobial activities against planktonic cells, since peptides with preferential activity against one or the other bacterial growth states were identified [118]. Critically antibiofilm peptides exhibit broad-spectrum activity versus Gram-positive and Gram-negative bacteria [114,115,118]. Since these initial observations, a number of other naturally occurring HDPs have been found to exhibit antibiofilm properties as well as including phylloseptin-1 from waxy monkey frogs [119,120], pleurocidin from winter flounder, and human β-defensin 3 [121]. Recently synthetic HDPs have also been identified with even more potent broad-spectrum antibiofilm activity often inhibiting or killing organisms in biofilms at concentrations as low as ~ 1 μg/mL. For example, IDR-1018 and the D-enantiomeric peptides DJK-5 and DJK-6, have all been shown to possess potent broad-spectrum activity against preformed biofilms from *P. aeruginosa* and other antibiotic-resistant species [114,115]. Consistent with the lack of relationship between antibiofilm and antimicrobial activity, *Burkholderia cepacia* is completely resistant to the antimicrobial effects of HDPs when growing planktonically but highly susceptible to antibiofilm peptides when growing as biofilms [114,115]. Mechanistically, de la Fuente-Núñez et al. demonstrated that IDR-1018 and the DJK peptides work by intracellular targeting and promoting degradation of the stringent response signaling molecule, ppGpp, which plays an important role in bacterial biofilm formation [114,115].

Recently, synthetic HDPs have also been demonstrated to effectively reduce abscess size in a murine model of chronic, high-density bacterial infections [122]. Though bacteria within an abscess are generally considered to be physiologically different from those found within biofilms, both are strongly dependent on the stringent response, suggesting a strong mechanistic connection between the two growth phenotypes [122]. Importantly, DJK-5 [114], previously shown to possess potent antibiofilm activity in vitro, reduced the severity of abscesses formed by MRSA and *P. aeruginosa* in mice by significantly reducing tissue injury and lesion size by up to fourfold compared to controls [122]. Though this model is not precisely a direct mimic of a biofilm, peptides that can eradicate high-density abscess infections have also been shown to work in nonmammalian biofilm infection models [114].

Although the examples describing the antibiofilm activity of HDPs presented here are encouraging, their relatively high costs have spurred researchers to explore combination therapies of peptides with conventional antibiotics. Synergistic application of HDPs in combination with conventional antibiotics has been successfully demonstrated with the synthetic peptides DJK-5, DJK-6, and IDR-1018 in vitro. In these cases, the application of the synthetic peptide synergistically decreased the concentration of antibiotics such as ciprofloxacin, gentamicin, ceftazidime, and imipenem by up to 64-fold in treating biofilms formed by several clinically relevant pathogens [123,124]. Natural peptides, such as LL-37 and a cecropin–melittin hybrid (CAMA), have also shown synergy in vitro against *P. aeruginosa* biofilms, where combinations with ciprofloxacin decreased biofilm-eradication concentrations by four- and eightfold compared to single antibiotic treatments [125]. Taken together, these findings highlight the potential for HDPs to be used as standalone or adjunctive therapies to conventional antibiotics. Overall, the potent antibiofilm activity exerted by many synthetic HDPs is rapidly emerging as a treatment option for recalcitrant biofilm-associated infections and chronic abscesses.

10.6.1 Biofilm prevention using peptide-coated surfaces

Implanted medical devices, such as catheters or prosthetic valves, are often colonized by bacteria and this can lead to the formation of biofilms that require the removal of these devices and/or can cause severe infections recalcitrant to treatment by conventional antibiotics [126]. In fact, nearly 100,000 deaths per year in the United States are due to medical device-related infections largely due to the notorious (adaptive) antibiotic resistance of biofilms [127]. One possible method of preventing biofilm formation on implanted medical devices is to attach peptides to the surface of these devices either physically through adsorption or chemically through covalent bonding [126,127]. As an example, Lim et al. established CWR11-coated Foley catheters with effective antimicrobial and antibiofilm activities against *S. aureus*, *E. coli*, and *P. aeruginosa* which could prevent infections commonly associated with indwelling devices [126]. Tethering approaches to coat biomedical plastics with peptides have also been used to prevent bacterial adhesion of Gram-negative and Gram-positive organisms [128]. Recently, Yu et al. [128] used branched polymers with multiple peptide covalent-attachment sites to increase the density of peptides on catheter surfaces. The catheters were tested in a mouse urinary catheter model, and the AMP-conjugated catheters significantly reduced the formation of *P. aeruginosa* biofilms as compared to uncoated catheters [128]. In addition, the peptide, hLf1-11 (Table 10.1), currently being clinically evaluated against nosocomial bacterial and *Candida* infections has been shown to exhibit effective antibiofilm activity against MRSA when covalently tethered to chitosan ultrathin film [151]. Chitosan is a naturally occurring polymer used in several medical devices used for gene therapy and as drug-delivery carriers [152]. Since biofilms play a major role in medical device-related infections, the use of tethered peptides is a promising new application to prevent medical device-related infections.

Ultimately, the immunomodulatory, antimicrobial, and antibiofilm functions of HDPs have the potential to be exploited for many applications to combat bacterial infections and biofilms, as well as modulate serious inflammatory diseases. Furthermore, the ability of many natural and synthetic HDPs to exhibit multiple activities allows them to have a multifaceted effect in many diseases, or synergy with conventional treatments, which makes them ideal candidates for antiinfective therapies. In this context, to improve on current HDP activities and increase future applications, research is focused on actively designing and developing novel synthetic HDPs with improved clinical potential.

10.7 Designing novel HDPs

Synthetic peptide design is of increasing interest as progressively more information on peptide structure and function becomes available. To be therapeutically relevant, the design of a synthetic peptide has to enhance a specific desired activity while exhibiting low cytotoxicity at the therapeutic dose as well as retaining favorable properties such as high solubility and stability. To design such optimized HDPs, three approaches are generally employed: template-based design, biophysical design, and computational design methods [9].

10.7.1 Template-based design

Template-based peptide synthesis relies on a template peptide with known activity and sequence [153]. This template peptide serves as the starting point to design novel peptide sequences in which various biophysical properties are altered such as charge, hydrophobicity, or amphipathicity. Researchers have been able to examine the importance of specific amino acids and residue positions on peptide activity by synthesizing derivatives with altered residues or functional motifs, truncated or deleted regions, or even scrambling the peptide sequence to determine functional residues and structural characteristics that contribute to the desired biological activity. Many peptide derivatives with enhanced antimicrobial activity have been made

Table 10.1 HDPs previously and currently in clinical trials and their proposed application in the context of specific diseases

Peptide name (company)	Clinical indication seeking approval	Clinical status	Most recent clinical trial ID and/or information	References
Omiganan (Cutanea Life Sciences)	Treatment against acne vulgaris, rosacea, genital warts, and vulvar intraepithelial neoplasia	Phase II/III		[129,130]
Surotomycin (MK-4261/ CB-315) (Merck)	*Clostridium difficile*-associated diarrhea	Phase III	Clinical trial ID NCT01597505	[131,132]
Brilacidin or PMX-30063 (PolyMedix)	Acute *Staphylococcus aureus* skin infections	Phase II	Clinical Trial ID: NCT01211470	[133,134]
Novexatin/NP213 (NovaBiotics)	Fungal nail infections	Phase IIb		[135,136]
P-113/PAC-113 (Pacgen Biopharmaceuticals)	Gingivitis and oral candidiasis	Phase II	Clinical Trial ID: NCT00659971	[137]
DPK-060 (DermaGen)	Atopic dermatitis and acute external otitis	Phase II	Clinical Trial ID: NCT01447017	[138,139]
LTX-109 (Lytix Biopharma)	Topical treatment of bacterial impetigo	Phase II	Clinical Trial ID: NCT01803035	[140,141]
OP-145 (OctoPlus)	Otitis media: chronic middle ear infection	Phase II	Acquired by Dr Reddys Laboratories in 2013 Clinical Trial ID: ISRCTN84220089; ISRCTN12149720	[19,142]

(CKPV)2/CZEN-002 (Abiogen Pharma)	Treatment of urogenital conditions (vulvovaginal candidiasis)	Phase II	Clinical Trial ID: CN 1867349 A	[143]
Dusquetide/SGX-942/ IMX-942 (Soligenix)	Treatment of oral mucositis	Phase II	Recently passed phase II	[133,144,145]
Synthetic LL-37 (Lipopeptide AB)	Venous leg ulcers	Phase I/II	EU Clinical Trials Register: 2012-002100-41	[22]
Iseganan or IB-367 (Ardea Biosciences)	Oral mucositis for patients undergoing radiation therapy	Phase III	Discontinued as of 2004 due to health risks Clinical Trial ID: NCT0002233	[146,147]
hLf1-11 (Am-Pharma)	Nosocomial infections; systemic *Candida* infections; bone marrow transplantation patients	Phase II	Company suspended trials for strategic reasons Clinical Trial ID: NCT00509938	[68,148,149]
Plectasin NZ2114 (Novozymes)	Endocarditis associated with MRSA	Phase I	Discontinued in 2011	[74,150]

from natural HDPs using this approach, including derivatives of cecropins, magainins, protegrins, bactenicins, and cathelicidins [154–156].

This type of design strategy is often limited by the number of peptides that can be synthesized and screened for activity, however, this problem has been somewhat alleviated through the use of SPOT-synthesized peptide libraries synthesized on cellulose sheets [157]. This technology has allowed researchers to simultaneously evaluate the activity of hundreds of synthetic peptides at a fraction of the cost and has made it feasible to perform complete amino acid substitution screens on known HDP sequences. For example, 228 single amino acid variants of the linear HDP bactenecin, Bac2a, were SPOT-synthesized on cellulose peptide arrays and their antimicrobial activity was assessed against a luminescent *P. aeruginosa* lux strain. This approach permitted the identification of residues that were favored in the Bac2a sequence and that when combined together in next-generation HDP sequences resulted in peptides with improved antibacterial potency [158].

Synthetic HDPs with immunomodulatory activity have also served as templates for novel IDR peptides with potent immunomodulatory properties. IDR-1, -1018, and -1002 were developed from the Bac2a template using substitution methods [14,16,21]. Originally, IDR-1 was designed to contain sequence features that were incompatible with direct antimicrobial activity [14]. IDR-1018 and IDR-1002 were subsequently discovered to have enhanced immunomodulatory functions compared to IDR-1, as well as enhanced antibiofilm properties [16,21,116]. Derivatives of IDR-1002 and IDR-HH2 have also been optimized using SPOT-synthesized peptide arrays and high-throughput screening methods for various biological activities [159], demonstrating that such an approach could be used to further develop these IDR peptides to treat biofilm-associated chronic infections.

10.7.2 Structure-guided design

A second method of peptide design is the structure-guided method which uses models based on peptide structures determined in hydrophobic environments (similar to the structures that would be formed in membranes) and biophysical measurements, rather than properties associated with the primary amino acid sequence. Bactenecin, indolicidin, and protegrin have been investigated using structure-based design, molecular modeling, and biophysical studies in order to increase their broad-spectrum activity, stability, and elucidate mechanistic membrane interactions to improve their bacterial killing [59,160].

Molecular dynamic modeling can extend structure-guided approaches by using simulations to computationally represent the atoms in an HDP molecule and evaluate their interactions with the solvent conditions, membranes, and each other [9,161]. This information can then serve to inform next-generation peptide sequences that maximize these important biophysical interactions. As an example, molecular dynamic simulations were used to design indolicidin analogues with enhanced antimicrobial activity towards *E. coli* by increasing the charge density at the membrane interface by replacing proline with a lysine residue. At the same time, derivatives with decreased hemolytic activity were generated by replacing tryptophan with phenylalanine to retain rigid but smaller aromatic rings as side chains, and decrease the overall disruption of the eukaryotic membrane interface [162].

10.7.3 Computational modeling of HDPs

Beyond simple structure and template-based modeling, complex computational peptide analysis is being used to model peptide activity based on quantified biological activities and features of the peptide structure. Such an approach is known as quantitative structure–activity relationship (QSAR) studies and uses large numbers of molecular descriptors that describe each individual peptide sequence and tries to model experimentally measured activities such as antibacterial potency using sophisticated machine learning techniques [9]. Feature selection to describe peptide activity and structure is a crucial step in computational modeling and is typically done automatically based on peptide data sets and statistical models of variable selection [163]. Ultimately, with any model used, feature selection is a tradeoff between predictability of the model and minimizing the necessary descriptors used. An ideal model has equal predictability with fewer descriptors to lower the computational workload and make it easier to interpret [164].

Virtual screening and random design of peptides rely on using numerical methods to determine quantifiable peptide descriptors to design and test peptide structures without the need for large high-throughput screening experiments. In 2009, Cherkasov et al. used available chemical biology information of small broad-spectrum peptides and test sets of randomly generated peptides using previously developed QSAR descriptor preferences to create models of antibiotic activity with artificial neural networks (based on previous Bac2a screens) [165]. A library of 100,000 virtual peptide sequences was scored and classified based on the QSAR models and the top 200 peptides with predicted activity were synthesized using SPOT technology and screened against the lux-*Pseudomonas*. Compared to the Bac2a control, 98% of the peptides predicted to have increased activity actually did, and two lead peptides, HHC-10 and HHC-36, were selected for further analysis revealing significant activity against a wide range of multidrug-resistant bacterial strains [165]. Current research is focused on using similar computation approaches to design novel peptides with specific antibiofilm or immunomodulatory properties and to associate these activities with various HDP structural characteristics.

10.8 The future of HDPs: From the bench to the clinic

Numerous preclinical studies, as discussed in this chapter, reveal the overwhelming potential and efficacy associated with the use of HDPs to treat microbial infections either by directly targeting planktonic cells, stimulating the immune system to mediate clearance of the infection, or by targeting complex growth adaptations like biofilms which exhibit broad-spectrum adaptive resistance against antibiotics. From all the studies and examples presented in this chapter, we can appreciate the prospects of using HDPs in a clinical setting. Table 10.1 presents a nonexhaustive list of synthetic HDPs that are currently being tested for clinical use. Many peptides such as IMX-942 and a synthetic LL-37 peptide are in early stages of the development pipeline, completing phase I or II trials as antiinfectives [22,133,144]. Similarly there are multiple immunomodulatory peptides such as DPK-060 [138] and OP-145 [19] that are currently being tested (phase II) for the treatment of inflammatory ear infections known as otitis media. Although many of the peptides currently undergoing trials are being tested for topical use (e.g., DPK-060, Novexatin, Brilacidin, Omiganan [129]), HDP Surotomycin has reached phase III clinical trials for systemic treatment of *C. difficile* infections in order to prevent and alleviate associated bowel irritability [131]. The promising clinical applications of peptides are highlighted by the many peptides currently in the pipeline. However, challenges still arise that prevent peptides from continuing clinical testing (see notes on Iseganan, hLf1-11, and Plectasin). These challenges and corresponding solutions are discussed below.

10.8.1 Current commercialization challenges and potential solutions

One of the major limitations associated with HDP use is their inherent low stability in serum and their susceptibility to degradation by host proteases. The use of D-amino acids makes the peptides impervious to proteolytic degradation by bacterial or host proteases [60]. Such a strategy has successfully generated protease-resistant LL-37 derivatives as well as the highly active antibiofilm peptides, DJK-5 and DJK-6, and the angiogenic peptide SR-0379 [39,114].

Another major hurdle facing commercialization of peptides is their relatively short shelf-life and potential for degradation during storage. Several solutions for this have been proposed including chemical modification of functional groups within the peptide [166]. In addition, various formulation strategies have been proposed including loading peptides into various nanoparticles or encapsulating peptides into lipid vesicles [167].

To reduce the high costs of synthetic peptidse (up to \$50–400 mg^{-1}), researchers have favored the development of shorter peptides that are typically easier and cheaper to make. Truncation studies on peptides can reduce the number of residues, and often tryptophanyl substitutions at the hydrophobic–hydrophilic interface of the amphipathic helix are incorporated to maximize activity in shorter peptides [6,168]. Additionally, the use of recombinant fusion peptides allows higher yields of soluble proteins and is potentially a more cost-efficient alternative to solid-phase synthesis chemistry [154].

Peptide toxicity, a major issue in the development of HDPs clinically, has been poorly studied but might have limited peptides to being administered topically as opposed to systemically. Indeed, many of the peptides presently in clinical development are seeking approval for infections of the skin or other topical applications (Table 10.1). As an example, Iseganan was a synthetic protegrin analogue that was shown to be effective as an oral rinse to treat oral mucositis [169] but this peptide failed phase III of clinical trials due to high systemic toxicity when used to treat ventricular-associated pneumonia [170]. Unfortunately, the relatively high levels of peptide needed to exhibit antiinfective activity also resulted in high eukaryotic cell toxicity and hemolysis [171]. Numerous studies have tried to address the issue of toxicity using the design approaches described above and our group previously established that there exists a positive correlation between peptide hydrophobicity and toxicity [172]. Therefore, by replacing amino acids with less hydrophobic residues or interrupting hydrophobic patches with basic groups, it may be possible to alleviate some of the prospective issues associated with nonspecific peptide toxicity [172], although it is unlikely that this generalization holds true for all HDP sequences. Alternatively, peptide toxicity could also be reduced by using specialized drug-delivery systems such as liposome-encapsulated peptides [168].

10.8.2 Final thoughts

Due to rising antibiotic resistance in pathogenic microbes around the world and a significant lack of antibiotics for treatment of multidrug-resistant infections, new treatments are needed that can replace and improve on current antibiotic therapies. HDPs have certain advantages over antibiotics because they have broad-spectrum activity while also employing multiple mechanisms of action, which makes them less likely to induce resistance in bacteria. Furthermore, HDPs have potent immunomodulatory activities, which is a feature that gives peptides a distinct edge over conventional antibiotics in terms of fighting infections and promoting healing [6,14,116]. Given that infections often trigger local inflammatory responses and can occur in the context of wounds, such activities are highly relevant to the treatment of chronic and biofilm infections. Similarly, the ability of these peptides to act against multiple bacterial species represents a potential advantage against mixed infections [6]. Furthermore, studies have found that combinations of HDPs and antibiotics can be used synergistically to treat infections such as biofilms, which allows for lower quantities of both agents to be used and exhibit increased efficacy against the infections being treated [116,123,128]. With improved screening methods and in silico testing, it is apparent that we are making great strides to identify and understand the properties of HDPs that allow them to exert their diverse range of biological activities. Armed with this information, the design of next-generation HDPs with enhanced biological activity profiles is a feasible goal and will permit future applications of HDPs to improve human health.

Acknowledgments

Our peptide research has been generously supported by grants from the Canadian Institutes for Health Research (CIHR) (Funding reference MOP-74493) and by the National Institute of Allergy and Infectious Diseases of the National Institutes of Health. C.R.B. received a CIHR Canada Graduate Master's Scholarship and a Cystic Fibrosis Canada Doctoral Studentship Award. R.E.W.H. holds a Canada Research Chair in Health and Genomics and a UBC Killam Professorship.

Abbreviations

ABP antibiofilm peptide
AMP antimicrobial peptide
COPD chronic obstructive pulmonary disease
CRAMP cathelin-related antimicrobial peptide
DC dendritic cells
GlcNAc β-1,4-linked *N*-acetylglucosamine
HDP host defense peptide
IDR innate defense regulator
IL interleukin
LPS lipopolysaccharide
MAPK modulating mitogen-activated protein kinase
MRSA methicillin resistant *Staphylococcus aureus*
MurNAc *N*-acetylmuramic acid
ODN oligodeoxynucleotides
PBMC peripheral blood mononuclear cells
PG peptidoglycan
PTd pertussis toxoid
QSAR quantitative structure−activity relationship.
ROS reactive oxygen species
STAT signal transducer and activator of transcription
TLR Toll-like receptor
TNF tumor necrosis factor

References

[1] WHO, Antimicrobial Resistance: Global Report on Surveillance, 2014, WHO, Geneva, Switzerland, 2014. <http://www.who.int/drugresistance/documents/surveillancereport/en/> (accessed 17.03.17).
[2] General Assembly of the United Nations, High-level meeting on antimicrobial resistance, September 21, 2016. <http://www.un.org/pga/71/2016/09/21/press-release-hl-meeting-on-antimicrobial-resistance/> (accessed 20.03.17).
[3] R.E.W. Hancock, E.F. Haney, E.E. Gill, The immunology of host defence peptides: beyond antimicrobial activity, Nat. Rev. Immunol. 16 (2016) 321−334.
[4] A.L. Hilchie, K. Wuerth, R.E.W. Hancock, Immune modulation by multifaceted cationic host defense (antimicrobial) peptides, Nat. Chem. Biol. 9 (2013) 761−768.
[5] D. Yang, O. Chertov, J.J. Oppenheim, Participation of mammalian defensins and cathelicidins in anti-microbial immunity: receptors and activities of human defensins and cathelicidin (LL-37), J. Leukoc. Biol. 69 (2001) 691−697.
[6] S.C. Mansour, O.M. Pena, R.E.W. Hancock, Host defense peptides: front-line immunomodulators, Trends Immunol. 35 (2014) 443−450.
[7] A.A. Bahar, D. Ren, Antimicrobial peptides, Pharmaceuticals (Basel) 6 (2013) 1543−1575.
[8] L.T. Nguyen, E.F. Haney, H.J. Vogel, The expanding scope of antimicrobial peptide structures and their modes of action, Trends Biotechnol. 29 (2011) 464−472.
[9] C.D. Fjell, J.A. Hiss, R.E.W. Hancock, et al., Designing antimicrobial peptides: form follows function, Nat. Rev. Drug Discov. 11 (2011) 37−51.

[10] D.M.E. Bowdish, D.J. Davidson, Y.E. Lau, et al., Impact of LL-37 on anti-infective immunity, J. Leukoc. Biol. 77 (2005) 451–459.
[11] A. Nijnik, R. Hancock, Host defence peptides: antimicrobial and immunomodulatory activity and potential applications for tackling antibiotic-resistant infections, Emerg. Health Threats J. 2 (2009) e1. Available from: http://dx.doi.org/10.3134/ehtj.09.001.
[12] N. Mookherjee, K.L. Brown, D.M.E. Bowdish, et al., Modulation of the TLR-mediated inflammatory response by the endogenous human host defense peptide LL-37, J. Immunol. 176 (2006) 2455–2464.
[13] S.M. Walters, V.S. Dubey, N.R. Jeffrey, et al., Antibiotic-induced *Porphyromonas gingivalis* LPS release and inhibition of LPS-stimulated cytokines by antimicrobial peptides, Peptides 31 (2010) 1649–1653.
[14] M.G. Scott, E. Dullaghan, N. Mookherjee, et al., An anti-infective peptide that selectively modulates the innate immune response, Nat. Biotechnol. 25 (2007) 465–472.
[15] A.H. Achtman, S. Pilat, C.W. Law, et al., Effective adjunctive therapy by an innate defense regulatory peptide in a preclinical model of severe malaria, Sci. Transl. Med. 4 (2012) 135ra64.
[16] A. Nijnik, L. Madera, S. Ma, et al., Synthetic cationic peptide IDR-1002 provides protection against bacterial infections through chemokine induction and enhanced leukocyte recruitment, J. Immunol. 184 (2010) 2539–2550.
[17] J. Vane, R. Botting, Inflammation and the mechanism of action of anti-inflammatory drugs, FASEB J. 1 (1987) 89–96.
[18] K.A. Brogden, A.M. Bates, C.L. Fischer, Antimicrobial peptides in host defense: functions beyond antimicrobial activity, in: J. Harder, J.-M. Schröder (Eds.), Antimicrobial Peptides, Springer International Publishing, Cham, 2016, pp. 129–146.
[19] N. Malanovic, R. Leber, M. Schmuck, et al., Phospholipid-driven differences determine the action of the synthetic antimicrobial peptide OP-145 on Gram-positive bacterial and mammalian membrane model systems, Biochim. Biophys. Acta 1848 (2015) 2437–2447.
[20] B. Rivas-Santiago, J.E. Castañeda-Delgado, C.E. Rivas Santiago, et al., Ability of innate defence regulator peptides IDR-1002, IDR-HH2 and IDR-1018 to protect against *Mycobacterium tuberculosis* infections in animal models, PLoS One 8 (2013) e59119.
[21] M. Wieczorek, H. Jenssen, J. Kindrachuk, et al., Structural studies of a peptide with immune modulating and direct antimicrobial activity, Chem. Biol. 17 (2010) 970–980.
[22] A. Grönberg, M. Mahlapuu, M. Ståhle, et al., Treatment with LL-37 is safe and effective in enhancing healing of hard-to-heal venous leg ulcers: a randomized, placebo-controlled clinical trial, Wound Repair Regen. 22 (2014) 613–621.
[23] P.H.A. Lee, J.A. Rudisill, K.H. Lin, et al., HB-107, a nonbacteriostatic fragment of the antimicrobial peptide cecropin B, accelerates murine wound repair, Wound Repair Regen. 12 (2004) 351–358.
[24] L. Steinstraesser, T. Koehler, F. Jacobsen, et al., Host defense peptides in wound healing, Mol. Med. 14 (2008) 528–537.
[25] W. Xiao, Y.-P. Hsu, A. Ishizaka, et al., Sputum cathelicidin, urokinase plasminogen activation system components, and cytokines discriminate cystic fibrosis, copd, and asthma inflammation, Chest 128 (2005) 2316–2326.
[26] L.N.Y. Chow, K.-Y. Choi, H. Piyadasa, et al., Human cathelicidin LL-37-derived peptide IG-19 confers protection in a murine model of collagen-induced arthritis, Mol. Immunol. 57 (2014) 86–92.
[27] D. Yang, Q. Chen, A.P. Schmidt, et al., LL-37, the neutrophil granule- and epithelial cell-derived cathelicidin, utilizes formyl peptide receptor-like 1 (FPRL1) as a receptor

to chemoattract human peripheral blood neutrophils, monocytes, and T cells, J. Exp. Med. 192 (2000) 1069–1074.
[28] M. Dürr, A. Peschel, Chemokines meet defensins: the merging concepts of chemoattractants and antimicrobial peptides in host defense, Infect. Immun. 70 (2002) 6515–6517.
[29] G.S. Tjabringa, D.K. Ninaber, J.W. Drijfhout, et al., Human cathelicidin LL-37 is a chemoattractant for eosinophils and neutrophils that acts via formyl-peptide receptors, Int. Arch. Allergy Immunol. 140 (2006) 103–112.
[30] S.-K. Zhang, J.-W. Song, F. Gong, et al., Design of an α-helical antimicrobial peptide with improved cell-selective and potent anti-biofilm activity, Sci. Rep. 6 (2016) 27394.
[31] L.T. Nguyen, H.J. Vogel, Structural perspectives on antimicrobial chemokines, Front. Immunol. 3 (2012) 384. Available from: http://dx.doi.org/10.3389/fimmu.2012.00384.
[32] J.M. Pérez-Cañadillas, A. Zaballos, J. Gutiérrez, et al., NMR solution structure of murine CCL20/MIP-3alpha, a chemokine that specifically chemoattracts immature dendritic cells and lymphocytes through its highly specific interaction with the beta-chemokine receptor CCR6, J. Biol. Chem. 276 (2001) 28372–28379.
[33] F. Niyonsaba, H. Ushio, N. Nakano, et al., Antimicrobial peptides human beta-defensins stimulate epidermal keratinocyte migration, proliferation and production of proinflammatory cytokines and chemokines, J. Invest. Dermatol. 127 (2007) 594–604.
[34] T. Velnar, T. Bailey, V. Smrkolj, The wound healing process: an overview of the cellular and molecular mechanisms, J. Int. Med. Res. 37 (2009) 1528–1542.
[35] M.P. Caley, V.L.C. Martins, E.A. O'Toole, Metalloproteinases and wound healing, Adv. Wound Care 4 (2015) 225–234.
[36] R. Ramos, J.P. Silva, A.C. Rodrigues, et al., Wound healing activity of the human antimicrobial peptide LL37, Peptides 32 (2011) 1469–1476.
[37] M. Carretero, M.J. Escámez, M. García, et al., In vitro and in vivo wound healing-promoting activities of human cathelicidin LL-37, J. Invest. Dermatol. 128 (2008) 223–236.
[38] K. Kanazawa, K. Okumura, H. Ogawa, et al., An antimicrobial peptide with angiogenic properties, AG-30/5C, activates human mast cells through the MAPK and NF-κB pathways, Immunol. Res. 64 (2016) 594–603.
[39] H. Tomioka, H. Nakagami, A. Tenma, et al., Novel anti-microbial peptide SR-0379 accelerates wound healing via the PI3 kinase/Akt/mTOR pathway, PLoS One 9 (2014) e92597.
[40] M.L. Mangoni, A.M. McDermott, M. Zasloff, Antimicrobial peptides and wound healing: biological and therapeutic considerations, Exp. Dermatol. 25 (2016) 167–173.
[41] A. Thorburn, Apoptosis and autophagy: regulatory connections between two supposedly different processes, Apoptosis 13 (2008) 1–9.
[42] Z. Hu, T. Murakami, K. Suzuki, et al., Antimicrobial cathelicidin peptide LL-37 inhibits the LPS/ATP-induced pyroptosis of macrophages by dual mechanism, PLoS One 9 (2014) e85765.
[43] J.-M. Yuk, D.-M. Shin, H.-M. Lee, et al., Vitamin D3 induces autophagy in human monocytes/macrophages via cathelicidin, Cell Host Microbe 6 (2009) 231–243.
[44] Y.E. Lau, D.M.E. Bowdish, C. Cosseau, et al., Apoptosis of airway epithelial cells: human serum sensitive induction by the cathelicidin LL-37, Am. J. Respir. Cell Mol. Biol. 34 (2006) 399–409.
[45] C.I. Chamorro, G. Weber, A. Grönberg, et al., The human antimicrobial peptide LL-37 suppresses apoptosis in keratinocytes, J. Invest. Dermatol. 129 (2009) 937–944.

[46] P.G. Barlow, Y. Li, T.S. Wilkinson, et al., The human cationic host defense peptide LL-37 mediates contrasting effects on apoptotic pathways in different primary cells of the innate immune system, J. Leukoc. Biol. 80 (2006) 509–520.
[47] S.M. Alalwani, J. Sierigk, C. Herr, et al., The antimicrobial peptide LL-37 modulates the inflammatory and host defense response of human neutrophils, Eur. J. Immunol. 40 (2010) 1118–1126.
[48] J.R. Dunkelberger, W.-C. Song, Complement and its role in innate and adaptive immune responses, Cell Res. 20 (2010) 34–50.
[49] X.-T. Ma, B. Xu, L.-L. An, et al., Vaccine with beta-defensin 2-transduced leukemic cells activates innate and adaptive immunity to elicit potent antileukemia responses, Cancer Res. 66 (2006) 1169–1176.
[50] J.W. Lillard, P.N. Boyaka, O. Chertov, et al., Mechanisms for induction of acquired host immunity by neutrophil peptide defensins, Proc. Natl. Acad. Sci. U S A. 96 (1999) 651–656.
[51] J.H. Fritz, S. Brunner, M.L. Birnstiel, et al., The artificial antimicrobial peptide KLKLLLLLKLK induces predominantly a TH2-type immune response to co-injected antigens, Vaccine 22 (2004) 3274–3284.
[52] M. Li, D.-H. Yu, H. Cai, The synthetic antimicrobial peptide KLKL5KLK enhances the protection and efficacy of the combined DNA vaccine against *Mycobacterium tuberculosis*, DNA Cell Biol. 27 (2008) 405–413.
[53] L.-L. An, Y.-H. Yang, X.-T. Ma, et al., LL-37 enhances adaptive antitumor immune response in a murine model when genetically fused with M-CSFR (J6-1) DNA vaccine, Leuk. Res. 29 (2005) 535–543.
[54] V.W. Bramwell, S. Somavarapu, I. Outschoorn, et al., Adjuvant action of melittin following intranasal immunisation with tetanus and diphtheria toxoids, J. Drug Target. 11 (2003) 525–530.
[55] K. Kurosaka, Q. Chen, F. Yarovinsky, et al., Mouse cathelin-related antimicrobial peptide chemoattracts leukocytes using formyl peptide receptor-like 1/mouse formyl peptide receptor-like 2 as the receptor and acts as an immune adjuvant, J. Immunol. 174 (2005) 6257–6265.
[56] S.V. Vemula, O.A.K. Amen, J.M. Katz, et al., Beta-defensin 2 enhances immunogenicity and protection of an adenovirus-based H5N1 influenza vaccine at an early time, Virus Res. 178 (2013) 398–403.
[57] J. Kovacs-Nolan, J.W. Mapletoft, L. Latimer, et al., CpG oligonucleotide, host defense peptide and polyphosphazene act synergistically, inducing long-lasting, balanced immune responses in cattle, Vaccine 27 (2009) 2048–2054.
[58] A. Gracia, M. Polewicz, S.A. Halperin, et al., Antibody responses in adult and neonatal BALB/c mice to immunization with novel *Bordetella pertussis* vaccine formulations, Vaccine 29 (2011) 1595–1604.
[59] M. Wu, R.E. Hancock, Interaction of the cyclic antimicrobial cationic peptide bactenecin with the outer and cytoplasmic membrane, J. Biol. Chem. 274 (1999) 29–35.
[60] R.E.W. Hancock, H.-G. Sahl, Antimicrobial and host-defense peptides as new anti-infective therapeutic strategies, Nat. Biotechnol. 24 (2006) 1551–1557.
[61] D. Münch, H.-G. Sahl, Structural variations of the cell wall precursor lipid II in Gram-positive bacteria: impact on binding and efficacy of antimicrobial peptides, Biochim. Biophys. Acta 1848 (2015) 3062–3071.

[62] V.M. Hernández-Rocamora, C. Alfonso, W. Margolin, et al., Evidence that bacteriophage λ kil peptide inhibits bacterial cell division by disrupting FtsZ protofilaments and sequestering protein subunits, J. Biol. Chem. 290 (2015) 20325–20335.
[63] C.-F. Le, C.-M. Fang, S.D. Sekaran, Intracellular targeting mechanisms by antimicrobial peptides, Antimicrob. Agents Chemother. 61 (2017) e02340-16.
[64] K. Boesze-Battaglia, R. Schimmel, Cell membrane lipid composition and distribution: implications for cell function and lessons learned from photoreceptors and platelets, J. Exp. Biol. 200 (1997) 2927–2936.
[65] B. Leitgeb, A. Szekeres, L. Manczinger, et al., The history of alamethicin: a review of the most extensively studied peptaibol, Chem. Biodivers. 4 (2007) 1027–1051.
[66] T. Lazaridis, Y. He, L. Prieto, Membrane interactions and pore formation by the antimicrobial peptide protegrin, Biophys. J. 104 (2013) 633–642.
[67] Y. Tamba, M. Yamazaki, Magainin 2-induced pore formation in the lipid membranes depends on its concentration in the membrane interface, J. Phys. Chem. B. 113 (2009) 4846–4852.
[68] G. van den Bogaart, J.V. Guzmán, J.T. Mika, et al., On the mechanism of pore formation by melittin, J. Biol. Chem. 283 (2008) 33854–33857.
[69] D. Sengupta, H. Leontiadou, A.E. Mark, et al., Toroidal pores formed by antimicrobial peptides show significant disorder, Biochim. Biophys. Acta 1778 (2008) 2308–2317.
[70] F. Guilhelmelli, N. Vilela, P. Albuquerque, et al., Antibiotic development challenges: the various mechanisms of action of antimicrobial peptides and of bacterial resistance, Front. Microbiol. 4 (2013) 353.
[71] W.C. Wimley, Describing the mechanism of antimicrobial peptide action with the interfacial activity model, ACS Chem. Biol. 5 (2010) 905–917.
[72] X. Zeng, J. Lin, Beta-lactamase induction and cell wall metabolism in Gram-negative bacteria, Front. Microbiol. 4 (2013) 128. Available from: http://dx.doi.org/10.3389/fmicb.2013.00128.
[73] D.-J. Scheffers, M.G. Pinho, Bacterial cell wall synthesis: new insights from localization studies, Microbiol. Mol. Biol. Rev. 69 (2005) 585–607.
[74] T. Schneider, T. Kruse, R. Wimmer, et al., Plectasin, a fungal defensin, targets the bacterial cell wall precursor lipid II, Science 328 (2010) 1168–1172.
[75] S.-T. Hsu, E. Breukink, E. Tischenko, et al., The nisin-lipid II complex reveals a pyrophosphate cage that provides a blueprint for novel antibiotics, Nat. Struct. Mol. Biol. 11 (2004) 963–967.
[76] I. Wiedemann, E. Breukink, C. van Kraaij, et al., Specific binding of nisin to the peptidoglycan precursor lipid II combines pore formation and inhibition of cell wall biosynthesis for potent antibiotic activity, J. Biol. Chem. 276 (2001) 1772–1779.
[77] A. Müller, H. Ulm, K. Reder-Christ, et al., Interaction of type A lantibiotics with undecaprenol-bound cell envelope precursors, Microb. Drug Resist. 18 (2012) 261–270.
[78] J. Cleveland, T.J. Montville, I.F. Nes, et al., Bacteriocins: safe, natural antimicrobials for food preservation, Int. J. Food Microbiol. 71 (2001) 1–20.
[79] G. Gallo, G. Renzone, E. Palazzotto, et al., Elucidating the molecular physiology of lantibiotic NAI-107 production in Microbispora ATCC-PTA-5024, BMC Genomics 17 (2016) 42.
[80] NAICONS, Nai-107 (and related lantibiotics). <http://www.naicons.com/products/nai-107.html> (accessed 17.03.17).
[81] O.G. Ghobrial, H. Derendorf, J.D. Hillman, Pharmacodynamic activity of the lantibiotic MU1140, Int. J. Antimicrob. Agents 33 (2009) 70–74.

[82] OGEN, OG253: Oragenics, Inc. <http://www.oragenics.com/technology-pipeline/lantibiotics/og253> (accessed 17.03.17).
[83] M. Wenzel, A.I. Chiriac, A. Otto, et al., Small cationic antimicrobial peptides delocalize peripheral membrane proteins, Proc. Natl. Acad. Sci. U S A. 111 (2014) E1409–1418.
[84] C.L. Friedrich, D. Moyles, T.J. Beveridge, et al., Antibacterial action of structurally diverse cationic peptides on gram-positive bacteria, Antimicrob. Agents Chemother. 44 (2000) 2086–2092.
[85] J.D.F. Hale, R.E.W. Hancock, Alternative mechanisms of action of cationic antimicrobial peptides on bacteria, Expert Rev. Anti Infect. Ther. 5 (2007) 951–959.
[86] M. Mattiuzzo, A. Bandiera, R. Gennaro, et al., Role of the *Escherichia coli* SbmA in the antimicrobial activity of proline-rich peptides, Mol. Microbiol. 66 (2007) 151–163.
[87] M. Scocchi, C. Lüthy, P. Decarli, et al., The proline-rich antibacterial peptide Bac7 binds to and inhibits *in vitro* the molecular chaperone dnak, Int. J. Pept. Res. Ther. 15 (2009) 147–155.
[88] C.W. Gunderson, J.L. Boldt, R.N. Authement, et al., Peptide wrwycr inhibits the excision of several prophages and traps holliday junctions inside bacteria, J. Bacteriol. 191 (2009) 2169–2176.
[89] C.B. Park, H.S. Kim, S.C. Kim, Mechanism of action of the antimicrobial peptide buforin II: buforin II kills microorganisms by penetrating the cell membrane and inhibiting cellular functions, Biochem. Biophys. Res. Commun. 244 (1998) 253–257.
[90] J.-P.S. Powers, M.M. Martin, D.L. Goosney, et al., The antimicrobial peptide polyphemusin localizes to the cytoplasm of *Escherichia coli* following treatment, Antimicrob. Agents Chemother. 50 (2006) 1522–1524.
[91] A. Ghosh, R.K. Kar, J. Jana, et al., Indolicidin targets duplex DNA: structural and mechanistic insight through a combination of spectroscopy and microscopy, ChemMedChem 9 (2014) 2052–2058.
[92] Y.H. Nan, K.H. Park, Y. Park, et al., Investigating the effects of positive charge and hydrophobicity on the cell selectivity, mechanism of action and anti-inflammatory activity of a Trp-rich antimicrobial peptide indolicidin, FEMS Microbiol. Lett. 292 (2009) 134–140.
[93] M. Castle, A. Nazarian, S.S. Yi, et al., Lethal effects of apidaecin on *Escherichia coli* involve sequential molecular interactions with diverse targets, J. Biol. Chem. 274 (1999) 32555–32564.
[94] K. Konopka, B. Dorocka-Bobkowska, S. Gebremedhin, et al., Susceptibility of *Candida* biofilms to histatin 5 and fluconazole, Antonie Van Leeuwenhoek 97 (2010) 413–417.
[95] A. Peschel, C. Vuong, M. Otto, et al., The D-alanine residues of *Staphylococcus aureus* teichoic acids alter the susceptibility to vancomycin and the activity of autolytic enzymes, Antimicrob. Agents Chemother. 44 (2000) 2845–2847.
[96] C.M. Ernst, P. Staubitz, N.N. Mishra, et al., The bacterial defensin resistance protein MprF consists of separable domains for lipid lysinylation and antimicrobial peptide repulsion, PLoS Pathog. 5 (2009) e1000660.
[97] R.K. Ernst, K.N. Adams, S.M. Moskowitz, et al., The *Pseudomonas aeruginosa* Lipid a deacylase: selection for expression and loss within the cystic fibrosis airway, J. Bacteriol. 188 (2006) 191–201.
[98] K. Kawasaki, R.K. Ernst, S.I. Miller, Inhibition of *Salmonella enterica* serovar typhimurium lipopolysaccharide deacylation by aminoarabinose membrane modification, J. Bacteriol. 187 (2005) 2448–2457.

[99] A.J. McCoy, H. Liu, T.J. Falla, et al., Identification of *Proteus mirabilis* mutants with increased sensitivity to antimicrobial peptides, Antimicrob. Agents Chemother. 45 (2001) 2030–2037.

[100] A.E. Jerse, N.D. Sharma, A.N. Simms, et al., A gonococcal efflux pump system enhances bacterial survival in a female mouse model of genital tract infection, Infect. Immun. 71 (2003) 5576–5582.

[101] L.I. Kupferwasser, R.A. Skurray, M.H. Brown, et al., Plasmid-mediated resistance to thrombin-induced platelet microbicidal protein in staphylococci: role of the qacA locus, Antimicrob. Agents Chemother. 43 (1999) 2395–2399.

[102] V. Nizet, Antimicrobial peptide resistance mechanisms of human bacterial pathogens, Curr. Issues Mol. Biol. 8 (2006) 11–26.

[103] A. Schmidtchen, I.-M. Frick, E. Andersson, et al., Proteinases of common pathogenic bacteria degrade and inactivate the antibacterial peptide LL-37, Mol. Microbiol. 46 (2002) 157–168.

[104] T. Guina, E.C. Yi, H. Wang, et al., A PhoP-regulated outer membrane protease of *Salmonella enterica* serovar typhimurium promotes resistance to alpha-helical antimicrobial peptides, J. Bacteriol. 182 (2000) 4077–4086.

[105] K.M. Mason, F.K. Raffel, W.C. Ray, et al., Heme utilization by nontypeable *Haemophilus influenzae* is essential and dependent on sap transporter function, J. Bacteriol. 193 (2011) 2527–2535.

[106] T. Jin, M. Bokarewa, T. Foster, et al., *Staphylococcus aureus* resists human defensins by production of staphylokinase, a novel bacterial evasion mechanism, J. Immunol. 172 (2004) 1169–1176.

[107] D. Islam, L. Bandholtz, J. Nilsson, et al., Downregulation of bactericidal peptides in enteric infections: a novel immune escape mechanism with bacterial DNA as a potential regulator, Nat. Med. 7 (2001) 180–185.

[108] C.C. Taggart, C.M. Greene, S.G. Smith, et al., Inactivation of human beta-defensins 2 and 3 by elastolytic cathepsins, J. Immunol. 171 (2003) 931–937.

[109] C. de la Fuente-Núñez, F. Reffuveille, L. Fernández, et al., Bacterial biofilm development as a multicellular adaptation: antibiotic resistance and new therapeutic strategies, Curr. Opin. Microbiol. 16 (2013) 580–589.

[110] K. Poole, Stress responses as determinants of antimicrobial resistance in Gram-negative bacteria, Trends Microbiol. 20 (2012) 227–234.

[111] N. Høiby, O. Ciofu, T. Bjarnsholt, *Pseudomonas aeruginosa* biofilms in cystic fibrosis, Future Microbiol. 5 (2010) 1663–1674.

[112] L.O. Bakaletz, Bacterial biofilms in otitis media: evidence and relevance, Pediatr. Infect. Dis. J. 26 (2007) S17–S19.

[113] R.M. Donlan, Biofilm formation: a clinically relevant microbiological process, Clin. Infect. Dis. 33 (2001) 1387–1392.

[114] C. de la Fuente-Núñez, F. Reffuveille, S.C. Mansour, et al., D-enantiomeric peptides that eradicate wild-type and multidrug-resistant biofilms and protect against lethal *Pseudomonas aeruginosa* infections, Chem. Biol. 22 (2015) 196–205.

[115] C. de la Fuente-Núñez, S.C. Mansour, Z. Wang, et al., Anti-biofilm and immunomodulatory activities of peptides that inhibit biofilms formed by pathogens isolated from cystic fibrosis patients, Antibiotics (Basel) 3 (2014) 509–526.

[116] S.C. Mansour, C. de la Fuente-Núñez, R.E.W. Hancock, Peptide IDR-1018: modulating the immune system and targeting bacterial biofilms to treat antibiotic-resistant bacterial infections, J. Pept. Sci. 21 (2015) 323–329.

[117] J. Overhage, A. Campisano, M. Bains, et al., Human host defense peptide LL-37 prevents bacterial biofilm formation, Infect. Immun. 76 (2008) 4176–4182.
[118] C. de la Fuente-Núñez, V. Korolik, M. Bains, et al., Inhibition of bacterial biofilm formation and swarming motility by a small synthetic cationic peptide, Antimicrob. Agents Chemother. 56 (2012) 2696–2704.
[119] J.-K. Lee, S.W. Chang, H. Perinpanayagam, et al., Antibacterial efficacy of a human β-defensin-3 peptide on multispecies biofilms, J. Endod. 39 (2013) 1625–1629.
[120] R. Zhang, M. Zhou, L. Wang, et al., Phylloseptin-1 (PSN-1) from *Phyllomedusa sauvagei* skin secretion: a novel broad-spectrum antimicrobial peptide with antibiofilm activity, Mol. Immunol. 47 (2010) 2030–2037.
[121] R. Tao, Z. Tong, Y. Lin, et al., Antimicrobial and antibiofilm activity of pleurocidin against cariogenic microorganisms, Peptides 32 (2011) 1748–1754.
[122] S.C. Mansour, D. Pletzer, C. de la Fuente-Núñez, et al., Bacterial abscess formation is controlled by the stringent stress response and can be targeted therapeutically, EBioMedicine 12 (2016) 219–226.
[123] F. Reffuveille, C. de la Fuente-Núñez, S. Mansour, et al., A broad-spectrum antibiofilm peptide enhances antibiotic action against bacterial biofilms, Antimicrob. Agents Chemother. 58 (2014) 5363–5371.
[124] S.M. Ribeiro, C. de la Fuente-Núñez, B. Baquir, et al., Antibiofilm peptides increase the susceptibility of carbapenemase-producing *Klebsiella pneumoniae* clinical isolates to β-lactam antibiotics, Antimicrob. Agents Chemother. 59 (2015) 3906–3912.
[125] S. Dosler, E. Karaaslan, Inhibition and destruction of *Pseudomonas aeruginosa* biofilms by antibiotics and antimicrobial peptides, Peptides 62 (2014) 32–37.
[126] K. Lim, R.R.Y. Chua, H. Bow, et al., Development of a catheter functionalized by a polydopamine peptide coating with antimicrobial and antibiofilm properties, Acta Biomater. 15 (2015) 127–138.
[127] S.A. Onaizi, S.S.J. Leong, Tethering antimicrobial peptides: current status and potential challenges, Biotechnol. Adv. 29 (2011) 67–74.
[128] J. Yu, N. Mookherjee, K. Wee, et al., Host defense peptide LL-37, in synergy with inflammatory mediator IL-1beta, augments immune responses by multiple pathways, J. Immunol. 179 (2007) 7684–7691.
[129] E. Rubinchik, D. Dugourd, T. Algara, et al., Antimicrobial and antifungal activities of a novel cationic antimicrobial peptide, omiganan, in experimental skin colonisation models, Int. J. Antimicrob. Agents 34 (2009) 457–461.
[130] T.R. Fritsche, P.R. Rhomberg, H.S. Sader, et al., Antimicrobial activity of omiganan pentahydrochloride against contemporary fungal pathogens responsible for catheter-associated infections, Antimicrob. Agents Chemother. 52 (2008) 1187–1189.
[131] C.T.M. Mascio, L.I. Mortin, K.T. Howland, et al., *In vitro* and *in vivo*characterization of cb-183,315, a novel lipopeptide antibiotic for treatment of *Clostridium difficile*, Antimicrob. Agents Chemother. 56 (2012) 5023–5030.
[132] BioCentury, Surotomycin (MK-4261) (formerly CB-183,315, CB-315). Product Profile. <http://bciq.biocentury.com/products/cb-183315> (accessed 16.03.17).
[133] R.E.W. Hancock, A. Nijnik, D.J. Philpott, Modulating immunity as a therapy for bacterial infections, Nat. Rev. Microbiol. 10 (2012) 243–254.
[134] B. Mensa, G.L. Howell, R. Scott, et al., Comparative mechanistic studies of brilacidin, daptomycin, and the antimicrobial peptide LL16, Antimicrob. Agents Chemother. 58 (2014) 5136–5145.
[135] S. Li, R.R. Breaker, Fluoride enhances the activity of fungicides that destabilize cell membranes, Bioorg. Med. Chem. Lett. 22 (2012) 3317–3322.

[136] NovaBiotics, NP213 (Novexatin®) is a novel cationic antifungal peptide. <http://www.novabiotics.co.uk/pipeline/novexatin-np213> (accessed 17.03.17).
[137] D.M. Rothstein, P. Spacciapoli, L.T. Tran, et al., Anticandida activity is retained in P-113, a 12-amino-acid fragment of histatin 5, Antimicrob. Agents Chemother. 45 (2001) 1367–1373.
[138] L. Boge, H. Bysell, L. Ringstad, et al., Lipid-based liquid crystals as carriers for antimicrobial peptides: phase behavior and antimicrobial effect, Langmuir 32 (2016) 4217–4228.
[139] ClinicalTrials.gov, A study of DPK-060 to investigate clinical safety and efficacy in patients with acute external otitis – full text view – clinicaltrials.gov. <https://clinicaltrials.gov/ct2/show/NCT01447017>, 2017 (accessed 17.03.17).
[140] J. Isaksson, B.O. Brandsdal, M. Engqvist, et al., A synthetic antimicrobial peptidomimetic (LTX 109): stereochemical impact on membrane disruption, J. Med. Chem. 54 (2011) 5786–5795.
[141] A.C. Nilsson, H. Janson, H. Wold, et al., LTX-109 is a novel agent for nasal decolonization of methicillin-resistant and -sensitive *Staphylococcus aureus*, Antimicrob. Agents Chemother. 59 (2015) 145–151.
[142] R. Roukema, OctoPlus proves efficacy of OP-145 in phase II ear infection study, *Marketwire*. <http://www.marketwired.com/press-release/octoplus-proves-efficacy-of-op-145-in-phase-ii-ear-infection-study-883010.htm>, 2008 (accessed 15.03.17).
[143] H. Ji, Y. Zou, J. Duan, et al., The synthetic melanocortin (CKPV)2 exerts anti-fungal and anti-inflammatory effects against *Candida albicans* vaginitis via inducing macrophage M2 polarization, PLoS One 8 (2013) e56004.
[144] A.A. Miller, P.F. Miller, Emerging Trends in Antibacterial Discovery: Answering the Call to Arms, Caister Academic Press, Norfolk, 2011.
[145] Soligenix, SGX942 oral mucositis. <http://www.soligenix.com/pipeline/biotherapeutics/sgx942-oral-mucositis/> (accessed 27.03.17).
[146] O. Cirioni, C. Silvestri, E. Pierpaoli, et al., IB-367 pre-treatment improves the in vivo efficacy of teicoplanin and daptomycin in an animal model of wounds infected with meticillin-resistant *Staphylococcus aureus*, J. Med. Microbiol. 62 (2013) 1552–1558.
[147] D.A. Mosca, M.A. Hurst, W. So, et al., IB-367, a protegrin peptide with in vitro and *in vivo* activities against the microflora associated with oral mucositis, Antimicrob. Agents Chemother. 44 (2000) 1803–1808.
[148] M. Godoy-Gallardo, C. Mas-Moruno, K. Yu, et al., Antibacterial properties of hLf1-11 peptide onto titanium surfaces: a comparison study between silanization and surface initiated polymerization, Biomacromolecules 16 (2015) 483–496.
[149] AM-Pharma, AM-Pharma announces €2.5 million financing | Blog, August 6, 2007. <http://www.am-pharma.com/blog/2007/08/am-pharma-announces-25-million-financing> (accessed 17.03.17).
[150] Y.Q. Xiong, W.A. Hady, A. Deslandes, et al., Efficacy of NZ2114, a novel plectasin-derived cationic antimicrobial peptide antibiotic, in experimental endocarditis due to methicillin-resistant *Staphylococcus aureus*, Antimicrob. Agents Chemother. 55 (2011) 5325–5330.
[151] F. Costa, S. Maia, J. Gomes, et al., Characterization of hLf1-11 immobilization onto chitosan ultrathin films, and its effects on antimicrobial activity, Acta Biomater. 10 (2014) 3513–3521.
[152] M. Bulwan, S. Zapotoczny, M. Nowakowska, Robust "one-component" chitosan-based ultrathin films fabricated using layer-by-layer technique, Soft Matter 5 (2009) 4726–4732.

[153] J.A. Robinson, Protein epitope mimetics as anti-infectives, Curr. Opin. Chem. Biol. 15 (2011) 379–386.
[154] E.F. Haney, R.E.W. Hancock, Peptide design for antimicrobial and immunomodulatory applications, Biopolymers 100 (2013) 572–583.
[155] Y. Huang, J. Huang, Y. Chen, Alpha-helical cationic antimicrobial peptides: relationships of structure and function, Protein Cell 1 (2010) 143–152.
[156] N. Wiradharma, U. Khoe, C.A.E. Hauser, et al., Synthetic cationic amphiphilic α-helical peptides as antimicrobial agents, Biomaterials 32 (2011) 2204–2212.
[157] K. Hilpert, D.F. Winkler, R.E. Hancock, Peptide arrays on cellulose support: SPOT synthesis, a time and cost efficient method for synthesis of large numbers of peptides in a parallel and addressable fashion, Nat. Protoc. 2 (2007) 1333–1349.
[158] K. Hilpert, R. Volkmer-Engert, T. Walter, et al., High-throughput generation of small antibacterial peptides with improved activity, Nat. Biotechnol. 23 (2005) 1008–1012.
[159] E.F. Haney, S. Mansour, A.L. Hilchie, et al., High throughput screening methods for assessing antibiofilm and immunomodulatory activities of synthetic peptides, Peptides 71 (2015) 276–285.
[160] D.S. Bolintineanu, Y.N. Kaznessis, Computational studies of protegrin antimicrobial peptides: a review, Peptides 32 (2011) 188–201.
[161] T.D. Romo, L.A. Bradney, D.V. Greathouse, et al., Membrane binding of an acyl-lactoferricin B antimicrobial peptide from solid-state NMR experiments and molecular dynamics simulations, Biochim. Biophys. Acta 1808 (2011) 2019–2030.
[162] C.-W. Tsai, N.-Y. Hsu, C.-H. Wang, et al., Coupling molecular dynamics simulations with experiments for the rational design of indolicidin-analogous antimicrobial peptides, J. Mol. Biol. 392 (2009) 837–854.
[163] M.P. Gonzalez, C. Teran, L. Saiz-Urra, et al., Variable selection methods in QSAR: an overview, Curr. Top. Med. Chem. 8 (2008) 1606–1627.
[164] G. Maccari, M.D. Luca, R. Nifosí, et al., Antimicrobial peptides design by evolutionary multiobjective optimization, PLoS Comput. Biol. 9 (2013) e1003212.
[165] A. Cherkasov, K. Hilpert, H. Jenssen, et al., Use of artificial intelligence in the design of small peptide antibiotics effective against a broad spectrum of highly antibiotic-resistant superbugs, ACS Chem. Biol. 4 (2009) 65–74.
[166] R. Eckert, Road to clinical efficacy: challenges and novel strategies for antimicrobial peptide development, Future Microbiol. 6 (2011) 635–651.
[167] J.P. da Costa, M. Cova, R. Ferreira, et al., Antimicrobial peptides: an alternative for innovative medicines? Appl. Microbiol. Biotechnol. 99 (2015) 2023–2040.
[168] M.-D. Seo, H.-S. Won, J.-H. Kim, et al., Antimicrobial peptides for therapeutic applications: a review, Molecules 17 (2012) 12276–12286.
[169] A. Trotti, A. Garden, P. Warde, et al., A multinational, randomized phase III trial of iseganan HCl oral solution for reducing the severity of oral mucositis in patients receiving radiotherapy for head-and-neck malignancy, Int. J. Radiat. Oncol. Biol. Phys. 58 (2004) 674–681.
[170] M. Kollef, D. Pittet, M. Sánchez García, et al., A randomized double-blind trial of iseganan in prevention of ventilator-associated pneumonia, Am. J. Respir. Crit. Care Med. 173 (2006) 91–97.
[171] A.K. Marr, W.J. Gooderham, R.E. Hancock, Antibacterial peptides for therapeutic use: obstacles and realistic outlook, Curr. Opin. Pharmacol. 6 (2006) 468–472.
[172] Y. Chen, C.T. Mant, S.W. Farmer, et al., Rational design of alpha-helical antimicrobial peptides with enhanced activities and specificity/therapeutic index, J. Biol. Chem. 280 (2005) 12316–12329.

Peptides in immunoengineering

11

John C. Barrett, Handan Acar, Michael J. Mellas and Matthew V. Tirrell
University of Chicago, Chicago, IL, United States

11.1 Introduction

11.1.1 Progress and challenges in engineering immunity

The immune system plays a critical role in our health. When functioning correctly, the immune system continually protects us from the most common to the most deadly and mysterious maladies: common influenza to HIV and cancer. When functioning incorrectly, autoimmune diseases such as diabetes and multiple sclerosis ensue. Understanding how the immune system functions and learning how to rationally control it to protect against or treat disease has been a challenge historically burdened by the immunology and medical fields. Recently, exciting strides in immunological discoveries and investigational tools have allowed outsiders to enter the fold, chief among them being the biomaterials field [1]. With biomaterials as an engineering tool, rational manipulation of the immune system at the cellular and molecular levels has expanded, shedding new light on the functioning of the immune system while also showing promising preclinical and clinical data for more effective and safer treatments [2]. As a flourishing subset of the biomaterials field, peptides have been used as a major design component of the immune engineering strategy. Peptides, because of their chemically defined nature, engineerability, and range of complexity, provide useful functionalities when designing strategies for vaccination, cancer immunotherapy, and the treatment of autoimmune disorders. This chapter will examine the peptide-based strategies for immune modulation in detail and discuss the promise and outstanding challenges facing the field of immune engineering.

11.1.1.1 Key cellular actors in the immune system

A brief review of the immune system as a whole and the activities of central immune cells during an immune response is important to discuss before delving into the role peptides play in the immunoengineering field today. The immune system is a collection of specialized cells that collaborate to protect against invading pathogens [3]. The cells belong to one of two arms of the immune system, the innate immune system or adaptive immune system. Innate immune cells, such as neutrophils and macrophages, are the rapid response arm of the immune system. These cells rapidly respond to pathogen invasions through the receptors they express that recognize conserved molecular motifs (i.e., flagella) characteristic of bacteria, viruses, and fungi, to quickly phagocytose (internalize) pathogens and

Peptide Applications in Biomedicine, Biotechnology and Bioengineering.
DOI: http://dx.doi.org/10.1016/B978-0-08-100736-5.00011-9

secrete reactive oxygen species or cytokines (soluble cell-signaling molecules that aid in cell–cell communication in immune responses).

While the innate immune system is the first, immediate response to a pathogen, the adaptive immune system is a secondary response which follows the processing and recognition of an antigen. The adaptive immune system is comprised of T cells and B cells. More specifically, there are two types of T cells. First, $CD4^+$ helper T cells are cells that secrete cytokines to direct the function of other immune cells, including innate cells, natural killer cells, and B cells. The second type of T cell is the $CD8^+$ killer T cell that recognizes and destroys infected or transformed cells. B cells are responsible for producing antibodies that bind to and neutralize the ability of pathogens to invade host cells and/or promote their phagocytosis. The adaptive immune response is a very specific response because of the clonal development of the immune system where each T cell and B cell expresses a unique T cell receptor (TCR) or B cell receptor (BCR), respectively. These receptors are formed in part by a process of DNA recombination, creating a wide variety of receptors that can recognize any microbial antigen (an antigen is essentially any biological molecule from a pathogen that is recognized by a TCR or BCR) [4]. When a T cell or B cell binds an antigen, this specific cell undergoes massive proliferation, generating a pool of effector cells within ~7 days following exposure, all specific to that antigen. During these 7 days, the innate immune system is attempting to clear the pathogen until the adaptive immune system steps up to finish clearing out the invasion. Following pathogen clearance, a majority of these effector B and T cells (~90%) enter a phase of programmed cell death, leaving behind a small group of long-lived memory cells that provide fast memory protection if the same antigen or pathogen ever invades again [5].

A final important group of immune cells are the antigen-presenting cells (APCs), which connect the innate arm to the adaptive arm of the immune system. APCs, including the most famous dendritic cell (DC), are responsible for activating naïve T cells, and in some cases B cells [6,7]. DCs act like innate immune cells because they reside in all peripheral tissues, constantly collecting antigens from the surrounding fluid. Similar to innate cells, DCs keep an eye out for and respond to pathogen invasions or tissue damage through the receptors they express (most studied are Toll-like receptors) that recognize conserved molecular motifs [8]. Unlike innate cells that remain at the site and fight the initial stages of the infection, once activated by a danger signal, DCs migrate from their peripheral tissue through the lymphatic vessels to local draining lymph nodes, where they physically present antigen to T cells and B cells. T-cell activation is achieved through the loading of short (8–15 amino acids) peptide fragments of antigens into the cleft of major histocompatibility complex (MHC) molecules displayed on the DC surface. DCs and other APCs have MHC class II, but all other nucleated cells in the body can also present peptides with MHC class I. These peptides are surveyed by the TCRs of T cells, and on finding a cognate peptide, T cells become activated by the DC to proliferate and carry out their adaptive immune response. In this way, DCs are the connection between the innate arm and the adaptive arm of the immune system.

The complex activities of individual immune cells and their interactions that make up the immune system summarized above (greatly oversimplified) are viewed as a network of potential for the immunoengineering field. In this chapter, we aim to summarize the numerous ways in which bioengineers, chemical engineers, material scientists, chemists, and physicists (often in collaboration with immunologists) use peptides as tools to probe or manipulate immune responses for therapeutic ends.

11.1.2 Peptides in immunoengineering

The purpose of this chapter is to highlight the current ways in which peptides are being used in immunology as well as the future ways which peptides can be used to assist in or directly tackle the challenges of vaccine development, immunotherapy, and the broader immune engineering field as a whole. Peptides are used for many purposes, as has been described many times throughout this textbook already. In cell biology, peptides are usually considered intermediate degradation products on their way to full degradation. The resulting free amino acids can then be used to build new proteins. Until 1986, it was never considered that peptides could have an essential role during an immune response: as specific antigens that DCs present to T cells and B cells for activation [9]. Over the past three decades, not only have we uncovered the natural function of peptides in immunology, but we have also come to identify and utilize other functions for peptides in immunology.

Peptides encompass a wide spectrum of complexity (Fig. 11.1). In the first level of complexity, peptides act as relatively simple, passive antigens. Moving up in complexity, peptides can possess functional/active properties. In the highest level of complexity, peptides can form supramolecular/assembling structures. As complexity increases, both breadth and strength of available functions expand for these systems. The range of engineered complexity allows for the selection of optimal system benefits for each application. While a majority of the immune engineering strategies currently in the clinic today are based on traditional drug development approaches—antibodies, recombinant proteins, small molecule drugs—simple peptides, which act mainly as antigens, also make an appearance in treatments. Many peptide antigens in commercially available drugs come from the first, lower tranche of complexity. These peptides are chosen for exhibiting the general, natural properties of peptides as a category: highly selective and efficacious, as well as, safe and well tolerated. These properties are the primary differentiating factor of peptides compared with small molecules and provide peptides with an excellent starting point for the design of immune therapies.

While opportunities still exist for the design and use of peptide antigens in their simplest form, a range of peptide technologies has been emerging that represents the opportunities and future directions within the peptide immunoengineering field. Beyond a peptide's use as an antigen, immunoengineering opportunities exist with the next level of peptide complexity: functional or "active" peptides. Peptides can possess a diversity of functions, including targeting (cell binding, integrin binding, surface binding), penetrating, antiinflammatory, proteolyzable, and combinations thereof [14]. Moreover, the tailorable nature and chemically defined properties of peptides leaves much room for the discovery and development of additional

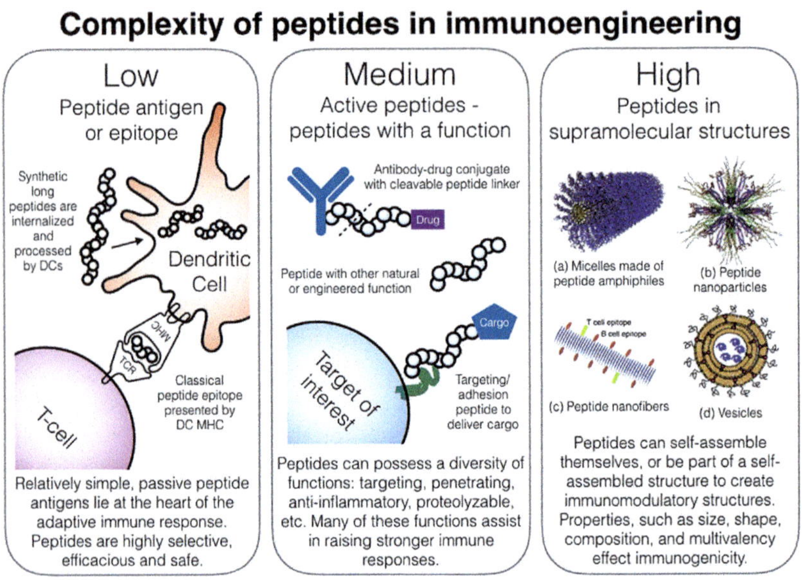

Figure 11.1 Peptide-based biomaterials for immunoengineering. As complexity increases, both breadth and strength of available functions expand for these systems. The range of engineered complexity allows for the selection of optimal system benefits for each application.
Source: (a) Reprinted from M. Black, A. Trent, Y. Kostenko, J.S. Lee, C. Olive, M. Tirrell, Self-assembled peptide amphiphile micelles containing a cytotoxic T-cell epitope promote a protective immune response in vivo, Adv. Mater. 24 (28) (2012) 3845–3849, doi:10.1002/adma.201200209 [10]; Copyright (2012), with permission from Wiley-VCH Verlag GmbH & Co. KGaA, Weinheim. (b) Reprinted from S.A. Kaba, M.E. McCoy, T.A.P.F. Doll, C. Brando, Q. Guo, D. Dasgupta, et al., Protective antibody and CD8 + T-cell responses to the Plasmodium falciparum circumsporozoite protein induced by a nanoparticle vaccine, PLoS One 7 (10) (2012) e48304–e48311, doi:10.1371/journal.pone.0048304 [11]. (c) Reprinted from Y. Wen, J.H. Collier, Supramolecular peptide vaccines: tuning adaptive immunity, Curr. Opin. Immunol. 35 (2015) 73–79, doi:10.1016/j.coi.2015.06.007 [12]; Copyright (2015), with permission from Elsevier. (d) Reprinted from J.J. Moon, H. Suh, A. Bershteyn, M.T. Stephan, H. Liu, B. Huang, et al., Interbilayer-crosslinked multilamellar vesicles as synthetic vaccines for potent humoral and cellular immune responses, Nat. Mater. 10 (3) (2011) 243–251, doi:10.1038/NMAT2960 [13]; Copyright (2011), with permission from Nature Publishing Group.

functions in the future. Many of these functional peptides assist in the basic design principle that immunomodulatory compounds much reach their target cell types to exert their effects. Using peptides to target, attach, or enzymatically release drugs in target lymphoid tissues or within specific immune cells can dramatically increase their efficiency and potency. With these enhanced functions and directed tissue and/or cellular targeting, engineered peptides can also lead to safer vaccines and therapies [15–18]. Immunomodulatory drugs (such as small molecules or cytokines) typically act on a broad range of cell types, causing them to have severe toxicity side effects resulting in their failure as a drug [19,20]. Functional peptide incorporation to the engineered system has the potential to modulate the right target cells at the right location and avert typical toxicity.

Continuing along the complexity scale, peptides can self-assemble by themselves or be a part of a self-assembled supramolecular structure. These peptide particles can possess immunomodulatory functions, acting as antigens, adjuvants (immune response booster), immune potentiators, or a combination thereof. These engineered platforms acquire their immunological properties from many controllable features, including size, shape, multivalency, and tightly controlled molecular content of different functional components.

In the end, a successfully engineered immunotherapy involves orchestrating the activities of a broad range of immune cells to elicit a specific immune response. Peptides, with the range of complexities they possess, are ideal for engineering these immune responses. Promising data discussed in this chapter suggest that peptides will have an important role to play in the future of the burgeoning field of immune engineering.

11.2 Peptides as antigens: Immunogenic peptides to engineer immune responses

Peptides fulfill many roles in immunology, yet none are more important than their role as immunogenic epitopes driving the adaptive immune response. Epitopes—T-cell epitopes, B-cell epitopes, or any other part of an antigen that is recognized by the immune system—lie at the heart of the adaptive immune response. It has been known for over 30 years now that peptides are critical factors for instructing the immune system to mobilize against foreign invaders. With our knowledge of how peptides are used to alarm the immune system (their processing and presentation on MHC), the question is how to use this information to modulate the immune system to treat or prevent infections, cancer, and autoimmunity.

11.2.1 Considerations for the design of peptide antigens

11.2.1.1 Choosing the peptide antigen

A variety of considerations need to be made during the design of a peptide antigen, depending on the context within which the peptide will be used. Chief among the considerations is the identification of immuno-dominant domains of epitopes that are capable of inducing an immune response [21]. The immune phenotype required to treat or prevent a disease will determine the epitopes chosen. For example, extracellular pathogens or infections may be effectively neutralized by an antibody response, which could dictate the selection of an immuno-dominant B-cell epitope. On the other hand, intracellular pathogens or cancers may require the identification of epitopes that induce cytotoxic T-cell responses [22,23]. In addition to the effector B-cell or cytotoxic T-cell responses, a helper T-cell response may also be crucial, requiring the identification of an activation solution for that cell as well [23,24]. From among the epitopes that are found to induce the desired immune response, the next challenge is to identify the epitope(s) or peptide(s) that activate the immune response to the correct magnitude that can provide protective immunity. The processing, presentation, and association of the candidate peptide antigen by APCs (DCs) in such a highly MHC heterogeneous human population is another issue that requires consideration.

11.2.1.2 Improving the peptide antigen

A peptide antigen should be chosen first for its immunogenic characteristics, as described above. However, when deploying the peptide in vivo, some peptide characteristics such as stability, affinity, and delivery should be improved upon in order to obtain sustained immunological effects. As described in earlier chapters, one major drawback of peptides is their short half-life. Peptides have a short half-life in circulation, are rapidly cleared through the kidney by filtration, and are also destroyed by many extracellular peptidases [25]. Over the years, however, many strategies have been devised to mitigate this problem. A number of these strategies are discussed in more detail elsewhere in the text, so we will briefly touch on the strategies most pertinent to immune engineering.

Solving peptide stability through chemical modification employs one of the most advantageous properties of peptides. Prodrugs or propeptides are stable, chemically altered peptides that are processed into the desired antigen on location. Propeptides are designed to have the antigenic sequence flanked by additional amino acid sequences preferred by dominant endopeptidases. Combining this with blocking N-or C-terminal groups may further protect from degradation and allow the antigen sequence to arrive intact and perform its immunological function [26]. Stability may also be increased by delivering the peptide with an adjuvant [27]. Many adjuvants rely on creating a depot, in which peptides are protected from immediate degradation, and to which immune cells will be attracted and activated by simultaneous stimulation of peptides and adjuvant molecular motifs [28–30]. Peptide antigen stability can also be increased through incorporation into larger structures such as antibody–drug conjugates, supramolecular peptide structures, liposomes, or other assemblies, many of which will be discussed later in this chapter and other chapters.

Among all the peptide antigens that produce an immune response, many will bind with relatively high affinity to MHC molecules, biased by the selection criteria. But even if the peptide antigen can induce an immune response and bind to MHC molecules, the binding may still have room for improvement. In other words, a peptide's natural affinity or on/off rates may not be ideal and this may provide an area of opportunity to improve the immune response. These parameters are important to the development of an immune response because it stands to reason that such peptides that make more stable MHC molecules will have more time to activate the immune system [31,32]. Improving peptide affinity to MHC molecules can be achieved in a few different ways. One way is with a chemical biology approach to improve the interactions of the MHC with the peptide cargo, primarily through modifying the N- and C-termini and anchoring residues of the peptide. For peptide side chains, small modifications may sometimes improve affinity, but care must be taken to not affect the MHC–peptide complex surface for fear of altering the ability of a T-cell receptor to recognize the complex. Finally, a more direct way to improve affinity is to just directly conjugate the peptide to the MHC through molecular biology/protein engineering techniques, as some researchers have done [33].

11.2.2 Peptide antigens modulating immune responses

There are many ways an immune response may need to be modulated, including inducing or suppressing a particular response. Because of a peptide's specificity and chemical definition, it can be engineered to be useful in almost any case required. Autoimmune diseases (such as type I diabetes and multiple sclerosis) arise when T cells that recognize self-proteins are not completely eliminated before leaving the thymus [34]. It is possible to control these autoreactive cells through mechanisms of tolerance in the peripheral tissues, but if these mechanisms fail, disease ensues. In type I diabetes, one reason why tolerance mechanisms fail is because the patient happens to be homozygous for a particular MHC molecule. 45% of type I diabetes patients are homozygous for HLA-ER3 or HLA-DR4, although this is present in only 3% of the population [35]. These MHC molecules selectively present a peptide from a tissue-specific protein, which are then recognized by self-recognizing T cells that were not deleted in T-cell maturation. If there are a set of dominant peptides in this interaction, modifying these peptides to become less able to associate with T cells [36] or to selectively activate regulatory T cells (T_{reg}s), could stifle the autoimmune response [37]. Researchers have successfully used this approach in mouse models of defined and homogeneous T-cell responses. However, the strategy becomes more complicated if multiple populations of T cells or T_{reg}s need to be antagonized or activated, as is the case in human translation. To realize a strategy for human type I diabetes, more studies are currently underway.

Another clever peptide approach to treating type I diabetes involves tolerizing polyspecific $CD8^+$ T cells that recognize multiple self-epitopes expressed by pancreatic islet cells [33]. Tolerization in this approach is attempted with iron oxide nanoparticles conjugated with autoantigen peptide–MHC complexes (pMHC-NPs). In this antigen-specific strategy, it was found that pMHC-NPs stimulated self-antigen-specific $CD8^+$ T cells and led to the expansion of autoregulatory memory-like T cells. These autoregulatory T cells kill APCs that present autoantigens, thereby preventing the activation of autoreactive $CD8^+$ T cells. If the activation of polyspecific autoreactive $CD8^+$ T cells is also inhibited, normal glycemic levels in nonobese diabetic mice can be established (Fig. 11.2). It is thought that the expansion of the autoregulatory memory-like T cells may be a result of TCR crosslinking by the autoantigen peptide-loaded pMHC-NPs [33].

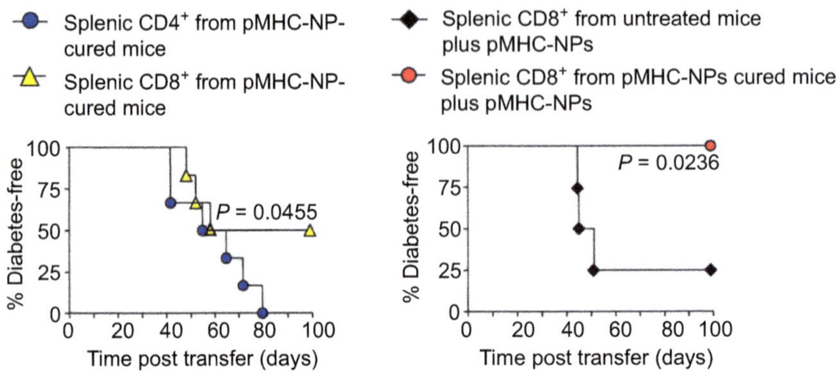

Figure 11.2 Peptide–MHC nanoparticles (pMHC-NPs) restore normoglycemia in diabetic NOD mice. Left: pMHC-NP-expanded $CD8^+$ cells from cured NOD mice ($n = 6$; 5×10^6 $CD8^+$ cells, containing $<5 \times 10^4$ pMHC-NP cells) suppress the adoptive transfer of diabetes by splenic T cells from prediabetic NOD mice into NOD.*scid* females compared to total $CD4^+$ cells (5×10^6) from cured donors ($n = 6$). Right: pMHC-NP treatment (twice/week) potentiates the transfer of protection by splenic $CD8^+$ cells from pMHC-NP-cured mice into T-cell-reconstituted NOD.*scid* hosts ($n = 3$) but cannot induce protective activity in splenic $CD8^+$ T cells from untreated NOD donors ($n = 4$).
Source: Reprinted from S. Tsai, A. Shameli, J. Yamanouchi, X. Clemente-Casares, J. Wang, P. Serra, et al., Reversal of autoimmunity by boosting memory-like autoregulatory T cells, Immunity 32 (4) (2010) 568–580, doi:10.1016/j.immuni.2010.03.015; Copyright (2010), with permission from Elsevier.

Cancer therapy with peptide antigen is a conceptually simpler task than autoimmune disease. If a tumor peptide antigen is defined, injecting the peptide to activate a T-cell response against this antigen could result in an immune response against the tumor. This strategy has been applied in the treatment of premalignant lesions of cervical cancer in human patients, in melanoma tumors, and many more cancer mouse models [38–42]. In general, tumor microenvironments weaken the local immune system's ability to develop a strong immune response against the tumor. By injecting a peptide vaccine and raising an immune response elsewhere in the body, this can help to induce a local immune response to clear the tumor. Of course, a major player in this response is the DC. When presented with the correct peptide antigen, DCs can activate a strong and specific immune response against tumors. In mouse models, it is enough to provide the minimal 9–11-amino-acid peptide sequence recognized by $CD8^+$ T cells in order to get an immune response [38]. But, by delivering a longer peptide which includes both MHC class I and MHC class II sequences, a stronger immune response can generally be raised [38]. Delivering this longer, combined peptide resulted in a T-cell response against human papillomavirus (HPV) proteins and clearance of the premalignant cervix tissue in patients [38,43]. A longer peptide could perform better in vaccines for multiple reasons; longer peptides are retained longer in the circulation, and combining both MHC class I and MHC class II peptides increases the probability of a DC processing and presenting both peptides on its surface to activate a strong $CD8^+$ T-cell response [44–46]. Additionally, delivering a longer peptide allows the immune system of each individual to process and trim the peptide in slightly different ways. This is important because human tumor microenvironments are complex and are all slightly different, which means the peptide antigens required for an immune response are all slightly different between patients. Therefore, delivering a longer peptide allows the peptides to be trimmed more naturally.

11.3 "Active" peptides: Peptides with a function

While a peptide's basic function as an antigen in immunology is exceptionally important, peptides can offer much more functionality beyond that role. Due in large part to a peptide's chemical definition and tailorable nature, peptides can possess a diversity of functions, including targeting (cell binding, integrin binding, surface binding), penetrating, antiinflammatory, proteolyzable, and combinations thereof. Moreover, the tailorable nature and chemically defined properties of peptides provide opportunities for the discovery and development of additional functions in the future.

11.3.1 Targeting peptides

Peptides that target specific cell types involved in the immune systems have several key uses, including antigen delivery for vaccine development, delivery of immunosuppressant agents for the treatment of autoimmune diseases, and redirection of an immune response to eradicate diseased cells. Unlike antibodies, peptides are easy to synthesize in large quantities, and their smaller size improves tissue penetration [47]. Additionally, peptides can be chemically modified to alter affinity, charge, hydrophobicity, stability, and solubility. Because of these tailorable qualities, peptides can display antibody-like affinities for their receptors, making them effective targeting agents [48]. Actively targeting immune cells in order to modulate immune responses is a heavily and actively explored area, and here we will discuss promising examples of this strategy.

11.3.1.1 Targeting the innate immune system

Neutrophils are the "first responders" at sites of inflammation and play an important role in providing the initial defense against invading pathogens through phagocytosis of microbes and secretion of cytokines and reactive oxygen species [49]. Inflammation caused by neutrophils is important for the initiation of a complete immune response, however, prolonged neutrophil-mediated inflammation can lead to tissue damage and the pathogenesis of diseases such as arthritis, cancer, and chronic obstructive pulmonary disease (COPD) [50]. Targeting neutrophils to mediate excessive inflammation is an attractive strategy, so long as the beneficial roles of neutrophils and inflammation are not completely inhibited. One strategy begins by using collagen IV-targeting peptides. Collagen IV is a useful target for tissue

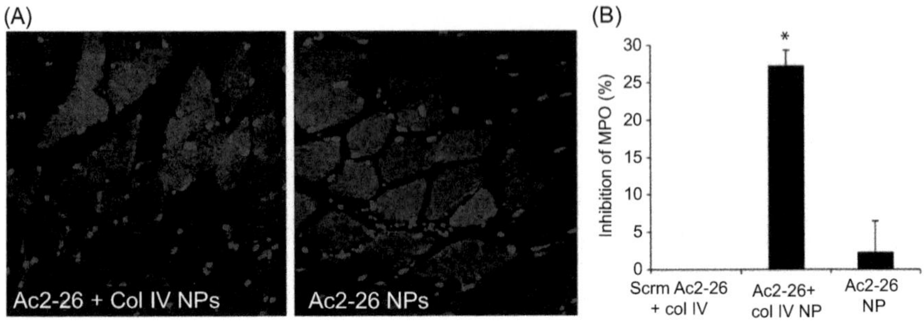

Figure 11.3 Collagen-IV-targeted Ac2-26 NPs (Col IV NPs) limit polymorphonuclear neutrophil (PMN) infiltration into injured tissue. Ischemia was induced by placing a tourniquet around the hindlimb for 1 h. After 1 h, the tourniquet was released and vehicle, Scrm-Ac2-26 Col-IV NPs, Ac2-26 Col-IV NPs, or Ac2-26 NPs were injected IV. Reperfusion was carried out for 1 h. The gastrocnemius muscle tissue was harvested and (A) sectioned for confocal imaging using a Nikon A1R microscope, 20 × magnification. Images are representative of $n = 3$. (B) Gastrocnemius tissue was lysed and homogenized to assess PMNs using a myeloperoxidase (MPO) ELISA ($n = 3$; mean ± SEM). The data are plotted as inhibition of tissue MPO. $^*P < .05$ Col-IV for Ac2-26 NPs versus Ac2-26 NPs or versus Scrm-Ac2-26 Col-IV-targeted NPs.
Source: Reprinted from N. Kamaly, G. Fredman, M. Subramanian, S. Gadde, A. Pesic, L. Cheung, et al., Development and in vivo efficacy of targeted polymeric inflammation-resolving nanoparticles, Proc. Natl. Acad. Sci. U. S. A. 110 (16) (2013) 6506–6511, doi:10.1073/pnas.1303377110 [55]; with permission from PNAS.

injury applications because this type of collagen is abundant on the basal lamina—a layer of extracellular matrix secreted by epithelial cells. These collagen IV-targeting peptides were used to deliver polymer nanoparticles carrying the antiinflammatory peptide Ac2-26. Isolated from annexin A1 N-terminus, Ac2-26 (Ac-AMVSEFLKQAWFIENEEQEYVQTVK) was discovered for its sequence similarity to uteroglobin, and functions as a mimetic peptide that acts on the G-protein-coupled formyl peptide receptor, ALX/FPR2 [51–54]. After intravenous administration, the collagen IV peptide targets and delivers the encapsulated Ac2-26 peptides to sites of tissue injury. In a hindlimb ischemia–reperfusion tissue injury mouse model, this resulted in a 30% reduction in neutrophil recruitment to the site of injury (Fig. 11.3) [55]. Targeting collagen IV during an injury is one way to deliver cargo to neutrophils, though it may be delivered to other cells as well. To have even more specific delivery to a neutrophil, the peptide WAWVWLTETAV can be used. This peptide was isolated against the FcγRIIA receptor, a receptor only expressed by neutrophils and mononuclear phagocytes, and was shown to mediate cell internalization and degradation of an OVA model antigen after binding [56].

11.3.1.2 Targeting the adaptive immune system

Human immunodeficiency virus (HIV) is a terrible infection for which there is no cure. During infection, $CD4^+$ T cells are a primary target of the virus. This has

Table 11.1 Intracellular indinavir concentrations after incubation with soluble or lipid-associated indinavir nanoparticles

	Intracellular indinavir (ng/10^5 CEM × 174 cells)			
	Day 0		Day 4	
Indinavir formulation	**6.25 μM**	**25 μM**	**6.25 μM**	**25 μM**
Soluble	2774 ± 53	3772 ± 1066	53 ± 1	124 ± 4
DSPC:DSPE-mPEG				
Untargeted	34 ± 10[a]	102 ± 3[a]	7 ± 0.1[a]	5 ± 0.4[a]
CD4-BP2 LNP	3105 ± 164	3584 ± 62	93 ± 1[a]	148 ± 1[a]
CD4-BP2 LNP	3584 ± 62[a]	6586 ± 328[a]	162 ± 4[a]	282 ± 5[a]

[a]Significant at $P < .05$ compared with soluble drug.
*To quantify indinavir levels after enhanced delivery to cells with LNPs, triplicate 10^5 CEM × 174 cells were incubated with 6.25 or 25 μM indinavir formulations, for 30 min at 37°C, washed, pelleted, then lysed in 100 μL of acetoniltrile and 100 μL water was added to each sample. 10 μL of each sample in duplicate was injected onto a Zorbax SB-C18 column with mobile phase consisting of a 0.1% acetic acid solution in methanol. The mass spectrometer was operated in API-ESI$^+$ mode, and the analytes were detected using selected ion monitoring at m/z 603.7–623.7 to detect indinavir.
API-ESI$^+$, atmospheric pressure ionization-electrospray ionization; DSPC, 1,2-distearoyl-sn-glycero-3phosphocholine.
Source: Reprinted from A.N. Endsley, R.J.Y. Ho, Enhanced anti-HIV efficacy of indinavir after inclusion in CD4-targeted lipid nanoparticles, J. Acquir. Immune Defic. Syndr. 61 (4) (2012) 417–424, doi:10.1097/QAI.0b013e3182653c1f [57]; Copyright (2012), with permission from Lippincott Williams & Wilkins, Inc.

motivated a number of immunoengineering strategies to target therapeutic agents to CD4$^+$ T cells to block HIV replication, and peptides are frequently used in the targeting strategy. For example, lipid nanoparticles containing an encapsulated antiretroviral drug, idinavir, used specific CD4 coreceptor targeting peptides to target T cells. Pretreatment of CD4$^+$ T cells with the peptide-targeted lipid nanoparticles resulted in a reduced number of infected cells compared to nontargeted lipid nanoparticles in vitro (Table 11.1) [57].

11.3.1.3 Targeting antigen-presenting cells

Dendritic cells (DCs), as mentioned above, are the cells that connect the innate immune system to the adaptive immune system. DCs are constantly on the lookout for foreign antigens, and after capturing an antigen, present peptides to activate naïve T cells, B cells, natural killer cells, and natural killer T cells [58]. Due to their ability to activate all of these different immune pathways, DCs are ideal targets for immune engineering. Because of this, several DC-specific targeting peptides have been isolated for the purposes of specifically delivering antigens to dendritic cells. One such peptide, the p30 (CGRWSGWPADLC) peptide, was isolated against the DC-specific marker CD11c/CD18 and was later used for vaccination [59,60]. Chicken ovalbumin (OVA) is a T-cell-dependent antigen commonly used as a model antigen for studying antigen-specific immune responses. In this vaccination, peptides on the surface of an OVA antigen-loaded liposome targeted DCs with the p30 peptide and produced a much stronger OVA-specific antibody response compared to the control OVA liposomes without the targeting peptide. In another example, the p30 DC-targeting peptide was grafted onto plasma membrane vehicles derived from B16-OVA cells, a metastatic murine melanoma cell line that secretes OVA. Vaccinating mice with these peptide membrane vehicles before introducing B16-OVA cells into the mice significantly reduced the number of lung metastases. Additionally, mice bearing existing B16-OVA tumors that underwent subsequent vaccination with the peptide membrane vehicles induced antigen-specific T-cell priming and antibody production, followed by dramatic antitumor responses (Fig. 11.4) [60].

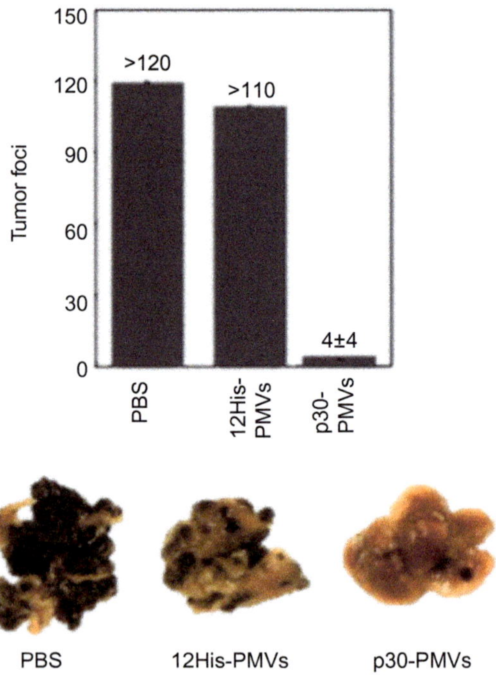

Figure 11.4 Ag-bearing liposomes engrafted with peptides that interact with CD11c/CD18 induce potent Ag-specific and antitumor immunity. Vaccination with plasma membrane vehicles modified with the dendritic-cell-specific peptide CGRWSGWPADLC (p30) leads to antitumor activity. Naïve mice were injected IV with B16-OVA cells on day 0. At days 2, 8, and 14 different groups of mice (five mice per group) were vaccinated with PBS or B16-OVA-derived plasma membrane vehicles modified with the control peptide 12His or the dendritic-cell-specific peptide p30. At day 21, the lungs were removed from the mice, and tumor foci were counted via microscopy. (Top) Bars indicate the mean number of tumor foci for each vaccination group, and this number is indicated above each bar. (Bottom) Representative lung images from each vaccination group.
Source: Reprinted from A. Faham, J.G. Altin, Ag-bearing liposomes engrafted with peptides that interact with CD11c/CD18 induce potent Ag-specific and antitumor immunity, Int. J. Cancer 129 (6) (2011) 1391–1403, doi:10.1002/ijc.25810; Copyright (2010), with permission from John Wiley & Sons, Inc.

In recent years, microbiome research has made many promising advances. Because of the microbiome's wide-reaching influence throughout the body, it has garnered interest from many different fields, including immunology. Immunoengineering offers the opportunity to expand applications in this exciting new field of research. The intestines are lined with a single layer of epithelium for protection, making it difficult for most things to pass through, or out of, the gut [61]. Microfold cells (M cells) are immune cells located in Peyer's patches of the intestinal epithelium that sample antigens in the gut and transport them a short distance to a binding pocket in which immune cells such as lymphocytes or macrophages dock [61]. Because M cells can carry antigens out of the gut to other immune cells in order to facilitate an immune response, M cells are very desirable targets for orally available vaccines. Engineering delivery directly to the M cells for initiation of an immune response resulted in the identification of the Col, P8, and P25 peptides (Col: SFHQLPARSPLP; P8: LETTCASLCYPS; P25: VPPHPMTYSCQY) [62,63]. Kim et al. [63] elucidated the delivery of Col peptide by fusing it to the fluorescent protein EGFP. These peptides, and in particular the Col peptide, are able to specifically bind M cells and transport across the intestine in vivo when injected into mouse intestine, as well as bind to tissue sections of human intestine as evidenced by immunohistochemistry. After oral administration, the targeting peptide-fused antigen (Col-EGFP) enhanced immune responses by raising two- to threefold higher antibodies against the fused antigen compared with those of the control antigen without targeting peptide (Fig. 11.5). In addition, the use of the targeting peptide supported a skewed Th2-type immune response against the fused antigen. Th2 immune responses are typically raised against extracellular infections or parasites, which makes sense in this context because the EGFP represents an extracellular infection. Because of Col's ability to raise strong immune responses and show binding to human M-like cells in culture, Col peptide has high potential for clinical translation in the effort to develop an oral vaccine.

11.3.1.4 Other targeting peptides in immune engineering

One of the most heavily studied peptides in the context of biomaterials and immunoengineering has been RGD (Arg-Gly-Asp). RGD was originally developed as a small molecule capable of functioning analogously to the 10th type III domain of fibronectin, which is a high-molecular-weight protein of the extracellular matrix (ECM) that binds to special cell membrane receptors called integrins. RGD has undergone exhaustive exploration and is now known to bind promiscuously to many different integrin subtypes and many different ECM molecules beyond fibronectin [64,65]. RGD peptide has been studied in many contexts, but one example with single-walled carbon nanotubes (SWNTs) is especially relevant to immunoengineering. In this example, SWNTs were coated with lipid-tailed poly(ethylene glycol) and terminated with RGD. These particles were shown to accumulate at tumor sites in mouse models of cancer [66]. This was first attributed to the preferential accumulation of SWNTs in tumors via the enhanced permeability and retention (EPR) effect and the ability of the RGD peptide to bind to integrins expressed on

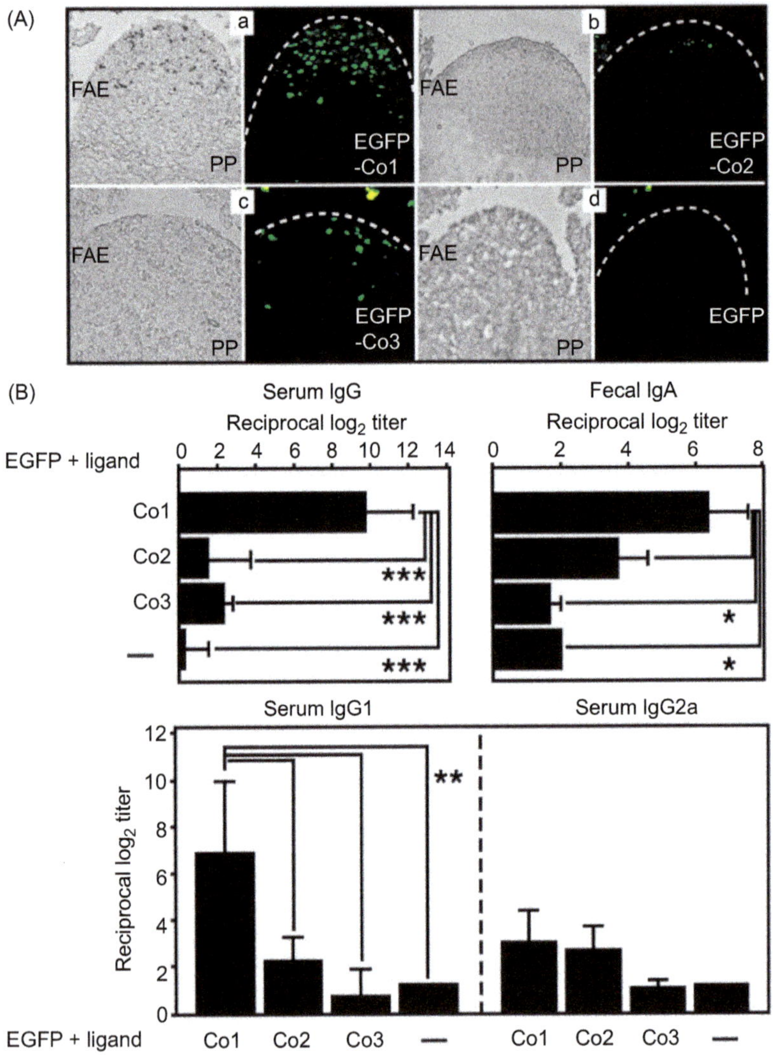

Figure 11.5 Targeting of ligand-fused EGFP into Peyer's patches (PPs) and induction of EGFP-specific mucosal and systemic immune responses after oral administration of the ligand-fused antigens. (A) Immunohistochemical analysis of mouse gut loops incubated with (a) EGFP-Co1, (b) EGFP-Co2, (c) EGFP- Co3, and (d) EGFP (green) alone using the ex vivo loop assay. Left and right columns of each panel represent the results from light and fluorescence microscopy, respectively (original magnification × 310). (B) Levels of EGFP-specific serum IgG and fecal IgA (upper panel) and EGFP-specific IgGs of different subclasses (lower panel) in the sera after oral administration of ligand-fused EGFP or EGFP alone. Results were calculated as the reciprocal of the geometric mean log_2 titer. *Source*: Reprinted from S.-H. Kim, K.-W. Seo, J. Kim, K.-Y. Lee, Y.-S. Jang, The M cell-targeting ligand promotes antigen delivery and induces antigen-specific immune responses in mucosal vaccination, J. Immunol. 185 (10) (2010) 5787–5795, doi:10.4049/jimmunol.0903184; Copyright (2010), with permission from the American Association of Immunology.

tumor vasculature on the surface of tumor cells [67,68]. EPR is the property by which molecules of certain sizes tend to accumulate in tumor tissue much more than they do in normal tissues [69]. Upon further investigation using intravital microscopy, it was shown that in addition to accumulation in the tumor due to EPR, 25% of intravenously injected SWNTs were taken up preferentially by $Ly6C^{hi}$ monocytes in the circulation, which were then recruited to the site of the tumor in response to inflammation. The RGD peptide played a key role in that RGD conjugation to SWNTs increased the recruitment of $Ly6C^{hi}$ monocytes into the tumor interstitium and resulted in increased accumulation of SWNTs at the tumor site (Fig. 11.6) [70]. Although RGD is powerfully useful in some contexts, caution must be taken because of its broad integrin-binding properties. For example, if cell-specific behaviors are required, or if a defined set of integrins need to be activated and bound, the broad RGD integrin binding may not be desired [65].

Another promising, more recently developed adhesion peptide is the erythrocyte-binding peptide ERY1 (H_2N-WMVLPWLPGTLDGGSGCRG-$CONH_2$) [71]. As mentioned earlier, researchers are exploring new approaches based on induction of antigen-specific immunological tolerance for the treatment of autoimmune diseases. In this case, the antigen-specific tolerization approach involves using erythrocyte (red blood cell)-binding peptides, with the premise that as erythrocytes circulate, age, and are cleared, the erythrocyte surface-bound antigen payload will be cleared tolerogenically along with the eryptotic debris. ERY1 targets and attaches to the erythrocyte-specific cell surface marker glycophorin-A after IV injection (Fig. 11.7). In an ovalbumin (OVA) model antigen proof of concept experiment, it was shown that erythrocyte-binding antigen is collected much more efficiently than free antigen by splenic and hepatic immune cell populations and hepatocytes, and that it induces antigen-specific deletional responses in $CD4^+$ and $CD8^+$ T cells [72,73]. In a separate application for the same ERY1 system, a tolerance induction strategy was applied to prevent antigen-specific humoral (antibody) immune responses to therapeutic proteins. In this tolerance induction strategy, they demonstrate that binding the clinical therapeutic enzyme *Escherichia coli* L-asparaginase to erythrocytes in situ antigen-specifically abrogates development of antibody titers by >1000-fold and extends the pharmacodynamics effect of the drug 10-fold in mice, tolerizing mice to multiple subsequent doses of the wild-type enzyme [74]. This tolerance strategy, led by the erythrocyte-binding peptide ERY1, shows strong translational promise to enable more effective and safer treatment with therapeutic proteins and drug candidates that are hampered by immunogenicity, as well as in the treatment of autoimmune diseases such as type I diabetes and multiple sclerosis.

11.3.2 Enzyme cleavable peptides for immunoengineering

As described above, peptides in immunoengineering have been used to bind very specific cells and cell surface receptors. Another fantastic feature of a peptide is that its specificity extends beyond just the cell surface. Peptides can also interact with specific enzymes and subcellular compartments. Immunoengineering strategies can use peptide sequences with distinctive binding constants and degradation

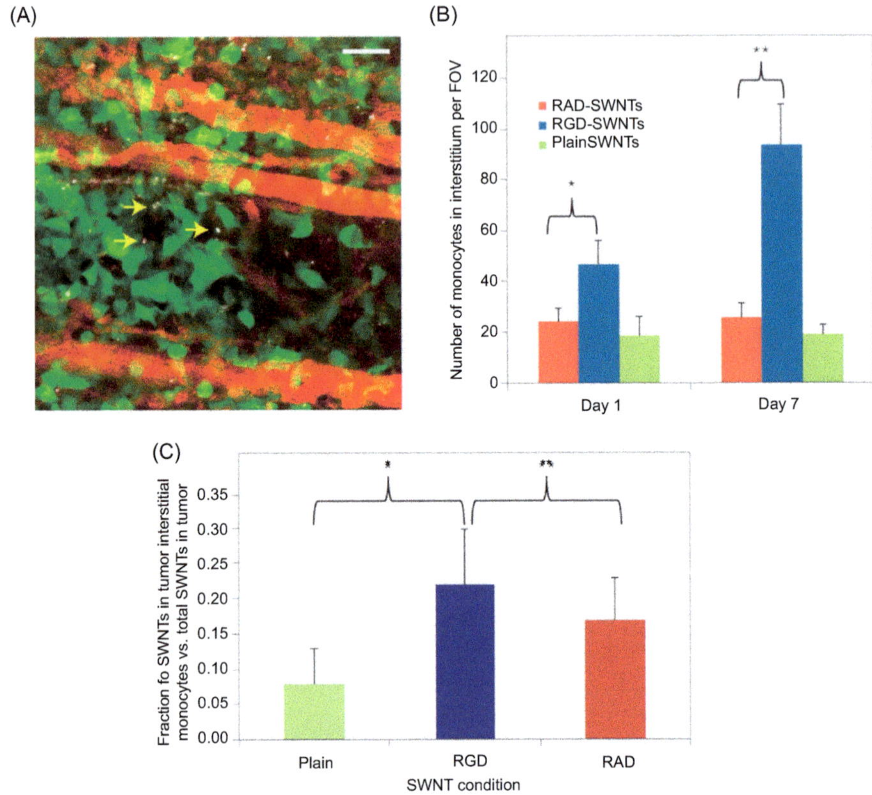

Figure 11.6 SWNT-laden monocytes enter the tumor interstitium in a peptide-dependent manner. (A) Representative intravital micrograph of a tumor region (tumor cells, green; blood vessels, red; SWNTs, grayscale). Yellow arrows point to several SWNT-laden monocytes within the tumor interstitium. Scale bar, 50 μm. (B) Bar graph showing that significantly more monocytes carrying RGD−SWNTs accumulate in tumor interstitium than monocytes carrying RAD−SWNTs and plain SWNTs on days 1 and 7 post-injection of SWNTs. Moreover, significantly more RGD−SWNT-laden monocytes are in the interstitium per field of view (FOV) on day 7 than on day 1 ($P < .001$). $^{*}P < .05$; $^{**}P < .0005$. Error bars represent standard error of the mean (s.e.m.) (C) SWNTs can enter the tumor via a variety of mechanisms, such as leakage through blood vessel pores. This graph shows the relative amounts of SWNTs in the tumor interstitium that were ferried in via the Trojan horse monocytes compared with all SWNTs within the tumor interstitium as a function of peptide on day 1 post-injection More than 20% of SWNTs in the tumor interstitium in the RGD−SWNT condition are carried in via monocytes. $^{*}P < .0001$; $^{**}P < .05$. Error bars represent standard deviation.
Source: Reprinted by permission from B.R. Smith, E.E.B. Ghosn, H. Rallapalli, J.A. Prescher, T. Larson, L.A. Herzenberg, et al., Selective uptake of single-walled carbon nanotubes by circulating monocytes for enhanced tumour delivery, Nat. Nanotechnol. 9 (6) (2014) 481−487, doi:10.1038/NNANO.2014.62 [70]; Copyright (2017), Macmillan Publishers Ltd: Nature Nanotechnology.

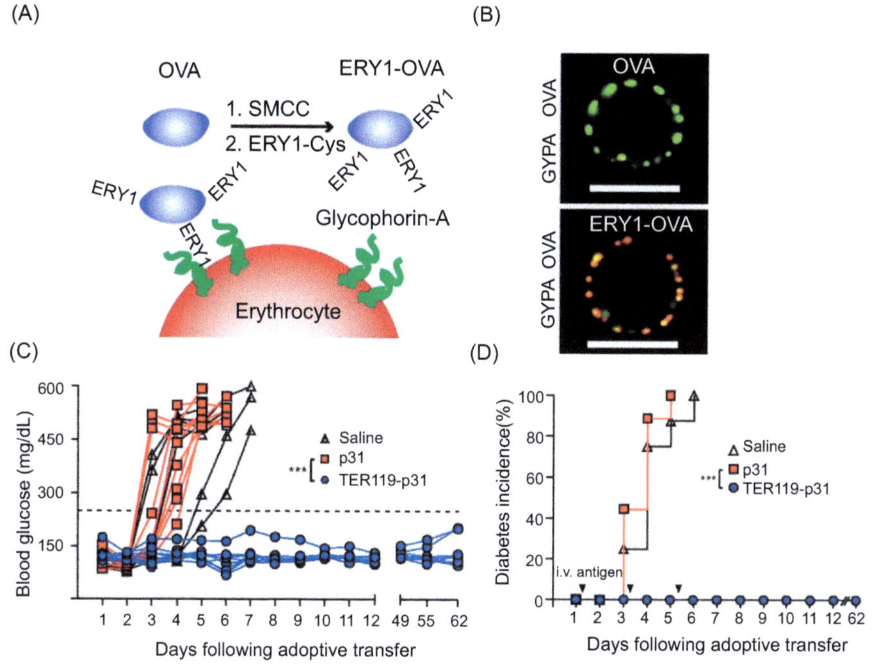

Figure 11.7 ERY1-OVA binds the equatorial periphery of mouse erythrocytes with high affinity and protects mice from T-cell-induced autoimmune type I diabetes. (A) Schematic of conjugation of ERY1 peptide to OVA, resulting in binding to erythrocyte surface glycophorin-A. (B) High-resolution confocal microscopy images of mouse erythrocytes labeled ex vivo with (green) antimouse glycophorin-A (GYPA) and (red) either OVA (upper) or ERY1-OVA (lower) (Scale bar = 5 μm). (C) Glycemia monitoring as measured by daily blood glucose measurements following adoptive transfer of diabetogenic BDC2.5 $CD4^+$ T cells and a tolerogenic treatment regimen of either saline, p31, or TER119-p31 ($n = 8$, $n = 9$, and $n = 9$, respectively). $^{***}P < .0001$. (D) Diabetes incidence rate quantified by measurements in (C); arrows indicate antigen administration time points. $^{***}P < .0001$. *Source*: Reprinted from S. Kontos, I.C. Kourtis, K.Y. Dane, J.A. Hubbell, Engineering antigens for in situ erythrocyte binding induces T-cell deletion, Proc. Natl. Acad. Sci. U. S. A. 110 (1) (2013) E60–E68, doi:10.1073/pnas.1216353110/-/DCSupplemental [72]; with permission from PNAS.

kinetics to various proteases in order to direct the most desired immune response or other outcome for a particular application.

As a comparison to targeting peptides, monoclonal antibodies (mAb) are also an attractive immunoengineering tool, thanks to their extracellular target specificity [75–78]. However, the effectiveness of directly triggering cell death through extracellular activation of intracellular protein pathways can be complicated and highly cancer cell/tissue-type dependent. Instead of relying on these complicated pathways to death, mAbs can be used to kill cells more directly by simply carrying and delivering a toxic payload to the target. These combined particles, termed antibody–drug conjugates (ADCs), have undergone considerable optimization since they were first used more than three decades ago, and have recently shown major success [79–82].

The delayed success of antibody–drug conjugates was mainly due to the complex, detailed design of their conjugation chemistry; linking antibody to cargo (the linker). Design parameters for this critical segment are that: (1) the conjugation of antibody to cargo should occur in mild medium so as to not do damage to the antibody; (2) the linker must be resistant to proteases in the serum during circulation in order to deliver the cargo to the target intact; and (3) once the cargo is delivered to the target, it is commonly desired that the cargo be internalized by the cell and released from the antibody so that the cargo can perform its function. Luckily, when an mAb binds to its target, endocytosis is frequently triggered and the ADC is taken into an acidic environment where the cargo is released from the antibody through the acid labile linker.

The development of acid labile linkers significantly advanced ADC technologies, and improved their efficacy and tolerability. Some examples of the first acid labile linker chemistries to be used include hydrazones and disulfide bonds. The advantage of hydrazone is that this linker is highly unstable at pH 5, largely allowing for the complete intracellular release of active drug. Unfortunately, hydrazones are only moderately stable at neutral pH in circulation (Fig. 11.8) [83]. Disulfide bonded linkers, on the other hand, such as glutathione and cysteine, show higher stability than hydrazones in circulation because of steric hindrance and the simple fact that thiols are generally not present outside the cell (Fig. 11.8) [84]. However, this strong stability can persist even within the lysosome, leaving disulfides incompletely degraded, thus preventing the cargo from being released [85]. One added limitation to the disulfide linker technology is the extensive and complicated purification procedures required to produce homogeneous product. In order to produce

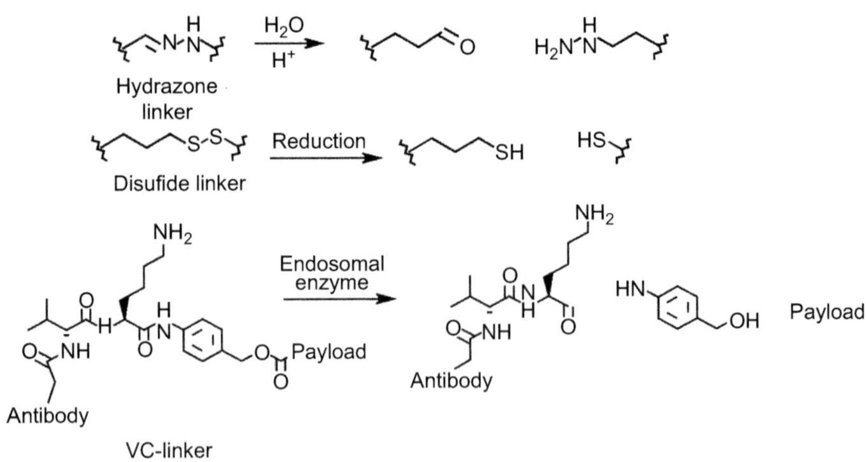

Figure 11.8 Antibody–drug conjugate linkers and their cleavage mechanisms.

the desired molecule, solvent chemicals to disrupt excess disulfide bonds may damage the stability of mAbs.

Since each of these linkers had limitations, the search for an improvement continued. Ultimately, cleavable peptides emerged as the lead candidate to advance ADCs. Peptide linker technologies presented significant advantages over those nonpeptide linkers [86]. Chief among the advantages is that the hydrolysis of peptide-based linkers is enzymatic. This allows the cleavage sequence within the peptide linker to be selected based upon which enzymes are preferentially expressed by the target tumor cells or solid tumor masses. Therefore, the cargo will only be cleaved and released from the mAb within a target cell, and be shuttled out of nontarget cells by the continued progression of lysosomes to exosomes. In addition, the peptides will not be cleaved in the systemic circulation because the peptides can only be cleaved by the intracellular tumor-associated proteases. Peptide linkers are also biocompatible, producing safe byproducts after the cargo is delivered.

Cathepsin B enzyme was found overexpressed in a wide variety of human cancers, often associated with the tumor cell membrane [87]. Thus, cathepsin B enzyme cleavable linkers are among the most advanced linkers used with ADCs for the treatment of cancer. As mentioned before, the internalization of mAbs occurs through endocytosis. Endolysosomes degrade their contents using proteases, such as cathepsins, in acidic pH. Cathepsins recognize and cleave specific amino-acid sequences. These sequences have been extensively studied with cathepsin B being the most popular endocytotic cysteine-based protease [88,89]. Cathepsin B is rarely found in the extracellular matrix or elsewhere outside the cell, and therefore conjugates produced with cathepsin B cleavable linkers are likely to be stable in circulation [90].

Initially, Dubowchik et al. studied a library of enzyme cleavable dipeptide linkers and measured the rate of drug (doxorubicin) release by enzyme activity [90,91]. According to Dubowchik's studies, in the presence of cathepsin enzyme, phenylalanine-lysine (FK), valine-lysine (VK), and valine-citrulline (VC) showed the fastest cleavage kinetics. With the addition of a spacer such as para-(amino benzoyl) (PAB), the cleavage reaction rates were improved even further, to half-life times of 8, 9, and 240 min, respectively. While in vitro studies were a great place to start, in vivo experiments revealed that cleavage kinetics were vastly different between the three sequences. The cleavage kinetics of VC with PAB spacer stood out amongst the rest, showing impressively high cleavage as identified through drug efficacy. The cleavage mechanism of VC linkers is shown in Fig. 11.8. In this reaction, enzyme interacts with VC dipeptide and cleaves through the N-terminus of the peptide, which triggers the 1,6-elimination of carbon dioxide on PAB and contaminant release of the free drug in parent amine form [92]. According to extensive studies, the VC peptide linker shows superior stability in human plasma, robust cleavage after endocytosis, and potent, antigen-specific cytotoxicity when compared to nonpeptide linkers [90].

Following the initial studies on enzyme cleavable peptide linkers, many fields saw their potential and quickly adopted the linkers. The immunoengineering field applied the linkers to ADCs because of their superior properties compared to previously used acid-liable linkers. One of the first ADC studies to use these peptide linkers was performed by Doronina et al., who wanted to deliver monomethyl auristatin E (MMAE) [93]. MMAE cannot be delivered systemically because of its severe toxicity and adverse effects. Therefore, Doronina wanted to test whether delivering MMAE as part of an ADC could circumvent the toxicity and potentially improve MMAE safety and efficacy. To test this, three linkers (FK, VC, and hydrazone) in an otherwise unchanged ADC system, were compared against one another for their ability to kill cancer cells and prevent toxic effects by MMAE. The results showed that hydrazone was far less stable in plasma compared to the peptide linkers. Moreover, comparison of the dipeptide linkers showed that FK and VC both activated the enzyme trigger robustly; however, FK was found to be less stable in plasma. In vitro experiments of the same study demonstrated that peptide-linked ADCs were highly potent; inducing 10- to 100-fold greater immunologically dependent cell death compared to the corresponding hydrazone linker ADCs [93].

Of course, with such dramatic success brought on by essentially only a change to a peptide linker, enzymatically cleavable peptide applications have been greatly repeated and extended. Burris et al. switched the antibody to a fully-human mAb (CR011) to target breast cancer [94]. More specifically, the antibody targets antiglycoprotein NMB (GPNMB), also known as osteoactivin, that is expressed in many breast cancer types and associated with an increased risk of cancer recurrence. The ADC was prepared with MMAE and a VC peptide linker, and phase II studies have met the criteria for advancement (Table 11.2). Other examples of ADCs under clinical trials that use the VC peptide linker are shown in Table 11.2.

ADCs have also been prepared against the CD30 antigen on Hodgkin's disease and the LeY antigen on carcinomas. In vivo tests of anti-CD-30-MMAE (SGN-35), prepared with the VC peptide linker, showed no sign of toxicity in healthy mice up

Table 11.2 Examples of antibody–drug conjugates prepared with cleavable VC dipeptide

Antibody–drug conjugate	Antibody	Cancer type	Status	Reference
Brentuximab vedotin (SGN-35)	Anti-CD30	Lymphomas	Approved	[95]
Glembatummumab vedotin (CDX-011)	Anti-GPNMB	Breast cancer	Phase II	[94]
SGN-75	Anti-CD70	Renal cell carcoinoma	Phase I	[96]
AGS-22M6E	Anti-Nectin fully human IgG	Solid tumors	Phase I	[97]
PSMA ADC	Anti-PSMA fully human IgG1	Prostate cancer	Phase II	[98]

to a 30 mg/kg concentration even though a 1 mg/kg dose is known to be therapeutic to a xenografted mice model [99]. The human trials of SGN-35 produced high response rates; 75% for Hodgkin's lymphoma and 87% for anaplastic large cell lymphoma, and it was approved by the Food and Drug Association (FDA) (Table 11.2) [95]. Another interesting example of an ADC using the VC peptide linker for Hodgkin's lymphoma is SGN-75. SGN-75 is a drug conjugated version of naked antibody SGN-70; a humanized anti-CD70 IgG1 mAb. CD70 is a member of a TNF superfamily that is transiently expressed on activated T- and B-lymphocytes, natural killer cells, and mature dendritic cells [100–103]. In normal tissues, CD70 has very limited expression. However, in several malignancies, such as renal cell carcinoma and non-Hodgkin lymphoma cells, CD70 overexpression is observed [104,105]. SGN-70 demonstrated antitumor activity, which was associated with dose-dependent antibody-dependent cell-mediated cytotoxicity [106]. Addition of a drug to this mAb, like MMAE through an enzyme-cleavable VC dipeptide linker, SGN-75, enhanced the antitumor activity in phase I studies [96] (Table 11.2).

Biologic conjugate drug design, simplified by the knowledge and use of such effective peptide linkers, has seen an explosion of success. With such robust and specific enzymatically cleavable peptide linkers, the bottleneck for development of new ADCs has shifted away from linker design. Instead, discovery of new biomarkers specific to various cancer cells is now backlogging the future pipeline. Examples of ADCs on deck for success include ADCs against nectin-4 and PMSA. Nectin-4 is a transmembrane type I protein that has been discovered on epithelial cancer cells. The ADC prepared with VC linker of enfortumab vedotin binding to nectin-4 showed high antitumor efficacy in xenografted mice models of several solid tumors [97].

The use of cancer-associated enzyme-responsive peptide linkers for the delivery of anticancer agents significantly improved over the course of the last decade. ADCs, many of which are under clinical trials or already FDA approved, showed impressive success in cancer treatment. The specificity of peptides only to the cancer and prompt response potential, make these linkers superior linkers for the delivery of the drugs to carcinogenic cells, while not damaging healthy cells. The studies on exploiting specific biomarkers and antigens for different types of cancer can improve the fight of humanity against cancer, with great help provided by enzyme-specific cleavable peptide linkers (Table 11.3).

11.4 Peptides in supramolecular structures

A peptide's role in immunology and immunoengineering is not only critical, but also practical. Peptide antigens are an essential part of the immunologic response, and functional "active" peptides make new technologies such as ADCs possible, as described above.

Nonetheless, even with the previously described examples of successfully engineered systems, many complex diseases still remain, for which engineered immunotherapies do not exist. These diseases will require the activation of a complex

Table 11.3 **Example peptides used in immune engineering**

Peptide application	Name of peptide: sequence	References
Antiinflammatory peptide	Ac2-26: Ac-AMVSEFLKQAWFIENEEQEYVQTVK	[51–54]
Neutrophil receptor (FcγRIIA) binding peptide	FcγRIIA binding peptide: WAWVWLTETAV	[56]
Dendritic cell (CD11c/CD18) binding peptide	p30: CGRWSGWPADLC	[59,60]
M cell binding peptides	Co1: SFHQLPARSPLP; P8: LETTCASLCYPS; P25: VPPHPMTYSCQY	[62,63]
Erythrocyte binding peptide	ERY1: H_2N-WMVLPWLPGTLDGGSGCRG-$CONH_2$	[71]
Beta-sheet fiber forming peptide	Q11: Ac-QQKFQFQFEQQ-am	[107]
Tail added to Q11 in order to incorporate antigens and slow assembly kinetics	β-tail: MALKVELEKLKSELVVLHSELHKLKSEL	[108]
Secondary structure transitioning peptide	W3K: C16-WAAAAKAAAAKAAAAKA	[109]
Chicken egg model antigen	OVA(253–266): EQLESIINFEKLTE	[10]
Group A *Streptococcus* epitope	J8: QAEDKVKQSREAKKQVEKALKQLEDKVQK	[110]
Group A *Streptococcus* epitope	J14: KQAEDKVKASREAKKQVEKALEQLEDKVK	[111]
Trimeric coiled-coil peptide derived from respiratory syncytial virus	HR1(153–202): AVSKVLHLEGEVNKIKSALLSTNKAVVSLSNGVSVLTSKVLDLKNYIDKQ	[112]

and specific immune response not easily induced. Complementary to their antigenic and functional roles, peptides can also be a part of larger structures designed to present antigen or provide immunity-promoting secondary signals. Peptides themselves can self-assemble or be part of a self-assembled supramolecular particle through specific and engineered supramolecular interactions. These peptides in higher complexity structures can be used to instruct the immune system via a different approach than normal peptide antigen. These structures can provide signals to the immune system by (1) the compounds they present/encapsulate or (2) by virtue of the structure or composition of the structure itself. While a more general overview of the uses of self-assembling peptides is discussed elsewhere in this textbook, here we would like to highlight in particular the role that peptides in supramolecular structures play in immunological applications.

11.4.1 Advantages of supramolecular assembly for immunological applications

In the field of immunoengineering, the antigenic function of peptides has inspired their incorporation into supramolecular assemblies. The resulting assemblies have been incredibly diverse and elegant in design, employing peptides for a wide array of functions ranging from coatings to encapsulation to driving the assembly itself, among others. Because peptide supramolecular assemblies in particular can simultaneously incorporate immunologically active peptides while inducing an immunological response by virtue of the structure or composition of the assembly itself, these systems have commanded great interest from the research community.

The immune system is highly sensitive to inputs, requiring very specific signals to raise the correct immune response. The way an antigen is presented matters just as much as, if not more than, the antigen itself. Supramolecular assemblies offer the ability to provide this fine-tuned immunological response by rational modification of the moieties responsible for driving assembly in tandem with modification of the immunologically active peptides themselves. Through rational design, it is possible to modify the size, shape, and composition of the supramolecular assembly, as well as any cargos carried in the core or on the surface of the assembly. The fine control of the assembly's structure allows for the precise functioning of the assembly required to achieve the desired immune response. For example, because assemblies are composed of many monomers, supramolecular assemblies can be designed to act as potent adjuvants for their cargo or even exhibit self-adjuvanting properties to promote cellular or humoral response due to the local density of immunogenic moieties present on the surface of the assembly [10,107,113]. Furthermore, these multifunctional assemblies can be designed to precisely control the relative amount of each functional component of the assembly to fine-tune the resulting immune response [114].

Through rational design, supramolecular assemblies are also capable of controlled, targeted localization, increasing the bioavailability of both the enclosed cargo and the assembly itself [115]. Localization is particularly crucial for immunological applications, as "on-target" effects exhibited at undesired locations have been associated with severe toxicity [19,20]. The use of targeting moieties in combination with nanoparticles can therefore mitigate the toxicity of immunomodulatory compounds while improving their ability to elicit desirable immune responses.

11.4.2 Peptide assemblies

Peptides themselves are capable of self-assembly into supramolecular structures. In the field of immunoengineering, self-assembling peptides have been used to create platforms capable of highly controlled presentation of antigen. Due to the variety of amino acid building blocks available, peptide assemblies in particular are highly modular and have extensive design capabilities. The assembling peptides can be designed to contain specific secondary structure to control the interactions between peptides, allowing for tuning of the shape and size of the assembly as well as the density of antigen presented at the surface. As a result, it is possible to achieve a more specific immune response, making peptide–peptide assemblies potentially useful as vaccines.

11.4.2.1 Assembly of peptides into fibers

One clever vaccine system relies on coiled-coil self-assembly of peptides. In this case, self-assembly is achieved through the detailed design of a linear peptide monomer. This monomer has three domains: a central "self-assembly" domain with two coiled-coil oligomerization sequences, flanked on the N- and C-termini by the antigenic domains (antigen domain 1–self-assembly domain–antigen 2 domain). The self-assembly sequence contains a pentameric and a trimeric coiled-coil sequence so that the monomer will not coil on itself, but coil only in the presence of other monomers under the correct conditions. The antigen domains are highly modular and can each accommodate multiple epitopes, allowing for great control of overall epitope presentation. In a study utilizing this system as a malaria vaccine, by attaching a B-cell epitope, a T-helper epitope and a cytotoxic T-cell epitope to this system, the assembly was able to generate a persistent humoral and cellular response to provide protection from malaria parasites for up to 1 year [11].

Vaccines utilizing the self-assembly of β-sheet-forming peptides into fibers have also been shown to provoke a strong immune response [15]. The strong hydrogen-bonding residues responsible for β-sheet formation permit the attachment of a wide variety of antigens to the assembling domains without disrupting supramolecular assembly of the β-sheets into fibers [11]. Fibrillar structures formed by self-assembling peptides seem to be potent adjuvants without exhibiting inflammatory responses on their own, making fiber assemblies a particularly useful delivery mechanism of vaccines [113].

A prominent example of a β-sheet fiber-forming peptide tested as a potential vaccine platform is the "Q11" peptide reported by Rudra et al. The peptide (Ac-QQKFQFQFEQQ-am; Fig. 11.9A) forms fibers 5–10 nm wide and up to 1 μm

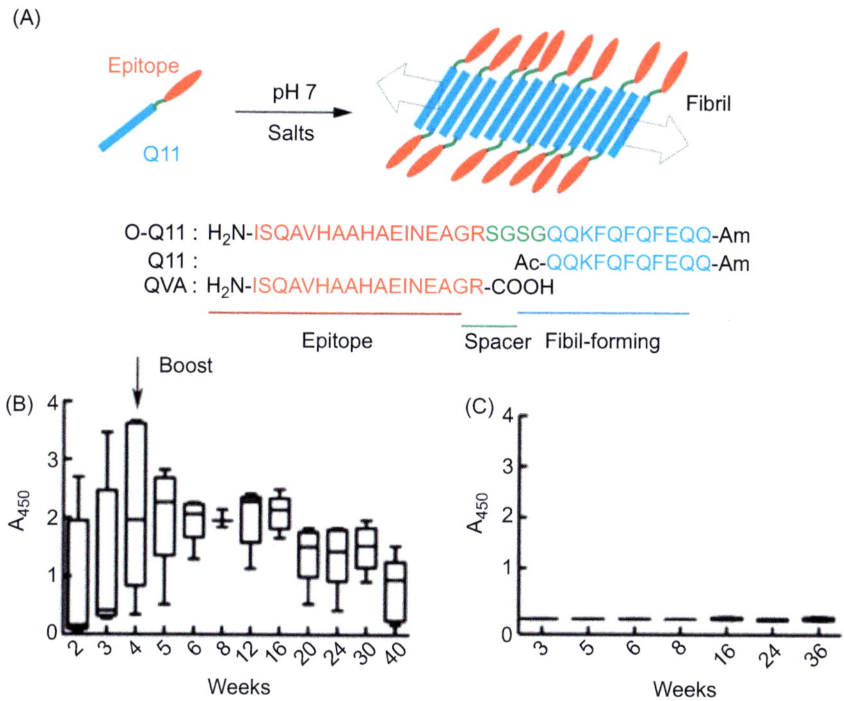

Figure 11.9 (A) Schematic representation of formation of Q11 fiber assembly and sequence of Q11-OVA, Q11, and ovalbumin epitope ($OVA_{323-339}$). (B) and (C) Antibody response to $(NANP)_3$-Q11 (B) and soluble Q11 (C) in B6 mice.
Source: Reprinted from J.S. Rudra, Y.F. Tian, J.P. Jung, J.H. Collier, A self-assembling peptide acting as an immune adjuvant, Proc. Natl. Acad. Sci. U. S. A. 107 (2) (2010) 622–627, doi:10.1073/pnas.0912124107; with permission from PNAS.

in length [107]. The fibers were nonimmunogenic, even when coadministered with mycobacterial complete Freund's adjuvant. However, when ovalbumin B- and T-cell epitopes ($OVA_{323-339}$) were attached to Q11 fibers and administered to mice, high levels of IgG1, IgG2a, and IgG3 antibodies against OVA developed without the use of adjuvant. Interestingly, when the hydrogen bonds driving Q11 self-assembly were disrupted, no immune response was induced. Without the self-assembly, the system becomes merely an injection of soluble Q11 peptide, which is known to have weak immunogenic properties. This suggests that the assembly itself was essential to the function of the vaccine [113]. The vaccine was further shown to induce a strong humoral response against malaria parasite when the malaria peptide $(NANP)_3$ was coupled to Q11 (Fig. 11.9B,C) [116]. More recently, Rudra has shown that $OVA_{257-264}$-Q11 vaccine was capable of eliciting a cytotoxic T-cell response in mice, raising the possibility of using the Q11 system in response to viral infections or tumors [117]. The modularity and tunability of Q11 peptide was demonstrated in an experiment in which fibers were formed via coassembly of the

B-cell epitope E214-Q11 and the T-cell epitope PADRE-Q11 [114]. Modulating the concentrations of each epitope changed the antibody and helper T-cell immune response to the vaccine in a manner dependent on the respective concentrations of each epitope. Furthermore, Chen et al. have demonstrated the safety of the OVA-Q11 vaccine, by showing that $OVA_{323-339}$-Q11 fibers did not induce inflammation at the site of injection nor cell death [118].

One potential challenge facing the use of Q11 is that its fast fiber formation kinetics can lead to aggregation, which can reduce the potency of the Q11 constructs. To address this, Hudalla et al. report the use of a peptide (MALKVELEKLKSELVVLHSELHKLKSEL) whose kinetics permit slow formation of a β-sheet, which they called a "β-tail" (Fig. 11.10A–C). The slow assembly kinetics of the β-tail peptides was shown to improve their incorporation into Q11 fibers and reduce aggregate formation, providing a platform by which to incorporate antigens into the system. Hudalla proceeded to demonstrate that β-tails attached

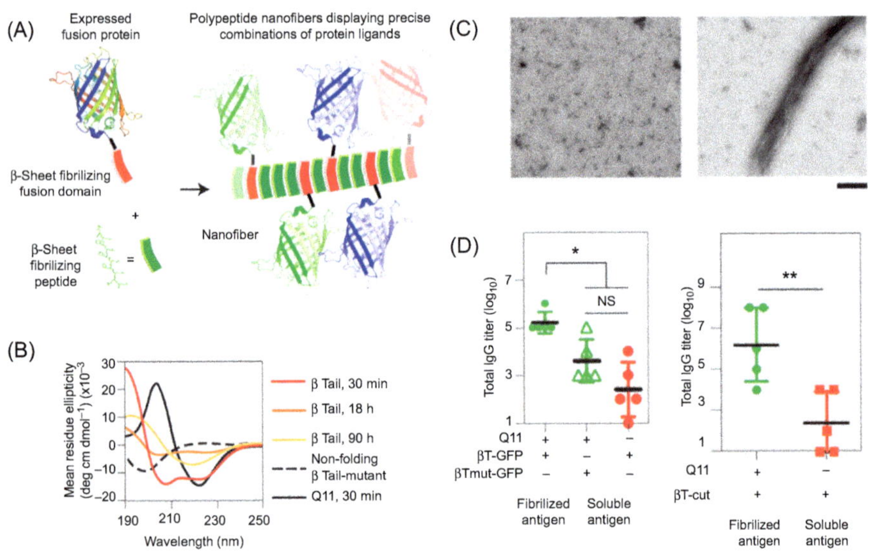

Figure 11.10 (A) β-Tail formation from a fusion of a β-sheet fibrillizing peptide and an antigenic peptide. (B) Circular dichroism shows that β-tails undergo a slow transition to β-sheet structure, while Q11 transitions quickly and a nonfunctional β-tail mutant retains a random coil structure. (C) TEM images of β-tail peptide at 30 min (left) and 100 h (right). Scale bar = 200 nm. (D) Antibody titers of C57BL/6 mice injected with β-tail-GFP (left) or β-tail-cutinase (right) incorporated into Q11 fibers are greatly increased compared to freely soluble β-tail-antigen (red) or mutated β-tail/Q11 assemblies (left, empty triangles). $^{*}P < .05$, $^{**}P < .01$.

Source: Reprinted by permission from G.A. Hudalla, T. Sun, J.Z. Gasiorowski, H. Han, Y.F. Tian, A.S. Chong, et al., Gradated assembly of multiple proteins into supramolecular nanomaterials, Nat. Mater. 13 (8) (2014) 829–836, doi:10.1038/nmat3998; Copyright (2014), Macmillan Publishers Ltd: Nature Materials.

to fungal cutinase, when coassembled with Q11, could raise antibody titers against cutinase. Furthermore, assemblies incorporating β-tail-cutinase and β-tail-GFP were able to raise titers against both proteins (Fig. 11.10D). These observations suggest that the β-tail system could be used to incorporate multiple types of antigen, modifying the relative ratios of β-tail-antigen constructs to modulate the resulting immune response [108].

11.4.2.2 Conjugated peptide assemblies

While hydrogen bonding is an important force that can drive self-assembly, it is not the only force at our disposal. The hydrophobic force is an excellent means of driving self-assembly of hydrophilic peptides. To promote self-assembly via the hydrophobic force, the peptide should contain a hydrophobic component, which could be a peptide sequence composed of hydrophobic amino acids, a hydrophobic polymer, or a lipid or acyl chain. The resulting molecule is capable of self-assembly into a monolayered spherical or cylindrical micelle, or a bilayered vesicle or liposome, depending on the respective properties of the hydrophilic peptide "head" and hydrophobic carbon "tail." For example, the peptide conjugate amphiphile W3K (C_{16}-WAAAAKAAAAKAAAAKA), designed by Shimada et al., self-assembled into spherical micelles in aqueous solution, but as the peptide's secondary structure transitioned from α-helix to β-sheet morphology over the course of days, the micelles gradually elongated to worm-like micelles [109].

One advantage of the peptide–conjugate assembly strategy is the stabilizing effect it can have on the secondary structure of the peptide epitope. A peptide's secondary structure is a vital parameter of the specific binding of the peptide to its target, so this stabilizing effect generally strengthens the immunogenic function of the peptide. A platform reported by Tirrell et al. utilizes a peptide–dialkyl lipid (diC_{16}) conjugate with self-adjuvanting properties (Fig. 11.11A). Using an elongated form of the OVA cytotoxic T-cell epitope ($OVA_{253-266}$, EQLESIINFEKLTE-diC16), the assembly induced a cytotoxic T-cell response that inhibited the growth of E.G7-OVA tumors in mice (Fig. 11.11B) [10,119]. Another platform utilizing the J8 B-cell epitope against group A *Streptococcus* (QAEDKVKQSREAKKQVEKALKQLEDKVQK-diC16) exhibited strong IgG and IgM antibody responses (Fig. 11.11C) [110,120]. Notably, the J8-diC16 construct did not require a helper T-cell epitope, whereas the Q11 systems did, though the cause of this difference is unknown [113].

Peptide amphiphiles are also highly modular. Because an antigenic peptide needs only to be hydrophilic and attachable to the hydrophobic component, a wide variety of antigens may be used. Furthermore, it is possible to synthesize a branched molecule capable of binding multiple antigens, allowing for mixed-antigen systems capable of a more comprehensive response. A related platform utilizes self-assembling lipoamino acids, comprised of a fatty acid linked to two peptide sequences through the amine moieties of a lysine linker unit. Azmi et al. report the use of a lipoamino acid system utilizing the J14 peptide epitope (KQAEDKVKASREAKKQVEKALEQLEDKVK) [111]. Building on this platform,

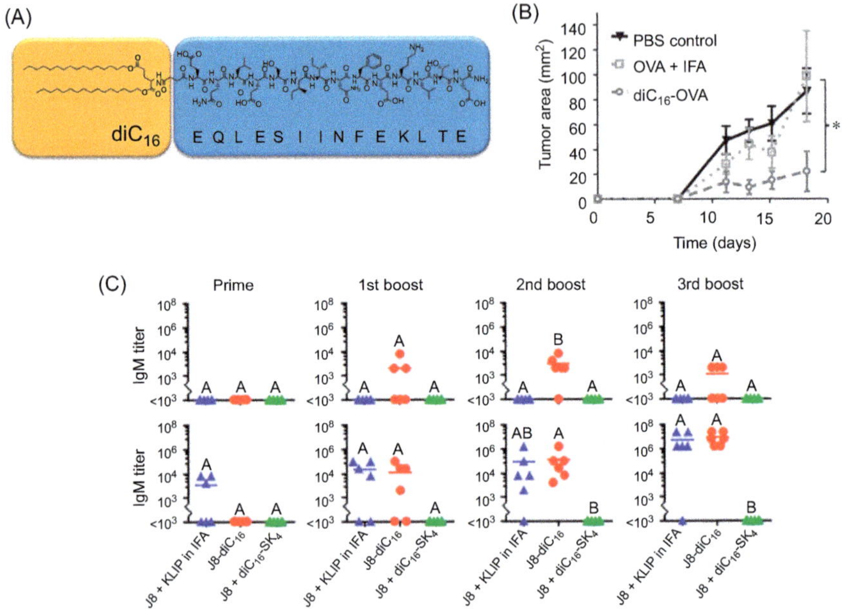

Figure 11.11 (A) Structure of $OVA_{253-266}$-diC_{16}. (B) $OVA_{253-266}$-diC_{16} inhibits tumor growth in mice. IFA, incomplete Freund's adjuvant. $^*P < .05$. (C) J8-diC_{16} raises IgG and IgM antibody titers against group A *Streptococcus*. Within each graph, different letters indicate statistically significant differences in mean ($P < .05$), whereas a shared letter indicates no significance ($P > .05$).

Source: (A) and (B) Reprinted from M. Black, A. Trent, Y. Kostenko, J.S. Lee, C. Olive, M. Tirrell, Self-assembled peptide amphiphile micelles containing a cytotoxic T-cell epitope promote a protective immune response in vivo, Adv. Mater. 24 (28) (2012) 3845–3849, doi:10.1002/adma.201200209; Copyright (2012), with permission from Wiley-VCH Verlag GmbH & Co. KGaA, Weinheim. (C) Reprinted from A. Trent, B.D. Ulery, M.J. Black, J.C. Barrett, S. Liang, Y. Kostenko, et al., Peptide amphiphile micelles self-Aajuvant group A streptococcal vaccination, AAPS J. 17 (2) (2015), 380–388, doi:10.1208/s12248-014-9707-3; Copyright (2014), with permission from Springer.

Zaman et al. designed a lipopeptide vaccine against *Streptococcus* utilizing a C_{16} lipid linked to P25 epitope and J14 [121].

11.4.3 Further delivery strategies

In addition to the use of engineered peptides to drive assembly, other supramolecular systems exist to stimulate the immune response. These systems use structures such as viral proteins (as in virus-like particles (VLPs)) or synthetic polymers to form particles. The immunogenic cargo is then delivered attached to the surface or encapsulated within the particles. This section will briefly discuss the strategies utilizing VLPs and peptide encapsulation.

11.4.3.1 Virus-like particles

VLPs are a class of supramolecular assemblies composed of an underlying self-assembling structure coated with an antigen of interest. The goal of a VLP is to raise an immune response against the coated antigen, which can be almost anything (peptides, whole proteins, etc.). The self-assembly drivers of a VLP are viral envelope proteins known as capsids. Capsids do not contain any viral genetic material and are therefore noninfectious. Even though the capsid shell is noninfectious, it resembles a real pathogen and still contains conserved pattern-associated molecular patterns (PAMPs) recognized by the innate immune system. This allows the VLP to induce strong innate and adaptive immune responses against the conjugated antigen without the need for additional adjuvants [122]. In a sense, the VLP itself acts as the adjuvant. However, an unintended drawback of a VLP's innate immunogenicity is that an immune response will be raised against the VLP capsid proteins in addition to the coated antigen that is delivered with it. Attaching antigens to VLPs therefore enhances the strength of the immune response, but the immunogenicity of the capsids themselves reduces the capacity for fine control of the resulting response. This contrasts with the supramolecular systems mentioned previously where an immune response is raised against the antigen only, not the self-assembling platform it employs (though as mentioned the self-assembled structure is required to raise an immune response). Currently, two VLP vaccines are used in the market: hepatitis B and human papillomavirus.

By modifying nonassembling viral proteins with functionalities that drive self-assembly into a virus-sized particle, one can construct synthetic virus-like particles (SVLPs). Lipids are a common functionality to add to drive hydrophobic assembly, but the final structural shape often varies. Boato et al. synthesized SVLPs utilizing a sequence derived from respiratory syncytial virus, $HR1_{153-202}$, that organized into trimeric coiled-coils which in turn formed micelles (Fig. 11.12) [112]. These SVLPs were fused to HIV-1 peptide antigen, and subsequently induced antibody production without adjuvants in rabbits. Another recent study, by Robinson et al., used a similar SVLP to deliver B-cell and helper T-cell epitopes against malaria parasite subcutaneously to mice and rabbits, which were met with strong antibody responses in the animals [124].

11.4.3.2 Encapsulation of peptides in nanoparticle assemblies

An alternative strategy that uses self-assembled particles to deliver immunomodulatory agents is encapsulation. Encapsulation is a thoroughly developed field that utilizes many strategies to deliver antigens of interest (peptides, proteins, cytokines, etc.), and has been thoroughly reviewed elsewhere [125,126]. One primary advantage to encapsulation of antigens, including peptides, is their extended release profile, as freely soluble antigens generally have shorter lifetimes. For example, nanoparticles composed of the biodegradable synthetic diblock copolymer poly(lactic-co-glycolic acid)-*b*-poly(ethylene glycol) (PLGA-PEG) targeted to tissue injury sites via collagen IV-targeting peptide and loaded with the antiinflammatory

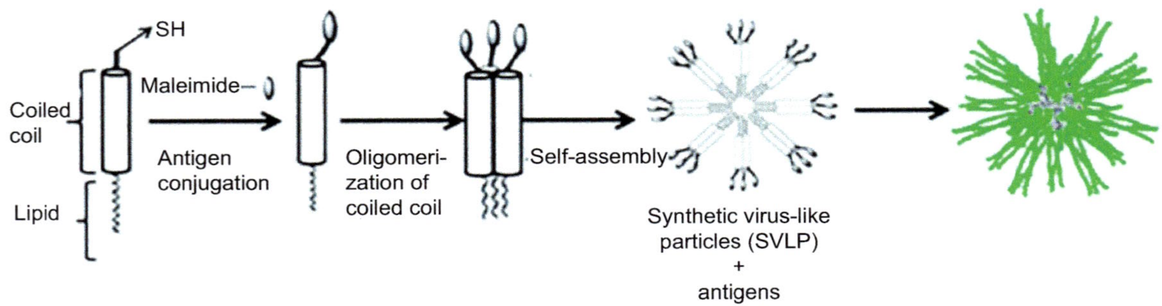

Figure 11.12 Formation of synthetic VLPs by attaching antigens to self-assembling peptide–lipid conjugates. The resulting structure resembles a virus in size and shape.
Source: Reprinted from L. Yang, W. Li, M. Kirberger, W. Liao, J. Ren, Design of nanomaterial based systems for novel vaccine development, Biomater. Sci. 4 (5) (2016) 785–802, doi:10.1039/C5BM00507H [123]; Copyright (2016), with permission from The Royal Society of Chemistry.

peptide Ac2-26 (AMVSEFLKQAWFIENEEQEYVQTVK) preferentially targeted sites of tissue injury and reduced neutrophil recruitment at those sites to promote healing via sustained release on long timescales (Fig. 11.13) [55].

A particular class of drug-encapsulating self-assembling particle is the interbilayer-crosslinked multilamellar vesicle (ICMV). ICMVs are multilamellar lipid vesicles that have been stabilized by covalently crosslinking the headgroups of the lipid assembly to the peptide adjuvant loaded within. The covalent bonds increase the density of loaded peptides while also prolonging the release time of encapsulated cargo. In a system reported by Moon et al., the malaria antigen VMP001 was crosslinked with phospholipid vesicles to form ICMVs. The increased density of VMP001 in the ICMVs, in combination with a long release time of

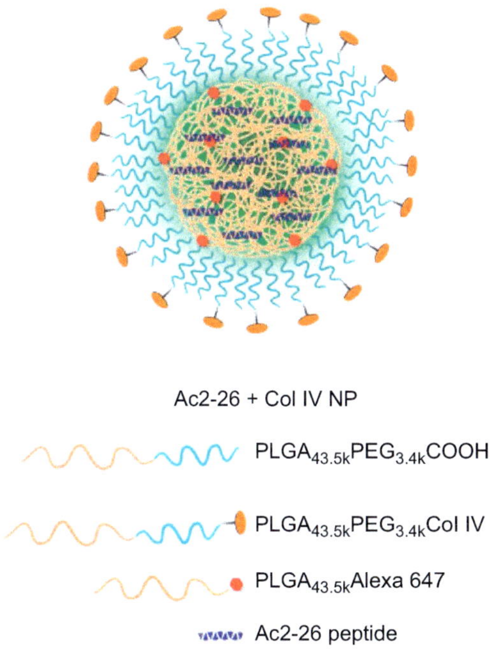

Figure 11.13 Targeted (Col IV) NPs encapsulating the Ac2-26 peptide were developed using biodegradable polymers via a single-step nanoprecipitation method. The synthesized polymers and Ac2-26 peptide were dissolved in acetonitrile (total polymer 3 mg/mL), and 2% (wt/wt) of the fluorescent PLGA-Alexa 647 was added to all formulations. NPs contained 4% (wt/wt) peptide and 5% (wt/wt) of the Col IV peptide-conjugated targeting polymer. The organic mixture containing the polymers and peptide was then added dropwise to nuclease-free water (10 mL). The solution was stirred for 2–4 h, and the particles were filtered, washed, and resuspended in water or PBS.
Source: Reprinted from N. Kamaly, G. Fredman, M. Subramanian, S. Gadde, A. Pesic, L. Cheung, et al., Development and in vivo efficacy of targeted polymeric inflammation-resolving nanoparticles, Proc. Natl. Acad. Sci. U. S. A. 110 (16) (2013) 6506–6511, doi:10.1073/pnas.1303377110; with permission from PNAS.

2 weeks, produced robust immune responses against VMP001 in mice that lasted for over 1 year [127].

11.4.4 Future directions

One major challenge remaining for fiber and cylindrical assemblies is control of their length. While spherical assemblies tend to have well-defined monomer numbers as a result of the properties of the monomer subunits, cylindrical assemblies have a defined cross-radius but poorly defined length. However, overall, supramolecular assemblies present great opportunities in immunomodulation due to their ability to aggregate epitopes, antigens, and other immunomodulatory agents into one system in modular fashion. This modularity lends precision to the future engineering of supramolecular assemblies to modulate immune response.

References

[1] N.A. Hotaling, L. Tang, D.J. Irvine, J.E. Babensee, Biomaterial strategies for immunomodulation, Annu. Rev. Biomed. Eng. 17 (1) (2015) 317–349. Available from: http://dx.doi.org/10.1146/annurev-bioeng-071813-104814.

[2] D.J. Irvine, M.C. Hanson, K. Rakhra, T. Tokatlian, Synthetic nanoparticles for vaccines and immunotherapy, Chem. Rev. 115 (19) (2015) 11109–11146. Available from: http://dx.doi.org/10.1021/acs.chemrev.5b00109.

[3] B.S. Sheridan, L. Lefrancois, Regional and mucosal memory T cells, Nat. Immunol. 12 (6) (2011) 485–491. Available from: http://dx.doi.org/10.1038/ni.2029.

[4] C.A. Janeway Jr, P. Travers, M. Walport, M.J. Shlomchik, Principles of Innate and Adaptive Immunity, Garland Science, New York, 2001.

[5] E. Cha, M. Klinger, Y. Hou, C. Cummings, A. Ribas, M. Faham, et al., Improved survival with T cell clonotype stability after anti-CTLA-4 treatment in cancer patients, Sci. Transl. Med. 6 (238) (2014) 238ra70. Available from: http://dx.doi.org/10.1126/scitranslmed.3008211.

[6] R.M. Steinman, J. Banchereau, Taking dendritic cells into medicine, Nature 449 (7161) (2007) 419–426. Available from: http://dx.doi.org/10.1038/nature06175.

[7] I. Mellman, R.M. Steinman, Dendritic cells: specialized and regulated antigen processing machines, Cell 106 (3) (2001) 255–258, doi:10.1016/S0092-8674(01)00449-4.

[8] T. Kawai, S. Akira, Toll-like receptors and their crosstalk with other innate receptors in infection and immunity, Immunity 34 (5) (2011) 637–650. Available from: http://dx.doi.org/10.1016/j.immuni.2011.05.006.

[9] A. Townsend, J. Rothbard, F.M. Gotch, G. Bahadur, D. Wraith, A.J. McMichael, The epitopes of influenza nucleoprotein recognized by cytotoxic T-lymphocytes can be defined with short synthetic peptides, Cell 44 (6) (1986) 959–968, doi:10.1016/0092-8674(86)90019-X.

[10] M. Black, A. Trent, Y. Kostenko, J.S. Lee, C. Olive, M. Tirrell, Self-assembled peptide amphiphile micelles containing a cytotoxic T-cell epitope promote a protective immune response in vivo, Adv. Mater. 24 (28) (2012) 3845–3849. Available from: http://dx.doi.org/10.1002/adma.201200209.

[11] S.A. Kaba, M.E. McCoy, T.A.P.F. Doll, C. Brando, Q. Guo, D. Dasgupta, et al., Protective antibody and CD8 + T-cell responses to the *Plasmodium falciparum* circumsporozoite protein induced by a nanoparticle vaccine, PLoS One 7 (10) (2012) e48304–e48311. Available from: http://dx.doi.org/10.1371/journal.pone.0048304.

[12] Y. Wen, J.H. Collier, Supramolecular peptide vaccines: tuning adaptive immunity, Curr. Opin. Immunol. 35 (2015) 73–79. Available from: http://dx.doi.org/10.1016/j.coi.2015.06.007.

[13] J.J. Moon, H. Suh, A. Bershteyn, M.T. Stephan, H. Liu, B. Huang, et al., Interbilayer-crosslinked multilamellar vesicles as synthetic vaccines for potent humoral and cellular immune responses, Nat. Mater. 10 (3) (2011) 243–251. Available from: http://dx.doi.org/10.1038/NMAT2960.

[14] J.H. Collier, T. Segura, Evolving the use of peptides as components of biomaterials, Biomaterials 32 (18) (2011) 4198–4204. Available from: http://dx.doi.org/10.1016/j.biomaterials.2011.02.030.

[15] T.Y.H. Wu, M. Singh, A.T. Miller, E. De Gregorio, F. Doro, U. D'Oro, et al., Rational design of small molecules as vaccine adjuvants, Sci. Transl. Med. 6 (263) (2014) 263ra160. Available from: http://dx.doi.org/10.1126/scitranslmed.3009980.

[16] B. Kwong, S.A. Gai, J. Elkhader, K.D. Wittrup, D.J. Irvine, Localized immunotherapy via liposome-anchored anti-CD137 + IL-2 prevents lethal toxicity and elicits local and systemic antitumor immunity, Cancer Res. 73 (5) (2013) 1547–1558. Available from: http://dx.doi.org/10.1158/0008-5472.CAN-12-3343.

[17] B. Kwong, H. Liu, D.J. Irvine, Induction of potent anti-tumor responses while eliminating systemic side effects via liposome-anchored combinatorial immunotherapy, Biomaterials 32 (22) (2011) 5134–5147. Available from: http://dx.doi.org/10.1016/j.biomaterials.2011.03.067.

[18] H. Liu, K.D. Moynihan, Y. Zheng, G.L. Szeto, A.V. Li, Structure-based programming of lymph-node targeting in molecular vaccines, Nature 507 (7493) (2014) 519–522. Available from: http://dx.doi.org/10.1038/nature12978.

[19] L. Niu, S. Strahotin, B. Hewes, B. Zhang, Y. Zhang, D. Archer, et al., Cytokine-mediated disruption of lymphocyte trafficking, hemopoiesis, and induction of lymphopenia, anemia, and thrombocytopenia in anti-CD137-treated mice, J. Immunol. 178 (7) (2007) 4194–4213.

[20] J.P. Leonard, M.L. Sherman, G.L. Fisher, L.J. Buchanan, G. Larsen, M.B. Atkins, et al., Effects of single-dose interleukin-12 exposure on interleukin-12-associated toxicity and interferon-gamma production, Blood 90 (7) (1997) 2541–2548.

[21] S.L. Demento, A.L. Siefert, A. Bandyopadhyay, F.A. Sharp, T.M. Fahmy, Pathogen-associated molecular patterns on biomaterials: a paradigm for engineering new vaccines, Trends Biotechnol. 29 (6) (2011) 294–306. Available from: http://dx.doi.org/10.1016/j.tibtech.2011.02.004.

[22] P.F. Brunsvig, S. Aamdal, M.K. Gjertsen, G. Kvalheim, C.J. Markowski-Grimsrud, I. Sve, et al., Telomerase peptide vaccination: a phase I/II study in patients with non-small cell lung cancer, Cancer Immunol. Immunother. 55 (12) (2006) 1553–1564. Available from: http://dx.doi.org/10.1007/s00262-006-0145-7.

[23] J.S. Testa, R. Philip, Role of T-cell epitope-based vaccine in prophylactic and therapeutic applications, Future Virol. 7 (11) (2012) 1077–1088. Available from: http://dx.doi.org/10.2217/FVL.12.108.

[24] J.G. Lanier, M.J. Newman, E.M. Lee, A. Sette, R. Ahmed, Peptide vaccination using nonionic block copolymers induces protective anti-viral CTL responses, Vaccine 18 (5–6) (1999) 549–557.

[25] J. Roelse, M. Gromme, F. Momburg, G. Hammerling, J. Neefjes, Trimming of tap-translocated peptides in the endoplasmic-reticulum and in the cytosol during recycling, J. Exp. Med. 180 (5) (1994) 1591–1597. Available from: http://dx.doi.org/10.1084/jem.180.5.1591.

[26] E. Reits, A. Griekspoor, J. Neijssen, T. Groothuis, K. Jalink, P. van Veelen, et al., Peptide diffusion, protection, and degradation in nuclear and cytoplasmic compartments before antigen presentation by MHC class I, Immunity 18 (1) (2003) 97–108, doi:10.1016/S1074-7613(02)00511-3.

[27] R.L. Coffman, A. Sher, R.A. Seder, Vaccine adjuvants: putting innate immunity to work, Immunity 33 (4) (2010) 492–503. Available from: http://dx.doi.org/10.1016/j.immuni.2010.10.002.

[28] N. Petrovsky, J.C. Aguilar, Vaccine adjuvants: current state and future trends, Immunol. Cell. Biol. 82 (2004) 488–496. Available from: http://dx.doi.org/10.1111/j.1440-1711.2004.01272.x.

[29] S.G. Reed, M.T. Orr, C.B. Fox, Key roles of adjuvants in modern vaccines, Nat. Med. 19 (12) (2013) 1597–1608. Available from: http://dx.doi.org/10.1038/nm.3409.

[30] M. Black, A. Trent, M. Tirrell, C. Olive, Advances in the design and delivery of peptide subunit vaccines with a focus on Toll-like receptor agonists, Expert Rev. Vaccines 9 (2) (2010) 157–173. Available from: http://dx.doi.org/10.1586/erv.09.160.

[31] B. Engels, V.H. Engelhard, J. Sidney, A. Sette, D.C. Binder, R.B. Liu, et al., Relapse or eradication of cancer is predicted by peptide–major histocompatibility complex affinity, Cancer Cell 23 (4) (2013) 516–526. Available from: http://dx.doi.org/10.1016/j.ccr.2013.03.018.

[32] E. Valeur, M. Bradley, Amide bond formation: beyond the myth of coupling reagents, Chem. Soc. Rev. 38 (2) (2009) 606–631. Available from: http://dx.doi.org/10.1039/b701677h.

[33] S. Tsai, A. Shameli, J. Yamanouchi, X. Clemente-Casares, J. Wang, P. Serra, et al., Reversal of autoimmunity by boosting memory-like autoregulatory T cells, Immunity 32 (4) (2010) 568–580. Available from: http://dx.doi.org/10.1016/j.immuni.2010.03.015.

[34] G.L. Stritesky, S.C. Jameson, K.A. Hogquist, Selection of self-reactive T cells in the thymus, Annu. Rev. Immunol. 30 (1) (2012) 95–114. Available from: http://dx.doi.org/10.1146/annurev-immunol-020711-075035.

[35] G. Thomson, W.P. Robinson, M.K. Kuhner, S. Joe, M.J. Macdonald, J.L. Gottschall, et al., Genetic-heterogeneity, modes of inheritance, and risk estimates for a joint study of Caucasians with insulin-dependent diabetes-mellitus, Am. J. Hum. Genet. 43 (6) (1988) 799–816.

[36] D.S. Lyons, S.A. Lieberman, J. Hampl, J.J. Boniface, Y.H. Chien, L.J. Berg, et al., TCR binds to antagonist ligands with lower affinities and faster dissociation rates than to agonists, Immunity 5 (1) (1996) 53–61, doi:10.1016/S1074-7613(00)80309-X.

[37] M. Larche, D.C. Wraith, Peptide-based therapeutic vaccines for allergic and autoimmune diseases, Nat. Med. 11 (4) (2005) S69–S76. Available from: http://dx.doi.org/10.1038/nm1226.

[38] G.G. Kenter, M.J.P. Welters, A.R.P.M. Valentijn, M.J.G. Lowik, D.M.A. Berends-van der Meer, A.P.G. Vloon, et al., Vaccination against HPV-16 oncoproteins for vulvar intraepithelial neoplasia, N. Engl. J. Med. 361 (19) (2009) 1838–1847. Available from: http://dx.doi.org/10.1056/NEJMoa0810097.

[39] D. Sesardic, Synthetic peptide vaccines, J. Med. Microbiol. 39 (1993) 241–242.

[40] T.F. Greten, A. Forner, F. Korangy, G. N'Kontchou, N. Barget, C. Ayuso, et al., A phase II open label trial evaluating safety and efficacy of a telomerase peptide

vaccination in patients with advanced hepatocellular carcinoma, BMC Cancer 10 (1) (2010) 1862–1867. Available from: http://dx.doi.org/10.1186/1471-2407-10-209.
[41] M.D. Joshi, W.J. Unger, G. Storm, Y. van Kooyk, E. Mastrobattista, Targeting tumor antigens to dendritic cells using particulate carriers, J. Control. Release 161 (1) (2012) 25–37. Available from: http://dx.doi.org/10.1016/j.jconrel.2012.05.010.
[42] Z. Xu, S. Ramishetti, Y.-C. Tseng, S. Guo, Y. Wang, L. Huang, Multifunctional nanoparticles co-delivering Trp2 peptide and CpG adjuvant induce potent cytotoxic T-lymphocyte response against melanoma and its lung metastasis, J. Control. Release 172 (1) (2013) 259–265. Available from: http://dx.doi.org/10.1016/j.jconrel.2013.08.021.
[43] A.A. Moĭsa, E.F. Kolesanova, Synthetic peptide vaccines, Biomed. Khim. 57 (1) (2011) 14–30.
[44] M.S. Bijker, C.J. Melief, R. Offringa, S.H. van der Burg, Design and development of synthetic peptide vaccines: past, present and future, Expert Rev. Vaccines 6 (4) (2007) 591–603. Available from: http://dx.doi.org/10.1586/14760584.6.4.591.
[45] E.D. Quakkelaar, E.M. Bunnik, F.P.J. van Alphen, B.D.M. Boeser-Nunnink, A.C. van Nuenen, H. Schuitemaker, Escape of human immunodeficiency virus type 1 from broadly neutralizing antibodies is not associated with a reduction of viral replicative capacity in vitro, Virology 363 (2) (2007) 447–453. Available from: http://dx.doi.org/10.1016/j.virol.2007.02.011.
[46] E.D. Quakkelaar, C.J.M. Melief, Experience with synthetic vaccines for cancer and persistent virus infections in nonhuman primates and patients, Adv. Immunol. 114 (2012) 77–106. Available from: http://dx.doi.org/10.1016/B978-0-12-396548-6.00004-4.
[47] B.L. Bray, Large-scale manufacture of peptide therapeutics by chemical synthesis, Nat. Rev. Drug Discov. 2 (2003) 587–593.
[48] B.P. Gray, K.C. Brown, Combinatorial peptide libraries: mining for cell-binding peptides, Chem. Rev. 114 (2) (2014) 1020–1081. Available from: http://dx.doi.org/10.1021/cr400166n.
[49] E. Kolaczkowska, P. Kubes, Neutrophil recruitment and function in health and inflammation, Nat. Rev. Immunol. 13 (3) (2013) 159–175. Available from: http://dx.doi.org/10.1038/nri3399.
[50] H.L. Wright, R.J. Moots, R.C. Bucknall, S.W. Edwards, Neutrophil function in inflammation and inflammatory diseases, Rheumatology 49 (9) (2010) 1618–1631. Available from: http://dx.doi.org/10.1093/rheumatology/keq045.
[51] M. Perretti, F. D'Acquisto, Annexin A1 and glucocorticoids as effectors of the resolution of inflammation, Nat. Rev. Immunol. 9 (1) (2009) 62–70. Available from: http://dx.doi.org/10.1038/nri2470.
[52] N. Dufton, R. Hannon, V. Brancaleone, J. Dalli, H.B. Patel, M. Gray, et al., Anti-inflammatory role of the murine formyl-peptide receptor 2: ligand-specific effects on leukocyte responses and experimental inflammation, J. Immunol. 184 (5) (2010) 2611–2619. Available from: http://dx.doi.org/10.4049/jimmunol.0903526.
[53] M. Perretti, N. Chiang, M. La, I.M. Fierro, S. Marullo, S.J. Getting, et al., Endogenous lipid- and peptide-derived anti-inflammatory pathways generated with glucocorticoid and aspirin treatment activate the lipoxin A4 receptor, Nat. Med. 8 (11) (2002) 1296–1302. Available from: http://dx.doi.org/10.1038/nm786.
[54] L. Miele, E. Cordella-Miele, A. Facchiano, A.B. Mukherjee, Novel anti-inflammatory peptides from the region of highest similarity between uteroglobin and lipocortin I, Nature 335 (6192) (1988) 726–730. Available from: http://dx.doi.org/10.1038/335726a0.
[55] N. Kamaly, G. Fredman, M. Subramanian, S. Gadde, A. Pesic, L. Cheung, et al., Development and in vivo efficacy of targeted polymeric inflammation-resolving

nanoparticles, Proc. Natl. Acad. Sci. U. S. A. 110 (16) (2013) 6506–6511. Available from: http://dx.doi.org/10.1073/pnas.1303377110.
[56] G. Berntzen, J.T. Andersen, K. Ustgard, T.E. Michaelsen, S.A. Mousavi, J.D. Qian, et al., Identification of a high affinity FcγRIIA-binding peptide that distinguishes FcγRIIA from FcγRIIB and exploits FcγRIIA-mediated phagocytosis and degradation, J. Biol. Chem. 284 (2) (2009) 1126–1135. Available from: http://dx.doi.org/10.1074/jbc.M803584200.
[57] A.N. Endsley, R.J.Y. Ho, Enhanced anti-HIV efficacy of indinavir after inclusion in CD4-targeted lipid nanoparticles, J. Acquir. Immune Defic. Syndr. 61 (4) (2012) 417–424. Available from: http://dx.doi.org/10.1097/QAI.0b013e3182653c1f.
[58] J. Banchereau, A.K. Palucka, Dendritic cells as therapeutic vaccines against cancer, Nat. Rev. Immunol. 5 (4) (2005) 296–306. Available from: http://dx.doi.org/10.1038/nri1592.
[59] C. Frick, A. Odermatt, K. Zen, K.J. Mandell, H. Edens, R. Portmann, et al., Interaction of ICAM-1 with β2-integrin CD11c/CD18: characterization of a peptide ligand that mimics a putative binding site on domain D4 of ICAM-1, Eur. J. Immunol. 35 (12) (2005) 3610–3621. Available from: http://dx.doi.org/10.1002/eji.200425914.
[60] A. Faham, J.G. Altin, Ag-bearing liposomes engrafted with peptides that interact with CD11c/CD18 induce potent Ag-specific and antitumor immunity, Int. J. Cancer 129 (6) (2011) 1391–1403. Available from: http://dx.doi.org/10.1002/ijc.25810.
[61] M.A. Clark, M.A. Jepson, B.H. Hirst, Exploiting M cells for drug and vaccine delivery, Adv. Drug Deliv. Rev. 50 (2001) 81–106.
[62] L.M. Higgins, I. Lambkin, G. Donnelly, D. Byrne, C. Wilson, J. Dee, et al., In vivo phage display to identify M cell-targeting ligands, Pharm. Res. 21 (4) (2004) 1–11.
[63] S.-H. Kim, K.-W. Seo, J. Kim, K.-Y. Lee, Y.-S. Jang, The M cell-targeting ligand promotes antigen delivery and induces antigen-specific immune responses in mucosal vaccination, J. Immunol. 185 (10) (2010) 5787–5795. Available from: http://dx.doi.org/10.4049/jimmunol.0903184.
[64] M.D. Pierschbacher, E. Ruoslahti, Cell attachment activity of fibronectin can be duplicated by small synthetic fragments of the molecule, Nature 309 (5963) (1984) 30–33.
[65] E. Ruoslahti, RGD and other recognition sequences for integrins, Annu. Rev. Cell Dev. Biol. 12 (1) (1996) 697–715. Available from: http://dx.doi.org/10.1146/annurev.cellbio.12.1.697.
[66] Z. Liu, W. Cai, L. He, N. Nakayama, K. Chen, X. Sun, et al., In vivo biodistribution and highly efficient tumour targeting of carbon nanotubes in mice, Nat. Nanotechnol. 2 (1) (2007) 47–52. Available from: http://dx.doi.org/10.1038/nnano.2006.170.
[67] H. Maeda, The enhanced permeability and retention (EPR) effect in tumor vasculature: the key role of tumor-selective macromolecular drug targeting, Adv. Enzyme Regul. 41 (1) (2001) 189–207, doi:10.1016/S0065-2571(00)00013-3.
[68] B.R. Smith, C. Zavaleta, J. Rosenberg, R. Tong, J. Ramunas, Z. Liu, et al., High-resolution, serial intravital microscopic imaging of nanoparticle delivery and targeting in a small animal tumor model, Nano Today 8 (2) (2013) 126–137. Available from: http://dx.doi.org/10.1016/j.nantod.2013.02.004.
[69] K. Greish, Enhanced permeability and retention (EPR) effect for anticancer nanomedicine drug targeting, Vaccine Adjuvants: Methods in Molecular Biology, vol. 624, Humana Press, Totowa, NJ, 2010, pp. 25–37.
[70] B.R. Smith, E.E.B. Ghosn, H. Rallapalli, J.A. Prescher, T. Larson, L.A. Herzenberg, et al., Selective uptake of single-walled carbon nanotubes by circulating monocytes for enhanced tumour delivery, Nat. Nanotechnol. 9 (6) (2014) 481–487. Available from: http://dx.doi.org/10.1038/NNANO.2014.62.

[71] S. Kontos, J.A. Hubbell, Improving protein pharmacokinetics by engineering erythrocyte affinity, Mol. Pharm. 7 (6) (2010) 2141–2147. Available from: http://dx.doi.org/10.1021/mp1001697.
[72] S. Kontos, I.C. Kourtis, K.Y. Dane, J.A. Hubbell, Engineering antigens for in situ erythrocyte binding induces T-cell deletion, Proc. Natl. Acad. Sci. U. S. A. 110 (1) (2013) E60–E68. Available from: http://dx.doi.org/10.1073/pnas.1216353110/-/DCSupplemental.
[73] A.J. Grimm, S. Kontos, G. Diaceri, X. Quaglia-Thermes, J.A. Hubbell, Memory of tolerance and induction of regulatory T cells by erythrocyte-targeted antigens, Sci. Rep. 5 (2015) 1–11. Available from: http://dx.doi.org/10.1038/srep15907.
[74] K.M. Lorentz, S. Kontos, G. Diaceri, H. Henry, J.A. Hubbell, Engineered binding to erythrocytes induces immunological tolerance to *E. coli* asparaginase, Sci. Adv. 1 (6) (2015) e1500112. Available from: http://dx.doi.org/10.1126/sciadv.1500112.
[75] K.A. Gelderman, S. Tomlinson, G.D. Ross, A. Gorter, Complement function in mAb-mediated cancer immunotherapy, Trends Immunol. 25 (3) (2004) 158–164. Available from: http://dx.doi.org/10.1016/j.it.2004.01.008.
[76] M. Stern, R. Herrmann, Overview of monoclonal antibodies in cancer therapy: present and promise, Crit. Rev. Oncol. Hematol. 54 (1) (2005) 11–29. Available from: http://dx.doi.org/10.1016/j.critrevonc.2004.10.011.
[77] G.P. Adams, L.M. Weiner, Monoclonal antibody therapy of cancer, Nat. Biotechnol. 23 (9) (2005) 1147–1157. Available from: http://dx.doi.org/10.1038/nbt1137.
[78] J.M. Reichert, V.E. Valge-Archer, Development trends for monoclonal antibody cancer therapeutics, Nat. Rev. Drug Discov. 6 (5) (2007) 349–356. Available from: http://dx.doi.org/10.1038/nrd2241.
[79] G.F. Rowland, G.J. O'Neill, D.A.L. Davies, Suppression of tumour growth in mice by a drug–antibody conjugate using a novel approach to linkage, Nature 255 (5508) (1975) 487–488. Available from: http://dx.doi.org/10.1038/255487a0.
[80] C.H. Ford, C.E. Newman, J.R. Johnson, C.S. Woodhouse, T.A. Reeder, G.F. Rowland, et al., Localisation and toxicity study of a vindesine-anti-CEA conjugate in patients with advanced cancer, Br. J. Cancer 47 (1) (1983) 35–42. Available from: http://dx.doi.org/10.1038/bjc.1983.4.
[81] P.J. Carter, P.D. Senter, Antibody-drug conjugates for cancer therapy, Cancer J. 14 (3) (2008) 154–169. Available from: http://dx.doi.org/10.1097/PPO.0b013e318172d704.
[82] H. Bouchard, C. Viskov, C. Garcia-Echeverria, Antibody–drug conjugates: a new wave of cancer drugs, Bioorg. Med. Chem. Lett. 24 (23) (2014) 5357–5363. Available from: http://dx.doi.org/10.1016/j.bmcl.2014.10.021.
[83] P.R. Hamann, L.M. Hinman, I. Hollander, C.F. Beyer, D. Lindh, R. Holcomb, et al., Gemtuzumab ozogamicin, a potent and selective anti-CD33 antibody–calicheamicin conjugate for treatment of acute myeloid leukemia, Bioconjug. Chem. 13 (1) (2002) 47–58. Available from: http://dx.doi.org/10.1021/bc010021y.
[84] P.E. Thorpe, P.M. Wallace, P.P. Knowles, M.G. Relf, New coupling agents for the synthesis of immunotoxins containing a hindered disulfide bond with improved stability in vivo, Cancer Res. 47 (22) (1987) 5924–5931.
[85] R.S. Zolot, S. Basu, R.P. Million, Antibody–drug conjugates, Nat. Rev. Drug Discov. 12 (4) (2013) 259–260. Available from: http://dx.doi.org/10.1038/nrd3980.
[86] J.R. McCombs, S.C. Owen, Antibody drug conjugates: design and selection of linker, payload and conjugation chemistry, AAPS J. 17 (2) (2015) 339–351. Available from: http://dx.doi.org/10.1208/s12248-014-9710-8.
[87] N. Aggarwal, B.F. Sloane, Cathepsin B: multiple roles in cancer, Proteomics Clin. Appl. 8 (5–6) (2014) 427–437. Available from: http://dx.doi.org/10.1002/prca.201300105.

[88] Y. Li, Cathepsin B-cleavable doxorubicin prodrugs for targeted cancer therapy (review), Int. J. Oncol. 42 (2) (2012) 373–383. Available from: http://dx.doi.org/10.3892/ijo.2012.1754.

[89] C.S. Gondi, J.S. Rao, Cathepsin B as a cancer target, Expert Opin. Ther. Targets 17 (3) (2013) 281–291. Available from: http://dx.doi.org/10.1517/14728222.2013.740461.

[90] G.M. Dubowchik, R.A. Firestone, L. Padilla, D. Willner, S.J. Hofstead, K. Mosure, et al., Cathepsin B-labile dipeptide linkers for lysosomal release of doxorubicin from internalizing immunoconjugates: model studies of enzymatic drug release and antigen-specific in vitro anticancer activity, Bioconjug. Chem. 13 (4) (2002) 855–869. Available from: http://dx.doi.org/10.1021/bc025536j.

[91] G.M. Dubowchik, R.A. Firestone, Cathepsin B-sensitive dipeptide prodrugs. 1. A model study of structural requirements for efficient release of doxorubicin, Bioorg. Med. Chem. Lett. 8 (23) (1998) 3341–3346, doi:10.1016/S0960-894X(98)00609-X.

[92] P.J. Burke, P.D. Senter, D.W. Meyer, J.B. Miyamoto, M. Anderson, B.E. Toki, et al., Design, synthesis, and biological evaluation of antibody–drug conjugates comprised of potent camptothecin analogues, Bioconjug. Chem. 20 (6) (2009) 1242–1250. Available from: http://dx.doi.org/10.1021/bc9001097.

[93] S.O. Doronina, B.E. Toki, M.Y. Torgov, B.A. Mendelsohn, C.G. Cerveny, D.F. Chace, et al., Development of potent monoclonal antibody auristatin conjugates for cancer therapy, Nat. Biotechnol. 21 (7) (2003) 778–784. Available from: http://dx.doi.org/10.1038/nbt832.

[94] H. Burris, M. Saleh, J. Bendell, L. Hart, A. Rose, Z. Dong, et al., A phase (Ph) I/II study of CR011-VcMMAE, an antibody-drug conjugate, in patients (Pts) with locally advanced or metastatic breast cancer (MBC), Cancer Res. 69 (24 Supplement) (2009) 6096–7096. Available from: http://dx.doi.org/10.1158/0008-5472.SABCS-09-6096.

[95] J. Katz, J.E. Janik, A. Younes, Brentuximab vedotin (SGN-35), Clin. Cancer Res. 17 (20) (2011) 6428–6436. Available from: http://dx.doi.org/10.1158/1078-0432.CCR-11-0488.

[96] N.M. Tannir, A. Forero-Torres, R. Ramchandren, S.K. Pal, S.M. Ansell, J.R. Infante, et al., Phase I dose-escalation study of SGN-75 in patients with CD70-positive relapsed/refractory non-Hodgkin lymphoma or metastatic renal cell carcinoma, Invest. New Drugs 32 (6) (2014) 1246–1257. Available from: http://dx.doi.org/10.1007/s10637-014-0151-0.

[97] P.M. Challita-Eid, D. Satpayev, P. Yang, Z. An, K. Morrison, Y. Shostak, et al., Enfortumab vedotin antibody-drug conjugate targeting nectin-4 is a highly potent therapeutic agent in multiple preclinical cancer models, Cancer Res. 76 (10) (2016) 3003–3013. Available from: http://dx.doi.org/10.1158/0008-5472.CAN-15-1313.

[98] V.A. DiPippo, W.C. Olson, H.M. Nguyen, L.G. Brown, R.L. Vessella, E. Corey, Efficacy studies of an antibody-drug conjugate PSMA-ADC in patient-derived prostate cancer xenografts, Prostate 75 (3) (2015) 303–313. Available from: http://dx.doi.org/10.1002/pros.22916.

[99] J.A. Francisco, C.G. Cerveny, D.L. Meyer, B.J. Mixan, K. Klussman, D.F. Chace, et al., cAC10-vcMMAE, an anti-CD30-monomethyl auristatin E conjugate with potent and selective antitumor activity, Blood 102 (4) (2003) 1458–1465. Available from: http://dx.doi.org/10.1182/blood-2003-01-0039.

[100] M.C. Peitsch, J. Tschopp, Comparative molecular modelling of the Fas-ligand and other members of the TNF family, Mol. Immunol. 32 (10) (1995) 761–772, doi:10.1016/0161-5890(95)00016-8.

[101] T.N.J. Bullock, H. Yagita, Induction of CD70 on dendritic cells through CD40 or TLR stimulation contributes to the development of CD8 + T cell responses in the absence of CD4 + T cells, J. Immunol. 174 (2) (2005) 710–717.
[102] S.M.A. Lens, R. De Jong, B. Hooibrink, G. Koopman, S.T. Pals, M.H.J. van Oers, et al., Phenotype and function of human B cells expressing CD70 (CD27 ligand), Eur. J. Immunol. 26 (12) (1996) 2964–2971. Available from: http://dx.doi.org/10.1002/eji.1830261223.
[103] A.M. Orengo, C. Cantoni, F. Neglia, R. Biassoni, S. Ferrini, Reciprocal expression of CD70 and of its receptor, CD27, in human long term-activated T and natural killer (NK) cells: inverse regulation by cytokines and role in induction of cytotoxicity, Clin. Exp. Immunol. 107 (3) (1997) 608–613.
[104] J. Diegmann, K. Junker, B. Gerstmayer, A. Bosio, W. Hindermann, J. Rosenhahn, et al., Identification of CD70 as a diagnostic biomarker for clear cell renal cell carcinoma by gene expression profiling, real-time RT-PCR and immunohistochemistry, Eur. J. Cancer 41 (12) (2005) 1794–1801. Available from: http://dx.doi.org/10.1016/j.ejca.2005.05.005.
[105] S.M. Lens, P. Drillenburg, B.F. den Drijver, G. van Schijndel, S.T. Pals, R.A. van Lier, et al., Aberrant expression and reverse signalling of CD70 on malignant B cells, Br. J. Haematol. 106 (2) (1999) 491–503.
[106] J.A. McEarchern, E. Oflazoglu, L. Francisco, C.F. McDonagh, K.A. Gordon, I. Stone, et al., Engineered anti-CD70 antibody with multiple effector functions exhibits in vitro and in vivo antitumor activities, Blood 109 (3) (2006) 1185–1192. Available from: http://dx.doi.org/10.1182/blood-2006-07-034017.
[107] J.S. Rudra, Y.F. Tian, J.P. Jung, J.H. Collier, A self-assembling peptide acting as an immune adjuvant, Proc. Natl. Acad. Sci. U. S. A. 107 (2) (2010) 622–627. Available from: http://dx.doi.org/10.1073/pnas.0912124107.
[108] G.A. Hudalla, T. Sun, J.Z. Gasiorowski, H. Han, Y.F. Tian, A.S. Chong, et al., Gradated assembly of multiple proteins into supramolecular nanomaterials, Nat. Mater. 13 (8) (2014) 829–836. Available from: http://dx.doi.org/10.1038/nmat3998.
[109] T. Shimada, S. Lee, F.S. Bates, A. Hotta, M. Tirrell, Wormlike micelle formation in peptide-lipid conjugates driven by secondary structure transformation of the headgroups, J. Phys. Chem. B. 113 (42) (2009) 13711–13714. Available from: http://dx.doi.org/10.1021/jp901727q.
[110] A. Trent, B.D. Ulery, M.J. Black, J.C. Barrett, S. Liang, Y. Kostenko, et al., Peptide amphiphile micelles self-Aajuvant group A streptococcal vaccination, AAPS J. 17 (2) (2015) 380–388. Available from: http://dx.doi.org/10.1208/s12248-014-9707-3.
[111] F. Azmi, A.A. Ahmad Fuaad, A.K. Giddam, M.R. Batzloff, M.F. Good, M. Skwarczynski, et al., Self-adjuvanting vaccine against group A *Streptococcus*: application of fibrillized peptide and immunostimulatory lipid as adjuvant, Bioorg. Med. Chem. 22 (22) (2014) 6401–6408. Available from: http://dx.doi.org/10.1016/j.bmc.2014.09.042.
[112] F. Boato, R.M. Thomas, A. Ghasparian, A. Freund-Renard, K. Moehle, J.A. Robinson, Synthetic virus-like particles from self-assembling coiled-coil lipopeptides and their use in antigen display to the immune system, Angew. Chem. Int. Ed. 46 (47) (2007) 9015–9018. Available from: http://dx.doi.org/10.1002/anie.200702805.
[113] J.S. Rudra, T. Sun, K.C. Bird, M.D. Daniels, J.Z. Gasiorowski, A.S. Chong, et al., Modulating adaptive immune responses to peptide self-assemblies, ACS Nano 6 (2) (2012) 1557–1564. Available from: http://dx.doi.org/10.1021/nn204530r.

[114] R.R. Pompano, J. Chen, E.A. Verbus, H. Han, A. Fridman, T. McNeely, et al., Titrating T-cell epitopes within self-assembled vaccines optimizes CD4+ helper T cell and antibody outputs, Adv. Healthc. Mater. 3 (11) (2014) 1898–1908. Available from: http://dx.doi.org/10.1002/adhm.201400137.

[115] D.J. Irvine, M.A. Swartz, G.L. Szeto, Engineering synthetic vaccines using cues from natural immunity, Nat. Mater. 12 (11) (2013) 978–990. Available from: http://dx.doi.org/10.1038/nmat3775.

[116] J.S. Rudra, S. Mishra, A.S. Chong, R.A. Mitchell, E.H. Nardin, V. Nussenzweig, et al., Self-assembled peptide nanofibers raising durable antibody responses against a malaria epitope, Biomaterials 33 (27) (2012) 6476–6484. Available from: http://dx.doi.org/10.1016/j.biomaterials.2012.05.041.

[117] C.B. Chesson, E.J. Huelsmann, A.T. Lacek, F.J. Kohlhapp, M.F. Webb, A. Nabatiyan, et al., Antigenic peptide nanofibers elicit adjuvant-free CD8+ T cell responses, Vaccine 32 (10) (2014) 1174–1180. Available from: http://dx.doi.org/10.1016/j.vaccine.2013.11.047.

[118] J. Chen, R.R. Pompano, F.W. Santiago, L. Maillat, R. Sciammas, T. Sun, et al., The use of self-adjuvanting nanofiber vaccines to elicit high-affinity B cell responses to peptide antigens without inflammation, Biomaterials 34 (34) (2013) 8776–8785. Available from: http://dx.doi.org/10.1016/j.biomaterials.2013.07.063.

[119] A. Trent, R. Marullo, B. Lin, M. Black, M. Tirrell, Structural properties of soluble peptide amphiphile micelles, Soft Matter 7 (20) (2011) 9572–9582. Available from: http://dx.doi.org/10.1039/c1sm05862b.

[120] J.C. Barrett, B.D. Ulery, A. Trent, S. Liang, N.A. David, M.V. Tirrell, Modular peptide amphiphile micelles improving an antibody-mediated immune response to group A *Streptococcus*, ACS Biomater. Sci. Eng. 3 (2) (2016) 144–152. Available from: http://dx.doi.org/10.1021/acsbiomaterials.6b00422.

[121] M. Zaman, A.-B. Abdel-Aal, Y. Fujita, K.S.M. Phillipps, M.R. Batzloff, M.F. Good, et al., Immunological evaluation of lipopeptide group A *Streptococcus* (GAS) vaccine: structure–activity relationship, PLoS One 7 (1) (2012) e30146–e30147. Available from: http://dx.doi.org/10.1371/journal.pone.0030146.

[122] R. Noad, P. Roy, Virus-like particles as immunogens, Trends Microbiol. 11 (9) (2003) 438–444.

[123] L. Yang, W. Li, M. Kirberger, W. Liao, J. Ren, Design of nanomaterial based systems for novel vaccine development, Biomater. Sci. 4 (5) (2016) 785–802. Available from: http://dx.doi.org/10.1039/C5BM00507H.

[124] A. Ghasparian, T. Riedel, J. Koomullil, K. Moehle, C. Gorba, D.I. Svergun, et al., Engineered synthetic virus-like particles and their use in vaccine delivery, ChemBioChem 12 (1) (2010) 100–109. Available from: http://dx.doi.org/10.1002/cbic.201000536.

[125] N. Matougui, L. Boge, A.-C. Groo, A. Umerska, L. Ringstad, H. Bysell, et al., Lipid-based nanoformulations for peptide delivery, Int. J. Pharm. 502 (1–2) (2016) 80–97. Available from: http://dx.doi.org/10.1016/j.ijpharm.2016.02.019.

[126] R. Narayanaswamy, T. Wang, V.P. Torchilin, Improving peptide applications using nanotechnology, Curr. Top. Med. Chem. 16 (3) (2016) 253–270.

[127] J.J. Moon, H. Suh, A.V. Li, C.F. Ockenhouse, A. Yadava, D.J. Irvine, Enhancing humoral responses to a malaria antigen with nanoparticle vaccines that expand Tfh cells and promote germinal center induction, Proc. Natl. Acad. Sci. U. S. A. 109 (4) (2012) 1080–1085. Available from: http://dx.doi.org/10.1073/pnas.1112648109.

Peptide-based vaccines

12

Reshma J. Nevagi, Istvan Toth and Mariusz Skwarczynski
The University of Queensland, Brisbane, QLD, Australia

12.1 Types of vaccines

Since the discovery of the first vaccine by Edward Jenner in 1796, vaccination has been described as the most powerful approach to fight against infectious diseases and cancer [1]. During the following two centuries, many new vaccines have been developed against diseases such as tuberculosis, measles, typhoid fever, mumps, rubella, and influenza [2]. Although many more infectious diseases can potentially be treated or controlled by vaccination, vaccines against these pathogens have not yet been successfully introduced for clinical use.

Traditionally, vaccines consist of live attenuated or killed (inactivated) pathogens [3]. Live vaccines are usually highly immunogenic, do not require a boost, and can produce humoral and cellular immune responses. However, these vaccines are not stable and require storage at a consistent cold temperature—"cold chain." Attenuated pathogens can undergo secondary mutation and revert to their virulent form leading to severe complications, especially in the case of immunocompromised patients. Killed or inactivated vaccines are more stable and safer; but generate a relatively weaker immune response and require additional boosts to induce the desired protection. All of these vaccines also possess a questionable safety profile due to the potential risk of inducing autoimmunity or strong allergic reactions. The lack of technology to produce some of the pathogens in large quantities and problems associated with purification of biological materials limit the use of whole organisms in the development of vaccines. These limitations have led to the generation of highly purified subunit vaccines that contain the minimum part of the pathogen necessary to evoke an effective immune response [4].

Subunit vaccines can be composed of polysaccharides (e.g., meningococcal vaccines Mencevax ACWY and Menomune), proteins (e.g., acellular pertussis (aP), and hepatitis B vaccines), or their peptide fragments. These vaccines are safer because they lack any live components and therefore cannot cause disease. They can induce a controlled immune response and the risk of autoimmunity and allergic responses is greatly reduced.

12.2 Peptide-based vaccines

Although protein- and polysaccharide-based vaccines have been developed, further improvements in vaccine stability, selectivity, and reduction of any undesired immune response could be achieved using peptide-based vaccines. As the name

Peptide Applications in Biomedicine, Biotechnology and Bioengineering.
DOI: http://dx.doi.org/10.1016/B978-0-08-100736-5.00012-0

suggests, these vaccines are based on peptide epitopes, which are minimum components necessary for pathogen recognition [5]. The human immune system is equipped with pattern-recognition receptors (PRRs) to recognize and detect invading pathogens. These PRRs are expressed by antigen-presenting cells (APCs) such as dendritic cells (DCs) and macrophages [6]. As shown in Fig. 12.1, the immature

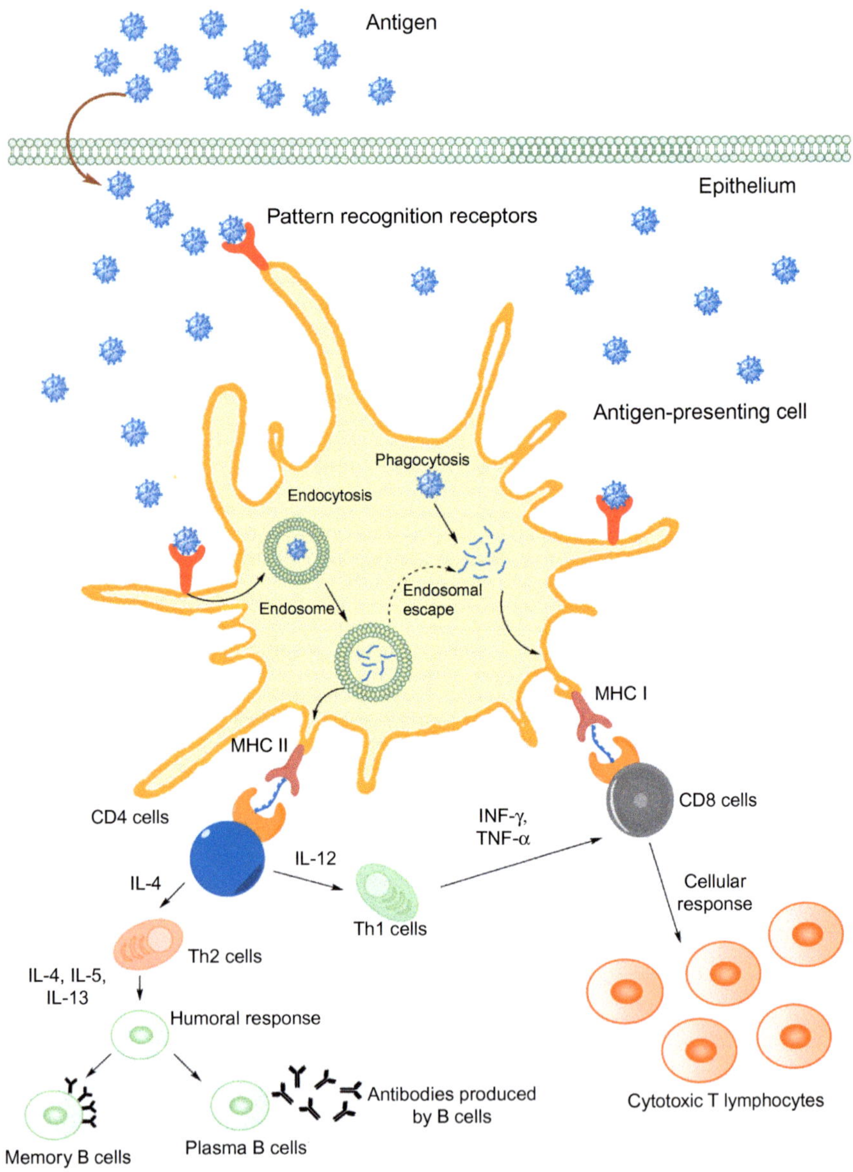

Figure 12.1 Antigen presentation, processing, and generation of an immune response.

DCs detect the entry of foreign antigens via various PRRs (mostly present on the DCs surface) and capture the pathogens either by phagocytosis or receptor-mediated endocytosis. During migration, APCs undergo maturation in which they digest endo/phagocytized pathogens and express the antigen in the form of peptide epitopes on their MHC complexes. These peptide–MHC complexes present the peptide antigen to naïve T cells in order to stimulate humoral or cellular immunity [7,8]. Thus, it is very clear that the immune system needs to process the pathogen or their fragments into small peptides to induce an immune response. Therefore, the use of peptide epitopes instead of whole protein or pathogens in vaccine development is a logical extension of the molecular mechanism behind the generation of an immune response. This approach allows exclusion of deleterious sequences from full-length antigens and other pathogen-derived molecules, and therefore reduces the risk of generating antibodies against ineffective epitopes that would lead to poor vaccine performance, or an allergic response.

Some other key features of peptide vaccines include:

1. Peptide-based vaccines are safer than classical vaccines because they exclude unnecessary biological material, reducing the potential to cause an unwanted host immune response, including allergic shock and autoimmunity [5].
2. They overcome immune evasion, when immunodominant epitopes on a protein trigger non-neutralizing and nonopsonic antibody responses, by directing humoral immunity against epitopes (cryptic epitopes), which are recognized only when the more immunogenic parts of the protein are absent.
3. Peptides can be obtained in highly pure form and are easily characterized using analytical techniques such as liquid chromatography and mass spectrometry.
4. They can be lyophilized and stored in a powder form, eliminating the need for cold chain during transport and storage.
5. The properties and size of the peptide can easily be controlled or tailored to generate the desired immune effect.
6. Peptides can be covalently conjugated to small molecules or a protein carrier to modulate their immunogenicity in a highly controllable manner.
7. Peptide-based vaccines can be designed to include multiple epitopes from one or more pathogens.
8. The advances in automated microwave peptide synthesis and purification have enabled large-scale manufacturing of peptide vaccines with lot-to-lot consistency. This also facilitates quality control and regulatory approval by the authorities.

12.3 Peptide-based vaccine development

The first step in the development of the peptide vaccines is the identification of cryptic peptide epitopes [9]. The second step may include the optimization of peptide epitope structure to achieve effective and long-term immunity against the pathogen. The final step focuses on improving the immunogenicity of peptide epitopes by supporting the epitope with adjuvants or enhancing its delivery by using carriers or delivery systems.

12.3.1 Epitope identification

The choice of epitope is a crucial factor when designing peptide-based vaccines [10]. Generally, B-cell epitopes are used to target vaccine against extracellular pathogens (humoral immunity) and CD8 T-cell epitopes are used against intracellular pathogens (cellular immunity). The approaches used to identify the epitope that binds to the antibody include computational docking, combinatorial approaches, mutagenesis, biophysical methods, and binding analyses. In the computational docking method, two separate crystal structures of antigen and antibody are produced and one structure is docked onto another using a series of computer algorithms [11]. Various databases are available to obtain a crystal structure of the specific antibody, however, generating a crystal structure for antigen of interest is a serious challenge. A general model of antigen structure can be produced either experimentally or by a computer model. Combinatorial methods include the use of phage display libraries of random peptides that are constructed using phage display vectors [12]. This approach is widely used to map B-cell epitopes. The monoclonal antibody (mAb) of interest is immobilized on the solid surface and reacted with a library of random peptides. Then, the affinity phages are collected and the epitope is identified by matching the peptide sequences of affinity phages to areas of the antigen against which the antibody was generated.

Another useful approach is the generation of a series of overlapping linear peptides that cover the entire sequence of the antigen of interest. The series of peptides is reacted with the corresponding antibody. The sequences of the peptide that binds to the antibody are determined using binding analyses such as ELISA, western blot, and surface plasmon resonance (SPR). The epitopes can also be identified in a protein using biophysical methods such as mass spectrometry [13], nuclear magnetic resonance (NMR), and X-ray crystallography using cocrystallized antigen–antibody complex [9].

12.3.2 Modification of a peptide epitope

The epitopes can be categorized into two types: linear and conformational. Linear epitopes are usually composed of sequential amino acids that participate in antibody binding and their interaction is based on the primary structure of the epitope. CD8 T-cell epitopes presented by MHC I molecules are typically linear peptides with length that ranges from 8 to 11 amino acids. The conformational epitopes are defined by their primary amino acid sequence as well as their secondary structure. Conformational epitopes should maintain native conformation in order to generate antibodies that are able to recognize parent protein. For example, a B-cell epitope derived from aspartic protease of *Necator americanus* (Na-APR-1) that did not display native β-sheet conformation failed to induce the production of APR-1 selective antibodies [14]. B-cell epitopes are usually 5–17 amino acids in length but can be extended without adverse effects on the immune function.

The conformational epitopes used for vaccine purposes are usually constrained minimal B-cell epitopes, which are either chemically modified or elongated significantly in chain length to maintain the native conformation. One technique used to lock the peptide into an α-helical conformation is stapling [5]. It is a process of introducing a chemical linker between distant side chains of two amino acids that lie on the same face of the peptide helix. This process reinforces the peptide into the desired α-helical conformation by inducing structural rigidity and it also protects the peptide from proteolytic degradation. Stapled peptides can be obtained by lactamization between Lys and Glu/Asp residues, cycloaddition reactions, hydrocarbon stapling, ring closing metathesis, thioether formation, a double-click approach, cysteine crosslinking, etc. [15]. Another method to induce desired peptide conformation is flanking the minimal epitopes with sequences that are known to promote the chosen conformation. For example, when the short random coiled minimal epitope (J_I8, SREAKKQVEKAL) was flanked with sequences derived from GCN4 yeast protein, it was able to form an α-helical structure (J8, QAEDKVKQ**SREAKKQVEKAL**KQLEDKVQ). The helical peptide induced the production of antibodies that bind to and opsonize group A *Streptococcus* (GAS) pathogens [16] while the random-coiled peptide did not.

Along with B-cell or CD8 T-cell epitopes, CD4 T-cell (T-helper, Th) epitopes are also needed to induce a universal long-term immune response in a diverse population. In some cases, the Th epitope may be present within the sequences of B-cell or CD8 T-cell epitopes [17,18]. However, the Th epitopes present in B-cell or CD8 T-cell epitopes sometimes may have deleterious effects. For example, the Th epitope present in the N-terminal region of the p145 sequence (GAS B-cell epitope, LRRDLDASREAKKQVEKALE), has the potential to stimulate antibodies that can crossreact with cardiac myosin [19]. Therefore, some vaccines need to incorporate a distinct Th epitope to induce the T-cell help necessary to provide an optimum B-cell antibody response [20]. For example, the carrier protein diphtheria toxoid (DT) can be conjugated to a B-cell epitope as a source of Th epitopes [21]. However, conjugating a large protein to a peptide antigen may induce an immune response against the carrier protein instead of against the incorporated epitope [22]. These problems led to the identification of universal promiscuous Th epitopes that can bind to many different MHC II molecules [23]. Compared to large protein carriers, these helper peptides do not stimulate a crossreactive response because they do not carry any B-cell or CD8 T-cell epitope in their sequence [24]. For example, universal Pan HLA-DR-binding epitope (PADRE) (AKFVAAWTLKAAA) was incorporated into several experimental vaccines to stimulate an antibody-mediated response against B-cell epitopes [25,26]. Moreover, this epitope showed high efficiency and safety in clinical trials [27].

12.3.3 Improving immunogenicity of peptide epitopes

12.3.3.1 Adjuvants (immunostimulants) for peptide vaccines

Since peptide epitopes do not possess the danger signals that are necessary to evoke an efficient immune response, the administration of adjuvants (immunostimulants) is required to induce an immune response against peptide antigens [28]. Adjuvants are nonimmunogenic agents that improve the potency and quality of an immune response against an antigen [29]. The word "adjuvant" is derived from the Latin word *adjuvare*, which means “to help.”

The first adjuvant approved by the US Food and Drug Administration (FDA) for human use was alum, and it has been the most common adjuvant used in licensed vaccines since its introduction in the 1930s [30]. It is widely used in prophylactic vaccines for meningococcal and pneumococcal infections, hepatitis A, hepatitis B, anthrax, diphtheria-tetanus-pertussis (DTaP, Tdap), *Haemophilus influenzae* type B, and cervical cancer. However, it can lead to local reactions such as erythema, subcutaneous nodules, contact hypersensitivity, and granulomatous inflammation. Generally, alum cannot induce cellular immunity and is too weak to stimulate an immune response against peptide epitopes.

In the 1990s, a hepatitis A vaccine that contained a virosome adjuvant system as an alternative to alum was approved. Virosomes are spherical vesicles made up of phospholipid layers that carry viral envelope proteins or inactivated strains of pathogen. In recent years, squalene-based emulsions such as MF59, AS03, and AF03 have also been included in licensed vaccines [31]. However, AF03, which was initially approved for use in influenza vaccine in Europe, was later withdrawn by the European Medical Agency. Unlike alum, the emulsions induce both humoral and cellular immune responses. In 2009, adjuvant system 04 (AS04), which combines the toll-like receptor (TLR)-4 agonist 3-O-desacyl-4′-monophosphoryl lipid A (MPL) and aluminum salt, was approved for human use [32]. In addition, a synthetic derivative of MPL, RC-529, was included as an adjuvant in a prophylactic hepatitis B vaccine approved in Argentina [33]. A montanide ISA51 adjuvanted cancer vaccine was recently registered in Chile and Cuba [34].

The development of effective and safe adjuvants is a delicate balance between adjuvanticity and safety. Despite extensive research to develop new adjuvants for vaccines, only a few adjuvants (Table 12.1) were approved for use in licensed vaccines. Moreover, these adjuvants were approved only for specific vaccines in a limited number of countries. This is in part due to the undesired activation of the host immune system and induction of inflammatory responses associated with many experimental adjuvants. Also, these approved adjuvants have predominantly been used for whole pathogen, protein-based vaccines; and information about their ability to adjuvant peptide-based vaccines is limited.

Including adjuvants such as complete Freund's adjuvant (CFA), incomplete Freund's adjuvant (IFA), and lipid A [47] with peptide-based vaccines has been widely studied in animal models. Despite their excellent immunostimulating properties, they are too toxic for human use. Hence, there is an urgent need to develop novel adjuvants or discover a new combination of adjuvants that are safe for use in peptide-based vaccines for human use.

Typically, adjuvants have the properties of pathogen-associated molecular patterns (PAMPs), which resemble the molecular and structural features of the pathogen and are recognized by PRRs (Fig. 12.1) [48,49]. Adjuvants are usually agonists of toll-like receptors (TLRs), nucleotide-binding oligomerization domain (NOD)-like receptors (NLRs), or C-type lectin receptors (CLRs). Different PRRs recognize different adjuvants and the selection of an appropriate adjuvant for peptide vaccine design is based on the required type of immunity to be induced.

Table 12.1 Adjuvants in use in approved vaccines

Adjuvant	Type	Vaccine marketed names	Disease	Company	Adjuvant composition	Location licensed	Types of vaccine	References
Alum	Mineral salt	Variety of vaccines available	DTaP/Tdap, hepatitis A and B, human papillomavirus (HPV), meningococcal and pneumococcal infections, etc.	–	Aluminum phosphate or aluminum hydroxide	Worldwide	Inactivated or killed, toxoid, subunit, conjugate vaccines	[30]
Virosome	Liposome	Epaxal	Hepatitis A	Crucell, Solvay	Inactivated strain of hepatitis A virus adsorbed to the surface of liposomes	~45 countries	Inactivated and protein-based vaccine	[35,36]
		Invivac, Inflexal	Influenza		Liposomes carrying viral envelope proteins			
MF59	Oil-in-water emulsion	Fluad, Aflunov, Focetria, Celtura	Influenza	Novartis	Squalene, Polysorbate 80, Span 85	Canada, Europe	Inactivated vaccine	[37]
AS04	Lipopoly-saccharide with alum salts	Fendrix	Hepatitis B	GSK	MPL and alum	USA	Recombinant DNA vaccine	[38,39]
		Cervarix	HPV infection				Protein subunit vaccine	

(*Continued*)

Table 12.1 (Continued)

Adjuvant	Type	Vaccine marketed names	Disease	Company	Adjuvant composition	Location licensed	Types of vaccine	References
RC-529	Synthetic MPL	Supervax	Hepatitis B	Dynavax	Modified MPL	Argentina	Recombinant DNA vaccine	[40]
AS01	Liposome	Mosquirix	Malaria	GSK	Liposomal adjuvant system containing MPL and QS-21	Europe	Recombinant protein vaccine	[41,42]
AS03	Oil-in-water emulsion	Prepandrix, Pandemrix, Arepanrix	Influenza	GSK	Squalene oil and α-tocopherol	Europe	Split virion, inactivated vaccine	[31,43]
AF03	Water-in-oil emulsion	Humenza	Influenza	Sanofi Pasteur	Squalene oil-based emulsion	Europe (later withdrawn by European Medical Agency)	Split virion, inactivated vaccine	[44]
ISA-51	Montanide	Cimavax	Non-small cell lung cancer (NSCLC)	Bioven	Mineral oil with a mannide monooleate emulsifier	Cuba and Chile	Protein-based vaccine	[45]
Recombinant CTB	Toxin subunit	Dukoral	Cholera	Crucell	B subunit of cholera toxin	Europe and Canada	Inactivated vaccine	[46]

Among a variety of chemical structures able to act as an adjuvant, peptidic adjuvants or immunostimulatory peptides have also been able to trigger humoral and cellular immunity against antigens. Muramyl dipeptide (MDP) is a synthetic adjuvant composed of *N*-acetylmuramic acid linked by its lactic acid moiety to the N-terminus of an L-alanine D-isoglutamine dipeptide. Derivatives of MDP such as adamantylamide dipeptide (AdDP), threonyl muramyl dipeptide (TMDP), and glucosaminyl muramyl dipeptide (GMDP) have been researched for their adjuvanting properties [50]. In addition, many potential peptide adjuvants have been examined in combination with the antigen of choice. For example, a 15-mer RN peptide (Table 12.2) was identified as a GMDP peptide mimic based on its interaction with anti-GMDP monoclonal antibodies in a phage display combinatorial library [62]. Other immunostimulatory peptides identified by phage display peptide library are WKYMVm, RS09, p3, p40, and p15 (Table 12.2). Recently, high-mobility group box 1 (HMGB1) protein-derived peptide, Hp91, was found to potentiate cellular immune responses to ovalbumin (OVA) protein-derived peptide and cellular and humoral immune responses to OVA protein in vivo [52]. Hp91 promoted the in vivo production of the immunomodulatory cytokines as well as antigen-specific activation of CD8 T cells. However, not all of the peptide-based adjuvants are recognized by PRRs. For example, Q11 is a short peptide that can self-assemble into a network of β-sheet nanofibers. When conjugated to Q11, peptide antigens are displayed on the surface of the fibrils [66]. Malaria peptide antigen (NANP)3 (NANPNANPNANP) conjugated to Q11 and self-assembled into nanofibers, generated a strong antibody response that could be detected for up to 40 weeks in C57BL/6 mice [69]. Similarly, a self-assembling peptide, h9e, formed a hydrogel in Ca^{2+} solution that has been used as an adjuvant [68]. The immunostimulatory peptides, along with their sequence, source, and mechanism of action are summarized in Table 12.2.

The method of incorporating an adjuvant within a peptide vaccine also plays a role in their potency to induce an immune response. The adjuvants can either be mixed with the peptide or chemically conjugated to the peptides. The hydrophobic and electrostatic interactions between the oppositely charged peptide and adjuvant can help to form complexes that in turn govern their interactions with receptors. Chemical conjugation ensures that the adjuvant and peptide antigen are both taken up by the same APCs, therefore a strong immune response can be induced and nonspecific stimulation of immune cells can be avoided [70]. However, conformational changes in the secondary structure of the peptide epitope during its modification or conjugation are areas that need careful consideration. The antigen and adjuvant can also be delivered together with the help of a delivery system; the importance of delivery systems is discussed in Section 12.4.

12.3.3.2 Peptide vaccine delivery

Route of administration

The immunization route used to deliver peptide vaccine can dramatically influence the magnitude and type of an immune response generated. Historically, vaccines were administered by the parenteral route. However, mucosal routes, such as the

Table 12.2 Immunostimulatory peptides

Peptide	Amino acid sequence (peptide source)	Mechanism of action	References
P28	KFLTTAKDKQRWEDPGKQLYQVEATSYA (complement protein C3d)	Acts through complement receptor 2 (CR2) binding	[51]
Hp91	DPNAPKRPPSAFFLFCSE (HMGB1 protein-derived peptide)	Induces secretion of IL-6 which is mediated through a MyD88/TLR4-dependent pathway involving p38 MAPK and NFκB	[52,53]
RS09	APPHALS (phage display peptide library)	Interacts with TLR-4 receptors	[54,55]
CEL-1000	DGQEEKAGVVSTGLIGGG (human MHC II β chain)	Binds to CD4 T-cell receptor and cell surface protein present on some DCs	[56,57]
WKYMVm	WKYMVm where m is D-methionine (phage display peptide library)	Acts through formyl peptide receptor (FPR) and FPRL1 to stimulate phagocytes. It possesses higher efficacy for FPRL1	[58]
EP54	YSFKPMPLaR (complement protein C5a)	C5a agonist	[59]
EP67	YSFKDMP(MeL)aR (complement protein C5a)	C5a agonist	[59]
Ii-Key	LRMK (immunoregulatory Ii protein)	Facilitates antigen binding to MHC II molecules and enhances Th activation	[60]
KLK	KLKLLLLLKLK (cationic antimicrobial peptide)	Enhances the association of antigen to APCs and forms a depot of antigen at the injection site	[61]
RN	RVPPRYHAKISPMVN (phage display peptide library)	Enhances expression of MHC II antigens on the surface of macrophages	[62]
p3	VWRLLAPPFSNRLLP (phage display peptide library)	Ganglioside GM1-binding	[63]

p40	NPPSPLSVSHRT (phage display peptide library)	Ganglioside GM1-binding	[64]
p15	RSTKPPLSPLG (phage display peptide library)	Ganglioside GM1-binding	[64]
TxA_{C314}	$TcdA_{2394-2706}$ (C-terminal region of *Clostridium difficile* toxin A)	Binds to epithelial cells in intestine and induces release of regulatory cytokines in DCs	[65]
Peptide-based self-adjuvanting delivery systems			
Q11	QQKFQFQFEQQ	Self-assembles into networks of nanofibers when conjugated to peptide antigen	[66,67]
h9e	FLIVIGSIIGPGGDGPGGD (combination of native sequences from an elastic segment of spider silk and transmembrane segment of human muscle L-type calcium channel)	The turning segment GSII of h9e promotes hydrogel formation with Ca^{2+} in solution	[68]

oral and intranasal routes offer many advantages, especially for peptide-based vaccines [71]. The typical primary site of entry for pathogens is through mucosal surfaces, therefore, targeting the immune cells on these surfaces is an added benefit. Mucosal administration of vaccine should induce the release of antigen-specific IgA antibodies, and this local immune response is required for the prevention of infection. Along with IgA, mucosal vaccination also induces systemic IgG responses, which lead to further protection against invading pathogens. Along with mucosal and systemic responses, mucosal immunization can stimulate CD8 T cells, which are important in combating intracellular pathogens. Furthermore, and importantly, mucosal vaccines are easier to administer compared to parenteral vaccines.

Oral and intranasal delivery is considered to be the best choice for vaccine development. The oral route is the most patient-friendly and perfect for self-administration. However, peptides are subjected to degradation due to proteolytic enzymes of the gastrointestinal tract (GIT) [72]. The peptide vaccines are often encapsulated within the delivery system to improve stability. The antigen dilution due to the large surface area of the GIT is also an obstacle to oral vaccine delivery. It leads to need for administration of a higher dose that ultimately increases the cost of production. Additionally, large doses of orally administered antigen may cause oral tolerance, leading to altered delayed-type hypersensitivity (DTH) reactions and decreased T-cell proliferation. Thus, the nasal route is often recommended over the oral route, because it bypasses the first-pass metabolism and usually does not induce tolerance. The nasal epithelium is also rich in immune cells [73]. Beneficially, the nasally administered peptide vaccine is stable because nasal mucosa is low on proteolytic enzymes. Absorption of the peptide antigen could be low if administered as a simple aqueous solution, hence formulation of a particulate delivery system is required to improve absorption and bioavailability of the peptide vaccine. The retention of a mucosally delivered peptide antigen formulation also depends on various factors such as the size, charge, shape, and surface properties, which are discussed in the next section.

Influence of size, charge, shape, and surface properties

Usually, to activate the immune system, a peptide-based vaccine is first recognized by PRRs and then taken up by APCs [74]. If delivered by nanosized carriers, the peptide vaccine can mimic the size of bacterial and viral pathogens and disseminate within body tissues and the lymphatics [75]. Moreover, such peptide nanovaccines are often able to induce an immune response without an additional adjuvant. The degree and mechanisms of their uptake by APCs are highly dependent on the size, shape, charge, and surface properties of peptide vaccine particles. Thus, the strength of the immune response usually depends on the size of peptide vaccine particles, with smaller nanoparticles of 10–50 nm often reported as preferable [76]. Smaller nanoparticles (<50 nm) can penetrate the barrier of the lymphatic system; however, very small particles (<10 nm) are reported to cause cytotoxicity [77–79]. Optimal particle size is also dependent on the vaccine administration route. In the case of intranasal administration, the mucosal clearance of smaller particles should be taken into consideration. Therefore, particle size is one parameter that needs serious consideration in peptide vaccine design to elicit the desired immune response with no toxicity [80].

Both surface charge and particle shape also play essential roles in the interaction between peptide vaccine particles and APCs [81]. Surface charge affects in vivo distribution and the APCs' uptake of peptide vaccine particles. Owing to the anionic nature of the cell membrane, cationic particles tend to enter cells more readily than anionic particles. For example, negatively (carboxymethyl chitosan nanoparticles, −46 mV), neutrally (chitosan-coated carboxymethyl chitosan nanoparticles, 0.5 mV), and positively (trimethyl chitosan (TMC) nanoparticles, +39 mV) charged nanoparticles were compared to evaluate the effect of surface charge on cellular uptake and intracellular trafficking [82]. All three nanoparticles possessed uniform size (~200 nm), as well as similar physicochemical properties including matrix, shape, and mechanical characteristics. The positively charged nanoparticles were taken up faster and in a larger amount. Considering intracellular trafficking of the same particles, negatively and neutrally charged nanoparticles were localized in lysosomes, whereas positively charged nanoparticles escaped from lysosomes and showed perinuclear localization. Similarly, many reports have shown that positively charged or negatively charged particles coated with cationic polymer were preferentially taken up by APCs, which in turn directly influence the generation of humoral and cellular immune responses [83,84]. In addition, spherical particles were more readily endocytosed compared to rod-shaped particles [85−87].

Surface properties are also important for interactions between APCs and peptide vaccine particles [88]. Cell membranes are usually made up of a lipid bilayer; therefore, increasing the surface hydrophobicity of peptide vaccine helps to intensify the physical interaction between the vaccine construct and cell membranes, thus allowing the peptide vaccine to enter the APC. In a study by Marasini et al., poly(D,L-lactic-co-glycolic) acid (PLGA)/dextran nanoparticles were coated with the cationic polymers TMC and polylysine. TMC-coated nanoparticles induced an efficient immune response, while polylysine-coated nanoparticles did not, despite bearing a similar positive charge [89,90]. This shows that the surface properties, not only the charge, of materials influence the effectiveness of peptide-based vaccines.

12.4 Importance of delivery systems in peptide-based vaccines

Biodegradable, biocompatible, nontoxic, and possibly mucoadhesive delivery systems would be an excellent choice to improve the efficacy and safety of peptide-based vaccines. These delivery systems would ideally serve as carriers that protect the peptide antigens from degradation, and synchronously present both peptide antigens and adjuvants to APCs. Modern delivery systems, such as liposomes, nano/microparticles, hydrogels, and immune-stimulating complexes (ISCOMs) (Fig. 12.2) have been extensively studied in peptide-based vaccines to achieve these goals. Since peptide antigens are poorly immunogenic, the use of delivery systems

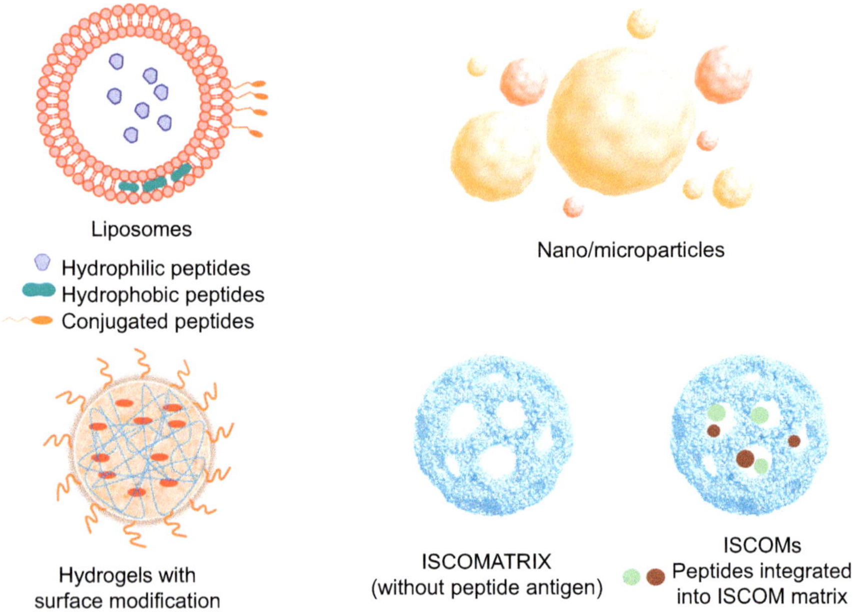

Figure 12.2 Overview of delivery systems for particulate peptide-based vaccines.

would have a big influence in boosting the immune response against peptide antigens, and this could possibly be achieved without external adjuvant because some delivery systems have self-adjuvanting properties. The importance of the most common peptide vaccine delivery systems is discussed below.

12.4.1 Liposomes

Liposomes are closed membrane vesicles made up of lipid bilayers and have been widely used for peptide vaccine delivery [91]. Liposomes have received extensive attention in peptide-based vaccine development due to their biocompatibility, biodegradability, good encapsulation efficiency, and excellent safety profile. The properties and size of the liposomes can be tailored by changing the lipid composition and method of manufacture, and they offer flexibility in terms of the properties of peptide/adjuvant to be loaded. Most hydrophilic peptides or adjuvants can be incorporated in the aqueous core of the liposomes, while lipophilic peptides can be intercalated between lipid layers or coupled onto the surface of the liposomes.

Peptides derived from OVA protein (found in egg white) are often used as model antigens to study the ability of a delivery system to help induce the desired immune response. One such example is OVA24 (DEVSGLEQLESIINFEKLAAAAAK), which contains amino acids 247–264 of the OVA protein. This peptide includes CD8 T-cell epitope OVA8 (SINFEKL, 257–264 amino acid sequences), which is an immunodominant MHC I epitope. When the OVA24 peptide was coencapsulated with poly I:C adjuvant within cationic liposomes (140 nm, +26 mV), the liposomes efficiently delivered the peptide to DCs and induced an OVA24-specific CTL immune response in mice more efficiently than by poly(I:C)-adjuvanted soluble OVA24 [92]. Another example of an OVA-derived peptide is the MHC II peptide epitope OVA17 (ISQAVHAAHAEINEAGR), which contains amino acids 323–339 of the OVA protein and includes the Th epitope AAHAEINEA. OVA17 and OVA24 were encapsulated into TLR ligand-adjuvanted cationic liposomes and evaluated for cancer immunotherapy [17]. The liposomal formulations enhanced APC uptake and induced activation of OVA17-specific T cells better than that of the emulsion-based formulations Montanide ISA-51 and squalene.

Coupling peptide antigens to the surface of the liposomes has become a popular strategy in vaccine delivery. Uchida et al. initially investigated the effectiveness of the conjugates of OVA8 peptide with oleayl liposomes and reported the complete eradication of tumors in 12 days in E.G7 tumor-bearing mice [93]. Also, $NP_{366-374}$ (ASNENMDAM), a nucleoprotein peptide derived from H3N2 influenza virus, conjugated to an oleoyl liposome in the presence of CpG ODNs, also protected mice against influenza infection [94].

Liposomes can be targeted to specific APC subsets by covalently linking or anchoring targeting moieties (e.g., mannose) to their outer membranes. Influenza hemagglutinin (HA)-derived promiscuous Th epitope HA307–319 (PKYVKQNTLKLAT) and ErbB2 protein-derived CD8 T-cell epitope p63–71 (TYLPTNASL), were anchored onto a mannosylated liposomal surface. A TLR2/1 agonist (Pam3CAG) or TLR2/6 agonist (Pam2CAG\Pam2CGD) was separately embedded into the vesicles [95]. These vaccine constructs were evaluated for the generation of specific antitumor cellular immunity in mice that bore renal tumors that overexpressed the ErbB2 protein. Mannose-targeted liposomes displayed higher therapeutic efficiency against tumors, allowing treatment with decreased quantities of peptide epitopes.

The incorporation of self-adjuvanting lipopeptides (immunostimulatory lipids attached to a peptide) into liposomes can generate long lasting immunity. Ghaffar et al. produced liposomes that anchored a lipopeptide that bore the B-cell epitope J14 (KQAEDKVKASREAKKQVEKALEQLEDKVK) from GAS M-protein, universal Th epitope p25 (GKLIPNASLIENCTKAEL), and two lipidic moieties [96]. This lipopeptide vaccine induced high titers of IgA and IgG, even 5 months after the last intranasal immunization. This lipopeptide was further incorporated into liposomes layered with oppositely charged polymers (alginate-TMC) and evaluated for oral immunization in mice against GAS infection. High levels of systemic antibody titers were observed, even 185 days after the primary immunization [97]. The most successful application of liposomes in lipopeptide vaccine delivery is Tecemotide (formerly known as Stimuvax or BLP25) which reached phase III trials [98]. Tecemotide is a synthetic palmitoylated (Pal) mucin I peptide (STAPPAHGVTSAPDTRPAPGSTAPP-K(Pal)G) that was used as an antigen in an investigational therapeutic anticancer vaccine along with MPL [99].

12.4.2 Nano/microparticles

Nano/microparticles have gained tremendous attention in the delivery of peptide-based vaccines [100]. It is reported that stronger immunity is developed when the peptide epitope is delivered in the form of nano/microparticles, compared with a physical mixture or soluble peptide [76]. In these systems, peptide epitopes are either encapsulated within a shell or conjugated on the surface of the nano/microparticles [101]. Microparticles can act as a depot to improve peptide vaccine efficacy through prolonged presentation and gradual release of peptide antigens at the administration site, thus providing more effective uptake and delivery to lymph nodes by peripheral DCs [102]. In contrast, nanoparticles can efficiently reach the lymphatic system and travel to lymph nodes on their own to deliver the peptide vaccine. Nanoparticles also serve as a vehicle for mucosal immunization, protect encapsulated peptide antigen from proteolytic enzymes when delivered orally, and promote M-cell uptake of peptide in nasal-associated lymphoid tissue (NALT) on nasal administration.

Different types of materials are used to formulate nano/microparticles, depending on properties of peptide antigen. Many polymeric nano/microparticles are known for their biocompatible and nontoxic properties [103]. A variety of natural and synthetic polymers are available, which can be used for particulate peptide vaccine delivery. Mucoadhesive polymers are preferred in the formulation of nano/microparticles for peptide vaccines aiming for mucosal immunity; however, very small and very large particles are usually not retained on mucosal layers long enough to be effective [104].

The two most extensively used polymers in the preparation of particulate peptide vaccines are chitosan and PLGA [105,106]. For example, MAGE-3 peptide (YEEYYPLIFLDNDQETMETSEEEEYEEYYPLIF), was incorporated into chitosan–deoxycholic acid conjugate nanoparticles (30–50 nm) and evaluated for antitumor effect [107]. This peptide vaccine generated an antigen-specific CTL immune response and was able to kill tumor cells. Hamdy et al. encapsulated a melanocyte differentiation antigen tyrosinase-related protein 2 (TRP2)-derived $TRP2_{180-188}$ peptide (SVYDFFVWL) and 7-acyl lipid A adjuvant into PLGA nanoparticles and evaluated the efficacy of these nanoparticles as a cancer vaccine [108]. Mice treated with $TRP2_{180-188}$/lipid A nanoparticles showed an immunostimulatory milieu at the tumor microenvironment compared to mice treated with nanoparticles that contained only peptide. Other polymers, such as alginate and dextran, are mostly used in combination with other oppositely charged polymers to form coating layers to prevent degradation of peptide antigen before internalization into APCs, facilitating the release of antigenic cargo at the site of action. For example, dextran nanoparticles (145 nm, −38 mV) that bore lipopeptides comprised of GAS B-cell epitope J14 and T-helper P25 epitope were coated with TMC (303 nm, +38 mV), PLGA (193 nm, −31 mV), and PLGA-TMC (186 nm, +10 mV) [90]. Of these, lipopeptides enclosed in dextran/TMC-based nanoparticles induced maturation of DCs in the most effective manner, and produced opsonic antibodies against clinical GAS isolates. While DC-uptake was not related to the charge of the nanoparticles, antibody responses were highly correlated with strong positive charge.

Polystyrene nanobeads or nanoparticles have also been widely explored as a peptide vaccine delivery system because polystyrene is readily available and nanoparticles can be synthesized in a wide size range. OVA8 peptide was conjugated to 0.02, 0.04, 0.1, 0.5, 1.0, or 2.0 μm sized polystyrene beads, and 0.04 μm nanoparticles generated significantly stronger antibody and CTL responses, compared with other sized nanobeads and conventional adjuvants such as alum, MPL, and Quil A [109]. Furthermore, OVA8 peptide was conjugated to narrow-sized beads of 20, 40, 49, 67, 93, 101, and 123 nm, and it was found that IFN-γ induction from CD8 T cells was achieved by 40–49 nm beads, whereas CD4 T-cell activation and IL-4 induction were limited to 93–123 nm OVA8-beads [110]. Wilson et al. conjugated an immunodominant CD8 T-cell peptide epitope from circumsporozoite protein (CSP) of *Plasmodium berghei*, SYIPSAEKI (referred to as KI) to 50 nm polystyrene nanoparticles, and suggested that chemical conjugation of minimal epitope to polystyrene nanoparticles is a requirement for in vivo delivery of the peptide and induction of a CTL response [111]. However, polystyrene is nonbiodegradable, which poses a serious problem with the use of polystyrene nanoparticles for vaccine delivery [112].

Controlled-release nanoparticles are of interest in peptide-based vaccines because the release of peptide antigen over a period of time may increase the period of antigen interaction with the immune cells, and thus lead to the induction of a more effective immune response. These nanoparticles can be tailored to degrade over variable rates and can serve as a booster, eliminating the need for a prime-boost immunization scheme. However, in the case of cancer vaccines, persistent release of vaccine antigen and adjuvant at the administration site may lead to localization of CD8 T cells at the vaccination site rather than accumulating at the tumor [113]. A single dose of oral hepatitis B vaccine prepared by incorporating B-cell peptide epitope (CTTPAQGNSMFPSCCCTKPTDGNCY) from hepatitis B surface antigen (HBsAg) into PLGA microparticles (10 μm) released the peptide in sustained form and produced significant IgG and IgM antibodies [114]. Mice also showed rapid production of antibodies against HBsAg in a challenge study.

Self-assembled nanoparticles are a versatile tool for peptide-based vaccine delivery. Amphiphilic materials are known to self-assemble into nanoparticles, forming a hydrophobic core and hydrophilic surface. For example, branched hydrophobic synthetic polyacrylates possess the ability to self-assemble into nanoparticles when hydrophilic peptide epitopes are attached to their periphery. Using this approach, Skwarczynski et al. synthesized polyacrylate–J14 peptide conjugates and self-assembled these conjugates into 20 nm nanoparticles [115]. Mice injected subcutaneously with these nanoparticles produced high titers of J14-specific antibodies, equal to those stimulated by the J14 peptide emulsified with powerful, but toxic, CFA. Furthermore, when polyacrylate–J14 peptide conjugates were self-assembled into small (~20 nm) and large (~500 nm) nanoparticles, it was found that the smaller-sized nanoparticles produced significantly stronger immune response, compared to the larger ones [116]. Similarly, four copies of T-cell epitopes derived from HPV-16 E7 protein were conjugated to four-arm star polyacrylate polymer and self-assembled to form microparticles ($\sim 10\ \mu m$) which were evaluated for the generation of cellular immunity [18]. The conjugate efficiently activated CD4 and CD8 T cells and eradicated tumor in a mouse challenge study [117]. Among the examined branched and linear polymeric structures, E7 epitope conjugated to the most branched dendritic polyacrylate was superior, even when compared with peptide epitope emulsified with commercial ASA51 adjuvant [118,119].

Gold nanoparticles are attractive nanocarriers because they are inert, nontoxic, and their particle sizes can be well controlled (generally ranging from 1 to 100 nm in diameter) [120]. Gold particles are readily endocytosed by DCs and other phagocytic mononuclear cells. The thiol group possesses a high binding affinity for the gold surface; therefore, the peptides can be directly conjugated onto the gold surface via thiol linkers. For example, gold nanoparticles (80 nm) were coated by a self-assembled monolayer of carboxyl-polyethylene glycol (PEG)-thiol, and MHC I peptides, such as OVA or gp100 (KVPRNQDWL) or $TRP2_{180-188}$, were subsequently conjugated on the surface of the nanoparticles [121]. Gp100 is a shorter 25–33-amino-acid fragment of human melanoma antigen gp100 that is recognized by T cells. The peptide–gold nanoparticle complexes were readily taken up by DCs and were far more effective at stimulating CTL immune response compared to free peptides. In a study by Zhang et al., gold nanoparticles were used as templates to build polyelectrolyte multilayers of oppositely charged adjuvants and antigens [122]. Polyanionic TLR agonist poly I:C and cationic OVA8-R_9 (OVA8 modified with nine arginine residues) were alternatively deposited on a gold nanoparticle template. The bilayered gold nanoparticles (~ 90 nm) induced high levels of antigen-specific CD8 T cells 1 week after last immunization in mice, compared to a mixture of peptide antigen and adjuvant.

In contrast to some polymers and gold carriers, lipid-based nanoparticles are biodegradable. Lipopeptides are amphiphiles, which possess the ability to self-assemble into nanoparticles in an aqueous environment [123]. The size that the nanoparticles form depends on the nature of lipidic moieties and peptides. For example, lipopeptides that bore self-adjuvanting units built of two lipoamino acids (with 16 carbon atoms) were conjugated to a variety of GAS peptide epitopes [124]. The size of the formed particles was inversely proportional to the hydrophilicity of the attached peptides, and varied from completely soluble material, to small nanoparticles, all the way to micrometer-sized agglomerates. This shows that the balance between hydrophilicity and hydrophobicity can be used to control the size of self-assembling nanoparticles. Following this study, similar lipopeptides that bore GAS epitopes were synthesized and self-assembled into 10 and 100 nm particles [125]. The smaller nanoparticles induced significantly higher IgG titers than the larger ones, demonstrating once again that size plays an important role in nanoparticle immunogenicity. In another study, lipopeptides that incorporated GAS B-cell epitope J14 and T-helper P25 epitope, were synthesized by altering the spatial arrangement (site of conjugation) and length of lipoamino acids conjugated to the antigen [126]. These lipopeptides were self-assembled into nanoparticles (7–25 nm) and the longer lipid-based nanoparticles generated the highest antibody responses, even 5 months after primary immunization, while the arrangement of peptide epitopes and lipids in the conjugate did not significantly influence the immune response.

12.4.3 Hydrogels

A hydrogel is a type of gel that swells in water, and is made up of a three-dimensional crosslinked configuration of amphiphilic or hydrophilic polyionic polymers. Hydrogel-forming vesicles in the nanometer size (1–1000 nm) are often termed as nanogels, or hydrogel nanoparticles [127]. Hydrogels could be designed to show phase transition (and therefore cargo release) in response to temperature, pH, molecular species, and ionic strength. Nanogels represent an excellent platform for the protection of peptides against degradation and their delivery to APCs to stimulate the desired immune response. For example, OVA8 and OVA17 encapsulated within polymethacrylate hydrogel capsules were internalized within DCs and induced an improved proliferation of OVA-specific CD4 and CD8 T cells in both in vitro and in vivo studies, compared to OVA alone [128]. Cancer gp-100 peptide-loaded nanogels (68 nm) were used to immunize mice via the transdermal route using the iontophoresis technique. The peptide-loaded nanogels were accumulated in the epidermis and suppressed tumor growth, whereas peptide alone was not effective [129].

Bioreducible hydrogels trigger the release of peptide antigen in intracellular reductive environments. Reduction-sensitive PEG hydrogels co-conjugated with OVA8 and TLR9 agonist CpG ODNs were prepared for intracellular delivery of vaccine components [130]. Bone-marrow-derived DCs treated with OVA8-conjugated nanogel (260–270 nm) efficiently crosspresented the OVA8 peptide via MHC I surface receptor and induced the proliferation of CD8 T cells in OT-I mice. CpG ODN-conjugated PEG nanogel (240–290 nm) led to the maturation of BMDCs, as shown by the overexpression of CD80 and CD40 costimulatory receptors. Moreover, OVA8/CpG ODN-conjugated nanogel significantly increased the frequency of IFN-γ producing CD8 effector T cells in mice, compared to soluble peptide adjuvanted with CpG ODN.

12.4.4 Immune-stimulating complexes (ISCOMs)

ISCOMs are spherical, hollow, cage-like self-assembled particles (40–60 nm) consisting of saponin Quil A adjuvant, antigen, phospholipids, and cholesterol [131]. The particle without antigen is known as an ISCOM matrix. Quil A possesses an affinity for cholesterol which induces stability in the ISCOM matrix [132]. ISCOMs allow multimeric representation of the incorporated peptide antigen and can be very stable during lyophilization, depending on the stability of the peptide antigen. ISCOMs are designed to carry peptide antigen and adjuvant to APCs and the lymphatic system. Normally, hydrophobic molecules can be only loaded within these complexes. ISCOMs can also be used to deliver hydrophilic peptide antigens, provided that the peptides are conjugated to a hydrophobic payload [133]. Anionic peptide antigens can also be loaded onto cationic ISCOMs through electrostatic interactions. On immunization with ISCOMs containing various antigenic peptides, such as vesicular stomatitis virus nucleoprotein 52−59 (RGYVYQGL), listeriolysin peptide 91−99 (GYKDGNEYI), cytomegalovirus pp89 168−176 (YPHFMPTNL), and OVA8, it was found that the CTL responses generated were peptide-specific and Quil-A was necessary for this activity [134]. ISCOMs have also been used to deliver the mutant V3 loop peptide of human immunodeficiency virus (HIV)-1 (CKRKIHIGPGQAFYT) [135] and nonameric peptides I ($VP1_{26-34}$, RRQHTDVSF), II ($VP1_{157-165}$, RTLPTSFNY), and III ($VP1_{45-53}$, KEQVNVLDL) from the VP1 capsid protein of FMDV, among others [136].

12.5 Current status, unmet needs, and future directions

Though there is tremendous interest and investment in developing a peptide-based vaccine, no such vaccine is currently on the market. The development of safe and effective peptide vaccines needs to address a number of issues. Most importantly, the identification of minimal peptide epitope possessing no detrimental effect is the foremost criteria towards the successful design of peptide-based vaccine. The maintenance of the desired conformation of peptide during formulation and delivery is also vital in evoking an optimal humoral immune response. Multiantigenic peptide vaccines have significant advantages over single epitope-based constructs; however, the combination of various B-cell and T-cell peptide epitopes into single vaccines can biasedly produce immune response towards a dominant epitope.

Another challenge in peptide-based vaccines is sourcing safe and effective adjuvants. Only a few adjuvants, other than traditional alum, were introduced into the market; however, they were only licensed regionally, and for specific vaccines, and their mechanisms of action are not fully understood. Even in the case of alum, multiple adjuvanting mechanisms were reported, and the picture is still not complete, despite decades of research. Without such knowledge, it is difficult to develop a safe vaccine and predict the potential adverse effects associated with immunization.

Similar to adjuvant-related complications, obstacles are still present in the development of an efficient peptide vaccine delivery systems. Various delivery systems, such as liposomes, nano/microparticles, hydrogels, ISCOMs, etc., have been investigated; however, the exact pathway of cellular transportation and uptake of these nanocarriers are not completely understood. Also, appropriate attention needs to be given to peptide loading during formulation. If the peptide antigen is on the surface of the carrier, then degradation of the antigen and premature release are common areas of concern. If the antigen is within the core, then the effective release of cargo and the prevention of antigen leakage prior to reaching APCs, needs careful consideration. Also, the stability of the particulate system, especially when formulated by complexation through electrostatic interaction, may be questionable. The selection of materials for formulating delivery systems is highly dependent on the route of administration. Batch-to-batch consistency and reproducibility could be a major constraint for polymer−peptide

Table 12.3 **Examples of peptide vaccines that have reached clinical trials**

Peptide name	Adjuvant/ carrier	Peptide sequence	Therapeutic indication	References
J8	DT	QAEDKVKQSREAKKQVEKALKQLEDKVQ	GAS infections	[137]
$(T_1B)_4$ MAP	Alum/QS-21	B-cell epitope $(NANP)_3$ and Th epitope $(DPNANPNV)_2$	Malaria	[138]
MSP3 SLP	Aluminum hydroxide	RKTKEYAEKAKNAYEKAKNAYQKANQA VLKAKEASSYDYILGWEFGGGVPEHKK EENMLSHLYVSSKDKENISKENDDVLD EKEEEAEETEEEELE	Malaria	[139]
AFO-18	CAF01	15 CD8 cell peptides, 2 Th epitopes (KRWIILGLNKIVRMY and VWGIKQLQARVLAVERYLKD) and PADRE	HIV	[140]
Peptides derived from B-cell epitopes of HA	Ii-key	Peptide sequences are not revealed	Influenza	[141]
HerpV	QS-21	32 synthetic 35mer peptides noncovalently linked to heat shock protein	Herpes simplex virus	[142]
E75, HER2/neu peptide	GM-CSF	KIFGSLAFL	Breast cancer	[143]
E6/E7 peptides	Pegylated INF-α	13 synthetic peptides (25–35 amino acids long) derived from the E6 and E7 proteins of HPV 16	Cervical cancer	[144]
CDCA1, CDH3 and KIF-20A	ISA51	VYGIRLEHF, DYLNEWGSRF, KVYLRVRPLL	Advanced biliary tract cancer	[145]
SurVaxM (SVN53–67/ M57-KLH)	ISA 51	DLAQCFFMFKELEGW	Recurrent malignant glioma	[146]

(Continued)

Table 12.3 (Continued)

Peptide name	Adjuvant/ carrier	Peptide sequence	Therapeutic indication	References
VEGFR1-770/1084 peptides	ISA-51VG	TLFWLLLTL (VEGFR1 770), SYGVLLWEI (VEGFR1 1084)	Renal cell carcinoma	[147]
WT1 Peptides	GM-CSF or CpG-ODN	CYTWNQMNL or SLGEQQYSV	Solid malignancy	[148]
R3 peptide	ISA-51 and GM-CSF	ILSLELMKL	Acute myeloid leukemia, myelodysplastic syndrome, multiple myeloma	[149]
gp100:209–217 (210M)	ISA-51 or interleukin-2	IMDQVPFSV	Metastatic melanoma	[150]
GV1001	GM-CSF	EARPALLTSRLRFIPK	Locally advanced and/or metastatic pancreatic cancer	[151]
Tecemotide	Liposome and MPL	(STAPPAHGVTSAPDTRPAPGSTAPP-K(Pal)G)	NSCLC	[99]
Affitope AD02	–	6 amino acid sequence of B-cell epitope of Aβ42	Alzheimer's disease	https://clinicaltrials.gov

CAF01, cationic adjuvant formulation consisting of trehalose 6,6-dibehenate (TDB) incorporated into dimethyldioctadecyl-ammonium (DDA) liposomes; *DT*, diphtheria toxoid; *GM-CSF*, granulocyte-macrophage colony-stimulating factor; *IFN*, interferon-α; *KLH*, keyhole limpet hemocyanin; *MAP*, multiple antigen peptide; *QS-21*, a saponin adjuvant; *SLP*, synthetic long peptide; *SurVaxM*, survivin peptide mimic; *VEGFR1*, vascular endothelial growth factor receptor 1; *WT1*, Wilms' tumor 1.

conjugate vaccines. The failure to obtain a consistent substitution ratio in peptide–polymer conjugates may have unpredictable effects on the generation of an immune response.

Although there are many areas that demand further investigation, advances in organic and medicinal chemistry, biotechnology, and immunology have enabled the progression of peptide-based vaccines from the laboratory bench to clinical trials. Therapeutic cancer vaccines often use specific peptides that can mark a tumor as dangerous and further activate CD8 T cells to kill the original cancer cells. Progress in reverse immunology and gene expression has enabled the identification of tumor-associated specific antigenic peptides to develop therapeutic cancer vaccines for many types of cancers. There are currently several peptide-based prophylactic and therapeutic vaccines under investigation in clinical trials and, hopefully, one will make it to the clinic soon. Table 12.3 lists some of the peptides that have reached clinical trials.

12.6 Concluding remarks

Traditional whole organism vaccines have been approved and used in large quantities for many decades. However, the safety of these vaccines is questionable, due to the limited purification of biological material and their ability to induce an undesired immune response. As such, different novel strategies have been investigated for vaccine production in recent years. Indeed, subunit protein-based vaccines have started to dominate the market. Furthermore, interest in the development of peptide-based vaccines has been growing rapidly over the last decades. Peptide-based vaccines are inexpensive, highly defined and safer, due to the elimination of redundant components that can cause adverse effects. However, these peptides are not highly immunogenic and this makes the development of peptide vaccines challenging. The use of the right adjuvant, antigen, and delivery system as a single vesicle may solve these problems. Again, the safety of the adjuvant and the identification of an optimal delivery system, preferably delivered by a single immunization, is a real challenge in peptide-based vaccine development. Therefore, an advanced understanding of the immune system and the smart design of a particulate nanovaccine that incorporates one or more peptide epitopes and adjuvants would definitely benefit peptide vaccine technology.

References

[1] S. Riedel, Edward Jenner and the history of smallpox and vaccination, Proc. (Bayl. Univ. Med. Cent.) 18 (2005) 21–25.

[2] S.A. Plotkin, History of Vaccine Development, Springer, New York, 2011.

[3] R.S. Kallerup, C. Foged, Classification of vaccines, in: C. Foged, T. Rades, Y. Perrie, S. Hook (Eds.), Subunit Vaccine Delivery, Springer, New York, 2015.

[4] J. Maurice, S. Davey, State of the World's Vaccines and Immunization, World Health Organization, UNICEF, World Bank, Geneva, Switzerland, 2009.
[5] M. Skwarczynski, I. Toth, Peptide-based synthetic vaccines, Chem. Sci. 7 (2016) 842–854.
[6] M.-L. De Temmerman, J. Rejman, J. Demeester, D.J. Irvine, B. Gander, S.C. De Smedt, Particulate vaccines: on the quest for optimal delivery and immune response, Drug Discov. Today 16 (2011) 569–582.
[7] B. Spellberg, J.E. Edwards, Type 1/type 2 immunity in infectious diseases, Clin. Infect. Dis. 32 (2001) 76–102.
[8] J. Neefjes, M.L. Jongsma, P. Paul, O. Bakke, Towards a systems understanding of MHC class I and MHC class II antigen presentation, Nat. Rev. Immunol. 11 (2011) 823–836.
[9] J.M. Gershoni, A. Roitburd-Berman, D.D. Siman-Tov, N.T. Freund, Y. Weiss, Epitope mapping, BioDrugs 21 (2007) 145–156.
[10] A. Sette, J. Fikes, Epitope-based vaccines: an update on epitope identification, vaccine design and delivery, Curr. Opin. Immunol. 15 (2003) 461–470.
[11] M.N. Davies, D.R. Flower, Computational epitope mapping, in: V. Sintchenko (Ed.), Infectious Disease Informatics, Springer, New York, 2010.
[12] L.-F. Wang, M. Yu, Epitope identification and discovery using phage display libraries: applications in vaccine development and diagnostics, Curr. Drug Targets 5 (2004) 1–15.
[13] C. Hager-Braun, K.B. Tomer, Determination of protein-derived epitopes by mass spectrometry, Expert Rev. Proteomics 2 (2005) 745–756.
[14] A.A. Fuaad, M.S. Pearson, D.A. Pickering, L. Becker, G. Zhao, A.C. Loukas, et al., Lipopeptide nanoparticles: development of vaccines against hookworm parasite, ChemMedChem 10 (2015) 1647–1654.
[15] Y.H. Lau, P. De Andrade, Y. Wu, D.R. Spring, Peptide stapling techniques based on different macrocyclisation chemistries, Chem. Soc. Rev. 44 (2015) 91–102.
[16] W.A. Hayman, E.R. Brandt, W.A. Relf, J. Cooper, A. Saul, M.F. Good, Mapping the minimal murine T cell and B cell epitopes within a peptide vaccine candidate from the conserved region of the M protein of group A *Streptococcus*, Int. Immunol. 9 (1997) 1723–1733.
[17] E.M. Varypataki, A.L. Silva, C. Barnier-Quer, N. Collin, F. Ossendorp, W. Jiskoot, Synthetic long peptide-based vaccine formulations for induction of cell mediated immunity: a comparative study of cationic liposomes and PLGA nanoparticles, J. Control. Release 226 (2016) 98–106.
[18] T.-Y. Liu, W.M. Hussein, Z. Jia, Z.M. Ziora, N.A. McMillan, M.J. Monteiro, et al., Self-adjuvanting polymer–peptide conjugates as therapeutic vaccine candidates against cervical cancer, Biomacromolecules 14 (2013) 2798–2806.
[19] S. Pruksakorn, B. Currie, E. Brandt, C. Phornphutkul, S. Hunsakunachal, A. Manmontri, et al., Identification of T cell autoepitopes that cross-react with the C-terminal segment of the M protein of group A streptococci, Int. Immunol. 6 (1994) 1235–1244.
[20] G. Sharma, R.A. Holt, T-cell epitope discovery technologies, Hum. Immunol. 75 (2014) 514–519.
[21] M.R. Batzloff, W.A. Hayman, M.R. Davies, M. Zeng, S. Pruksakorn, E.R. Brandt, et al., Protection against group A *Streptococcus* by immunization with J8-diphtheria toxoid: contribution of J8-and diphtheria toxoid-specific antibodies to protection, J. Infect. Dis. 187 (2003) 1598–1608.
[22] A.R. Lussow, M.T. Aguado, G. Del Giudice, P.-H. Lambert, Towards vaccine optimisation, Immunol. Lett. 25 (1990) 255–263.

[23] D.S. Rosa, S.P. Ribeiro, E. Cunha-Neto, CD4 + T cell epitope discovery and rational vaccine design, Arch. Immunol. Ther. Exp. (Warsz). 58 (2010) 121–130.
[24] K.S. Rosenthal, Immune peptide enhancement of peptide-based vaccines, Front. Biosci. 10 (2005) 478–482.
[25] C.C. Wilson, M.J. Newman, B.D. Livingston, S. MaWhinney, J.E. Forster, J. Scott, et al., Clinical phase 1 testing of the safety and immunogenicity of an epitope-based DNA vaccine in human immunodeficiency virus type 1-infected subjects receiving highly active antiretroviral therapy, Clin. Vaccine Immunol. 15 (2008) 986–994.
[26] J. Alexander, M.-F. del Guercio, A. Maewal, L. Qiao, J. Fikes, R.W. Chesnut, et al., Linear PADRE T helper epitope and carbohydrate B cell epitope conjugates induce specific high titer IgG antibody responses, J. Immunol. 164 (2000) 1625–1633.
[27] C. La Rosa, J. Longmate, S.F. Lacey, T. Kaltcheva, R. Sharan, D. Marsano, et al., Clinical evaluation of safety and immunogenicity of PADRE-cytomegalovirus (CMV) and tetanus-CMV fusion peptide vaccines with or without PF03512676 adjuvant, J. Infect. Dis. 205 (2012) 1294–1304.
[28] A.D. Pasquale, S. Preiss, F.T.D. Silva, N. Garçon, Vaccine adjuvants: from 1920 to 2015 and beyond, Vaccines 3 (2015) 320–343.
[29] C.J. Hackett, D.A. Harn, Vaccine Adjuvants: Immunological and Clinical Principles, Humana Press, Totowa, NJ, 2006.
[30] R.K. Gupta, Aluminum compounds as vaccine adjuvants, Adv. Drug Deliv. Rev. 32 (1998) 155–172.
[31] R.R. Shah, L.A. Brito, D.T. O'Hagan, M.M. Amiji, Emulsions as vaccine adjuvants, in: C. Foged, T. Rades, Y. Perrie, S. Hooks. (Eds.), Subunit Vaccine Delivery, Springer, New York, 2015.
[32] N. Garçon, L. Segal, F. Tavares, M. Van Mechelen, The safety evaluation of adjuvants during vaccine development: the AS04 experience, Vaccine 29 (2011) 4453–4459.
[33] J. Dupont, J. Altclas, A. Lepetic, M. Lombardo, V. Vázquez, C. Salgueira, et al., A controlled clinical trial comparing the safety and immunogenicity of a new adjuvanted hepatitis B vaccine with a standard hepatitis B vaccine, Vaccine 24 (2006) 7167–7174.
[34] N. Garçon, G. Leroux-Roels, W.-F. Cheng, Vaccine adjuvants, Perspect. Vaccinol. 1 (2011) 89–113.
[35] R. Mischler, I.C. Metcalfe, Inflexal® V a trivalent virosome subunit influenza vaccine: production, Vaccine 20 (2002) B17–B23.
[36] P.A. Bovier, Epaxal®: a virosomal vaccine to prevent hepatitis A infection, Expert Rev. Vaccines 7 (2008) 1141–1150.
[37] D.T. O'Hagan, MF59 is a safe and potent vaccine adjuvant that enhances protection against influenza virus infection, Expert Rev. Vaccines 6 (2007) 699–710.
[38] N. Garçon, Preclinical development of AS04, in: G. Davies (Ed.), Vaccine Adjuvants: Methods and Protocols, Springer, New York, 2010.
[39] A. Monie, C.-F. Hung, R. Roden, T.C. Wu, Cervarix™: a vaccine for the prevention of HPV 16, 18-associated cervical cancer, Biologics 2 (2008) 107–113.
[40] N. Garcon, S. Hem, M. Friede, Evolution of adjuvants across the centuries, in: S. Plotkin, P. Offit, W. Orenstein (Eds.), Vaccines, sixth ed, Saunders Elsevier, Philadelphia, PA, 2013.
[41] S. Rts, Efficacy and safety of RTS, S/AS01 malaria vaccine with or without a booster dose in infants and children in Africa: final results of a phase 3, individually randomised, controlled trial, Lancet 386 (2015) 31–45.

[42] A.M. Didierlaurent, B. Laupèze, A. Di Pasquale, N. Hergli, C. Collignon, N. Garçon, Adjuvant system AS01: helping to overcome the challenges of modern vaccines, Expert Rev. Vaccines 16 (2017) 55–63.
[43] W.T. Walker, S.N. Faust, Monovalent inactivated split-virion AS03-adjuvanted pandemic influenza A (H1N1) vaccine, Expert Rev. Vaccines 9 (2010) 1385–1398.
[44] C. Caillet, F. Piras, M.-C. Bernard, A. de Montfort, F. Boudet, F.R. Vogel, et al., AF03-adjuvanted and non-adjuvanted pandemic influenza A (H1N1) 2009 vaccines induce strong antibody responses in seasonal influenza vaccine-primed and unprimed mice, Vaccine 28 (2010) 3076–3079.
[45] P.C. Rodríguez, G. Rodríguez, G. González, A. Lage, Clinical development and perspectives of CIMAvax EGF, Cuban vaccine for non-small-cell lung cancer therapy, MEDICC Rev. 12 (2010) 17–23.
[46] T. Stratmann, Cholera toxin subunit B as adjuvantm—an accelerator in protective immunity and a break in autoimmunity, Vaccines 3 (2015) 579–596.
[47] S.G. Reed, S. Bertholet, R.N. Coler, M. Friede, New horizons in adjuvants for vaccine development, Trends Immunol. 30 (2009) 23–32.
[48] F. Steinhagen, T. Kinjo, C. Bode, D.M. Klinman, TLR-based immune adjuvants, Vaccine 29 (2011) 3341–3355.
[49] W.M. Hussein, T.-Y. Liu, M. Skwarczynski, I. Toth, Toll-like receptor agonists: a patent review (2011–2013), Expert Opin. Ther. Pat. 24 (2014) 453–470.
[50] C. Ogawa, Y.-J. Liu, K.S. Kobayashi, Muramyl dipeptide and its derivatives: peptide adjuvant in immunological disorders and cancer therapy, Curr. Bioact. Compd. 7 (2011) 180–197.
[51] J.F. Bower, T.M. Ross, A minimum CR2 binding domain of C3d enhances immunity following vaccination, in: J.D. Lambris (Ed.), Current Topics in Complement, Springer, New York, 2006.
[52] R. Saenz, C. da Silva Souza, C.-T. Huang, M. Larsson, S. Esener, D. Messmer, HMGB1-derived peptide acts as adjuvant inducing immune responses to peptide and protein antigen, Vaccine 28 (2010) 7556–7562.
[53] R. Saenz, D. Futalan, L. Leutenez, F. Eekhout, J.F. Fecteau, S. Sundelius, et al., TLR4-dependent activation of dendritic cells by an HMGB1-derived peptide adjuvant, J. Transl. Med. 12 (2014) 211.
[54] A. Shanmugam, S. Rajoria, A.L. George, A. Mittelman, R. Suriano, R.K. Tiwari, Synthetic Toll like receptor-4 (TLR-4) agonist peptides as a novel class of adjuvants, PLoS One 7 (2012) e30839.
[55] R. Tiwari, R. Suriano, A. Mittelman, J. Geliebter, Synthetic toll-like receptor-4 (TLR-4) agonist peptides, US patent application 14/377,782, 2013.
[56] Y. Charoenvit, N. Goel, M. Whelan, K.S. Rosenthal, D.H. Zimmerman, CEL-1000—a peptide with adjuvant activity for Th1 immune responses, Vaccine 22 (2004) 2368–2373.
[57] D.H. Zimmerman, Method for inducing an immune response against avian, swine, spanish, h1n1, h5n9 influenza viruses and formulations thereof, US patent application 13/994,092, 2011.
[58] C.G. Lee, S.Y. Choi, S.-H. Park, K.S. Park, S.H. Ryu, Y.C. Sung, The synthetic peptide Trp-Lys-Tyr-Met-Val-D-Met as a novel adjuvant for DNA vaccine, Vaccine 23 (2005) 4703–4710.
[59] G.J. Kollessery, T.M. Nordgren, A.K. Mittal, G.V. Hegde, S.D. Sanderson, S.S. Joshi, Development of vaccine-based immunotherapy for murine large B-cell lymphoma, Blood 112 (2008) 2619–3619.

[60] R. Humphreys, S. Adams, G. Koldzic, B. Nedelescu, E. von Hofe, M. Xu, Increasing the potency of MHC class II-presented epitopes by linkage to Ii-Key peptide, Vaccine 18 (2000) 2693–2697.
[61] J.H. Fritz, S. Brunner, M.L. Birnstiel, M. Buschle, Av Gabain, F. Mattner, et al., The artificial antimicrobial peptide KLKLLLLLKLK induces predominantly a TH2-type immune response to co-injected antigens, Vaccine 22 (2004) 3274–3284.
[62] A.G. Laman, A.O. Shepelyakovskaya, I.A. Berezin, K.M. Boziev, I.L. Rodionov, I.A. Chulina, et al., Identification of pentadecapeptide mimicking muramyl peptide, Vaccine 25 (2007) 2900–2906.
[63] T. Matsubara, D. Ishikawa, T. Taki, Y. Okahata, T. Sato, Selection of ganglioside GM1-binding peptides by using a phage library, FEBS Lett. 456 (1999) 253–256.
[64] A.D. Montaner, A. De Nichilo, F. Elias, J.M. Rodríguez, J.M. Fló, R.A. López, et al., Ganglioside GM1-binding peptides as adjuvants of antigens inoculated by the intranasal route, Vaccine 24 (2006) 1889–1896.
[65] I. Castagliuolo, M. Sardina, P. Brun, C. DeRos, C. Mastrotto, L. Lovato, et al., *Clostridium difficile* toxin A carboxyl-terminus peptide lacking ADP-ribosyltransferase activity acts as a mucosal adjuvant, Infect. Immun. 72 (2004) 2827–2836.
[66] J.S. Rudra, Y.F. Tian, J.P. Jung, J.H. Collier, A self-assembling peptide acting as an immune adjuvant, Proc. Natl. Acad. Sci. U. S. A. 107 (2010) 622–627.
[67] J.P. Jung, A.K. Nagaraj, E.K. Fox, J.S. Rudra, J.M. Devgun, J.H. Collier, Co-assembling peptides as defined matrices for endothelial cells, Biomaterials 30 (2009) 2400–2410.
[68] H. Huang, J. Shi, J. Laskin, Z. Liu, D.S. McVey, X.S. Sun, Design of a shear-thinning recoverable peptide hydrogel from native sequences and application for influenza H1N1 vaccine adjuvant, Soft Matter 7 (2011) 8905–8912.
[69] J.S. Rudra, S. Mishra, A.S. Chong, R.A. Mitchell, E.H. Nardin, V. Nussenzweig, et al., Self-assembled peptide nanofibers raising durable antibody responses against a malaria epitope, Biomaterials 33 (2012) 6476–6484.
[70] D.T. O'Hagan, M. Singh, Microparticles as vaccine adjuvants and delivery systems, Expert Rev. Vaccines 2 (2003) 269–283.
[71] M.M. Levine, Can needle-free administration of vaccines become the norm in global immunization? Nat. Med. 9 (2003) 99–103.
[72] N. Marasini, M. Skwarczynski, I. Toth, Oral delivery of nanoparticle-based vaccines, Expert Rev. Vaccines 13 (2014) 1361–1376.
[73] A. Almeida, H. Alpar, Nasal delivery of vaccines, J. Drug Target. 3 (1996) 455–467.
[74] R.R. Shah, D.T. O'Hagan, M.M. Amiji, L.A. Brito, The impact of size on particulate vaccine adjuvants, Nanomedicine 9 (2014) 2671–2681.
[75] M.F. Bachmann, G.T. Jennings, Vaccine delivery: a matter of size, geometry, kinetics and molecular patterns, Nat. Rev. Immunol. 10 (2010) 787–796.
[76] M. Skwarczynski, I. Toth, Recent advances in peptide-based subunit nanovaccines, Nanomedicine 9 (2014) 2657–2669.
[77] Z.J. Deng, M. Liang, M. Monteiro, I. Toth, R.F. Minchin, Nanoparticle-induced unfolding of fibrinogen promotes Mac-1 receptor activation and inflammation, Nat. Nanotechnol. 6 (2011) 39–44.
[78] Y. Pan, A. Leifert, D. Ruau, S. Neuss, J. Bornemann, G. Schmid, et al., Gold nanoparticles of diameter 1.4 nm trigger necrosis by oxidative stress and mitochondrial damage, Small 5 (2009) 2067–2076.
[79] S.T. Reddy, A.J. van der Vlies, E. Simeoni, V. Angeli, G.J. Randolph, C.P. O'Neil, et al., Exploiting lymphatic transport and complement activation in nanoparticle vaccines, Nat. Biotechnol. 25 (2007) 1159–1164.

[80] M. Skwarczynski, I. Toth, Peptide-based subunit nanovaccines, Curr. Drug Deliv. 8 (2011) 282–289.
[81] Y. Tabata, Y. Ikada, Effect of the size and surface charge of polymer microspheres on their phagocytosis by macrophage, Biomaterials 9 (1988) 356–362.
[82] Z.-G. Yue, W. Wei, P.-P. Lv, H. Yue, L.-Y. Wang, Z.-G. Su, et al., Surface charge affects cellular uptake and intracellular trafficking of chitosan-based nanoparticles, Biomacromolecules 12 (2011) 2440–2446.
[83] C. Foged, B. Brodin, S. Frokjaer, A. Sundblad, Particle size and surface charge affect particle uptake by human dendritic cells in an in vitro model, Int. J. Pharm. 298 (2005) 315–322.
[84] M. Lundqvist, J. Stigler, G. Elia, I. Lynch, T. Cedervall, K.A. Dawson, Nanoparticle size and surface properties determine the protein corona with possible implications for biological impacts, Proc. Natl. Acad. Sci. U. S. A. 105 (2008) 14265–14270.
[85] J.A. Champion, S. Mitragotri, Shape induced inhibition of phagocytosis of polymer particles, Pharm. Res. 26 (2009) 244–249.
[86] B.D. Chithrani, W.C. Chan, Elucidating the mechanism of cellular uptake and removal of protein-coated gold nanoparticles of different sizes and shapes, Nano Lett. 7 (2007) 1542–1550.
[87] B.D. Chithrani, A.A. Ghazani, W.C. Chan, Determining the size and shape dependence of gold nanoparticle uptake into mammalian cells, Nano Lett. 6 (2006) 662–668.
[88] A. Verma, F. Stellacci, Effect of surface properties on nanoparticle–cell interactions, Small 6 (2010) 12–21.
[89] N. Marasini, A.K. Giddam, M.R. Batzloff, M.F. Good, M. Skwarczynski, I. Toth, Poly-L-lysine-coated nanoparticles are ineffective in inducing mucosal immunity against group A *Streptococcus*, Biochem. Compd. 5 (2017) 1.1–1.6.
[90] N. Marasini, A.K. Giddam, Z.G. Khalil, W.M. Hussein, R.J. Capon, M.R. Batzloff, et al., Double adjuvanting strategy for peptide-based vaccines: trimethyl chitosan nanoparticles for lipopeptide delivery, Nanomedicine 11 (2016) 3223–3235.
[91] M. Henriksen-Lacey, K.S. Korsholm, P. Andersen, Y. Perrie, D. Christensen, Liposomal vaccine delivery systems, Expert Opin. Drug Deliv. 8 (2011) 505–519.
[92] E.M. Varypataki, K. van der Maaden, J. Bouwstra, F. Ossendorp, W. Jiskoot, Cationic liposomes loaded with a synthetic long peptide and poly (I: C): a defined adjuvanted vaccine for induction of antigen-specific T cell cytotoxicity, AAPS J. 17 (2015) 216–226.
[93] M. Taneichi, H. Ishida, K. Kajino, K. Ogasawara, Y. Tanaka, M. Kasai, et al., Antigen chemically coupled to the surface of liposomes are cross-presented to CD8+ T cells and induce potent antitumor immunity, J. Immunol. 177 (2006) 2324–2330.
[94] T. Nagata, T. Toyota, H. Ishigaki, T. Ichihashi, K. Kajino, Y. Kashima, et al., Peptides coupled to the surface of a kind of liposome protect infection of influenza viruses, Vaccine 25 (2007) 4914–4921.
[95] J.-S. Thomann, B. Heurtault, S. Weidner, M. Brayé, J. Beyrath, S. Fournel, et al., Antitumor activity of liposomal ErbB2/HER2 epitope peptide-based vaccine constructs incorporating TLR agonists and mannose receptor targeting, Biomaterials 32 (2011) 4574–4583.
[96] K.A. Ghaffar, N. Marasini, A.K. Giddam, M.R. Batzloff, M.F. Good, M. Skwarczynski, et al., Liposome-based intranasal delivery of lipopeptide vaccine candidates against group A *Streptococcus*, Acta Biomater. 41 (2016) 161–168.

[97] N. Marasini, A.K. Giddam, K.A. Ghaffar, M.R. Batzloff, M.F. Good, M. Skwarczynski, et al., Multilayer engineered nanoliposomes as a novel tool for oral delivery of lipopeptide-based vaccines against group A *Streptococcus*, Nanomedicine 11 (2016) 1223–1236.

[98] G. Goss, H. Anderson, A. Maksymiuk, D. Vergidis, D. Soulieres, Y. Cormier, et al., BLP25 liposome vaccine (L-BLP25): long-term safety results from a phase IIb study of patients with stage IIIB and IV non-small-cell lung cancer (NSCLC) treated for 2 years or more, J. Thorac. Oncol., 2009, Lippincott Williams & Wilkins 4 (2009) S681.

[99] Y.-L. Wu, K. Park, R.A. Soo, Y. Sun, K. Tyroller, D. Wages, et al., INSPIRE: a phase III study of the BLP25 liposome vaccine (L-BLP25) in Asian patients with unresectable stage III non-small cell lung cancer, BMC Cancer 11 (2011) 430.

[100] M. Singh, A. Chakrapani, D. O'Hagan, Nanoparticles and microparticles as vaccine-delivery systems, Expert Rev. Vaccines 6 (2007) 797–808.

[101] M.-G. Kim, J.Y. Park, Y. Shon, G. Kim, G. Shim, Y.-K. Oh, Nanotechnology and vaccine development, Asian J. Pharm. Sci. 9 (2014) 227–235.

[102] A. Salvador, M. Igartua, R.M. Hernández, J.L. Pedraz, An overview on the field of micro- and nanotechnologies for synthetic peptide-based vaccines, J. Drug Deliv. 2011 (2011) 18.

[103] A.C. Rice-Ficht, A.M. Arenas-Gamboa, M.M. Kahl-McDonagh, T.A. Ficht, Polymeric particles in vaccine delivery, Curr. Opin. Microbiol. 13 (2010) 106–112.

[104] S.P. Vyas, P.N. Gupta, Implication of nanoparticles/microparticles in mucosal vaccine delivery, Expert Rev. Vaccines 6 (2007) 401–418.

[105] M. Allahyari, E. Mohit, Peptide/protein vaccine delivery system based on PLGA particles, Hum. Vaccin. Immunother. 12 (2016) 806–828.

[106] L. Illum, I. Jabbal-Gill, M. Hinchcliffe, A. Fisher, S. Davis, Chitosan as a novel nasal delivery system for vaccines, Adv. Drug Deliv. Rev. 51 (2001) 81–96.

[107] J. Yang, Z. Li, J. Zhou, R. Chen, L. Cheng, Q. Zhou, et al., Preparation and antitumor effects of nanovaccines with MAGE-3 peptides in transplanted gastric cancer in mice, Chin. J. Cancer 29 (2010) 359–364.

[108] S. Hamdy, O. Molavi, Z. Ma, A. Haddadi, A. Alshamsan, Z. Gobti, et al., Co-delivery of cancer-associated antigen and Toll-like receptor 4 ligand in PLGA nanoparticles induces potent CD8+ T cell-mediated anti-tumor immunity, Vaccine 26 (2008) 5046–5057.

[109] T. Fifis, A. Gamvrellis, B. Crimeen-Irwin, G.A. Pietersz, J. Li, P.L. Mottram, et al., Size-dependent immunogenicity: therapeutic and protective properties of nano-vaccines against tumors, J. Immunol. 173 (2004) 3148–3154.

[110] P.L. Mottram, D. Leong, B. Crimeen-Irwin, S. Gloster, S.D. Xiang, J. Meanger, et al., Type 1 and 2 immunity following vaccination is influenced by nanoparticle size: formulation of a model vaccine for respiratory syncytial virus, Mol. Pharm. 4 (2007) 73–84.

[111] K.L. Wilson, S.D. Xiang, M. Plebanski, Montanide, poly I: C and nanoparticle based vaccines promote differential suppressor and effector cell expansion: a study of induction of CD8 T cells to a minimal *Plasmodium berghei* epitope, Front. Microbiol. 6 (2016) 29.

[112] C. Loos, T. Syrovets, A. Musyanovych, V. Mailänder, K. Landfester, G.U. Nienhaus, et al., Functionalized polystyrene nanoparticles as a platform for studying bio–nano interactions, Beilstein J. Nanotechnol. 5 (2014) 2403–2412.

[113] Y. Hailemichael, Z. Dai, N. Jaffarzad, Y. Ye, M.A. Medina, X.-F. Huang, et al., Persistent antigen at vaccination sites induces tumor-specific CD8+ T cell sequestration, dysfunction and deletion, Nat. Med. 19 (2013) 465–472.

[114] R. Rajkannan, M. Dhanaraju, D. Gopinath, D. Selvaraj, R. Jayakumar, Development of hepatitis B oral vaccine using B-cell epitope loaded PLG microparticles, Vaccine 24 (2006) 5149–5157.
[115] M. Skwarczynski, M. Zaman, C.N. Urbani, I. Lin, Z. Jia, M.R. Batzloff, et al., Polyacrylate dendrimer nanoparticles: a self-adjuvanting vaccine delivery system, Angew. Chem. Int. Ed. Engl. 49 (2010) 5742–5745.
[116] A.A.A. Fuaad, Z. Jia, M. Zaman, J. Hartas, Z.M. Ziora, I.-C. Lin, et al., Polymer–peptide hybrids as a highly immunogenic single-dose nanovaccine, Nanomedicine 9 (2014) 35–43.
[117] T.-Y. Liu, A.K. Giddam, W.M. Hussein, Z. Jia, N.A. McMillan, M.J. Monteiro, et al., Self-adjuvanting therapeutic peptide-based vaccine induce CD8+ cytotoxic T lymphocyte responses in a murine human papillomavirus tumor model, Curr. Drug Deliv. 12 (2015) 3–8.
[118] W.M. Hussein, T.-Y. Liu, Z. Jia, N.A. McMillan, M.J. Monteiro, I. Toth, et al., Multiantigenic peptide–polymer conjugates as therapeutic vaccines against cervical cancer, Bioorg. Med. Chem. 24 (2016) 4372–4380.
[119] T.-Y. Liu, W.M. Hussein, A.K. Giddam, Z. Jia, J.M. Reiman, M. Zaman, et al., Polyacrylate-based delivery system for self-adjuvanting anticancer peptide vaccine, J. Med. Chem. 58 (2014) 888–896.
[120] Y. Fujita, H. Taguchi, Current status of multiple antigen-presenting peptide vaccine systems: application of organic and inorganic nanoparticles, Chem. Cent. J. 5 (2011) 48.
[121] A.Y. Lin, J. Lunsford, A.S. Bear, J.K. Young, P. Eckels, L. Luo, et al., High-density sub-100-nm peptide-gold nanoparticle complexes improve vaccine presentation by dendritic cells in vitro, Nanoscale Res. Lett. 8 (2013) 1–11.
[122] P. Zhang, Y.-C. Chiu, L.H. Tostanoski, C.M. Jewell, Polyelectrolyte multilayers assembled entirely from immune signals on gold nanoparticle templates promote antigen-specific T cell response, ACS Nano 9 (2015) 6465–6477.
[123] M. Skwarczynski, M. Zaman, I. Toth, Lipo-peptides/saccharides in peptide vaccine delivery, in: A. Kastin (Ed.), Handbook of the Biologically Active Peptides, second ed., Academic Press, Cambridge, MA, 2013.
[124] M. Skwarczynski, B.H. Parhiz, F. Soltani, S. Srinivasan, K.A. Kamaruzaman, I.-C. Lin, et al., Lipid peptide core nanoparticles as multivalent vaccine candidates against *Streptococcus pyogenes*, Aust. J. Chem. 65 (2012) 35–39.
[125] M. Zaman, S. Chandrudu, A.K. Giddam, J. Reiman, M. Skwarczynski, V. McPhun, et al., Group A streptococcal vaccine candidate: contribution of epitope to size, antigen presenting cell interaction and immunogenicity, Nanomedicine 9 (2014) 2613–2624.
[126] A. Chan, W.M. Hussein, K.A. Ghaffar, N. Marasini, A. Mostafa, S. Eskandari, et al., Structure–activity relationship of lipid core peptide-based group A *Streptococcus* vaccine candidates, Bioorg. Med. Chem. 24 (2016) 3095–3101.
[127] S. Salatin, J. Barar, M. Barzegar-Jalali, K. Adibkia, M.A. Milani, M. Jelvehgari, Hydrogel nanoparticles and nanocomposites for nasal drug/vaccine delivery, Arch. Pharm. Res. 39 (2016) 1181–1192.
[128] A. Sexton, P.G. Whitney, S.-F. Chong, A.N. Zelikin, A.P. Johnston, R. De Rose, et al., A protective vaccine delivery system for in vivo T cell stimulation using nanoengineered polymer hydrogel capsules, ACS Nano 3 (2009) 3391–3400.
[129] M. Toyoda, S. Hama, Y. Ikeda, Y. Nagasaki, K. Kogure, Anti-cancer vaccination by transdermal delivery of antigen peptide-loaded nanogels via iontophoresis, Int. J. Pharm. 483 (2015) 110–114.

[130] C.H. Kapadia, S. Tian, J.L. Perry, J.C. Luft, J.M. DeSimone, Reduction sensitive PEG hydrogels for codelivery of antigen and adjuvant to induce potent CTLs, Mol. Pharm. 13 (2016) 3381–3394.
[131] H.-X. Sun, Y. Xie, Y.-P. Ye, ISCOMs and ISCOMATRIX™, Vaccine 27 (2009) 4388–4401.
[132] S. Hook, T. Rades, Immune stimulating complexes (ISCOMs) and Quil-A containing particulate formulations as vaccine delivery systems, in: D. Flower, Y. Perrie (Eds.), Immunomic Discovery of Adjuvants and Candidate Subunit Vaccines, Springer, New York, 2013.
[133] H.M. Nielsen, H.B. Hübschmann, T. Rades, ISCOMs as a vaccine delivery system, in: C. Foged, T. Rades, Y. Perrie, S. Hook (Eds.), Subunit Vaccine Delivery, Springer, New York, 2015.
[134] G.B. Lipford, H. Wagner, K. Heeg, Vaccination with immunodominant peptides encapsulated in Quil A-containing liposomes induces peptide-specific primary CD8+ cytotoxic T cells, Vaccine 12 (1994) 73–80.
[135] A. Ahluwalia, K. Gokulan, I. Nath, D.N. Rao, Modification of delivery system enhances MHC nonrestricted immunogenicity of V3 loop region of HIV-1 gp120, Microbiol. Immunol. 41 (1997) 779–784.
[136] F.-S. Gao, L. Feng, Q. Zhang, R.-Q. Yan, Y.-G. Li, X.-S. Li, Immunogenicity of two FMDV nonameric peptides encapsulated in liposomes in mice and the protective efficacy in guinea pigs, PLoS One 8 (2013) e68658.
[137] A.C. Steer, J.R. Carapetis, J.B. Dale, J.D. Fraser, M.F. Good, L. Guilherme, et al., Status of research and development of vaccines for *Streptococcus pyogenes*, Vaccine 34 (2016) 2953–2958.
[138] E.H. Nardin, G.A. Oliveira, J.M. Calvo-Calle, Z.R. Castro, R.S. Nussenzweig, B. Schmeckpeper, et al., Synthetic malaria peptide vaccine elicits high levels of antibodies in vaccinees of defined HLA genotypes, J. Infect. Dis. 182 (2000) 1486–1496.
[139] S.B. Sirima, A.B. Tiono, A. Ouédraogo, A. Diarra, A.L. Ouédraogo, J.B. Yaro, et al., Safety and immunogenicity of the malaria vaccine candidate MSP3 long synthetic peptide in 12–24 months-old Burkinabe children, PLoS One 4 (2009) e7549.
[140] I. Karlsson, L. Brandt, L. Vinner, I. Kromann, L.V. Andreasen, P. Andersen, et al., Adjuvanted HLA-supertype restricted subdominant peptides induce new T-cell immunity during untreated HIV-1-infection, Clin. Immunol. 146 (2013) 120–130.
[141] F.B. Scorza, V. Tsvetnitsky, J.J. Donnelly, Universal influenza vaccines: shifting to better vaccines, Vaccine 34 (2016) 2926–2933.
[142] A. Wald, D.M. Koelle, K. Fife, T. Warren, K. LeClair, R.M. Chicz, et al., Safety and immunogenicity of long HSV-2 peptides complexed with rhHsc70 in HSV-2 seropositive persons, Vaccine 29 (2011) 8520–8529.
[143] E.A. Mittendorf, J.P. Holmes, S. Ponniah, G.E. Peoples, The E75 HER2/neu peptide vaccine, Cancer Immunol. Immunother. 57 (2008) 1511–1521.
[144] P. Vici, L. Pizzuti, L. Mariani, G. Zampa, D. Santini, L. Di Lauro, et al., Targeting immune response with therapeutic vaccines in premalignant lesions and cervical cancer: hope or reality from clinical studies, Expert Rev. Vaccines 15 (2016) 1327–1336.
[145] A. Aruga, N. Takeshita, Y. Kotera, R. Okuyama, N. Matsushita, T. Ohta, et al., Phase I clinical trial of multiple-peptide vaccination for patients with advanced biliary tract cancer, J. Transl. Med. 12 (2014) 61.
[146] R.A. Fenstermaker, M.J. Ciesielski, J. Qiu, N. Yang, C.L. Frank, K.P. Lee, et al., Clinical study of a survivin long peptide vaccine (SurVaxM) in patients with recurrent malignant glioma, Cancer Immunol. Immunother. 65 (2016) 1339–1352.

[147] K. Yoshimura, T. Minami, M. Nozawa, H. Uemura, Phase I clinical trial of human vascular endothelial growth factor receptor 1 peptide vaccines for patients with metastatic renal cell carcinoma, Br. J. Cancer 108 (2013) 1260–1266.

[148] S. Ohno, R. Okuyama, A. Aruga, H. Sugiyama, M. Yamamoto, Phase I trial of Wilms' tumor 1 (WT1) peptide vaccine with GM-CSF or CpG in patients with solid malignancy, Anticancer Res. 32 (2012) 2263–2269.

[149] M. Schmitt, A. Schmitt, M.T. Rojewski, J. Chen, K. Giannopoulos, F. Fei, et al., RHAMM-R3 peptide vaccination in patients with acute myeloid leukemia, myelodysplastic syndrome, and multiple myeloma elicits immunologic and clinical responses, Blood 111 (2008) 1357–1365.

[150] D.J. Schwartzentruber, D.H. Lawson, J.M. Richards, R.M. Conry, D.M. Miller, J. Treisman, et al., gp100 peptide vaccine and interleukin-2 in patients with advanced melanoma, N. Engl. J. Med. 364 (2011) 2119–2127.

[151] C. Staff, F. Mozaffari, J.-E. Frödin, H. Mellstedt, M. Liljefors, Telomerase (GV1001) vaccination together with gemcitabine in advanced pancreatic cancer patients, Int. J. Oncol. 45 (2014) 1293–1303.

Cell-penetrating peptides for medical theranostics and targeted drug delivery

13

Yue-Wern Huang[1] *and Han-Jung Lee*[2]
[1]Missouri University of Science and Technology, Rolla, MO, United States,
[2]National Dong Hwa University, Hualien, Taiwan

13.1 Introduction of cell-penetrating peptides

Intracellular delivery of biologically active molecules is essential in treating diseases. However, mucous layer, cell membrane, nuclear membrane, and tightly protective vasculature barriers (e.g., blood–brain barrier, blood–eye barrier, and blood–testis barrier) impede cellular entry of certain molecules (i.e., cargoes) such as nucleic acids, peptides, nanomaterials, and small synthetic diagnostics or therapeutics [1,2]. Thus, there is a need to develop effective means to facilitate cellular uptake of cargoes. Mechanical and electrical transfection techniques and chemical/biochemical methods have been developed to achieve this. The mechanical and electrical transfection techniques include microinjection, bioballistics, hydrodynamic force, ultrasonic nebulization, and electroporation. The chemical/biochemical methods include calcium phosphate coprecipitation, membrane fusion catalyzed by artificial lipids, peptides and small proteins, dendrimers and nanomaterials, and adenovirus-associated virus vectors [1]. Each has its advantages and drawbacks. Some are suitable for in vitro manipulations, while others are potential tools for delivery at the organism level.

As safety concerns over viral-based delivery grow, nonviral-based tools have become more attractive. Cell-penetrating peptides (CPPs) (also known as protein transduction domains (PTDs)), a nonviral-based tool, have been studied intensively over nearly three decades for in vitro and in vivo delivery as shown in Fig. 13.1. The first CPP was identified within the transactivator of transcription (Tat) PTD in HIV-1 [3,4]. It was originally recognized to consist of 11 amino acids (HIV Tat 47–57: YGRKKRRQRRR) or 13 amino acids (HIV Tat 48–60: GRKKRRQRRRPPQ). Penetratin (also known as Antennapedia) of the *Drosophila* [5] and structural protein VP22 of the herpes simplex virus [6] are two other types of early discoveries. Based upon the origin of CPPs, they can be classified into protein-derived peptides (e.g., Tat-PTD), chimeric peptides which consist of two or more motifs from different peptides (e.g., transportan) [7], and synthetic peptides (e.g., polyarginines) [8]. Most CPPs are of nonhuman origin. To date, approximately 1855 CPPs have been characterized in the CPPsite 2.0 database (http://crdd.osdd.net/raghava/cppsite/). In general, CPPs are rich in basic amino acids,

Peptide Applications in Biomedicine, Biotechnology and Bioengineering.
DOI: http://dx.doi.org/10.1016/B978-0-08-100736-5.00013-2

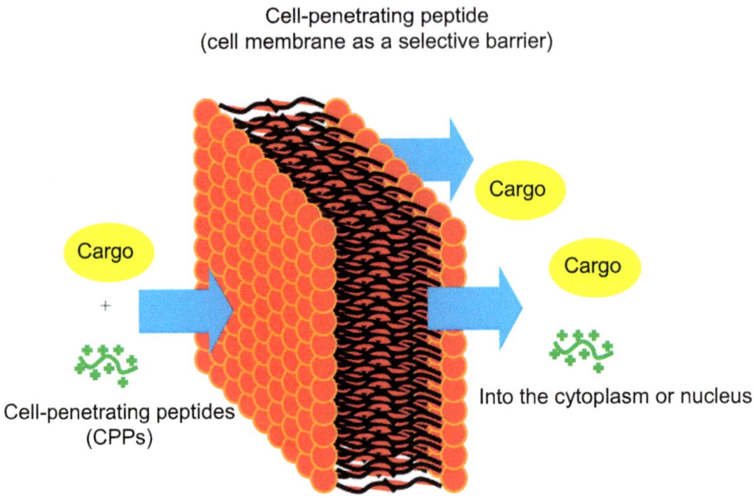

Figure 13.1 Cell-penetrating peptides (CPPs) as a tool to deliver biologically active molecules into cells.

positively charged or amphipathic, with less than 30 amino acids in length. The merits of CPPs as carriers are easy preparation, lack of cytotoxicity, and high efficiency of penetration. CPPs have been shown to effectively deliver cargoes such as nucleic acids, peptides, proteins, antibodies, enzymes, peptide nucleic acids, nanomaterials, liposomes, and small synthetic diagnostics or therapeutics.

13.2 Cellular uptake mechanisms

13.2.1 Methods to study mechanisms of CPP-mediated cellular uptake

Pharmacological inhibitors are common tools to study mechanisms of CPP-directed cellular uptake of biologically active molecules. We have described extensively proper usage of inhibitors in a previous publication [1]. Pharmacological inhibitors are nonspecific, meaning that an inhibitor may inhibit several biomolecules that are involved in multiple uptake pathways. For instance, hypertonic sucrose inhibition is evidence of clathrin-dependent endocytosis; however, it also interferes with caveolin-dependent endocytosis and macropinocytosis [9]. Although monodansylcadaverine (MDC) can block clathrin-mediated uptake via stabilizing nascent clathrin-coated vesicles thus limiting formation of new clathrin lattice, it also suppresses the transglutaminase family (activator Rho GTPases) which plays a key role in actin-dependent macropinocytosis [9,10]. Thus, it is prudent to apply more than one inhibitor to study a specific entry pathway.

We suggest that complementary tools are needed to gain a more comprehensive and precise picture in terms of involvement of specific uptake mechanism(s). One such tool is the small interfering RNA (siRNA) knockdown technique. siRNA (also known as short interfering RNA or silencing RNA) is a class of double-stranded RNA molecules of 20–25 base pairs in length. It interferes with the expression of specific genes with complementary nucleotide sequences by degrading mRNA leading to no translation [11]. Previously, we found that although pharmacological inhibitors could reduce CPP-mediated cellular uptake, siRNA knockdown of cellular components such as clathrin and caveolin did not. The discrepancy may arise from nonspecificity of inhibitors and the extent of siRNA knockdown. Therefore, interpretation of internalization pathways becomes complicated. Another issue with pharmacological inhibitors is the concentration to be applied. Though manufacturers recommend a certain concentration (or a range of concentrations) of a specific inhibitor to be used, caution needs to be taken as sensitivity of inhibitors is cell type-specific. Too high a concentration of an inhibitor may compromise cellular physiological functions leading to wrong interpretation of uptake mechanisms, while too low concentrations of an inhibitor may not be effective [1].

13.2.2 Diverse cellular uptake routes

Investigation on mechanisms of CPP-directed cellular uptake has been vigorous, but the level of understanding is still incomplete. Not only will the understanding of uptake mechanisms enrich basic science in molecular biology, but it may also inform design of more effective delivery platforms to treat diseases. Commonly known pathways of CPP-directed cellular uptake include direct membrane translocation (also known as direct cell penetration), macropinocytosis, clathrin-dependent endocytosis, caveolin-dependent endocytosis, or a combination of multiple pathways. Reuter et al. provided empirical evidence that direct membrane translocation involves binding of CPPs to the phosphate groups of the phospholipids on the membrane surface leading to rearrangement of lipid molecules. Side chains of arginines translocate through the distal layer and form a water pore. Eventually, CPPs diffuse through the pore, followed by closure of the pore [12]. An example of CPP-mediated direct membrane translocation is depicted in Fig. 13.2.

13.2.3 Improved prediction of potential CPPs

Effectiveness of a CPP's cellular penetration is highly influenced by types of cargo, concentration of the CPP, and its total electric charge and amino acid sequence (these factors dictate its three-dimensional structure and its interactions with membrane molecules). Secondary amphipathicity is an especially important determinant of uptake efficiency [14]. CPPred-RF (http://server.malab.cn/CPPred-RF/), C2Pred (http://lin.uestc.edu.cn/server/C2Pred), CPPpred (http://bioware.ucd.ie/~compass/biowareweb/Server_pages/cpppred.php), and CellPPD (http://crdd.osdd.net/raghava/cellppd/submission.php) websites are four powerful tools that allow researchers to predict the effectiveness of CPPs.

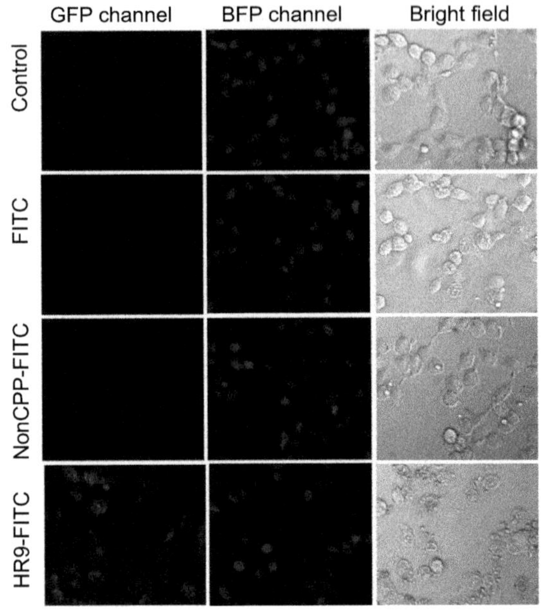

Figure 13.2 Cellular internalization of CPPs. Human bronchoalveolar carcinoma A549 cells were treated with 20 μM of HR9-fluorescein isothiocyanate (FITC) peptide [13] at 37°C for 1 h, followed by staining with Hoechst 33342, and then analyzed using fluorescent microscopy. Green and blue fluorescent protein (GFP and BFP) channels and bright fields revealed the distributions of HR9, nuclei, and cell morphologies, respectively. All images are obtained using a Motic AE31 fluorescent microscope with a magnification of 200x.

13.2.4 Release from lysosomal entrapment

One issue with endocytosis-mediated cellular entry is entrapment of the CPP/cargo complex inside the acidic lysosome, which may lead to undesired degradation of the cargo. Several strategies have evolved to preserve intended functions of endocytosed cargoes. For instance, we and others have employed endosome-disruptive peptides such as hemagglutinin-2 (HA2; a subunit of hemagglutinin which is an envelope glycoprotein of human influenza virus) [15] and penetration accelerating sequence (PAS) to circumvent the problem [16,17].

13.3 Medical imaging and drug delivery

Conventional CPPs are limited by nontargeting that results in binding to both normal and diseased tissues. Roger Tsien's team has developed a theranostic system using an activatable cell-penetrating peptide (ACPP) or ACPP with dendrimer (ACPPD) polyamidoamine (PAMAM) system for cancer detection, staging, and improvement of surgical precision [18,19]. The merits of the ACPP are enhanced targeted delivery to the tumor site with higher local cellular uptake, longer circulation time, and less toxicity. The ACPP system possesses a polycationic domain whose nonspecific adsorption and cellular uptake are minimized by a covalently attached polyanionic inhibitory domain. A linker, a substrate sequence for proteases MMP-2 and MMP-9 overexpressed in tumors, exists between the two domains. Upon reaching the tumor, proteases cleave the linker, releasing the polyanionic domain and leaving the highly positive domain to enter the cell. The system can be complexed with dendrimer PAMAM and loaded with cyanine 5 (Cy5) and gadolinium (Gd) for dual imaging purposes. The most significant contribution of the ACPPD system to disease treatment is its ability to improve surgical margin accuracy and survival. Conventional approaches to aid surgical accuracy include preoperative placement of a physical marker or intraoperative margin evaluation. For the former approach, a guide-wire with mammographic stereotactic guidance or with ultrasound guidance can be used to distinguish breast tumor from adjacent normal tissue. The disadvantages of these techniques are their limited spatial resolution and the difficulty of translating two-dimensional information to the three-dimensional surgical field. For the latter approach, small samples from the surgical bed are subject to histological examinations before wound closure. The disadvantages include extended anesthesia-related risks, time-consuming, increased healthcare system costs, intensive labor, and possible misdiagnosis. The ACPPD system provides an alternative means of surgical improvement and achieves (1) a 90% reduction of residual cancer cells, (2) 93% specificity in identifying malignant cells, and (3) improved tumor-free survival [20].

Ruoslahti's laboratory used the distinct biochemical signatures in the vasculature, namely the "vascular zip codes," to establish tumor-penetrating peptides [21–23]. Using the phage display technique, integrin-binding RGDs were identified and tested to deliver drug and imaging agents [24,25]. However, crossing the vascular wall and penetrating into the tumor parenchyma against the elevated interstitial pressure in the tumor remained a major challenge. The research team later identified that the C-terminal arginine (or lysine) in R/KXXR/K is required for tissue penetration. This is called C-end Rule (CendR). Adding another amino acid or blocking the free carboxyl group of the arginine residue would lose the binding and penetration capacity [26,27]. The RGDs with the R/KXXR/K motif are dubbed as iRGD (internalizing RGD). Penetration of iRGDs into tumor and tissues involves the following processes. The iRGD peptides accumulate at the surface of αv integrin-expressing endothelial and other cells in tumors. The iRGD motif mediates binding to the integrin. Upon cleavage by proteases, the CendR element is exposed and then mediates binding to neuropilin-1, which leads to penetration of cells and tissues. The iRGD system has been successfully tested in various in vivo cancer models for tissue binding. Importantly, iRGD-coated iron oxide nanoworms are capable of penetrating tumor tissues of high interstitial pressure and delineating tumors more effective than conventional RGD peptides do. Application of abraxane-loaded iRGD peptides in delivering anticancer drugs results (1) inhibition of the proliferation of 22Rv1 cells more efficiently than abraxane-loaded RGD peptides, (2) eightfold more abraxane accumulation in the tumor than the untargeted abraxane, and (3) more significant reduction of tumor growth than other control groups [26]. Collectively, iRGD peptides are a powerful tool for theranostics.

13.4 Gene delivery and editing

13.4.1 Introduction of the CRISPR-Cas9 system in genome editing

Genome editing is the introduction of a desired change to the sequence of genomic DNA of a living organism using engineered nucleases, and is driving a revolution in the biomedical sciences [28]. There were four major programmable genome-editing platforms in use: meganucleases, zinc-finger nucleases (ZFNs), transcription activator-like effector nucleases (TALENs), and the clustered regularly interspaced short palindromic repeats (CRISPR)-associated (Cas) protein 9 system [29]. In principle, it is possible to correct many genetic diseases with genome-editing technologies. However, the biggest challenge is the safe and efficient delivery of genome-editing biomacromolecules [29]. Because only programmable RNA is required to generate sequence specificity, the CRISPR-Cas9 system was easily applicable and developed quickly over the past few years.

The prokaryotic type II CRISPR-Cas9 system was originally derived from a natural adaptive immune system in bacteria [30], and has been developed by enabling fast and accurate alternations of the target genome in mammalian cells and model organisms [31]. This system requires a ribonucleoprotein complex formed by a single endonuclease protein Cas9, a short CRISPR RNA (crRNA), and a *trans*-acting crRNA (tracrRNA) to achieve efficient crRNA-directed recognition and site-specific cleavage of targeted DNA. Recognition of cleavage sites is determined by crRNA-DNA base pairing and a protospacer-adjacent motif (PAM) located downstream of target sequences [30]. This system has been further simplified with the development of a chimeric single-guide RNA (sgRNA) and a Cas9 protein [32]. Together, both sgRNA and Cas9 are sufficient for targeted DNA binding and DNA double-strand breaks in an RNA-guided manner [31].

13.4.2 Applications of CPP-mediated delivery of the CRISPR-Cas9 system

The first evidence of CPP-mediated delivery of the site-specific recombinase technology Cre-loxP system in genome editing was documented by Dowdy's laboratory [2,33]. The Cre-loxP system is one of the most popular genome-editing tools in both eukaryotic and prokaryotic systems [34]. This system requires a Cre enzyme that recombines a pair of short target *LoxP* sites in the DNA of cells. The Tat-Cre fusion protein and a loxP-STOP-loxP-EGFP reporter gene were established to measure protein transduction in mouse T cells [2]. Tat-Cre-mediated excision of transcriptional STOP region flanked by *loxP* sequences showed constitutive enhanced green fluorescent protein (EGFP) expression. Using this transducible reporter system, they concluded that the interaction of Tat-PTD with cell membrane leads to the internalization of Tat-fusion proteins by lipid raft-dependent macropinocytosis.

The first innovative application of CPP-mediated delivery of the CRISPR-Cas9 system in genome editing was reported recently by Ramakrishna et al. [35]. Covalent conjugation of Cas9-CPP by a thioether bond and noncovalent CPP/sgRNA complex led to efficient gene disruptions in several human cell types with lower off-target mutation frequencies than plasmid DNA transfection. The CPP/sgRNA complex resulted in about 500-nm particles with a net positive charge. This study demonstrated CPP-enabled direct delivery of both Cas9 protein and sgRNA into cultured mammalian cells with highly efficient genome editing. In contrast, Oude Blenke et al. argued that genome editing frequency of this Cas9-CPP and CPP/sgRNA treatment was low (<15% after three rounds of treatment) [36]. Moreover, the efficiency of CPP-based methods often depends on the type and status of the cell [29].

Genome editing using CPP-mediated delivery of the CRISPR-Cas9 system has several advantages as the approach: (1) may avoid potential problems associated with plasmid-based genome-editing methods, such as uncontrolled integration of the plasmid DNA into the host genome and unwanted immune responses caused by bacterial sequences on plasmid DNA; (2) involves a simple and cloning-free process; (3) reduces off-target mutations related to plasmid-mediated delivery; (4) implies potential development of in vivo genome editing; and (5) shortens exposure time at the cellular level, leading to lower toxicity compared to plasmid- or viral vector-mediated delivery [37]. However, the application of this approach in vivo was questioned to be less likely, because the individual components need to be delivered to the same cell simultaneously [36]. Furthermore, the relatively large and highly cationic CPP/sgRNA complex is unlikely to have favorable circulation kinetics or biodistribution. It remains to be investigated whether incorporation of the ACPPD concept into the CPP/sgRNA complex will improve such a disadvantage.

13.5 Safety and biocompatibility

Immune responses to CPPs and cytotoxicity have been vigorously tested in vitro and in vivo. Studies have shown that Tat, penetratin, or transportan and its derivatives do not either cause cytotoxicity (as measured by cell viability assays or apoptosis) nor illicit innate immune response (e.g., IL-1β, IL-6, IL-8, IL-18, TNF-α) in various types of epithelial cells and animal models (up to 5 mg/kg) [38,39]. In a different study of intratracheal administration of siRNA to knock down p38 mitogen-activated protein (MAP) kinase mRNA, while penetratin alone does not induce the innate immune response (e.g., TNF-α, IL-12p40), penetratin–siRNA conjugate does [40]. As for cytotoxicity, the above-mentioned studies provided no evidence of CPP-induced toxicity. Additionally, in our extensive studies with R9 (nona-arginines), HR9 (CH5-R9-CH5C), IR9 (INF7-R9), PR9 (FFLIPKG-R9), and lactoferricin-derived CPPs that addressed molecular mechanisms of CPP-mediated cellular uptake was as shown in Table 13.1, CPPs are not cytotoxic [13,41–45]. Other studies using a single bioassay or a metabolomics analysis pointed to less

Table 13.1 Summary of cell-penetrating peptides reported in the chapter

Type of peptide	Type of application	References
Protein-derived peptides		
Tat-PTD (GRKKRRQRRRPPQ)	Translocate to cytoplasm and nucleus	[3,4]
Penetratin; Antennapedia (RQIKIWFQNRRMKWKK)	Deliver protein and DNA to cytoplasm and nucleus	[5]
VP22 (DAATATRGRSAASRPTERPRAPARSASRPRRPVD)	Translocate to nucleus	[6]
Chimeric peptides		
Tat-Cre	Translocate to nucleus and recombine DNA	[2]
Transportan (CLIKKALAALAKLNIKLLYGASNLTWG)	Translocate to cytoplasm, nucleus, and nucleolus	[7]
CPP-HA2-RFP (R9-HA2-mCherry, R9-HA2-NLS-mCherry)	Endocytosed into endosomes, and release from endosomes into cytoplam and nucleus	[15]
Cas9-m9R (Cas9 linked GGGGRRRRRRRRRLLLL) sgRNA:9R (CGGGRRRRRRRRRLLLLC)	Translocate to nucleus for genome editing	[35,37]
CPP-PBase (R9-*piggyBac*)	Translocate to nucleus for gene therapy	[41]
R9 containing peptide	Deliver protein to cyanobacteria	[42]
Synthetic peptides		
SR9 (RRRRRRRRR)	Deliver quantum dots to cytoplasm and nucleus	[8]
PLA (poly-L-arginines)	Translocate through cell membrane	[12]
PasR8 (FFLIPKGRRRRRRRRGC)	Endocytosed into endosomes, and release from endosomes into cytoplam and nucleus	[16]
PR9 (FFLIPKGRRRRRRRRR)	Deliver quantum dots and DNA through the escape from endosomes into cytoplasm and nucleus	[17]

(*Continued*)

Table 13.1 (Continued)

Type of peptide	Type of application	References
ACPP (EEEEEEEEPLGLAGRRRRRRRRRC)	Endocytosed into endosomes of targeted tumor cells	[18]
ACPPD (EEEEEEEEPLGCAGRRRRRRRRRC)	Deliver quantum dots to reveal targeted tumor imaging	[19,20]
L5a (RRWQW)	Deliver DNA to nucleus	[43]
L6 (RRWQWR)	Deliver DNA to nucleus	[43]
HR9 (CHHHHHRRRRRRRRRHHHHHC)	Deliver quantum dots to cytoplasm and nucleus	[13,44]
IR9 (GLFEAIEGFIENGWEGMIDGWYGRRRRRRRRR)	Deliver quantum dots and DNA to cytoplasm and nucleus	[44,45]

toxic cationic CPPs than amphipathic CPPs [46,47]. Caution needs to be made that studies of toxicity were usually conducted with different assays, different cell lines, and different concentrations of CPPs. These factors may have influences on the observed outcomes.

Collectively, current data suggest that it is impossible to generalize potential immune response to CPPs and cytotoxicity, in vitro or in vivo, as the consequence depends on the sequence of a CPP, the nature of the cargo and its attachment to a CPP, dose of the CPP, and types of cells and animal models. A systematic approach using the quantitative structure and activity relationship (QSAR) principle may help reveal the relationship between structures of CPPs and the immune response and cytotoxicity [48].

Acknowledgments

This work was supported by the Center for Biomedical Science and Engineering at Missouri University of Science and Technology to Yue-Wern Huang and the Ministry of Science and Technology, Taiwan, to Han-Jung Lee (Grant No. MOST 104-2320-B-259-002-MY3).

Abbreviations

ACPP activatable cell-penetrating peptide
ACPPD activatable cell-penetrating peptide with dendrimer
Cas9 CRISPR-associated protein 9

CendR C-end rule
CPP cell-penetrating peptide
CRISPR clustered regularly interspaced short palindromic repeats
iRGD internalizing Arg-Gly-Asp
PAMAM polyamidoamine
PTD protein transduction domain
sgRNA single-guide RNA
siRNA small interfering RNA
Tat transactivator of transcription

References

[1] Y.W. Huang, H.J. Lee, L.M. Tolliver, R.S. Aronstam, Delivery of nucleic acids and nanomaterials by cell-penetrating peptides: opportunities and challenges, Biomed. Res. Int. 2015 (2015) 834079.

[2] J.S. Wadia, R.V. Stan, S.F. Dowdy, Transducible TAT-HA fusogenic peptide enhances escape of TAT-fusion proteins after lipid raft macropinocytosis, Nat. Med. 10 (3) (2004) 310–315.

[3] M. Green, P.M. Loewenstein, Autonomous functional domains of chemically synthesized human immunodeficiency virus tat *trans*-activator protein, Cell 55 (6) (1988) 1179–1188.

[4] A.D. Frankel, C.O. Pabo, Cellular uptake of the tat protein from human immunodeficiency virus, Cell 55 (6) (1988) 1189–1193.

[5] D. Derossi, G. Chassaing, A. Prochiantz, Trojan peptides: the penetratin system for intracellular delivery, Trends Cell. Biol. 8 (2) (1998) 84–87.

[6] G. Elliott, P. O'Hare, Intercellular trafficking and protein delivery by a herpesvirus structural protein, Cell 88 (2) (1997) 223–233.

[7] M. Pooga, M. Hallbrink, M. Zorko, U. Langel, Cell penetration by transportan, FASEB J. 12 (1) (1998) 67–77.

[8] Y. Xu, B.R. Liu, H.J. Lee, K.B. Shannon, J.G. Winiarz, T.C. Wang, et al., Nona-arginine facilitates delivery of quantum dots into cells via multiple pathways, J. Biomed. Biotechnol. 2010 (2010) 948543.

[9] A.I. Ivanov, Exocytosis and Endocytosis, Humana, Totowa, NJ, 2008.

[10] Y. Phonphok, K.S. Rosenthal, Stabilization of clathrin coated vesicles by amantadine, tromantadine and other hydrophobic amines, FEBS Lett. 281 (1–2) (1991) 188–190.

[11] A. Fire, S. Xu, M.K. Montgomery, S.A. Kostas, S.E. Driver, C.C. Mello, Potent and specific genetic interference by double-stranded RNA in *Caenorhabditis elegans*, Nature 391 (6669) (1998) 806–811.

[12] M. Reuter, C. Schwieger, A. Meister, G. Karlsson, A. Blume, Poly-L-lysines and poly-L-arginines induce leakage of negatively charged phospholipid vesicles and translocate through the lipid bilayer upon electrostatic binding to the membrane, Biophys. Chem. 144 (1–2) (2009) 27–37.

[13] B.R. Liu, Y.W. Huang, J.G. Winiarz, H.J. Chiang, H.J. Lee, Intracellular delivery of quantum dots mediated by a histidine- and arginine-rich HR9 cell-penetrating peptide through the direct membrane translocation mechanism, Biomaterials 32 (13) (2011) 3520–3537.

[14] E.D. Karagiannis, A.M. Urbanska, G. Sahay, J.M. Pelet, S. Jhunjhunwala, R. Langer, et al., Rational design of a biomimetic cell penetrating peptide library, ACS Nano 7 (10) (2013) 8616–8626.
[15] J.S. Liou, B.R. Liu, A.L. Martin, Y.W. Huang, H.J. Chiang, H.J. Lee, Protein transduction in human cells is enhanced by cell-penetrating peptides fused with an endosomolytic HA2 sequence, Peptides 37 (2) (2012) 273–284.
[16] K. Takayama, I. Nakase, H. Michiue, T. Takeuchi, K. Tomizawa, H. Matsui, et al., Enhanced intracellular delivery using arginine-rich peptides by the addition of penetration accelerating sequences (Pas), J. Control. Release 138 (2) (2009) 128–133.
[17] B.R. Liu, S.Y. Lo, C.C. Liu, C.L. Chyan, Y.W. Huang, R.S. Aronstam, et al., Endocytic trafficking of nanoparticles delivered by cell-penetrating peptides comprised of nona-arginine and a penetration accelerating sequence, PLoS One 8 (6) (2013) e67100.
[18] T.A. Aguilera, E.S. Olson, M.M. Timmers, T. Jiang, R.Y. Tsien, Systemic *in vivo* distribution of activatable cell penetrating peptides is superior to that of cell penetrating peptides, Integr. Biol. (Camb.) 1 (5–6) (2009) 371–381.
[19] E.S. Olson, T. Jiang, T.A. Aguilera, Q.T. Nguyen, L.G. Ellies, M. Scadeng, et al., Activatable cell penetrating peptides linked to nanoparticles as dual probes for *in vivo* fluorescence and MR imaging of proteases, Proc. Natl. Acad. Sci. U. S. A. 107 (9) (2010) 4311–4316.
[20] Q.T. Nguyen, E.S. Olson, T.A. Aguilera, T. Jiang, M. Scadeng, L.G. Ellies, et al., Surgery with molecular fluorescence imaging using activatable cell-penetrating peptides decreases residual cancer and improves survival, Proc. Natl. Acad. Sci. U. S. A. 107 (9) (2010) 4317–4322.
[21] E. Ruoslahti, RGD and other recognition sequences for integrins, Annu. Rev. Cell Dev. Biol. 12 (1996) 697–715.
[22] E. Ruoslahti, Specialization of tumour vasculature, Nat. Rev. Cancer 2 (2) (2002) 83–90.
[23] E. Ruoslahti, D. Rajotte, An address system in the vasculature of normal tissues and tumors, Annu. Rev. Immunol. 18 (2000) 813–827.
[24] W. Arap, R. Pasqualini, E. Ruoslahti, Cancer treatment by targeted drug delivery to tumor vasculature in a mouse model, Science 279 (5349) (1998) 377–380.
[25] F. Curnis, A. Gasparri, A. Sacchi, R. Longhi, A. Corti, Coupling tumor necrosis factor-alpha with alphaV integrin ligands improves its antineoplastic activity, Cancer Res. 64 (2) (2004) 565–571.
[26] K.N. Sugahara, T. Teesalu, P.P. Karmali, V.R. Kotamraju, L. Agemy, O.M. Girard, et al., Tissue-penetrating delivery of compounds and nanoparticles into tumors, Cancer Cell 16 (6) (2009) 510–520.
[27] T. Teesalu, K.N. Sugahara, V.R. Kotamraju, E. Ruoslahti, C-end rule peptides mediate neuropilin-1-dependent cell, vascular, and tissue penetration, Proc. Natl. Acad. Sci. U. S. A. 106 (38) (2009) 16157–16162.
[28] A.C. Komor, A.H. Badran, D.R. Liu, CRISPR-based technologies for the manipulation of eukaryotic genomes, Cell 168 (1–2) (2017) 20–36.
[29] H. Yin, K.J. Kauffman, D.G. Anderson, Delivery technologies for genome editing, Nat. Rev. Drug Discov. 16 (6) (2017) 387–399.
[30] J.E. Garneau, M.E. Dupuis, M. Villion, D.A. Romero, R. Barrangou, P. Boyaval, et al., The CRISPR/Cas bacterial immune system cleaves bacteriophage and plasmid DNA, Nature 468 (7320) (2010) 67–71.
[31] C. Fellmann, B.G. Gowen, P.C. Lin, J.A. Doudna, J.E. Corn, Cornerstones of CRISPR-Cas in drug discovery and therapy, Nat. Rev. Drug Discov. 16 (2) (2017) 89–100.

[32] M. Jinek, K. Chylinski, I. Fonfara, M. Hauer, J.A. Doudna, E. Charpentier, A programmable dual-RNA-guided DNA endonuclease in adaptive bacterial immunity, Science 337 (6096) (2012) 816–821.

[33] H. Liu, F. Zeng, M. Zhang, F. Huang, J. Wang, J. Guo, et al., Emerging landscape of cell penetrating peptide in reprogramming and gene editing, J. Control. Release 226 (2016) 124–137.

[34] M.A. McLellan, N.A. Rosenthal, A.R. Pinto, Cre-loxP-mediated recombination: general principles and experimental considerations, Curr. Protoc. Mouse Biol. 7 (1) (2017) 1–12.

[35] S. Ramakrishna, A.B. Kwaku Dad, J. Beloor, R. Gopalappa, S.K. Lee, H. Kim, Gene disruption by cell-penetrating peptide-mediated delivery of Cas9 protein and guide RNA, Genome Res. 24 (6) (2014) 1020–1027.

[36] E. Oude Blenke, M.J. Evers, E. Mastrobattista, J. van der Oost, CRISPR-Cas9 gene editing: delivery aspects and therapeutic potential, J. Control. Release 244 (Pt B) (2016) 139–148.

[37] B. Suresh, S. Ramakrishna, H. Kim, Cell-penetrating peptide-mediated delivery of Cas9 protein and guide RNA for genome editing, Methods Mol. Biol. 1507 (2017) 81–94.

[38] E. Carter, C.Y. Lau, D. Tosh, S.G. Ward, R.J. Mrsny, Cell penetrating peptides fail to induce an innate immune response in epithelial cells in vitro: implications for continued therapeutic use, Eur. J. Pharm. Biopharm. 85 (1) (2013) 12–19.

[39] J. Suhorutsenko, N. Oskolkov, P. Arukuusk, K. Kurrikoff, E. Eriste, D.M. Copolovici, et al., Cell-penetrating peptides, PepFects, show no evidence of toxicity and immunogenicity *in vitro* and *in vivo*, Bioconjug. Chem. 22 (11) (2011) 2255–2262.

[40] S.A. Moschos, S.W. Jones, M.M. Perry, A.E. Williams, J.S. Erjefalt, J.J. Turner, et al., Lung delivery studies using siRNA conjugated to TAT(48-60) and penetratin reveal peptide induced reduction in gene expression and induction of innate immunity, Bioconjug. Chem. 18 (5) (2007) 1450–1459.

[41] C.Y. Lee, J.F. Li, J.S. Liou, Y.C. Charng, Y.W. Huang, H.J. Lee, A gene delivery system for human cells mediated by both a cell-penetrating peptide and a *piggyBac* transposase, Biomaterials 32 (26) (2011) 6264–6276.

[42] B.R. Liu, Y.W. Huang, H.J. Lee, Mechanistic studies of intracellular delivery of proteins by cell-penetrating peptides in cyanobacteria, BMC Microbiol. 13 (2013) 57.

[43] B.R. Liu, Y.W. Huang, R.S. Aronstam, H.J. Lee, Identification of a short cell-penetrating peptide from bovine lactoferricin for intracellular delivery of DNA in human A549 cells, PLoS One 11 (3) (2016) e0150439.

[44] B.R. Liu, J.S. Liou, Y.J. Chen, Y.W. Huang, H.J. Lee, Delivery of nucleic acids, proteins, and nanoparticles by arginine-rich cell-penetrating peptides in rotifers, Mar. Biotechnol. (NY). 15 (5) (2013) 584–595.

[45] B.R. Liu, J.S. Liou, Y.W. Huang, R.S. Aronstam, H.J. Lee, Intracellular delivery of nanoparticles and DNAs by IR9 cell-penetrating peptides, PLoS One 8 (5) (2013) e64205.

[46] K. Kilk, R. Mahlapuu, U. Soomets, U. Langel, Analysis of *in vitro* toxicity of five cell-penetrating peptides by metabolic profiling, Toxicology 265 (3) (2009) 87–95.

[47] K. Saar, M. Lindgren, M. Hansen, E. Eiriksdottir, Y. Jiang, K. Rosenthal-Aizman, et al., Cell-penetrating peptides: a comparative membrane toxicity study, Anal. Biochem. 345 (1) (2005) 55–65.

[48] J. Regberg, A. Srimanee, M. Erlandsson, R. Sillard, D.A. Dobchev, M. Karelson, et al., Rational design of a series of novel amphipathic cell-penetrating peptides, Int. J. Pharm. 464 (1–2) (2014) 111–116.

Tumor-penetrating peptides

14

Ruirui Qiao[*,1], *Kun Wang*[2] *and Jian Zhong*[*,3]
[1]Monash University, Parkville, VIC, Australia, [2]Chinese Academy of Sciences, Beijing, China, [3]Shanghai Ocean University, Shanghai, China

14.1 Introduction

Tumor targeting is attractive in anticancer therapy since it largely improves the therapeutic efficacy from increased local accumulation and lower systemic exposure. Tumor cells express molecules that can be markers for differentiating them from normal cells. Historically, such molecules were recognized and detected by antibodies but certain peptides have shown great potential through targeting and penetrating the tumor cell [1–4]. Cargo molecules can be attached to such peptides and then delivered into tumor tissue and enable enhanced drug delivery.

The major obstacle for targeting of solid tumor is that drugs can only penetrate a few cell diameters into the extravascular tumor tissue from blood vessels [5]. Limited drug distribution within tumors is mainly due to the poorly organized vascular architecture, irregular perfused blood vessels, and high interstitial pressure which causes tissue fluid to flow out of the tumor [6–8]. The poor penetration from the leakiness of tumor vessels gives rise to the so-called enhanced permeability and retention (EPR) effect for passive targeting, but the EPR effect still suffers from variability from tumor to tumor and efficiency is still very low [9,10]. The discovery of cell-penetrating peptides (CPPs) provided enhanced permeability for passive targeting of payloads due to the capability of transporting the varying cargo across cell membranes. However, the nonspecific targeting to normal tissues can result in severe toxicity. Targeting the specific biology of tumor tissue by using different ligands could improve their efficacy due to more extensive penetration of the tumor tissue [2]. The targeted receptors can be on the surface of tumor cells, vascular cells, and stromal cells within tumors. Peptides that can specifically target tumor and tumor-associated microenvironments are often called "tumor homing peptides" (THPs) which can largely improve the selectivity of drug delivery. In 2002, Ruoslahti and collaborators discovered a type of peptide that can specifically recognize the endothelium of tumor vessels, extravagate, and penetrate deep into the extravascular tumor tissue and this type of peptide has been called a tumor-penetrating peptide (TPP). This chapter will focus on the discovery of TPPs, the mechanism for tumor targeting and the biomedical applications of TPPs for imaging and/or drug-delivery applications.

* Corresponding authors: ruirui.qiao@monash.edu; jzhong@shou.du.cn

Peptide Applications in Biomedicine, Biotechnology and Bioengineering.
DOI: http://dx.doi.org/10.1016/B978-0-08-100736-5.00014-4

14.2 Discovery of TPPs

14.2.1 Phage display

In 1985, the discovery of a technology that enabled filamentous bacteriophages to be genetically engineered to express peptide sequences at the surface of phage particles lead to the development of libraries of phages which contains up to 10^9 different peptides [4,11]. Phage screening libraries can be used in vitro to isolate clones displaying peptides with the desired activity and selectivity to the target. In vivo phage screening allows the sequencing and identification of peptides in live animals, therefore, more specific binding at the target tissue can be achieved [12]. Since being introduced by Smith [11], phage display peptide libraries and the screening technique have uncovered many specific peptides for tumor targeting.

Phage display is particularly suitable for the discovery of markers on the surface of tumor cells and tumor endothelial cells since the method relies on recognition of the accessible targets rather than overall expression levels. Tumor blood vessels express various cell surface and extracellular matrix proteins which are different from those expressed on the surface of normal vessels. The expressed proteins in cancer cells mostly result from neoangiogenesis, and the proteins are functionally important in this distinctive biological process [13,14]. In vivo phage screening mostly probes the molecular heterogeneity in the vasculature of normal and tumor tissues due to the reason that the phage particles (40–600 nm) do not readily penetrate the vascular wall [15]. Using this technique, Ruoslahti and collaborators have discovered a tumor-homing peptide, LyP-1, which was specifically accumulated in tumor lymphatics [16]. The receptor of LyP-1 is a p32 protein (gC1q receptor, hyaluronic acid-binding protein), expressed at the cell surface of lymphatic, myeloid, and cancer cells in tumors [17]. In vivo studies revealed that the peptide primarily targets myeloid/macrophages in tumors after intravenous injection into tumor-bearing mice [17].

14.2.2 Markers in tumor vasculature

Specific activations are present in vascular cells during specific responses and processes such as tumor growth and inflammation. For example, expression of a set of proteins including P-selectin, E-selectin, ICAM-1, and VCAM-1 can be turned on during the process of inflammation in venular endothelial cells allowing for leukocyte rolling and adhesion/migration in response to inflammation [18,19]. Another set of molecules including integrins, growth factor receptors, extracellular proteases, and extracellular matrix proteins, are expressed during angiogenesis, which makes tumor vasculature distinguishable from vasculature in normal tissues [18]. Among these, integrins have been intensively studied and are considered to be prime targets for tumor targeting since they are readily expressed on the surface of the endothelium and are available for the binding of bloodborne compounds. Integrins are the primary cell–matrix adhesion molecules and each integrin is composed of noncovalently associated α and β subunits that combine to activate signaling pathways by coclustering with one or several extracellular matrix (ECM) components [20]. In particular, $\alpha_v\beta_3$ and $\alpha_v\beta_5$ integrins, which are highly expressed in the vessels of most malignant tumors, recognize the Arg-Gly-Asp (RGD) sequence in their respective substrates including fibronectin, vitronectin, von Willebrand factor, tenascin, osteopontin, fibrillin, fibrinogen, and thrombospondin [21,22]. Integrin-binding RGD motif-containing peptides, which have been used for enhanced drug delivery to tumors, have strongly stimulated the field of tumor targeting of diagnostics and therapeutic strategies [23].

14.3 Molecular structure of TPPs

14.3.1 CendR motif

By using phage display technology, Ruoslahti and collaborators discovered that a peptide that combines the RGD sequence with a peptide motif for neuropilin-1 (NRP-1) can lead to both selective tumor vascular targeting and facilitate penetration through the vascular wall into extravascular tumor tissue [24,25]. The discovery of the CendR motif (arginine/lysine-X-X-arginine/lysine [R/KXXR/K]) by the same group revealed the mechanism for spreading within tumors.

Two different peptides, LyP-1 and iRGD, both contain three independent modules including a vascular homing motif, a CendR tissue penetration motif, and a protease recognition site, were identified through phage display. The amino acid sequence and structure in the peptides allows for multistep penetration into tumors: First, the RGD motif recognizes and binds to $\alpha_v\beta_3$ and $\alpha_v\beta_5$ integrins; second, the CendR motif is cut off by a protease and the internal R/KXXR/K motif at the C-terminus of the truncated peptide is exposed; third, the CendR motif binds to neuropilin-1 (NRP-1) or neuropilin-2 (NRP-2) followed by the endocytotic/exocytotic transport, leading to the enhanced transport of payloads into the tumor parenchyma (Fig. 14.1). This mechanism is called the CendR pathway. NRP1 is a cell-surface protein identified as a receptor for both the collapsin/semaphorin family of receptors and vascular endothelial growth factor 165 (VEGF165) (a key regulator of angiogenesis) [26]. The presence of free C-terminal arginine or lysine residues is required for the binding and internalization of the R/KXXR/K motif since the addition of other amino acid residues to the motif or amidation of the carboxyl terminus resulted in loss of activity [24]. Furthermore, the internalization of the NRP1 receptor after binding with CendR peptide can stimulate cell penetration [24].

Molecular modeling studies show that peptides with a C-terminal CendR motif, such as RPARPAR, fit well to the binding pocket at the b1b2 domain of NRP-1, but the role of the penultimate arginine residue outside the binding pocket is unclear (Fig. 14.2) [27,28]. Whereas the role of NRP-1 is to act as a coreceptor for various receptors to be recruited and presented to their receptors, overexpression of NRP-1 in tumor cells and stroma is believed to be implicated in the development and maintenance of tumor vessels and in tumor growth and progression [29]. Meanwhile, NRP-1 has been reported to be a target of anticancer therapy using

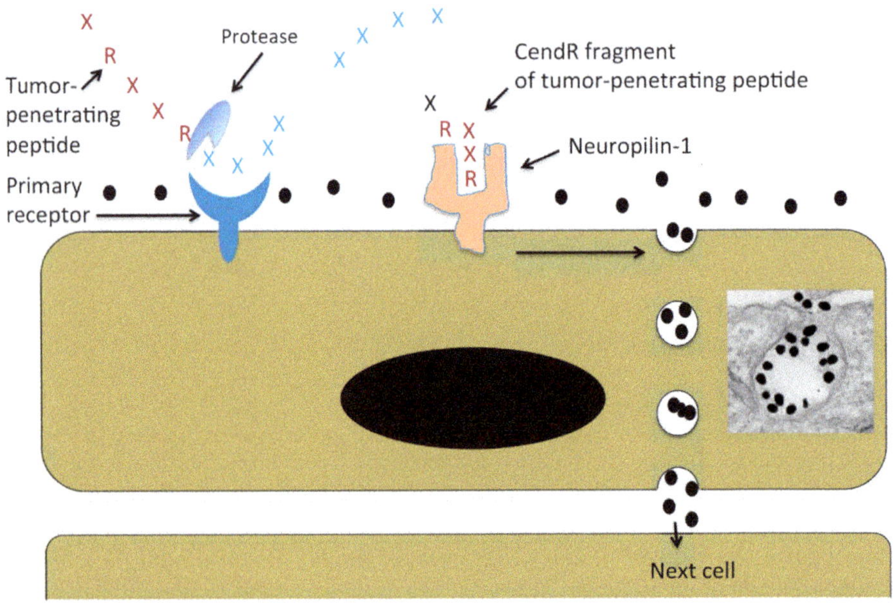

Figure 14.1 Schematic representation of the CendR transport pathway [4]. The process of tumor penetration with TPPs: (1) binding with a primary receptor on tumor endothelial cells; (2) peptide cleavage by protease to expose the CendR motif; and (3) binding with neuropilin-1 (NRP-1) and mediating vascular and tissue permeability [4].

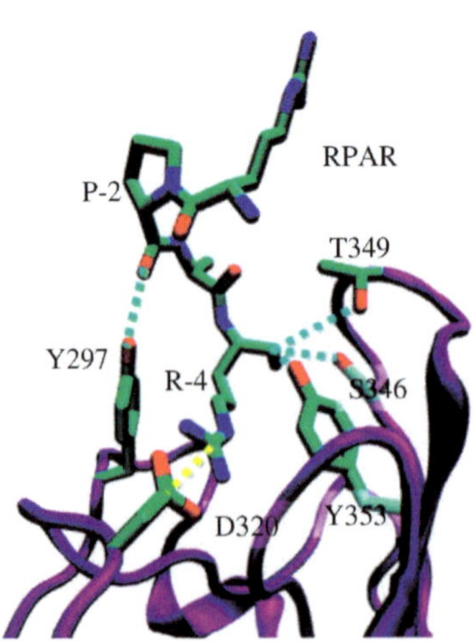

Figure 14.2 Ribbon representation of the NRP-1–RPAR complex [27].

antibodies or peptides as targeting ligands [30–36]. However, targeting of NRP-1 suffers from limited tumor specificity due to expression in normal vessels and the plasma protease carboxypeptidases in the bloodstream can remove C-terminal arginine residues. Therefore, the efficacy of systemically active CendR peptides in tumor drug delivery is largely limited. Benefiting from the localized tumor-specific proteolytic activation, the exposed CendR motif in TPPs provides tumor-specific activation of cell and tissue penetration (Table 14.1).

14.3.2 Tumor-penetrating CendR peptides

Another design of TPPs is based on a concept with two motifs for proteolytic activation and tumor-homing. The CendR motif containing the RGD peptide, iRGD (CRGDKGPDC), has been demonstrated to be able to carry 10 times more drug cargo into a tumor than the conventional RGD peptide [25]. Apart from the iRGD, other tumor-homing CendR peptides, such as F3 [64] and LyP-1 [16], have been shown to increase the extravasation of cargo loading and drug uptake to tumors [65,66]. The NGR tumor-homing motif, which binds to endothelial CD13 in angiogenic tumor vessels, was converted to a CendR motif (RNGR) and embedded in the iRGD framework to form a new peptide (CRNGRGPDC, iNGR). In comparison to NRG, the iNGR showed greater tumor penetration of coupled nanoparticles and coadministered compounds [59]. It has been demonstrated that the tumor penetration of CendR peptides is through an endosomal route and the process of tumor penetrating has been identified to be energy dependent, indicating an active transport pathway [18]. The CendR pathway has been characterized as starting with an endocytosis step and it is different from other endocytosis pathways. Initiated by an NRP receptor, the CendR endocytic vesicles resemble macropinocytotic vesicles with an average diameter of about 200 nm. A considerable volume of extracellular fluid and payloads within this size can be swept through into the pathway, leading to the remarkable feature of a coadministered effect. Under this mechanism, the therapeutic agent can be transported into the tumor parenchyma without conjugating to the CendR peptide [43]. A slight advantage of the coadministered modality was observed, whereas the conjugated modality gained better selective tumor penetration for drugs ranging in size from low-molecular-weight chemicals to nanoparticles and antibodies [43].

14.3.3 Cell-penetrating peptides

Different from TPPs, CPPs generally refers to peptides that are internalized into cells. CPPs are small polypeptides that contain several positively charged amino acids, such as lysine or arginine, or having sequences that contain a pattern of alternating polar and nonpolar, hydrophobic amino acids [1]. There are two main types of structures: polycationic and amphipathic. The mechanism of cell uptake of CPPs still remains controversial since the methods employed by different labs are often not comparable. However, it has been shown that positively charged CPPs can target and penetrate into tumor cells in a receptor-independent manner [3,67].

Table 14.1 Tumor-penetrating peptides

Type of peptide	Type of application	References
iRGD	Cisplatin in nanoparticles (coadministration drug delivery)	[37]
iRGD	Gemcitabine (coadministration drug delivery)	[38,39]
iRGD	Doxorubicin, sorafenib (coadministration drug delivery)	[40,41]
iRGD	Nanoparticles containing doxorubicin (coadministration drug delivery)	[42]
iRGD	Nab-paclitaxel (conjugates drug delivery)	[25,43]
iRGD	Paclitaxel in nanoparticles (conjugates drug delivery)	[44]
iRGD	Thymopentin fused to iRGD (conjugates drug delivery)	[45]
iRGD	Doxorubicin in chitosan nanoparticles (conjugates drug delivery)	[46]
iRGD	Doxorubicin liposomes (conjugates drug delivery)	[47]
iRGD	Doxorubicin multilamellar liposomes (conjugates drug delivery)	[48]
iRGD	Paclitaxel nanoparticles (conjugates drug delivery)	[48]
iRGD	Oncolytic viruis (conjugates drug delivery)	[49]
iRGD	Paclitaxel liposomes (conjugates drug delivery)	[50]
iRGD	Paclitaxel and surviving siRNA in nanoparticles (conjugates drug delivery)	[51]
iRGD	Doxorubicin-containing exosomes (conjugates drug delivery)	[52]
iRGD	ATAP mitochondrial targeting peptide (conjugates drug delivery)	[53]
iRGD	Anti-EGFR (conjugates drug delivery)	[41]
iRGD	Nanomicelles containing salinomycin (conjugates drug delivery)	[54]
iRGD	Microbubbles for ultrasound imaging and photothermal/ photodynamic therapy	[55]
iRGD	Liposomal doxorubicin (conjugates drug delivery)	[56]
iRGD	Paclitaxel nanocrystallite (conjugates drug delivery)	[57]
iRGD	Sorafenib in porous silicon nanoparticles (conjugates drug delivery)	[58]
iNGR	Doxorubicin (coadministration drug delivery)	[59]
tLyP-1	Paclitaxel nanoparticles (coadministration drug delivery)	[60]
tLyp-1	Tumor targeted nanoparticles (coadministration drug delivery)	[61]
RPARPAR	Doxorubicin liposomes (conjugates drug delivery)	[62]
RGERPPR	Doxorubicin liposomes (conjugates drug delivery)	[63]

Although CPPs have been widely applied as delivery vehicles in tumor diagnosis and therapy studies, in vivo application is still limited for two reasons. Firstly, the positively charged peptides can be easily taken up by the reticuloendothelial system (RES), leading to fast clearance from the blood circulation. Secondly, CPPs suffer from low specificity for targeting tumor cells versus normal cells, resulting in toxicity and insufficient therapy. To solve this problem, multistage nanocomplexes which carry targeting ligands that respond to the tumor microenvironment were designed including the activatable cell-penetrating peptide (ACPP) systems, deshielding systems, "pop-up" systems, and trojan-horse targeting systems.

The cationic CPPs may also contain the CendR sequence motifs and it is conceivable that they may use the CendR pathway to enter cells. One such example is a herpes virus protein, VP22, which has been shown to have similar tumor-penetrating function [68,69]. However, the tissue-penetrating function in other CPPs is unclear. The mechanism of cell uptake can occur through different endocytosis routes including caveolae [70], micropinocytosis [71], or clathrin-dependent pathway [72], cholesterol-dependent clathrin-mediated pathway [73], or the trans-Golgi network [74]. The Tat peptide, which is a prototypic CPP, is commonly applied in delivering payloads into cells through the endocytic pathway [71,75]. Generally, the cell uptake is considered to be heparan sulfate-dependent and it can be inhibited with an excess of free heparan sulfate [76]. Different from CPPs, typical CendR peptides do not bind to heparan sulfate but only use the CendR pathway, even though the endocytic vesicles containing the two different pathway cargoes are morphologically similar. The distinctive feature of the two different pathways is that the CendR receptor, i.e., NRP1, is overexpressed in tumors while the heparan sulfate is widely expressed in other cell surfaces. Therefore, higher tumor specificity can be achieved with TPPs.

14.4 Tumor-penetrating CendR peptides in drug delivery

14.4.1 The drug penetration problem

Drugs have to cross the vascular barrier and penetrate into the tumor stroma to reach tumor and tumor-associated parenchymal cells to effectively kill cancer cells. However, it has been demonstrated in many cases that drugs penetrate the equivalent of only a few cell diameters from blood vessels into the extravascular tumor tissue [77]. Due to the highly heterogeneous features of the cancerous tissue including (1) regional differences in tumor structure such as leaky vasculature and (2) defective lymphatics, and physiology such as inflammation, fibrosis, acidity, and hypoxia, the uptake and distribution of drugs can be entirely different in different regions. Small-molecule drugs, such as doxorubicin, may distribute up to a distance of about 40 μm [77], whereas larger drug molecules such as trastuzumab (herceptin antibody) are not detected inside many tumors [78]. As a result, low concentrations of drugs often cause drug resistance and the efficacy of chemotherapy to be low.

The reasons for the poor penetration of drugs into tumor are: First, the tumor microenvironment constitutes a dense extracellular matrix which creates a physical barrier to the penetration of anticancer drugs into tumor tissue; and second, the abnormal structure and function of tumor vessels and the poorly functional lymphatic drainage leading to high interstitial pressure [77,78]. Dissolving extracellular matrix with enzymes such as collagenase or hyaluronidase and inhibiting the activity of tumor-associated fibroblasts can partially destroy the physical barrier and increase drug delivery [79,80]. To lower the interstitial pressure, vasodilatory compounds such as bradykinin, endothelin, and calcium channel antagonists can be used to allow better tumor perfusion and therefore increase the drug concentration at the tumor site [81]. However, tumor penetration still remains a challenge in cancer drug delivery, especially with nanoparticle drugs [4].

14.4.2 TPP-enhanced drug delivery

Drug delivery with tumor-homing CendR peptides may overcome the tumor penetration problem. Drugs attached to CendR peptides can be delivered to the tumor tissue more effectively than the drug alone. Tumor penetration with fluorescein (FAM)-labeled peptides showed a fast distribution in tumor parenchyma within several minutes after intravenous administration in comparison to FAM-labeled control peptides. Meanwhile, the THPs that lack a CendR motif only showed binding to blood vessels [16,25,59,82]. Studies on iRGD and conventional RGD peptides, such as CRGDC and cRGDfK, demonstrated that iRGD peptides can deliver the payloads to tumor parenchyma, whereas the conventional RGD peptides only target tumor vessels [25,43].

The coadministration of CendR peptides with drugs can also overcome drug resistance in tumors. Ruoslahti et al. reported a study in a breast cancer xenograft model (BT474), which is essentially resistant to paclitaxel, showing effectiveness after the coadministration of nab-paclitaxel (Abraxane) nanoparticles together with iRGD or with iRGD coupled to the nanoparticle surface [43]. This finding may be explained by the fact that the drugs with the iRGD tumor endothelium-targeting domain enter cells through the CendR pathway rather than directly through the cell membrane [83,84]. Apart from iRGD, other CendR peptides also showed enhanced tumor penetration for drug delivery. For example, the LyP-1 peptide coupled with Abraxane nanoparticles, which is a clinically approved paclitaxel-albumin nanoparticle, have been demonstrated to produce more significant inhibition of tumor growth in mice bearing MDA-MB-435 human cancer model compared to the Abraxane nanoparticles without the tumor-targeting peptide [66]. Another CendR peptide, iNGR, possesses the same targeting ability as iRGD for delivery of doxorubicin in a mouse model [59].

Interestingly, drug delivery using CendR peptides can be achieved under the condition that the therapeutic agent is not linked to the peptides and can be enhanced by the peptides' stimulatory effect (bystander effect). This strategy offers a number of advantages. First, a faster route for clinical translation can be provided since no new chemical entities are synthesized; second, there is no limitation to the number of available receptors, which is usually quite low on the surface of cells. Therefore, significant work has been performed to allow for the clinical translation of iRGD peptides. However, the bystander effect of intravenously injected iRGD can only last for a very short time since the blood half-life of the peptide is only about 8 min [75]. Obviously, extending the iRGD half-life will offer better effect for drug delivery.

A more recent study by Jindřich Kopeček et al. demonstrated that a stimuli-responsive drug-delivery system with iRGD conjugated to a HPMA copolymer−DOX conjugate via a MMP-2 cleavable spacer can enhance cell accumulation and penetration of DOX in both monolayer and multicellular spheroid models of prostate cancer [85]. The main goal of this design was to develop a smart drug-delivery system to conquer the problem that the conjugated form of iRGD may lose its cell-penetrating/targeting ability from lack of exposure of the CendR motif. iRGD in the resulting HPMA copolymer was linked with a peptide sequence that can be released into the tumor microenvironment. Therefore, drug delivery and penetration were achieved through the coadministration effect from the free iRGD [86]. Moreover, this nanosystem can also prolong the half-life of iRGD as well as preserve its penetrating/targeting ability.

14.5 Tumor-penetrating CendR peptides in tumor imaging

CendR peptides can also be used to enhance tumor diagnosis using techniques such as magnetic resonance imaging (MRI) and positron emission tomography (PET). Superparamagnetic iron oxide nanoworms with sizes of about 80 nm long and 30 nm thick were conjugated with iRGD peptide and injected into a tumor-bearing mouse model. The results showed enhanced T2-weighted contrast in MR images 3 h postinjection (Fig. 14.3) [25]. CRGDC nanoworms caused decreased signal intensity in the tumor vasculature, whereas the untargeted nanoworms showed no significant contrast enhancement at the tumor site (Fig. 14.3). To some extent, the successful delivery of molecular imaging probes offers a promising tool for validating the drug-delivery efficacy of CendR peptides.

A similar diagnostic platform was also developed using LyP-1 peptide as a targeting regent and it was tested in an atherosclerotic plaque model. After intravenous injection of the LyP-1-labeled superparamagnetic iron-oxide nanoworms, the plaques showed decreased signals in the interior of plaques which was also validated by the colocalization of p32 and CD68 markers in the plaques [87]. More sensitive detection was also achieved with PET imaging, which showed a specific accumulation of the LyP-1-labeled probe within the aortic plaque. Xiaoyuan Chen et al. investigated the imaging ability of iRGD peptide using a novel design of an optical coupling structure [88]. Significant tumor-targeting imaging was observed 4 h post-injection in an orthotopic MDA-MB-435 xenograft model, suggesting the importance of the iRGD targeting peptide for tumor imaging [88].

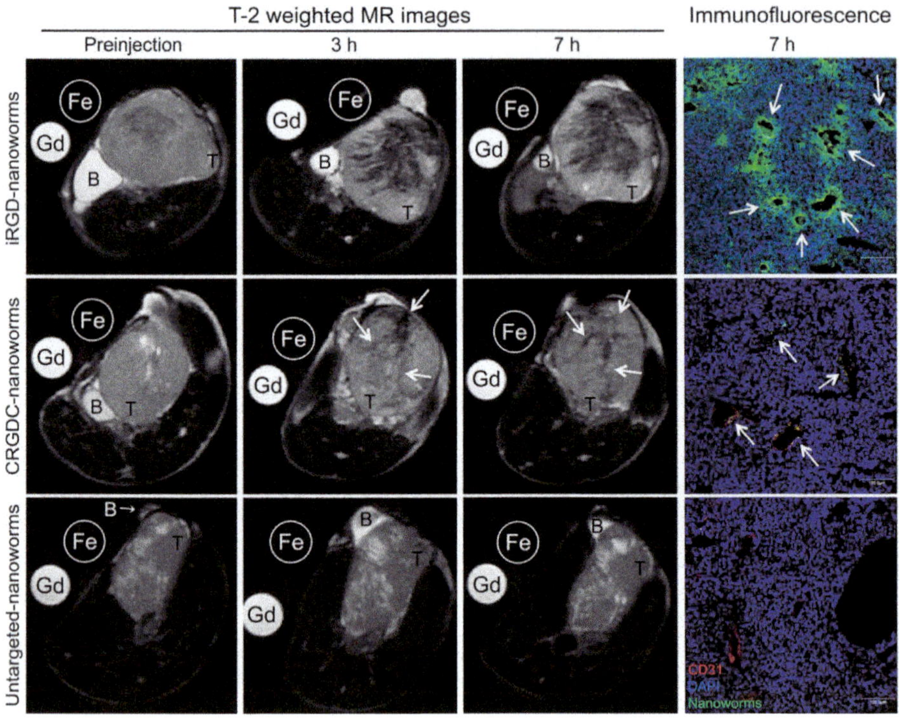

Figure 14.3 Tumor imaging with iRGD-coated iron oxide nanoworms [25].

Recently, Zheng et al. constructed a multifunctional near-infrared (NIR) probe by using iRGD-modified indocyanine green (ICG) liposomes (iRGD–ICG-LPs) for molecular imaging-guided photothermal therapy (PTT) and photodynamic therapy (PDT) against breast tumors. The results from optical imaging suggested that iRGD–ICG-LPs can target breast tumor with sensitive detection of breast tumor through NIR fluorescence molecular imaging. In vivo studies in a 4T1 tumor-bearing model showed significant tumor inhibition and almost complete tumor regression upon treatment of the animals with the iRGD–ICG-LP formulations in comparison to control animals. Therefore, theranostic application for both tumor diagnosis and therapy can be achieved by integrating molecular imaging into PTT/PDT therapy [55].

14.6 Conclusion

TPPs have shown significant efficiency in delivering both covalently coupled and coadministered drug payloads deep into tumor tissues. Evidence from in vivo animal studies demonstrated that drug distribution in tumor tissue is significantly increased upon the use of such TPPs. Activated by the CendR motif with the NRP-

1 receptor on the vasculature surface, TPPs can activate the process for the form of resembling micropinocytosis that can uptake the drug link to the peptide or be coadministered with it. Due to overexpression of the NRP-1 receptor in tumor tissues, CendR peptides are generally more specific than traditional CPP or vascular targeting peptides. However, the short blood half-life of the peptide largely reduces its overall efficacy. Therefore, future improvements will involve the rational design of TPP with increased blood circulation. Nevertheless, TPPs, such as iRGD, have been poised for clinical studies. Their applications in molecular imaging or tumors and in theranostics clearly suggest that the system is very promising not only for the functionalization of nanoparticle-based diagnostic tools but also for the treatment of several types of cancer.

Acknowledgments

This work has been supported by research grants from the Special National Key Research and Development Plan (No. 2016YFD0400202-8) and the National Natural Science Foundation of China (No. 81571746).

References

[1] V. Kersemans, B. Cornelissen, Targeting the tumour: cell penetrating peptides for molecular imaging and radiotherapy, Pharmaceuticals 3 (3) (2010) 600–620.
[2] O. Feron, Tumor-penetrating peptides: a shift from magic bullets to magic guns, Sci. Transl. Med. 2 (34) (2010) 34ps26.
[3] M. Zorko, U. Langel, Cell-penetrating peptides: mechanism and kinetics of cargo delivery, Adv. Drug Deliv. Rev. 57 (4) (2005) 529–545.
[4] E. Ruoslahti, Tumor penetrating peptides for improved drug delivery, Adv. Drug Deliv. Rev. 110–111 (2017) 3–12.
[5] T.W. Hambley, Is anticancer drug development heading in the right direction? Cancer Res. 69 (4) (2009) 1259–1261.
[6] R.K. Jain, Transport of molecules, particles, and cells in solid tumors, Ann. Rev. Biomed. Eng. 1 (1999) 241–263.
[7] I.F. Tannock, A. Primeau, R. Grantab, et al., Limited penetration of anticancer drugs to cells in solid tumors: a neglected and modifiable cause of drug resistance, J. Clin. Oncol. 23 (16) (2005), 852s-s.
[8] A.I. Minchinton, I.F. Tannock, Drug penetration in solid tumours, Nat. Rev. Cancer 6 (8) (2006) 583–592.
[9] A.K. Iyer, G. Khaled, J. Fang, et al., Exploiting the enhanced permeability and retention effect for tumor targeting, Drug Discov. Today 11 (17–18) (2006) 812–818.
[10] H. Maeda, J. Wu, T. Sawa, et al., Tumor vascular permeability and the EPR effect in macromolecular therapeutics: a review, J. Control. Release 65 (1–2) (2000) 271–284.
[11] G.P. Smith, Filamentous susion phage – novel expression vectors that display cloned antigens on the virion surface, Science 228 (4705) (1985) 1315–1317.

[12] J. Babickova, L. Tothova, P. Boor, et al., In vivo phage display – a discovery tool in molecular biomedicine, Biotechnol. Adv. 31 (8) (2013) 1247–1259.

[13] D. Hanahan, J. Folkman, Patterns and emerging mechanisms of the angiogenic switch during tumorigenesis, Cell 86 (3) (1996) 353–364.

[14] K. Alitalo, P. Carmeliet, Molecular mechanisms of lymphangiogenesis in health and disease, Cancer Cell 1 (3) (2002) 219–227.

[15] E. Ruoslahti, Peptides as targeting elements and tissue penetration devices for nanoparticles, Adv. Mater. 24 (28) (2012) 3747–3756.

[16] P. Laakkonen, K. Porkka, J.A. Hoffman, et al., A tumor-homing peptide with a targeting specificity related to lymphatic vessels, Nat. Med. 8 (7) (2002) 751–755.

[17] V. Fogal, L. Zhang, S. Krajewski, et al., Mitochondrial/cell-surface protein p32/gC1qR as a molecular target in tumor cells and tumor stroma, Cancer Res. 68 (17) (2008) 7210–7218.

[18] T. Teesalu, K.N. Sugahara, E. Ruoslahti, Tumor-penetrating peptides, Front. Oncol. 3 (2013) 216.

[19] P. Libby, M. DiCarli, R. Weissleder, The vascular biology of atherosclerosis and imaging targets, J. Nucl. Med. 51 (2010) 33s–37ss.

[20] S.M. Weis, D.A. Cheresh, Tumor angiogenesis: molecular pathways and therapeutic targets, Nat. Med. 17 (11) (2011) 1359–1370.

[21] J.D. Humphries, A. Byron, M.J. Humphries, Integrin ligands at a glance, J. Cell Sci. 119 (19) (2006) 3901–3903.

[22] A. Erdreich-Epstein, H. Shimada, S. Groshen, et al., Integrins alpha(v)beta3 and alpha(v) beta5 are expressed by endothelium of high-risk neuroblastoma and their inhibition is associated with increased endogenous ceramide, Cancer Res. 60 (3) (2000) 712–721.

[23] W. Arap, R. Pasqualini, E. Ruoslahti, Cancer treatment by targeted drug delivery to tumor vasculature in a mouse model, Science 279 (5349) (1998) 377–380.

[24] T. Teesalu, K.N. Sugahara, V.R. Kotamraju, et al., C-end rule peptides mediate neuropilin-1-dependent cell, vascular, and tissue penetration, Proc. Natl. Acad. Sci. U. S. A. 106 (38) (2009) 16157–16162.

[25] K.N. Sugahara, T. Teesalu, P.P. Karmali, et al., Tissue-penetrating delivery of compounds and nanoparticles into tumors, Cancer Cell 16 (6) (2009) 510–520.

[26] S. Soker, S. Takashima, H.Q. Miao, et al., Neuropilin-1 is expressed by endothelial and tumor cells as an isoform-specific receptor for vascular endothelial growth factor, Cell 92 (6) (1998) 735–745.

[27] N. Haspel, D. Zanuy, R. Nussinov, et al., Binding of a C-end rule peptide to the neuropilin-1 receptor: a molecular modeling approach, Biochemistry 50 (10) (2011) 1755–1762.

[28] D. Zanuy, R. Kotla, R. Nussinov, et al., Sequence dependence of C-end rule peptides in binding and activation of neuropilin-1 receptor, J. Struct. Biol. 182 (2) (2013) 78–86.

[29] A. Bagri, M. Tessier-Lavigne, R.J. Watts, Neuropilins in tumor biology, Clin. Cancer Res. 15 (6) (2009) 1860–1864.

[30] T.M. Hong, Y.L. Chen, Y.Y. Wu, et al., Targeting neuropilin 1 as an antitumor strategy in lung cancer, Clin. Cancer Res. 13 (16) (2007) 4759–4768.

[31] W.C. Liang, M.S. Dennis, S. Stawicki, et al., Function blocking antibodies to neuropilin-1 generated from a designed human synthetic antibody phage library, J. Mol. Biol. 366 (3) (2007) 815–829.

[32] J. Sun, L. Wang, W.H. Lou, et al., Construction and identification of small hairpin RNA plasmids targeting neuropilin-1 gene and their inhibitory effect on human nasopharyngeal carcinoma CNE-2Z cell proliferation in vitro and in vivo, Eur. Arch. Otorhinolaryngol. 273 (9) (2016) 2541–2547.

[33] K. Karjalainen, D.E. Jaalouk, C.E. Bueso-Ramos, et al., Targeting neuropilin-1 in human leukemia and lymphoma, Blood 117 (3) (2011) 920–927.
[34] D. Bechet, L. Tirand, F. Plenat, et al., Targeting of neuropilin-1 to improve the anti-vascular effect of photodynamic therapy in xenograft human malignant glioma, EJC Suppl. 6 (9) (2008) 25–26.
[35] L. Tirand, C. Frochot, R. Vanderesse, et al., A peptide competing with VEGF(165) binding on neuropilin-1 mediates targeting of a chlorin-type photosensitizer and potentiates its photodynamic activity in human endothelial cells, J. Control. Release 111 (1–2) (2006) 153–164.
[36] C. Nasarre, M. Roth, L. Jacob, et al., Peptide-based interference of the transmembrane domain of neuropilin-1 inhibits glioma growth in vivo, Oncogene 29 (16) (2010) 2381–2392.
[37] W.T. Song, M.Q. Li, Z.H. Tang, et al., Methoxypoly(ethylene glycol)-block-poly(L-glutamic acid)-loaded cisplatin and a combination with iRGD for the treatment of non-small-cell lung cancers, Macromol. Biosci. 12 (11) (2012) 1514–1523.
[38] Y. Akashi, T. Oda, Y. Ohara, et al., Anticancer effects of gemcitabine are enhanced by co-administered iRGD peptide in murine pancreatic cancer models that overexpressed neuropilin-1, Br. J. Cancer 110 (6) (2014) 1481–1487.
[39] Q. Zhang, Y. Zhang, K. Li, et al., A novel strategy to improve the therapeutic efficacy of gemcitabine for non-small cell lung cancer by the tumor-penetrating peptide iRGD, PLoS One 10 (6) (2015) e0129865.
[40] C. Schmithals, V. Koberle, H. Korkusuz, et al., Improving drug penetrability with iRGD leverages the therapeutic response to sorafenib and doxorubicin in hepatocellular carcinoma, Cancer Res. 75 (15) (2015) 3147–3154.
[41] H.Z. Sha, Z.Y. Zou, K. Xin, et al., Tumor-penetrating peptide fused EGFR single-domain antibody enhances cancer drug penetration into 3D multicellular spheroids and facilitates effective gastric cancer therapy, J. Control. Release 200 (2015) 188–200.
[42] X.L. Cun, J.T. Chen, S.B. Ruan, et al., A novel strategy through combining iRGD peptide with tumor-microenvironment-responsive and multistage nanoparticles for deep tumor penetration, ACS Appl. Mater. Interfaces 7 (49) (2015) 27458–27466.
[43] K.N. Sugahara, T. Teesalu, P.P. Karmali, et al., Coadministration of a tumor-penetrating peptide enhances the efficacy of cancer drugs, Science 328 (5981) (2010) 1031–1035.
[44] Z.S. Zhu, C. Xie, Q. Liu, et al., The effect of hydrophilic chain length and iRGD on drug delivery from poly(epsilon-caprolactone)-poly(N-vinylpyrrolidone) nanoparticles, Biomaterials 32 (35) (2011) 9525–9535.
[45] X.Z. Lao, M. Liu, J. Chen, et al., A tumor-penetrating peptide modification enhances the antitumor activity of thymosin alpha 1, PLoS One 8 (8) (2013) e72242.
[46] X. Wang, X. Zhen, J. Wang, et al., Doxorubicin delivery to 3D multicellular spheroids and tumors based on boronic acid-rich chitosan nanoparticles, Biomaterials 34 (19) (2013) 4667–4679.
[47] K.F. Yu, W.Q. Zhang, L.M. Luo, et al., The antitumor activity of a doxorubicin loaded, iRGD-modified sterically-stabilized liposome on B16-F10 melanoma cells: in vitro and in vivo evaluation, Int. J. Nanomedicine 8 (2013) 2473–2485.
[48] G.Z. Gu, X.L. Gao, Q.Y. Hu, et al., The influence of the penetrating peptide iRGD on the effect of paclitaxel-loaded MT1-AF7p-conjugated nanoparticles on glioma cells, Biomaterials 34 (21) (2013) 5138–5148.
[49] C. Puig-Saus, L.A. Rojas, E. Laborda, et al., iRGD tumor-penetrating peptide-modified oncolytic adenovirus shows enhanced tumor transduction, intratumoral dissemination and antitumor efficacy, Gene Ther. 21 (8) (2014) 767–774.

[50] R. Du, T. Zhong, W.Q. Zhang, et al., Antitumor effect of iRGD-modified liposomes containing conjugated linoleic acid-paclitaxel (CLA-PTX) on B16-F10 melanoma, Int. J. Nanomedicine 9 (2014) 3091–3105.
[51] J.N. Shen, Q.S. Meng, H.P. Sui, et al., iRGD conjugated TPGS mediates codelivery of paclitaxel and survivin shRNA for the reversal of lung cancer resistance, Mol. Pharm. 11 (8) (2014) 2579–2591.
[52] Y.H. Tian, S.P. Li, J. Song, et al., A doxorubicin delivery platform using engineered natural membrane vesicle exosomes for targeted tumor therapy, Biomaterials 35 (7) (2014) 2383–2390.
[53] G.J. De, J.K. Ko, T. Tan, et al., Amphipathic tail-anchoring peptide is a promising therapeutic agent for prostate cancer treatment, Oncotarget 5 (17) (2014) 7734–7747.
[54] X.L. Mao, J.J. Liu, Z.R. Gong, et al., iRGD-conjugated DSPE-PEG2000 nanomicelles for targeted delivery of salinomycin for treatment of both liver cancer cells and cancer stem cells, Nanomedicine (Lond.) 10 (17) (2015) 2677–2695.
[55] F. Yan, H. Wu, H.M. Liu, et al., Molecular imaging-guided photothermal/photodynamic therapy against tumor by iRGD-modified indocyanine green nanoparticles, J. Control. Release 224 (2016) 217–228.
[56] W.B. Dai, Y.C. Fan, H. Zhang, et al., A comprehensive study of iRGD-modified liposomes with improved chemotherapeutic efficacy on B16 melanoma, Drug Deliv. 22 (1) (2015) 10–20.
[57] D.Z. Ni, H. Ding, S. Liu, et al., Superior intratumoral penetration of paclitaxel nanodots strengthens tumor restriction and metastasis prevention, Small 11 (21) (2015) 2518–2526.
[58] M.Q. Li, Z.H. Tang, D.W. Zhang, et al., Doxorubicin-loaded polysaccharide nanoparticles suppress the growth of murine colorectal carcinoma and inhibit the metastasis of murine mammary carcinoma in rodent models, Biomaterials 51 (2015) 161–172.
[59] L. Alberici, L. Roth, K.N. Sugahara, et al., De novo design of a tumor-penetrating peptide, Cancer Res. 73 (2) (2013) 804–812.
[60] D.Y. Miao, M.Y. Jiang, Z.Y. Liu, et al., Co-administration of dual-targeting nanoparticles with penetration enhancement peptide for antiglioblastoma therapy, Mol. Pharm. 11 (1) (2014) 90–101.
[61] Q.Y. Hu, G.Z. Gu, Z.Y. Liu, et al., F3 peptide-functionalized PEG-PLA nanoparticles co-administrated with tLyp-1 peptide for anti-glioma drug delivery, Biomaterials 34 (4) (2013) 1135–1145.
[62] Z.Q. Yan, Y.Y. Yang, X.L. Wei, et al., Tumor-penetrating peptide mediation: an effective strategy for improving the transport of liposomes in tumor tissue, Mol. Pharm. 11 (1) (2014) 218–225.
[63] Y.Y. Yang, Z.Q. Yan, D.X. Wei, et al., Tumor-penetrating peptide functionalization enhances the anti-glioblastoma effect of doxorubicin liposomes, Nanotechnology 24 (40) (2013) 405101.
[64] V. Fogal, K.N. Sugahara, E. Ruoslahti, et al., Cell surface nucleolin antagonist causes endothelial cell apoptosis and normalization of tumor vasculature, Angiogenesis 12 (1) (2009) 91–100.
[65] K. Porkka, P. Laakkonen, J.A. Hoffman, et al., A fragment of the HMGN2 protein homes to the nuclei of tumor cells and tumor endothelial cells in vivo, Proc. Natl. Acad. Sci. U. S. A. 99 (11) (2002) 7444–7449.
[66] P.P. Karmali, V.R. Kotamraju, M. Kastantin, et al., Targeting of albumin-embedded paclitaxel nanoparticles to tumors, Nanomedicine 5 (1) (2009) 73–82.

[67] J.Q. Fu, C.M. Yu, L. Li, et al., Intracellular delivery of functional proteins and native drugs by cell-penetrating poly(disulfide)s, J. Am. Chem. Soc. 137 (37) (2015) 12153–12160.

[68] N.D. Brewis, A. Phelan, N. Normand, et al., Particle assembly incorporating a VP22-BH3 fusion protein, facilitating intracellular delivery, regulated release, and apoptosis, Mol. Ther. 7 (2) (2003) 262–270.

[69] X.D. Xue, J.H. Huang, H.S. Wang, The study of the intercellular trafficking of the fusion proteins of herpes simplex virus protein VP22, PLoS One 9 (6) (2014) e100840.

[70] A. Fittipaldi, A. Ferrari, M. Zoppe, et al., Cell membrane lipid rafts mediate caveolar endocytosis of HIV-1 Tat fusion proteins, J. Biol. Chem. 278 (36) (2003) 34141–34149.

[71] I.M. Kaplan, J.S. Wadia, S.F. Dowdy, Cationic TAT peptide transduction domain enters cells by macropinocytosis, J. Control. Release 102 (1) (2005) 247–253.

[72] J.P. Richard, K. Melikov, H. Brooks, et al., Cellular uptake of unconjugated TAT peptide involves clathrin-dependent endocytosis and heparan sulfate receptors, J. Biol. Chem. 280 (15) (2005) 15300–15306.

[73] R.E. Vandenbroucke, S.C. De Smedt, J. Demeester, et al., Cellular entry pathway and gene transfer capacity of TAT-modified lipoplexes, Biochim. Biophys. Acta 1768 (3) (2007) 571–579.

[74] R. Fischer, K. Kohler, M. Fotin-Mleczek, et al., A stepwise dissection of the intracellular fate of cationic cell-penetrating peptides, J. Biol. Chem. 279 (13) (2004) 12625–12635.

[75] H.B. Pang, G.B. Braun, T. Friman, et al., An endocytosis pathway initiated through neuropilin-1 and regulated by nutrient availability, Nat. Commun. 5 (2014) 4904.

[76] J.M. Gump, S.F. Dowdy, TAT transduction: the molecular mechanism and therapeutic prospects, Trends Mol. Med. 13 (10) (2007) 443–448.

[77] A.J. Primeau, A. Rendon, D. Hedley, et al., The distribution of the anticancer drug doxorubicin in relation to blood vessels in solid tumors, Clin. Cancer Res. 11 (24) (2005) 8782–8788.

[78] C.H. Heldin, K. Rubin, K. Pietras, et al., High interstitial fluid pressure: an obstacle in cancer therapy, Nat. Rev. Cancer 4 (10) (2004) 806–813.

[79] I. Tufto, R. Hansen, D. Byberg, et al., The effect of collagenase and hyaluronidase on transient perfusion in human osteosarcoma xenografts grown orthotopically and in dorsal skinfold chambers, Anticancer Res. 27 (3b) (2007) 1475–1481.

[80] F. Xing, J. Saidou, K. Watabe, Cancer associated fibroblasts (CAFs) in tumor microenvironment, Front. Biosci. (Landmark Ed.) 15 (2010) 166–179.

[81] O. Feron, Targeting the tumor vascular compartment to improve conventional cancer therapy, Trends Pharm. Sci. 25 (10) (2004) 536–542.

[82] P. Laakkonen, M.E. Akerman, H. Biliran, et al., Antitumor activity of a homing peptide that targets tumor lymphatics and tumor cells, Proc. Natl. Acad. Sci. U. S. A. 101 (25) (2004) 9381–9386.

[83] C.M.J. Hu, L.F. Zhang, Therapeutic nanoparticles to combat cancer drug resistance, Curr. Drug Metab. 10 (8) (2009) 836–841.

[84] J. Huwyler, A. Cerletti, G. Fricker, et al., By-passing of P-glycoprotein using immunoliposomes, J. Drug Target. 10 (1) (2002) 73–79.

[85] Z.H. Peng, J. Kopecek, Enhancing accumulation and penetration of HPMA copolymer-doxorubicin conjugates in 2D and 3D prostate cancer cells via iRGD conjugation with an MMP-2 cleavable spacer, J. Am. Chem. Soc. 137 (21) (2015) 6726–6729.

[86] Z.H. Peng, J. Kopecek, Synthesis and activity of tumor-homing peptide iRGD and histone deacetylase inhibitor valproic acid conjugate, Bioorg. Med. Chem. Lett. 24 (8) (2014) 1928–1933.

[87] J. Hamzah, V.R. Kotamraju, J.W. Seo, et al., Specific penetration and accumulation of a homing peptide within atherosclerotic plaques of apolipoprotein E-deficient mice, Proc. Natl. Acad. Sci. U. S. A. 108 (17) (2011) 7154–7159.

[88] Y.P. Ye, L. Zhu, Y. Ma, et al., Synthesis and evaluation of new iRGD peptide analogs for tumor optical imaging, Bioorg. Med. Chem. Lett. 21 (4) (2011) 1146–1150.

Self-assembling peptides in biomedicine and bioengineering: Tissue engineering, regenerative medicine, drug delivery, and biotechnology*

15

Sotirios Koutsopoulos
Massachusetts Institute of Technology, Cambridge, MA, United States

15.1 Introduction

Self-assembly is a spontaneous and reversible process in which disordered components are hierarchically organized into structures or patterns due to specific, noncovalent interactions with the components themselves. At the molecular level, self-assembly is the organization of molecules without outside guidance. This chapter focuses on the intermolecular association of peptides and the formation of structurally defined, stable organizations through local interactions. These interactions include hydrogen bonds, ionic bonds, van der Waals' bonds, hydrophobic interactions, and water-mediated hydrogen bonds. Although weak, these bonds in combination control the final structure and stability of the self-assembled system.

It is beyond the scope of this chapter to detail all the self-assembling peptide systems that have been discovered during the last 30 years after the pioneer work in the groups of Abraham and Beachey [1], Glabe [2], Imanishi and Kimura [3], Ghadiri [4,5], and Zhang [6]. Instead, this chapter will focus on important representatives of ionic-complementary, self-assembling peptide hydrogels [6,7], and other interesting cases of self-assembling peptides will be briefly discussed. The observation of the unusual behavior of a peptide fragment from the yeast protein zuotin opened new research paths resulting in the design and study of various peptide sequences. Self-organization of these peptides and those that were designed based on the first generation of ionic-complementary, self-assembling peptides provided a platform for research in diverse fields including biotechnology, bioengineering, biomedical engineering, biotechnology, and bioelectronics [7–10]. The most important of these applications with relevance to biomedicine will be discussed in this chapter.

* The author declares no conflict of interest.

Peptide Applications in Biomedicine, Biotechnology and Bioengineering.
DOI: http://dx.doi.org/10.1016/B978-0-08-100736-5.00015-6

15.2 Self-assembly of ionic-complementary peptides into hydrogels consisting of nanofibers

Inspired by the tandem repeats of the self-assembling zuotin protein fragment, the first designer ionic-complementary self-assembling peptide was Ac-AEAEAKAKAEAEAKAK-$CONH_2$ (i.e., Ac-$(AEAEAKAK)_2$-$CONH_2$) [6]. Several other self-assembling peptide variants were designed based on this sequence including Ac-$(RARADADA)_2$-$CONH_2$, Ac-$(RADA)_4$-$CONH_2$, and Ac-$(KLDL)_3$-$CONH_2$ [16–18]. Composed of 12–16 amino acids, these peptides form stable β-sheet structures in aqueous media. On one side of the β-sheet are the charged amino acid side groups with the hydrophobic side groups on the other side. Ionic-complementarity and chemical compatibility of the peptides monomers is critical for allowing the interaction and spontaneous assembly into nanofibers (Fig. 15.1) [7,19].

The design principles for self-assembling peptides have been established and optimized over the years. Self-assembling peptides are characterized by periodic repeats of alternating charged hydrophilic and hydrophobic amino acids. Depending on the charge arrangements on the hydrophilic side of the β-sheet, the complementary ionic sides have been classified into several moduli, e.g., modulus I, II, III, IV, etc., and mixed moduli [7]. The charge distribution for modulus I is "− + − + − + − +"; for modulus II "− − + + − − + +"; for modulus III "− − − + + +"; and for modulus IV "− − − − + + + +". Reversing the order of the charge orientation results in peptide molecules with new self-assembling features and properties. Stable nanofibers result from peptide alignment and association in the presence of an electrolyte. These nanofibers have a charged surface with a hydrophobic interior (Fig. 15.1) [20]. On the hydrophilic surface, alternating positively and negatively charged amino acid residues with regular repeats facilitate complementary ionic interactions and form checkerboard-like structures. Shielded from interaction with water, the hydrophobic residues self-assemble in the interior of the nanofiber.

The presence of electrolytes in the solution drives peptide self-assembly. A small net positive or negative charge present in the self-assembling peptides results in

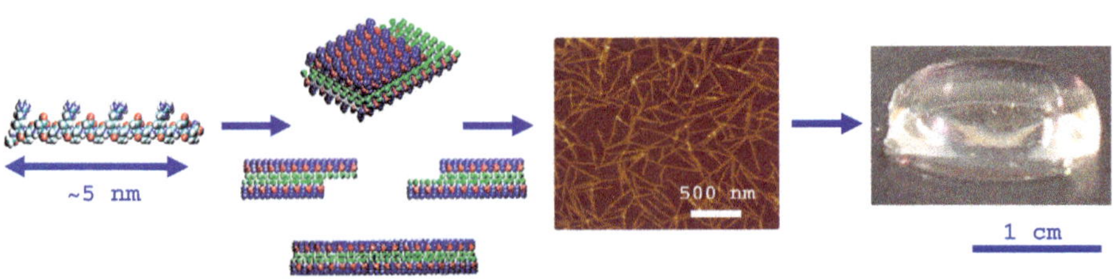

Figure 15.1 Self-assembly of peptides into nanofibers and the formation of the peptide hydrogel. At physiological conditions (e.g., 0.15 M NaCl, pH 7.4) individual peptides of ~5 nm each self-assemble and form nanofibers which are further organized to form a hydrogel which contains water up to 99.5%.
Source: Reprinted with permission from S.G. Zhang, Fabrication of novel biomaterials through molecular self-assembly, Nat. Biotechnol. 21 (2003) 1171–1178.

small electrostatic repulsions of the peptide monomers. Relatively high concentrations of these peptides remain dissolved in water for prolonged periods. Addition of an electrolyte initiates self-assembly instantly. Although the phenomenon has not been thoroughly understood, peptide self-assembly is thought to occur through a decrease of the electrostatic repulsions of the like-charged peptide monomers in the presence of the electrolyte ions. By bringing the peptides closer to each other, hydrophobic interactions become significant, leading to the formation of nanofibers [9,21]. These nanofibers are further organized into 3D architectures that retain up to 99.5% water, forming a hydrogel with pore diameter of 5–200 nm [22].

The first generation of self-assembling peptides was designed to have amino acid sequences that self-assemble forming a nanofiber hydrogel (Fig. 15.1). The second generation was developed to provide internal stimuli that promote cell attachment, proliferation, and differentiation (Fig. 15.2). ECM proteins provided a template to choose functional amino acid sequences that promote biological compatibility between cells and the hydrogel scaffold [23]. These functional motifs (e.g., RGD domain) were synthesized and attached to the C- or N-terminus of the self-assembling peptide unit (i.e., -$(RADA)_4$-). The addition of a two-glycine residue linker was found to be necessary to ensure sufficient flexibility and exposure of the functional motifs to cell membrane receptors. By this approach several tailor-made self-assembling peptide sequences were obtained which showed increased cell attachment, proliferation, migration, differentiation, and expression of cell-specific biological markers.

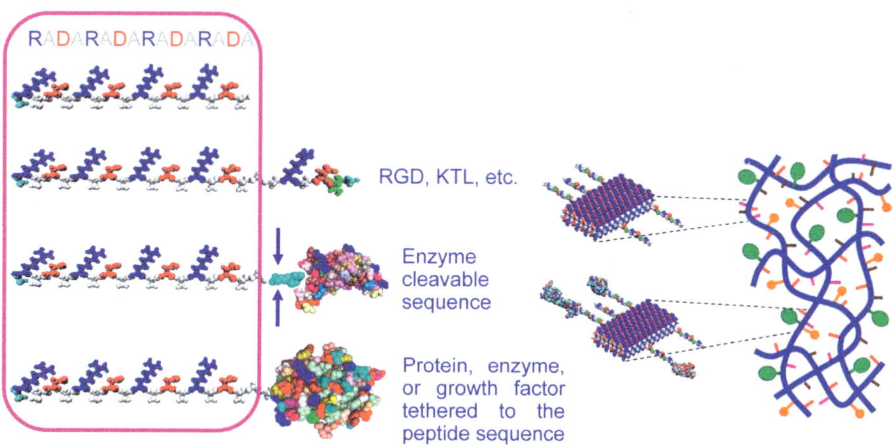

Figure 15.2 Designer peptide scaffolds carrying proteins, enzymes, and growth factors tethered directly or through the biotin–avidin complex to the self-assembling peptide sequence forming peptide nanofiber hydrogels that provide constant stimuli to cells or tissues.
Source: Reprinted with permission from S. Koutsopoulos, Self-assembling peptide nanofiber hydrogels in tissue engineering and regenerative medicine: progress, design guidelines, and applications, J. Biomed. Mater. Res. A. 104 (2016) 1002–1016.

The third generation of ionic-complementary, self-assembling peptides involves the tethering of proteins to the self-assembling peptide unit (Fig. 15.2). Soluble proteins such as cytokines and growth factors are important effectors in regenerative medicine and tissue engineering. These proteins may be tethered to the N- or C-terminus of the peptides to provide stimuli to cells interacting with the peptide nanofiber hydrogel. Chau et al. utilized this approach to tether a matrix metalloproteinase-2 (MMP-2) cleavable hexapeptide sequence to the Ac-$(RADA)_4$-self-assembling unit [24]. This hexapeptide sequence and other sequences can be used as linkers between molecules with therapeutic properties against cancer cells and the self-assembling peptide sequence. When MMP-2 interacts with such peptide hydrogels, the linker is cleaved and the bioactive molecule is released to kill cancer cells. Lee and colleagues immobilized a bioactive protein to the self-assembling peptide nanofibers of the peptide hydrogel through the biotin–avidin binding system [25].

15.2.1 Advantages of ionic-complementary self-assembling peptides

Ionic-complementary peptide hydrogels are ideal biomaterials because they do not use toxic crosslinkers or other harmful chemicals to initiate the sol-gel transformation and their mechanical properties can be fine-tuned. Degradation of such hydrogels results in amino acids which are naturally metabolized. Furthermore, the sol-gel transition occurs at physiological conditions and results in high internal hydration. The hydrogel consists of nanoscale peptide fibers allowing for the encapsulation of cells. They are easy to use, nontoxic, nonimmunogenic, nonthrombogenic, biodegradable, and applicable to localized therapies [21,22,26].

Tissue culture studies using the Ac-$(AEAEAKAK)_2$-$CONH_2$, Ac-$(RARADADA)_2$-$CONH_2$, Ac-$(RADA)_4$-$CONH_2$, and Ac-$(KLDL)_3$-$CONH_2$ self-assembling peptide hydrogels showed excellent compatibility (Table 15.1) for various cell types, including neural cells, mesenchymal stem cells (MSCs), chondrocytes, hepatocytes, cardiomyocytes, skin epithelial cells, and endothelial cells [18,30,31,33,35,36,40–42,56,57]. Compared with many synthetic polymer matrices or animal-derived materials commonly used, the results of these studies suggest that self-assembling peptides are better biomaterials for tissue cell cultures [58]. Composed of nanofibers built with natural amino acids, self-assembling peptide hydrogels resemble the composition and size scale of the extracellular matrix (ECM) fibers surrounding cells. Most of the polymer matrices used in past decades for tissue engineering consist of ~10–50-μm microfibers that are the same size or larger than mammalian cells. Utilizing such type of materials for cell cultures leads to cell spreading on the microfiber surface and tissue cultures that do not mimic the in vivo 3D microenvironment of the cell. On the other hand, tissue culture studies are generally not reproducible when using biologically derived materials such as collagen and Matrigel. These materials contain undefined concentrations of impurities, growth factors, and cytokines because the matrix composition varies from lot to lot.

Table 15.1 Types of ionic-complementary, self-assembling peptides and some of their applications in tissue engineering, tissue regeneration, and drug delivery

Peptide	Application	References
Ac-(AEAEAKAK)$_2$-CONH$_2$	Bone, cartilage, and myocardium regeneration	[27]
	Drug delivery	[28,29]
Ac-(RARADADA)$_2$-CONH$_2$	Angiogenesis, heart tissue regeneration, neural tissue regeneration	[25,30–32]
Ac-(RADA)$_4$-CONH$_2$	Angiogenesis, bone regeneration, cartilage regeneration, ligament regeneration, neural tissue regeneration, optical nerve regeneration, wound healing, liver regeneration, cochlea regeneration, ear mucosal regeneration, tooth regeneration	[32–40, 42–48, 66]
	Drug delivery of encapsulated small molecules, proteins, and cleavable groups (drugs, diagnostic molecules, or proteins)	[21,22,26,76,77]
Ac-(KLDL)$_3$-CONH$_2$	Cartilage regeneration	[18,51–53]
	Drug delivery of encapsulated small molecules, peptides, proteins, and cleavable groups (drugs, diagnostic molecules, or proteins)	[54,55]

Furthermore, the biochemical pathways observed may be due to unknown cell-signaling factors present in the matrix [59].

15.3 Applications of ionic-complementary, self-assembling peptide hydrogels in tissue engineering and tissue regeneration

15.3.1 Bone regeneration

Compared to bone tissue, hydrogel scaffolds have poor mechanical properties. However, to their advantage, hydrogels are compatible with bone cells allowing for cell–cell interactions, cell proliferation, and ECM production, which promotes tissue regeneration. Investigating bone tissue regeneration in vivo, Misawa et al. showed that bone defects in mice calvaria healed significantly faster when they were treated with the injectable Ac-$(RADA)_4$-$CONH_2$ self-assembling peptide hydrogel compared with mice treated with Matrigel [56].

In another study, Bokhari et al. examined the formation of bone nodules. They created a composite material by mixing PolyHIPE with the Ac-$(RADA)_4$-$CONH_2$ self-assembling peptide hydrogel. PolyHIPE is a porous hydrophobic polymer material with good mechanical properties, whereas the hydrogel provides a suitable microenvironment for osteoblast cell growth, differentiation, and biomineralization in vitro [35]. A 35-day study showed that mineralized bone nodules were extensively formed inside the peptide hydrogel–PolyHIPE composite, but were not observed in the PolyHIPE scaffold alone.

In a similar strategy, Kim et al. used a poly(L-lactide) (PLA) polymer scaffold coated with β-tricalcium phosphate (β-TCP). β-TCP is a calcium phosphate and precursor salt of hydroxyapatite, the thermodynamically most stable calcium phosphate in the bone [60]. The PLA scaffold was filled with a Ac-$(KLDL)_3$-$CONH_2$ peptide hydrogel encapsulating substance P, resulting in the facilitation of bone tissue formation and regeneration due to chemoattraction for MSCs [61].

Horii et al. designed several self-assembling peptides containing biologically active motifs to promote cell adhesion and performed 2D tissue culture studies of preosteoblasts seeded on the surface of the peptide hydrogels [36]. Biologically active motifs that were added to the self-assembling peptide Ac-$(RADA)_4$-$CONH_2$ included ALKRQGRTLYGF from the osteogenic growth peptide, DGRGDSVAYG from the osteopontin cell adhesion motif, and PRGDSGYRGDS from fibronectin and other ECM proteins. Compared with tissue culture studies using Ac-$(RADA)_4$-$CONH_2$ alone, the addition of the motifs resulted in increased mouse preosteoblast MC3T3-E1 cell proliferation and osteocalcin secretion, which is a marker for osteoblastic differentiation.

Following-up on the previous work in which bone cells were cultured on the surface of peptide hydrogels, Hamada et al. performed 3D studies in 5-mm-thick tissue cultures of MSCs encapsulated inside the synthetic Ac-$(RADA)_4$-$CONH_2$ self-assembling peptide hydrogel [37]. Increased concentration and gene expression levels for alkaline phosphatase activity and osteocalcin were observed during the culture period of 2–4 weeks. Within the peptide hydrogel scaffold, MSCs differentiated into mature osteoblasts and formed mineralized matrices.

In another work, Zamuner et al. tested the $(AEAEAKAK)_2$-$CONH_2$ ionic-complementary self-assembling peptide as a scaffold for tissue cultures of human bone marrow MSCs [27]. MSCs are immunomodulatory, multipotent, and can be used for a wide range of applications including bone, cartilage, and myocardium regeneration [62]. The authors of this work tethered to the $(AEAEAKAK)_2$-$CONH_2$ self-assembling peptide unit (i) the $(GRGDSP)_4K$ peptide (which belongs to the class of the RGD-containing peptides), (ii) the PRHRNRKGK peptide (which contains a cell-adhesive amino acid sequence), and (iii) the IGF-1 protein. They observed that these three modifications promote MSC adhesion and proliferation compared with the (AEAEAKAK)$_2$-CONH$_2$ peptide hydrogels alone.

Further, Dettin et al. investigated peptide self-assembly during electrospinning while the solvent was evaporated and the fibers were formed [63]. The enrichment of poly-ε-caprolactone (PCL) with self-assembling peptides produced materials with different performances in tensile modulus, maximum stress and maximum strain depending on the concentration and the sequence of the self-assembling peptide. Matrices containing Ac-RGD-(AEAEAKAK)$_2$-CONH$_2$ 10% w/w, Ac-(AbuEAbuEAbuKAbuK)$_2$-CONH$_2$ 5% w/w (where Abu is 2-aminobutyric acid), or Ac-(AEAEAKAK)$_2$-CONH$_2$ 15% w/w promoted osteoblast vitality, supported calcium deposition, and increased osteoblast-related gene expression in the electrospun hybrid polymer–peptide hydrogel, suggesting that this material may have potential applications in bone regeneration.

15.3.2 Cartilage regeneration and repair

Various synthetic polymers and biologically derived materials have been utilized in cartilage tissue repair [64,65]. Self-assembling peptide hydrogel scaffolds composed of the Ac-(KLDL)$_3$-CONH$_2$ peptide were used to encapsulate chondrocytes by Kisiday et al. [18,51]. In this work, cartilage-like ECM rich in proteoglycans and type II collagen were produced by the encapsulated chondrocytes after 4 weeks of culture in vitro, suggesting that the self-assembling peptide hydrogel can be used to repair injured cartilage tissue. In another study, bone marrow MSCs and adipose-derived progenitor cells from skeletally mature horses were mixed with the Ac-(KLDL)$_3$-CONH$_2$ peptide hydrogel to test for cartilage resurfacing [53]. The chondrogenic potential of these cells was significantly higher in the peptide hydrogel compared with cells inside agarose gels. For the cells inside the peptide hydrogel, there was extensive formation of cartilage-like ECM. Further work in Grodzinsky's lab showed that the Ac-(KLDL)$_3$-CONH$_2$ self-assembling peptide hydrogel may be used in cartilage tissue repair through coencapsulation of chondrocytes and insulin-like growth factor-1 (IGF-1) [52]. In a recent work, Florine et al. showed that 3D tissue cultures of human (HMSCs) and bovine bone marrow stromal cells (BMSCs) in Ac-(RADA)$_4$-CONH$_2$ peptide hydrogels were characterized by ECM accumulation, aggrecanase activity, and proteoglycan synthesis and accumulation [38].

15.3.3 Ligament regeneration

To fabricate a composite material with good mechanical properties, Chen et al. coated silk-based microfibrous scaffold with the Ac-(RADA)$_4$-CONH$_2$ self-assembling peptide hydrogel to provide an ECM-like network of peptide nanofibers for ligament regeneration [39]. After seeding this composite material with BMSCs, the peptide-coated scaffolds showed enhanced BMSC proliferation, metabolism, and fibroblastic differentiation during 3 weeks of culture compared with the uncoated control. The results indicate that the Ac-(RADA)$_4$-CONH$_2$–silk hybrid scaffold system has potential for use for the reconstruction of injured ligament tissues.

15.3.4 Neural tissue formation and regeneration

Extensive neurite outgrowth and the formation of active synapses resulted from seeding neural cells on the surface of the Ac-$(RADA)_4$-$CONH_2$ and Ac-$(RARADADA)_2$-$CONH_2$ self-assembling peptide hydrogels (Fig. 15.3) [57]. Neurons grown on the surface of the peptide hydrogels and Matrigel had similar levels of synapse formation and synaptic activities. In this early work, Holmes et al. suggested that the peptide hydrogels are a suitable matrix for neural tissue engineering and neural tissue regeneration.

The migration of neural cells from sliced postnatal hippocampal tissue to the Ac-$(RADA)_4$-$CONH_2$ peptide nanofiber scaffold was investigated by Semino and colleagues [40]. For this study, tissue sections were brought in contact with the peptide hydrogel to monitor the migration of cells across the tissue/hydrogel interface. After 1 week, glial and neuronal cells including progenitor neural cells progressively invaded the peptide hydrogel scaffold up to 500 μm from the tissue/hydrogel interface.

In another study, neural stem cells from the subventricular zone of adult mice were encapsulated inside the peptide hydrogel to create 3D neural tissue cultures [66]. Such tissue cultures were truly biomimetic because they resemble the in vivo mammalian brain. When cultured in serum-free medium, neural stem cells inside the peptide hydrogel proliferated and differentiated into progenitor neural cells, neurons, astrocytes, and oligodendrocytes. The neural cells in the peptide hydrogels thrived for at least 5 months in vitro. This work was compelling because better cell survival rates were observed in long-term studies for neural tissue cultured in the peptide hydrogels compared to the decreasing population of cells in Matrigel tissue cultures. Additionally, superior cell survival and differentiation properties were observed when self-assembling peptides functionalized with amino acid sequences from proteins promoting cell adhesion and cell viability were used compared to nonmodified peptide hydrogels (Fig. 15.3). The well-defined chemical and physical properties of the peptide nanofiber hydrogels and the use of serum-free medium results in more realistic biological studies of neural cells in a biomimetic 3D environment.

In another study, Ellis-Behnke et al. demonstrated vision restoration by using Ac-$(RADA)_4$-$CONH_2$ self-assembling peptide hydrogel to fill in the gap of the injured optic tract in the brain of adult hamsters. The result was axonal regeneration to reconnect the disconnected parts of the CNS with sufficient density to allow for functional return of vision [41].

15.3.5 Angiogenesis

Narmoneva et al. tested the Ac-$(RARADADA)_2$-$CONH_2$ self-assembling peptide scaffold for 3D cultures of human microvascular endothelial cells. The peptide

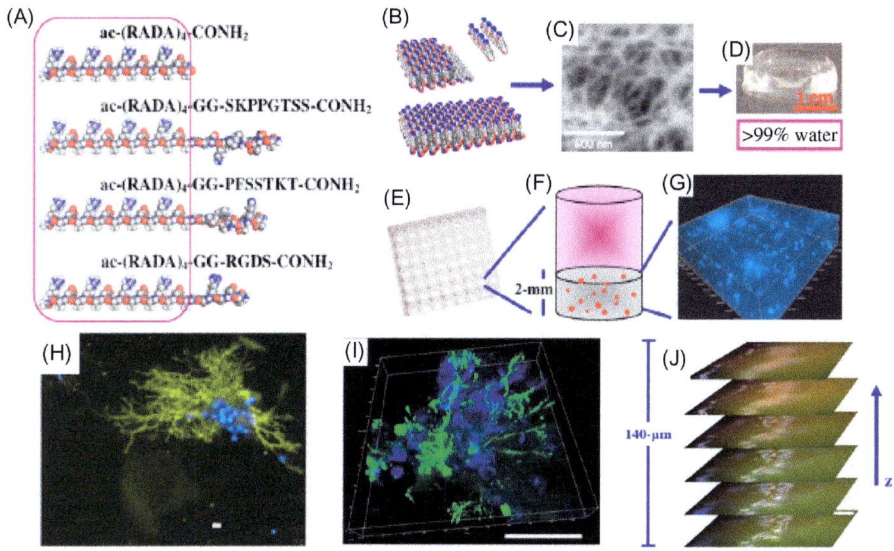

Figure 15.3 (A) Molecular models of the self-assembling peptide Ac-$(RADA)_4$-$CONH_2$ and of the modified self-assembling peptides carrying a diglycine linker and the functional motifs SKPPGTSS, PFSSTKT, and RGDS [66]. Color code: carbon, gray; oxygen, red; nitrogen, blue; hydrogen, white. (B) Molecular model of the self-assembling peptide and of the nanofiber formed upon assembly of the peptide monomers. (C) SEM picture of the peptide nanofibers inside the hydrogel. (D) Picture of the peptide hydrogel. (E and F) Ninety-six-well plate and graphical representation of a plate well in which the cells (red spheres) are encapsulated inside the peptide hydrogel (gray shaded volume at the bottom of the well); the well is filled with cell growth medium (red shaded volume). (G) Microscopy image volume representation of neural cells encapsulated in the peptide hydrogel; series of images were taken at different z-planes of cells cultured in the 3D environment of a 2-mm-thick peptide hydrogel (cell nuclei were stained blue with DAPI). Neural stem cells were encapsulated in the peptide nanofiber hydrogel scaffold containing the Ac-$(RADA)_4$-GG-SKPPGTSS-$CONH_2$ peptide and proliferated and differentiated therein. (H and I) Oligodendrocytes were stained with anti-GalC (green). (J) Images of 2-week tissue culture in which nestin(+) neural progenitors, cells (green), and Tuj1(+) neurons (red) appear at different z-planes. Images were collected using an inverted fluorescence microscope (G, H, J) and a confocal microscope (I). Scale bar is 200 μm in (A) and 20 μm in (H and I).
Source: Figure modified with permission from S. Koutsopoulos, S. Zhang, Long-term three-dimensional neural tissue cultures in functionalized self-assembling peptide hydrogels, Matrigel and Collagen I, Acta Biomater. 9 (2013) 5162–5169.

hydrogel formed an angiogenic environment which promoted long-term cell survival and capillary-like formation [31]. In contrast to the collagen scaffold, the peptide hydrogel not only prevented endothelial cell apoptosis without the addition of angiogenic factors but also showed expression of the angiogenic factor vascular endothelial growth factor (VEGF). Injection of the self-assembling peptides into the left ventricle of a mouse heart showed that the peptide nanofiber microenvironment facilitated angiogenesis by recruiting endothelial cells from the surrounding tissue and by promoting their survival and organization [30].

Genové et al. modified the N-terminal of the Ac-*X*-$(RADA)_4$-$CONH_2$ self-assembling peptide with functional motifs, represented by *X*, to examine the effect of these peptide hydrogels on human aortic endothelial cell function and tubular formation. The functional motifs chosen are present in two major proteins of the basement membrane (i.e., *X* = YIGSR and RYVVLPR are from laminin 1 and TAGSCLRKFSTM is from collagen IV) [33]. Therefore, the peptide hydrogel system was designed to be a synthetic basement analogue to enhance endothelial cell phenotype and to promote basement membrane deposition. Human aortic endothelial cell monolayers obtained on these functionalized peptide hydrogel scaffolds maintained LDL uptake activity, had enhanced nitric oxide release, and elevated laminin 1 and collagen IV deposition. This study suggested that functionalized peptide hydrogels provide a good physiological substrate for endothelial cell culture and angiogenesis, allowing for their application in the development of viable and engineered tissues.

Further work by Sieminski et al. showed that human umbilical vein endothelial cells (HUVEC) cultured in Ac-$(RADA)_4$-$CONH_2$ or Ac-$(RARADADA)_2$-$CONH_2$ hydrogels formed interconnected capillary-like networks [34].

Wang et al. tested the amino acid sequence PRGDSGYRGDS (PRG) which contains two repeats of the RGD domain that facilitates cell attachment and the sequence KLTWQELYQLKYKGI (KLT) which contains an active motif that induces endothelial cell proliferation. Tethering these sequences to the self-assembling peptide Ac-$(RADA)_4$-$CONH_2$ and plating HUVECs on the surface of the functionalized peptide hydrogels resulted in cell growth, cell migration, and capillary-like network formation [32].

15.3.6 Heart regeneration

With a goal of heart tissue regeneration, Lee and colleagues coseeded human microvascular endothelial cells and cardiac myocytes in the Ac-$(RARADADA)_2$-$CONH_2$ self-assembling peptide scaffold. This resulted in increased cell survival and optimal spatial reorganization of cardiac myocytes adjacent to endothelial cell microvascular structures [25]. The cardiomyocytes inside the peptide hydrogel presented synchronized, spontaneous contractions and prolonged survival allowing for heart tissue regeneration.

In another study, a solution of the Ac-$(RARADADA)_2$-$CONH_2$ self-assembling peptide was mixed with endothelial and smooth muscle cells and the mixture was injected into the heart of a mouse model of myocardial infarction. Upon injection, the peptide solution gelled and encapsulated the cells [30]. Cell injection techniques suffer from low cell survival. However, mixing the peptide solution with endothelial and smooth muscle cells prior to injection resulted in prolonged cell viability in the heart tissue environment and facilitated the recruitment of migrating exogenous cells into the hydrogel, promoting regeneration of the injured heart muscle. In contrast, there was little penetration of endothelial cells and cardiac myocyte precursors inside the injected matrix in control experiments using Matrigel.

15.3.7 Skin wound healing

Wound healing requires the orchestrated action of a number of cells to rebuild the tissue, including the recruitment and proliferation of keratinocytes on the wound site. Schneider et al. [42] found that wound reepithelialization using the Ac-$(RADA)_4$-$CONH_2$ self-assembling peptide hydrogel supplemented with epidermal growth factor (EGF) accelerated the rate of wound coverage. Compared to the controls (i.e., saline solution mixed with EGF and peptide scaffold without EGF), the rate was increased 5- and 3.5-fold, respectively. This suggests that self-assembling peptide hydrogels are good candidates for wound-healing applications and in vitro engineering of skin grafts.

15.3.8 Liver regeneration

To foster tissue-like function, Semino et al. used the Ac-$(RADA)_4$-$CONH_2$ self-assembling peptide hydrogel to culture a putative adult rat hepatocyte progenitor cell line in vitro [43]. In the 3D peptide hydrogel, the hepatocytes acquired a spheroidal morphology and produced progeny cells with the properties of mature hepatocytes, including the expression of several indicators such as albumin and cytochrome P450s.

In another study, Yarmush and colleagues, cultured freshly isolated primary rat hepatocytes in Ac-$(RADA)_4$-$CONH_2$ self-assembling peptide hydrogel [45]. After 3 days of culture inside the peptide hydrogel, the authors observed hepatocyte attachment, migration, formation of 3D spheroids of 80 mm of mean diameter, and expression of the apical membrane marker dipeptidyl peptidase IV at cell–cell contacts. Compared to tissue cultures performed in collagen, tissue cultures in the peptide hydrogel resulted in higher albumin and urea secretion for the first week and higher cytochrome P450IA1 activity for the entire study compared to tissue cultures in collagen. Albumin and urea production were similar in the peptide hydrogel and in Matrigel. Since Matrigel is derived from mouse cancerous tissues, the peptide hydrogel is a significantly more attractive system for generating biologically active hepatocyte spheroids.

In another study, Ac-$(RADA)_4$-$CONH_2$ self-assembling peptide hydrogel was functionalized with the GRGDSP sequence from the RGD integrin-binding sequence, the YIGSR sequence from the 67-kDa laminin receptor and the TAGSCLRKFSTM sequence from collagen IV that is responsible for binding to heparin [44]. When hepatocytes were cultured in these functionalized peptide hydrogels, they acquired parenchymal morphology, formed functional bile canaliculi-like structures, produced urea, and secreted proteins normally produced by hepatocytes. These results represent a significant advance in culturing hepatocytes in vitro using a biomimetic platform.

15.3.9 Middle-ear mucosal regeneration and cochlea cell tissue cultures for hearing organ regeneration

Using immunosuppressed rats which had undergone surgical removal of their middle-ear mucosa, Akiyama et al. tested the peptide hydrogel system for regeneration of the hearing organ [47]. The authors isolated and encapsulated mucosal cells from Sprague-Dawley transgenic rats within the Ac-$(RADA)_4$-$CONH_2$ self-assembling peptide hydrogel followed by transplantation. There was a significant number of donor mucosal cells in the recipient middle ear after transplantation compared with the control (i.e., donor cells in culture medium transplanted by injection). With normal morphology and function, these donor cells proliferated to form new epithelial and subepithelial layers and the basement membrane.

Using an avian hair cell system, Spencer et al. tested the effect of numerous biomaterials, peptide hydrogel, native matrix cultures, collagen I, collagen I/chondroitin-sulfate, and Matrigel on the ability to culture hair cells. In contrast to mammalian hair cells, avain hair cells have self-regenerating properties [46]. Hair cell tissue cultures in the Ac-$(RADA)_4$-$CONH_2$ hydrogel were superior compared with the other biomaterials and despite the absence of the basement membrane and accessory structures, cell density was comparable to that of the native matrix cultures.

15.3.10 Tooth regeneration

Dissanayaka et al. investigated the potential of the peptide hydrogel system for tooth regeneration [48]. They cocultured dental pulp stem cells and HUVECs inside the Ac-$(RADA)_4$-$CONH_2$ self-assembling peptide hydrogel for 4 weeks. In the absence of added growth factors, 3D tissue cultures inside the peptide nanofiber hydrogel supported cell survival, migration, and capillary network formation. Additionally, HUVEC migration and VEGF expression were promoted by the dental pulp stem cells, resulting in increased vascular network formation, significant ECM production, extended vascularization, and hydrogel mineralization in vivo. The peptide nanofiber hydrogel microenvironment is important for the support of cell–cell interactions and cell migration patterns resembling the in vivo situation, thus allowing for dental pulp regeneration.

15.4 Other notable self-assembling peptide systems in tissue engineering and regenerative medicine

Self-assembling peptides are biocompatible and tunable biomaterials. This is advantageous for applications in the field of tissue engineering and tissue regeneration. Several case studies exist in the literature in which self-assembling peptide hydrogels are tested using a broad range of mammalian cells. Ultrashort, Fmoc dipeptides containing glycine, alanine, leucine, and phenylalanine self-assemble to form hydrogel scaffolds consisting of nanofibers. Tissue cultures of bovine chondrocytes encapsulated inside such hydrogels showed proliferation and retention of chondrocyte phenotype as shown in the work of Ulijn and colleagues [11]. In other research, Fmoc-FF mixed with Fmoc-RGD peptides formed nanofiber hydrogels which were used to culture human dermal fibroblasts in 3D [12]. This hydrogel promotes adhesion of encapsulated dermal fibroblasts through specific RGD–integrin binding, cell spreading and proliferation as demonstrated by F-actin staining, live/dead and proliferation/viability assays.

In a different application, in situ enamel regeneration was promoted by the facilitation of crystal growth of hydroxyapatite and biomineralization in dental lesions by self-assembling CH_3CO-QQRFEWEFEQQ-$CONH_2$ (P_{11}-4) peptide hydrogels [13]. In vitro and in vivo clinical trials suggested that the P_{11}-4 peptide hydrogel assembly takes place within the lesion, allowing the scaffold to act as nucleator for hydroxyapatite, resulting in tissue regeneration from within [14].

Gough and colleagues investigated the injectable EFEFKFK self-assembling peptide hydrogel system for its potential as a cell carrier and scaffold for nucleus pulposus tissue engineering [15]. The authors of this work adjusted the mechanical properties of the hydrogel to match those of the native tissue in which the cultured cells grow in the body. In vitro 3D tissue culture studies showed upregulation of nucleus pulposus cell-specific genes, high cell viability, and increased aggrecan and type II collagen deposition, which are two important nucleus pulposus extracellular matrix components.

15.5 Ionic-complementary, self-assembling peptides for drug delivery

Self-assembling peptide nanofibers form highly hydrated hydrogels of up to ~99.5% w/v water. Inside the hydrogel, there is a network of pores with sizes between 5 and 200 nm [22]. Because the hydrogel is composed of such a high percentage of water, the system allows for 100% loading efficiency of the therapeutic compound. The maximum amount of drug or antibody loading into the hydrogel depends solely on its solubility in water.

The physical properties of this hydrogel make it ideal for drug delivery applications [18,21,22,26]. The sol-gel transition occurs at physiological conditions and results in high internal hydration. Self-assembling peptide hydrogels are biocompatible and biodegradable. Hydrophilic, self-assembling peptide nanofiber hydrogels provide a mild, toxic-free, nondenaturing environment for encapsulation and subsequent release of small drug molecules [26] and proteins with therapeutic properties including growth factors and antibodies [21,22]. The liquid-to-gel transition occurs during interaction of the peptide solution with biological fluids containing electrolytes, making this drug delivery system injectable (Fig. 15.1). The release profile may be fine tuned to last from days to months and it depends on the size of the protein, the size/charge of the drug compound, and the nanofiber density of the hydrogel. For example, the release kinetics for human immunoglobulin (IgG) could be controlled by changing the hydrogel nanofiber density [21,22]. Injectable, multilayered hydrogel structures were also created with an "onion-like" architecture consisting of concentric spheres with a ac-$(RADA)_4$-$CONH_2$ core and ac-$(KLDL)_3$-$CONH_2$ shell. Release through this architecture can be adjusted for more than 3 months. Notably, even after 3 months inside the hydrogel, the encapsulation and release of IgG did not affect the conformation of the antibody and the biological activity.

Liu et al. designed self-assembling peptides carrying a net positive charge by the addition of arginines to the self-assembling unit of the ac-$(KLDL)_3$-$CONH_2$ peptide (i.e., ac-$(KLDL)_3$-R_x-$CONH_2$, where x is between 1 and 3) [74]. In vitro and in vivo data showed that sustained release of the antiinflammatory protein hepatocyte growth factor (HGF) through the ac-$(KLDL)_3$-R_x-$CONH_2$ peptide hydrogel, which contained preadsorbed heparin, significantly enhanced β-cell survival and insulin secretion by inhibiting NFκB/p38 pathways, inflammatory cytokine expression, and lymphocyte infiltration in grafts.

Prior to injection, the peptide solution may also be mixed with cells to stimulate the release of growth factors and other cytokines to the injured site for the reconstruction of the tissue. Lee and colleagues showed that injection of the peptide solution mixed with endothelial cells into infracted myocardium facilitates heart muscle regeneration [25].

Peptide hydrogels are amenable to molecular design and can carry multiple drugs tethered to the peptide nanofibers by cleavable linkers. Law et al. tethered a protease cleavable cytotoxic peptide to the self-assembling unit of the ac-$(KLDL)_3$-$CONH_2$ [54]. Drug release was mediated by enzymatic treatment of the peptide hydrogel carrying the therapeutic peptide incorporated into the peptide nanofibers in a manner similar to that described in Fig. 15.2. Furthermore, the ac-$(KLDL)_3$-$CONH_2$ self-assembling peptide hydrogel has been tested for the prolonged delivery of doxorubicin and the Smac-derived proapoptotic peptide for the treatment of cancer [55]. These results suggest that modified peptide hydrogels have applications for targeted drug delivery.

In another application, Chen and colleagues investigated the Ac-$(AEAEAKAK)_2$-$CONH_2$ ionic-complementary, self-assembling peptide hydrogel for encapsulation and delivery of the anticancer drug ellipticine using an in vitro platform and rat models [28,75]. The authors observed significant differences in the main pharmacokinetic parameters between ellipticine in nanoparticle form and ellipticine encapsulated in peptide hydrogel, suggesting that the formulation of ellipticine in peptide hydrogel prolongs the residence time of the drug and enlarges the area under the concentration–time curve (AUC). These results show that the hydrogel formulation of the drug can increase the bioavailability of ellipticine in the body.

15.6 Other notable cases of self-assembling peptides for drug delivery

Self-assembling peptide formulations have long been recognized as being well suited for drug delivery applications. Upon self-assembly, such peptides may form supramolecular architectures including hydrogels, nanoparticles, nanovesicles, and nanotubes which have been tested for encapsulation and delivery of active pharmaceutical ingredients (APIs) in a controlled, sustainable, or targeted manner. There are several case studies for self-assembling peptide platforms for drug delivery applications. Dipeptides, such as diphenylglycine self-assemble into well-ordered closed-caged nanospheres. LeuΔPhe (containing an α,β-dehydrophenylalanine) self-assembles into emulsions. Fmoc-dipeptide derivatives (i.e., YL, YA, and YS) self-assemble into hydrogels consisting of nanofibers, and diphenylalanine peptides self-assemble into microtubules which have been tested for drug delivery applications [67–70].

Lipid-like or surfactant-like peptides self-assemble into nanovesicles similar to liposomes (Fig. 15.4). These may encapsulate hydrophilic molecules or integrate hydrophobic molecules into the peptide bilayer [71]. Similar to liposomes, such nanovesicular formulations may be used for the sustained and controlled release of therapeutic molecules. Interestingly, compelling evidence suggests that lipid-like peptide nanovesicle formulations can carry drugs through biological membranes such as the intestinal barrier and, therefore, such formulations may be used for oral drug delivery [72].

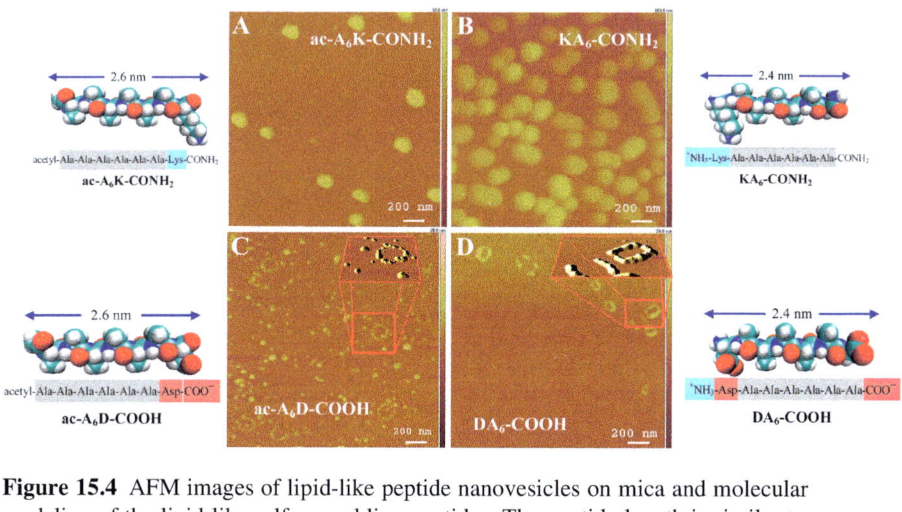

Figure 15.4 AFM images of lipid-like peptide nanovesicles on mica and molecular modeling of the lipid-like self-assembling peptides. The peptide length is similar to biological phospholipids. The hydrophobic domain of the peptides consists of six alanines. Color code: cyan, carbon; red, oxygen; blue, nitrogen; white, hydrogen. Illustrations were generated by VMD. Highlighted domains represent amino acids with positive charge (blue), negative charge (red) and hydrophobic side chains (gray). The net charge of the peptides at pH 7.4 was calculated to be ac-A_6K-$CONH_2$ (+1), KA_6-$CONH_2$ (+2), ac-A_6D-COOH (−2), DA_6-COOH (−1).
Source: Reprinted with permission from D.G. Fatouros, D.A. Lamprou, A.J. Urquhart, S.N. Yannopoulos, I.S. Vizirianakis, S. Zhang, et al., Lipid-like self-assembling peptide nanovesicles for drug delivery, ACS Appl. Mater. Interfaces 6 (2014) 8184–8189.

In other work, Nisbet and colleagues, showed that hydrogels consisting of self-assembling peptides carrying the Fmoc group, the laminin sequence IKVAV, and lysine in the peptide's C-terminal, are effective as biocompatible gene carriers for localized delivery of viral vectors in vivo [73]. The authors showed that the addition of the lysine residue resulted in increased electrostatic interactions between the peptide nanofibers and the viral particles allowing for effective immobilization of the lentiviral vector particles, thereby limiting their activity exclusively to the site of injection and enabling focal gene delivery in vivo in a tissue-specific manner.

15.7 Other notable applications of ionic-complementary, self-assembling peptides in biotechnology and bioengineering

Ionic-complementary self-assembling peptides have also been studied in other applications that are relevant to biomedicine. In an early work, Zhang et al. synthesized the Ac-$(RADS)_2$-AAAC and the Ac-$(RADS)_3$-AAAC self-assembling peptides which contain the cell adhesion motif $(RADS)_n$ where $n = 2$ or 3, at the N-terminus, followed by an alanine linker and a cysteine residue at the C-terminus. The thiol group of the cysteines is the anchor that covalently binds to gold atoms on the gold-coated surface of a substrate allowing for the peptides to form monolayers [77]. These peptides can be used as ink for an inkjet printer to print patterns on a suitable surface, allowing proteins and cells to adsorb and grow in specific directions or to microfabricate systems in which cells align in a well-defined manner to form specific cell arrays.

Ellis-Behnke and colleagues discovered that the Ac-$(RADA)_4$-$CONH_2$ self-assembling peptide was effective in bleeding models [41,78]. Their experiments showed that the self-assembling peptide solution when injected into the bleeding tissue (including brain, spinal cord, femoral artery, liver, or skin of mammals) gels and establishes a nanofiber barrier to achieve complete hemostasis in less than 15 s. Using this approach they were able to stop bleeding without the use of pressure, cauterization, vasoconstriction, coagulation, or crosslinked adhesives.

Luo et al. designed self-assembling peptides containing D-amino acids and showed that the Ac-$(\text{dAdEdAdEdAdKdAdK})_2$-$CONH_2$ peptide self-assembles into a hydrogel, while in solution it forms stable β-sheets with the exact inverted CD spectrum as the Ac-$(AEAEAKAK)_2$-$CONH_2$ which contains L-amino acids [49]. Furthermore, AFM analysis showed that nanofibers containing the D- and L-amino acid peptides were indistinguishable. The D-peptide in solution was also tested in hemostasis models and showed that upon interaction with biological fluids it forms a hydrogel material which is an efficient hemostatic agent [50] like the respective L-amino acids counterparts Ac-$(AEAEAKAK)_2$-$CONH_2$ and Ac-$(RADA)_4$-$CONH_2$. This suggests that the hemostatic properties of peptide hydrogels are likely due to the gelation process, the viscoelastic properties of the hydrogel, and the interaction of the hydrogel with tissues and membranes.

15.8 Future perspectives

Peptides represent a unique system for the design of biomaterials with controllable nanoscale features and biological resemblance to the body's microenvironment. The versatility of the chemical design of the amino acids which form the peptides allows for the fabrication of tailor-made biocompatible and biodegradable supramolecular structures and matrices with numerous applications in tissue engineering, regenerative medicine, pharmaceutical engineering biotechnology, and bioengineering.

During the last 30 years, research on a variety of cells cultured inside self-assembling peptide hydrogel scaffolds was characterized by physiological cellular behaviors, including proliferation, differentiation, migration, and production of ECM components. These results were significantly better than those observed in other biomaterials or in some cases better than those observed in Matrigel [66]. Self-assembling peptides, due to their synthetic nature, do not contain toxic impurities or pathogens or evoke immune/inflammatory responses. Several advantageous properties include inherent biocompatibility, biodegradability, functionalization options with biologically recognizable moieties to support cellular activities, and the ability to induce vascularization, a crucial step toward an effective tissue-regenerative therapy. Growth factors, important for attracting cells or promoting growth, may be released from the hydrogel. Because self-assembling peptide hydrogels mimic the real in vivo microenvironment of cells in mammalian tissues, the peptide hydrogel system is useful not only for furthering our understanding of cell biology but also in the future of healthcare technology. Personalized therapies are possible because the injectable peptide hydrogel allows encapsulation and implantation of the patient's own cells into the injured tissue, promoting tissue regeneration [58]. As time progresses, more and more laboratory studies and clinical trials using the peptide hydrogels will likely become available, providing more evidence for their potential use in biomedicine.

References

[1] S.N. Abraham, E.H. Beachey, Assembly of a chemically synthesized peptide of *Escherichia coli* type 1 fimbriae into fimbria-like antigenic structures, J. Bacteriol. 69 (1987) 2460–2465.

[2] D. Burdick, B. Soreghan, M. Kwon, J. Kosmoski, M. Knauer, A. Henschen, et al., Assembly and aggregation properties of synthetic Alzheimer's A4/beta amyloid peptide analogs, J. Biol. Chem. 267 (1992) 546–554.

[3] Y. Imanishi, S. Kimura, Peptide self-assembly in phospholipid bilayer membrane, Proc. Jpn. Acad. 68 (1992) 121–126.

[4] M.R. Ghadiri, C. Soares, C. Choi, Design of an artificial four-helix bundle metalloprotein via a novel ruthenium (II)-assisted self-assembly process, J. Am. Chem. Soc. 114 (1992) 4000–4002.

[5] M.R. Ghadiri, J.R. Granja, R.A. Milligan, D.E. McRee, N. Khazanovich, Self-assembling organic nanotubes based on a cyclic peptide architecture, Nature 366 (1993) 324–327.

[6] S. Zhang, T. Holmes, C. Lockshin, A. Rich, Spontaneous assembly of a self-complementary oligopeptide to form a stable macroscopic membrane, Proc. Natl. Acad. Sci. U.S.A. 90 (1993) 3334–3338.
[7] S.G. Zhang, Fabrication of novel biomaterials through molecular self-assembly, Nat. Biotechnol. 21 (2003) 1171–1178.
[8] S. Koutsopoulos, L. Kaiser, H.M. Eriksson, S. Zhang, Designer peptide surfactants stabilize diverse functional membrane proteins, Chem. Soc. Rev. 41 (2012) 1721–1728.
[9] S. Koutsopoulos, Self-assembling peptide nanofiber hydrogels in tissue engineering and regenerative medicine: progress, design guidelines, and applications, J. Biomed. Mater. Res. A. 104 (2016) 1002–1016.
[10] K. Matsumoto, M. Vaughn, B.D. Bruce, S. Koutsopoulos, S. Zhang, Designer peptide surfactants stabilize functional Photosystem-I membrane complex in aqueous solution for extended time, J. Phys. Chem. B. 113 (2008) 75–83.
[11] V. Jayawarna, M. Ali, T.A. Jowitt, A.F. Miller, A. Saiani, J.E. Gough, et al., Nanostructured hydrogels for three-dimensional cell culture through self-assembly of fluorenylmethoxycarbonyl-dipeptides, Adv. Mater. 18 (5) (2006) 611–614.
[12] M. Zhou, A.M. Smith, A.K. Das, N.W. Hodson, R.F. Collins, R.V. Ulijn, et al., Self-assembled peptide-based hydrogels as scaffolds for anchorage-dependent cells, Biomaterials 30 (13) (2009) 2523–2530.
[13] P.A. Brunton, R.P.W. Davies, J.L. Burke, A. Smith, A. Aggeli, S.J. Brookes, et al., Treatment of early caries lesions using biomimetic self-assembling peptides—a clinical safety trial, Br. Dent. J. 215 (2013) E6.
[14] L. Kind, S. Stevanovic, S. Wuttig, S. Wimberger, J. Hofer, B. Müller, et al., Biomimetic remineralization of carious lesions by self-assembling peptide, J. Dent. Res. 96 (7) (2017) 790–797.
[15] S. Wan, S. Borland, S.M. Richardson, C.L. Merry, A. Saiani, J.E. Gough, Self-assembling peptide hydrogel for intervertebral disc tissue engineering, Acta Biomater. 46 (2016) 29–40.
[16] S. Zhang, T.C. Holmes, C.M. DiPersio, R.O. Hynes, X. Su, A. Rich, Self-complementary oligopeptide matrices support mammalian cell attachment, Biomaterials 16 (1995) 1385–1393.
[17] S. Zhang, M. Altman, Peptide self-assembly in functional polymer science and engineering, React. Funct. Polym. 41 (1999) 91–102.
[18] J. Kisiday, M. Jin, B. Kurz, H. Hung, C. Semino, S. Zhang, et al., Self-assembling peptide hydrogel fosters chondrocyte extracellular matrix production and cell division implications for cartilage tissue repair, Proc. Natl. Acad. Sci. U.S.A. 99 (2002) 9996–10001.
[19] S.G. Zhang, D.M. Marini, W. Hwang, S. Santoso, Design of nanostructured biological materials through self-assembly of peptides and proteins, Curr. Opin. Chem. Biol. 6 (2002) 865–871.
[20] S.G. Zhang, C. Lockshin, R. Cook, A. Rich, Unusually stable beta-sheet formation of an ionic self-complementary oligopeptide, Biopolymers 34 (1994) 663–672.
[21] S. Koutsopoulos, S.G. Zhang, Two-layered injectable self-assembling peptide scaffold hydrogels for long-term sustained release of human antibodies, J. Control. Release 160 (2012) 451–458.
[22] S. Koutsopoulos, L.D. Unsworth, Y. Nagai, S.G. Zhang, Controlled release of functional proteins through designer self-assembling peptide nanofiber hydrogel scaffold, Proc. Natl. Acad. Sci. U.S.A. 106 (2009) 4623–4628.

[23] S. Ayad, R.P. Boot-Handford, M.J. Humphreise, K.E. Kadler, C.A. Shuttleworth, The Extracellular Matrix: Facts Book, second ed, Academic Press, San Diego, CA, 1998.
[24] Y. Chau, Y. Luo, A.C. Cheung, Y. Nagai, S. Zhang, J.B. Kobler, et al., Incorporation of a matrix metalloproteinase-sensitive substrate into self-assembling peptides—a model for biofunctional scaffolds, Biomaterials 29 (11) (2008) 1713–1719.
[25] D.A. Narmoneva, R. Vukmirovic, M.E. Davis, R.D. Kamm, R.T. Lee, Endothelial cells promote cardiac myocyte survival and spatial reorganization – implications for cardiac regeneration, Circulation 110 (2004) 962–968.
[26] Y. Nagai, L.D. Unsworth, S. Koutsopoulos, S.G. Zhang, Slow release of molecules in self-assembling peptide nanofiber scaffold, J. Control. Release 115 (2006) 18–25.
[27] A. Zamuner, M. Cavo, S. Scaglione, G.M.L. Messina, T. Russo, A. Gloria, et al., Design of decorated self-assembling peptide hydrogels as architecture for mesenchymal stem cells, Materials 9 (2016) 727.
[28] S.Y. Fung, H. Yang, P. Chen, Sequence effect of self-assembling peptides on the complexation and in vitro delicery of the hydrophobic anticancer drug ellipticine, PLoS One 3 (2008) e1956.
[29] J. Liu, L. Zhang, Z. Yang, X. Zhao, Controlled release of paclitaxel from a self-assembling peptide hydrogel formed in situ and antitumor study in vitro, Int. J. Nanomedicine 6 (2011) 2143–2153.
[30] M.E. Davis, J.P.M. Motion, D.A. Narmoneva, T. Takahashi, D. Hakuno, R.D. Kamm, et al., Injectable self-assembling peptide nanofibers create intramyocardial microenvironments for endothelial cells, Circulation 111 (2005) 442–445.
[31] D.A. Narmoneva, O. Oni, A.L. Sieminski, S.G. Zhang, J.P. Gertler, R.D. Kamm, et al., Self-assembling short oligopeptides and the promotion of angiogenesis, Biomaterials 26 (2005) 4837–4846.
[32] X.M. Wang, A. Horii, S. Zhang, Designer functional self-assembling peptide nanofiber scaffolds for the growth, migration, and tubulogenesis of human umbilical vein endothelial cells, Soft Matter 4 (2008) 2388–2395.
[33] E. Genové, C. Shen, S.G. Zhang, C.E. Semino, The effect of functionalized self-assembling peptide scaffolds on human aortic endothelial cell function, Biomaterials 26 (2005) 3341–3351.
[34] A.L. Sieminski, A.S. Was, G. Kim, H. Gong, R.D. Kamm, The stiffness of three-dimensional ionic self-assembling peptide gels affects the extent of capillary-like network formation, Cell Biochem. Biophys. 49 (2007) 73–83.
[35] M.A. Bokhari, G. Akay, S. Zhang, M.A. Birch, The enhancement of osteoblast growth and differentiation in vitro on a peptide hydrogel-polyHIPE polymer hybrid material, Biomaterials 26 (2005) 5198–5208.
[36] A. Horii, X. Wang, F. Gelain, S. Zhang, Biological designer self-assembling peptide nanofiber scaffolds significantly enhance osteoblast proliferation, differentiation and 3-D migration, PLoS One 2 (2007) e190.
[37] K. Hamada, M. Hirose, T. Yamashita, H. Ohgushi, Spatial distribution of mineralized bone matrix produced by marrow mesenchymal stem cells in self-assembling peptide hydrogel scaffold, J. Biomed. Mater. Res. A. 84 (2008) 128–136.
[38] E.M. Florine, R.E. Miller, R.M. Porter, C.H. Evans, B. Kurz, A.J. Grodzinsky, Effects of dexamethasone on mesenchymal stromal cell chondrogenesis and aggrecanase activity: comparison of agarose and self-assembling peptide scaffolds, Cartilage 4 (2013) 63–74.

[39] K. Chen, S. Sahoo, P. He, K.S. Ng, S.L. Toh, J.C. Goh, A hybrid silk/RADA-based fibrous scaffold with triple hierarchy for ligament regeneration, Tissue Eng. Part A. 18 (2012) 1399–1409.
[40] C.E. Semino, J. Kasahara, Y. Hayashi, S.G. Zhang, Entrapment of migrating hippocampal neural cells in three-dimensional peptide nanofiber scaffold, Tissue Eng. 10 (2004) 643–655.
[41] R.G. Ellis-Behnke, Y.X. Liang, S.W. You, D.K.C. Tay, S.G. Zhang, K.F. So, et al., Nano neuro knitting: peptide nanofiber scaffold for brain repair and axon regeneration with functional return of vision, Proc. Natl. Acad. Sci. U.S.A. 103 (2006) 5054–5059.
[42] A. Schneider, J.A. Garlick, C. Egles, Self-assembling peptide nanofiber scaffolds accelerate wound healing, PLoS One 3 (2008) e1410.
[43] C.E. Semino, J.R. Merok, G.G. Crane, G. Panagiotakos, S. Zhang, Functional differentiation of hepatocyte-like spheroid structures from putative liver progenitor cells in three-dimensional peptide scaffolds, Differentiation 71 (2003) 262–270.
[44] E. Genove, S. Schmitmeier, A. Sala, S. Borros, A. Bader, L.G. Griffith, et al., Functionalized self-assembling peptide hydrogel enhance maintenance of hepatocyte activity in vitro, J. Cell. Mol. Med. 13 (2009) 3387–3397.
[45] S. Wang, D. Nagrath, P.C. Chen, F. Berthiaume, M.L. Yarmush, Three-dimensional primary hepatocyte culture in synthetic self-assembling peptide hydrogel, Tissue Eng. Part A. 14 (2008) 227–236.
[46] N.J. Spencer, D.A. Cotanche, C.M. Klapperich, Peptide- and collagen-based hydrogel substrates for in vitro culture of chick cochleae, Biomaterials 29 (2008) 1028–1042.
[47] N. Akiyama, T. Yamamoto-Fukuda, H. Takahashi, T. Koji, In situ tissue engineering with synthetic self-assembling peptide nanofiber scaffolds, PuraMatrix, for mucosal regeneration in the rat middle-ear, Int. J. Nanomed. 8 (2013) 2629–2640.
[48] W.L. Dissanayaka, K.M. Hargreaves, L. Jin, L.P. Samaranayake, C. Zhang, The interplay of dental pulp stem cells and endothelial cells in an injectable peptide hydrogel on angiogenesis and pulp regeneration in vivo, Tissue Eng. Part A. 21 (2015) 550–563.
[49] Z. Luo, X. Zhao, S. Zhang, Structural dynamic of a self-assembling peptide d-EAK16 made of only D-amino acids, PLoS One 3 (5) (2008) e2364.
[50] Z. Luo, S. Wang, S. Zhang, Fabrication of self-assembling D-form peptide nanofiber scaffold D-EAK16 for rapid hemostasis, Biomaterials 32 (2011) 2013–2020.
[51] J.D. Kisiday, M.S. Jin, M.A. DiMicco, B. Kurz, A.J. Grodzinsky, Effects of dynamic compressive loading on chondrocyte biosynthesis in self-assembling peptide scaffolds, J. Biomech. 37 (2004) 595–604.
[52] R.E. Miller, P.W. Kopesky, A.J. Grodzinsky, Growth factor delivery through self-assembling peptide scaffolds, Clin. Orthop. Relat. Res. 469 (2011) 2716–2724.
[53] J.D. Kisiday, P.W. Kopesky, C.H. Evans, A.J. Grodzinsky, C.W. McIlwraith, D.D. Frisbie, Evaluation of adult equine bone marrow- and adipose-derived progenitor cell chondrogenesis in hydrogel cultures, J. Orthop. Res. 26 (2007) 322–331.
[54] B. Law, R. Weissleder, C.H. Tung, Peptide-based biomaterials for protease-enhanced drug delivery, Biomacromolecules 7 (4) (2006) 1261–1265.
[55] E. Yishay-Safranchik, M. Golan, A. David, Controlled release of doxorubicin and Smac-derived pro-apoptotic peptide from self-assembled KLD-based peptide hydrogels, Polym. Adv. Technol. 25 (5) (2014) 539–544.
[56] H. Misawa, N. Kobayashi, A. Soto-Gutierrez, Y. Chen, A. Yoshida, J.D. Rivas-Carrillo, et al., PuraMatrix (TM) facilitates bone regeneration in bone defects of calvaria in mice, Cell Transplant. 15 (2006) 903–910.

[57] T.C. Holmes, S. de Lacalle, X. Su, G.S. Liu, A. Rich, S.G. Zhang, Extensive neurite outgrowth and active synapse formation on self-assembling peptide scaffolds, Proc. Natl. Acad. Sci. U.S.A. 97 (2000) 6728–6733.
[58] S. Koutsopoulos, Molecular fabrications of smart nanobiomaterials and applications in personalized medicine, Adv. Drug Deliv. Rev. 64 (2012) 1459–1476.
[59] C.S. Hughes, L.M. Postovit, G.A. Lajoie, Matrigel: a complex protein mixture required for optimal growth of cell culture, Proteomics 10 (2010) 1886–1890.
[60] S. Koutsopoulos, Synthesis and characterization of hydroxyapatite crystals: a review study on the analytical methods, J. Biomed. Mater. Res. 62 (2002) 600–612.
[61] S.H. Kim, W. Hur, J.E. Kim, H.J. Min, S. Kim, H.S. Min, et al., Self-assembling peptide nanofibers coupled with neuropeptide substance P for bone tissue engineering, Tissue Eng. Part A. 21 (2015) 1237–1246.
[62] P. Bianco, P.G. Robey, Stem cells in tissue engineering, Nature 414 (2001) 118–121.
[63] M. Dettin, A. Zamuner, M. Roso, A. Gloria, G. Iucci, G.M.L. Messina, et al., Electrospun scaffolds for osteoblast cells: peptide-induced concentration-dependent improvements of polycaprolactone, PLoS One 10 (9) (2015) e0137505.
[64] L.E. Freed, J.C. Marquis, A. Nohria, J. Emmanual, A.G. Mikos, R. Langer, Neocartilage formation in vitro and in vivo using cells cultured on synthetic biodegradable polymers, J. Biomed. Mater. Res. 27 (1993) 11–23.
[65] C. Perka, R.S. Spitzer, K. Lindenhayn, M. Sittinger, O. Schultz, Matrix-mixed culture: new methodology for chondrocyte culture and preparation of cartilage transplants., J. Biomed. Mater. Res. 49 (2000) 305–311.
[66] S. Koutsopoulos, S. Zhang, Long-term three-dimensional neural tissue cultures in functionalized self-assembling peptide hydrogels, Matrigel and Collagen I, Acta Biomater. 9 (2013) 5162–5169.
[67] M. Reches, E. Gazit, Formation of closed-cage nanostructures by self-assembly of aromatic dipeptides, Nano Lett. 4 (2004) 581–585.
[68] C.K. Thota, N. Yadav, V.S. Chauhan, A novel highly stable and injectable hydrogel based on a conformationally restricted ultrashort peptide, Sci. Rep. 6 (2016) 31167.
[69] C. Bai, S. Pappas, P.W.J.M. Debnath, J. Frederix, S. Leckie, Fleming, et al., Stable emulsions formed by self-assembly of interfacial networks of dipeptide derivatives, ACS Nano 8 (2014) 7005–7013.
[70] R.F. Silva, D.R. Araujo, E.R. Silva, R.A. Ando, W.A. Alves, L-diphenylalanine microtubes as a potential drug-delivery system: characterization, release kinetics, and cytotoxicity, Langmuir 29 (32) (2013) 10205–10212.
[71] D.G. Fatouros, D.A. Lamprou, A.J. Urquhart, S.N. Yannopoulos, I.S. Vizirianakis, S. Zhang, et al., Lipid-like self-assembling peptide nanovesicles for drug delivery, ACS Appl. Mater. Interfaces 6 (2014) 8184–8189.
[72] C. Karavasili, M. Spanakis, D. Papagiannopoulou, I.S. Vizirianakis, D.G. Fatouros, S. Koutsopoulos, Bioactive self-assembling lipid-like peptides as permeation enhancers for oral drug delivery, J. Pharm. Sci. 104 (2015) 2304–2311.
[73] A.L. Rodriguez, T.Y. Wang, K.F. Bruggeman, R. Li, R.J. Williams, C.L. Parish, et al., Tailoring minimalist self-assembling peptides for localized viral vector gene delivery, Nano Res. 9 (3) (2016) 674–684.
[74] S. Liu, L. Zhang, J. Cheng, Y. Lu, J. Liu, Sustained release of hepatocyte growth factor by cationic self-assembling peptide/heparin hybrid hydrogel improves β-cell survival and function through modulating inflammatory response., Int. J. Nanomedicine 11 (2016) 4875–4890.

[75] W. Ma, S. Lu, P. Pan, P. Sadatmousavi, Y. Yuan, P. Chen, Pharmacokinetics of peptide mediated delivery of anticancer drug ellipticine, PLoS One 7 (8) (2012) e43684.
[76] A. Zhou, S. Chen, B. He, W. Zhao, X. Chen, D. Jiang, Controlled release of TGF-beta 1 from RADA self-assembling peptide hydrogel scaffolds, Drug Des. Devel. Ther. 10 (2016) 3043–3051.
[77] S. Zhang, L. Yan, M. Altman, M. Lässle, H. Nugent, F. Frankel, et al., Biological surface engineering: a simple system for cell pattern formation, Biomaterials 20 (1999) 1213–1220.
[78] R.G. Ellis-Behnke, Y.X. Liang, D.K. Tay, P.W. Kau, G.E. Schneider, S. Zhang, et al., Nano hemostat solution: immediate hemostasis at the nanoscale, Nanomedicine 2 (4) (2006) 207–215.

Peptide-based systems for drug delivery

16

Dana M. Copolovici and Andreea I. Lupitu
Aurel Vlaicu University of Arad, Arad, Romania

16.1 Introduction

Peptides are short sequences of up to 40 amino acids that are connected through amide (peptide) bonds. Peptides present important biological functions including biocompatibility and affinity to specific cells [1]. They are naturally produced because they are integral components of living organisms and they may also be biosynthetically obtained through recombinant microbial fermentation or chemically synthesized by methods which allow for the conjugation of small molecules or the incorporation of natural and non-natural amino acids.

During the last 4 decades many different delivery platforms and methods have been developed to transport and deliver therapeutic substances into specific cells and tissues. These methods include the use of viral vectors [2], physical methods such as microinjection or electroporation [3], physicochemical approaches such as liposome encapsulation [4–6], and proteins [7]. However, these approaches suffer from various drawbacks such as inefficient drug delivery, cellular damage, toxicity, immunogenicity, limited applicability in vivo, and restrictions due to the properties of the drug molecule and the cell type. In this respect, the use of peptides as delivery agents has attracted research interest.

Peptides are used in nanotechnologies as they are suitable moieties that may lead to biocompatible nanostructures, such as hydrogels, nanovesicles, and nanoparticles, which are able to load and deliver chemical drugs and bomolecules.

16.2 Peptide-based systems as delivery vectors of anticancer drugs

Recent data revealed that in the USA alone there were nearly 14.5 million people diagnosed with diverse types of cancer in 2014, and about new 1.6 million new cases were expected to be reported in 2016 [8]. Breast, lung, and prostate cancers represent more than 55% of all cancer cases reported in economically developed countries followed by stomach, liver, and colorectal cancers.

Peptides are used to generate therapeutics to enhance cellular uptake, drug targeting, or/and vaccination. Researchers are now given the challenge to seek

Peptide Applications in Biomedicine, Biotechnology and Bioengineering.
DOI: http://dx.doi.org/10.1016/B978-0-08-100736-5.00017-X

effective therapeutic approaches that eliminate cancerous cells while sparing the normal, healthy tissue. Proteins and peptides have effectiveness, activity, specificity, and relatively lower toxicity, which are important factors for the treatment of numerous diseases. Due to their biocompatibility and affinity toward target cells, peptides have been investigated for possible medical and clinical applications. Controlled drug-delivery technology represents one of the broader areas of science, involving a multidisciplinary scientific approach, contributing to human health care. The concept of controlled drug delivery is used in an attempt to improve the therapeutic index of drugs by increasing their localization to specific organs, tissues, or cells and by decreasing their potential toxic side effects. Therefore, peptides as drug carriers of bioactive agents have great potential in treating or detecting various types of cancer including breast cancer [9–13], prostate cancer [14–17], pancreatic cancer [18–23], ovarian cancer [24–28], and lung cancer [14,29–32].

Chemotherapeutic agents have toxic side effects for tumor cells as well as for normal cells. The controlled delivery of these agents to diseased sites would enable the use of higher doses to increase therapeutic efficacy while minimizing the drug's adverse effects to healthy tissues and cells [33]. Controlled drug delivery involves the association of a drug with a carrier system, thereby allowing modulation of the pharmacokinetic properties and biodistribution of the drug. Different nano-sized peptide-containing carriers for delivering drugs, such as hydrogels, nanovesicles, and nanoparticles which have been developed and suggested for achieving these goals, are depicted in Table 16.1 [33,59–63].

16.2.1 Hydrogels

A hydrogel contains mostly water, usually over 95% by volume which allows for the encapsulation of therapeutic molecules. Peptide-based hydrogel represents a promising alternative to current drug-delivery approaches [64].

The encapsulation of doxorubicin and 5-fluorouracil within the hydrogel matrix containing the aromatic peptide building block, *N*-fluorenylmethoxycarbonyl-diphenylalanine (Fmoc-FF), shows release kinetics of the drugs depending on their chemical structure, molecular weight, and hydrophobicity. The results indicate that the Fmoc-FF-based hydrogel has the potential to be used as encapsulation and delivery system for various drugs and bioactive molecules [62].

A self-assembling peptide, RADA16, which can spontaneously form a hydrogel under physiological conditions, was used for the controlled release of paclitaxel. Studies performed in vitro indicated that RDA16 peptide hydrogels encapsulating paclitaxel could effectively inhibit the growth of the breast cancer cell line, MDA-MB-435S. The hydrogel was effective at controlling the release of paclitaxel and inhibiting tumor cell growth [35].

The peptide hydrogel system was also shown to be an effective vehicle for the localized delivery of curcumin over for prolonged periods of time. β-Hairpin hydrogels loaded with curcumin were analyzed and the in vitro studies with a medulloblastoma cell line confirm that the encapsulation of the curcumin within the hydrogel does not have an adverse effect on its bioactivity. The rate of curcumin

Table 16.1 Sequences of peptides included in peptide-based delivery systems of drugs/therapeutic molecules

Name and/or sequence of the peptide included in the nanoplatform delivery system	Drug/molecule delivered	Type of application	References
MAX8 (VKVKVKVK-V^{D}PPT-KVEVKVKV-NH_2)	Curcumin	Medulloblastoma cell line	[34]
RADA16(RADARADARADARADA-NH_2)	Paclitaxel	Breast cancer cells, MDA-MB-435S	[35]
RAD16-I, AcN-RADARADARA DARADA-$CONH_2$	Collagen	HP75 cells	[36]
CRGDK and GPLGVRGK	Irinotecan	HT-29 cells	[37]
AVNTANSTGSLVHC	Topotecan	Mouse breast cancer cells	[38]
Cholesterol_PEG200_cyclo-RGD	(DiD) fluorescent dye	Endothelial cells, HMEC-1	[39]
ac-A_6K$CONH_2$, KA_6-$CONH_2$, ac-A_6D-COOH, DA_6-COOH	Carboxyfluorescein, Nile red dye, FITC-dextran	Human colorectal adenocarcinoma Caco-2 cells and rat jejunum	[40,41]
Ac-FLIVIK(Ac-FLIVI)KKKK-$CONH_2$, Ac-FLIVIGSIIK(Ac-FLIVIGSII)KKKK-$CONH_2$	5(6)-Carboxyfluorescein, tryptophan and rhodamine 6G	N/N 1003A lens epithelial cells	[42]
NH_2-YHWYGYTPQNVIeGGGSGGGS-Cys-COOH	Doxorubicin	Hepatocellular carcinoma tumors	[43]
Poly (L-lysine) to the hydrophobic poly (L-leucine)	Doxorubicin	HepG2 cells	[44]
K237 (HTMYYHHYQHHL)	Paclitaxel	Tumor neovasculature targeting drug delivery	[45]
SP204 (KQFSALPFNFYT)	Doxorubicin and vinorelbine	Decreased tumor angiogenesis in prostate cancer	[46]
pVEC-CREKA (LLIILRRRIRKQAHAHSKCREKA)	Chlorambucil	Breast cancer cells	[47]

(*Continued*)

Table 16.1 (Continued)

Name and/or sequence of the peptide ineluded in the nanoplatform delivery system	Drug/molecule delivered	Type of application	References
pVEC-gHo (LLIILRRRIRKQAHAHSKNHQQQNPHQPPM-NH_2)	Doxorubicin	Glioma tumor	[48]
ADDA-TCPP C12-CGRKKRRQRRRPPQRGDS	Doxorubicin	HeLa cells	[49]
iRGD (Ac-CCRGDKGPDC) and TAT (RKKRRQRRRC)	Nile red	Choroidal flat mount	[50]
$QD(TAT)_{10}$, $QD(TAT)_{20}$, $QD(TAT)_{40}$, $QD(TAT)_{75}$	Quantum dots	Target choroidal neovascularization	[51]
CD3ac-Ac-LK(Ac)-LK(Ac)-LK(Ac)-LW-DL-LW-DL-LW-DL-LW-NH_2	Rose bengal and 5-carboxy-fluorescein	Targeted drug delivery	[52]
MNP–DGL–RGD-GX1 RGD: cyclo[-Arg-Gly-Asp-Phe-Lys], GX1: cyclo[-Cys-Gly-Asn-Ser-Asn-Pro-Lys-Ser-Cys]	Doxorubicin	Early diagnosis and therapy in hepatocellular carcinoma (HepG2)	[53]
H-Tyr-D-Arg-Phe-Lys-NH_2 (SS-01), H-Dmt-D-Arg-Phe-Lys-NH_2 (SS-02), H-Phe-D-Arg-Phe-Lys-NH_2 (SS-20), H-D-Arg-Dmt-Lys-Phe-NH_2 (SS-31)	Fluorescent probes	Cancer cells, mitochondria	[54]
$F_xrF_xrF_xr$ (F_x = cyclohexylalanine, r = D-arginine)	Chlorambucil	HeLa cells	[55]
$F_xrF_xrF_xr$ (F_x = cyclohexylalanine, r = D-arginine)	Doxorubicin	HeLa cells	[56]
KCFQWQRNMRKVRGPPVSCIKR	Streptavidin	HeLa, IEC-6	[57]
RΔF: arginine-α,β-dehydrophenylalanine; VΔF: valine-α,β-dehydrophenylalanine; MΔF: methonine-α,β-dehydrophenylalanine	Curcumin	Antimalaria	[58]

Ac, acetyl; DiD, 1,1′-dioctadecyl-3,3,3′,3′-tetramethylindodicarbocyanine perchlorate; Dmt, 2′6′-dimethyltyrosine; FITC, fluorescein isothiocyanate; MNP, magnetic nanoparticle.

release and its consequent therapeutic efficacy can be conveniently modulated as a function of the concentration of the MAX8 peptide [34].

A self-assembling peptide hydrogel (RAD16-I, AcN-RADARADARA DARADA-$CONH_2$), also called PuraMatrix, was used to investigate the discoid in domain receptor (DDR) 1-mediated protein interactions between collagen type I and matrix metalloproteinase (MMP)-2 or -9, as well as cell invasion. Results obtained revealed that the dynamic cell culture technique produced highly stable harvesting of cells throughout the constructs with HP-75, human pituitary adenoma cell line, when compared to the traditional seeding methods. The results showed that incorporation of this peptide with each ECM component provides a more permissive environment to elucidate ECM to cell signal interaction [36].

16.2.2 Nanovesicles

Peptide-based nanovesicles or nanospheres are an important category of vesicles due to their small size which permits their suspension in fluids and their capacity to load and deliver intracellularly therapeutic agents, including drugs. These nanovesicles can enhance therapeutic efficiency and specificity and may induce less severe side effects from the delivered drug (Figs. 16.1 and 16.2).

Irinotecan hydrochloride (ITC) was encapsulated into a peptide-based nanovesicle system, named ITC$\subset$N-G-C [37]. Nanovesicles (N) were prepared by using L-α-phosphatidylcholine, cholesterol, 1,2-dioleoyloxy-3-trimethylammonium propane chloride (DOTAP), and 1,2-distearoyl-snglycero-3-phosphoethanolamine-polyethyleneglycol-maleimide (DSPE-PEG 2000-MAL). DSPE-PEG-NH_2 was added to the lipid mixture before formation of the lipid film. After that, the tumor homing peptide CRGDK was covalently attached to the maleimide group of the nanovesicle to lead to (N-C) system, then MMP-2 responsive peptide, GPLGVRGK, was used to form the protective shell around the cores to obtain (N-C-G). Irinotecan hydrochloride (ITC) was incorporated into the nanovesicles and ITC$\subset$N-G-C formulation was obtained. These drug-loaded nanovesicles were able to penetrate into the

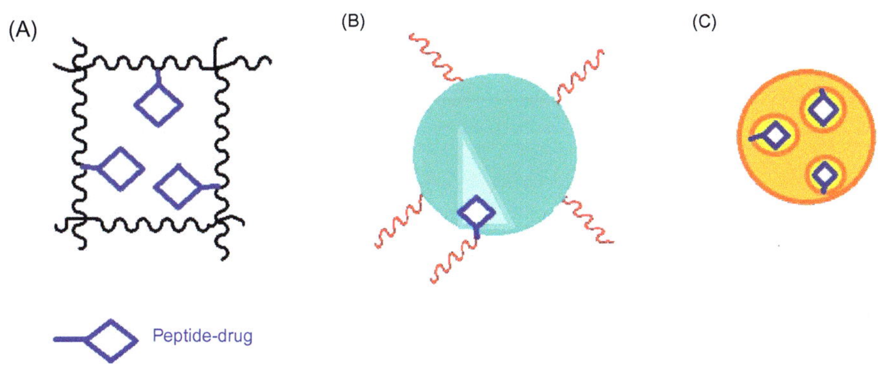

Figure 16.1 Schematic representation of peptide-based hydrogels (A), peptide-based nanovesicles (B), and peptide-based nanoparticles (C) that encapsulated drug molecules.

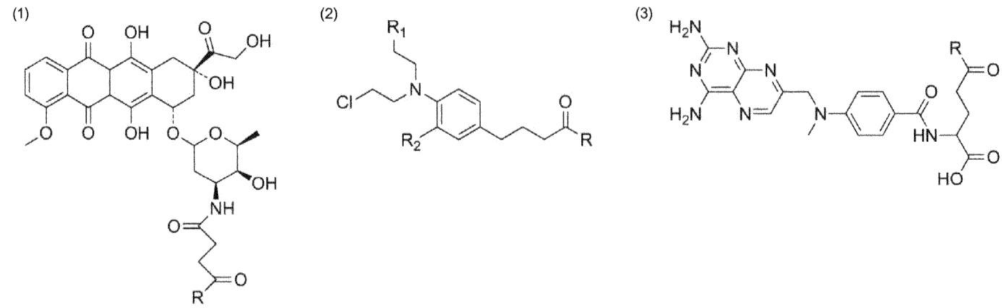

Figure 16.2 Anticancer drugs frequently used in chemotherapy: doxorubicin (1), chlorambucil (2), methotrexate (3). All R groups represent covalently bonded peptide.

multicellular spheroids of HT-29 cells in a larger area of the tumor tissue compared to nanovesicles containing irinotecan (ITC⊂N) [37].

Sortino et al. developed nanoassemblies, diverse morphological vesicles, at different pH, containing chlorogenic acid, a natural polyphenol, and then covalently bonded the tumor-targeting peptide sequence AVNTANSTGSLVHC [38]. This peptide contains two important sequences: GLS, which has been shown to home to neovasculature of breast carcinoma, melanoma, and a variety of other types of cancer tissues, and VNTANST, which binds to vimentin receptors and inhibited lung and skin cancer cells. Then the encapsulation efficiency with the anticancer drug Topotecan as well the release of the drug were evaluated. These assemblies successfully delivered Topotecan into mouse breast cancer cells.

Cebrera et al. developed another type of nanovesicle based on a single-step fluid-based method named DELOS-SUSP (depressurization of an expanded liquid organic solution-suspension) and prepared small unilamellar vesicles (SUVs) of about 100 nm by using cholesterol and the phospholipid 1,2-dipalmitoyl-sn-glycero-3-phosphocholine (DPPC) in the presence of ethanol and compressed CO_2 [39]. This technique is simple, fast, environmentally friendly, and scale-up may be easily developed to obtain diverse types of SUVs for diverse applications. Then, a PEG2000 unit was attached to the SUV through an ether bond and a cRGDfK peptide was coupled to this moiety through a carbamate bond leading to a cholesterol_PEG200_RGD molecule. This liposome–RGD conjugate with 1,1′-dioctadecyl-3,3,3′,3′-tetramethylindodicarbocyanine perchlorate (DiD) fluorescent dye was able to transfect HMEC-1 cells more efficiently than plain liposomes, probably via $\alpha_v\beta_3$ integrin-mediated endocytosis. In vitro assays on endothelial cells HMEC-1, demonstrated that these nanocarriers were nontoxic, nonhemolytic, and sterile.

A class of lipid-like peptides with an aspartic acid or lysine hydrophilic head and a hydrophobic tail composed of six alanines (i.e., ac-A_6KCONH_2, KA_6-$CONH_2$, ac-A_6D-COOH, and DA_6-COOH) that formed nanovesicles was obtained [40]. These peptide-based nanovesicles were nontoxic, and were able to encapsulate and slowly release fluorescent molecules (carboxifluorescein and FITC-dextran) [40,41]. From this series ac-A_6D-COOH was identified as the most suitable for further development of a future drug-delivery system (DDS). Nanovesicles containing ac-A_6D-COOH peptide showed an increased mucosal absorption of FITC-dextran in rat jejunum compared with control and other lipid-like peptides from the series.

Branched amphiphilic peptides capable of forming supramolecular self-assemblies that are filled with solvent were obtained. The peptide assemblies obtained, namely Ac-FLIVIK(Ac-FLIVI)KKKK-$CONH_2$ and Ac-FLIVIGSIIK (Ac-FLIVIGSII)KKKK-$CONH_2$, were able to efficiently encapsulate dyes, such as 5(6)-carboxyfluorescein, tryptophan, and rhodamine 6G and to translocate them through plasma membranes of N/N 1003A lens epithelial cells [42].

GE11 (NH_2-YHWYGYTPQNVIeGGGSGGGS-Cys-COOH), an EGFR peptide ligand, was used to obtain self-assembled peptide nanovesicles with various sizes and surface characteristics that were loaded with diverse cargoes (fluorescent dyes, siRNA, doxorubicin) [43]. The peptide-based nanovesicles exhibited good thermal stability and an affinity for EGFR-overexpressing cancer cells, hepatocellular carcinoma cells (SMMC-7721), and Eca 109 cells. These vesicles were able to encapsulate and deliver both AChE gene and the antitumor drug doxorubicin into mice bearing subcutaneous hepatocellular carcinoma tumors [43].

A self-assembly of artificial supramolecular peptide amphiphiles was developed by Xu et al. that contained covalently bonded the hydrophilic poly(L-lysine) to the hydrophobic poly(L-leucine) [44]. These core–shell nanostructures, which were formed in water at pH 7.4, were able to encapsulate and release doxorubicin into HepG2 cells.

16.2.3 Nanoparticles

Nanoparticles (NPs) can encapsulate poorly soluble anticancer drugs, protect unstable therapeutic molecules, and may modify the drug's blood circulation and tissue distribution. Peptide-containing NPs can be easily prepared and scaled-up during manufacture. Owing to multiple functional groups in the primary sequences of polypeptides, peptide NPs can be exploited to create different interactions with therapeutic compounds and subsequently form three-dimensional networks offering a variety of possibilities for reversible binding of active molecules, protecting them in a matrix, as well as specific targeting to the site of action.

There are a number of criteria that must be met when creating NPs: they must effectively bind and carry the drug(s), they must be stable in the circulation and biocompatible in the body, they should have high bioavailability so as to optimize their anticancer effects on tumor cells, they must be able to target cancer cells while avoiding healthy cells, and they must only release the drug(s) once inside the tumor [65].

Aldehyde-PEG–PLA block copolymer nanoparticles conjugated with K237-(HTMYYHHYQHHL) peptides were used for tumor neovasculature-targeting drug delivery. Paclitaxel, a chemotherapeutic agent with potent antiangiogenic activity, was used as a prototype drug. Conjugation of the K237 with paclitaxel-loaded nanoparticles (K237-PTX-NP) allowed for long-term, accurate in vivo tumor targeting, and significantly higher apoptosis of tumor neovasculature endothelial cells and necrosis of tumor tissues compared to cells treated with the commercial formulation of Taxol and with PTX-NP [45].

Peptide SP204 was obtained from screening of a peptide library using a phage display method. Peptide SP204 is capable of binding specifically to the cell surface of prostate cancer cells both in vitro and in vivo and can be conjugated to liposomes containing doxorubicin and vinorelbine. Such conjugation leads to increased therapeutic efficacy of these drugs in mice bearing human prostate cancer xenografts through enhanced tumor apoptosis and decreased tumor angiogenesis. These results suggest that it might be possible to use the SP204 peptide for specific targeting of tumor cells in the treatment of prostate cancer as well as for imaging of this malignancy [46].

Tumor-targeting peptides (TTPs), that are specific for tumor-related surface markers, such as membrane receptors, can be used to deliver cytotoxic cargo (i.e., drugs or a cytotoxic peptide) specifically to tumor tissue or vasculature. These peptides interact with receptors which are overexpressed on tumor cells.

The maximum therapeutic effect and minimal off-target side effects can be achieved through effective strategies for targeted delivery of anticancer peptides, and also restricted release of the drug payload to the tumor cell or at least to its local environment. The ability to specifically deliver therapeutics to tumor cells decreases unwanted side effects and increases the therapeutic effect.

RGD (Arg-Gly-Asp) and NGR (Asn-Gly-Arg) are two of the original TTPs, discovered in the early 1980s and are the most extensively studied. Besides those, more than 740 TTPs are known to target various types of tumors in hepatocellular carcinoma, colon, prostate, breast, and lung cancers [66–68]. Several RGD motif peptides are available which can be used for synthesizing peptide–drug conjugates which include c(N-Me-VRGDf)–cyclo(MeVal-Arg-Gly-Asp-D-Phe), RGD-4C–Cys-Phe-Cys-Asp-Gly-Arg-Cys-Asp-Cys, c(RGDyK)–cyclo(Arg-Gly-Asp-D-Tyr-Lys), c(RGDfC)–cyclo(Arg-Gly-Asp-D-Phe-Cys), and c(RGDfK)–cyclo(Arg-Gly-Asp-D-Phe-Lys). Apart from the RGD sequence peptides, NGR sequence peptides have also shown targeted delivery to integrin-expressing vasculature and integrin-positive tumor cells which can be used as alternatives to RGD peptides [69].

Lipid nanocapsules (LNCs) modified with grafting cRGD peptides were assessed in vitro in tumor targeting of U87MG glioma cells. The results obtained revealed that cRGD grafting improved binding and internalization compared to negative control LNC-cRAD and blank LNC. The peptide-grafted LNC remained in the blood circulation to up to 3 h, with reduced capture by the RES organs. Tumor accumulation of LNC-cRGD with respect to LNC-cRAD was significantly higher at 1–3 h [70].

The peptide $H_7K(R_2)_2$ was used as a targeting ligand and in vitro studies revealed the peptide's efficiency for targeting of glioma tumor cells. Furthermore, $H_7K(R_2)_2$-modified pH-sensitive liposomes containing antitumor drugs demonstrated antitumor activity in mice bearing glioma tumors and revealed that this is a promising delivery system involving the response stimuli at the acidic pH of the glioma tumor microenvironment and that it is suitable for antitumor therapy [71].

The engineered production of a CTP (NGR)-containing C-1027 has been reported in order to evaluate its activity against breast cancer cell lines (MDA-MB-468, MDA-MB-231, SK-BR-3, MCF-7, and T47D for CD13). The results obtained demonstrate the feasibility to produce CTP-containing enediyne chromo proteins, and the possibility of CTP-containing drug-binding proteins for targeted delivery.

A 5′-carboxy-derived nucleolipid has been synthesized and coupled to the cytotoxic D-$(KLAKLAK)_2$ sequence in order to obtain a peptide–nucleolipid bioconjugate. The human A549 nonsmall-cell lung carcinoma cell line was used to assess the anticancer properties of a peptide–nucleolipid bioconjugate. The biophysical and structural properties of the peptide–nucleolipid bioconjugate were assessed and the results concluded that the amphiphilic peptides favored helical-type secondary structures and well-defined nanoparticle formulations that were found to be important towards their biological activity [72].

Two chimeric peptides were reported for the delivery of the covalently bonded DNA alkylating agents chlorambucil or doxorubicin. These chimeric peptides contain a cell-penetrating peptide pVEC (LLIILRRRIRKQAHAHSK-NH_2) and a tumor homing peptide: (1) CREKA [47] and (2) gHo, NHQQQNPHQPPM-NH_2 [48]. The pVEC-CREKA-chlorambucil derivative efficiently killed specifically the breast cancer cells [47], while doxorubicin-pVEC-gHo demonstrated efficacy in both in vitro and in vivo glioma [48].

A nanosystem can realize the controlled drug release and serve as a platform for designing multifunctional nanocarriers using diversified bioactive peptide-based amphiphiles. A tumor-targeting redox-responsive DDS with bioactive surface was constructed by immobilizing peptide-based amphiphile C12-CGRKKRRQRRRP PQRGDS (defined as ADDA-TCPP) onto the mesoporous silica nanoparticles (MSNs) as an end-capping nanovalve. Doxorubicin (DOX) was used as a model drug that was efficiently trapped in the pores of MSNs. An in vitro release experiment demonstrated that the mesopores of the resulting DOX-loaded MSNs (DOX@MSN-ss-ADDA-TCPP) could be sealed tightly with ADDATCPP self-assemblies. Investigations confirmed that incorporation of RGDS peptide facilitated the active targeting delivery of DOX to $\alpha_v\beta_3$ integrin overexpressed tumor cells [49].

A NP was developed to target choroidal neovascularization (CNV) via topical ocular administration. The NPs were obtained through conjugation of arginine-glycine-aspartic acid RGD (iRGD, Ac-CCRGDKGPDC) and transactivated transcription (TAT) (RKKRRQRRRC) peptides to polymerized ethylene glycol and lactic-co-glycolic acid. Choroidal flat mounts in Brown Norway rats showed that the red fluorescence intensity of NPs followed the order of iRGD and TAT NPs. iRGD and transactivated transcription dual-modified NPs have significantly displayed targeting and penetration ability both in vitro and in vivo, indicating that it is a promising DDS for managing choroidal neovascularization via topical ocular administration [50].

The functional role of nanoparticle multivalency in a cellular uptake mechanism and the importance of lower multivalency for efficient subcellular targeting were studied using TAT-peptide functionalized with quantum dot (QD) of 30–35 nm hydrodynamic diameter with varied multivalency from 10 to 75 (e.g., $QD(TAT)_{10}$, $QD(TAT)_{20}$, $QD(TAT)_{40}$, $QD(TAT)_{75}$). The role of multivalency in endocytosis and subcellular trafficking was studied. The results obtained showed that both low and high multivalent nanoparticles enter into cells predominantly via lipid-raft-mediated endocytosis but the higher multivalencies of 40 and 75 induce vesicular trapping followed by exocytosis within 12 h. In contrast, lower multivalencies of 10 and 20 offer efficient traffic toward the perinuclear region and Golgi apparatus [51].

A new 10-amino-acid peptide, CD3ac (Ac-LK(Ac)-LK(Ac)-LK(Ac)-LW-DL-LW-DL-LW-DL-LW-NH2) was synthesized, purified, and characterized. Peptide CD3ac forms solid particles in the nano- and micrometer size range and holds promise for drug-delivery applications. Rose bengal and 5-carboxy-fluorescein, two xanthene derivatives, were encapsulated into CD3ac. Their presence in the nanoparticles was confirmed by confocal fluorescence microscopy and the encapsulation efficiency was quantified by absorption measurements of dissolved RB-containing peptide bead suspensions [73]. A fully synthetic NP material was obtained by the self-assembly of three molecular components, Flutax-2 as cargo and Tfn-AF568, encapsulated in CD3ac. This fully synthetic NP material can be used as a targeted DDS [52].

NPs were used as a promising tool in early diagnosis and therapy in hepatocellular carcinoma. DOX was loaded into MNP–DGL–RGD-GX1–DOX NPs (where MNP is a magnetic nanoparticle, RGD is the pentapeptide cyclo[-Arg-Gly-Asp-Phe-Lys], and GX1 is the nonapeptide (cyclo[-Cys-Gly-Asn-Ser-Asn-Pro-Lys-Ser-Cys])) in order to target HepG2 cells. In vitro exploration of a molecular mechanism revealed that cell apoptosis was associated with the overexpression of Fas protein and the significant activation of caspase-3. NPs MNP–DGL–RGD-GX1–DOX possessed significant antitumor activity, with a tumor inhibitory rate of 78.5% [53].

16.3 Peptide-based systems as delivery vectors of drugs for neurodegenerative diseases

Neurodegenerative diseases are considered to be caused by mitochondrial dysfunctions with subsequent oxidative stress accumulation. The selective delivery of bioactive molecules, including drugs, into mitochondria is a new therapeutic approach to treat neurodegenerative diseases, and peptide-based technologies that target mitochondria have been developed.

Two strategies that target bioactive molecules to mitochondria have been developed: lipophilic cations and peptides. Lipophilic cations that are most commonly used are the derivatives of triphenylphosphonium (TPP) that can be covalently bonded to small, bioactive molecules to facilitate delivery into the mitochondria. An example from this category is the anticancer drug cisplatin complexed with rhodamine-123 [74]. Two peptide classes were the most developed for efficient delivery of covalently bonded bioactive molecules to mitochondria: Szeto-Schiller

(SS) peptides [75] and mitochondrial-penetrating peptides (MPPs) [76,77]. These peptides are usually amphipathic and are capable of forming complexes with diverse cargoes however the uptake mechanism is still under study. SS peptides are localized within the inner mitochondrial membrane.

The first small SS peptides (DMTrFK) were developed by H.H. Szeto and P.W. Schiller: SS-01 (H-Tyr-D-Arg-Phe-Lys-NH_2) and SS-02 (H-Dmt-D-ArgPhe-Lys-NH_2; Dmt = 2′6′-dimethyltyrosine) are highly polar water-soluble tetrapeptides that carry 3+ net charge at physiologic pH and have high-affinity for binding to opioid receptors. Thereafter, more SS peptides were designed and obtained that can readily cross the blood–brain barrier (BBB) [78], and they have demonstrated in vivo efficacy in animal models of Parkinson's disease, amyotrophic lateral sclerosis (ALS), and Alzheimer's disease [54,79]. The most important mode of action of mitochondria-targeted therapies is demonstrated to be that of antioxidants that block mitochondria oxidative injury. For example, the SS peptides present multiple protective functions on mitochondria, such as facilitation of mitochondrial respiration and ATP synthesis, reducing ROS generation, and inhibiting mitochondrial swelling. The SS31 peptide (H-D-Arg-Dmt-Lys-Phe-NH_2 [SS-31]) was demonstrated to be efficient in reducing myocardial ischemia–reperfusion injury, in neuroprotection, metabolic syndrome, and heart failure, and may prevent muscle atrophy [54,75]. The mitochondrial oxidative damage is an important therapeutic target and more bioactive and pharmacophore-containing complexes may be developed in the future to treat patients, especially after successful establishment of the mitochondrial function in humans.

S.O. Kelley et al. developed a new class of peptide-based vectors efficient for the delivery of bioactive molecules to mitochondria named mitochondria penetrating peptides (MPPs). MPPs can be easily modified for delivery of specific cargo as well as behaving as bioactive compounds [80–82]. MMPs present antioxidant, apoptosis-inducing and mitochondrial DNA-damaging properties. A library of MMPs was obtained by incorporating hydrophobic residues into cell-penetrating peptides. These peptides present elevated levels of intracellular uptake with different mitochondrial localization efficacy. The MMPs that contain cyclohexylalanine are the most efficient in mitochondrial localization and also induce a cytotoxic effect on cancer cells [83]. MMPs efficiently transported chlorambucil [55,84–86], doxorubicin [56,77,87–89], and platinum compounds [90] in cancer cells, induced apoptosis and necrosis. Cyclohexylalanine-containing MMPs exhibited the most elevated levels of mitochondrial localization [77]. Even if peptide vectors may be used to target mitochondria, more strategies have to be established in order to rationally design therapeutic-containing molecules to manipulate mitochondria.

16.4 Antimicrobial peptide-based drug delivery systems

Peptides with antimicrobial properties, named antimicrobial peptides (AMPs), behave as well as efficient delivery vectors of bioactive substances, including drugs, in many applications, such as cancer, genetic disorders, cardiovascular, infectious, and inflamatory diseases. Firstly, AMPs were isolated from natural sources, however many newly designed and synthetically obtained peptides were produced. An online database containing more than 2800 AMPs with diverse activities is available on the following link: http://aps.unmc.edu/AP/main.php.

The antimicrobial protein lactoferrin (hLF) is found in human milk, and is important as it exhibits antifungal, antimicrobial, and antiviral properties. Lactoferrin is cleaved in the gastrointestinal tract through pepsin digestion to a shorter peptide, named lactoferricin. Lactoferricin contains 49 amino acids (amino acids 20–68 of the parent sequence) and exhibits antimicrobial, antiviral, antitumor, and immunological activity [91]. Another peptide sequence derived from lactoferrin, named hLF peptide, which consists of amino acids 38–59 (19–40 of lactoferricin: KCFQWQRNMRKVRGPPVSCIKR), was shown to efficiently enter in different cell types (e.g., HeLa, IEC-6). Efficient protein (streptavidin) delivery mediated by hLF peptide has been described [57].

Also, SynB sequences from the antimicrobial peptide protegrin-1 (PG-1) can be used for delivery of therapeutic cargoes (doxorubicin [92–95], paclitaxel [96]). Protregrins are small cysteine- and arginine-rich antimicrobial peptides. PG-1 consists of 18 amino acids, it was originally derived from porcine leukocytes, and it is active against fungi, bacteria, and viruses [97]. Due to its uptake mechanism, the efficient delivery of the covalently bonded complex of the peptide with doxorubicin into the brain has been reported. The blood–brain barrier was traversed with high efficiency and without damaging its integrity [92,94].

Buforin II, a 21-amino-acid antimicrobial peptide, was identified in the stomach tissue of the Asian toad. Buforin II and its analogues are able to penetrate through the cell membrane without destroying it and can kill bacteria efficiently by binding to nucleic acids. BF2d, a modified analogue of buforin II, is able to deliver green fluorescent protein (GFP) to HeLa cells [98].

Neundorf et al. have developed a cathelicidin-derived peptide, sC18, that originates from the 18-kDa cationic antimicrobial protein (CAP18), isolated from rabbit leukocytes. CAP18 itself, and also shorter variants, exhibit antimicrobial properties. sC18 peptide consists of amino acids 106–121 of CAP18. This so-called sC18 peptide was able to internalize into several different cell lines without being toxic in the tested concentration range and also was efficiently used as a delivery transporter for cytostatic organometallic complexes [99–101].

The LL-37 peptide contains 37 amino acids and was the first identified α-helical antimicrobial peptide from a human source. It belongs to the cathelicidin family and it was demonstrated to present high activity toward a broad antimicrobial spectrum (Gram-positive and Gram-negative bacteria) [102] and also cytotoxic activity against eukaryotic cells [103]. LL-37 was reported to deliver noncovalently linked fluorescently labeled oligonucleotides [104] and also plasmid DNA into different cell lines [105].

Curcumin, an efficient antimalarian agent with low solubility and biocompatibility, was successfully incorporated into self-assembled and biocompatible nanovesicles containing dipeptides (RΔF: arginine-α,β-dehydrophenylalanine; VΔF: valine-α,β-dehydrophenylalanine; MΔF: methonine-α,β-dehydrophenylalanine) [58]. The in vitro and in vivo studies revealed that these dipeptide nanovesicles were able to deliver curcumin and presented a higher antimalarian efficacy compared with curcumin alone. These short peptide nanovesicles did not show evidence of any cytotoxity or hemolytic activities as confirmed by MTT and LDH assays.

Taxifolin 3-*O*-β-D-glucopyranoside (TXG: 3′,4′,5′,70-tetrahydroxyflavonol glucoside), a flavonoid isolated from *Rhododendron mucronulatum* and *Silybum marianum*, was shown to reduce serum levels of eosinophils, IgE, IL-4, and IL-5 in NC/Nga mice so it was proposed to be an ideal drug candidate for topical application in atopic dermatitis therapy. Due to its high hydrophilicity, TXG may be used only together with a delivery system, therefore the TXG-loaded Pep1-elastic liposome formulation was developed [106]. Pep-1 peptide (KETWWETWWTEWSQPKKKRKVC, 22mer) was conjugated via the thiolmaleimide reaction to the TXG liposome preparation. The in vivo studies revealed that the TXG-loaded Pep1-elastic liposome formulation was able to deliver TXG in large quantities into the skin and to amplify the immune-modulating and antiinflammatory activities of the drug.

16.5 Peptide-based delivery systems of drugs under clinical trial

Peptide-based formulations that are capable of delivering therapeutical cargoes into specific cell lines with low toxicity and immunogenicity have been developed in the last three decades. Clinical studies revealed that peptides are efficacious, specific, and safe for delivering therapeutic molecules into cells, as well as easy to produce and purify. The majority of the peptide-based formulations involved in clinical trials are efficient for parenteral routes of administration [107,108] but less are suitable for oral administration [108,109]. A phase Ib study was performed to verify the safety of the tumor homing NGR peptide-targeted hTNF plus doxorubicin in the treatment of refractory/resistant solid tumors [110]. The main goal of the peptide development is to be incorporated in an adequate formulation with proper excipients or to modify the known peptides/design new peptides for oral delivery purposes. More research is needed, with standardized methods and correlation of the structure–activity properties, in order to develop better peptide-based formulations for therapeutic purposes.

16.6 Conclusions

Significant progress has been made in identifying peptides with cell-penetrating properties, active and stable in diverse biological systems, capable of delivering therapeutic agents, including drugs, in targeted cells, and with low toxicity and immunogenicity. The standardization of the protocols and practices that are necessary for the faster development of peptide-based research is still in a developing phase. Therefore, new approaches are required for the determination of the uptake mechanism, pharmacodynamic parameters, toxicity and immunogenicity of peptide-based systems both in vitro and in vivo. Peptides are versatile molecules that can evade the immune system and cross biological barriers, including skin, blood–brain barrier, gastrointestinal walls, can enter into the cytoplasm or nucleus of the targeted cells, and avoid clearance by the mononuclear phagocyte system. Due to the fact that there is abundant information on peptide-based systems investigated under different conditions (e.g., purity of peptides, concentrations, pH, cell type, and density, etc.), the modest understanding of the peptide sequence–structure function relationships is a considerable impediment to progress in this field. Also, the mode of administration of the complex is important for patients and the potential for peptides to enhance the oral bioavailability of therapeutic molecules can clearly be adjusted by the formulation used. Although considerable improvements are being made in this regard, further optimization and development are needed. Chemical conjugation, physical mixtures, nanoparticles, or combinations of the above may lead toward the ideal system. About 140 peptides are undergoing clinical trials, but no peptide-based delivery system is currently in clinical use. Therefore, in order to accelerate the development of peptide-based DDSs it is necessary to design peptides that are able to penetrate the cell membrane or to cross the biological barriers and to specifically deliver cargoes into targeted cells, to possess increased circulation times, as well to be nontoxic and nonimmunogenic.

Acknowledgments

Funding for this study has been provided by the European Commission and the Romanian government (project POSCCE 621/2014).

References

[1] H. Feyzizarnagh, D.Y. Yoon, M. Goltz, D.S. Kim, Peptide nanostructures in biomedical technology, Wiley Interdiscip. Rev. Nanomed. Nanobiotechnol. 8 (2016) 730–743.
[2] W.F. Anderson, Human gene therapy, Nature 392 (1998) 25–30.
[3] T.K. Wong, E. Neumann, Electric-field mediated gene-transfer, Biochem. Biophys. Res. Commun. 107 (1982) 584–587.
[4] D.A. Eavarone, X.J. Yu, R.V. Bellamkonda, Targeted drug delivery to C6 glioma by transferrin-coupled liposomes, J. Biomed. Mater. Res. 51 (2000) 10–14.
[5] G. Kong, G. Anyarambhatla, W.P. Petros, R.D. Braun, O.M. Colvin, D. Needham, et al., Efficacy of liposomes and hyperthermia in a human tumor xenograft model: importance of triggered drug release, Cancer Res. 60 (2000) 6950–6957.
[6] M. Schwonzen, C.M. Kurbacher, P. Mallmann, Liposomal doxorubicin and weekly paclitaxel in the treatment of metastatic breast cancer, Anticancer Drugs 11 (2000) 681–685.
[7] L. Schoonen, J.C.M. van Hest, Functionalization of protein-based nanocages for drug delivery applications, Nanoscale 6 (2014) 7124–7141.

[8] Cancer Facts & Figures, American Cancer Society, American Cancer Society, Atlanta, GA, 2016.
[9] X.L. Wu, J.H. Kim, H. Koo, S.M. Bae, H. Shin, M.S. Kim, et al., Tumor-targeting peptide conjugated pH-responsive micelles as a potential drug carrier for cancer therapy, Bioconjug. Chem. 21 (2010) 208–213.
[10] M. Essler, E. Ruoslahti, Molecular specialization of breast vasculature: a breast-homing phage-displayed peptide binds to aminopeptidase P in breast vasculature, Proc. Natl. Acad. Sci. U.S.A. 99 (2002) 2252–2257.
[11] S. Aroui, S. Brahim, M. De Waard, J. Breard, A. Kenani, Efficient induction of apoptosis by doxorubicin coupled to cell-penetrating peptides compared to unconjugated doxorubicin in the human breast cancer cell line MDA-MB 231, Cancer Lett. 285 (2009) 28–38.
[12] S. Aroui, S. Brahim, J. Hamelin, M. De Waard, J. Breard, A. Kenani, Conjugation of doxorubicin to cell penetrating peptides sensitizes human breast MDA-MB 231 cancer cells to endogenous TRAIL-induced apoptosis, Apoptosis 14 (2009) 1352–1365.
[13] M. Lindgren, K. Rosenthal-Aizman, K. Saar, E. Eiriksdottir, Y. Jiang, M. Sassian, et al., Overcoming methotrexate resistance in breast cancer tumour cells by the use of a new cell-penetrating peptide, Biochem. Pharmacol. 71 (2006) 416–425.
[14] W.H. Yang, D.F. Luo, S.X. Wang, R. Wang, R. Chen, Y. Liu, et al., TMTP1, a novel tumor-homing peptide specifically targeting metastasis, Clin. Cancer Res. 14 (2008) 5494–5502.
[15] T. Teesalu, K.N. Sugahara, V.R. Kotamraju, E. Ruoslahti, C-end rule peptides mediate neuropilin-1-dependent cell, vascular, and tissue penetration, Proc. Natl. Acad. Sci. U.S.A. 106 (2009) 16157–16162.
[16] M. Brabant, L. Baux, R. Casimir, J.P. Briand, O. Chaloin, M. Porceddu, et al., A flavivirus protein M-derived peptide directly permeabilizes mitochondrial membranes, triggers cell death and reduces human tumor growth in nude mice, Apoptosis 14 (2009) 1190–1203.
[17] J. Hodoniczky, C.G. Sims, W.M. Best, J.M. Bentel, J.A. Wilce, The intracellular and nuclear-targeted delivery of an antiandrogen drug by carrier peptides, Biopolymers 90 (2008) 595–603.
[18] M. Hallbrink, K. Saar, C.G. Ostenson, U. Soomets, S. Efendic, J. Howl, et al., Effects of vasopressin-mastoparan chimeric peptides on insulin release and G-protein activity, Regul. Pept. 82 (1999) 45–51.
[19] C.G. Ostenson, S. Zaitsev, P.O. Berggren, S. Efendic, U. Langel, T. Bartfai, Galparan: a powerful insulin-releasing chimeric peptide acting at a novel site, Endocrinology 138 (1997) 3308–3313.
[20] P.O. Carlsson, C.G. Ostenson, S. Efendic, U. Langel, L. Jansson, Pituitary adenylate cyclase activating polypeptide (PACAP) redistributes the blood within the pancreas of anesthetized rats, Regul. Pept. 63 (1996) 123–128.
[21] K. Kask, M. Berthold, J. Bourne, S. Andell, U. Langel, T. Bartfai, Binding and agonist antagonist actions of M35, galanin(1-13)-bradykinin(2-9)amide chimeric peptide, in RIN-M-5F insulinoma cells, Regul. Pept. 59 (1995) 341–348.
[22] O.M. Faruque, D. Le-Nguyen, A.D. Lajoix, E. Vives, P. Petit, D. Bataille, et al., Cell-permeable peptide-based disruption of endogenous PKA-AKAP complexes: a tool for studying the molecular roles of AKAP-mediated PKA subcellular anchoring, Am. J. Physiol. Cell Physiol. 296 (2009) C306–C316.

[23] I. Massodi, E. Thomas, D. Raucher, Application of thermally responsive elastin-like polypeptide fused to a lactoferrin-derived peptide for treatment of pancreatic cancer, Molecules 14 (2009) 1999–2015.
[24] A. Bagnato, C. Moretti, J. Ohnishi, G. Frajese, K.J. Catt, Expression of the growth hormone-releasing hormone gene and its peptide product in the rat ovary, Endocrinology 130 (1992) 1097–1102.
[25] C. Palm, M. Jayamanne, M. Kjellander, M. Hallbrink, Peptide degradation is a critical determinant for cell-penetrating peptide uptake, Biochim. Biophys. Acta 1768 (2007) 1769–1776.
[26] A. Anas, T. Okuda, N. Kawashima, K. Nakayama, T. Itoh, M. Ishikawa, et al., Clathrin-mediated endocytosis of quantum dot-peptide conjugates in living cells, ACS Nano 3 (2009) 2419–2429.
[27] S. Aroui, S. Brahim, M. De Waard, A. Kenani, Cytotoxicity, intracellular distribution and uptake of doxorubicin and doxorubicin coupled to cell-penetrating peptides in different cell lines: a comparative study, Biochem. Biophys. Res. Commun. 391 (2010) 419–425.
[28] I. Massodi, S. Moktan, A. Rawat, G.L. Bidwell, D. Raucher, Inhibition of ovarian cancer cell proliferation by a cell cycle inhibitory peptide fused to a thermally responsive polypeptide carrier, Int. J. Cancer 126 (2010) 533–544.
[29] S.A. Moschos, S.W. Jones, M.M. Perry, A.E. Williams, J.S. Erjefalt, J.J. Turner, et al., Lung delivery studies using siRNA conjugated to TAT(48-60) and penetratin reveal peptide induced reduction in gene expression and induction of innate immunity, Bioconjug. Chem. 18 (2007) 1450–1459.
[30] M.N. Fukuda, C. Ohyama, K. Lowitz, O. Matsuo, R. Pasqualini, E. Ruoslahti, et al., A peptide mimic of E-selectin ligand inhibits sialyl Lewis X-dependent lung colonization of tumor cells, Cancer Res. 60 (2000) 450–456.
[31] S.A. Moschos, A.E. Williams, M.A. Lindsay, Cell-penetrating-peptide-mediated siRNA lung delivery, Biochem. Soc. Trans. 35 (2007) 807–810.
[32] J. Nguyen, X.L. Xie, M. Neu, R. Dumitrascu, R. Reul, J. Sitterberg, et al., Effects of cell-penetrating peptides and pegylation on transfection efficiency of polyethylenimine in mouse lungs, J. Gene Med. 10 (2008) 1236–1246.
[33] L. Crombez, M.C. Morris, S. Deshayes, F. Heitz, G. Divita, Peptide-based nanoparticle for ex vivo and in vivo dug delivery, Curr. Pharm. Des. 14 (2008) 3656–3665.
[34] A. Altunbas, S.J. Lee, S.A. Rajasekaran, J.P. Schneider, D.J. Pochan, Encapsulation of curcumin in self-assembling peptide hydrogels as injectable drug delivery vehicles, Biomaterials 32 (2011) 5906–5914.
[35] J. Liu, L. Zhang, Z. Yang, X. Zhao, Controlled release of paclitaxel from a self-assembling peptide hydrogel formed in situ and antitumor study in vitro, Int. J. Nanomedicine 6 (2011) 2143–2153.
[36] D. Yoshida, A. Teramoto, The use of 3-D culture in peptide hydrogel for analysis of discoidin domain receptor 1-collagen interaction, Cell Adh. Migr. 1 (2007) 92–98.
[37] Y. Liu, D. Zhang, Z.Y. Qiao, G.B. Qi, X.J. Liang, X.G. Chen, et al., A peptide-network weaved nanoplatform with tumor microenvironment responsiveness and deep tissue penetration capability for cancer therapy, Adv. Mater. 27 (2015) 5034–5042.
[38] R.M. Sortino, S.M. Romanelli, G.A. Knoll, I.A. Banerjee, Development of peptide conjugated chlorogenic acid nanoassemblies for targeting tumorigenic cells, Soft Mater. 13 (2015) 150–159.

[39] I. Cabrera, E. Elizondo, O. Esteban, J.L. Corchero, M. Melgarejo, D. Pulido, et al., Multifunctional nanovesicle-bioactive conjugates prepared by a one-step scalable method using CO_2-expanded solvents, Nano Lett. 13 (2013) 3766–3774.

[40] D.G. Fatouros, D.A. Lamprou, A.J. Urquhart, S.N. Yannopoulos, I.S. Vizirianakis, S.G. Zhang, et al., Lipid-like self-assembling peptide nanovesicles for drug delivery, ACS Appl. Mater. Interfaces 6 (2014) 8184–8189.

[41] C. Karavasili, M. Spanakis, D. Papagiannopoulou, I.S. Vizirianakis, D.G. Fatouros, S. Koutsopoulos, Bioactive self-assembling lipid-like peptides as permeation enhancers for oral drug delivery, J. Pharm. Sci. 104 (2015) 2304–2311.

[42] S. Gudlur, P. Sukthankar, J. Gao, L.A. Avila, Y. Hiromasa, J.H. Chen, et al., Peptide nanovesicles formed by the self-assembly of branched amphiphilic peptides, PLoS One 7 (2012) e45374.

[43] X.F. Liang, B.Z. Shi, K. Wang, M.L. Fan, D.J. Jiao, J.P. Ao, et al., Development of self-assembling peptide nanovesicle with bilayers for enhanced EGFR-targeted drug and gene delivery, Biomaterials 82 (2016) 194–207.

[44] X.H. Xu, Y.K. Li, H.P. Li, R. Liu, M.M. Sheng, B. He, et al., Smart nanovehicles based on pH-triggered disassembly of supramolecular peptide-amphiphiles for efficient intracellular drug delivery, Small 10 (2014) 1133–1140.

[45] D.-H. Yu, Q. Lu, J. Xie, C. Fang, H.-Z. Chen, Peptide-conjugated biodegradable nanoparticles as a carrier to target paclitaxel to tumor neovasculature, Biomaterials 31 (2010) 2278–2292.

[46] C.-Y. Yeh, J.-K. Hsiao, Y.-P. Wang, C.-H. Lan, H.-C. Wu, Peptide-conjugated nanoparticles for targeted imaging and therapy of prostate cancer, Biomaterials 99 (2016) 1–15.

[47] M. Mae, H. Myrberg, S. El-Andaloussi, U. Langel, Design of a tumor homing cell-penetrating peptide for drug delivery, Int. J. Pept. Res. Ther. 15 (2009) 11–15.

[48] E. Eriste, K. Kurrikoff, J. Suhorutsenko, N. Osokolkov, D.M. Copolovici, S. Jones, et al., Peptide-based glioma-targeted drug delivery vector gHoPe2, Bioconjug. Chem. 24 (2013) 305–313.

[49] Y.J. Cheng, A.Q. Zhang, J.J. Hu, F. He, X. Zeng, X.Z. Zhang, et al., Multifunctional Peptide-Amphiphile End-Capped Mesoporous Silica Nanoparticles for Tumor Targeting Drug Delivery, ACS Appl. Mater. Inter. 9 (2017) 2093–2103.

[50] Y.C. Chu, N. Chen, H.J. Yu, H.J. Mu, B. He, H.C. Hua, et al., Topical ocular delivery to laser-induced choroidal neovascularization by dual internalizing RGD and TAT peptide-modified nanoparticles, Int. J. Nanomedicine 12 (2017) 1353–1368.

[51] C. Dalal, N.R. Jana, Multivalency effect of TAT-peptide-functionalized nanoparticle in cellular endocytosis and subcellular trafficking, J. Phys. Chem. B 121 (2017) 2942–2951.

[52] C. Dittrich, C.J. Burckhardt, G. Danuser, Delivery of membrane impermeable cargo into CHO cells by peptide nanoparticles targeted by a protein corona, Biomaterials 33 (2012) 2746–2753.

[53] J.M. Shen, X.X. Li, L.L. Fan, X. Zhou, J.M. Han, M.K. Jia, et al., Heterogeneous dimer peptide-conjugated polylysine dendrimer-Fe_3O_4 composite as a novel nanoscale molecular probe for early diagnosis and therapy in hepatocellular carcinoma, Int. J. Nanomedicine 12 (2017) 1183–1200.

[54] H.H. Szeto, S.Y. Liu, Y. Soong, D.L. Wu, S.F. Darrah, F.Y. Cheng, et al., Mitochondria-targeted peptide accelerates ATP recovery and reduces ischemic kidney injury, J. Am. Soc. Nephrol. 22 (2011) 1041–1052.

[55] S.B. Fonseca, M.P. Pereira, R. Mourtada, M. Gronda, K.L. Horton, R. Hurren, et al., Rerouting chlorambucil to mitochondria combats drug deactivation and resistance in cancer cells, Chem. Biol. 18 (2011) 445–453.
[56] G.R. Chamberlain, D.V. Tulumello, S.O. Kelley, Targeted delivery of doxorubicin to mitochondria, ACS Chem. Biol. 8 (2013) 1389–1395.
[57] F. Duchardt, I.R. Ruttekolk, W.P.R. Verdurmen, H. Lortat-Jacob, J. Burck, H. Hufnagel, et al., Peptide derived from human lactoferrin with conformation-dependent uptake efficiency, J. Biol. Chem. 284 (2009) 36099–36108.
[58] S. Alam, J.J. Panda, T.K. Mukherjee, V.S. Chauhan, Short peptide based nanotubes capable of effective curcumin delivery for treating drug resistant malaria, J. Nanobiotechnology 14 (2016) 26.
[59] M.L. Becker, E.E. Remsen, D. Pan, K.L. Wooley, Peptide-derivatized shell-cross-linked nanoparticles. 1. Synthesis and characterization, Bioconjug. Chem. 15 (2004) 699–709.
[60] A. McDowell, B.J. McLeod, T. Rades, I.G. Tucker, Polymeric nanoparticles as an oral delivery system for biocontrol agents for the brushtail possum (Trichosurus vulpecula), N. Z. Vet. J. 57 (2009) 370–377.
[61] K. Basu, A. Baral, S. Basak, A. Dehsorkhi, J. Nanda, D. Bhunia, et al., Peptide based hydrogels for cancer drug release: modulation of stiffness, drug release and proteolytic stability of hydrogels by incorporating d-amino acid residue(s), Chem. Commun. 52 (2016) 5045–5048.
[62] R. Ischakov, L. Adler-Abramovich, L. Buzhansky, T. Shekhter, E. Gazit, Peptide-based hydrogel nanoparticles as effective drug delivery agents, Bioorg. Med. Chem. 21 (2013) 3517–3522.
[63] W. Xun, D.Q. Wu, Z.Y. Li, H.Y. Wang, F.W. Huang, S.X. Cheng, et al., Peptide-functionalized thermo-sensitive hydrogels for sustained drug delivery, Macromol. Biosci. 9 (2009) 1219–1226.
[64] P. Worthington, D.J. Pochan, S.A. Langhans, Peptide hydrogels – versatile matrices for 3D cell culture in cancer medicine, Front. Oncol. 5 (2015) 92.
[65] G. Seeta Rama Raju, L. Benton, E. Pavitra, J.S. Yu, Multifunctional nanoparticles: recent progress in cancer therapeutics, Chem. Commun. 51 (2015) 13248–13259.
[66] Z.C. Liu, J. Wang, P. Yin, J.H. Qiu, R.Z. Liu, W.Z. Li, et al., RGD-FasL induces apoptosis in hepatocellular carcinoma, Cell. Mol. Immunol. 6 (2009) 285–293.
[67] M. Amin, A. Badiee, M.R. Jaafari, Improvement of pharmacokinetic and antitumor activity of PEGylated liposomal doxorubicin by targeting with N-methylated cyclic RGD peptide in mice bearing C-26 colon carcinomas, Int. J. Pharm. 458 (2013) 324–333.
[68] T.J. Stott Reynolds, R. Schehr, D. Liu, J. Xu, Y. Miao, T.J. Hoffman, et al., Characterization and evaluation of DOTA-conjugated Bombesin/RGD-antagonists for prostate cancer tumor imaging and therapy, Nucl. Med. Biol. 42 (2015) 99–108.
[69] F. Araujo, N. Shrestha, M.J. Gomes, B. Herranz-Blanco, D. Liu, J.J. Hirvonen, et al., In vivo dual-delivery of glucagon like peptide-1 (GLP-1) and dipeptidyl peptidase-4 (DPP4) inhibitor through composites prepared by microfluidics for diabetes therapy, Nanoscale 8 (2016) 10706–10713.
[70] S. Hirsjärvi, C. Belloche, F. Hindré, E. Garcion, J.-P. Benoit, Tumour targeting of lipid nanocapsules grafted with cRGD peptides, Eur. J. Pharm. Biopharm. 87 (2014) 152–159.

[71] Y. Zhao, W. Ren, T. Zhong, S. Zhang, D. Huang, Y. Guo, et al., Tumor-specific pH-responsive peptide-modified pH-sensitive liposomes containing doxorubicin for enhancing glioma targeting and anti-tumor activity, J. Control. Release 222 (2016) 56–66.
[72] N. Rana, S. Huang, P. Patel, U. Samuni, D. Sabatino, Synthesis, characterization and anti-cancer activity of a peptide nucleolipid bioconjugate, Bioorg. Med. Chem. Lett. 26 (2016) 3567–3571.
[73] C. Dittrich, W. Meier, Solid peptide nanoparticles – structural characterization and quantification of cargo encapsulation, Macromol. Biosci. 10 (2010) 1406–1415.
[74] B.A. Teicher, S.A. Holden, K.N.S. Cathcart, Efficacy of $Pt(Rh\text{-}123)_2$ as a radiosensitizer with fractionated X-rays, Int. J. Radiat. Oncol. Biol. Phys. 13 (1987) 1217–1224.
[75] H.H. Szeto, P.W. Schiller, Novel therapies targeting inner mitochondrial membrane – from discovery to clinical development, Pharm. Res. 28 (2011) 2669–2679.
[76] L.F. Yousif, K.M. Stewart, S.O. Kelley, Targeting mitochondria with organelle-specific compounds: strategies and applications, Chembiochem 10 (2009) 1939–1950.
[77] S.R. Jean, M. Ahmed, E.K. Lei, S.P. Wisnovsky, S.O. Kelley, Peptide-mediated delivery of chemical probes and therapeutics to mitochondria, Acc. Chem. Res. 49 (2016) 1893–1902.
[78] G.M. Zhao, D.L. Wu, Y. Soong, M. Shimoyama, I. Berezowska, P.W. Schiller, et al., Profound spinal tolerance after repeated exposure to a highly selective mu-opioid peptide agonist: role of delta-opioid receptors, J. Pharmacol. Exp. Ther. 302 (2002) 188–196.
[79] M. Manczak, P.Z. Mao, M.J. Calkins, A. Cornea, A.P. Reddy, M.P. Murphy, et al., Mitochondria-targeted antioxidants protect against amyloid-beta toxicity in Alzheimer's disease neurons, J. Alzheimer Dis. 20 (2010) S609–S631.
[80] L.F. Yousif, K.M. Stewart, K.L. Horton, S.O. Kelley, Mitochondria-penetrating peptides: sequence effects and model cargo transport, Chembiochem 10 (2009) 2081–2088.
[81] H.H. Szeto, Cell-permeable, mitochondrial-targeted, peptide antioxidants, AAPS J. 8 (2006) E277–E283.
[82] H.H. Szeto, L.P. James, A.J. Atkinson, Mitochondrial pharmacology: its future is now, Clin. Pharmacol. Ther. 96 (2014) 629–633.
[83] K.L. Horton, M.P. Pereira, K.M. Stewart, S.B. Fonseca, S.O. Kelley, Tuning the activity of mitochondria-penetrating peptides for delivery or disruption, Chembiochem 13 (2012) 476–485.
[84] S.B. Fonseca, S.O. Kelley, Peptide-chlorambucil conjugates combat Pgp-dependent drug efflux, ACS Med. Chem. Lett. 2 (2011) 419–423.
[85] R. Mourtada, S.B. Fonseca, S.P. Wisnovsky, M.P. Pereira, X.M. Wang, R. Hurren, et al., Re-directing an alkylating agent to mitochondria alters drug target and cell death mechanism, PLoS One 8 (2013) e60253.
[86] S.R. Jean, M.P. Pereira, S.O. Kelley, Structural modifications of mitochondria-targeted chlorambucil alter cell death mechanism but preserve MDR evasion, Mol. Pharm. 11 (2014) 2675–2682.
[87] I. Buondonno, E. Gazzano, S.R. Jean, V. Audrito, J. Kopecka, M. Fanelli, et al., Mitochondria-targeted doxorubicin: a new therapeutic strategy against doxorubicin-resistant osteosarcoma, Mol. Cancer Ther. 15 (2016) 2640–2652.
[88] S.R. Jean, D.V. Tulumello, C. Riganti, S.U. Liyanage, A.D. Schimmer, S.O. Kelley, Mitochondrial targeting of doxorubicin eliminates nuclear effects associated with cardiotoxicity, ACS Chem. Biol. 10 (2015) 2007–2015.

[89] C. Riganti, B. Rolando, J. Kopecka, I. Campia, K. Chegaev, L. Lazzarato, et al., Mitochondrial-targeting nitrooxy-doxorubicin: a new approach to overcome drug resistance, Mol. Pharm. 10 (2013) 161–174.
[90] S.P. Wisnovsky, J.J. Wilson, R.J. Radford, M.P. Pereira, M.R. Chan, R.R. Laposa, et al., Targeting mitochondrial DNA with a platinum-based anticancer agent, Chem. Biol. 20 (2013) 1323–1328.
[91] J.L. Gifford, H.N. Hunter, H.J. Vogel, Lactoferricin: a lactoferrin-derived peptide with antimicrobial, antiviral, antitumor and immunological properties, Cell. Mol. Life Sci. 62 (2005) 2588–2598.
[92] C. Rousselle, P. Clair, J.M. Lefauconnier, M. Kaczorek, J.M. Scherrmann, J. Temsamani, New advances in the transport of doxorubicin through the blood-brain barrier by a peptide vector-mediated strategy, Mol. Pharmacol. 57 (2000) 679–686.
[93] C. Rousselle, M. Smirnova, P. Clair, M. Kaczorek, J.M. Scherrmann, J. Temsamani, The use of a new peptide vector-mediated strategy which bypasses brain P-gp for the transport of doxorubicin across the BBB, FASEB J. 14 (2000) A1480.
[94] C. Rousselle, M. Smirnova, P. Clair, J.M. Lefauconnier, A. Chavanieu, B. Calas, et al., Enhanced delivery of doxorubicin into the brain via a peptide-vector-mediated strategy: saturation kinetics and specificity, J. Pharmacol. Exp. Ther. 296 (2001) 124–131.
[95] X.H. Tian, F. Wei, T.X. Wang, P. Wang, X.N. Lin, J. Wang, et al., In vitro and in vivo studies on gelatin-siloxane nanoparticles conjugated with SynB peptide to increase drug delivery to the brain, Int. J. Nanomedicine 7 (2012) 1031–1041.
[96] E. Blanc, C. Bonnafous, P. Merida, S. Cisternino, P. Clair, J.M. Scherrmann, et al., Peptide-vector strategy bypasses P-glycoprotein efflux, and enhances brain transport and solubility of paclitaxel, Anticancer Drugs 15 (2004) 947–954.
[97] V.N. Kokryakov, S.S.L. Harwig, E.A. Panyutich, A.A. Shevchenko, G.M. Aleshina, O.V. Shamova, et al., Protegrins: leukocyte antimicrobial peptides that combine features of corticostatic defensins and tachyplesins, FEBS Lett. 327 (1993) 231–236.
[98] K. Takeshima, A. Chikushi, K.K. Lee, S. Yonehara, K. Matsuzaki, Translocation of analogues of the antimicrobial peptides magainin and buforin across human cell membranes, J. Biol. Chem. 278 (2003) 1310–1315.
[99] I. Neundorf, J. Hoyer, K. Splith, R. Rennert, H.W.P. N'Dongo, U. Schatzschneider, Cymantrene conjugation modulates the intracellular distribution and induces high cytotoxicity of a cell-penetrating peptide, Chem. Commun. (2008) 5604–5606.
[100] K. Splith, J. Hoyer, I. Neundorf, Cellular delivery and biological activity of metall complex-peptide conjugates, Drug Discov. Today 15 (2010) 1096–1097.
[101] K. Splith, W.N. Hu, U. Schatzschneider, R. Gust, I. Ott, L.A. Onambele, et al., Protease-activatable organometal-peptide bioconjugates with enhanced cytotoxicity on cancer cells, Bioconjug. Chem. 21 (2010) 1288–1296.
[102] Z. Oren, J.C. Lerman, G.H. Gudmundsson, B. Agerberth, Y. Shai, Structure and organization of the human antimicrobial peptide LL-37 in phospholipid membranes: relevance to the molecular basis for its non-cell-selective activity, Biochem. J. 341 (1999) 501–513.
[103] K.A.H. Wildman, D.K. Lee, A. Ramamoorthy, Mechanism of lipid bilayer disruption by the human antimicrobial peptide, LL-37, Biochemistry 42 (2003) 6545–6558.
[104] X.A. Zhang, K. Oglecka, S. Sandgren, M. Belting, E.K. Esbjorner, B. Norden, et al., Dual functions of the human antimicrobial peptide LL-37-target membrane perturbation and host cell cargo delivery, Biochim. Biophys. Acta 1798 (2010) 2201–2208.

[105] S. Sandgren, A. Wittrup, F. Cheng, M. Jonsson, E. Eklund, S. Busch, et al., The human antimicrobial peptide LL-37 transfers extracellular DNA plasmid to the nuclear compartment of mammalian cells via lipid rafts and proteoglycan-dependent endocytosis, J. Biol. Chem. 279 (2004) 17951–17956.

[106] M.J. Kang, J.Y. Eum, S.H. Park, M.H. Kang, K.H. Park, S.E. Choi, et al., Pep-1 peptide-conjugated elastic liposomal formulation of taxifolin glycoside for the treatment of atopic dermatitis in NC/Nga mice, Int. J. Pharm. 402 (2010) 198–204.

[107] J. Li, Y.J. Wang, R.J. Liang, X.J. An, K. Wang, G.X. Shen, et al., Recent advances in targeted nanoparticles drug delivery to melanoma, Nanomedicine 11 (2015) 769–794.

[108] Y. Gilad, M. Firer, G. Gellerman, Recent innovations in peptide based targeted drug delivery to cancer cells, Biomedicines 4 (2016) pii: E11.

[109] T.A.S. Aguirre, D. Teijeiro-Osorio, M. Rosa, I.S. Coulter, M.J. Alonso, D.J. Brayden, Current status of selected oral peptide technologies in advanced preclinical development and in clinical trials, Adv. Drug Deliv. Rev. 106 (2016) 223–241.

[110] V. Gregorc, A. Santoro, E. Bennicelli, C.J.A. Punt, G. Citterio, J.N.H. Timmer-Bonte, et al., Phase Ib study of NGR-hTNF, a selective vascular targeting agent, administered at low doses in combination with doxorubicin to patients with advanced solid tumours, Br. J. Cancer 101 (2009) 219–224.

Radiolabelled peptides in medical imaging

17

Subhani M. Okarvi[1] *and Helmut R. Maecke*[2]
[1]King Faisal Specialist Hospital and Research Centre, Riyadh, Saudi Arabia,
[2]University Hospital Freiburg, Freiburg, Germany

17.1 Introduction

Peptides are naturally gifted biomolecules as they physiologically play a crucial role in the human body and are responsible for many important biological functions. For the last several years, naturally occurring peptides and their synthetic derivatives have gained great clinical interest, especially in tumor diagnosis and peptide receptor radionuclide therapy in nuclear oncology [1]. This attribute is due to the fact that a wide variety of human cancers overexpress specific receptors for several types of peptides. Radiolabeled peptides that bind with high affinity and specificity to such receptors on tumor cells have shown enormous clinical potential for diagnostic molecular imaging and targeted radionuclide therapy. The biological action of peptides is usually mediated by their binding to specific, membrane-associated receptors. Many of these receptors belong to the family of G-protein-coupled receptors. G-protein-coupled receptors constitute a large family whose members are involved in numerous physiological functions and represent about 30%–40% of all modern pharmaceutical drugs targets. These receptors represent useful molecular targets for the detection and treatment of cancer since they are located on the plasma membrane and upon binding of a radiopeptide the receptor–ligand complex is internalized, allowing sufficiently long retention of radioactivity in tumor cells for efficient killing of cancer cells. It is worth mentioning that the presence of physiologic peptide receptors in the human body may appear as a weak prerequisite for specific and selective tumor targeting. However, compared with their relatively low density in physiological organs, many of these peptide receptors have been shown to be highly overexpressed in a variety of cancers. This is the basis for why these regulatory peptide receptors have become so attractive and valuable targets in cancer imaging and therapy in recent years [1–5]. The overexpression of peptides on various human cancers is summarized in Table 17.1.

Following intravenous injection of the radiolabeled receptor-specific peptide, it will extravasate from blood vessels and specifically bind to the sites with high peptide receptor density. If the patient has tumor-expressing peptide receptors (target) in large amounts, the radiopeptide will selectively bind to the respective peptide receptors and will actively be taken up by the tumor cells through a process called "receptor-ligand internalization." The cellular internalization will ultimately lead to high accumulation of radioactivity in the tumor lesions, compared with the rest of

Peptide Applications in Biomedicine, Biotechnology and Bioengineering.
DOI: http://dx.doi.org/10.1016/B978-0-08-100736-5.00019-3

Table 17.1 Expression of peptide receptors on various human tumors

Peptide	Receptors	Tumor type
Somatostatin	Somatostatin receptor subtypes sst_{1-5}	Neuroendocrine tumors, gastroenteropancreatic tumors, MTC, tumors of the nervous system, non-Hodgkin's lymphoma, carcinoids, melanoma, brain, SCLC.
VIP/PACAP	$VPAC_1$, $VPAC_2$, PAC_1 receptors	Various adenocarcinomas (stomach, colon, pancreas, lung, etc.)
CCK/gastrin	CCK_1, CCK_2 receptors	MTC, SCLC, stromal ovarian cancer, astrocytoma
LHRH	LHRH receptors	Breast, prostate, ovarian cancer
α-MSH	MSH receptors	Melanoma
Bombesin/GRP	BB_1, BB_2, BB_3, and BB_4 receptors	Prostate, breast, pancreas, gastric, SCLC, colorectal cancer, gastrointestinal stromal tumor
Neurotensin	NTR1, NTR2, and NTR3 receptors	Ewing sarcoma, meningioma, MTC, astrocytoma, SCLC, exocrine pancreatic cancer, breast, prostate
Opioid	Opioid receptors	SCLC, neuroblastoma, breast cancer
Substance P	NK1 receptors	Glioblastoma, astrocytoma, MTC, breast, peri- and intratumoral blood vessels
GLP-1/exendin	GLP-1 receptors	Insulinomas
Oxytocin	Oxytocin receptors	Endometrium, breast cancer
Neuropeptide Y	NPY receptors subtypes Y_1-Y_6	Breast, prostate, brain cancer
Chemokine (CXCR4)	CXCR4	Breast, prostate, lung, colon, multiple myeloma, lymphatic system

the organs; it will be amenable to detection through gamma camera imaging or positron emission tomography (PET). The detection sensitivity of this type of cancer imaging methodology is remarkable as tumors of a few millimeters in diameter are often visualized. This approach has been validated by numerous somatostatin receptor scintigraphy (SRS) studies, which have made this the preferred method for the efficient diagnosis of neuroendocrine tumors (NETs) [1,2,4–7].

The discovery of cell-surface receptors overexpressed in a wide range of human tumors has led to the development of many tumor-targeting radiopeptides for a variety of cancers. These radiopeptides are being used for both diagnostic imaging and therapy response monitoring as well as for tumor-targeted therapy. The first and most clinically successful registered peptide-based diagnostic imaging agent to date is primarily based on "octreotide" a synthetic and metabolically stable somatostatin analogue, ^{111}In-DTPA^{0}-octreotide (^{111}In-Octreoscan). Another important clinical application of radiolabeled octreotide and other somatostatin-targeting peptide analogues is peptide receptor-mediated radionuclide therapy (PRRT) [8−11]. The effectiveness of ^{111}In-DTPA^{0}-octreotide in localizing primary and metastatic tumors and in staging and treating these tumors has created great clinical interest in the development of radiopeptides to target other tumor-related peptide receptors, especially those overexpressed in the most commonly occurring cancers, such as prostate and breast cancers [12−14]. The continued research efforts have resulted in the discovery of bombesin peptide derivatives for targeting the gastrin-releasing peptide receptors (GRPRs) overexpressed in prostate and breast cancer, among others. Target bombesin/gastrin-releasing peptide receptors (BN/GRPR), which are overexpressed in many common human tumors including breast and prostate cancer [1,4,15,16], and chemokine receptor-targeting peptides [17−20], have been developed and are currently under preclinical or clinical evaluation to establish their applicability for the diagnosis or treatment of various cancers.

In this chapter, we present some of the recent developments of radiolabeled tumor receptor-binding peptides, particularly those based on somatostatin-derived peptides, and some of their applications in molecular imaging and nuclear oncology. In addition, a brief discussion on radiopeptides targeting bombesin, gastrin, chemokine receptor 4, and RGD is also included.

17.2 Molecular imaging in nuclear oncology

For better management of cancer, not only the detection of tumors but also their molecular characterization is necessary, as this collective information facilitates better diagnosis and prognosis of tumors. Molecular imaging is an emerging medical field that enables the visualization, characterization, and quantification of tumors and biological processes using probes known as molecular imaging agents (i.e., radiopeptides). Current advancements in the field of molecular imaging techniques have created several possibilities for more specific and effective tumor diagnosis and targeted therapy. Molecular imaging can be used for early tumor detection, characterization, and real-time monitoring of therapeutic responses. The principal goal of "molecular diagnostic imaging" is to visualize, characterize, and measure the biological processes at the cellular, subcellular, and molecular levels in humans and other living subjects using noninvasive procedures. Unlike the "classical diagnostic imaging," molecular imaging goes beyond structural assessment, and probes disease-specific abnormalities at the cellular and molecular levels and try to find the root cause. The main advantage of in vivo molecular imaging is that it provides information about the diseased tissues that other conventional methods cannot provide or else require more invasive procedures, such as biopsy or surgery. Additionally, its unique ability to characterize diseased tissues noninvasively allows the possibility of a more personalized treatment planning. Molecular imaging has the potential to identify disease (tumor) in its earliest stages and determine the exact location of tumor, often before symptoms occur or abnormalities can be detected with other methods, thus signaling a shifting paradigm from treating late disease to pinpointing and diagnosing disease sooner. Medical imaging has now become an essential tool in cancer management since imaging can determine the presence or spread of cancer in addition to visualizing a patient's response to radionuclide therapy [7,21,22].

Additional benefits of molecular imaging in oncology include the characterization of heterogeneity of tumor receptor expression, detection of molecular targets for personalized therapy, and the noninvasive nature of the field [1,7,21,22]. In tumor cells it is common for a particular receptor to be overexpressed and in abundance. It is important to consider however that even small concentrations of imaging agents may result in saturation of receptors, causing increased nonspecific background binding. Therefore, minute molar quantities (tracer amount) of an imaging agent are required. This has resulted in the radionuclide imaging methods PET and SPECT being preferred due to the micro- to picomolar concentrations used. Current molecular imaging approaches primarily utilize positron-emission tomography (PET) or single photon-emission computed tomography (SPECT)-based techniques. With the advent of hybrid technologies, such as PET/CT and SPECT/CT, information from the two different imaging modalities is precisely correlated together, allowing precise anatomic localization of functional abnormalities. The most commonly used radionuclide in SPECT is technetium-99m (^{99m}Tc) decay mainly by gamma-ray emission, whereas ^{18}F is the most commonly used radionuclide for PET imaging (see Table 17.2). The half-life of 109.8 min allows for synthesis and delivery from an external site to the PET center and finds a balance to allow for a minimized dose of radioactive material to a patient. It also has a low-energy positron emission, which decreases the risk for patients, which also leads to a lower linear range in tissue, enabling a higher spatial resolution image in PET. Furthermore, carbon–fluoride bonds are similar in length to carbon–oxygen bonds, which produce reasonable bioisosterism, improving the binding potential of radiotracers where hydroxyl functional groups are substituted for ^{18}F [2]. Another PET isotope gaining a great deal of clinical attention is gallium-68 (^{68}Ga) owing to its favorable characteristics and its availability through a generator system without the need for an on-site cyclotron system. With radionuclide imaging it is possible to make sure that radiation is delivered specifically to tumor sites, thereby killing cancer cells and avoiding significant damage to sensitive, healthy surrounding tissues. Overall, in vivo molecular imaging has shown great potential to impact medicine by detecting diseases in early stages (screening), identifying the extent of disease, selecting disease- and patient-specific treatment plans (personalized medicine), applying a directed or targeted therapy, and measuring molecular-specific effects of treatment. The research and development of new and effective imaging agents is the keystone for the attainment of this evolving technology [21–23].

Table 17.2 Characteristics of medically useful radionuclides and chelating agents for radiolabeling

Radionuclide	Half-life (h)	BFCA/ prosthetic group	Application	Production mode
Technetium-99m (^{99m}Tc)	6.02	HYNIC, MAG_3, GGC	Diagnosis	$^{99}Mo/^{99m}Tc$ generator
Fluorine-18 (^{18}F)	1.83	SFB	Diagnosis	Cyclotron
Iodine-123 (^{123}I)	13.2	SIB, SIPC	Diagnosis	Cyclotron
Gallium-68 (^{68}Ga)	1.13	NOTA, DOTA	Diagnosis	$^{68}Ge/^{68}Ga$ generator
Copper-64 (^{64}Cu)	12.7	TETA, DOTA, NOTA	Diagnosis/ therapy	Cyclotron
Indium-111 (^{111}In)	67.9	DTPA, DOTA	Diagnosis/ therapy	Cyclotron
Lutetium-177 (^{177}Lu)	160.8	DOTA	Therapy	Reactor
Yttrium-90 (^{90}Y)	64.1	DOTA	Therapy	$^{90}Sr/^{90}Y$ generator

BFCA, bifunctional chelating agent; DOTA, 1,4,7,10-tetraazacyclododecane-1,4,7,10-tetraacetic acid; DTPA, diethylenetriaminepentaacetic acid; GGC, Gly-Gly-Cys; HYNIC, 2-hydrazinonicotinic acid; MAG_3, mercaptoacetyltriglycine; NOTA, 1,4,7-triazacyclononane-1,4,7-triacetic acid; SFB, *N*-succinimidyl-4-[^{18}F] fluorobenzoate; SIB, *N*-succinimidyl-3-iodobenzoate; SIPC, *N*-succinimidyl-5-iodo-3-pyridinecarboxylate; TETA, 1,4,8,11-tetraazacyclotetradecane-1,4,8,11-tetraacetic acid.

17.3 Molecular imaging systems

A typical molecular imaging system usually consists of a clinically relevant molecular target or receptor, an imaging agent or probe, and an imaging modality or device. In nuclear oncology, the imaging agent is a radiotracer (e.g., radiopeptide), that contains in its structure a radioactive atom. Molecular imaging necessitates the interaction of the target with a "radiolabeed" agent that emits radiation, which can be detected externally by one or more imaging modalities. In agent—target interaction, the radiolabeled agent selectively binds to its target, for a long enough time to allow external detection by an appropriate imaging modality. Most molecular imaging approaches, however, rely on the use of exogenous probes to provide imaging signal. Many molecular imaging probes have more sophisticated structural design and include both a tumor-targeting component (e.g., peptide) and a signaling component (e.g., a PET- or SPECT-based radionuclide), and the resulting images reflect the function, rather than the structure of the organ, system, or metabolic process being evaluated. The increased use of multimodality imaging techniques (hybrid techniques), such as PET/CT, SPECT/CT, and PET/MRI allows better detection of molecular, pathophysiological, and anatomic changes in a single scan that involves simply the administration of radioactive probes and detection by suitable imaging modalities [3,21].

17.4 Characteristics of an ideal molecular imaging agent

Ideally, an imaging agent that has rapid and specific binding to its intended target, rapid clearance from the blood circulation, excretion mainly by the urinary pathway, high accumulation at the target site, and high in vivo stability is desirable. Although the factors that really influence the pharmacokinetic characteristics of an imaging agent are still somewhat unclear, it is generally believed that the size of a biomolecule plays an important role, and the charge and lipophilicity can also have a significant impact depending on the type of imaging agent. For proper visualization of a target site (receptor), an imaging agent must have high metabolic stability in vivo. Lipophilicity of an imaging agent to some degree is related to the extent of nonspecific binding as well as also influencing the route and rate of excretion; hence, it is important to consider lipophilicity while designing an imaging agent for potential clinical application. In addition, good safety profile (minimal toxicity), potential for clinical translation, and time- and cost-effective preparation are important considerations. Small peptides satisfy most if not all these requirements of an ideal molecular imaging agent [1–3].

17.5 Peptides in molecular imaging and radiotherapy

In recent years, small peptide-based radiopharmaceuticals have emerged as an important class of molecular imaging agent. The distinct advantages of small peptides as tumor-targeting agents over other bioactive macromolecules, such as proteins and antibodies, are summarized in Table 17.3.

Peptides offer a high degree of freedom and flexibility in terms of structural modifications. The formulation of a tumor-targeting peptide usually consists of identical molecules with a well-defined structure. Thus, peptides offer the advantage to be "as large as required, and as small as possible." Peptides are fairly small in size (up to 30 amino acids), can conveniently be prepared by modern solid-phase peptide synthesis, and the desired pharmacokinetic properties can be molecularly engineered by making appropriate changes in the peptide and/or by adding a biomodifying moiety to the sequence. Although peptides generally display a lower affinity compared to antibodies, they have a much higher stability at ambient temperature and a greater ability to penetrate tumor tissues due to their small size. The use of solid-phase peptide synthesis and phage display libraries has led to rapid high-throughput synthesis and screening of peptide-based molecular imaging agents [13,14,23].

Table 17.3 Key features of small peptides

- Low-molecular weight biomolecule
- Easy to synthesize chemically
- Synthetic in origin
- Well-defined structure
- Easy to radiolabel
- Flexible to chemical and molecular modifications
- Tolerate harsh radiolabeling conditions
- Favorable pharmacokinetics
- High specificity for target tissue
- High tumor penetration
- Low accumulation in healthy tissues
- Rapid clearance from blood and nontarget tissues
- Rate and route of excretion can be modified
- High receptor binding affinity
- Low toxicity
- Biological and chemical diversity
- Ability to attach chelator at the C- or N-terminus of the peptide
- Specific in vivo tissue distribution
- High expression on various human cancer cells
- Many clinically relevant targets (receptors)

Receptors for peptides are often found in higher density on tumor cells than in normal tissues; hence specifically designed receptor-binding radiopeptides could enable efficient visualization of tumors. Because of their small size, peptides usually exhibit rapid pharmacokinetics and good tumor-targeting properties, with the ability to penetrate into tumors more efficiently. The faster clearance of nontarget tissue leads to a better tumor-to-background ratio in vivo (better image contrast) and therefore less irradiation of nontarget organs. Additionally, peptides are less likely to cause immunogenic effects compared with antibodies and are more cost effective to produce (depending on the size of the peptide), as peptide synthesis is much cheaper than recombinant production techniques. Peptides can also withstand harsher conditions for modification and labeling than antibodies [13,14,24].

17.6 Developmental process of a peptide-based tumor-targeting agent

The development of peptide probes is based mainly on peptide chemistry and molecular biology techniques (solid-phase peptide synthesis, combinatorial peptide chemistry, phage display techniques), which permit rapid and high-throughput identification and isolation of high-affinity binders to the respective peptide receptor. Clinically, potential candidates emerging from this screening process are subject to further selection by their frequent very limited ability to tolerate harsh radiolabeling conditions without losing receptor affinity. From this point on, however, tracer characteristics beyond receptor affinity become the determining factors for the potential suitability for in vivo molecular imaging, i.e., in vivo stability, bioavailability, elimination kinetics, and accumulation in target and nontarget tissues. Optimization of these key parameters demands highly complex and time-consuming pharmaceutical development, and thus less than optimal performance of new peptide radiopharmaceuticals in animal biodistribution often leads to the ending of tracer optimization at this point. With the increasing availability of more sophisticated animal imaging modalities, such as small animal PET, PET/CT, SPECT, and SPECT/CT in recent years, this endpoint in tracer development has experienced a rapid shift towards "in vivo imaging in animal models" and ultimately translational radiopharmaceutical development. Although recent and improved technologies have yielded various strategies for designing "smart probes," utilization of peptide probes has been the most successful. To date, selected peptides that target a variety of disease-related receptors and biomarkers have been developed. These targeting peptides can be attached with the suitable imaging moieties by using facile radiolabeling techniques [2,3].

However, as mentioned before, there are several requirements for peptides to be used as clinically useful radiopharmaceuticals. Primarily, the corresponding peptide receptors have to be expressed on the tumor target in suitable amounts, overexpression or unique expression being desirable. The corresponding peptide ligand should retain the high affinity to the receptor as the natural compound. For therapeutic applications, internalization by endocytosis appears to be important because of the higher residence time in tumors. Furthermore, the preclinical pharmacokinetic findings should be encouraging. In addition to its metabolic stability, the strength of the metal–chelator complex is also crucial. Some other aspects that need to be considered when developing radiometal-labeled peptides are a high rate of complexation, ease of radiolabeling, and the availability of the clinically useful radionuclides. Undoubtedly, the very important role of peptide receptor imaging in clinical oncology is undeniable [3].

After a new imaging agent has been synthesized it must go through a series of in vitro and in vivo evaluations in order to determine its usefulness as a tumor targeting agent and its specificity and selectivity for the target of interest. These tests should only be performed after the identity and purity of the newly synthesized imaging agent has been confirmed via appropriate quality control methods. Depending on the type of imaging agent, quality control procedures might involve nuclear magnetic resonance, high-performance liquid chromatography, thin-layer chromatography, and transmission electron microscopy, amongst others [2,3].

17.7 Radiolabeling of peptides

With the availability of several new and efficient chelating agents and the recent development of facile methods for radiolabeling of peptides with various medically useful radionuclides there has been great acceleration in the development of peptide radiopharmaceuticals for various applications. Ideally, a radionuclide would be attached to a tumor-targeting peptide without changing the receptor binding and the biological properties of the peptide. Unfortunately, this is not always the case as different methods of radiolabeling have variable effects on binding and biological activity. The metal chelate can significantly influence the tumor uptake and biodistribution properties of a radiopharmaceutical based on small biomolecule (i.e., small peptide), due to the fact that in many cases the metal chelate contributes greatly to the overall size, charge, and molecular weight of the peptide radiopharmaceutical. Hence, the design and the selection of a bifunctional chelator (BFC) is important for the successful development of a clinically useful radiopharmaceutical [1,2,13,14].

The incorporation of a radionuclide to a peptide can be achieved using different radiolabeling approaches, of which the most common is the attachment of a radiometal to a peptide through pendant radiolabeling. This method requires a BFC to be incorporated to the peptide sequence either during peptide synthesis or later by manual conjugation. A BFC consists of a chelating agent to complex the radiometal and a functional group for the covalent attachment of the biomolecule. A variety of conjugation methods have been developed and applied successfully for the coupling of tumor-targeting peptides with the chelating agents. In postconjugation method, a BFC is attached to the peptide during peptide synthesis and the radiometal is introduced after the conjugation of the chelating moiety to the peptide molecule. Radiolabeling via a chelating agent proved to be most suitable for the preparation of peptide radiopharmaceuticals for clinical use, because of its convenience, well-defined labeling chemistry, and feasibility of instant kit formulation (where all ingredients are in "one vial" and only the radionuclide solution has to be added for labeling) [1,2,13].

Currently, there is no single radionuclide that satisfies all the requirements of an ideal radionuclide for the treatment of all types of cancers; consequently one BFC may not be suitable for radiolabeling with different therapeutic radionuclides. Thus several chelating agents, based on acyclic and macrocyclic ligand frameworks, have been developed for labeling with various radionuclides. These BFCs have their own advantages and limitations and thus require vigilant selection based on the specific application [13].

Bifunctional chelating agents can be divided into two broad categories: acyclic and macrocyclic chelators. Acyclic multidentate chelators based on the structure of DTPA have favorable labeling kinetics for most clinically useful radiometals. The first approved peptide-based tumor imaging agent was the DTPA-derivatized SST analogues labeled with ^{111}In (^{111}In-DTPA-octreotide) (Fig. 17.1). For this

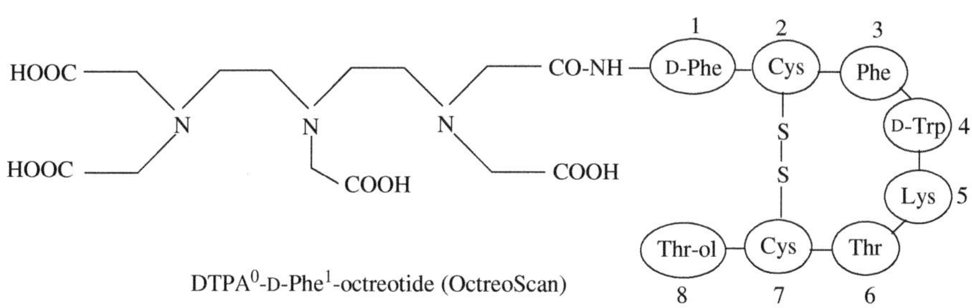

Figure 17.1 Structure of DTPA^0-D-Phe^1-octreotide.

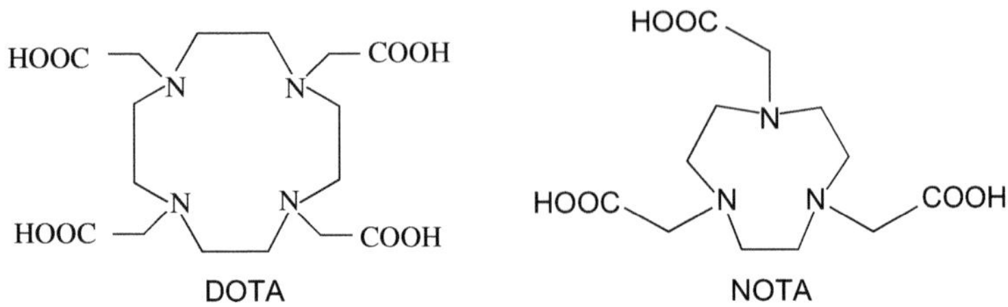

Figure 17.2 Chemical structures of DOTA and NOTA.

radionuclide, DTPA is an ideal chelator providing stable chelation and fast chelation kinetics, but not suitable for other useful radiometals. Because of the fast labeling kinetics, DTPA-coupled peptides can be readily and efficiently radiolabeled by incubation for 10–15 min at room temperature, using acetate or a citrate buffer (pH 5–7). The high labeling efficiency can be attributed to the linear chelator backbone of DTPA analogues. However, the kinetic stability of their metal complexes may prove to be problematic when they are used as BFCs for biomolecules having long retention in the blood. Several studies have shown the superiority of the new radiopeptides (i.e., DOTA- and NOTA-coupled) over DTPA-coupled Octreoscan and this was the basis for the introduction of the improved versions of somatostatin peptide analogues into clinical settings [1,2,13,14].

Macrocyclic chelators are rigid chemical structures designed to ensure the donor atoms are in the optimal location for radiometal chelation. Octadentate macrocyclic chelating agents, such as DOTA (1,4,7,10-tetraazacyclododecane-1,4,7,10-tetraacetic acid) (Fig. 17.2) derivatives, were developed for labeling with ^{90}Y, ^{177}Lu, and radiolanthanides and displayed better stability characteristics than acyclic chelating ligands. The slow-formation kinetics as compared to acyclic-based chelators can be overcome by heating the labeling reaction mixture, which usually does not destroy the integrity of peptides. DOTA-type chelators can be attached to the peptide either by solution-phase conjugation or by solid-phase peptide synthesis, thus allowing rapid and convenient preparation of a variety of DOTA-coupled peptides for both diagnosis and therapy. It is much more convenient to perform DOTA coupling to the targeting peptide sequence on the solid-phase support. While this synthetic approach offers the advantages of rapid solid-phase synthesis, it also poses some specific challenges for the bifunctional ligands. Most importantly, the chelator has to be compatible with the chemistry used for the peptide synthesis and should preferably be soluble under the conditions of solid-phase peptide synthesis. Despite a large number of DOTA–peptide conjugates having been developed for tumor-targeting applications, they are somewhat limited by their slow complex formation rate and require an undesirable heating step to enhance the complexation process [1,2,13]. Cyclic chelators, such as DOTA and 1,4,7-triazacyclononane-1,4,7-triacetic acid (NOTA), are commonly used for radiolabeling peptides with medically useful radiometals including ^{68}Ga, ^{64}Cu, and ^{111}In.

Another method designed for labeling peptides with lower atomic mass radioisotopes (i.e., ^{18}F, etc.) is by using a prosthetic group. A prosthetic group can be classified as a bifunctional agent since they contain the radionuclide of interest and a functional group facilitating bioconjugation to the tumor-targeting peptides. ^{18}F is first labeled on the prosthetic group before conjugating to the tumor-targeting peptide. This method involves multistep preparation in order to obtain pure radiolabeled peptide. The main challenge with this method is to obtain site-specific radiolabeling without disrupting the functionality of the side-chains, which may contain amines, carboxylic acids, or amides that are found in most peptide sequences. Prosthetic group labeling is the most common method for preparation of ^{18}F-labeled peptides. Conventional acylation prosthetic groups 4-nitrophenyl 2-[^{18}F] fluoropropionate ([^{18}F]NFP) and *N*-succinimidyl 4-[^{18}F]fluorobenzoate ([^{18}F]SFB) are still being exploited given their favorable biological character, in vivo stability, and ease of synthesis [25].

As mentioned before, it is somewhat a challenging task to incorporate chelating agents into tumor-targeting peptides since the radiolabeling may affect binding to the target receptor and the pharmacokinetics. For this reason, the choice of a radionuclide and labeling position of the radiolabel in the targeting peptide is important. There is a possibility that the attachment of a bulky chelating agent to a relatively small peptide molecule may lower the binding affinity and also the biological potency. To reduce this negative influence, a spacer group between the peptide and the chelator can often be introduced. The chelating agents are often placed at the N- or C-terminus of the peptide, on an amino acid side-chain such as lysine, or following an aliphatic spacer. While the nature of the peptide vector is determined by the cellular target (i.e., cell surface receptor), the spacer moiety is dependent on the pharmacokinetic requirements of the biomolecule as it often influences its hydrophilicity and improves the pharmacokinetic characteristics. The spacer group can also act as a separator between the chelating agent and the targeting peptide to prevent steric influence of the chelator on the binding affinity of the receptor-targeting part of the peptide. Additionally, chemical strategies, including modification of the amino acid sequence and the introduction of a specific linker/spacer group with different biological properties, have been explored with the aim of improvement of the overall performance of the radiopeptides in terms of high metabolic stability and favorable pharmacokinetics. The radiometal also has a significant influence on the overall properties, such as binding affinity, cellular internalization, and tissue biodistribution of the radiopharmaceutical [1,2,12–14].

17.8 Metabolic stability measurement

Human plasma stability measurements are well established and widely accepted methods for the evaluation of the metabolic stability of radiopeptide ligands, together with stability relevant to pH levels. Neither of these two in vitro analyses alone can mimic the challenging environment that a radiopeptide chelator complex encounters under in vivo conditions. However, both analyses together can be useful in determining the suitability of a chelator for use with the preferred radiometal [1,2,12–14].

A majority of the radiopeptides used in molecular imaging are structurally derived analogues of naturally occurring peptide hormones. These peptide hormones (e.g., SS-14 or SS-28, BN/GRP, etc.) showed short half-lives in human plasma of about 1−3 min. High contrast imaging of receptor-expressing tissues requires a sufficiently high bioavailability of the radiopeptide at the target tissue, and thus sufficient time to circulate within the blood pool, an increased in vivo resistance against degradation, e.g., via proteases, without affecting binding affinity was and still is a somewhat major issue in the development of peptide-based radiopharmaceuticals suitable for in vivo imaging.

17.9 Determination of binding affinity of radiolabeled peptides to their respective receptor targets

The tumor receptor-binding and pharmacokinetic characteristics are of major importance in the assessment of a receptor-targeting peptide. Once the stability of a radiolabeled peptide complex is determined, the next important in vitro test is the assessment of receptor binding affinity of the radiopeptide designed for tumor targeting. One of the initial steps in validating an imaging agent is the determination of its specificity and selectivity for the desire receptor target in vitro. This property can be examined through cell binding or cell accumulation studies, using the purified cellular target and/or cell assays. Assessment of new imaging agents in receptor-positive and receptor-negative cell lines can provide valuable information on binding specificity and also whether the imaging agent holds the ability to cross cell membranes. Most receptors have high affinities for their ligands and are active at low nanomolar concentrations of the ligands. Hence radiolabeled peptide preparation with high specific radioactivity is desirable. Even small molar quantities of imaging agents may saturate a receptor, limiting the ability to visualize receptor expression and increasing the background due to nonspecific binding [2,3,26,27].

The binding assay can be performed on intact cancer cells or on membrane preparations, either by direct or competitive binding assays, thereby assessing alterations in affinity and specificity of the radiopeptide analogues compared to the natural peptide. As shown in Table 17.1, there are several receptors overexpressed on different types of tumors so the knowledge of the affinity patterns for receptor subtypes is very important, since each subtype usually has a different expression profile on different tumor types. If the imaging agent displays high binding affinity (picomolar to nanomolar range) to the target of interest, with sufficiently high specificity and selectivity in vitro, the next step is to test the imaging agent in animal models in vivo.

17.10 In vivo evaluation of radiolabeled peptides in animal models

The in vivo evaluation of radiopeptides in animal models is of great importance, as it is the first real sign of the radiopeptide pharmacokinetics and merits the potential of radiopeptide for further development. The aim of biodistribution studies in animal models is the evaluation of radiopeptide uptake in different body organs and tissues, especially in the receptor-positive organs and in critical organs, such as the kidneys. For clinical applications of a peptides, kidney toxicity is the dose-limiting factor. Peptides are taken up by the tubular cells and radiometal chelates are trapped within the lysosomes; high retention of the radiolabel occurs in the kidneys, eventually causing nephrotoxicity [28,29]. Also, with animal biodistribution data, the rate and route of excretion can be determined, which is helpful in case any modification is needed to improve the biokinetics. Evaluation of an imaging agent in small animals can provide invaluable information regarding its in vivo binding properties, biodistribution, and pharmacokinetics. For instance, blood samples can be withdrawn at various time points following administration of imaging agent to evaluate its metabolism (by analyzing processed plasma samples via techniques such as HPLC). Ex vivo biodistribution studies are usually performed during the initial evaluation of imaging agents in living subjects. These studies involve sacrificing animals at predetermined time points postinjection of imaging agent, harvesting organs of interest, followed by an analysis of imaging signal in individual organ samples. Ex vivo samples (e.g., liver, brain, or other organs) can also be processed and analyzed for metabolites at various time points of interest. It is preferred that metabolites of an imaging agent be more polar than their parent compound and that they do not accumulate in the regions of interest [3].

For in vivo tumor-targeting studies, tumor cells for which the radiopeptide is being developed are inoculated from cell culture into animals (e.g., nude mice, rats, etc.) inducing the expression of a tumor. After administration of the radiopeptide into these experimentally grown tumor models, the potential of in vivo tumor targeting can be evaluated. From these studies, uptake in the tumors and other major organs, especially the receptor-positive organs and in the critical organs concerning toxicity effects can be determined. An important aspect of this is the proof of specificity of tumor targeting due to the possibility to block the receptors with excess cold (nonradioactive signal-producing) peptide. If necessary, the imaging agent can be evaluated in larger animals, depending on the application of the imaging agent. It is worth mentioning that imaging results obtained from rodent experiments are not always predictive of what will be observed in larger animals, or humans, due to potential species differences in binding and/or metabolism of the imaging agent. It is therefore important to assess imaging agents in a few different in vivo models if possible. Likewise, it is important to consider evaluating the imaging agent in humans (after obtaining regulatory approval) even if unfavorable metabolism is observed in animals.

The effectiveness of any tumor-targeting agent relies mainly on its pharmacokinetic characteristics. Undesirable pharmacokinetics leads to faster degradation and clearance of the peptide radiopharmaceutical. Elimination of radiopeptides mainly by renal excretion is advantageous for creating high-quality images with low background activity, however, unfortunately radiopeptides are often trapped in the kidneys due to tubular reabsorption. This is due to reabsorption in the renal proximal tubule and subsequent retention of the radiopeptide and its lysosomal degradation products and may lead to reduced contrast and quality of diagnostic peptide receptor imaging. Retention in the kidneys not only causes high background noise, but also delivers high radiation doses. High kidney uptake is of concern, particularly in peptide receptor radionuclide therapy, where high radiation doses to the kidneys may exert severe nephrotoxic effects. Nevertheless procedures have been developed and applied successfully in clinical settings to manage nephrotoxicity. Renal uptake of various radiolabeled peptides has been reduced by coadministration of cationic amino acids, such as lysine and arginine, yet these methods come with undesired physical effects such as nausea and arrhythmias. Nephrotoxicity has also been reported with side effects such as elevated electrolytes and difficulty urinating. Reducing renal uptake has also been achieved by coadministration of albumin, a megalin substrate. In addition, a careful selection of the chelator and biomodifying molecule in the peptide sequence and overall net charge on the peptide molecule are important considerations when designing a peptide probe for tumor targeting. All these in vivo studies together with the in vitro test described above are beneficial in designing and developing new and improved versions of peptide radiopharmaceuticals. Nonetheless, compounds that look promising in animal models do not necessarily behave similarly in humans.

If the imaging agent exhibited favorable preclinical results and seems to have potential for clinical trials, the imaging agent then must be tested for safety prior to pilot testing in humans. This process usually involves submission of animal data for approval of the imaging agent for human use by the FDA. For certain imaging agents, one can obtain exploratory investigational new drug (eIND) approval from the FDA to proceed with pilot studies in human volunteers or patients. Imaging agent development is a complex and time-consuming process. In order to accelerate the developmental process of effective imaging agents, there is a need for better collaboration between academic groups and industries, who have better resources to accelerate clinical translation and perform multicenter clinical trials [3,13,14].

17.11 Potential molecular targets for tumor diagnosis and therapy

One of the major challenges in the development of tumor receptor-targeting agents is to identify specific, clinically relevant, easily accessible, and highly expressed targets. They can be situated inside the cell (DNA, m-RNA, enzymes, proteins, etc.), on the cell surface membrane (receptors), or in the cell environment (pH, pO_2, neovasculature, etc.). Proteins as targets on the plasma membrane will allow the optimal selection of disease markers and more accessible drug targets. G-protein-coupled receptors are an important receptor target family; these are internalized upon peptide ligand binding (i.e., agonists). G-protein-coupled receptors represent more than 30%–40% of all modern pharmaceutical drug targets [3].

When identifying suitable molecular targets, one needs to keep a number of points in mind. In the case of imaging breast cancer, it is usually important that the potential targets are not present in high levels in "normal" breast tissue (e.g., a two-fold higher target level in cancerous breast tissue compared with normal breast tissue is likely insufficient for obtaining a high contrast image). Furthermore, the target needs to be "imageable" via the use of available or proposed techniques.

Molecular biology studies that precede successful receptor targeting with radiopeptides include in vitro identification and analysis of receptors with biochemical, biomolecular, and immunological techniques. For instance, radioligand binding analysis and bioassays with cells or membrane preparations are the means to characterize high-affinity binding sites and the pharmacological profile of a given tumor-targeting peptide. Anatomical information about the distribution of receptors in tissues is obtained by a quantitative receptor autoradiography technique, which measures radioligand binding on tissue sections and thus enables the localization of receptors at the microscopic level [30]. Only a minority of the useful regulatory peptides and peptide families has been reasonably thoroughly investigated so far and future work will probably reveal a multitude of clinically useful peptide radiopharmaceuticals.

After the identification of the suitable molecular target, it is then important to select the appropriate imaging modality, or combination thereof, to achieve the best possible results. External monitoring of radiolabeled targeting peptides can be achieved via sophisticated molecular imaging modalities used in nuclear medicine. The utilization of multimodality imaging techniques (hybrid techniques) including PET/CT, PET/MRI, and SPECT/CT facilitates better detection of molecular, pathophysiological, and anatomic changes in a single scan involves the administration of radiolabeled peptide probes and detection by appropriate imaging modalities.

Both PET and SPECT nuclear imaging modalities require a radionuclide to emit photons in the form of radiation, which can then be externally detected and processed into an image. There are many radionuclide options for both imaging modalities. The choice of radionuclide is dependent on the half-life, availability, method of incorporation, and method of radioactive decay. The half-life must be long enough to withstand synthesis, administration, and distribution of the peptide probe while maintaining enough radioactivity to be detectable by the imaging modality. The availability and proximity of a cyclotron limits the choice of radionuclides to generator-produced isotopes that can be produced on-site. As described before, there are different methods for incorporating a radioisotope to a tumor-targeting peptide. Optimal SPECT radionuclides decay mainly by gamma emission with little residual alpha and beta decay, as these forms of decay are detrimental to cellular processes. The most prevalent SPECT radionuclide is technetium-99m (^{99m}Tc). Through the use of ^{99}Mo/^{99m}Tc generators, ^{99m}Tc can be produced on-site and used in commercially available "instant labeling kits," allowing quick formulation of radiopharmaceuticals for patients. ^{99m}Tc, with a half-life of 6 h, has many clinical applications and has been used to radiolabel a wide variety of tumor-targeting peptides, such as neurotensin, somatostatin, bombesin, and other biomolecules. The easy availability, ideal gamma energy combined with short half-life makes this radioisotope ideal for diagnostic imaging requiring low radiation exposure. Another common SPECT radioisotope is indium-111 (^{111}In), which is a cyclotron-produced radiometal, with a half-life of 2.83 days and can be used for both diagnosis and treatment [1–3].

Similar to SPECT-based radioisotopes, several PET isotopes have been used for radiolabeling of tumor-targeting peptides for PET imaging. One of the major challenges in the development of PET radiopeptides is the need for more efficient synthetic methods for labeling short-lived positron-emitting radionuclides into the tumor-targeting peptides. In addition to the conventional positron-emitting radionuclides used in peptide radiolabeling, such as ^{18}F, a number of nonconventional metallic radionuclides are also being used recently for radiolabeling of tumor-targeting peptides. These metallic PET radionuclides are usually characterized by relatively longer half-life than ^{18}F, thus facilitating longer-term studies on the patient with a single injection of radiopeptides. Some of the metallic radionuclides used for peptide radiolabeling include ^{64}Cu, ^{68}Ga, ^{86}Y, and ^{89}Zr. Most of these radiometals labeled efficiently with well-known bifunctional chelators, such as DOTA and NOTA.

The use of a medical generator system is a less expensive and more convenient alternative to an on-site cyclotron system for the production of short-lived radionuclides. Among the PET radionuclides, gallium-68 (^{68}Ga) with a 68-min half-life is gaining enormous clinical acceptance as a PET isotope. The parent isotope, germanium-68 (^{68}Ge), has a 271-day half-life, allowing it to be packaged into a $^{68}Ge/^{68}Ga$ generator that functions similarly to the $^{99}Mo/^{99m}Tc$ generators. Since Ge-68 has a long half-life, the shelf-life of these generators is usually over a year. In recent years, several ^{68}Ga-labeled somatostatin analogues have been prepared and have shown great clinical potential in targeting somatostatin receptors expressing NETs [1–3].

Copper-64 (^{64}Cu) is another promising radiometal for site-directed PET imaging and therapy. The half-life of 12.7 h is long enough for radiopharmaceutical preparation, quality control, imaging, and therapy. ^{64}Cu-labeled peptides have attracted considerable interest in molecular imaging since it contains both $\beta+$ and $\beta-$ emissions, allowing both PET imaging and radionuclide therapy.

17.12 Theranostics

The concept of "theranostic" signifies the combination of the two medical terms "therapeutic" and "diagnostic." Theranostics demonstrates the close link between diagnosis and therapy for personalized management of disease. The aim of theranostics in relation to nuclear oncology is to identify the appropriate molecular targets in cancer, so that the optimal ligands and radionuclides with favorable labeling chemistry can be selected that enables direct diagnosis and treatment specifically directed and targeted towards the individual tumor of the patient. The effective use of ^{68}Ga for diagnosis, and of ^{177}Lu and ^{90}Y for radionuclide therapy using the same tumor-targeting peptide probe for targeting somatostatin receptor-positive NETs, has demonstrated the great clinical potential of theranostic-based radiopharmaceuticals in modern nuclear oncology [7].

The most effective theranostic approach would be a combined in vivo diagnostic with in vivo therapeutic strategy. Receptor-targeting radiopeptides are being developed as single agents for molecular imaging and treatment of tumors that overexpress peptide receptors. To balance the clinical benefits and risks of radionuclide therapy, the biodistribution, dosimetry, and toxicity of the therapeutic agents must be carefully monitored. Imaging with targeted radiopeptides can allow detection of metastases, monitoring of treatment efficacy, and detection of progression or recurrence. The ultimate goal of personalized medicine is to adjust the treatment plan for each patient based on the specific diagnostic information gathered on that patient's disease.

17.13 Limitations of peptides in tumor targeting

Peptides as tumor-targeting agents offer many advantages over other biomolecules, but also suffer with their own set of limitations. Natural peptides are known to have poor oral bioavailability as well as low metabolic stability in vivo. Poor oral bioavailability is insignificant for imaging agents as opposed to therapeutic agents, since peptide radiopharmaceuticals are usually administered intravenously, while poor metabolic stability can be minimized using structural modifications designed to inhibit enzymatic degradation [2,3].

One major problem with natural peptides is their often short biological half-life due to rapid proteolytic degradation in plasma. As native peptides are degraded rapidly by endogenous peptidases and proteases before reaching the intended target, most peptides have to be modified synthetically to minimize rapid enzymatic degradation in vivo [13,30].

Tremendous efforts have been focused towards developing metabolically stable peptides suitable for clinical use by performing appropriate molecular modifications. Endopeptidases that are capable of hydrolyzing peptide bonds within a peptide sequence are also of concern. Endopeptidases are only able to recognize natural L-amino acids; therefore replacing positions of hydrolysis with suitable D-amino acid or unnatural amino acid residues causes the peptides to become unrecognizable to the peptidase. Peptides are often degradated by exopeptidases, enzymes that specifically hydrolyze the C- and N-termini of a linear peptide. In order to resist exopeptidase degradation, the functionality of the termini can be altered. The simplest approach is to have the C-terminus synthesized as an amide and the N-terminus acetylated [1,2,13].

Degradation of peptides by exopeptidases can also be minimized by head-to-tail cyclization, which fully removes the termini. A somewhat more complex method for enhancing in vivo stability is to utilize the pseudo-peptide approach. Pseudopeptides resemble the natural peptide structure, but contain chemical modifications to the backbone that render them unrecognizable to peptidases. In addition, the inclusion of amino alcohols and the insertion of amino acid residues with modified side-chains without compromising the receptor-binding affinity and biological activity of the peptide [2,13].

These small structural modifications result in higher in vivo stability, while the integrity of peptide sequence remains virtually unchanged, allowing it to maintain target affinity. However, each approach has its own advantages and limitations depending on the specific natural peptide, target, and mode of action. Finding the optimal peptide analogues can require various permutations in peptide structure and the preparation and analysis of large libraries of peptide analogues is advantageous for inventing the most suitable peptide candidate.

It is the specific amino acid sequence of a peptide and usually the nature and type of particular amino acid side-chains that determine resistance to enzymatic degradation. For example, native somatostatin has a plasma half-life of 2–3 min in circulation, but its molecularly modified synthetic peptide derivative, "octreotide," has a half-life of 1.5–2 h, making it more suitable for clinical applications [31]. It is worth mentioning that it is not the absolute criteria that metabolically stable peptide radiopharmaceuticals always exhibit the most favorable tumor-targeting properties.

17.14 Agonist and antagonist peptide probes

It is meaningful to mention that common to all receptors is the interaction of a ligand and the receptor, in which specific binding of the ligand to the receptor results in downstream biochemical or physiological changes. Ligands that cause physiological functional changes upon binding to receptor, typically the naturally occurring ligands, are called "agonists." Then again ligands that bind to the receptor and block the binding of agonists but that do not activate changes are called "antagonists" [31]. It is generally believed that receptor agonist radioligands are more suitable for tumor targeting as they display a good degree of receptor-mediated internalization into the tumor cells upon binding to the respective cell-surface receptors, thus enabling active accumulation in the target site, resulting in optimal visualization [24].

Significant development of receptor-targeting peptides has focused on receptor agonists. Agonists are known to possess high binding affinities for their receptors that trigger internalization of the ligand–receptor complex. It was rationalized that the internalization and accumulation of the radioligand and in the cell over time would lead to better target-to-background ratios and a better radiopharmaceutical overall in terms of better chemical stability and longer duration of action.

Due to the inability of antagonists to internalize into tumor cells, as agonists readily do, accumulation has been assumed to be limited for antagonists. Agonistic activity has thus been assumed to be a prerequisite for high contrast in vivo receptor imaging and efficient peptide receptor therapy. However, despite being full antagonists towards sst_2 and sst_3, respectively, ^{111}In-DOTA-sst_2-ANT and ^{111}In-DOTA-sst_3-ODN-8 showed extremely high tumor accumulation (29.1% and 61.3% ID/g in HEK-sst_2 and HEK-sst_3 tumor-bearing nude mice, respectively, at 1 h post-injection [p.i.]) and retention ($>$48 h), resulting in unsurpassed imaging characteristics [32]. As revealed in vitro by Scatchard analysis, these findings can be attributed to a more than 15-fold increased number of accessible binding sites per cell for the antagonists as compared to their agonistic analogues, as well as very slow ligand dissociation from the receptor. Subsequently, these encouraging results have initiated the development of further antagonists with outstanding sst_2-binding properties [33], leading to the conclusion to reassess the current paradigm.

One recent study describes the influence of the chelate on the binding affinity and also on the pharmacokinetics of the somatostatin-based radiolabeled antagonists [34]. The somatostatin antagonist LM3 was attached to three different macrocyclic chelating agents, CB-TE2A, NODAGA, and DOTA. The CB-TE2A and NODAGA-coupled somatostatin conjugates were radiolabeled with the longer-living PET isotope ^{64}Cu while NODAGA- and DOTA- conjugate with the short-lived PET isotope ^{68}Ga. In vitro and in vivo studies exhibited the significant dependence of the affinity and pharmacokinetics on the chelator and the radiometal. For instance, ^{68}Ga-NODAGA-LM3 has a 10-fold higher sst_2-affinity than the corresponding DOTA-conjugate (^{68}Ga-DOTA-LM3) and a fivefold higher sst_2-affinity than the corresponding ^{64}Cu-complex (^{64}Cu-NODAGA-LM3), indicating that there is a much stronger influence of the appended chelate than found in somatostatin-based agonists [35].

Significant differences were observed in the in vivo behavior of two ^{64}Cu-labeled peptides: ^{64}Cu-CB-TE2A-LM3 and ^{64}Cu-NODAGA-LM3. Although ^{64}Cu-CB-TE2A-LM3 has 1.6-fold higher affinity than ^{64}Cu-NODAGA-LM3 this is not reflected in a higher tumor uptake. In contrast, ^{64}Cu-NODAGA-LM3 shows a distinctly higher uptake than ^{64}Cu-CB-TE2A-LM3 at 1 h (35.5% vs 19.3% ID/g) and 4 h (37.9% vs 26.9% ID/g). A significantly high amount of radioactivity was accumulated in the kidneys for both radiolabeled antagonists, but two to ninefold higher for ^{64}Cu-CB-TE2A-LM3, with time dependence, probably due to the positive charge of ^{64}Cu-CB-TE2A-LM3 versus the neutral charge of ^{64}Cu-NODAGA-LM3.

One important point in tumor targeting and especially for radiotherapy is the sufficient retention of the radionuclide in the tumors. Several studies have shown that low or absence of internalization correlates with very short tumor retention [36,37], and the converse; there is a good correlation of tumor uptake and uptake in somatostatin receptor-positive organs (e.g., pancreas) and the rate of internalization into sst_2-expressing cells in culture [38]. For ^{64}Cu-NODAGA-LM3 and ^{64}Cu-CB-TE2A-LM3 both showed low internalization capacity. Thus, it is notable that ^{64}Cu-NODAGA-LM3 has somewhat slow but distinctly faster washout from the tumor ($\sim$15% ID/g, 24 h p.i.) compared with ^{64}Cu-CB-TE2A-LM3, which shows almost no washout between 4 h (26.9% ID/g) and 24 h p.i. (21.6% ID/g). This may be of high significance if peptide receptor therapy is considered with the β-emitter ^{67}Cu ($t_{1/2} = 64$ h) as the area under the curve may be larger for the $^{64/67}$Cu-CB-TE2A conjugate. In spite of the negligible washout of ^{64}Cu-CB-TE2A-LM3 from the tumors, the tumor-to-normal tissue ratios are significantly higher for ^{64}Cu-NODAGA-LM3 as compared with ^{64}Cu-CB-TE2A-LM3 (e.g., tumor-to-kidney: 12.8 vs 1.7 and tumor-to-muscle: 1342 vs 75.2, at 24 h), an important parameter for obtaining good image contrast. These differences were striking particularly for the tumor-to-kidney ratios, which considerably increase with time and are much higher than usually seen with somatostatin-based radiolabeled agonists. MicroPET imaging showed clear tumor localization and very high image contrast, especially for ^{64}Cu-/^{68}Ga-NODAGA-LM3 [3].

Studies have shown that the chelated somatostatin-based agonists are sensitive to N-terminal radiometal modifications, with Ga-DOTA-agonists having significantly higher binding affinity than their Lu-, In-, and Y-DOTA-corresponding species [27]. For instance, Ga(III)-DOTA-OC has three times higher affinity for sst_2 than Y(III)-DOTA-OC, Ga(III)-DOTA-TOC has six times higher affinity than Y(III)-DOTA-TOC and Ga(III)-DOTA-TATE has eight times higher affinity than Y(III)-DOTA-TATE [39]. These improved binding affinities also translated into improved internalization rates and eventually higher uptake in the tumors. In those studies, it seemed acceptable to generalize that $^{67/68}$Ga is a radiometal that systematically improved the sst_2 affinity of somatostatin agonists as well as their pharmacokinetics. It was further confirmed in vivo in patients with NETs that ^{68}Ga-DOTA-NOC (NOC = [1-Nal^3]octreotide), ^{68}Ga-DOTA-TOC, or ^{68}Ga-DOTA-TATE were better imaging agents than the ^{111}In-DOTA congeners [39–46]. Recently, a paper was published on whether chelated somatostatin antagonists are also sensitive to radiometal modifications. In this study three different somatostatin antagonists, namely JR10, JR11, and LM3 (Table 17.4) were coupled to the chelators DOTA and NODAGA and various (radio)metals including In(III), Y(III), Lu(III), Cu(II), and Ga(III) were added [47]. Surprisingly, in all three resulting antagonists, the Ga-DOTA analogues were the lowest-affinity radioligands, with a sst_2 binding affinity up to 60 times lower than the respective Y(III)-DOTA, Lu(III)-DOTA, or In(III)-DOTA compounds (Table 17.5). Interestingly however, substitution of DOTA by NODAGA chelator was able to increase greatly its binding affinity in contrast to the Ga(III)-DOTA analogues. One important finding of this study is the eightfold difference between In(III)-DOTA-JR11 ($IC_{50} = 3.8 \pm 0.7$ nM) and Y(III)-DOTA-JR11 ($IC_{50} = 0.47 \pm 0.05$ nM). ^{111}In is commonly used as a surrogate for ^{90}Y for imaging the biodistribution of ^{90}Y-labeled compounds and for dosimetry (e.g., ^{111}In/^{90}Y-DOTATOC). The recent results suggest that ^{111}In may not be a suitable surrogate of ^{90}Y, and therefore may not be very reliable for dosimetric studies of the therapeutic radiopeptide based on somatostatin antagonists. Moreover, this may be the case for other classes of peptide and this required more careful investigation. One striking finding from these studies is the comparison of the antagonists with clinically used agonists. The side-by-side in vivo comparison between ^{68}Ga-DOTA-TATE and ^{68}Ga-DOTA-JR11 elucidates the great potential of the antagonists. ^{68}Ga-DOTA-JR11, having a considerably lower affinity for the sst_2 ($\sim$150-fold) than ^{68}Ga-DOTA-TATE (Table 17.5), showed a 1.3-fold higher tumor uptake (23.8% vs 17.8% IA/g, 1 h p.i.), while ^{68}Ga-NODAGA-JR11, with a sixfold lower affinity (Table 17.5), showed an up to 1.7-fold higher tumor uptake (30.7% vs 17.8% IA/g, 1 h p.i.) [47]. The low-affinity receptor antagonist is slightly superior, which can be explained by the higher number of binding sites for antagonists vs. agonists [34], outweighing the affinity differences. This was also nicely demonstrated with in vitro autoradiographic studies of human tumor specimens [48]. PET images showed the higher uptake of the

Table 17.4 Somatostatin-based ligands (other than fully agonists) that have been developed as potential tumor-targeting agents

Ligand code	Chemical structure	Chelating agent	(Radio)metal
KE88	(DOTA-)cyclo(DAB-Arg-Phe-Phe-D-Trp-Lys-Thr-Phe)	DOTA	^{nat}Y, $^{nat/67}Ga$ ^{111}In
AM3	(DOTA-)Tyr-cyclo(DAB-Arg-cyclo(Cys-Phe-D-Trp-Lys-Thr-Cys))	DOTA	^{177}Lu ^{68}Ga
sst_3-ODN-8	NH_2-CO-c(D-Cys-Phe-Tyr-D-Agl^8(Me,2-naphthoyl)-Lys-Thr-Phe-Cys)-OH	DOTA	^{111}In
sst_2-ANT (BASS)	*p*-NO_2-Phe-cyclo(D-Cys-Tyr-D-Trp-Lys-Thr-Cys)-D-Tyr-NH_2	DOTA CB-TE2A	^{111}In ^{177}Lu ^{64}Cu
LM3	*p*-Cl-Phe-cyclo(D-Cys-Tyr-D-Aph (Cbm)-Lys-Thr-Cys)-D-Tyr-NH_2	DOTA NODAGA CB-TE2A	^{68}Ga ^{64}Cu ^{nat}In
JR10	*p*-NO_2-Phe-c[D-Cys-Tyr-D-Aph (Cbm)-Lys-Thr-Cys]-D-Tyr-NH_2	DOTA NODAGA	$^{nat/68}Ga$, ^{nat}In, ^{nat}Lu
JR11	Cpa-c[D-Cys-Aph(Hor)-D-Aph (Cbm)-Lys-Thr-Cys]-D-Tyr-NH_2	DOTA NODAGA	$^{nat/68}Ga$, ^{nat}In, $^{nat/177}Lu$, ^{nat}Cu, ^{nat}Y

antagonists compared to agonists and the good background clearance already at 1 and 2 h p.i. [3].

There are already some published studies in animal tumor models where a comparison of somatostatin receptor antagonist versus agonist demonstrated that the antagonists bind to sst_2- and sst_3-expressing tumors in vivo better than agonists, with comparable or even higher affinity [37,47]. This was proven also to be the case in comparative in vitro binding studies on several different, well-characterized human tumor samples of the antagonist ^{177}Lu-DOTA-BASS versus the agonist ^{177}Lu-DOTA-TATE [52]. About 50 sst_2-positive human tumor tissue samples were analyzed by in vitro receptor autoradiography for the expression of sst_2, comparing the binding capacity of ^{177}Lu-DOTA-BASS and ^{177}Lu-DOTA-TATE in successive tissue sections. Both radioligands efficiently bind to the sst_2 receptor on different tumor tissues. However, in all cases, the radiolabeled antagonist ^{177}Lu-DOTA-BASS bound to more sst_2-receptor sites in the tumors than did the agonist ^{177}Lu-DOTA-TATE. The mean ratios of the antagonist ^{177}Lu-DOTA-BASS to the agonist ^{177}Lu-DOTA-TATE ranged from 4.2 up to 12.3. This significantly increased binding may increase the localization accuracy for tumors and metastases but also

Table 17.5 Chemical structures of three sst_2 antagonists based on the JR10, JR11, and LM3 family and their metallated conjugates and IC_{50} values for sst_2 expressed in nmol/L (mean ± SEM; $n \geq 3$)

Code	Chemical structure	IC_{50} (nM)
DOTA-JR11	DOTA-Cpa-c[D-Cys-Aph(Hor)-D-Aph(Cbm)-Lys-Thr-Cys]-D-Tyr-NH_2	0.72 ± 0.12
Ga-DOTA-JR11	[^{nat}Ga]-DOTA-Cpa-c[D-Cys-Aph(Hor)-D-Aph(Cbm)-Lys-Thr-Cys]-D-Tyr-NH_2	29 ± 2.7
Y-DOTA-JR11	[^{nat}Y]-DOTA-Cpa-c[D-Cys-Aph(Hor)-D-Aph(Cbm)-Lys-Thr-Cys]-D-Tyr-NH_2	0.47 ± 0.05
Lu-DOTA-JR11	[^{nat}Lu]-DOTA-Cpa-c[D-Cys-Aph(Hor)-D-Aph(Cbm)-Lys-Thr-Cys]-D-Tyr-NH_2	0.73 ± 0.15
Cu-DOTA-JR11	[^{nat}Cu]-DOTA-Cpa-c[D-Cys-Aph(Hor)-D-Aph(Cbm)-Lys-Thr-Cys]-D-Tyr-NH_2	16 ± 1.2
In-DOTA-JR11	[^{nat}In]-DOTA-Cpa-c[D-Cys-Aph(Hor)-D-Aph(Cbm)-Lys-Thr-Cys]-D-Tyr-NH_2	3.8 ± 0.7
NODAGA-JR11	NODAGA-Cpa-c[D-Cys-Aph(Hor)-D-Aph(Cbm)-Lys-Thr-Cys]-D-Tyr-NH_2	4.1 ± 0.2
Ga-NODAGA-JR11	[^{nat}Ga]-NODAGA-Cpa-c[D-Cys-Aph(Hor)-D-Aph(Cbm)-Lys-Thr-Cys]-D-Tyr-NH_2	1.2 ± 0.2
DOTA-JR10	DOTA-*p*-NO_2-Phe-c[D-Cys-Tyr-D-Aph(Cbm)-Lys-Thr-Cys]-D-Tyr-NH_2	0.62 ± 0.21
Ga-DOTA-JR10	[^{nat}Ga]-DOTA-*p*-NO_2-Phe-c[D-Cys-Tyr-D-Aph(Cbm)-Lys-Thr-Cys]-D-Tyr-NH_2	8.9 ± 2.2
In-DOTA-JR10	[^{nat}In]-DOTA-*p*-NO_2-Phe-c[D-Cys-Tyr-D-Aph(Cbm)-Lys-Thr-Cys]-D-Tyr-NH_2	2.3 ± 0.5
Lu-DOTA-JR10	[^{nat}Lu]-DOTA-*p*-NO_2-Phe-c[D-Cys-Tyr-D-Aph(Cbm)-Lys-Thr-Cys]-D-Tyr-NH_2	1.2 ± 0.2
NODAGA-JR10	NODAGA-*p*-NO_2-Phe-c[D-Cys-Tyr-D-Aph(Cbm)-Lys-Thr-Cys]-D-Tyr-NH_2	23.0 ± 1.5
Ga-NODAGA-JR10	[^{nat}Ga]-NODAGA-*p*-NO_2-Phe-c[D-Cys-Tyr-D-Aph(Cbm)-Lys-Thr-Cys]-D-Tyr-NH_2	6.5 ± 0.5
DOTA-LM3	DOTA-*p*-Cl-Phe-c[D-Cys-Tyr-D-Aph(Cbm)-Lys-Thr-Cys]-D-Tyr-NH_2	0.39 ± 0.05
Ga-DOTA-LM3	[^{nat}Ga]-DOTA-*p*-Cl-Phe-c[D-Cys-Tyr-D-Aph(Cbm)-Lys-Thr-Cys]-D-Tyr-NH_2	12.5 ± 4.3
In-DOTA-LM3	[^{nat}In]-DOTA-*p*-Cl-Phe-c[D-Cys-Tyr-D-Aph(Cbm)-Lys-Thr-Cys]-D-Tyr-NH_2	1.3 ± 0.1
Ga-NODAGA-LM3	[^{nat}Ga]-NODAGA-*p*-Cl-Phe-c[D-Cys-Tyr-D-Aph(Cbm)-Lys-Thr-Cys]-D-Tyr-NH_2	1.3 ± 0.2
Reference agonist		
Ga-DOTA-TATE	[^{nat}Ga]-DOTA-D-Phe-c[Cys-Tyr-D-Trp-Lys-Thr-Cys]-Thr	0.2 ± 0.04

increase the efficacy of radiotherapeutic intervention with such antagonist radiotracers. Of particular interest is the fact that tumors other than neuroendocrine, such as tumors that normally express a low density of receptors, for instance breast cancers, renal cell cancers, or non-Hodgkin lymphomas showed the same results. Those tumors are not currently among those being routinely investigated with somatostatin receptor imaging. It is therefore impressive to see that by using the antagonist radioligand, it is possible to increase, at least in vitro, the number of binding sites in breast carcinomas and renal cell cancers 11.4- and 5.1-fold, respectively. In addition, non-Hodgkin lymphomas which normally express a low amount of sst_2 showed a 4.8-fold increase in receptor number with the antagonist radioligand, reaching levels that may be detected more easily in vivo than with current agonists. These in vitro human data, together with the in vivo animal tumor data, are strong arguments indicating that sst_2 antagonists may be worth testing in patients in a wide range of tumors including non-NETs. Such sst_2 antagonist radioligands should be useful not only for diagnostic tumor imaging but also for peptide receptor-targeted tumor radiotherapy [3].

The first pilot study of somatostatin-based antagonists demonstrates the feasibility of imaging somatostatin-expressing tumors in patients using radiolabeled antagonists [53]. This pilot study provides the first clinical evidence that radiolabeled sst_2 antagonists not only detect sst_2-expressing NETs but also may even be superior to agonists for imaging and therapy of NETs. The studied somatostatin receptor antagonist ^{111}In-DOTA-BASS had a favorable biodistribution profile (higher tumor uptake and lower normal organ/tissue uptake) than the agonist ^{111}In-DTPA-octreotide, resulting in better tumor detection [3].

The two radiopeptides, ^{111}In-DOTA-BASS (^{111}In-DOTA-pNO$_2$-Phe-c(DCys-Tyr-DTrp-Lys-Thr-Cys)-DTyr-NH$_2$) and ^{111}In-DTPA-OC, were compared in five patients with metastatic thyroid carcinoma or NETs. ^{111}In-DOTA-BASS detected 25 of 28 lesions, whereas ^{111}In-DTPA-OC detected only 17 of 28 lesions. All lesions visible on ^{111}In-DTPA-OC scans were also detected by ^{111}In-DOTA-BASS, whereas there were eight lesions on ^{111}In-DOTA-BASS that were not visible on ^{111}In-DTPA-OC. There were additionally three bone lesions that were negative on both ^{111}In-DOTA-BASS and ^{111}In-DTPA-OC scans, but bone metastases were confirmed by serial CT. The longest residence times of ^{111}In-DOTA-BASS were observed in the tumor and kidneys [3].

The quantitative analysis showed relevant differences in the biodistribution between the antagonist ^{111}In-DOTA-BASS and agonist ^{111}In-DTPA-OC. The antagonist showed up to 4.1 times higher uptake in the tumor (^{111}In-DOTA-BASS, $3.5\% \pm 2.8\%$ IA and ^{111}In-DTPA-OC, $1.0\% \pm 0.99\%$ IA) 4 h after injection. At the same time, the uptake in the kidneys was lower for ^{111}In-DOTA-BASS (1.5% vs 2.3% IA), resulting in up to 5.2 times higher tumor-to-kidney uptake ratios in favor of the antagonist. The antagonist also showed lower uptake than the agonist in organs such as the liver ($1.4 \pm 0.2\%$ vs $1.7 \pm 0.6\%$ IA) and spleen ($1.0 \pm 0.2\%$ vs $2.3 \pm 1.0\%$ IA) 4 h after injection.

For radiotherapy, it is encouraging that tumor-to-kidney ratios were up to 5.2 times higher for the antagonist than for the agonist as the kidneys are the major dose-limiting organ in peptide receptor therapy. The several-fold higher tumor-to-kidney ratio could significantly improve the efficacy and toxicity profile of radiotherapy. To further evaluate the potential of somatostatin antagonists for radiotherapy the organ dose of ^{90}Y-DOTA-BASS was calculated based on the assumption that ^{90}Y-DOTA-BASS shows the same biodistribution as ^{111}In-DOTA-BASS. Compared with literature data for ^{90}Y-DOTA-TOC [59,92], ^{90}Y-DOTA-BASS showed lower renal, hepatic, and splenic radiation doses [53].

The pharmacokinetics of the first radiolabeled antagonist in humans, ^{111}In-DOTA-BASS, has been studied more thoroughly in a preclinical study to examine the potency of these new radiolabeled antagonists as radiotherapeuticals. The same antagonist was studied with ^{177}Lu to clarify if the radiometal has an influence on the pharmacokinetics [48]. The tumor uptake was found to be nearly identical for the two radiopeptides ^{111}In-DOTA-BASS and ^{177}Lu-DOTA-BASS (30%–35% IA/g, 4 h p.i.), with relatively slow tumor washout for both; about 50% still remained in the tumor at 48 h p.i. This shows no influence of radiometal on this DOTA-coupled antagonist making ^{177}Lu-DOTA-BASS a promising radiotherapeutic. This result is somewhat in conflict with what was shown with the radiolabeled somatostatin antagonist LM3 [34] as mentioned above, where distinct differences in pharmacology depending on radiometal and chelator were found. The blood clearance of both radiopeptides was fast resulting in high tumor-to-blood and tumor-to-normal organ ratios at early time points, which increased with time for most organs. Attempts to block the tubular cell-mediated kidney uptake of antagonistic radiopeptides with agents which are successful in the clinical use of corresponding radiolabeled agonists, showed that lysine and gelofusine were both effective in kidney blocking of ^{111}In-DOTA-BASS. Neither agent blocked tumor uptake but both blocked about 50% of the kidney uptake, increasing the tumor-to-kidney ratios about twofold. Finally, and importantly, this study excluded species differences, since the biodistribution results of ^{111}In-DOTA-BASS in mice bearing human sst_2-expressing tumors (sst_2) in comparison with mice bearing rat sst_2-expressing tumors ($rsst_2$) were found to be very similar. The uptake in the sst_2-tumors was the same as in the $rsst_2$-tumors; reaching a peak at 4 h (30% IA/g) and showing the same washout rate as described above. This study demonstrated the relevance of radiolabeled somatostatin-based antagonist not only in imaging but also in targeted radiotherapy of somatostatin receptor-positive tumors [3].

All the described radiopeptides here displayed antagonist potency in immunofluorescence microscopy-based internalization experiments [32,34,47,48]. They did not trigger sst_2 receptor internalization and antagonized the receptor internalization mediated by an agonist. Nevertheless, in the radioligand internalization studies the corresponding radiotracers (e.g., ^{68}Ga-NODAGA-LM3, ^{68}Ga-DOTA-LM3, or ^{111}In-DOTA-BASS, but also other radiolabeled antagonists, such as ^{64}Cu-CB-TE2A-BASS [56,57]) showed significant internalization in sst_2-positive cell lines; much lower than the internalization of corresponding radiolabeled agonists. Internalization studies of ^{111}In-DOTA-BASS in cells not expressing sst_2 showed no internalization of the radioligand, while pretreatment of the sst_2-expressing cells with hypertonic sucrose, known to prevent receptor endocytosis by preventing clathrin-coated pit formation [54] distinctly reduced radioligand internalization. These results are strong indications of a specific receptor-mediated internalization process of the radioligands. It is still not clear why these antagonists internalize in a radioligand experiment but are not able to stimulate sst_2 internalization. It should be considered that in the immunofluorescence microscopy experiments, the sst_2 receptor trafficking itself is monitored where ligand trafficking is monitored in the radioligand experiment. Furthermore, the ligand is used in high excess in the immunofluorescence microscopy experiments while in the radioligand internalization experiments the concentration of the radioligand is far below tumor saturation. Regarding in vivo experiments, the first in vivo data indicate receptor internalization triggered by agonists but apparently not by antagonists [93].

Some preclinical and clinical studies have shown that somatostatin-receptor expressing tumors exhibit higher uptake of radiolabeled somatostatin-antagonists than of somatostatin-agonists. In four consecutive patients with advanced NETs an assessment was made to determine whether treatment with ^{177}Lu-labeled somatostatin-antagonists is feasible. After the injection of ^{177}Lu-DOTA-[Cpa-c(DCys-Aph(Hor)-DAph(Cbm)-Lys-Thr-Cys)-DTyr-NH_2] (^{177}Lu-DOTA-JR11) and [^{177}Lu-DOTA^0,Tyr^3,Thr^8]-octreotide (^{177}Lu-DOTA-TATE) 3D voxel dosimetry analysis based on SPECT/CT was performed. A higher tumor-to-organ dose ratio of ^{177}Lu-DOTA-JR11 over ^{177}Lu-DOTA-TATE was the prerequisite for treatment with ^{177}Lu-DOTA-JR11. The sst_2-receptor antagonist ^{177}Lu-DOTA-JR11 exhibited a favorable pharmacokinetic and biodistribution profile (longer intratumoral residence time and higher tumor uptake) than the agonist ^{177}Lu-DOTA-TATE, resulting in 1.7–10.6 times higher tumor doses than ^{177}Lu-DOTA-TATE. The higher tumor uptake of ^{177}Lu-DOTA-JR11 is consistent with in vitro studies that have indicated that somatostatin-antagonists bind to more receptor sites on NETs than ^{177}Lu-DOTA-TATE. It is particularly encouraging that tumor-to-kidney and tumor-to-bone-marrow dose ratios were up to 6.2 and 7.2 times higher for the antagonist than for the agonist. The kidney and the bone marrow are the major dose-limiting organs in peptide radiotherapy; so several-fold higher tumor-to-kidney and tumor-to-bone marrow dose ratios could significantly improve the efficacy and toxicity profile of radionuclide therapy. All patients were treated with ^{177}Lu-DOTA-JR11 resulting in partial remission in two patients and stable disease/mixed response in one patient each. This pilot study provides clinical evidence that treatment of NETs with radiolabeled somatostatin-antagonists is clinically feasible. If the favorable tumor-to-organ dose ratios are confirmed in larger studies radiopeptide treatment with somatostatin-antagonist may improve the results of peptide receptor therapy in NETs [3].

The potential application of antagonistic peptide ligands as receptor-targeted agents has a particular impact on those G-protein-coupled receptor systems, in which agonist-binding stimulates tumor growth and thus exerts unwanted pharmacological actions. Recent studies demonstrate that the change of paradigm observed for somatostatin-based radiotracers may also be extended to radiolabeled BN/GRP, CXCR4, and possibly other peptide analogues.

Bombesin antagonist, ^{99m}Tc-Demobesin 1, displayed fourfold higher tumor uptake in PC-3 xenograft-bearing mice than the agonist ^{99m}Tc-Demobesin 4 at 4 h p.i. as well as prolonged retention of tumor-associated activity [58]. However, another radiolabeled GRP-antagonist, ^{111}In-DOTA-Bomproamide, showed 10-fold lower tumor uptake compared to ^{99m}Tc-Demobesin 1 in the same tumor model despite comparable receptor affinity, but nevertheless very high tumor-to-background ratios. The particularly high tumor-to-liver and tumor-to-intestine ratios (13 and 6 at 1 h p.i.) are notable, since they make this radiopeptide a potential candidate for prostate cancer imaging.

17.15 Prototypical radiolabeled peptides for diagnostic imaging and targeted radiotherapy

17.15.1 Radiolabeled somatostatin peptide analogues

Somatostatin (SS) is a naturally occurring cyclic disulfide-containing peptide. There are two endogenous biologically active forms of somatostatin comprising of 14 (SS-14) and 28 (SS-28) amino acids. Both natural SS-14 and SS-28 bind with high affinity to five different somatostatin receptor subtypes (sst_{1-5}), but have a short plasma half-life ($\sim$3 min), owing to rapid enzymatic degradation by endogenous peptidases, preventing their in vivo use. Both SS-14 and SS-28 contain in their structure a Cys–Cys disulfide bond forming a conformational constraint, which is important for receptor affinity. Somatostatin exerts a wide variety of functions in many tissues throughout the body. It binds with high affinity to somatostatin receptors expressed on target tissues exerting a large number of biological effects. There are five human somatostatin receptor subtypes (sst_1–sst_5). The most clinically relevant receptor subtype is sst_2, but all the other receptor subtypes are to some degree also overexpressed in human tumors [1–3,13,14,94].

Many synthetic somatostatin analogues that are more resistant to enzymatic degradation have been prepared by molecular modifications preserving most of the biological activity of the original somatostatin. Introduction of the D-amino acids, and decreasing the ring size to the bioactive core sequence, resulted in an eight-amino-acid-containing clinically approved somatostatin analogue, "octreotide" (Fig. 17.3), that preserves the four-amino-acid motif (Phe-D-Trp-Lys-Thr) of native SS-14 involved in receptor binding and has a significantly longer plasma half-life ($\sim$2 h)

Ala^1-Gly^2-Cys^3-Lys^4-Asn^5-Phe^6-Phe^7-Trp^8-Lys^9-Thr^{10}-Phe^{11}-Thr^{12}-Ser^{13}-Cys^{14}

Somatostatin-14

D-Phe^1-Cys^2-Phe^3-D-Trp^4-Lys^5-Thr^6-Cys^7-Thr(ol)^8

Octreotide

Figure 17.3 Peptide sequence of: (A) somatostatin-14; (B) octreotide.

as compared to endogenous somatostatin. These analogues preserved the cyclic form via a disulfide bond. These stabilized somatostatin analogues have different binding profiles for somatostatin receptor subtypes, but all show high affinity for sst_2, which is the most prominent somatostatin receptor in human tumors. The discoveries of these cyclic derivatives of somatostatin for diagnostic application are the crucial inventions that enabled the subsequent development of molecular imaging agents targeting the somatostatin receptors [2,13,14].

The presence of a high expression of sst_2 receptors in primary tumors and metastases has led to the development of molecular imaging agents based on the somatostatin structure. To date these imaging agents have found clinical use primarily for the detection of primary and metastatic NETs, as well as acromegaly, small-cell lung cancer, and pituitary adenomas [1,2].

The first imaging agent for targeting somatostatin receptor-positive tumors was reported in 1989 [8] was a radioiodinated version of the cyclic peptide octreotide, with a Tyr^3 modification allowing the attachment of ^{123}I. Due to major stability concerns and uptake in nontarget tissues, a DTPA-coupled octreotide was prepared and radiolabeled with ^{111}In. This was achieved by the addition of a D-Phe residue to the N-terminus of the peptide and subsequent addition of DTPA chelator. Evaluation in humans demonstrated the ability to noninvasively image small endocrine gastroenteropancreatic tumors and carcinoid tumors. Later this agent (octreoscan) was approved by the FDA in 1994 for the imaging of patients with NETs (Fig. 17.1) [2,3,13,14].

To facilitate radiolabeling with other medically useful radionuclides, such as ^{68}Ga, ^{177}Lu, and ^{90}Y, a further modification in the structure of somatostatin analogues has been performed by replacement of DTPA with DOTA. For instance, DOTA-TOC is a modified Tyr^3 octreotide, with DOTA at the N-terminus and is capable of coordinating diagnostic radionuclides, such as ^{111}In and ^{68}Ga, as well as therapeutic radionuclides ^{90}Y and ^{177}Lu. A further modification involved the replacement of the threoninol C-terminus with the natural amino acid threonine to form DOTA-TATE, a somatostatin analogue with improved affinity for the sst_2 receptor. Another modification, with a 1-Nal^3 unnatural amino acid, resulted in the formation of DOTA-NOC that has also been radiolabeled with various radiometals. This ability to modify the N-terminus of octreotide and octreotate with various metal chelating agents have been explored by many groups in order to formulate somatostatin analogues for both SPECT and PET imaging. For example, ^{99m}Tc analogues have been created using various chelating agents and linkers to separate the cyclic peptide from the metal complex. [Tyr^3]octreotide (TOC) was conjugated to HYNIC, a tetraamine chelator, and other tetra-coordinated ^{99m}Tc systems. The chemical structures of somatostatin and the octreotide are shown in Fig. 17.3 [2].

Somatostatin receptors have been identified on the surface of many NETs and represent a valid target for in vivo tumor imaging. SPECT- and PET-based agents have been applied successfully in clinics to image and quantify the uptake of receptor-specific radiopeptides in somatostatin receptor-positive tumors for cancer staging, treatment planning, as well as follow-up response to therapy. Also, somatostatin receptor-binding peptides allow the detection of small metastatic sites, which underline the importance and strength of molecular imaging in nuclear oncology [14,32]. In addition to its usefulness as a diagnostic imaging agent, nonradioactive somatostatin analogues have been used for symptomatic treatment of hormone-secreting NETs [2,11,27]. It has been shown that somatostatin analogues reduce the symptoms associated with excess hormone secretion and may also have direct anti-tumor effects [11,37]. Several somatostatin derivatives labeled with radiohalogens, such as ^{18}F via prosthetic groups, or linked to the chelating agents to facilitate labeling with radiometals, such as ^{99m}Tc, ^{111}In, ^{64}Cu, and ^{68}Ga have been used for in vivo imaging of somatostatin receptor-expressing tumors. When radiolabeled with β-emitters (^{90}Y or ^{177}Lu), these somatostatin analogues can be utilized for radionuclide therapy [1,2,11–14].

17.15.2 Imaging of neuroendocrine tumors with somatostatin-based agents

Despite massive research activities in the field of radiolabeled peptides in the last two decades, only radiopeptides targeting the somatostatin receptor family have a real impact in patient care, diagnostically and therapeutically. NETs have presented a diagnostic challenge for the clinician since their slow metabolic rate, small lesion size, and variable anatomical localization have limited their detection using various imaging procedures. The traditional routine diagnostic work up of NETs included conventional morphological imaging procedures such as computed tomography (CT), ultrasound (US), and magnetic resonance imaging (MRI) [92,95]. Based on the successful tumor-binding studies by Reubi and Landolt [59] using [3-^{125}I-Tyr^{3}] octreotide the idea was born to develop a tracer based on the γ-emitter [3-^{123}I-Tyr^{3}] octreotide for SRS.

In the late 1980s the first imaging data on planar images of a variety of tumors using [3-^{123}I-Tyr^{3}]octreotide were reported [8]. Due to the somewhat complicated radiolabeling procedure and high lipophilicity of the radiopeptide it was soon replaced by [^{111}In-DTPA]octreotide (Octreoscan), which was the first registered radiopeptide ahich has been used for almost 20 years in the diagnosis and staging of somatostatin-positive tumors and became the gold standard for the detection of NETs [96]. In addition it was shown to be an independent prognostic factor of survival in well-differentiated malignant endocrine tumors [97]. [^{111}In-DTPA]octreotide SPECT and SPECT/CT imaging were shown to outperform anatomic imaging [98] but have drawbacks as well: the sst_2 affinity is low ($IC_{50} = 22 \pm 3.6$ nM) and the affinities to the other somatostatin subtypes are at least 10-fold lower (>200 nM). In addition, ^{111}In is not the ideal radionuclide because of the nuclear physical properties, availability, and price.

Early on, conjugates were developed for ^{99m}Tc-labeling. A first compound published was a bifunctional analogues of HMPAO (hexamethylpropyleneamine oxime) [99]. Later, conjugates of HYNIC (6-hydrazino-nicotinamide) [62,63] and of the tetraamine derivative 1,4,8,11-tetraazaundecane carboxy-functionalized at 6-position were developed and studied preclinically and in patients. Both families of compounds made it to the clinic and showed exquisite pharmacokinetics [60,61]. The ^{99m}Tc-labeled depreotide (^{99m}Tc-NeoTect) is also registered in Europe; it has obtained registration for solitary pulmonary nodule imaging and showed good diagnostic accuracy for this indication but failed in comparisons with [^{111}In-DTPA] octreotide in NETs because of its relative lipophilicity. It has been shown that more tumors can be detected with [^{111}In-DTPA]octreotide than ^{99m}Tc-depreotide, particularly in the liver. The study concluded that the detection rate of ^{99m}Tc-depreotide is much lower than that of [^{111}In-DTPA]octreotide [55]. Furthermore, [^{99m}Tc-EDDA/HYNIC,Tyr^{3}]octreotide (^{99m}Tc-HYNIC-TOC) is now available in different European countries and is registered in Poland (Tektrotyd, Polatom, Otwock-Swierk, Poland). The corresponding ^{99m}Tc-EDDA/HYNIC-TATE (^{99m}Tc-HYNIC-TATE) analogue is also widely used. Another study [100] demonstrated the superiority of ^{99m}Tc-HYNIC-TATE over [^{111}In-DTPA]octreotide in 75 patients. They found high image quality of ^{99m}Tc-HYNIC-TATE and detected distinctly more lesions in comparison with [^{111}In-DTPA]octreotide. The superiority was confirmed by other groups as well. The same group [64] also showed the usefulness of ^{99m}Tc-HYNIC-TATE in radioguided surgery. One clinical study [65] compare site-by-site the two radiotracers ^{99m}Tc-HYNIC-TOC and ^{99m}Tc-HYNIC-TATE in the same 12 patients. They concluded that both radiotracers give high-quality images at low cost, with low patient dose and ideal availability. They depicted a similar number of metastatic sites. ^{99m}Tc-HYNIC-TATE seems to have some advantage in lymph node and liver metastases; the latter is due to its somewhat higher hydrophilicity. In summary, ^{99m}Tc-labeled somatostatin analogues have major advantages over [^{111}In-DTPA]octreotide, such as better image quality, higher sensitivity, availability, price, patient compliance, and lower mean effective dose.

A further important step ahead in the diagnostic work-up of NETs was the introduction of novel somatostatin-based PET tracers labeled with the generator produced radiometal ^{68}Ga. So far only DOTA-conjugated analogues have been studied in the clinic. There are many practical and economic advantages to using these tracers. Several ^{68}Ge/^{68}Ga-generators are now commercially available; the long half-life of the mother (270.8 days) allows the use of the generator for approximately 9–12 months which makes its use very cost-effective and the synthesis of these compounds can be done relatively easily also in centers without an on-site cyclotron. These factors resulted in an increasing use of this technique for the assessment of NETs in the past 5–8 years and several clinical studies reported the higher accuracy for NET lesion detection as compared to both morphological imaging procedures and SRS. An example of the superior image quality is shown in Fig. 17.4. Moreover, receptor-binding tracers offer the advantage of noninvasively providing data on receptor expression on NET cells with direct therapeutic implications.

^{68}Ga-labeled somatostatin analogues used in the clinic are based on the three octapeptides [Tyr^{3}]octreotide (TOC), [Tyr^{3},Thr^{8}]octreotide (TATE), and [1-Nal^{3}] octreotide (NOC) conjugated to DOTA. Interestingly, Ga-complexation to these DOTA-peptides results in an improved sst_2 binding affinity and increased internalization rate compared to other radiometals (e.g., ^{177}Lu, ^{90}Y, ^{111}In) labeled congeners [40]. This has to be considered if dosimetric calculation and prediction of targeted

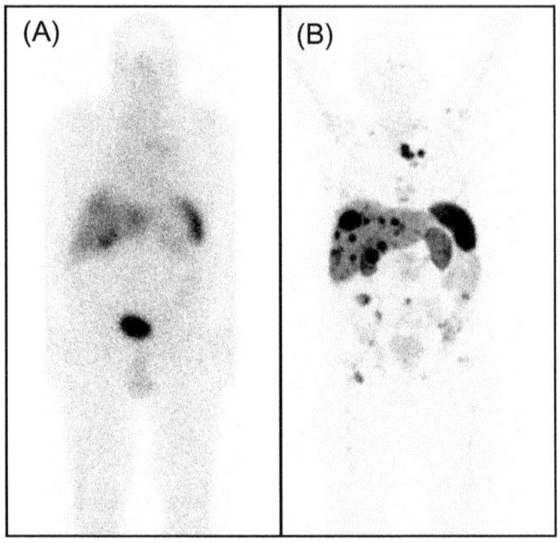

Figure 17.4 Planar scintigraphy 4 h p.i. of octreoscan (A) and ^{68}Ga-DOTA-TATE PET images 1 h p.i. (B) of a patient with a metastasized NET studied 3 weeks apart.
Source: Courtesy of Dr. D. Wild, University Hospital Freiburg.

therapy outcome with ^{90}Y- or ^{177}Lu-labeled DOTA-peptides are intended. The three PET radiotracers differ in regard to their receptor subtype binding profile. Whereas ^{68}Ga-DOTA-TATE is sst_2-selective with the highest binding affinity of any sst_2 binding peptides, ^{68}Ga-DOTA-TOC binds to sst_2 with high affinity and to sst_5 with reasonable affinity and ^{68}Ga-DOTA-NOC has high affinity to sst_2, sst_3, and sst_5.

The majority of the literature on these tracers employed ^{68}Ga-DOTA-TOC: this was, in fact, the first tracer to be employed in NET imaging. Many studies reported a high tumor to nontumor contrast and a higher sensitivity of ^{68}Ga-DOTA-TOC PET/CT as compared to SRS [45,46]. The study with the largest patient population (84 pts with NET) reported a superior sensitivity (97%) for DOTA-TOC compared to CT (61%) and SRS (52%) for the detection of NET lesions, especially in the case of small tumors at node or bone level [41]. Moreover, the accuracy of ^{68}Ga-DOTA-TOC for the detection of bone lesions was further investigated in detail in a recent paper that compared PET data with conventional imaging findings. In 51 patients with well differentiated NET, PET with ^{68}Ga-DOTA-TOC performed better than CT and SRS for the early detection of bone NET secondary lesions (sensitivity of 97%, specificity of 92%) [50].

One distinct feature of ^{68}Ga-DOTA-TATE is the selective and very high affinity for sst_2 [39,40]. In a recent paper, 51 patients with either negative (35/51) or equivocal (16/51) SRS, were studied by ^{68}Ga-DOTA-TATE. PET identified significantly more lesions than SRS and changed management in the majority of cases (70.6%), who were subsequently deemed suitable for PRRT [42].

^{68}Ga-DOTA-NOC is also increasingly used in clinical nuclear medicine and was reported to have a more favorable dosimetry [49] while it was also reported to present higher accuracy for NET lesion detection as compared to both conventional imaging and SRS. ^{68}Ga-DOTA-NOC showed good sensitivity for the visualization of small lesions, particularly at node and bone levels [84], or in cases with an unusual anatomical localization and it provided valuable data in the identification of the primary tumor site in patients with confirmed secondary NET lesions [101]. The clinical impact of ^{68}Ga-DOTA-NOC PET/CT was evaluated in a population of 90 NET cases [44] and it was shown that PET/CT findings had an impact on either modification of therapy or stage in half of the patients, such as initiation/continuation of "hot" or "cold" somatostatin analogues (36 pts), surgical indication/exclusion (6/6 pts), and other less frequent impacts. Considering the specific uptake of ^{68}Ga-DOTA-NOC and the observation that well differentiated NETs show a slow growth rate, higher somatostatin expression, higher response rate to somatostatin analogues and better outcome, has raised the issue of whether it is possible to define PET/CT parameters that can predict patient outcome. Recently, in a population of 44 NET patients, ^{68}Ga-DOTA-NOC PET/CT was reported to provide prognostic information [43]. In fact, the SUVmax was significantly higher in cases with stable disease or partial response at follow-up of peptide receptor radionuclide radiopeptide therapy and a SUVmax cut-off allowed differentiation between patients with stable disease/partial response at follow-up from those with progressive disease.

The observed differences in receptor-binding affinities of ^{68}Ga-DOTA-TOC, -TATE, and -NOC [40] have not yet found a direct clinical correlate but at present there is some indication that such differences are associated with advantages in clinical employment. Wild et al. [51] studied 21 consecutive patients within 2 weeks in a crossover design with ^{68}Ga-DOTA-TATE and ^{68}Ga-DOTA-NOC and found a distinct advantage of ^{68}Ga-DOTA-NOC which localized more lesions, in particular in the liver and pancreas. The authors concluded that the higher sensitivity may be due to the broader affinity profile of ^{68}Ga-DOTA-NOC. More clinical studies are needed along with in vitro receptor autoradiography to confirm this finding. Currently the employment of the three ^{68}Ga-labeled somatostatin analogues is limited to specialized centers as part of clinical trials and each center selects its preferred peptide. One can estimate that at least 10,000 scans are performed annually in Europe in about 100 centers.

Although ^{68}Ga-labeled somatostatin analogues have only recently been introduced into the clinic, the number of studies reporting better results as compared to other imaging procedures and comparative studies with other PET tracers (^{18}F-FDG and ^{18}F-DOPA) have increased exponentially. In a direct comparison between ^{68}Ga-DOTA-NOC and ^{18}F-DOPA in a limited patient population (13 pts), ^{68}Ga-DOTA-NOC showed a higher number of lesions and in more cases identified the site of the occult primary [84]. A comparison of ^{68}Ga-DOTA-TATE with ^{18}F-DOPA in 25 patients with well differentiated NET [66] showed higher sensitivity of ^{68}Ga-DOTA-TATE (96% vs 56%). It is interesting to note that since metabolic and receptor-binding PET tracers can investigate different tumor features, their employment in the same patient may offer more detailed information of the biological aspects of the tumor. Different patterns of tracer uptake (^{68}Ga-somatostatin analogues vs ^{18}F-FDG) have been described within areas of the same tumor lesion or in different lesions within the same patient [85]. This is particularly relevant since the identification of areas with high FDG uptake identifies the presence of less differentiated tumor areas that are associated with a worse prognosis and require systemic chemotherapy. In fact, a different pattern of tracer uptake was also documented in high- and low-grade NET [102]; while higher ^{68}Ga-DOTA-TATE uptake was reported in low-grade versus high-grade NET, FDG uptake was significantly higher in high-grade versus low-grade lesions. ^{68}Ga-DOTA-TOC was also reported to provide more accurate information in well differentiated NET as compared to FDG [86].

^{68}Ga-DOTA-somatostatin analogue PET/CT is an accurate imaging modality to evaluate patients with well differentiated NET lesions. All described compounds have agonistic properties, which were considered mandatory because of their ability to induce the internalization of the peptide−receptor complex. Recently, it was shown that ^{111}In-labeled somatostatin-based antagonists with affinity to $sst_{2,3}$ which very poorly internalize may show even better tumor visualization than agonists [32]. A first report describing ^{68}Ga- and ^{64}Cu-labeled sst_2-antagonists indicated the high potential of these radiopeptides in PET/CT [34]. This new family of tracers may even present a better imaging option.

PET/CT with ^{68}Ga-DOTA peptides not only accurately visualizes NET lesions but also noninvasively provides valuable information on somatostatin expression pattern on tumor cells, with direct therapeutic implications. In fact, patients with somatostatin-expressing lesions are candidates for targeted therapy with either hot or cold somatostatin analogues.

17.15.3 Radiolabeled bombesin/GRP analogues

Bombesin (BN) is a 14-amino-acid peptide analogue of the 27-amino-acid mammalian gastrin-releasing peptide (GRP) (Fig. 17.5). Bombesin-like peptide receptors also belong to the family of G-protein-coupled receptors. Four bombesin receptor subtypes are known, among them three mammalian: BB1 (neuromedin B receptor), BB2 (GRP receptor (GRPR)), and BB3 (orphan bombesin receptor), and one amphibian: BB4. These receptors have been found in different tumor types. Bombesin has high affinity for the human GRPR, a G-protein-coupled receptor which is overexpressed by a variety of cancers, including prostate, breast, pancreas, gastrointestinal, and small-cell lung cancer, making radiolabeled bombesin analogues attractive candidates for targeting GRPR. The natural BN has a plasma half-

Ala^{1}-Pro^{2}-Val^{3}-Ser^{4}-Val^{5}-Gly^{6}-Gly^{7}-Gly^{8}-Thr^{9}-Val^{10}-Leu^{11}-Ala^{12}-Lys^{13}-Met^{14}-Tyr^{15}-Pro^{16}-
Arg^{17}-Gly^{18}-Asn^{19}-His^{20}-Trp^{21}-Ala^{22}-Val^{23}-Gly^{24}-His^{25}-Leu^{26}-Met^{27}-NH_2

Gastrin-releasing peptide (GRP)

$pGlu^{1}$-Gln^{2}-Arg^{3}-Leu^{4}-Gly^{5}-Asn^{6}-Gln^{7}-Trp^{8}-Ala^{9}-Val^{10}-Gly^{11}-His^{12}-Leu^{13}-Met^{14}-NH_2

Bombesin (BN)

Tyr^{6}-Gln^{7}-Trp^{8}-Ala^{9}-Val^{10}-βAla^{11}-His^{12}-Phe^{13}-Nle^{14}-NH_2

Figure 17.5 Structures of gastrin-releasing peptide and bombesin.

life of about 2–3 min and therefore many more metabolically stable analogues of BN have been developed [1,2,12–14].

The discovery that BN/GRP receptors are overexpressed in many human cancers has created the opportunity of using BN analogues to image and treat these cancers. The mitogenic properties of BN/GRP agonists prompted researchers to investigate the potential use of antagonists as anticancer agents. GRPR antagonists can exert antiproliferative effects on several experimental cancers in vitro and in vivo, inhibiting tumor growth through a number of different mechanisms that are involved in tumor inhibition by reduction of epidermal growth factor receptor levels and the inhibition of neovascularization [103]. Diagnostic imaging with radiolabeled BN analogues might be used to identify tumors and metastases, to plan radiation treatment, to predict the potential of targeted radionuclide therapy in patients, and to estimate the dose to tumors and normal organs.

DOTA-coupled BN antagonists have shown great potential for targeting GRPR expressing tumors. Mansi et al. [67] compared directly the potent agonist ^{111}In-AMBA with the antagonist ^{111}In-RM1 in PC3 prostate cancer cells and in PC3-bearing nude mice. They demonstrated superior in vivo properties of the antagonist in terms of higher tumor uptake, lower uptake by the pancreas and kidneys, and better tumor retention compared with the agonist counterpart. Later, it was demonstrated that the use of a positively charged spacer improves the binding affinity and stability of the conjugate, rendering the antagonist RM2 an even better candidate for clinical translation. This new GRPR antagonist, labeled with ^{68}Ga (named BAY 86-7548), is currently being assessed in a phase II clinical study of 11 patients with prostate cancer for the treatment of both primary tumors and lymph node metastases [68]. A PEG_4-based bombesin antagonist derivative has been studied in combination with four different chelators and radiometals. The ^{64}Cu conjugate was found to have excellent in vivo pharmacokinetics and is currently under investigation in a phase I clinical study showing high potential in the visualization of primary prostate cancer using PET or PET/CT.

The currently available data for potential use of bombesin analogues as PET tracers are mainly preclinical. Several bombesin analogues and bombesin antagonists have been developed [104–106] and radiolabeled with different PET radioisotopes (^{68}Ga, ^{64}Cu, ^{18}F) and employing different chelators [107]. Overall, good-quality images were obtained in preclinical models using bombesin-like peptides as tracers to image bombesin-receptor-expressing tumors. A series of ^{18}F-labeled BN receptor antagonists has also been developed for PET imaging. Honer et al. [108] reported the first ^{18}F-labeled BN antagonist, which showed high and specific tumor uptake as well as high tumor-to-normal tissue ratios and is now in a phase I clinical study of patients with GRPR-expressing cancers. Radiolabeled GRPR antagonists show great promise as imaging agents for the different indications in prostate cancer.

Only a few human studies have employed positron-emitting bombesin analogues. Seventeen patients with GIST treated with imatinib (because of either unresectable primary or recurrent GIST or metastatic disease) were studied by dynamic PET scans using both ^{68}Ga-BZH3 (a pan-bombesin analogue binding to NMBr, GRPr, and BB3) [69] and ^{18}F-FDG. ^{18}F-FDG studies were superior: 14/17 patients were ^{18}F-FDG-positive while only 7/17 patients showed ^{68}Ga-BZH3 uptake. The authors concluded that only a subgroup of GIST patients showed significant ^{68}Ga-BZH3 uptake and therefore did not recommend the routine use of bombesin analogues in GIST patients. Interestingly, the authors reported a single case of recurrent gastric GIST that was ^{68}Ga-BZH3 positive while FDG was falsely negative: the authors suggested that the low proliferation activity within the tumor might explain the differential pattern of tracer uptake. Dimitrakopoulou-Strauss and coworkers [109] recently described the use of ^{68}Ga-BZH3 in a small patient population (15 patients) with recurrent gliomas. All patients were studied with ^{18}F-FDG PET/CT for metabolic staging and with ^{68}Ga-BZH3 PET/CT to detect bombesin receptor expression. In fact, available treatment options for gliomas are largely dependent on tumor grade, therefore the possibility to noninvasively assess tumor grade by PET is very interesting. The authors observed that ^{68}Ga-BZH3 was helpful for the differentiation between low- and high-grade gliomas, in particular for the identification of high-grade tumors. In fact, in their small patient population, ^{18}F-FDG correctly identified three of six patients with WHO II, three of six patients with WHO III, and none of three patients with WHO IV. On the contrary, ^{68}Ga-BZH3 revealed an enhanced uptake in three of six patients with WHO II, four of six patients with WHO III, and three of three patients with WHO IV.

Recently, bombesin antagonists were studied clinically using PET in primary prostate cancer. A statin-based peptide labeled with ^{64}Cu [110] showed focal uptake in the primary tumor in three out of four patients. The pharmacokinetics is very promising as fast washout from the kidneys and the pancreas is observed. The lack of available literature data on the use of bombesin peptides for PET imaging does not allow generalization of the results obtained only in small patient populations, however the results obtained so far seem promising and further studies are needed to better ascertain the role of these compounds in clinical practice.

Radiolabeled GRPR antagonists show great promise as imaging agents for the different stages within prostate cancer. Furthermore, therapeutic particle-emitting radionuclides can be coupled to the same targeting molecules yielding theranostic pairs for personalized treatment planning.

17.15.4 The chemokine receptor 4 (CXCR4) targeting peptide probes

The chemokine receptor 4 (CXCR4) belongs to the chemokine receptor family, which also is a member of the G-protein-coupled receptor family, and has recently been identified as an attractive target receptor for in vivo imaging and therapy. The important role of CXCR4 and its endogenous ligand, stromal cell-derived factor (SDF1-a; also named CXCL12), in oncology relies on two major factors: first, on SDF1-a binding to CXCR4, the G protein complex is activated and triggers parallel signal transduction cascades, leading to cancer cell proliferation, migration, and survival. Furthermore, CXCR4 expression on primary tumor cells directs circulating cancer cells to those organs that express SDF1-a, such as lung, liver, and bone marrow, resulting in organ-specific metastasis. In addition to its extensive physiological roles, CXCR4 has also been found to be highly overexpressed on both primary tumors and metastases including breast, prostate, lung, colon, and multiple myeloma, and has been suggested to be involved and play an important role in the process of cancer cell metastasis. In addition its expression has been correlated with tumor aggressiveness, poor prognosis, and resistance to chemotherapy. It was reported that CXCR4 expression could be a powerful predictive factor for prognosis (recurrence, metastasis, or survival rate) in colorectal cancer, malignant melanoma, and osteosarcoma. Therefore, a CXCR4 imaging agent would be a new type of radiopharmaceutical for predicting the prognosis of cancer patients. CXCR4 also represents a novel target for tumor therapy, and some CXCR4 inhibitors have been investigated as antimetastatic agents. These agents showed positive effects in suppressing tumor metastasis; however, they would also have deleterious effects on normal physiological functions since CXCR4 plays a crucial role in numerous biological processes [17–19]. Therefore, in vivo imaging of CXCR4 expression could be a potential method for determining the dose of antimetastatic agents and for monitoring their therapeutic efficacy.

However, currently there is no capability to evaluate CXCR4 expression noninvasively and in whole tumors, and diagnostic tests are limited to biopsies, which are restricted to a specific site and might give false-negative results. As mentioned earlier, CXCR4 is a potential target of growing interest for the development of new therapeutic drugs, and imaging agents and several promising antagonists and peptides/antibodies for in vivo targeting of CXCR4 expression were developed in recent years and are under preclinical evaluation. Subsequently, radiolabeled CXCR4 ligands could be interesting probes to monitor metastasis and the metastatic potential of primaries.

One of the initial attempts to detect CXCR4 in tumors using a radiolabeled antagonist was conducted by Hanaoka et al. [70], who radiolabeled the peptide antagonist (Ac-TZ14011) with ^{111}In via the DTPA chelating agent. ^{111}In-DTPA-Ac-TZ14011 was able to inhibit CXCL12 binding to CXCR4 in a similar extent to its parent molecule, with an IC_{50} value of 7.9 nM. The accumulation in CXCR4-expressing tumor in an animal model, however, was paltry (0.51% ID/g), reduced over time with kinetics similar to those of blood, and no in vivo specificity was proven.

Various PET- and SPECT-based CXCR4 radioligands have been prepared and evaluated in the last few years. But still none of these agents are able to fulfill the requirements of a fully optimal clinical agent, in terms of ease of preparation, suitable pharmacokinetics, and specificity of the radiotracer for determining CXCR4 expression in vivo. The first PET-based CXCR4 antagonist radiolabeled with ^{64}Cu was AMD3100. It is a bicyclam ligand that can readily chelate metal ions so the preparation of ^{64}Cu-AMD3100 is quick and relatively simple, resulting in high radiochemical yield. ^{64}Cu-AMD3100 showed rapid blood clearance and accumulation in CXCR4-expressing bone marrow and spleen in normal mice. The radiotracer was later reported to specifically accumulate in CXCR4-expressing tumors [18,71]. The main downside of the radiotracer was high accumulation ($>$40% ID/g) in the liver, which was specific to the parent molecule, and masked some of the adjacent organs. This phenomenon is not CXCR4-specific binding in the liver because (1) high CXCR4-expressing spleens do not exhibit such high uptake of the radiotracer; (2) only limited liver cells express CXCR4; and (3) other radiotracers (e.g., T140) do not show very high accumulation in the liver. Another similar small biomolecule, AMD3465, containing one cyclam that can chelate ^{64}Cu, was also evaluated for imaging CXCR4 and showed very high accumulation in CXCR4-expressing tumor (~100% ID/g), and relatively lower but still significant (40% ID/g) uptake in the liver [18,111].

A peptide-based CXCR4 antagonist, similar to Ac-TZ14011 previously labeled with ^{111}In, but having a fluorobenzoyl group at the N-terminus, was also radiolabeled with ^{18}F, without changing the chemical structure of the peptide [72] (Fig. 17.6A). Although the peptide ^{18}F-T140 accumulated in CXCR4-positive tumors, it also bound to red blood cells (RBCs), and injection of additional unlabeled mass of the peptide was required to visualize CXCR4 in vivo. Another drawback was its long radiosynthesis time and low radiochemical yield. To simplify the synthesis of a T140-based ligand, DOTA was introduced on the lysine groups to allow radiolabeling with the longer-lived PET isotope ^{64}Cu. The ^{64}Cu-labeled

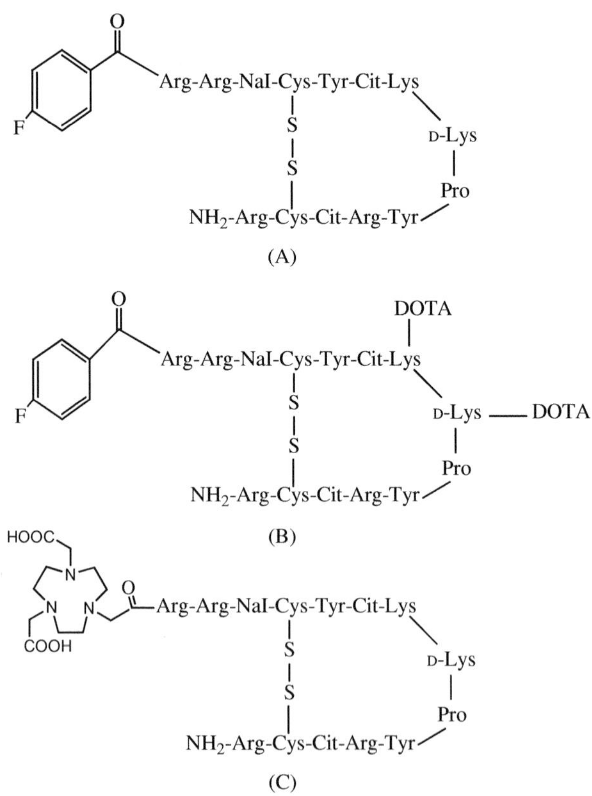

Figure 17.6 Structures of (A) 4-F-T140, (B) T140-2D, and (C) CCIC16 (NO2A-TN-14003).

T140-2D peptide (Fig. 17.6B) showed similar binding to RBCs as the corresponding ^{18}F-T140, and in low specific radioactivity form could be used for imaging CXCR4-positive tumors. Similar methodology was applied to label T140 with another PET isotope ^{68}Ga without significant differences. It is imperative to note that the accumulation of radioactivity in the liver was significantly higher when T140 was chelated with radiometals than when it was radiolabeled with ^{18}F, suggesting that some of the radioactivity in the liver may be due to transchelation of the radiometal [73]. Thus, it is important to improve the in vivo stability of a radiometal complex to minimize any transchelation occurrence.

In order to reduce undesired binding of radiolabeled T140 to RBCs, DOTA, or NOTA, chelators were attached at the N-terminus of the T140 peptide by replacing the fluorobenzoyl group and radiolabeling with the longer-lived PET isotope ^{64}Cu. It was found that replacing the fluorobenzoyl group with either DOTA or NOTA diminished the binding of radiolabeled T140 to RBCs, while only slightly reducing the binding to CXCR4 [18,73]. These two optimized radiotracers, used in high specific radioactivity, showed high accumulation in CXCR4-positive tumors but not in CXCR4-negative tumors, and high tumor-to-background ratio. Additional features of these radioligands include short radiosynthesis time, high radiochemical yield, and no nonspecific binding to RBCs.

For in vivo quantification of CXCR4 in cancer, 4-F-benzoyl-TN14003 (4-F-T140), a short peptide antagonist of CXCR4 with 1-(4,4-dimethyl-2,6-dioxocyclohexylidene)ethyl protecting groups on the ε-amino groups of the lysine residues, was prepared and radiolabeled with ^{18}F via *N*-succinimidyl-4-^{18}F-fluorobenzoate conjugation, followed by deprotection to give 4-^{18}F-T140 that was labeled on the α-amine at the N terminus. 4-^{18}F-T140 was found to bind specifically to RBCs in vitro and in vivo [74]. The binding of 4-^{18}F-T140 to RBCs was blocked with a small amount of cold 4-F-T140, which led to higher uptake of 4-^{18}FT140 by Chinese hamster ovarian (CHO)-CXCR4 tumors. Biodistribution of 4-^{18}F-T140 in tumor-bearing mice showed high and persistent retention of the radiopeptide in the blood, resulting in visualization of both CXCR4-negative and CXCR4-positive tumors, probably because of an enhanced but similar blood supply to the tumors. But when performed with the addition of 10 mg of nonradioactive 4-F-T140, a good uptake (3.03% ± 0.31% ID/g) in CHO-CXCR4 tumors was observed, with high tumor-to-blood and tumor-to-muscle ratios. It has been postulated that the binding of nonradioactive 4-F-T140 to RBCs allows 4-^{18}F-T140 to interact with CXCR4-expressing tumor cells but not with CXCR4-negative tumor cells, which enables visualization of the CXCR4-positive tumor with higher signal-to-noise ratio. Additionally, PET studies demonstrated clear visualization of CXCR4-transfected, but not CXCR4-negative, CHO tumors. These results suggest that 4-^{18}F-T140 can be used as a PET-tracer to determine tumor CXCR4 expression. Although 4-^{18}F-T140 binds to RBCs, this binding can be blocked by coinjection of the radiotracer with nonradioactive 4-F-T140 peptide, resulting in elevated accumulation in CXCR4-positive tumor, but not CXCR4-negative tumor, and high tumor-to-background ratios. Visualizing and quantifying CXCR4 expression by cancers could lead to more effective interventions adapted to the individual patient [72].

Another radiotracer that was developed for PET imaging of CXCR4 was based on a dimer of cyclic penta-peptide FC131 coupled with DOTA and radiolabeled with ^{68}Ga. The dimer radiotracer showed limited accumulation in the tumor, but very high accumulation in the liver ($\sim$44% ID/g). It appears that the high lipophilicity of the dimer compound contributed to its unfavorable biodistribution properties that results in the development of FC131 cyclic penta-peptide as a monomeric form (CPCR4-2). Even though the monomeric radiotracer hypothetically has lower polyvalency, it exhibited better biodistribution in vivo, with significantly lower accumulation in the liver, higher accumulation in the tumor, and good tumor-to-muscle ratios [18,74].

Significant efforts have been directed toward the development of clinically suitable CXCR4 imaging probes, most of which are derived from stromal cell-derived factor. The first downsized CXCR4 ligand, T140 (Arg^1-Arg^2-Nal^3-cyclo (Cys^4-Tyr^5-Arg^6-Lys^7-D-Lys^8-Pro^9-Tyr^{10}-Arg^{11}-Cit^{12}-Cys^{13})-Arg^{14}), was shown to exhibit strong antagonistic action, which is largely mediated by the four N- and C-terminal residues Nal^3, Tyr^5,Arg^{14}, and Arg^2. On the basis of this structure–activity relationship, cyclic pentapeptides, such as FC131 (cyclo(D-Tyr^1-Arg^2-Arg^3-2-Nal^4-Gly^5)), displaying improved metabolic stability as well as potent CXCR4 antagonism, have been developed. ^{124}I-FC131 was the first CXCR4-directed PET imaging agent in preclinical evaluation. It showed high accumulation in the CXCR4-positive tumor in an animal model and high focal uptake in lung micrometastases. Its applicability for whole-body CXCR4 imaging, however, is strictly limited because of its high lipophilicity and subsequently high liver and intestinal uptake. Based on the results obtained with ^{124}I-FC131 and encouraged by the convenient characteristics of ^{68}Ga as a generator produced PET radionuclide, ^{68}Ga-labeled cyclic CXCR4-binding pentapeptide, ^{68}Ga-CPCR4-2 (cyclo(D-Tyr^1-D-[NMe]Orn^2-[AMBS-DOTA]-Arg^3-2-Nal^4-Gly^5)), (AMBS: 4-(aminomethyl)benzoic acid) was prepared and evaluated [74]. However, it was found that the conjugation of the relatively bulky radiometal chelate to the small cyclic pentapeptide significantly affects the binding affinity, thus identification of the optimal site of attachment and optimization of suitable linker units were found to be necessary.

Among various cyclo(D-Tyr^1-Orn^2-Arg^3-2-Nal^4-Gly^5) pentapeptides that were examined for the flexibility of the CXCR4-binding pocket, the most promising candidate, CPCR4-2, the [NMe]-D-Orn^2 side chain was functionalized with DOTA via an AMBS ($HOOC$-C_6H_4-CH_2-NH_2) linker. As determined in CXCR4-expressing Jurkat cells, ^{nat}Ga-CPCR4-2 showed high affinity to CXCR4. In a competition binding assay, specific binding of ^{125}I-FC131 to Jurkat cells was inhibited by ^{nat}Ga-CPCR4-2 with a half maximal inhibitory concentration (IC_{50}) of 5 ± 1 nM. The radiolabeled peptide revealed CXCR4-specific tumor uptake, fast renal excretion, and high tumor-to-muscle ratios at 1 and 2 h p.i. in mice bearing CXCR4-overexpressing tumors [74].

Another study described the preparation of a new ^{68}Ga-labeled high-affinity cyclic CXCR4 ligand, ^{68}Ga-CPCR4-2 (cyclo(DTyr^1-[NMe]-D-Orn^2-[4-(aminomethyl) benzoic acid, ^{68}Ga-DOTA]- Arg^3-2-Nal^4-Gly^5)) for specific targeting of CXCR4. ^{nat}Ga-CPCR4-2 exhibited high CXCR4 affinity with an IC_{50} value in the low nanomolar range. ^{68}Ga-CPCR4-2 showed high in vivo stability and high and specific tumor accumulation, which was reduced by approximately 80% in competition studies with AMD3100. High CXCR4 expression in tumors was confirmed by immune histochemical staining. ^{68}Ga-CPCR4-2 showed low uptake in nontumor tissue and particularly low kidney uptake despite significant renal excretion, leading to high-contrast delineation of tumors in small-animal PET studies performed in nude mice bearing metastasizing OH1 human small cell lung cancer xenografts. ^{68}Ga-CPCR4-2 is the CXCR4-directed PET agent that seems to combine high CXCR4 affinity and thus high and persistent tumor uptake with suitable overall pharmacokinetics. These results demonstrate that ^{68}Ga-CPCR4-2 allows high-contrast PET of CXCR4-positive tumors and metastases even in the hepatic and abdominal region. It appears that ^{68}Ga-CPCR4-2 may be a promising radiotracer for imaging the CXCR4 receptor status in vivo with PET and deserves further investigation of ^{68}Ga-CPCR4-2 in a proof-of-concept study in humans [74].

A new compound based on CXCR4 peptidic antagonist TN14003, CCIC16, which is potentially amenable to radiolabeling by chelation with several clinically useful PET and SPECT radiometals, such as ^{68}Ga, ^{64}Cu, and ^{111}In, as well as ^{18}F (Al^{18}F), was designed and synthesized in order to develop a CXCR4-specific radiotracer [75] (Fig. 17.6C). Potent in vitro binding affinity and inhibition of signaling-dependent cell migration by unlabeled CCIC16 were confirmed by a threefold uptake in CXCR4-overexpressing cells compared to their isogenic counterparts. In mice injected with ^{68}Ga-CCIC16, the amount of radioactivity in the plasma remained constant through the time of the study showing good circulation of the radiotracer, and low nonspecific off-target uptake by RBCs. The major uptake was found in the liver, gall-bladder, kidneys, and urine, consistent with hepatic metabolism and urinary excretion. Uptake in the lung, spleen, and bone can be accounted for by the constitutive expression of CXCR4 in these tissues. In vivo evaluation of ^{68}Ga-CCIC16 in tumor-xenografted mice showed a significantly higher uptake into U87.CD4.CXCR4 compared to into U87.CD4 tumors. The coinjection of Ga-CCIC16 reduced the activity in the tumor insignificantly but it led to a general increase in uptake in several tissues, especially the kidneys and urine. This is consistent with the hypothesis of increased tracer availability following inhibition of specific binding and subsequent excretion. Dynamic PET imaging conducted on U87.CD4 and U87.CD4.CXCR4 tumor-bearing mice allowed good visualization of the CXCR4-expressing tumors by PET/CT. Together with the favorable PET imaging properties of ^{68}Ga, ^{68}Ga-CCIC16 shows essential features for the identification of CXCR4-expressing tumors in a clinical setting, with short time of residence of the tracer in the body. Furthermore, chelating groups may also be functionalized for cancer therapies by the incorporation of radionuclides suitable for radionuclide therapy of cancer, enabling theranostic dual-functionality to be envisaged. Overall, despite some favorable in vitro and in vivo properties, there is a need for optimization with regard to improving the tumor targeting characteristics and to make this radiotracer more suitable for further evaluation.

Substantial efforts are currently being directed on the development of small biomolecules that target CXCR4 (e.g., AMD3100) and antagonistic peptides (e.g.,

T140). Derivatives of the T140 antagonistic peptide, such as Ac-TZ14011, are more potent and biologically stable. Moreover, these peptides are very versatile imaging platforms as addition of various diagnostic labels is possible without interfering with the pharmacophore. An example of such a compound is ^{111}In-DTPA-Ac-TZ14011. Based on structure–activity relationships, a 14-residue peptidic CXCR4 inhibitor, Ac-TZ14011, was prepared, as a precursor for radiolabeled peptides. Considering the three residues on the restricted backbone (l-3-(2-naphthyl)alanine (Nal)^3, Tyr^5 and Arg^{14}) and the single residue in the flexible region (Arg^2) form the intrinsic pharmacophore of T140. This precursor contains the above four residues which are necessary for the inhibitory activity against CXCR4. Furthermore, for site-specific conjugation of radiolabels, Ac-TZ14011 has a single amino group (D-Lys^8), for the attachment of DTPA, which is distant from the pharmacophore, and the carboxyl group of Arg^{14} of Ac-TZ14011 is protected via amidation for stability in vivo [112].

17.15.5 Radiolabeled gastrin/CCK2 analogues

The gastrin receptor may be an attractive target for radiopeptide imaging and therapy of several human tumors. Reubi et al. [113] have identified by autoradiography that the gastrin receptor is overexpressed on MTC (92%), SCLC (57%), astrocytoma (65%), stromal ovarian tumors (100%), and gastroenteropancreatic tumors (22%). The first human studies were performed by Behr et al. [77]. Using a ^{111}In-labeled DTPA-conjugated minigastrin derivative they visualized medullary thyroid carcinomas.

From the same group [78], a study of 60 consecutive patients (51 carcinoids, 3 gastrinomas, 2 glucagonomas, 3 paragangliomas, and 1 insulinoma) was reported comparing gastrin receptor scintigraphy (GRS) with SRS. The detection rate was 74%. 11 patients had negative SRS, 6 of which were GRS positive. Based on the number of lesions GRS performed better in 13 patients, and SRS performed better in 24 cases. Overall GRS detected additional tumor sites in 20% of patients. There was no Pearson's correlation between the scan results and the grade of tumor differentiation; but this aspect appears to deserve more attention. In a similar comparative study of 29 patients with metastatic medullary thyroid cancer (MTC) Gotthardt et al. [79] found that tumor detection with GRS was 94.2%, whereas using SRS the detection rate was only 40.7% correlating with in vitro autography data.

Radiotherapeutic studies were also reported by the same group using ^{90}Y-DTPA-D-Glu-Glu$_5$-Ala-Tyr-Gly-Trp-Met-Asp-Phe-NH$_2$ in eight rapidly progressing, metastatic MTC patients [114]. Overall two partial remissions were found, four stable disease (SD) and two progressive disease (PD). The study had to be stopped because of serious kidney and hematologic toxicity. Recently, the development of new radioligands with the macrocyclic chelator DOTA was reported [76]. They show higher serum stability and improved tumor-to-kidney ratios. A study in MTC with these highly improved radiopeptides was initiated within a European project involving 10 hospitals.

17.15.6 Radiolabeled RGD peptides

PET imaging of $\alpha_{\nu}\beta_3$ integrin expression using radiolabeled RGD peptides made its way from bench to bedside. The first clinical data are on the glycosylated peptide [^{18}F]Galacto-RGD in a limited number of patients with malignant melanoma, sarcoma, osseous metastasis from renal cell carcinoma, and villonodular synovitis [80]. A great intra- as well as interindividual heterogeneity of the tracer uptake in the lesions was found, which indicates the great diversity in $\alpha_{\nu}\beta_3$ expression. Beer et al. studied if [^{18}F]Galacto-RGD uptake correlates with $\alpha_{\nu}\beta_3$ expression in patients with different solid tumors, such as musculoskeletal system, melanoma, head and neck cancer, glioblastoma, and breast cancer [81]. The patients were examined with [^{18}F]Galacto-RGD PET before surgical removal of the lesions. Sections from representative areas of the lesions with low as well as intense uptake were examined with immunohistochemistry. A significant correlation was found between the SUVs and the tumor-to-blood ratios with the immunohistochemical staining as well as with the microvessel density. Moreover, tumors negative on [^{18}F]Galacto-RGD PET showed lack of $\alpha_{\nu}\beta_3$ expression. Biodistribution of [^{18}F]Galacto-RGD in cancer patients showed specific receptor binding and rapid predominantly renal excretion resulting in low background activity in most regions of the body [82].

Side-by-side comparison of the [^{18}F]Galacto-RGD and [^{18}F]FDG uptake in patients with different primary or metastatic cancers, including nonsmall-cell lung, renal cell carcinoma, rectal cancer, and others, showed that there is no substantial correlation between glucose metabolism ([^{18}F]FDG uptake) and $\alpha_{\nu}\beta_3$ expression in malignant lesions [83]. Consequently, [^{18}F]FDG and [^{18}F]Galacto-RGD provide different information. Whereas [^{18}F]FDG PET was shown to be more sensitive for tumor staging, [^{18}F]Galacto-RGD PET has a potential value for appropriate selection of patients entering clinical trials with $\alpha_{\nu}\beta_3$-targeted antiangiogenic therapies, considering the correlation of the tracer uptake with $\alpha_{\nu}\beta_3$ expression.

The feasibility of using [^{18}F]Galacto-RGD PET in specific patient groups with respect to their $\alpha_{\nu}\beta_3$ expression patterns has been investigated in squamous cell carcinoma of the head and neck (SCCHN) [115], where immunohistochemistry confirmed $\alpha_{\nu}\beta_3$ expression mainly on tumor vessels, and in breast cancer where $\alpha_{\nu}\beta_3$ expression was predominantly on microvessels and, to a lesser extent, on tumor cells. In glioblastoma patients [116] the uptake was heterogeneous and substantially lower compared to tumors outside the central nervous system; probably related to the fact that [^{18}F]Galacto-RGD does not cross the intact blood–brain barrier. Higher accumulation was seen in the highly proliferating and infiltrating areas of tumors, while normal brain tissue did not show significant uptake. Immunohistochemical staining was prominent in microvessels as well as glial tumor cells and the tracer uptake correlated with $\alpha_{\nu}\beta_3$ expression of corresponding tumor samples.

The second clinically evaluated tracer was ^{99m}Tc-NC100692, introduced by GE Healthcare. In breast cancer patients the tracer was able to detect 19 of 22 histopathologically confirmed malignant lesions [87]. However, immunohistochemical correlation with $\alpha_{\nu}\beta_3$ expression was not reported.

A phase I trial has been performed with [^{18}F]AH111585 (also called [^{18}F]fluciclatide) in a small number of patients with metastasized breast cancer, where all tumors detected by CT were visible on the [^{18}F]AH111585 PET [88]. Metastases in the lung, pleura, bone, lymph node, and primary tumor showed increased uptake compared with background. However, liver metastases were indirectly identified as photopenic regions due to the high background activity in normal liver tissue. The tracer is being evaluated in a phase II study in patients with brain tumors, lung cancers, SCCHN, differentiated thyroid carcinoma, sarcoma, and melanoma to correlate [^{18}F]AH111585 PET imaging with histological parameters of angiogenesis (including $\alpha_v\beta_3$ expression) [117].

The radiotracer [^{18}F]RGD-K5 used in parallel with [^{18}F]FDG in a small number of patients with breast cancer showed high [^{18}F]RGD-K5 uptake in most lesions identified by [^{18}F]FDG [89]. The uptake of [^{18}F]RGD-K5 showed no correlation with the [^{18}F]FDG uptake; also seen for [^{18}F]Galacto-RGD. Although most lesions (80%) showed higher [^{18}F]FDG uptake, in a few cases (4%) [^{18}F]RGD-K5 uptake was higher than [^{18}F]FDG. [^{18}F]RGD-K5 is used in a phase II clinical study to assess the usefulness of predicting efficacy and early response monitoring in a combination therapy of the anti-VEGF antibody bevacizumab plus chemotherapy [90].

Recently, [^{68}Ga]NOTA-RGD has been used along with [^{18}F]FDG in patients with hepatic metastases of colorectal cancer before a combination therapy of FOLFOX and bevacizumab [91]. While all hypermetabolic liver lesions were visualized on [^{18}F]FDG PET, half of the patients had mild [^{68}Ga]NOTA-RGD uptake in these lesions and the other half no uptake. Interestingly, the patients with [^{68}Ga]NOTA-RGD uptake in the hepatic metastases showed a partial response to the antiangiogenic combination therapy, whereas the other patients had stable or progressive disease. These findings support the hypothesis that PET imaging of $\alpha_\nu\beta_3$ might help in the selection of the patient for antiangiogenic therapies (Table 17.6).

17.16 Conclusions

Radiolabeled peptides hold great clinical potential for diagnostic imaging and targeted radiopeptide receptor therapy in nuclear oncology. The advancement in the field of peptide synthesis chemistry and the availability of a wide variety of bifunctional chelating agents that allows radiolabeling with both diagnostic as well as therapeutic radionuclides will certainly accelerate the facile preparation of effective peptide-based radiopharmaceuticals for targeting a variety of human tumors.

There are several important areas in which molecular imaging will be expected to proceed, for example, development of newer and more effective imaging agents and imaging techniques and their use in standard practice in clinical environments and preclinical settings. It is important to realize that as with drug discovery and development only a few imaging agents that enter clinical trials will ever become truly clinically useful. The design and the development of new imaging agents is a complicated process and we will fail often in our efforts to design and implement

Table 17.6 **A summary of tumor receptor-targeting peptides evaluated in this study**

Peptide	Major application	References
^{111}In-DOTA-sst_2-ANT	Imaging of somatostatin receptor-positive tumors	[32]
^{111}In-DOTA-sst_3-ODN-8	Imaging of somatostatin receptor-positive tumors	[32,33]
^{68}Ga-NODAGA-LM3	Imaging of somatostatin receptor-positive tumors	[34,35]
^{68}Ga-DOTA-LM3	Imaging of somatostatin receptor-positive tumors	[3,35]
^{64}Cu-NODAGA-LM3	Imaging of somatostatin receptor-positive tumors	[35]
^{64}Cu-CB-TE2A-LM3	Imaging of somatostatin receptor-positive tumors	[34,35]
Ga(III)-DOTA-OC	Determination of somatostatin receptor affinity	[27]
Y(III)-DOTA-OC	Determination of somatostatin receptor affinity	[27]
Ga(III)-DOTA-TATE	Determination of somatostatin receptor affinity	[27]
Y(III)-DOTA-TATE	Determination of somatostatin receptor affinity	[27]
^{68}Ga-DOTA-NOC	Imaging of somatostatin receptor-positive tumors	[39–46,49]
^{68}Ga-DOTA-TOC	Imaging of somatostatin receptor-positive tumors	[39–46,50]
^{68}Ga-DOTA-TATE	Imaging of somatostatin receptor-positive tumors	[39–42,51]
In(III)-DOTA-JR11	Determination of somatostatin receptor affinity	[47]
$^{111}In/^{90}Y$-DOTATOC	Imaging and therapy of somatostatin receptor-positive tumors	[39–47]
^{68}Ga-DOTA-JR11	Imaging of somatostatin receptor-positive tumors	[47,48]
^{68}Ga-NODAGA-JR11	Imaging of somatostatin receptor-positive tumors	[47,48]

(*Continued*)

Table 17.6 (Continued)

Peptide	Major application	References
^{177}Lu-DOTA-BASS	Somatostatin receptor radionuclide therapy	[52]
^{177}Lu-DOTA-TATE	Somatostatin receptor radionuclide therapy	[52]
^{111}In-DOTA-BASS	Imaging of somatostatin receptor-positive tumors	[3,53]
^{111}In-DTPA-OC	Imaging of somatostatin receptor-positive tumors	[53–55]
^{90}Y-DOTA-BASS	Somatostatin receptor radionuclide therapy	[53]
^{177}Lu-DOTA-BASS	Somatostatin receptor radionuclide therapy	[48]
^{64}Cu-CB-TE2A-BASS	Imaging of somatostatin receptor-positive tumors	[56,57]
^{177}Lu-DOTA-JR11	Somatostatin receptor radionuclide therapy	[3]
^{99m}Tc-Demobesin 1	Imaging of somatostatin receptor-positive tumors	[58]
[3-^{123}I-Tyr^3]octreotide	Imaging of somatostatin receptor-positive tumors	[59]
^{99m}Tc-Dereotide (NeoTect)	Imaging of somatostatin receptor-positive tumors	[60,61]
^{99m}Tc-HYNIC-TOC	Imaging of somatostatin receptor-positive tumors	[55,62,63]
^{99m}Tc-HYNIC-TATE	Imaging of somatostatin receptor-positive tumors	[64,65]
^{18}F-DOPA	Imaging of somatostatin receptor-positive tumors	[66]
^{111}In-RM1	Imaging of somatostatin receptor-positive tumors	[67]
BAY 86-7548	Imaging of somatostatin receptor-positive tumors	[67,68]

^{68}Ga-BZH3	Imaging of somatostatin receptor-positive tumors	[69]
^{111}In-DTPA-Ac-TZ14011	Imaging of somatostatin receptor-positive tumors	[70]
^{64}Cu-AMD3100	Imaging of somatostatin receptor-positive tumors	[18,71]
^{18}F-T140	Imaging of somatostatin receptor-positive tumors	[72]
^{64}Cu-T140-2D	Imaging of somatostatin receptor-positive tumors	[72,73]
4-^{18}F-T140	Imaging of somatostatin receptor-positive tumors	[18,73]
^{124}I-FC131	Imaging of somatostatin receptor-positive tumors	[74]
^{68}Ga-CPCR4-2	Imaging of somatostatin receptor-positive tumors	[74]
^{68}Ga-CCIC16	Imaging of somatostatin receptor-positive tumors	[75]
^{90}Y-DTPA-gastrin	Somatostatin receptor radionuclide therapy	[76–79]
[^{18}F]Galacto-RGD	Imaging of somatostatin receptor-positive tumors	[80–83]
[^{18}F]FDG	Imaging of somatostatin receptor-positive tumors	[84–86]
^{99m}Tc-NC100692	Imaging of somatostatin receptor-positive tumors	[87]
^{18}F-AH111585 (^{18}F-fluciclatide)	Imaging of somatostatin receptor-positive tumors	[88]
[^{18}F]RGD-K5	Imaging of somatostatin receptor-positive tumors	[89,90]
[^{68}Ga]NOTA-RGD	Imaging of somatostatin receptor-positive tumors	[91]

new imaging strategies, but to go forward we need to learn from our mistakes and always have realistic expectations.

Another important point is to find more efficient screening and personalized medicine procedures. Both in vitro diagnostics and molecular imaging will likely permit screening of at-risk populations. Effective screening will result in early detection of disease, far before any clinical signs or symptoms manifest, ultimately leading to improved treatment strategies and patient outcomes. The merger of in vitro diagnostics and in vivo diagnostics will be critical to solving the limitations of each approach, and thus making personalized medicine a reality. Furthermore, the increased use of dual-modality and multimodality molecular imaging in clinical practice will provide better information regarding complex biological systems and disease states. Improvements in imaging agent design and development are the key element for the effectiveness of molecular imaging. Significant progress is also anticipated towards "smart" probe development and multimodality probes.

Even though many peptide receptor targets are available for investigation, it is the molecular targets, with relevance to a disease state, that are essential for further exploration in the framework of translational radiopharmaceuticals. It is important to identify a high clinically demanding target in order for the drug discovery process to permit the time and expense of developing a potential in vivo imaging agent. Validation of biological targets, in particular with an understanding of the relation of the target to the disease state, is an important step as part of the development pathway. Nonetheless, of equal importance are conclusive structure–activity relationship findings for the tumor-targeting peptide in order to provide the best possible clinical candidate. By performing all these vital steps, valuable SPECT- and PET-based peptide imaging agents with clinical potential will continue to be invented for better management of diseases. The future of molecular imaging is indeed very bright. The continual development in the field is likely to lead to major breakthroughs in our understanding of in vivo biology and also improvement in the clinical detection of disease and management of patients. The field of molecular imaging is still in its early stages, with many clinically important and innovative discoveries are imminent.

References

[1] S.M. Okarvi, Peptide-based radiopharmaceuticals and cytotoxic conjugates: potential tools against cancer, Cancer Treat. Rev. 34 (2008) 13–26.

[2] C.L. Charron, J.L. Hickey, T.K. Nasiama, D.R. Cruickshank, W.L. Turnbull, L.G. Luyt, Molecular imaging probes from natural peptides, Nat. Prod. Rep. 33 (2016) 761–800.

[3] S.M. Okarvi, H.R. Maecke, Radiometallo-Labeled Peptides in Tumor Diagnosis and Targeted Radionuclide Therapy, in: Rudi Van Eldik, Colin D. Hubbard (Eds.), Advances in Inorganic Chemistry, Vol. 68, Academic Press, Burlington, 2016, pp. 341–396.

[4] J.C. Reubi, Peptide receptors as molecular targets for cancer diagnosis and therapy, Endocr. Rev. 24 (2003) 389–427.

[5] G. Mariani, P.A. Erba, A. Signore, Receptor-mediated tumor targeting with radiolabeled peptides: there is more to it than somatostatin analogs, J. Nucl. Med. 47 (2006) 1904–1907.

[6] J.C. Reubi, H.R. Maecke, Peptide-based probes for cancer imaging, J Nucl Med 49 (2008) 1735–1738.

[7] R.P. Baum, H.R. Kulkarni, C. Carreras, Peptides and receptors in image-guided therapy: theranostics for neuroendocrine neoplasms, Semin. Nucl. Med. 42 (2012) 190–207.

[8] E.P. Krenning, E.P. Krenning, W.H. Bakker, W.A. Breeman, et al., Localisation of endocrine-related tumours with radioiodinated analogue of somatostatin, Lancet 1 (1989) 242–244.

[9] G.A. Kaltsas, D. Papadogias, P. Makras, et al., Treatment of advanced neuroendocrine tumours with radiolabelled somatostatin analogues, Endocr. Relat. Cancer 12 (2005) 683–699.

[10] L. Bodei, G. Paganelli, G. Mariani, Receptor radionuclide therapy of tumors: a road from basic research to clinical applications, J. Nucl. Med. 47 (2006) 375–377.

[11] J.J. Teunissen, D.J. Kwekkeboom, M. de Jong, et al., Endocrine tumours of the gastrointestinal tract. Peptide receptor radionuclide therapy, Best Pract. Res. Clin. Gastroenterol. 19 (2005) 595–616.

[12] S.M. Okarvi, Peptide-based radiopharmaceuticals: future tools for diagnostic imaging of cancers and other diseases, Med. Res. Rev. 24 (2004) 357–397.

[13] M. Fani, H.R. Maecke, S.M. Okarvi, Radiolabeled peptides: valuable tools for the detection and treatment of cancer, Theranostics 2 (2012) 481–501.

[14] M. Fani, H.R. Maecke, Radiopharmaceutical development of radiolabelled peptides, Eur. J. Nucl. Med. Mol. Imaging 39 (2012) S11–S30.

[15] N. Gonzalez, T.W. Moody, H. Igarashi, et al., Bombesin-related peptides and their receptors: recent advances in their role in physiology and disease states, Curr. Opin. Endocrinol. Diabetes Obes. 15 (2008) 58–64.

[16] H. Ohki-Hamazaki, M. Iwabuchi, F. Maekawa, Development and function of bombesin-like peptides and their receptors, Int. J. Dev. Biol. 49 (2005) 293–300.

[17] K. Philipp-Abbrederis, K. Herrmann, S. Knop, M. Schottelius, In vivo molecular imaging of chemokine receptor CXCR4 expression in patients with advanced multiple myeloma, EMBO Mol Med 7 (2015) 477–487.

[18] I.D. Weiss, O. Jacobson, Molecular imaging of chemokine receptor CXCR4, Theranostics 3 (2013) 76–84.

[19] J. Kuil, T. Buckle, F.W.B. van Leeuwen, Imaging agents for the chemokine receptor 4 (CXCR4), Chem Soc Rev 41 (2012) 5239–5261.

[20] O. Jacobson, I.D. Weiss, L.P. Szajek, et al., PET imaging of CXCR4 using copper-64 labeled peptide antagonist et al. PET imaging of CXCR4 using copper-64 labeled peptide antagonist, Theranostics 1 (2011) 251–262.

[21] M.L. James, S.S. Gambhir, A molecular imaging primer: modalities, imaging agents, and applications, Physiol Rev 92 (2012) 897–965.

[22] M.F. Kircher, H. Hricak, S.M. Larson, Molecular imaging for personalized cancer care, Molecular oncology 6 (2012) 182–195.

[23] M.A. Pysz, S.S. Gambhir, J.K. Willmann, Molecular imaging: current status and emerging strategies, Clinical Radiology 65 (2010) 500–516.

[24] M. Langer, A.G. Beck-Sickinger, Peptides as carrier for tumor diagnosis and treatment, Curr. Med. Chem. Anticancer Agents 1 (2001) 71–93.

[25] L. Carlie, A. Charron, L. Ashleigh, A. Farnsworth, D. Peter, B. Roselt, et al., Recent developments in radiolabelled peptides for PET imaging of cancer, Tetrahedron. Lett. 57 (2016) 4119–4127.

[26] J.C. Reubi, Neuropeptide receptors in health and disease: the molecular basis for in vivo imaging, J Nucl Med 36 (1995) 1825–1835.

[27] J.C. Reubi, J.C. Schaer, B. Waser, et al., Affinity profiles for human somatostatin receptor subtypes SST1-SST5 of somatostatin radiotracers selected for scintigraphic and radiotherapeutic use, Eur J Nucl Med 27 (2000) 273–282.
[28] R. Valkema, S.A. Pauwels, L.K. Kvols, et al., Long-term follow-up of renal function after peptide receptor radiation therapy with ^{90}Y-DOTA^{0},Tyr^{3}-octreotide and ^{177}Lu-DOTA^{0}, Tyr^{3}-octreotate, J. Nucl. Med. 46 (2005) 83S–91S.
[29] A. Imhof, P. Brunner, N. Marincek, et al., Response, survival, and long-term toxicity after therapy with the radiolabeled somatostatin analogue [^{90}Y-DOTA]-TOC in metastasized neuroendocrine cancers, J. Clin. Oncol. 29 (2011) 2416–2423.
[30] T.M. Behr, M. Gotthardt, A. Barth, et al., Imaging tumors with peptide-based radioligands, Q. J. Nucl. Med. 45 (2001) 189–200.
[31] D.A. Mankoff, J.M. Link, H.M. Linden, et al., Tumor receptor imaging, J. Nucl. Med. 49 (2008) 149S–163S.
[32] M. Ginj, H. Zhang, B. Waser, R. Cescato, D. Wild, X. Wang, et al., Radiolabeled somatostatin receptor antagonists are preferable to agonists for in vivo peptide receptor targeting of tumors, Proc. Natl. Acad. Sci. U.S.A. 103 (2006) 16436–16441.
[33] R. Cescato, J. Erchegyi, B. Waser, V. Piccand, H.R. Maecke, J.E. Rivier, et al., Design and in vitro characterization of highly sst2-selective somatostatin antagonists suitable for radiotargeting, J. Med. Chem. 51 (2008) 4030–4037.
[34] M. Fani, L. Del Pozzo, K. Abiraj, R. Mansi, M.L. Tamma, R. Cescato, et al., PET of somatostatin receptor-positive tumors using ^{64}Cu- and ^{68}Ga-somatostatin antagonists: the chelate makes the difference, J. Nucl. Med. 52 (2011) 1110–1111.
[35] K.P. Eisenwiener, M.I.M. Prata, I. Buschmann, H.W. Zhang, A.C. Santos, S. Wenger, et al., NODAGATOC, a new chelator-coupled somatostatin analogue labeled with [67/68Ga] and [111In] for SPECT, PET, and targeted therapeutic applications of somatostatin receptor (hsst2) expressing tumors, Bioconjug. Chem. 13 (2002) 530.
[36] M. Ginj, H. Zhang, K.P. Eisenwiener, D. Wild, S. Schulz, H. Rink, et al., New pansomatostatin ligands and their chelated versions: affinity profile, agonist activity, internalization, and tumor targeting, Clin. Cancer Res. 14 (2008) 2019–2027.
[37] M. Fani, A. Mueller, M.-L. Tamma, G. Nicolas, H.R. Rink, R. Cescato, et al., Radiolabeled bicyclic somatostatin-based analogs: a novel class of potential radiotracers for SPECT/PET of neuroendocrine tumors, J. Nucl. Med. 51 (2010) 1771–1779.
[38] D. Storch, M. Béhé, M.A. Walter, J. Chen, P. Powell, R. Mikolajczak, et al., Evaluation of [^{99m}Tc/EDDA/HYNIC^{0}]octreotide derivatives compared with [^{111}In-DOTA0,Tyr^{3}, Thr^{8}]octreotide and [^{111}In-DTPA^{0}]octreotide: does tumor or pancreas uptake correlate with the rate of internalization?, J. Nucl. Med. 46 (2005) 1561–1569.
[39] L. Wang, K. Tang, Q. Zhang, H. Li, H. Zhang, Somatostatin receptor-based molecular imaging and therapy for neuroendocrine tumors, Biomed. Res. Int. 2013 (2013) 102819.
[40] P. Antunes, M. Ginj, H. Zhang, B. Waser, R.P. Baum, J.C. Reubi, et al., Are radiogallium-labelled DOTA-conjugated somatostatin analogues superior to those labelled with other radiometals? Eur. J. Nucl. Med. Mol. Imaging 34 (2007) 982–993.
[41] M. Gabriel, C. Decristoforo, D. Kendler, G. Dobrozemsky, D. Heute, C. Uprimny, et al., ^{68}Ga-DOTA-Tyr^{3}-octreotide PET in neuroendocrine tumors: comparison with somatostatin receptor scintigraphy and CT, J. Nucl. Med. 48 (2007) 508–518.
[42] R. Srirajaskanthan, I. Kayani, A.M. Quigley, J. Soh, M.E. Caplin, J. Bomanji, The role of ^{68}Ga-DOTATATE PET in patients with neuroendocrine tumors and negative or equivocal findings on ^{111}In-DTPA-octreotide scintigraphy, J. Nucl. Med. 51 (2010) 875–881.

[43] D. Campana, V. Ambrosini, R. Pezzilli, S. Fanti, A.M. Labate, D. Santini, et al., Standardized uptake values of (68)Ga-DOTANOC PET: a promising prognostic tool in neuroendocrine tumors, J. Nucl. Med. 51 (2010) 353–359.

[44] V. Ambrosini, D. Campana, L. Bodei, C. Nanni, P. Castellucci, V. Allegri, et al., ^{68}Ga-DOTANOC PET/CT clinical impact in patients with neuroendocrine tumors, J. Nucl. Med. 51 (2010) 669.

[45] J. Kowalski, M. Henze, J. Schuhmacher, H.R. Mäcke, M. Hofmann, U. Haberkorn, Evaluation of positron emission tomography imaging using [^{68}Ga]-DOTA-D Phe(1)-Tyr(3)-Octreotide in comparison to [^{111}In]-DTPAOC SPECT. First results in patients with neuroendocrine tumors, Mol. Imaging Biol. 5 (2003) 42–48.

[46] M. Hofmann, H. Maecke, A. Borner, E. Weckesser, P. Schöffski, L. Oei, et al., Biokinetics and imaging with the somatostatin receptor PET radioligand (68)Ga-DOTATOC: preliminary data, Eur. J. Nucl. Med. 28 (2001) 1751–1757.

[47] M. Fani, F. Braun, B. Waser, K. Beetschen, R. Cescato, J. Erchegyi, et al., Unexpected sensitivity of sst2 antagonists to N-terminal radiometal modifications, J. Nucl. Med. 53 (2012) 1481–1489.

[48] X. Wang, M. Fani, S. Schulz, J. Rivier, J.C. Reubi, H.R. Maecke, Comprehensive evaluation of a somatostatin-based radiolabelled antagonist for diagnostic imaging and radionuclide therapy, Eur. J. Nucl. Med. Mol. Imaging 39 (2012) 1876–1885.

[49] C. Pettinato, A. Sarnelli, M. Di Donna, et al., 68Ga-DOTANOC: biodistribution and dosimetry in patients affected by neuroendocrine tumors, Eur. J. Nucl. Med. Mol. Imaging 35 (2008) 72–79.

[50] D. Putzer, M. Gabriel, B. Henninger, et al., Bone metastases in patients with neuroendocrine tumor: 68Ga-DOTA-Tyr3-octreotide PET in comparison to CT and bone scintigraphy, J. Nucl. Med. 50 (2009) 1214–1221.

[51] D. Wild, J.B. Bomanji, P. Benkert, H. Maecke, M.E. Caplin, P.J. Ell, Comparison of 68GaDOTA-NOC and 68Ga-DOTA-TATE PET/CT in the detection of GEP NETs, Eur. J. Nucl. Med. Mol. Imaging 36 (Suppl 2) (2009) S201.

[52] R. Cescato, B. Waser, M. Fani, J.C. Reubi, Evaluation of ^{177}Lu-DOTA-sst2 antagonist versus ^{177}Lu- DOTA-sst2 agonist binding in human cancers in vitro, J. Nucl. Med. 52 (2011) 1886–1890.

[53] D. Wild, M. Fani, M. Behe, I. Brink, J.E. Rivier, J.C. Reubi, et al., First clinical evidence that imaging with somatostatin receptor antagonists is feasible, J. Nucl. Med. 52 (2011) 1412–1417.

[54] J.E. Heuser, R.G. Anderson, Hypertonic media inhibit receptor-mediated endocytosis by blocking clathrin-coated pit formation, J. Cell. Biol. 108 (1989) 389–400.

[55] R. Lebtahi, J. Le Cloirec, C. Houzard, et al., Detection of neuroendocrine tumors: 99mTc-P829 scintigraphy compared with 111In-pentetreotide scintigraphy, J. Nucl. Med. 43 (2002) 889–895.

[56] T.J. Wadas, M. Eiblmaier, A. Zheleznyak, C.D. Sherman, R. Ferdani, K. Liang, et al., Preparation and biological evaluation of ^{64}Cu-CB-TE2A-sst2-ANT, a somatostatin antagonist for PET imaging of somatostatin receptor-positive tumors, J. Nucl. Med. 49 (2008) 1819–1827.

[57] K. Nguyen, J.J. Parry, B.E. Rogers, C.J. Anderson, Evaluation of copper-64-labeled somatostatin agonists and antagonist in SSTr2-transfected cell lines that are positive and negative for p53: implications for cancer therapy, Nucl. Med. Biol. 39 (2012) 187–197.

[58] R. Cescato, T. Maina, B. Nock, A. Nikolopoulou, D. Charalambidis, V. Piccand, et al., Bombesin receptor antagonists may be preferable to agonists for tumor targeting, J. Nucl. Med. 49 (2008) 318–326.

[59] J.C. Reubi, A.M. Landolt, High density of somatostatin receptors in pituitary tumors from acromegalic patients, J. Clin. Endocrinol. Metab. 59 (1984) 1148–1151.

[60] M. Bangard, M. Behe, S. Guhlke, et al., Detection of somatostatin receptor-positive tumours using the new 99mTc-tricine-HYNIC-D-Phe1-Tyr3-octreotide: first results in patients and comparison with 111In-DTPA-D-Phe1-octreotide, Eur. J. Nucl. Med. 27 (2000) 628–637.

[61] C. Decristoforo, T. Maina, B. Nock, M. Gabriel, P. Cordopatis, R. Moncayo, 99mTc-Demotate 1: first data in tumour patients-results of a pilot/phase I study, Eur. J. Nucl. Med. Mol. Imaging 30 (2003) 1211–1219.

[62] M. Béhé, H.R. Maecke, New somatostatin analogues labelled with technetium 99-m, Eur. J. Nucl. Med. 22 (1995) 791P.

[63] C. Decristoforo, S.J. Mather, W. Cholewinski, E. Donnemiller, G. Riccabona, R. Moncayo, 99mTc-EDDA/HYNIC-TOC: a new 99mTc-labelled radiopharmaceutical for imaging somatostatin receptor-positive tumours; first clinical results and intra-patient comparison with 111In-labelled octreotide derivatives, Eur. J. Nucl. Med. 27 (2000) 1318–1325.

[64] A. Hubalewska-Dydejczyk, J. Kulig, P. Szybinski, et al., Radio-guided surgery with the use of [99mTc-EDDA/HYNIC]octreotate in intra-operative detection of neuroendocrine tumours of the gastrointestinal tract, Eur. J. Nucl. Med. Mol. Imaging 34 (2007) 1545–1555.

[65] J.B. Cwikla, R. Mikolajczak, D. Pawlak, et al., Initial direct comparison of 99mTc-TOC and 99mTc-TATE in identifying sites of disease in patients with proven GEP NETs, J. Nucl. Med. 49 (2008) 1060–1065.

[66] A. Haug, C.J. Auernhammer, B. Wangler, et al., Intraindividual comparison of 68Ga-DOTA-TATE and 18F-DOPA PET in patients with well-differentiated metastatic neuroendocrine tumours, Eur. J. Nucl. Med. Mol. Imaging 36 (2009) 765–770.

[67] R. Mansi, X. Wang, F. Forrer, et al., Evaluation of a 1,4,7,10-tetraazacyclododecane-1,4,7,10-tetraacetic acid-conjugated bombesin-based radioantagonist for the labeling with single-photon emission computed tomography, positron emission tomography, and therapeutic radionuclides, Clin Cancer Res 15 (2009) 5240–5249.

[68] D. Wild, M. Frischknecht, Alpha- versus beta-particle radiopeptide therapy in a human prostate cancer model (^{213}Bi-DOTA-PESIN and ^{213}Bi-AMBA versus ^{177}Lu-DOTA-PESIN), Cancer Res. 71 (2011) 1009–1018.

[69] H. Zhang, J. Chen, C. Waldherr, et al., Synthesis and evaluation of bombesin derivatives on the basis of pan-bombesin peptides labeled with indium-111, lutetium-177, and yttrium-90 for targeting bombesin receptor-expressing tumors, Cancer Res. 64 (2004) 6707–6715.

[70] H. Hanaoka, T. Mukaic, H. Tamamura, et al., Development of a ^{111}In-labeled peptide derivative targeting a chemokine receptor, CXCR4, for imaging tumors, Nucl. Med. Biol. 33 (2006) 489–494.

[71] S. Nimmagadda, M. Pullambhatla, K. Stone, G. Green, Z.M. Bhujwalla, M.G. Pomper, Molecular imaging of CXCR4 receptor expression in human cancer xenografts with [^{64}Cu]AMD3100 positron emission tomography, Cancer Res. 70 (2010) 3935–3944.

[72] O. Jacobson, I.D. Weiss, D.O. Kiesewetter, J.M. Farber, X. Chen, PET of tumor CXCR4 expression with 4-F18-T140, J. Nucl. Med. 51 (2010) 1796–1804.

[73] O. Jacobson, I.D. Weiss, L.P. Szajek, G. Niu, Y. Ma, D.O. Kiesewetter, et al., Improvement of CXCR4 tracer specificity for PET imaging, J. Control Release 157 (2012) 216–223.

[74] E. Gourni, O. Demmer, M. Schottelius, C. D'Alessandria, S. Schulz, I. Dijkgraaf, et al., PET of CXCR4 expression by a (68)Ga-labeled highly specific targeted contrast agent, J. Nucl. Med. 52 (2011) 1803–1810.
[75] G.P.C. George, E. Stevens, O. Åberg, et al., Preclinical evaluation of a CXCR4-specific ^{68}Ga-labelled TN14003 derivative for cancer PET imaging, Bioorg Med Chem 22 (2014) 796–803.
[76] P. Laverman, L. Joosten, A. Eek, et al., Comparative biodistribution of 12 ^{111}In-labelled gastrin/CCK2 receptor-targeting peptides, Eur J Nucl Med Mol Imaging 38 (2011) 1410–1416.
[77] T.M. Behr, N. Jenner, M. Behe, et al., Radiolabeled peptides for targeting cholecystokinin-B/gastrin receptor-expressing tumors, J. Nucl. Med. 40 (1999) 1029–1044.
[78] M. Gotthardt, M.P. Behe, J. Grass, et al., Added value of gastrin receptor scintigraphy in comparison to somatostatin receptor scintigraphy in patients with carcinoids and other neuroendocrine tumours, Endocr. Relat. Cancer 13 (2006) 1203–1211.
[79] M. Gotthardt, M.P. Behe, D. Beuter, et al., Improved tumour detection by gastrin receptor scintigraphy in patients with metastasised medullary thyroid carcinoma, Eur. J. Nucl. Med. Mol. Imaging 33 (2006) 1273–1279.
[80] R. Haubner, W.A. Weber, A.J. Beer, et al., Noninvasive visualization of the activated alphavbeta3 integrin in cancer patients by positron emission tomography and [18F] Galacto-RGD, PLoS Med. 2 (2005) e70.
[81] A.J. Beer, R. Haubner, M. Sarbia, et al., Positron emission tomography using [18F] Galacto-RGD identifies the level of integrin alpha(v)beta3 expression in man, Clin. Cancer Res. 12 (2006) 3942–3949.
[82] A.J. Beer, R. Haubner, M. Goebel, et al., Biodistribution and pharmacokinetics of the alphavbeta3-selective tracer 18F-galacto-RGD in cancer patients, J. Nucl. Med. 46 (2005) 1333–1341.
[83] A.J. Beer, S. Lorenzen, S. Metz, et al., Comparison of integrin alphaVbeta3 expression and glucose metabolism in primary and metastatic lesions in cancer patients: a PET study using 18F-galacto-RGD and 18F-FDG, J. Nucl. Med. 49 (2008) 22–29.
[84] V. Ambrosini, P. Tomassetti, P. Castellucci, et al., Comparison between 68Ga-DOTA-NOC and 18F-DOPA PET for the detection of gastro-entero-pancreatic and lung neuro-endocrine tumours, Eur. J. Nucl. Med. Mol. Imaging 35 (2008) 1431–1438.
[85] C. von Falck, A.R. Boerner, M. Galanski, W.H. Knapp, Neuroendocrine tumour of the mediastinum: fusion of 18F-FDG and 68Ga-DOTATOC PET/CT datasets demonstrates different degrees of differentiation, Eur. J. Nucl. Med. Mol. Imaging 34 (2007) 812.
[86] S. Koukouraki, L.G. Strauss, V. Georgoulias, M. Eisenhut, U. Haberkorn, A. Dimitrakopoulou-Strauss, Comparison of the pharmacokinetics of 68Ga-DOTATOC and [18F]FDG in patients with metastatic neuroendocrine tumours scheduled for 90Y-DOTATOC therapy, Eur. J. Nucl. Med. Mol. Imaging 33 (2006) 1115–1122.
[87] T. Bach-Gansmo, R. Danielsson, A. Saracco, et al., Integrin receptor imaging of breast cancer: a proof-of-concept study to evaluate 99mTc-NC100692, J. Nucl. Med. 47 (2006) 1434–1439.
[88] L.M. Kenny, R.C. Coombes, I. Oulie, et al., Phase I trial of the positron-emitting Arg-Gly-Asp (RGD) peptide radioligand 18F-AH111585 in breast cancer patients, J. Nucl. Med. 49 (2008) 879–886.
[89] H.J. Cho, J.D. Lee, J.Y. Park, et al., First in human evaluation of a newly developed PET tracer, 18FRGD-K5 in patients with breast cancer: comparison with 18FFDG uptake pattern and microvessel density, J. Nucl. Med. 50 (Suppl 2) (2009) 1910.

[90] J. Zhang, Efficacy study of [18F]RGD-K5 positron emission tomography (PET) as a tool to monitor response to an antiangiogenic drug (K5-101), ClinicalTrialsgov, 2009 ed., US National Institutes of Health, 2009.

[91] R. Haubner, A.J. Beer, H. Wang, X. Chen, Positron emission tomography tracers for imaging angiogenesis, Eur. J. Nucl. Med. Mol. Imaging 37 (Suppl 1) (2010) S86−S103.

[92] J.K. Ramage, A.H. Davies, J. Ardill, et al., Guidelines for the management of gastroenteropancreatic neuroendocrine (including carcinoid) tumours, Gut 54 (Suppl 4) (2005) iv1-16.

[93] B. Waser, M.-L. Tamma, R. Cescato, R. Cescato, H.R. Maecke, J.C. Reubi, Highly efficient in vivo agonist-induced internalization of sst2 receptors in somatostatin target tissues, J. Nucl. Med. 50 (2009) 936−941.

[94] J.C. Schär, B. Waser, G. Mengod, J.C. Reubi, Somatostatin receptor subtypes sst1, sst2, sst3 and sst5 expression in human pituitary, gastroentero-pancreatic and mammary tumors: comparison of mRNA analysis with receptor autoradiography, Int. J. Cancer 70 (1997) 530−537.

[95] A. Sundin, U. Garske, H. Orlefors, Nuclear imaging of neuroendocrine tumours, Best Pract. Res. Clin. Endocrinol. Metab. 21 (2007) 69−85.

[96] E.P. Krenning, D.J. Kwekkeboom, W.H. Bakker, et al., Somatostatin receptor scintigraphy with [111In-DTPA-D-Phe1]- and [123I-Tyr3]-octreotide: the Rotterdam experience with more than 1000 patients, Eur. J. Nucl. Med. 20 (1993) 716−731.

[97] A. Asnacios, F. Courbon, P. Rochaix, et al., Indium-111-pentetreotide scintigraphy and somatostatin receptor subtype 2 expression: new prognostic factors for malignant well-differentiated endocrine tumors, J. Clin. Oncol. 26 (2008) 963−970.

[98] S.J. Goldsmith, Update on nuclear medicine imaging of neuroendocrine tumors, Future Oncol. 5 (2009) 75−84.

[99] T. Maina, B. Stolz, R. Albert, C. Bruns, P. Koch, H. Macke, Synthesis, radiochemistry and biological evaluation of a new somatostatin analogue (SDZ 219-387) labelled with technetium-99m, Eur. J. Nucl. Med. 21 (1994) 437−444.

[100] A. Hubalewska-Dydejczyk, K. Fross-Baron, R. Mikolajczak, et al., 99mTc-EDDA/HYNIC-octreotate scintigraphy, an efficient method for the detection and staging of carcinoid tumours: results of 3 years' experience, Eur. J. Nucl. Med. Mol. Imaging 33 (2006) 1123−1133.

[101] V. Prasad, V. Ambrosini, M. Hommann, D. Hoersch, S. Fanti, R.P. Baum, Detection of unknown primary neuroendocrine tumours (CUP-NET) using (68)Ga-DOTA-NOC receptor PET/CT, Eur. J. Nucl. Med. Mol. Imaging 37 (2010) 67−77.

[102] I. Kayani, J.B. Bomanji, A. Groves, et al., Functional imaging of neuroendocrine tumors with combined PET/CT using 68Ga-DOTATATE (DOTA-DPhe1,Tyr3-octreotate) and 18F-FDG, Cancer 112 (2008) 2447−2455.

[103] R. Mansi, A. Fleischmann, H.R. Mäcke, J.,C. Reubi, Targeting GRPR in urological cancers—from basic research to clinical application, Nat. Rev. Urol 10 (2013) 235−244.

[104] J.J. Parry, T.S. Kelly, R. Andrews, B.E. Rogers, In vitro and in vivo evaluation of 64Cu-labeled DOTA-linker-bombesin(7-14) analogues containing different amino acid linker moieties, Bioconjug. Chem. 18 (2007) 1110−1117.

[105] J.C. Garrison, T.L. Rold, G.L. Sieckman, et al., In vivo evaluation and small-animal PET/CT of a prostate cancer mouse model using 64Cu bombesin analogs: side-by-side comparison of the CB-TE2A and DOTA chelation systems, J. Nucl. Med. 48 (2007) 1327−1337.

[106] Z. Liu, Y. Yan, S. Liu, F. Wang, X. Chen, ^{18}F, ^{64}Cu, and ^{68}Ga labeled RGD-bombesin heterodimeric peptides for PET imaging of breast cancer, Bioconjugate Chem 20 (2009) 1016–1025.
[107] K.A. Lears, R. Ferdani, K. Liang, et al., In vitro and in vivo evaluation of 64Cu-labeled SarAr-bombesin analogs in gastrin-releasing peptide receptor-expressing prostate cancer, J. Nucl. Med. 52 (2011) 470–477.
[108] M. Honer, L. Mu, T. Stellfeld, K. Graham, M. Martic, C.R. Fischer, et al., ^{18}F-labeled bombesin analog for specific and effective targeting of prostate tumors expressing gastrin-releasing peptide receptors, J. Nucl. Med. 52 (2011) 270–278.
[109] A. Dimitrakopoulou-Strauss, M. Seiz, J. Tuettenberg, et al., Pharmacokinetic studies of Ga-labeled bombesin (Ga-BZH) and F-18 FDG PET in patients with recurrent gliomas and comparison to grading: preliminary results, Clin. Nucl. Med. 36 (2011) 101–108.
[110] G. Gornik, R. Mansi, K. Abiraj, et al., Evaluation of the GRPR radioantagonist 64Cu-CB-TE2A-AR-06 in mice and men, J. Nucl. Med. 52 (Suppl 1) (2011) 7P.
[111] R.A. De Silva, K. Peyre, M. Pullambhatla, J.J. Fox, M.G. Pomper, S. Nimmagadda, Imaging CXCR4 expression in human cancer xenografts: evaluation of monocyclam ^{64}Cu-AMD3465, J. Nucl.Med. 52 (2011) 986–993.
[112] T. Buckle, N.S. van den Berg, J. Kuil, A. Bunschoten, J. Oldenburg, A.D. Borowsky, et al., Non-invasive longitudinal imaging of tumor progression using an (111)indium labeled CXCR4 peptide antagonist, Am. J. Nucl. Med. Mol. Imaging 2 (2012) 99–109.
[113] J.C. Reubi, J.C. Schaer, B. Waser, Cholecystokinin(CCK)-A and CCK-B/gastrin receptors in human tumors, Cancer Res. 57 (1997) 1377–1386.
[114] T.M. Behr, M.P. Behe, Cholecystokinin-B/gastrin receptor-targeting peptides for staging and therapy of medullary thyroid cancer and other cholecystokinin-B receptor-expressing malignancies, Semin. Nucl. Med. 32 (2002) 97–109.
[115] A.J. Beer, A.L. Grosu, J. Carlsen, et al., [18F]galacto-RGD positron emission tomography for imaging of alphavbeta3 expression on the neovasculature in patients with squamous cell carcinoma of the head and neck, Clin. Cancer Res. 13 (2007) 6610–6616.
[116] O. Schnell, B. Krebs, J. Carlsen, et al., Imaging of integrin alpha(v)beta(3) expression in patients with malignant glioma by [18F] Galacto-RGD positron emission tomography, Neuro. Oncol. 11 (2009) 861–870.
[117] J. Winick, A proof-of-concept study to assess the ability of [18F]AH-111585 PET imaging to detect tumours and angiogenesis, ClinicalTrialsgov, 2007 ed., US National Institutes of Health, 2007.

Further reading

J.M. Palacios, R. Cortes, M. Dietl, A. Probst, Receptors in Human Brain Diseases: A use for Receptor Autoradiography in Neuropathology, J. Receptor Res 8 (1988) 509–520.

Peptide surfactants in membrane protein purification and stabilization

18

Shuguang Zhang[1] *and Karolina Corin*[2,3,4,5]
[1]Center for Bits and Atoms, Massachusetts Institute of Technology, Cambridge, MA, United States, [2]Biomedical Engineering Research Group in the School of Electrical and Information Engineering, University of the Witwatersrand, Johannesburg, South Africa, [3]Department of Molecular Medicine and Haematology, University of the Witwatersrand, Johannesburg, South Africa, [4]Department of Chemistry and Biochemistry, UCLA, Los Angeles, CA, United States, [5]Molecular Biology Institute, UCLA, Los Angeles, CA, United States

18.1 Introduction

Membrane proteins play vital roles in numerous biological processes, including sight, smell, taste, inflammation, cell signaling, cell migration, and ion transport. They account for ~30% of all proteins [1], are the targets of ~50% of pharmaceutical drugs [2,3], and have potential applications in the development of biosensors. In spite of their large numbers and importance, relatively little is known about their structure and function. Indeed, as of April 2017, of the 120,154 total protein structures deposited in the Protein Data Bank [4] (www.rcsb.org), only 686, or 0.6%, are unique membrane proteins [5].

This paucity of information is, in large part, due to the difficulty of solubilizing and stabilizing membrane proteins in aqueous solution. While chemical detergents have traditionally been used for this task, they present an array of problems. Extensive screens must usually be performed to find a suitable detergent, and the optimal detergent may depend on the specific application. Moreover, there is typically a trade-off between the ability of a detergent to stabilize a protein, and its ability to facilitate crystallization [6]. Unfortunately, both characteristics are required, as protein crystals can take weeks to grow.

The development of novel peptide surfactants offers potential solutions to the problems inherent with traditional detergents. These compounds can be designed to have desired properties, and promise to facilitate both protein stability and crystallization. Moreover, they show promise for numerous other applications as well. Four general types of peptide surfactants have been designed and studied: amphipathic helical peptides, lipopeptide complexes, short self-assembling lipid-like peptides, and beta-sheet peptides. While each type of peptide surfactant is designed with a hydrophobic region and a hydrophilic region, the implementation of this design varies greatly. This chapter discusses the specific design strategies for each type of peptide surfactant, as well as their properties, effects on membrane proteins, and potential applications.

Peptide Applications in Biomedicine, Biotechnology and Bioengineering.
DOI: http://dx.doi.org/10.1016/B978-0-08-100736-5.00020-X

18.2 Amphipathic helical peptide design

The idea of using peptides as surfactants in membrane protein studies was first proposed in 1993 [7]. PD_1, the first designed peptide detergent, was comprised of 24 residues in the sequence Ac-EELLKQALQQAQQLLQQAQELAKK-$CONH_2$. PD_1 was designed to fold into a helix. The residues were arranged to form a flat hydrophobic surface along the length of the helix with alanines in the middle and leucines extending from the sides (Fig. 18.1A). This hydrophobic face was sufficiently long to span the transmembrane region of a protein. To stabilize the termini and neutralize any dipole effects, glutamic acid and lysine were added to the termini to form salt bridges at positions i and $i+4$. To further stabilize the helix, the N-terminus was capped by acetylation, the C-terminus was capped by amidation, and an additional glutamic acid and lysine were added to the termini. Glutamine was primarily used for the hydrophilic surface because it has a high propensity to form helices.

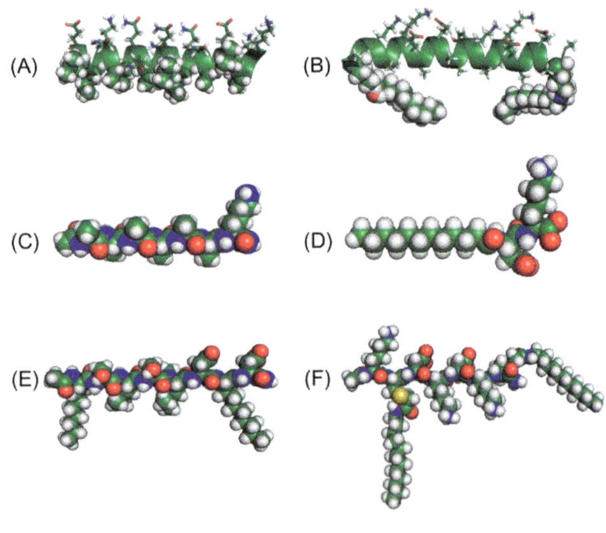

Figure 18.1 Proposed molecular models of surfactant peptides. (A) PD_1. Hydrophobic residues are shown as spheres and hydrophilic residues are shown as sticks. (B) LPD-12. This model is a rendering of a portion of the crystal structure from PDB entry 3CAY. (C) The short lipid-like peptide Ac-AAAAAAK-$CONH_2$ (A_6K). (D) The ultra-short lipopeptide C12DK. (E) The beta-sheet peptide BP-1. (F) The peptide gemini surfactant $DKDKC_{12}K$. Images were generated with PyMOL (*The PyMOL Molecular Graphics System, Version 1.8 Schrödinger, LLC*, https://www.pymol.org).

18.3 Lipopeptide detergent design

18.3.1 Cylindrical lipopeptides

Lipopeptide detergents (LPDs) are surfactants composed of both peptides and fatty acids. The first LPDs were developed in 2003 using the design of PD_1 as a basic template [8]. Like PD_1, an amphiphilic helical peptide sufficiently long to span the transmembrane region was incorporated into the surfactant design. However, unlike PD_1, these LPDs incorporate two alkyl chains using the peptide helix as a backbone, and it is these chains that facilitate both the formation of a helix and the self-assembly of LPD monomers [8–11] (Fig. 18.1B). The 25-residue backbone has the sequence Ac-AOAEAAEKAAKYAAEAAEKAAKAOA-NH_2, and uses ornithines at positions 2 and 24 as the sites of fatty acid coupling. Ornithines were chosen as the coupling residues because they may facilitate more favorable alkyl chain packing. The remaining design principles closely match those of PD_1. The parent peptide has a polar face designed to interact with aqueous solutions, and a hydrophobic face composed mainly of alanines. Residues with helix-forming propensities were chosen, and the N- and C-termini were capped to further assist with helix formation. Although bulky and not as hydrophilic as the other polar residues, the aromatic residue tyrosine was included for quantification by absorbance.

18.3.2 Ultra-short lipopeptides

Because it is expensive to produce LPDs with a 25-residue peptide head-group, subsequent designs used only 1–2 amino acids [12,13] (Fig. 18.1D). A single fatty acid chain was coupled to the N-terminus of this head group via an amide bond. Although these ultra-short LPDs are much simpler than their precursors, they can still be designed with a wide variety of physicochemical properties: the head groups can be designed to exhibit a wide variety of shapes, sizes, and charges, while different fatty acids can be used to manipulate the critical micelle concentration and hydrophobic character of the surfactant. Moreover, only three reaction steps are needed to produce these LPDs [13].

18.3.3 Peptide gemini surfactants

Peptide gemini surfactants (PGs) were inspired by previous successes using peptides for membrane protein studies, and were first tested on membrane proteins in 2013 [14,15]. PGs have a hydrophilic peptide head group comprised of 4–5 residues sandwiched between two cysteine residues. An alkyl chain is attached to each cysteine, and an N-terminal hydrophilic amino acid is often added to increase surfactant solubility (Fig. 18.1F). The peptide headgroup allows surfactants with a variety of different architectures and structures to be designed [14–16].

18.4 Short self-assembling lipid-like peptide design

Short lipid-like peptides are designed to mimic natural lipids. These peptides are ~2–3 nm in length, with a hydrophilic head group composed of 1–2 charged amino acids and a hydrophobic tail comprised of 3–6 hydrophobic residues (Fig. 18.1C). In aqueous solution, they self-assemble into organized nanostructures to sequester their hydrophobic tails from contact with water [17–19].

18.5 Beta-sheet peptide design

Beta-sheet peptides have recently been developed for membrane protein studies [20]. These peptides consist of the sequence acetyl-GSLSLDGD-NH_2 and employ some of the same design principles as the LPDs and short lipid-like peptides: the eight-residue sequence is sufficiently long to span the transmembrane section of a protein as a beta strand, and is coupled to alkyl side chains through the glycines at each end (Fig. 18.1E). The alternating polar and nonpolar residues facilitate beta-sheet formation via alternating hydrophobic and hydrophilic faces. *N*-methyl amino acids were incorporated to increase peptide solubility. Without these special residues, strong hydrogen-bond interactions between the peptides caused them to form an insoluble aggregate. The intent behind this design was to stabilize membrane proteins via a beta-barrel-like structure, as well as to minimize the dynamic instability in typical detergent micelles by the formation of hydrogen bonds.

18.6 Design principles and considerations

The idea to use peptides as surfactants was inspired by natural amphipathic helices that self-assemble into coiled-coil and helix-bundle structures in order to bury their hydrophobic regions. PD_1 was engineered to mimic key features of these natural proteins. It was designed to fold into an amphipathic helix with a hydrophobic side capable of interacting with the hydrophobic domains of membrane proteins. This hydrophobic face was long enough to cover the transmembrane region of a protein. Interdigitating leucines protruding from the sides of the hydrophobic surface were designed to allow PD_1 monomers to interact with themselves and to form higher-order structures, perhaps akin to micelles.

Although cylindrical LPDs utilize similar principles incorporated into the design of PD_1, the presence of fatty acid chains significantly alters their behavior. The unique design of cylindrical LPD monomers allows them to self-assemble into cylindrical micelles which are more stable than detergent micelles and more similar to a lipid bilayer. Traditional detergents self-associate into micelles with a roughly spherical or elliptical shape. Molecular dynamics simulations indicate that the alkyl tails pack in a compact yet disordered manner, and that many are likely exposed to the solvent. This packing conformation causes the micelle surface to be uneven, and may contribute to the instability of proteins [21]. In contrast, molecular modeling shows that cylindrical LPDs assemble into compact cylindrical micelles, and that such micelles are more efficient at packing the alkyl chains and shielding them from the bulk solvent [8,10]. The more uniform alkyl chain packing predicted for cylindrical LPD micelles is reminiscent of the packing in a lipid bilayer. Because of this similarity in structure, these micelles may provide a more stable environment for membrane proteins than traditional detergents. Moreover, because of their ability to form relatively small micelles, cylindrical LPDs may facilitate crystal formation.

In contrast to cylindrical LPDs, ultra-short LPDs were designed to mimic traditional surfactants and form roughly spherical micelles where the alkyl tails help solubilize the hydrophobic regions of integral membrane proteins. Because they are comprised of both lipids and peptides, they may provide an environment that more closely resembles the lipid bilayer relative to traditional detergents.

While cylindrical and ultra-short LPDs are designed to exhibit specific behavior, PGs can be designed to have conformations tailored to specific applications [16]. These potential architectures arise because the peptide head group is sufficiently long for PGs to form intra- and intermolecular bonds. Indeed, bilayers have been observed with some PGs, while others have been designed to form beta turns [15,16,22].

Each LPD design has many potential configurations and can be optimized for a given protein or application. For example, fatty acyl groups with 12–20 carbons have been attached and studied [8,13]. Head groups with various compositions have also been synthesized. Studies with ultra-short LPDs suggest that head groups with a larger size and more even charge distribution are better able to solubilize membrane proteins. Within a series of surfactants with the same tail length, the bulkier head groups DD and DK consistently solubilized more Photosystem I (PS-I) than the smaller head groups GD and GK. Moreover, the head group DK consistently solubilized more protein than DD or KK. This was attributed to the more even charge distribution of the DK moiety relative to the other head groups [13] (Fig. 18.2A). Studies with PGs suggest that the peptide sequence or secondary structure can influence the surfactant behavior. While several different PGs could keep PS-I and Photosystem II (PS-II) soluble and functional, only PGs with peptide heads designed to form beta turns were able to extract these proteins from lipid membranes [14,16]. Also, when a crosslink was introduced into the headgroup, PGs were able to form spherical bilayers capable of incorporating membrane proteins [15]. Finally, it has also been postulated that an aromatic group near the head group–alkyl tail interface can help stabilize membrane proteins [12].

Short lipid-like peptides were designed to behave like natural lipids while conferring advantages over traditional detergents. Traditional detergents and lipids pack together by sequestering their tails from contact with water. In addition to exhibiting these hydrophobic interactions, short lipid-like peptides can form protein–protein contacts and backbone hydrogen bonds, both between themselves and the target protein. Because of these extra interactions, short lipid-like peptides may provide a milder and more stable environment than traditional surfactants.

While it is likely that some receptors are naturally more amenable to experimental studies [23,25], peptide design can affect the expression, solubility, or stability

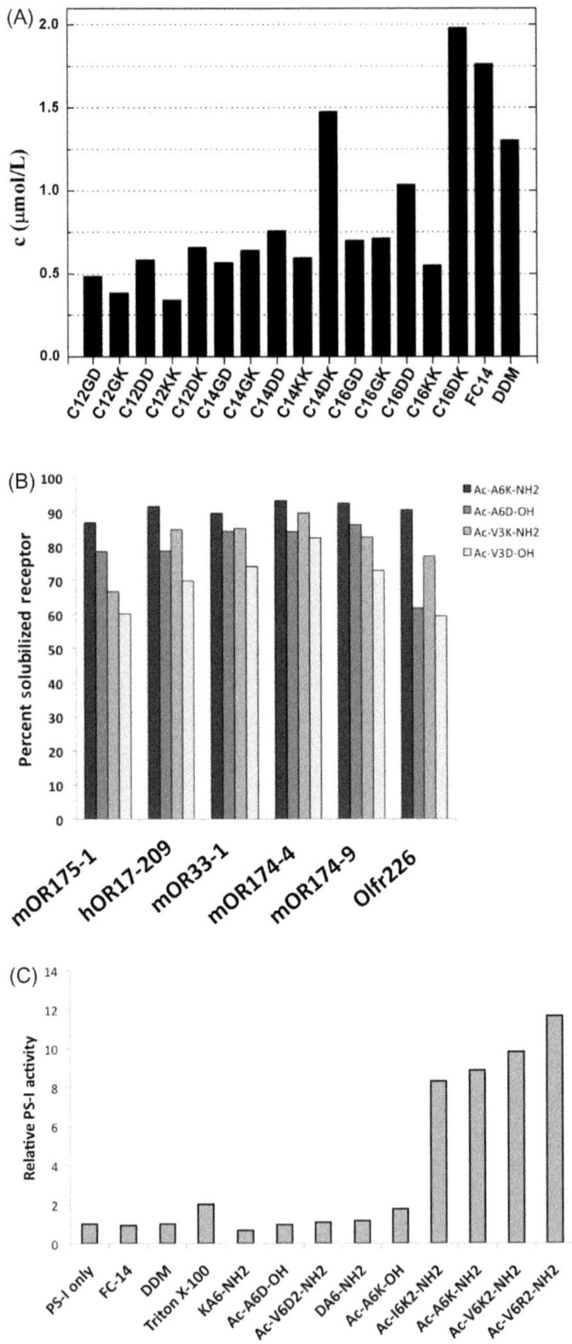

Figure 18.2 Surfactant peptide properties affect membrane protein solubility and activity. (A) Ultra-short LPDs with longer tails generally solubilized more Photosystem-I. Higher solubility

of a given receptor. The head group charge, charge distribution, tail length, and tail composition are all parameters that must be considered. The heterogeneity of a peptide solution (i.e., a single type of peptide vs a mix of several kinds of peptides) can also be a critical parameter.

Head group charge has consistently been shown to affect membrane protein behavior. Two studies investigating 14 GPCRs and 12 peptide surfactants found that the receptors were generally more soluble in peptides with positively charged head groups compared to their negatively charged counterparts [23,25] (Fig. 18.2B). A clear trend was not as obvious when total receptor expression was measured. However, some receptors like hOR17−10 and hFPR3 seemed to have higher yields in peptides with a negatively charged head group, while others like hTAAR5 and mOR103−15 had better yields in peptides with positively charged head groups [25]. Interestingly, the highest expression levels for most of the receptors were obtained using the peptide Ac-A_6D-COOH (A_6D), suggesting that some peptides may be more universal. Although extensive peptide screens were not performed with PS-I, it was again observed that the head group charge affected receptor stability. Indeed, a C-terminal positively charged head group was necessary to functionally stabilize PS-I [24,26]. Moreover, this configuration significantly enhanced the activity of PS-I. No increase in activity was seen when the head group was zwitterionic or anionic [24] (Fig. 18.2C).

The location of the head group can significantly affect membrane protein solubility, expression, or stability. Amongst eight different olfactory-related GPCRs, higher expression was consistently observed when the negative charge was on the C-terminus compared to the N-terminus (e.g., A_6D vs DA_6-CONH_2 [DA_6]) [23,25]. This trend was also often observed when the positive charge was on the C-terminus (e.g., Ac-A_6K-CONH_2 [A_6K] vs KA_6-CONH_2 [KA_6]). Interestingly, changing the location of the charge typically altered which type of charge was favored by the receptor. When the head group was on the C-terminus, most of the receptors had higher expression levels in the negatively charged peptide. When the head group was on the N-terminus, most receptors had higher expression in the positively charged peptide. It should be noted that A_6D has one more negative charge than

◀ was also attributed to larger headgroups with a more even charge distribution. (B) Short lipid-like peptides with a positive charge and longer tail generally solubilized more receptor than their negatively charged or shorter counterparts. (C) A C-terminal positive charge significantly enhances Photosystem-I activity. Surfactant peptides yielded better results than the traditional detergents FC-14, Triton-X, and DDM.
Source: (A) Adapted with permission from X. Wang, et al., Solubilization and stabilization of isolated photosystem I complex with lipopeptide detergents, PLoS One 8 (9) (2013) e76256 [13]; (B) Adapted with permission from K. Corin, et al., Designer lipid-like peptides: a class of detergents for studying functional olfactory receptors using commercial cell-free systems, PLoS One 6 (11) (2011) e25067 [23] (C) Adapted with permission from K. Matsumoto, M. Vaughn, B.D. Bruce, S. Koutsopoulos, S. Zhang, Designer peptide surfactants stabilize functional Photosystem-I membrane complex in aqueous solution for extended time, J. Phys. Chem. B. 113 (1) (2009) 75−83 [24], Copyright 2009 American Chemical Society.

DA_6, and that KA_6 has one more positive charge than A_6K. Thus, the effects of charge localization and total charge have not been completely isolated from one another. The charge distribution did not seem to have as much of an effect on receptor solubility. While solubility often decreased when the charge was on the N-terminus, most receptors were more soluble in the positively charged peptides, regardless of the head group location. As with the olfactory-related GPCRs, the amino acid sequence affected the stability of PS-I. While A_6K was able to increase the activity of PS-I ~9 times over that of PS-I alone, the peptide KA_6 had no significant effect on PS-I activity [24] (Fig. 18.2C).

Peptide tail length and composition can also affect membrane protein solubility and expression, but are unlikely to do so independent of the peptide head group charge and location. For example, Ac-I_5K-$CONH_2$ (I_5K) thermostabilized PS-I better than Ac-I_4K-$CONH_2$ and Ac-I_3K-$CONH_2$ (I_4K and I_3K, respectively). Likewise, Ac-I_6K_2-$CONH_2$ (I_6K_2) performed better than Ac-I_5K_2-$CONH_2$ (I_5K_2) and Ac-I_4K_2-$CONH_2$ (I_4K_2). An optimal number of head group lysines were required to maximize PS-I thermostability: I_5K_2 performed better than I_5K or Ac-I_5K_3-$CONH_2$ (I_5K_3) [27]. In a screen of 12 olfactory receptors, seven-residue peptides consistently solubilized a higher receptor fraction than four-residue peptides [23]. However, the positively charged peptides often performed better than the negatively charged ones (Fig. 18.2B). In a separate GPCR screen, four-residue peptides often solubilized a similar fraction of receptor as the seven-residue peptides [25]. Interestingly, although A_6K was often better at solubilizing the receptors, A_6D typically resulted in higher expression compared to the four-residue peptides. This trend was primarily observed among olfactory receptors; other GPCRs often had higher yields in the four-residue peptides [23,25]. In PS-I and in the GPCR screens, the hydrophobicity of the tail residues did not seem to affect protein solubility, stability, or expression [23–25].

Most membrane protein studies have been performed using a single lipid-like peptide with a tail composed of a repeating amino acid. However, solitary experiments suggest that it may be advantageous to use peptide mixtures, or to design tails composed of several amino acids. When A_6K or A_6D were used to extract GlpD from *Escherichia coli* membranes, they were able to extract ~25% of the protein when used alone, and ~60% of the protein when used as a mixture in a 60:40 ratio [28]. This mixture was also able to functionally stabilize GlpD and the enzyme Npx. While A_6D alone could stabilize Npx, the A_6K:A_6D mixture did so to a greater extent [28]. Only one study has examined peptides whose tails are composed of several amino acids [25]. The peptides Ac-GAVILEE-COOH (GAVILEE) and Ac-GAVILRR-NH_2 (GAVILRR) were designed to have a conical shape, and were screened with four GPCRs. Most of the GPCRs were less soluble and had lower yields in these peptides than in peptides with uniform tails. Because the head groups had two arginines or glutamic acids instead of one lysine or aspartic acid, it is not clear whether the poorer performance of these peptides is due to the tail or head composition. However, the highly increased activity of PS-I in Ac-V_6R_2-$CONH_2$ (V_6R_2) suggests that the tail shape or composition is the prime factor [24]. Indeed, the conical tails seem to drive peptide self-assembly into structures that are unique and morphologically distinct from structures formed from peptides with uniform tails [29].

18.7 Morphological, chemical, and physical properties of surfactant peptides

The crystal structure of PD_1 proved that it self-assembled as designed [7] (Fig. 18.3A). The peptide formed an antiparallel four-helix bundle where the four-helix bundles self-assembled to sequester a hydrophobic core from the aqueous solvent. Helices I and II interacted through their flat hydrophobic surfaces, as did helices III and IV. Helices I and IV interacted through the leucines on the sides of their hydrophobic surfaces, as did helices II and III. Although attempts were made to stabilize the helices via salt bridges, no evidence of such bridges could be seen in the crystal structure [7]. Interestingly, circular dichroism spectra showed that the peptide's structure increased in helical content in the presence of sodium dodecyl sulfate (SDS) and 1-stearoyl-2-oleoyl-glycero-3-phophocholine (SOPC) vesicles, presumably due to their charged nature [30]. Moreover, it was not possible to measure a critical micelle concentration for PD_1. While it was suggested that PD_1 should not be viewed as a typical detergent [30], the apparent lack of a critical micelle concentration (CMC) could be due to a low aggregation number, as evidenced by the crystal structure.

In direct contrast to PD_1, most LPDs form distinct micelles. The micelles formed from ultra-short LPDs have diameters that are ~2–3 nm [12], and critical aggregate concentrations (CACs) that typically range from ~0.03 to 5 mM and decrease by an order of magnitude for every two methylenes added to the tail of a series with a common head group [12,13]. The micelles formed from PGs display similar properties, with CACs ranging from ~0.003 to 11 mM and diameters ~5–6 nm. However, they can also form larger aggregates with diameters ~60–240 nm [14–16,22], as well as bilayer assemblies. The micelles formed from cylindrical LPDs exhibit strikingly different behavior: they are quite compact with small aggregation numbers [8–10]. This observation was consistent for alkyl chain lengths between 12 and 20 carbons, where ~8–20 monomers per micelle were estimated both computationally and experimentally [8,10]. Interestingly, free LPD monomers could not be measured by several techniques. This suggested that the entire series of cylindrical LPDs had a low CAC, which was estimated to be below 1 μM [8].

The crystal structure of LPD-12 verified that cylindrical LPDs assembled according to their intended design [9] (Fig. 18.3B). As predicted, eight monomers comprised each micelle. Interestingly, the structure showed that LPD monomers prefer to assemble into Alacoil antiparallel dimers. These dimers constitute the basic subunit from which micelles are formed: four pairs of dimers self-assemble into a cylindrical micelle with the acyl chains sequestered in the core. *n*-Dodecyl-β-D-maltopyranoside (DDM) was required to obtain a crystal, and the detergent molecules could be seen along the interface between dimers with their acyl chains extending into the hydrophobic interior. Because LPDs are capable of forming micelles without the presence of detergent [8–10], the micelles may assume a slightly different conformation in the absence of DDM.

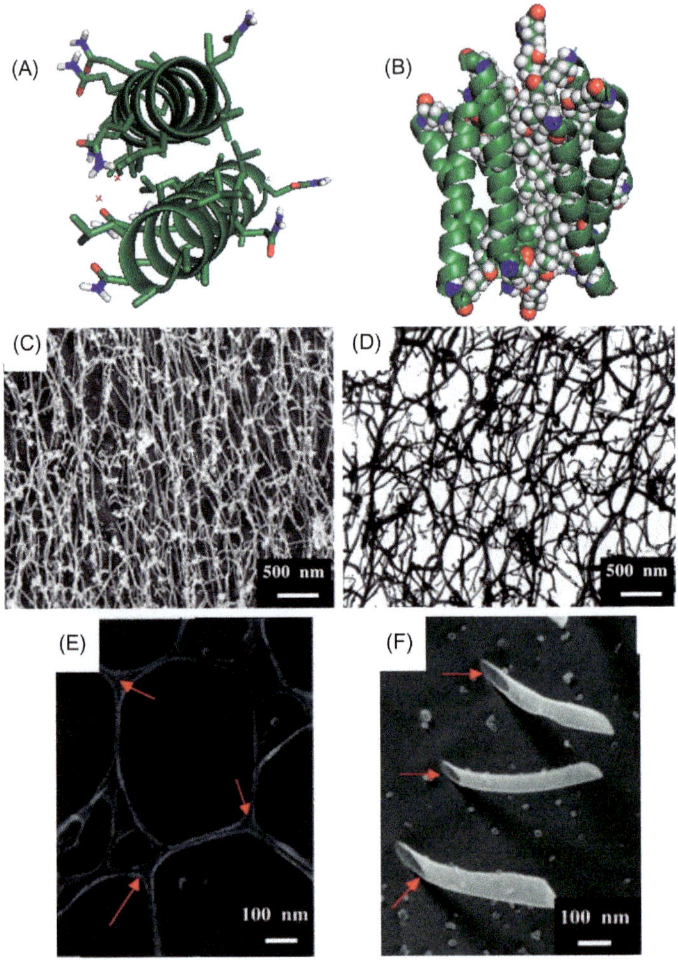

Figure 18.3 Morphological and physical properties of surfactant peptides. (A) Structure of two PD_1 monomers interacting through their hydrophobic faces (pdb 4HB1). (B) Structure of an LPD-12 micelle (pdb 3CAY). Eight LPD-12 molecules self-assemble into a cylindrical conformation with their alkyl tails sequestered in the core and the peptide helices facing the aqueous solution. Each micelle is comprised of four antiparallel dimer pairs. (C) Quick-freeze/deep-etch TEM image of A_6D. (D) Quick-freeze/deep-etch TEM image of V_6D. (E) Higher magnification of A_6D showing junctions. (F) Higher magnification of V_6D showing hollow tubes.
Source: (C–F) Adapted with permission from S. Vauthey, S. Santoso, H. Gong, N. Watson, S. Zhang, Molecular self-assembly of surfactant-like peptides to form nanotubes and nanovesicles, Proc. Natl. Acad. Sci. U. S. A. 99 (8) (2002) 5355–5360 [17], Copyright 2002 National Academy of Sciences.

Alkyl tails are critical for the detergent-like behavior of cylindrical LPDs. While LPDs with varying tail lengths could disrupt phosphatidylcholine liposomes to create mixed micelles, a control peptide with the same backbone as the LPDs, but without any alkyl chains (C-0), was unable to associate with or disrupt liposomes. Moreover, while cylindrical LPDs had circular dichroism spectra typical of alpha helices, the C-0 spectrum was reminiscent of a random coil. Taken together, this suggests that the alkyl tails are necessary for the peptide to properly fold into a helix, and that they are also necessary to confer surfactant-like properties to the LPDs [8].

The solubility of an LPD species decreases as the tail length is increased. Amongst cylindrical LPDs, the solubility decreases from >40 to >2 mM as the number of carbons in the tail is increased from 12 to 20 [8]. While the same trend is observed with traditional detergents, the solubility of cylindrical LPDs is far less dependent on the alkyl chain length. Indeed, the solubility of traditional detergents can decrease several orders of magnitude over a similar or smaller range of tail lengths. While the solubility of all ultra-short LPDs was not directly measured, it was observed that the longer-chained C16KK surfactant tended to precipitate, while the shorter-chained versions did not [13].

The head group composition of PGs likewise affects their solubility. PGs with a uniform aspartic acid head group were only soluble in basic solutions if there were at least two residues [22]. Similarly, PGs with repeating arginine, lysine, or serine residues were insoluble at neutral pH [14]. Addition of 1–2 hydrophilic K or D residues on the N-terminus was required to sufficiently solubilize them [14,16].

Molecular modeling and experimental evidence suggest that short self-assembling lipid-like peptides do indeed display behavior characteristic of phospholipids and commercial detergents. Namely, these peptides pack together tail-to-tail [31], and form ordered structures above a CAC [32,33].

Several studies demonstrated that the peptides formed hydrophobic regions when they self-assembled [33,34]. TEM images from initial experiments showing hollow nanotubes suggested that the peptides formed bilayers, with the hydrophilic heads forming two hydrophobic surfaces shielding a hydrophobic core [17–19]. This was proven experimentally [31], demonstrating that peptide self-assembly is driven by hydrophobic interactions. However, unlike the bilayer formed by traditional surfactants, in which the alkyl tails pack together so tightly as to exclude water, the peptide bilayer showed evidence of defects and the presence of vesicles. This may be due to hydrogen bonding with the peptide backbones [31,34].

The hydrophobic interactions driving peptide self-assembly were only seen above a critical concentration. The CACs of the peptides ranged from ~0.1 to 2 mM [24,32,33], but depended on the buffer in which the measurements were made. Lower CACs were measured in PBS than in water, presumably due to charge-shielding effects that allowed the peptide head groups to associate more easily [32]. Moreover, the CAC decreased as the hydrophobicity of the peptide increased [33]. This is expected, as the increased hydrophobicity of the tail lowers the energy required to overcome electrostatic repulsion of the head groups.

In spite of their diverse sequences, short self-assembling lipid-like peptides tended to form remarkably similar morphological structures. Networks of hollow nanotubes, micelles, and nanovesicles were consistently imaged for both cationic and anionic peptides with tail lengths ranging from 4 to 10 residues [17–19,32,34] (Fig. 18.3C–F). Experiments suggested that there might be an optimal tail length for vesicle formation [18], and that nanotubes might generally be the preferred self-assembled structure. Right-handed helical twists could be observed in the nanotubes, as well as branching and three-way junctions [17–19]. Dynamic light scattering showed that most structures were 30–50 nm in diameter, but that a small fraction of the peptides had structures 100–200 nm in diameter. Experiments with the peptide series Ac-G_nD_2-COOH ($n = 4, 6, 8, 10$) suggested that the fraction of structures with 100–200 nm diameters increased as the tail length increased [18]. In contrast, Ac-A_9K-$CONH_2$ (A_9K) formed smaller and more compact structures than A_6K [34]. While most studies measured nanofiber diameters ~30–50 nm in diameter, one study measured nanofibers and rods ~5–8 nm in diameter [34]. The reason for this difference is unclear, but could be related to the experimental peptide concentration: longer and wider nanotubes have been observed with increasing peptide concentrations [32]. The number of bulges visualized on nanotubes also increased with the peptide concentration [32].

Although less common, membrane structures were observed in addition to nanovesicles, nanotubes, and micelles. These structures seemed to form at longer peptide tail lengths, and could be seen connecting neighboring nanotubes [18]. The authors postulated that the emergence of membranes at longer tail lengths is expected, as the tight packing of the hydrophobic tail regions would preclude a curvature conducive to the formation of nanotubes and nanovesicles. The solution pH also likely affects whether membrane structures formed. TEM images suggested that cationic peptides formed membranous sheets at a pH above their pI, and nanotubes, micelles, and nanovesicles at a pH below their pI [19].

Interestingly, the self-assembled peptide structures are dynamic. DLS experiments showed that the peptide size distributions changed over time [17]. TEM images seemed to confirm this dynamic nature. Evidence of vesicle budding or fusion could be seen [18]. Further images suggested that nanovesicles could either bud from or fuse to nanotubes [17]. Moreover, AFM images of A_6K show the presence of small peptide discs and stacks after 1 h of incubation. After 6 h, short cylindrical fibrils were observed. These fibrils assembled into longer fibers, and an equilibrium was reached after 1 week, at which point only long fibers and no discs or stacks were observed [34].

Although most studies examined the tertiary structures formed by lipid-like peptides, several studies have used circular dichroism to examine the secondary structures of the peptides. Ge et al. [27] observed spectra for I_5K_2 which were characteristic of a beta-sheet. Vauthey et al. [17] observed a peak minimum at 220 nm, which was not characteristic of either an alpha helix or beta sheet. Zhuang and coworkers [35] observed that A_6K and A_6D had random coil conformations in water at a concentration of 0.4 mM. When the concentration was increased to 2.5 mM, A_6K adopted a helical structure. At 10 mM, both peptides formed β-sheets. Interestingly, the presence of ionic detergents could affect the secondary structure adopted by the peptides. In the presence of SDS micelles, A_6K formed helical structures while A_6D did not, suggesting that the positively charged lysine on A_6K interacted with the negatively charged SDS. At high SDS concentrations, a mixture of A_6K and A_6D formed antiparallel β-sheets that were stable in response to high temperatures and sonication.

While most studies have measured the properties of short lipid-like peptides with uniform tails, one study observed the characteristics of the conical peptide GAVILRR [29]. CAC measurements of GAVILRR were comparable to those of other peptides and surfactants. In contrast to the other peptides, AFM images suggested that GAVILRR self-assembled into nanodonuts. Because micelles and a small population of nanopipes could also be visualized, the authors postulated that the peptide formed nanodonuts from precursor micelles with the nanopipes as an intermediary structure [29].

18.8 Membrane protein–surfactant peptide interactions

In order to design optimal peptide detergents, it is necessary to understand how they interact with and stabilize membrane proteins. Unfortunately, the precise mechanisms by which they do so are unknown. A crystal structure of a membrane protein–peptide surfactant complex would help elucidate the nature of the protein–surfactant interaction. However, only structures of PD_1 and LPD-12 alone have been solved thus far.

Although the crystal structure of PD_1 in complex with itself has been solved, it is not sufficient for understanding how the peptide was able to solubilize bacteriorhodopsin and bovine rhodopsin. It has been theorized that the flat hydrophobic surfaces interact with and surround the hydrophobic surfaces of membrane proteins, thus protecting them from aqueous solutions [7]. However, this has not been proven. It should be noted that both bacteriorhodopsin and bovine rhodopsin are helical proteins, and that PD_1 was inspired by the ability of natural amphipathic helices to self-associate. The inability of PD_1 to solubilize the β-sheet protein PhoE porin [7] suggests that PD_1 may only be able to solubilize helical membrane proteins.

The crystal structure of an LPD-12 micelle yields clues as to how LPDs incorporating a helical peptide are able to solubilize and stabilize membrane proteins. As predicted, these LPDs formed cylindrical micelles with a tightly packed hydrophobic core (Fig. 18.3B). There is insufficient space for a protein in the core of a pure LPD-12 micelle. However, it has been postulated that the micelle comes apart at the loose interface between helix dimers in order to incorporate more LPD monomers and to accommodate a protein [9] (Fig. 18.4A). While there is typically a trade-off between the size of a micelle and its ability to stabilize a protein with traditional detergents [6], LPDs with a helical peptide seem capable of forming a small, compact, stabilizing environment [8]. The high resolution of the crystal

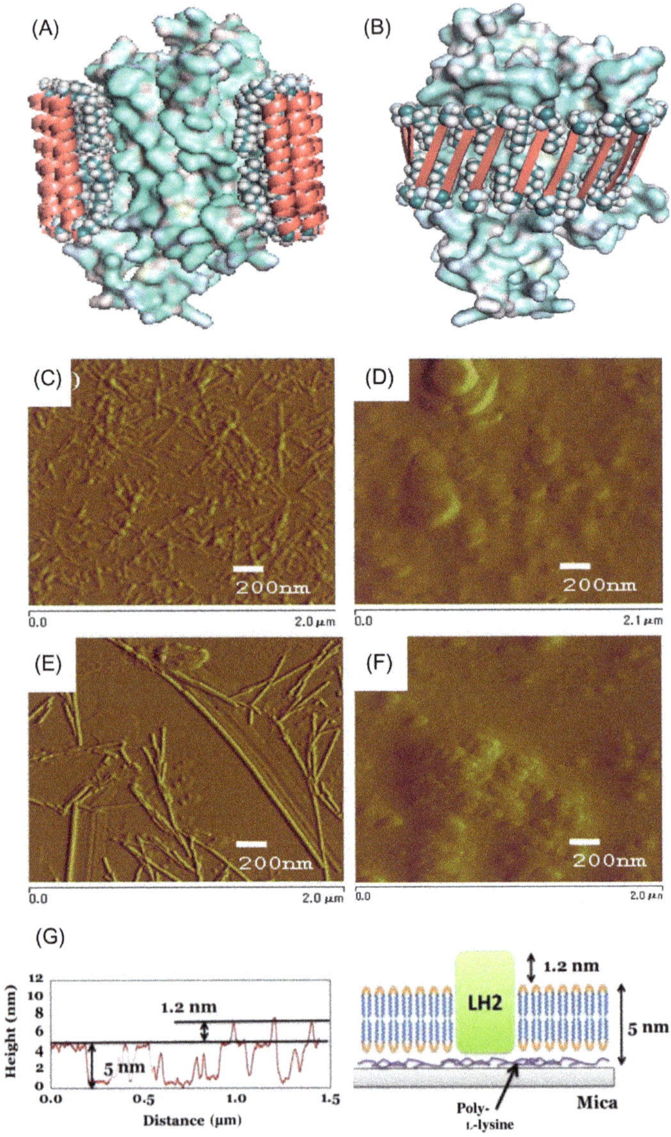

Figure 18.4 Surfactant peptide–membrane protein interactions. (A) Proposed model of the lipopeptide LPD-12 solubilizing a membrane protein. The alkyl chains surround the hydrophobic transmembrane region while the peptide helix interfaces with the aqueous solution. These peptides are proposed to form a cylinder around membrane proteins. (B) Proposed model of the beta-sheet peptide BP-1 solubilizing a membrane protein. (C) The peptide A_6K forms short nanotubes on its own. (D) When Photosystem-I is mixed with A_6K, large vesicles are formed. (E) The peptide A_6D forms long nanotubes on its own. (F) When Photosystem-I is mixed with A_6D, small vesicles are formed. (G) AFM evidence of light-harvesting antenna complex 2 (LH2) incorporation into PG Cr-D_2C_{12} bilayers.

structure suggests that the micelles form a very well-ordered structure with a rigid peptide exterior which is favorable for maintaining protein stability. This may be due in part to the acyl chain packing, which is thought to provide an environment that better mimics the membrane bilayer. The low aggregation numbers suggest that these micelles are small and compact, roughly half the size of the most stabilizing traditional detergents [11]. The small micelle size and stability are attributes which should favor both crystallization screens and NMR studies. It should be noted that DDM was required to obtain a crystal structure of LPD-12, indicating a degree of conformational dynamism of pure LPD micelles. Thus, while experimental and computational results indicate that pure LPD micelles are stable enough for membrane protein studies [8,10], the crystal structure suggests that the addition of detergent may confer additional benefits [9].

While the means by which ultra-short LPDs and PGs solubilize and stabilize membrane proteins is not known, it is thought that they form micelles which behave as traditional detergents: The alkyl tails are in contact with the hydrophobic protein domains, while the hydrophilic heads are in contact with the hydrophilic protein domains and the bulk solution [13,14]. The details of these interactions have not been elucidated, but AFM images of light-harvesting antenna complex 2 (LH2) in PG bilayers suggest that membrane proteins are indeed incorporated in this manner (Fig. 18.4G) [15]. Additional experiments with PS-I and PS-II suggest that ultra-short LPDs and PGs with an optimal head group provide a mild, yet stabilizing environment comparable to that of traditional detergents.

The mechanism by which short self-assembling lipid-like peptides stabilize membrane proteins is also postulated to be similar to that of commercial detergents. Namely, it is believed that the peptide tails surround and shield the hydrophobic regions of the proteins, protecting them from aqueous solutions and subsequent aggregation and loss of function [24,26,36]. AFM images of peptides alone and interacting with PS-I seem to confirm this hypothesis. A_6K and A_6D alone form nanotubes, while A_6K or A_6D that were incubated with PS-I formed vesicles [24] (Fig. 18.4C–F). Similarly, I_5K_2 formed nanotubes in the presence and absence of Triton X-100. When PS-I was included in an I_5K_2/Triton X-100 mixture, vesicles were observed [27]. Moreover, vesicles composed of peptides that best enhanced PS-I activity were more stable than vesicles formed from other peptides. This suggests that encapsulation of membrane proteins in peptide vesicles is indeed the mechanism by which the peptides stabilize the proteins [24].

Although evidence suggests that short lipid-like peptides stabilize membrane proteins by forming vesicles, additional mechanisms of stabilization have also been

◀ *Source*: (C–F) Adapted with permission from K. Matsumoto, M. Vaughn, B.D. Bruce, S. Koutsopoulos, S. Zhang, Designer peptide surfactants stabilize functional Photosystem-I membrane complex in aqueous solution for extended time, J. Phys. Chem. B. 113 (1) (2009) 75–83 [24]. Copyright 2009 American Chemical Society. (G) Adapted with permission from S. Koeda, et al., Creation of cross-linked bilayer membranes that incorporate membrane proteins from oligo-Asp-based peptide Gemini surfactants, Langmuir 29 (37) (2013) 11695–11704 [15]. Copyright 2013 American Chemical Society.

proposed. The peptides may have antioxidant properties that can aid in retaining enzymatic activity and function [28]. The small size of the peptides may allow them to stabilize membrane proteins via insertion into molecular gaps, as was proposed in the case of PS-I [26]. Finally, the large, ordered three-dimensional networks that the peptides are capable of forming may aid in stabilizing large multimeric proteins [24]. These mechanisms may all be modes via which the peptides interact with membrane proteins, and may depend on the specific peptide sequence and chemical properties, as well as the specific membrane protein.

The precise mechanism by which beta-sheet peptides solubilize and stabilize membrane proteins is unknown. Electron micrograph images show that pure peptide solutions assemble into long filaments, and that the addition of membrane proteins disrupts this fibrous organization. Circular dichroism spectra indicate the presence of beta-sheet secondary structures in both pure BP-1 and MsbA-BP-1 solutions. The authors postulate that it is feasible for the peptides to rearrange themselves from filamentous structures to beta-barrel structures to accommodate and stabilize membrane proteins (Fig. 18.4B), but no structural data currently exist to support this hypothesis [20].

18.9 Applications

18.9.1 Protein functional studies

The first step in any studies or projects involving membrane proteins is to synthesize adequate quantities of soluble, sufficiently stable, and functional proteins. However, it is often difficult to express membrane proteins in appropriate amounts. This difficulty is compounded by the fact that even once expressed, it is difficult to keep them soluble and stable for sufficient periods of time. Extensive detergent screens have traditionally been required. Surfactant peptides may address these problems, thus making it possible to study proteins that have previously remained elusive.

18.9.1.1 Expression

The peptitergent PD_1 may enable high expression of select proteins. When fused to the N-terminal region of plant and mammalian cytochrome P450s, milligram quantities of protein could be expressed in yeast and *E. coli* systems [37,38]. *Escherichia coli* expression even yielded four times more of the PD_1-P450 mutant than the native P450 protein [38].

Short lipid-like peptides also have the potential to facilitate the expression of milligram quantities of protein. In a cell-free expression system, $\sim 1-5$ mg of 8 different GPCRs could be expressed using 10 different peptides [23,25] (Fig. 18.5A). The peptides performed comparably to the best-performing traditional detergents, and in some cases even better.

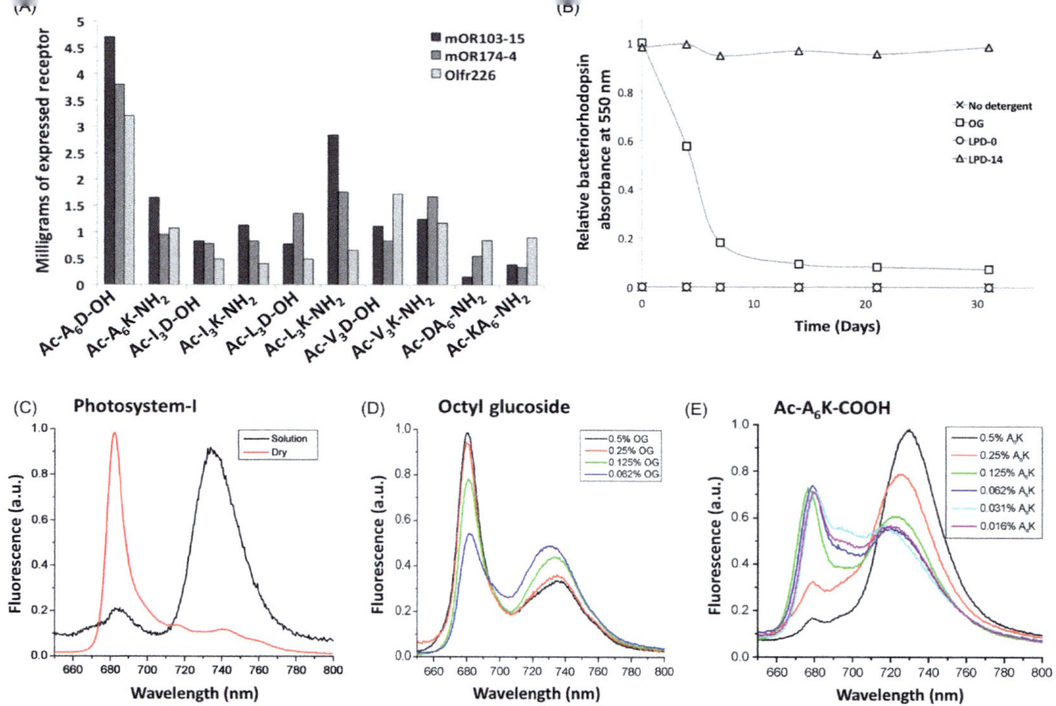

Figure 18.5 Surfactant peptides stabilize membrane proteins, enable their expression, and enhance their activity. (A) Milligrams of GPCRs can be expressed in a cell-free reaction using short lipid-like peptides. (B) LPD-14 stabilizes bacteriorhodopsin for >30 days, while OG does not. The alkyl chains are required for LPD-14 to exhibit surfactant behavior. (C) The emission spectrum of Photosystem-I alone exhibits a blueshift when the protein is dried. (D) The detergent OG is unable to stabilize Photosystem-I. (E) The peptide A_6K is able to stabilize Photosystem-I when it is dried. *Source*: (A) Adapted with permission from K. Corin, et al., Designer lipid-like peptides: a class of detergents for studying functional olfactory receptors using commercial cell-free systems, PLoS One 6 (11) (2011) e25067 [23]. (B) Data adapted with permission from C.-L. McGregor, et al., Lipopeptide detergents designed for the structural study of membrane proteins, Nat. Biotechnol. 21 (2) (2003) 171–176 [8]. (C–E) Adapted with permission from P. Kiley, et al., Self-assembling peptide detergents stabilize isolated Photosystem I on a dry surface for an extended time, PLoS Biol. 3 (7) (2005) e230 [26].

18.9.1.2 Solubilization

Surfactant peptides can be used to solubilize various classes of membrane proteins. They may even outperform the best traditional detergents.

Initial experiments on PD_1 had limited success, and merely proved the concept of using peptides as detergents [7,30,39]. However, when placed at the N-terminus of various P450 enzymes, PD_1 yielded proteins that were soluble without detergent. These engineered proteins were more stable than their native counterparts, and retained enzymatic activity [37,38].

LPDs were capable of solubilizing the helical proteins bacteriorhodopsin, halorhodopsin, lactose permease, MsbA, PS-I, and PS-II, as well as the *E. coli* β-barrel protein PagP, and the dual-morphology YidC [8,12–14,16]. Lactose permease is highly susceptible to aggregation, and of the traditional detergents, DDM has been the only one capable of being used. Surprisingly, cylindrical LPDs performed better than DDM. Significantly fewer aggregates were seen in LPD-treated samples, and the bulk of the sample remained soluble after 5 days of incubation at 37°C [8]. Likewise, the best ultra-short LPDs with two-residue-heads could solubilize PS-I to an equal or greater extent than the traditional detergents DDM and fos-choline 14 (FC-14) [13]. Ultra-short LPDs with just one residue in the head group could solubilize halorhodopsin, MsbA, and YidC directly from membranes comparably to or better than DDM or *N*-octyl-β-D-glucoside (OG) [12]. PGs were also able to solubilize PS-I and PS-II without loss of activity directly from membranes or after a detergent exchange [14,16]. The tested PGs typically performed comparably to DDM, but were able to solubilize the proteins into micelles without any observed aggregation at significantly lower surfactant concentrations (0.001% vs 0.1%) [14]. Moreover, when the headgroup was designed to include a crosslink, PGs were able to form bilayer structures capable of incorporating the membrane protein LH2 [15].

The effect of LPD fatty acyl chain length on membrane protein solubility was examined [8,13]. Cylindrical LPDs with fatty acyl chains 12–20 carbons long were equally capable of maintaining bacteriorhodopsin stable for 1 month at room temperature, though this does not preclude the possibility that specific chain lengths may be optimal for different proteins [8]. In contrast, amongst ultra-short LPDs with a given head group, those with longer alkyl tail lengths were better able to solubilize PS-I [13].

Like LPDs, short-lipid-like peptides could perform comparably to, and sometimes better than, the best detergents. A large number of peptide detergents and GPCRs were screened in two separate studies: one study examined four unique GPCRs in 12 different peptide detergents [25], while the other examined 12 olfactory receptors in 10 different peptide detergents [23]. The surfactant peptides performed comparably to Brij-35, which had empirically been determined to be the best detergent for these receptors. The peptides could solubilize up to 93% of receptors produced in an *E. coli* cell-free system. In a separate study, the peptides Ac-V_6D-COOH (V_6D), A_6K, and A_6D were successfully used to solubilize and stabilize glycerol-3-phosphate dehydrogenase (GlpD) [28]. An A_6D:A_6K mixture in a 40:60 ratio was able to solubilize about 60% of GlpD directly from *E. coli* membranes. While this was less than the 90% solubilized by OG, it was more than the 50% of protein solubilized by decyl maltoside (DM) and Triton X-100.

Although it is a new and emerging surfactant peptide, studies suggest that the beta-sheet peptide BP-1 may be compatible with a broad range of membrane proteins. It was able to solubilize and functionally stabilize the ABC exporter MsbA. BP-1 was also able to stabilize bacteriorhodopsin, the potassium channel KcsA, and the glucagon receptor [20].

18.9.1.3 Stability

Once membrane proteins have been expressed, it is necessary to keep them functionally stable for the duration of their intended usage. Long-term stability is especially crucial for crystallization screens or technological developments, as conditions under these circumstances can easily exceed those in typical laboratory procedures.

All of the LPDs conferred extra degrees of structural or thermal stability to the tested proteins. Absorbance spectra indicated that cylindrical LPDs could maintain bacteriorhodopsin stable at room temperature for 1 month at a pH of 7.4 (Fig. 18.5B). These conditions typically cause loss of receptor function [8]. Circular dichroism and fluorescence spectroscopy suggested that the structural and thermal stability of PS-I in ultra-short LPDs was comparable to or better than that in the traditional detergents DDM and FC-14 [13]. Moreover, absorbance spectra suggested that ultra-short LPDs could also help stabilize halorhodopsin [12]. Finally, PS-I solubilized in the PG $DKDKC_{12}D$ was more stable at 60°C than when it was solubilized in DDM. Interestingly, DDM afforded more thermal stability at 80°C, while PGs and DDM yielded similar results at 70°C [14].

Short lipid-like peptides have been used more extensively than other peptide surfactants, and show great promise in functionally stabilizing membrane proteins. The peptide A_6K could stabilize PS-I for over 2 months, keeping it ~5 times more active than DDM [24]. The peptide V_6D and an A_6D:A_6K mixture were able to maintain the specific activity of GlpD at ~77% of the value in a membrane bilayer, and retain ~90%−99% of this activity for 2 weeks. While the specific activity of GlpD was highest in OG, approaching that in the presolubilized membrane, about 40% of this was lost within 1 day [28]. The same authors also showed that V_6D, A_6D, and A_6D:A_6K were able to extend the half-life of the soluble NADH peroxidase 3−5 times its base value [28]. The peptide A_6D was successfully used to stabilize rhodopsin: a mixture of A_6D with OG increased the thermal half-life of rhodopsin ~4 times compared to OG alone [36]. When A_6D was used with DM, or without any detergent, rhodopsin had negligible denaturation, even at 55°C. A_6D was also able to increase the half-life of delipidated rhodopsin by a factor of 2.6. The peptide I_5K_2 improved the thermal stability and half-life of PS-I [27]. PS-I solubilized in I_5K_2 had a higher melting temperature than when it was solubilized in the most effective traditional detergents. At an optimal concentration, the half-life of PS-I increased from 67 to 173 min.

Initial experiments indicate that beta-sheet peptides can also functionally stabilize membrane proteins. In the presence of BP-1, high MsbA activity was maintained for 40 days at room temperature. In contrast, incubation in the detergent β-D-undecylmaltoside (UDM) resulted in precipitation and an activity loss of ~80% after just 1 day [20]. Similar results were seen with bacteriorhodopsin, KcsA, and the glucagon receptor. BP-1 could keep bacteriorhodopsin >70% active after 60 days at 37°C. Almost all activity was lost after 2.5 days in OG, and 10 days in DDM. BP-1 increased the thermal transition temperature of the glucagon receptor, and stabilized KcsA better than DM or DDM when KcsA was incubated for 5 days at 37°C.

18.9.1.4 Activity

In addition to keeping membrane proteins structurally and thermally stable, surfactant peptides have been able to keep them active. In several cases, they yield better results than the best commercially available detergents.

PS-I oxygen uptake in ultra-short LPDs was better or equivalent to the uptake in DDM. While the uptake in FC-14 was the highest, spectroscopy experiments suggested that FC-14-solubilized PS-I was not as structurally stable as PS-I solubilized by peptides [13]. Similarly, the electron transfer activity of PS-I and PS-II solubilized in PGs was comparable to that observed for DDM-solubilized proteins [14,16], and absorbance spectra of LH2 in PG bilayers exhibited similar behavior to LH2 solubilized in phospholipids [15].

Further experiments with PS-I demonstrated that short lipid-like peptides could also increase PS-I activity. The peptides I_6K_2, A_6K, Ac-V_6K_2- NH_2 (V_6K_2), and V_6R_2 could enhance the activity of PS-I 8–12 times that of commercial detergents or PS-I alone. A C-terminal positively charged head group was likely necessary to functionally stabilize PS-I. No increase in activity was seen when the head group was zwitterionic or anionic. Moreover, while A_6K was able to increase the activity of PS-I ~9 times that of PS-I alone, the peptide KA_6 had no significant effect on PS-I activity [24] (Fig. 18.2C).

In addition to PS-I, short lipid-like peptides could functionally solubilize four GPCRs. Circular dichroism and microscale thermophoresis measurements indicated that the receptors mOR103–15, mOR174–4, mOR174–9, and Olfr226 were properly folded and could bind their respective ligands. The results for receptors expressed with peptides were comparable to or better than those for receptors expressed in the detergent Brij-35. Indeed, Olfr226 expressed with A_6K had an EC_{50} value nearly 30 times lower than the same receptor expressed in Brij-35 [23].

18.9.2 Protein structural studies

Structural studies require milligram quantities of soluble, stable, and active receptors. Historically, each of these bottlenecks was difficult to overcome and typically required laborious detergent screens. Surfactant peptides not only address each of these issues, but may confer additional benefits over traditional detergents. For example, the stable and ordered structures formed by several surfactant peptides may decrease the problems inherent in crystal lattice formation. Indeed, by addressing each of these obstacles, surfactant peptides provide one more tool that can facilitate membrane protein structure determination.

Although no high-resolution structure of a peptide surfactant–membrane protein complex currently exists, several studies show promising results towards reaching that goal. An NMR ^{1}H-^{15}N HSQC spectrum could be obtained for PagP solubilized in LPD-14 [8]. The well-dispersed peaks suggested that the protein was properly folded. Moreover, they were reminiscent of those obtained for OG- and DPC-solubilized PagP, for which an NMR structure has been solved [40]. NMR data of various substrates in the active site of a PD_1-engineered P450 cytochrome could likewise be obtained [37]. Moreover, electron microscopy images of MsbA solubilized in BP-1 clearly showed single particles in various conformations [20]. Although UDM is a preferred detergent for MsbA, and has even been used to crystallize MsbA [41], individual MsbA particles could not be visualized in UDM. Unlike BP-1, UDM-solubilized protein had a high background, and tended to aggregate when the detergent was diluted [20]. Finally, unpublished reports of GlpD-V_6D crystals diffracting to 7 Å have been cited [26].

18.9.3 Technological applications

One of the biggest potential uses for membrane proteins is in the development of specialized sensors. Biological systems have capabilities unsurpassed by modern technologies. For example, PS-I efficiently transfers energy, and olfactory receptors can specifically detect particles in concentrations as small as parts per trillion [42,43]. If these receptors could be integrated with electronics, it might be possible to design sensors capable of generating electricity, detecting cancer at early stages, of fingerprinting individuals, and of detecting dangerous materials without risking human or animal lives. However, membrane proteins are designed to function in a lipid bilayer. Suitable methods must thus be developed to facilitate protein-based sensors.

Peptide surfactants can help integrate membrane proteins into electronic chips. PS-I in particular has been the focus of experiments with reasonable success [14,26,43,44]. Indeed, photocurrents have been observed when PS-I has been attached to electrodes [14,43,44].

The success in creating PS-I-based circuits partly stems from the surprising fact that short lipid-like peptides can functionally stabilize PS-I in the dry state (Fig. 18.5C–E). Minimal changes in PS-I fluorescence spectra demonstrated that the peptide A_6K could stabilize dried PS-I for 3 weeks [26,44]. The peptide V_6D stabilized PS-I to a lesser extent, but performed better than the detergents Triton X-100, *N*-dimethyldodecylamine *N*-oxide, *N*-dodecyl-β-D-maltoside (DM), or OG [26]. Moreover, photocurrents could be detected for PS-I dried with A_6K and V_6D, and a highly efficient yet low-cost PS-I-based photovoltaic device could be constructed with the help of A_6K [43,44]. Because most biological experiments must be conducted in aqueous solutions, the ability to use dried proteins constitutes a critical advancement towards the development of biologically based technologies.

Another advantage of using peptide-based detergents over traditional detergents is the ability to couple them to a variety of functional groups. This feature can be exploited in the development of novel technologies. For example, when the electron acceptor methyl viologen (MV) was coupled to the PG surfactant $DKDKC_{12}K$ (MV-$DKDKC_{12}K$) and used to solubilize PS-I, better electron transfer rates were seen at lower MV concentrations than when MV was in free solution with DDM-solubilized PS-I [14]. The MV-$DKDKC_{12}K$-solubilized PS-I could then be adsorbed onto an electrode and used to generate photocurrents [14].

18.10 Future directions

Surfactant peptides have proven that they can facilitate membrane protein studies and overcome the current major bottlenecks of expression, stability, activity, and solubility. Future work will likely focus on improving these areas, as well as understanding how peptides work, and designing the best peptide for the protein or application at-hand. Most of the peptides that we discussed in this chapter are summarized in Table 18.1.

Table 18.1 Examples of peptides that have been used for membrane protein purification, solubilization and stabilization

Peptide	Applications	References
Ac-EELLKQALQQAQQLLQQAQELAKK-$CONH_2$ (PD_1)	Solubilization of rhodopsin and bacteriorhodopsin (bR); expression, solubilization, stabilization, and NMR studies of P450	[7,37,38]
Ac-AOAEAAEKAAKYAAEAAEKAAKA-OA-$CONH_2$ (12–20 carbon alkyl chains attached to the ornithines)	Solubilization of bR, lactose permease, PagP; stabilization of bR and lactose permease; NMR of PagP	[8]
C12-X; X = G, A, V, I, S, or E	Extraction and solubilization of halorhodopsin (HR), MsbA, YidC; maintains HR activity	[12]
C12DK, C14DK, C16DK, C14DD	Functional solubilization and stabilization of PS-I	[13]
$DKDKC_{12}D$ $DKDKC_{12}K$ (peptide gemini surfactants, PG)	Functional solubilization and stabilization under 60°C of PS-I and PS-II, PS-I integration into electrodes ($DKDKC_{12}K$)	[14,16]

(Continued)

Table 18.1 (Continued)

Peptide	Applications	References
Cr-D_2C_{12}, Cr-D_3C_{12}	Functional incorporation of LH2 into bilayers	[15]
NPDGC_{12}KK DPDGC_{12}KK	Extraction and functional solubilization of PS-I and PS-II	[16]
$A^{\beta}Np^{(-)}Np^{(+)}A^{\beta}C_{12}K$	Functional solubilization of PS-I and PS-II	[16]
Ac-A_6D-COOH	Stabilization of rhodopsin; GPCR expression and solubilization	[23,25,36]
Ac-I_4K_2-$CONH_2$, Ac-I_5K_2-$CONH_2$ Ac-I_3K-$CONH_2$, Ac-I_4K-$CONH_2$ Ac-I_5K-$CONH_2$, Ac-I_5K_3-$CONH_2$	Thermal stabilization of PS-I	[27]
Ac-I_6K_2-$CONH_2$	Thermal stabilization of PS-I; enhances PS-I activity	[24,27]
Ac-V_6K_2-$CONH_2$	Enhances PS-I activity	[24]
Ac-V_6R_2-$CONH_2$	Enhances PS-I activity	[24]
Ac-A_6K-$CONH_2$	GPCR expression and solubilization; maintains or enhances GPCR and PS-I activity; stabilization of aqueous and dried PS-I, integration of PS-I on electrodes	[23–25,27,43]
Ac-A_6K-COOH	Stabilizes dried PS-I	[26]
Ac-V_6D-COOH	GlpD crystallization, stabilization, and maintenance of activity; stabilizes dried PS-I	[26,28]

(Continued)

Table 18.1 **(Continued)**

Peptide	Applications	References
Ac-A_6K-COOH: Ac-V_6D-COOH mixture	Integration of PS-I and bacterial reaction center (RC) on electrodes, stabilization of dried PS-I	[44]
Ac-V_3D-COOH, Ac-V_3K-$CONH_2$, Ac-I_3D-COOH, Ac-I_3K-$CONH_2$, Ac-L_3D-COOH, Ac-L_3K-$CONH_2$, DA_6-$CONH_2$, KA_6-$CONH_2$	Expression and solubilization of GPCRs	[23,25]
Ac-GAVILEE-COOH, Ac-GAVILRR-$CONH_2$	Expression and solubilization of GPCRs	[25]
Ac-A_6K-COOH: Ac-A_6D-COOH	GlpD extraction, maintenance of activity, and stabilization	[28]
Ac-GSLSLDGD-NH_2 (BP-1) (eight carbon alkyl chains attached to the glycines, second leucine has *N*-methyl group)	Solubilization and stabilization of bR, MsbA, KcsA, and the glucagon receptor; maintains MsbA activity	[20]

Surfactant peptides can help in the utilization of membrane proteins in technologies beneficial to society. It has already been shown that they can functionally stabilize membrane proteins in the dry state, and that they can be used to interface membrane proteins with solid-state electronics [26,43,44]. It has also been shown that they can be coupled to functional groups to enhance the functionality of membrane-protein-based chips [14]. Moreover, surfactant peptides can enhance the activity or stability of proteins [8,12–14,16,20,23,24,27,28,36], making it possible to design sensors with higher sensitivities and longer lifetimes. Future work will likely build on these successes. For example, it might be possible to design many types of sensors with the aid of membrane proteins. Olfactory-based sensors could potentially detect cancer, identify dangerous materials, help find missing people, or even detect life-threatening airborne allergens [45–48]. Other membrane protein-based sensors could be designed to detect various diseases, while chips made with PS-I or other related proteins could be used to generate energy or electricity.

The potential for surfactant peptides and membrane proteins extends beyond what has currently been done. For example, peptides can self-assemble into scaffolds that can aid in tissue regeneration [49,50]. It might be possible to incorporate membrane proteins into these scaffolds to better facilitate this process. Surfactant peptides have potential applications in numerous other areas as well, including controlled drug delivery, cancer therapeutics, antibiotic development, and even bioremediation [51–53]. Because membrane proteins are involved in many of these areas [54–56], it may be possible to incorporate them to create more stable and effective therapeutics and technologies.

18.11 Conclusions

One major bottleneck hindering membrane protein studies is the need to find a suitable detergent. While commercial detergents have been used, success has largely been due to luck, laborious detergent screens, or the robustness of the protein under study.

Peptide-based surfactants are an attractive alternative to traditional commercial detergents. They can easily be designed to have desired physical or chemical properties. They can confer equal or greater stability without compromising protein solubility, likely due to milder protein–peptide interactions. Most surfactant peptides are also inexpensive to produce. Moreover, the superior performance of peptide surfactants over commercial detergents in several studies, as well as their ability to functionally stabilize proteins in the dry state, can make them useful in the development of membrane protein-based technologies like photovoltaic cells and sensors.

In order to make peptide surfactants universally useful, there is a need to fine tune their properties and establish clear design principles. Some principles are beginning to emerge, though more studies are needed to ensure that current results are not limited to isolated cases or specific families of proteins. However, the results thus far have generated a versatile toolbox that is easily available to, and modifiable by, researchers. While still preliminary, this toolbox promises to open exciting new possibilities in membrane protein research.

Acknowledgments

The data showing bacteriorhodopsin stability in OG and LPDs were graciously shared by Dr. Gilbert G. Privé [8].

References

[1] E. Wallin, G. von Heijne, Genome-wide analysis of integral membrane proteins from eubacterial, archaean and eukaryotic organisms, Protein Sci. 7 (4) (1998) 1029–1038.
[2] D.R. Flower, Modelling G-protein coupled receptors for drug design, Biochim. Biophys. Acta 1422 (3) (1999) 207–234.

[3] J.P. Overington, B. Al-Lazikani, A.L. Hopkins, How many drug targets are there? Nat. Rev. Drug Discov. 5 (12) (2006) 993–996.
[4] H.M. Berman, et al., The protein data bank, Nucleid Acids Res. 28 (2000) 235–242, <www.rcsb.org>.
[5] Stephen White Laboratory, Membrane proteins of known structure run by the Stephen White Laboratory at UCI. http://blanco.biomol.uci.edu/mpstruc/.
[6] G.G. Prive, Detergents for the stabilization and crystallization of membrane proteins, Methods 41 (4) (2007) 388–397.
[7] C.E. Schafmeister, L.J.W. Miercke, R.M. Stroud, Structure at 2.5Å of a designed peptide that maintains solubility of membrane proteins, Science 262 (5134) (1993) 734–738.
[8] C.-L. McGregor, et al., Lipopeptide detergents designed for the structural study of membrane proteins, Nat. Biotechnol. 21 (2) (2003) 171–176.
[9] D.N. Ho, N.C. Pomroy, J.A. Cuesta-Seijo, G.G. Prive, Crystal structure of a self-assembling lipopeptide detergent at 1.20 Å, Proc. Natl. Acad. Sci. U.S.A. 105 (35) (2008) 12861–12866.
[10] E. Kelly, G.G. Prive, D.P. Tieleman, Molecular models of lipopeptide detergents: large coiled-coils with hydrocarbon interiors, J. Am. Chem. Soc. 127 (39) (2005) 13446–13447.
[11] G.G. Prive, Lipopeptide detergents for membrane protein studies, Curr. Opin. Struct. Biol. 19 (4) (2009) 379–385.
[12] R. Abe, J.M.M. Caaveiro, M. Kudou, K. Tsumoto, Solubilization of membrane proteins with novel *N*-acylamino acid detergents, Mol. Biosyst. 6 (4) (2010) 677–679.
[13] X. Wang, et al., Solubilization and stabilization of isolated photosystem I complex with lipopeptide detergents, PLoS One 8 (9) (2013) e76256.
[14] S. Koeda, et al., Application of peptide Gemini surfactants as novel solubilization surfactants for photosystems I and II of cyanobacteria, Langmuir 29 (37) (2013) 11667–11680.
[15] S. Koeda, et al., Creation of cross-linked bilayer membranes that incorporate membrane proteins from oligo-Asp-based peptide Gemini surfactants, Langmuir 29 (37) (2013) 11695–11704.
[16] M. Shibata, et al., Design of new extraction surfactants for membrane proteins from peptide Gemini surfactants, Bioconjug. Chem. 27 (10) (2016) 2469–2479.
[17] S. Vauthey, S. Santoso, H. Gong, N. Watson, S. Zhang, Molecular self-assembly of surfactant-like peptides to form nanotubes and nanovesicles, Proc. Natl. Acad. Sci. U.S.A. 99 (8) (2002) 5355–5360.
[18] S. Santoso, W. Hwang, H. Hartman, S. Zhang, Self-assembly of surfactant-like peptides with variable glycine tails to form nanotubes and nanovesicles, Nano Lett. 2 (7) (2002) 687–691.
[19] G. Von Maltzahn, S. Vauthey, S. Santoso, S. Zhang, Positively charged surfactant-like peptides self-assemble into nanostructures, Langmuir 19 (2003) 4332–4337.
[20] H. Tao, et al., Engineered nanostructured β-sheet peptides protect membrane proteins, Nat. Methods 10 (8) (2013) 759–761.
[21] R.M. Garavito, S. Ferguson-Miller, Detergents as tools in membrane biochemistry, J. Biol. Chem. 276 (35) (2001) 32403–32406.
[22] K. Umezaki, et al., Formation of planar bilayer membranes on solid supports using peptide gemini surfactants, Chem. Lett. 41 (2012) 1430–1432.

[23] K. Corin, et al., Designer lipid-like peptides: a class of detergents for studying functional olfactory receptors using commercial cell-free systems, PLoS One 6 (11) (2011) e25067.
[24] K. Matsumoto, M. Vaughn, B.D. Bruce, S. Koutsopoulos, S. Zhang, Designer peptide surfactants stabilize functional Photosystem-I membrane complex in aqueous solution for extended time, J. Phys. Chem. B. 113 (1) (2009) 75–83.
[25] X. Wang, et al., Peptide surfactants for cell-free production of functional G protein-coupled receptors, Proc. Natl. Acad. Sci. U.S.A. 108 (22) (2011) 9049–9054.
[26] P. Kiley, et al., Self-assembling peptide detergents stabilize isolated Photosystem I on a dry surface for an extended time, PLoS Biol. 3 (7) (2005) e230.
[27] B. Ge, F. Yang, D. Yu, S. Liu, H. Xu, Designer amphiphilic short peptides enhance thermal stability of isolated Photosystem-I, PLoS One 5 (4) (2010) e10233.
[28] J.I. Yeh, S. Du, A. Tortajada, J. Paulo, S. Zhang, Peptergents: peptide detergents that improve stability and functionality of a membrane protein, glycerol-3-phosphate dehydrogenase, Biochemistry 44 (51) (2005) 16912–16919.
[29] U. Khoe, Y. Yang, S. Zhang, Self-assembly of nanodonut structure from a cone-shaped designer lipid-like peptide surfactant, Langmuir 25 (7) (2009) 4111–4114.
[30] A. Bavec, A. Juréus, B. Cigic, U. Langel, M. Zorko, Peptitergent PD1 affects the GTPase activity of rat brain cortical membranes, Peptides 20 (2) (1999) 177–184.
[31] X.B. Zhao, et al., Interfacial assembly of cationic peptide surfactants, Soft Matter 5 (2009) 1630–1638.
[32] A. Nagai, Y. Nagai, H. Qu, S. Zhang, Dynamic behaviors of lipid-like self-assembling peptide A6D and A6K nanotubes, J. Nanosci. Nanotechnol. 7 (7) (2007) 2246–2252.
[33] S.J. Yang, S. Zhang, Self-assembling behavior of designer lipid-like peptides, Supramol. Chem. 18 (5) (2006) 389–396.
[34] J. Wang, et al., Dynamic self-assembly of surfactant-like peptides A_6K and A_9K, Soft Matter 5 (2009) 3870–3878.
[35] F. Zhuang, K. Oglecka, C.A. Hauser, Self-assembling peptide surfactants A6K and A6D adopt a-helical structures useful for membrane protein stabilization, Membranes (Basel) 1 (4) (2011) 314–326.
[36] X. Zhao, et al., Designer short peptide surfactants stabilize G protein coupled receptor bovine rhodopsin, Proc. Natl. Acad. Sci. U.S.A. 103 (47) (2006) 17707–17712.
[37] G.A. Schoch, R. Attias, M. Belghazi, P.M. Dansette, D. Werck-Reichhart, Engineering of a water-soluble plant cytochrome P450, CYP73A1, and NMR-based orientation of natural and alternate substrates in the active state, Plant Physiol. 133 (3) (2003) 1198–1208.
[38] T. Sueyoshi, L.J. Park, R. Moore, R.O. Juvonen, M. Negishi, Molecular engineering of a microsomal P450 2a-4 to a stable, water-soluble enzyme, Arch. Biochem. Biophys. 322 (1) (1995) 265–271.
[39] U. Soomets, C. Kairane, M. Zilmer, U. Langel, Attempt to solubilize Na + /K + -exchanging ATPase with amphiphilic peptide PD1, Acta Chem. Scand. 51 (3 Suppl) (1997) 403–406.
[40] P.M. Hwang, et al., Solution structure and dynamics of the outer membrane enzyme PagP by NMR, Proc. Natl. Acad. Sci. U.S.A. 99 (21) (2002) 13560–13565.
[41] A. Ward, C.L. Reyes, J. Yu, C.B. Roth, G. Chang, Flexibility in the ABC transporter MsbA: alternating access with a twist, Proc. Natl. Acad. Sci. U.S.A. 104 (48) (2007) 19005–19010.

[42] M. Laska, et al., Detecting danger—or just another odorant? Olfactory sensitivity for the fox odor component 2,4,5-trimethylthiazoline in four species of mammals, Physiol. Behav. 84 (2) (2005) 211–215.
[43] A. Mershin, et al., Self-assembled photosystem-I biophotovoltaics on nanostructured TiO(2) and ZnO, Sci. Rep. 2 (234) (2012). Available from: http://dx.doi.org/10.1038/srep00234.
[44] R. Das, et al., Integration of photosynthetic protein molecular complexes in solid-state electronic devices, Nano Lett. 4 (6) (2004) 1079–1083.
[45] G. Horvath, G.A. Jarverud, S. Jarverud, I. Horvath, Human ovarian carcinomas detected by specific odor, Integr. Cancer Ther. 7 (2) (2008) 76–80.
[46] F. Kuhn, A. Natsch, Body odour of monozygotic human twins: a common pattern of odorant carboxylic acids released by a bacterial aminoacylase from axilla secretions contributing to an inherited body odour type, J. R. Soc. Interface 6 (2009) 377–392.
[47] D.J. Penn, et al., Individual and gender fingerprints in human body odour, J. R. Soc. Interface 4 (2007) 331–340.
[48] M. McCulloch, et al., Diagnostic accuracy of canine scent detection in early- and late-stage lung and breast cancers, Integr. Cancer Ther. 5 (1) (2006) 30–39.
[49] H. Hosseinkhani, P.-D. Hong, D.-S. Yu, Self-assembled proteins and peptides for regenerative medicine, Chem. Rev. 113 (7) (2013) 4837–4861.
[50] R. Pugliese, F. Gelain, Peptidic biomaterials: from self-assembling to regenerative medicine, Trends Biotechnol. 35 (2) (2017) 145–158.
[51] R. Domalaon, B. Findlay, M. Ogunsina, G. Arthur, F. Schweizer, Ultrashort cationic lipopeptides and lipeptoids: evaluation and mechanistic insights against epithelial cancer cells, Peptides 84 (2016) 58–67.
[52] J. Liang, W. Wu, D. Lai, J. Li, C. Fang, Enhanced solubility and targeted delivery of curcumin by lipopeptide micelles, J. Biomater. Sci. Polym. Ed. 26 (6) (2015) 369–383.
[53] S. Patel, S. Ahmed, J.S. Eswari, Therapeutic cyclic lipopeptides mining from microbes: latest strides and hurdles, World J. Microbiol. Biotechnol. 31 (2015) 1177–1193.
[54] K.R. Kampen, Membrane proteins: the key players of a cancer cell, J. Membr. Biol. 242 (2) (2011) 69–74.
[55] S. Chatterjee, B.B. Azad, S. Nimmagadda, The intricate role of CXCR4 in cancer, Adv. Cancer Res. 124 (2014) 31–82.
[56] J.N. Varela, M.S. Izidoro Jr, L.M. de Hollanda, M. Lancellotti, Membrane proteins as novel targets for vaccine production in Haemophilus influenzae and Neisseria meningitidis, J. Vaccines Vaccin. 3 (6) (2012) 152.

Catalysis by peptides

19

Kengo Akagawa
University of Tokyo, Meguro, Tokyo, Japan

19.1 Introduction

In living organisms, enzymes catalyze the transformations of organic compounds efficiently and selectively. Chemists have been attracted to enzymatic reactions and have utilized them to produce useful materials in organic synthesis [1]. Meanwhile, enzymatic catalysis has some drawbacks, such as a narrow substrate scope and limited availability of enzymes; thus, the practical applications of enzymes in organic synthesis are still limited. The development of artificial peptide catalysts by mimicking enzymes or by designing sequences suitable for specific transformations enables the drawbacks of enzymatic reactions to be overcome. So far, there have been three major streams of the research on peptide catalysis: (1) the development of peptides with a multifunctional catalytic site for hydrolysis of esters and transesterification; (2) the use of peptides as ligands for metal catalysts; and (3) the use of peptides themselves as organocatalysts.

The first topic is related to the success in the X-ray crystallographic analyses of serine proteases such as α-chymotrypsin and lipase. In 1967, the X-ray structure of α-chymotrypsin revealed the presence of the "catalytic triad" consisting of His, Ser, and Asp [2]. These amino acid residues are appropriately arranged in the reaction center and collaboratively function for hydrolysis. Chemists have attempted to construct such a catalytic site with artificial peptides for efficient and stereoselective hydrolysis. Such peptides are potentially applicable for transesterification, in which an alcohol is used instead of water as a solvolysis reagent. The peptide catalysts for hydrolysis and transesterification are discussed in Section 19.2.

The second and the third topics have emerged along with the advancement of catalytic asymmetric synthesis. Since the catalytic production of chiral molecules was attained by chiral transition metal complexes in 1968 [3,4], the research field of asymmetric catalysis has flourished. For the development of asymmetric catalysts, amino acid derivatives have been widely used as abundant chiral sources. Oligomers of amino acids, peptides, are also candidates for capable chiral catalysts. In 1975, poly(amino acid)-mediated asymmetric Michael addition of a thiol was reported by Inoue and coworkers [5]. Since then, poly(amino acid)s, oligopeptides, and cyclic peptides have been employed for asymmetric synthesis. In relation to the progress on transition-metal-catalyzed reactions, peptides that coordinate to metal ions have been used as chiral ligands. Such applications of peptides for ligands are discussed in Section 19.3.

In 2000, amino acid proline was re-realized as a potent chiral aminocatalyst for an asymmetric aldol reaction [6]. In the same year, the concept of "organocatalysis"

Peptide Applications in Biomedicine, Biotechnology and Bioengineering.
DOI: http://dx.doi.org/10.1016/B978-0-08-100736-5.00021-1

was proposed in the report on the chiral-amine-catalyzed Deals–Alder reaction [7]. These amine-catalyzed reactions proceed through the formation of either an enamine [8] or iminium ion [9] intermediate. Because peptides have an amino group at the N-terminus, they potentially catalyze such reactions. Since *N*-prolyl peptides were applied to asymmetric aldol reactions in 2003 [10,11], N-terminal secondary and primary amino groups of peptides have been used as chiral aminocatalysts. Peptides with other functional groups in the side chains of amino acids such as histidine, aspartic acid, and artificially designed ones have also been used for asymmetric transformations. According to the concept of organocatalysis, the peptides promoting reactions without a metal ion can be classified as organocatalysts. The use of peptides as organocatalysts is discussed in Section 19.4.

19.2 Peptides for hydrolysis of esters and transesterification

Serine proteases catalyze efficient and selective hydrolysis of amides and esters. A catalytic triad, which typically consists of His, Ser, and Asp/Glu, plays a critical role for promoting reactions. The construction of such a reaction site has been attempted by designing artificial peptides for hydrolysis of esters and transesterification. In 1980, Ohkubo and coworkers reported dipeptide-catalyzed enantioselective hydrolysis of a racemic phenylalanine derivative (Fig. 19.1) [12]. The formation of micelles by a chiral surfactant and the reactants with long alkyl chains enhanced the selectivity. The ratio of the reaction rates for L- and D-enantiomers (k_L/k_D) was up to 5.7. Later, much higher selectivity was achieved by using a bilayer vesicular system with Z-Leu-His-OH ($k_L/k_D = 83.6$) [13] and Z-Phe-His-Leu-OH ($k_L/k_D = 71$) [14]. The NOE (nuclear Overhauser effect) measurement of the vesicular system indicated the presence of the strong hydrophobic interaction between the phenyl group of a substrate phenylalanine ester and the side chain of the peptide catalyst within a vesicle [15]. Nolte and coworkers also performed enantioselective hydrolysis of an ester with a His-containing dipeptide in a micellar system, and proposed a detailed mechanism for the selective reaction [16].

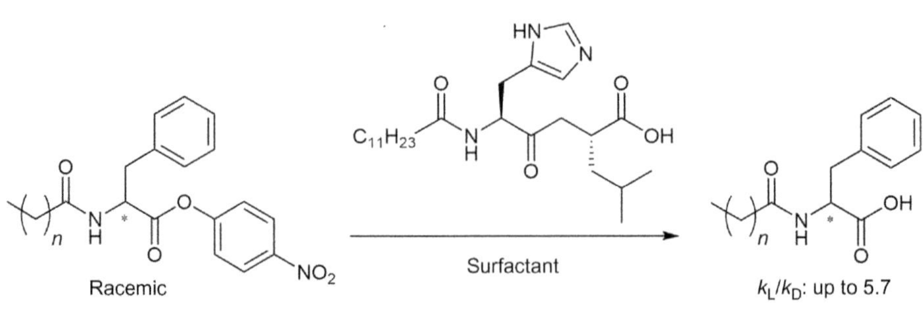

Figure 19.1 Enantioselective hydrolysis by a dipeptide catalyst in a micellar system.

In 1990, Stewart and coworkers reported a chymotrypsin-like peptide catalyst for efficient hydrolysis of esters (Fig. 19.2) [17]. To construct a reaction site similar to the catalytic triad of chymotrypsin, His, Ser, Asp, and Glu residues were attached to the N-termini of four α-helical peptide chains, respectively, and these helices were linked together to form a bundle. With this catalyst, hydrolysis of tyrosine-derived esters was examined. In addition to a reactive 4-nitrophenyl ester, relatively stable ethyl esters could be hydrolyzed, although the reaction rate was much lower than chymotrypsin.

The catalysis by four-helix bundle peptides was intensively studied by Baltzer and coworkers. They revealed that the peptide with a helix-loop-helix motif containing 42 amino acid residues (Fig. 19.3) forms an antiparallel dimer in solution [18,19]. The resulting four-helix bundle showed high catalytic activity for hydrolysis of esters and transesterification. It is proposed that a $HisH^+$-His pair acts as a cooperative nucleophilic and general-acid catalyst [20]. By changing some His residues of this sequence to other amino acids, more reactive peptides were obtained [21,22]. The peptides having Gln at 26 position, Lys or Arg at 30 position, and Lys at 34 position promoted enantioselective hydrolysis of a norleucine-derived ester. Related peptide catalysts based on the four-helix bundle template were applied to hydrolysis of RNA analogues, phosphodiesters [23–25]. As for the hydrolysis of

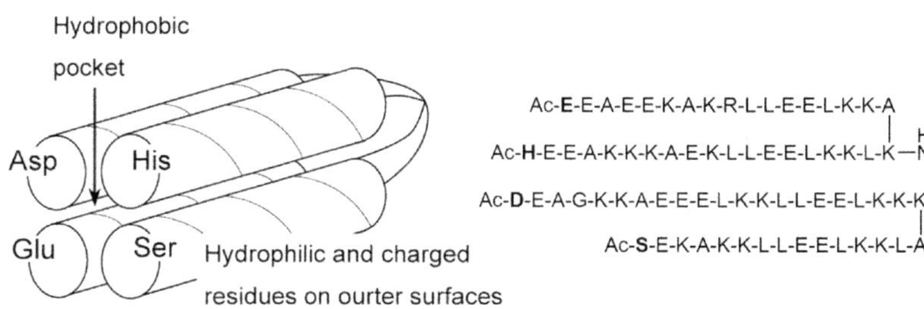

Figure 19.2 Design of a chymotrypsin-like peptide.

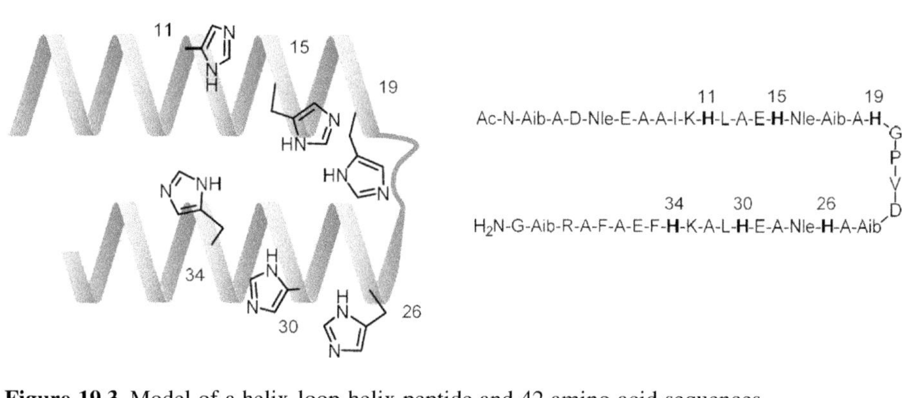

Figure 19.3 Model of a helix-loop-helix peptide and 42 amino acid sequences. *Aib*, 2-aminoisobutyric acid; *Nle*, norleucine.

oligoRNA, Brack and coworkers performed the reaction with poly(Leu-Lys) [26], a cyclic peptide [27], and α-helical poly(Leu-Lys-Lys-Leu) [28]. Schepartz and coworkers developed helix bundle catalysts consisting of β-peptides, and evaluated the catalytic activities for hydrolysis of esters [29]. They successfully obtained the high-resolution X-ray structure of the peptide, and revealed the detailed structure of the catalytically active site containing His and Arg residues.

Ueno and coworkers used a monomeric α-helical peptide having carboxylate, imidazole, and cyclodextrin functionalities at the same side of the helix (Fig. 19.4) [30]. It is assumed that the substrate captured by the cyclodextrin moiety is efficiently attacked by the imidazole group activated by the carboxylate. The peptide showed high enantioselectivity for hydrolysis of an alanine derivative.

Gagné, Waters, and coworkers developed a histidine-containing helical peptide for transesterification [31]. The efficient catalyst was found by a dye-tagging screening, in which resin-supported peptides were colored depending on their reactivity. By applying this method to a β-hairpin peptide library, they identified a more reactive peptide with multiple His residues (Fig. 19.5) [32]. The histidyl side chains are considered to function cooperatively for hydrolysis. Such an action of His residues is observed in the His-His dyad of paraoxonase, where the basicity of one His is increased by the other. Although the aggregating nature of the peptide used in their study is not clear, peptides with a hydrophobic β-sheet structure easily aggregate and form amyloid fibrils and nanofibers. The use of such peptide assemblies for hydrolysis was reported by other groups [33,34].

Sugiura and coworker showed that the modification of a zinc finger peptide, which contains 30 amino acids and folds into a ββα conformation through the coordination to a zinc ion, was effective for hydrolysis of an ester [35]. The coordinating residues of the wild-type zinc finger were replaced to other amino acids to convert a tetradentate-zinc site to a tridentate-zinc site (Fig. 19.6). This allowed the generation of a zinc-coordinated hydroxide ion for high catalytic performance. One

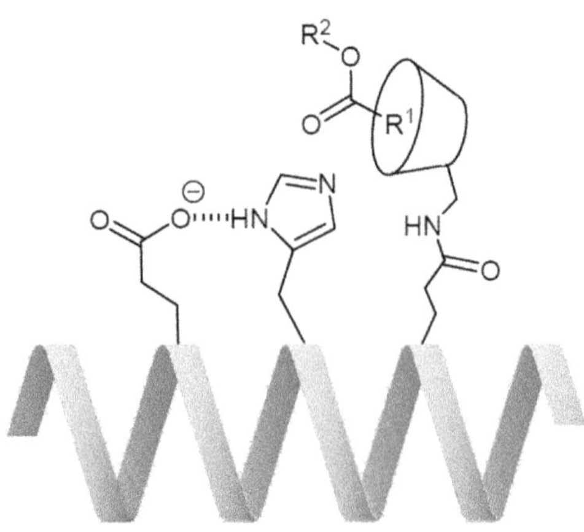

Figure 19.4 Model of a cyclodextrin–peptide hybrid and a captured substrate.

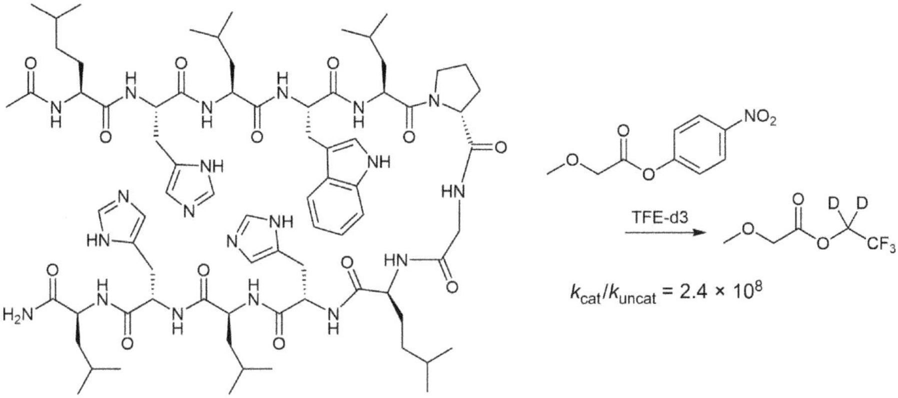

Figure 19.5 β-Hairpin peptide and transesterification.

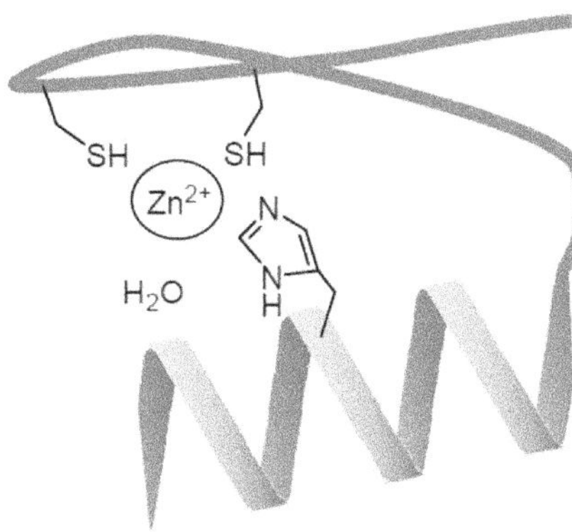

Figure 19.6 Model of a modified zinc finger peptide.

of the modified zinc fingers catalyzed highly enantioselective hydrolysis of a glutamine-derived ester.

A peptide dendrimer can provide the reaction site where multiple functional groups are placed in close proximity without the aid of a secondary structure such as an α-helix and β-sheet. Reymond and coworkers employed peptide dendrimers for hydrolysis of esters. A diamino acid was introduced at branching points, and two dendrimer units were connected by a disulfide bond (Fig. 19.7A) [36]. The dendrimer with His residues at the outermost position showed high activity, and a heterodimer of the dendrimer units was most reactive. In their study of a different type of dendrimer having an iterative sequence of His-Ser-Dap (Dap = 2,3-diaminopropionic acid), a positive dendritic effect was observed [37,38]. They also demonstrated that the screening of a resin-bound dendrimer library could be performed rapidly using a fluorogenic ester (Fig. 19.7B) [39,40]. Such a screening technique was employed by another group to identify an active linear peptide catalyst [41].

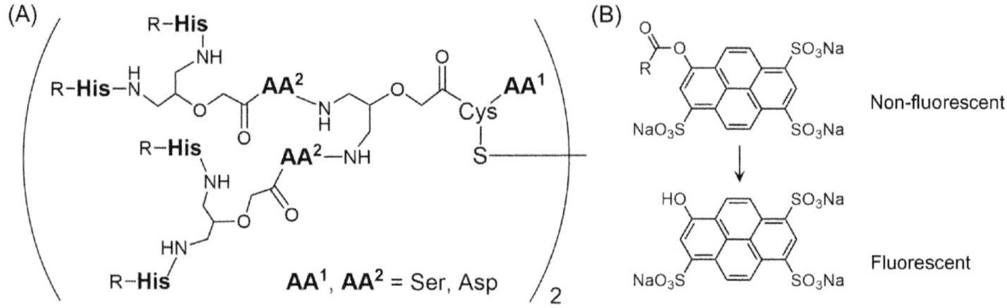

Figure 19.7 (A) Dimer of peptide dendrimer units. (B) Fluorogenic ester for library screening.

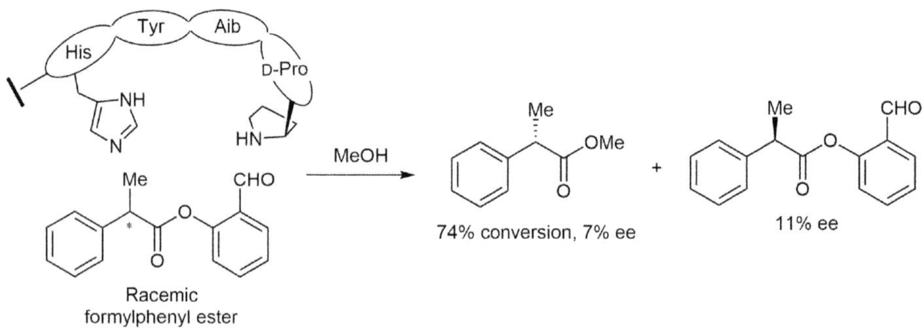

Figure 19.8 Solvolysis of a formylphenyl ester by an N-terminal prolyl peptide with a histidyl residue.

Kudo and coworker developed a bifunctional catalytic system for hydrolysis of an ester and transesterification (Fig. 19.8) [42]. A formylphenyl ester was efficiently hydrolyzed by the N-terminal prolyl peptide with a His residue. It is considered that N-terminal Pro forms an iminium ion intermediate with the formyl group of the substrate, and the histidyl side chain allocated nearby the ester bond promotes the reaction. Enantioselective transesterification of a racemic formyl ester was shown to be possible, although the selectivity was low.

19.3 Peptides for ligands of metal catalysts

In organic synthesis, various types of asymmetric transformations have been achieved by metal catalysts with chiral ligands such as phosphines, bisoxazolines, and salens [43]. To attain high enantioselectivity, it is important to construct an asymmetric reaction site around a metal center. Peptides with coordinating functionalities can be used as chiral ligands with tunable asymmetric environments. The attractive feature of peptide ligands is that the diversity of three-dimensional structures can be attained simply by changing amino acid sequences. The large conformational space of peptides ensures the potential for the development of a capable catalyst. It is also advantageous that peptides can be readily prepared by solid-phase synthesis, which enables the rapid evaluation of a number of peptide ligands.

In 1996, Hoveyda and coworkers reported a titanium-catalyzed enantioselective ring-opening reaction with a dipeptide-based ligand (Fig. 19.9) [44]. The ligand was synthesized by the formation of a Schiff base between an aromatic aldehyde and the N-terminal amino group of a peptide. A stepwise optimization of two amino acids and the aldehyde part gave an optimum dipeptide ligand. Similar types of ligands were applied to titanium-catalyzed Strecker reaction [45] and aluminum-catalyzed cyanosilylation of ketones [46].

The above Schiff base ligand can be easily modulated by changing its components (Fig. 19.10). The use of the aldehyde containing a pyridyl group was effective for copper-catalyzed alkylation of allylic phosphates [47]. By optimizing the peptide part, a highly enantioselective reaction was attained. The dipeptide ligand with a phosphine group was employed to copper-catalyzed asymmetric conjugate addition of alkyl zinc [48]. Tomioka and coworkers performed the same reaction with a different type of phosphine-containing peptide ligand [49]. They introduced the phosphine functionality at the C-terminus of the peptide through an amide bond formation.

In 1996, Gilbertson and coworker reported the synthesis of a β-turn peptide with two phosphine groups [50]. The peptide functioned as a bidentate ligand and

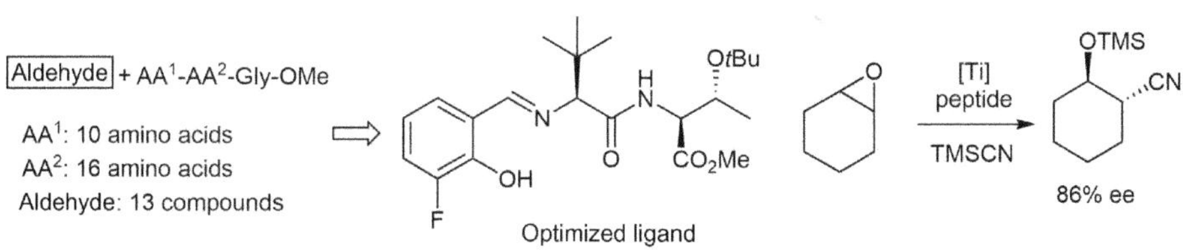

Figure 19.9 Optimization of a Schiff base ligand and titanium-catalyzed reaction.

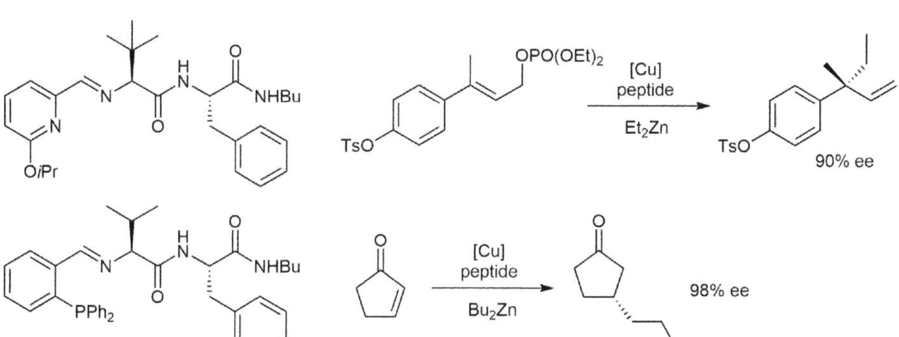

Figure 19.10 Other Schiff base ligands and copper-catalyzed reactions.

formed a complex with a rhodium(I) ion. Later, the same group used a related bisphosphine peptide ligand for palladium-catalyzed asymmetric allylic substitution (Fig. 19.11) [51]. They constructed a resin-supported peptide library and identified an optimum sequence for high enantioselectivity. The turn structure formed by a Pro-D-Val sequence was essential; the replacement of D-Val to Gly resulted in complete loss in the enantioselectivity. Other turn motifs and cyclic structures were examined, but no improvement for the selectivity was observed [52]. Cyclic bisphosphine peptide ligands based on a natural product were prepared and evaluated by Kamer, Overhand, and coworkers [53]. The enantioselectivities of their cyclic peptides for rhodium- and palladium-catalyzed reactions were moderated to low. Breit and coworkers synthesized supramolecular bidentate phosphine ligands with an antiparallel β-sheet peptide structure [54]. The supramolecule promoted rhodium-catalyzed asymmetric hydroformylation, although the enantioselectivity was low.

Meldal and coworkers applied other types of bidentate ligands (Fig. 19.12) for palladium-catalyzed asymmetric allylic substitution. In addition to an N-terminal diphenylphosphino group, the peptide had a 2-oxazoline ring derivatized from a Ser residue [55]. In another case, phosphine and sulfide groups were incorporated into the peptide [56]. These peptide ligands could be used for the reaction without detaching from resin beads after solid-phase synthesis.

Roelfes and coworkers developed an artificial peptide ligand based on the structure of a peptide hormone [57]. Bovine pancreatic polypeptide, a 36-residue peptide

Figure 19.11 Solid-supported β-turn peptide with phosphine side chains and palladium-catalyzed reaction.

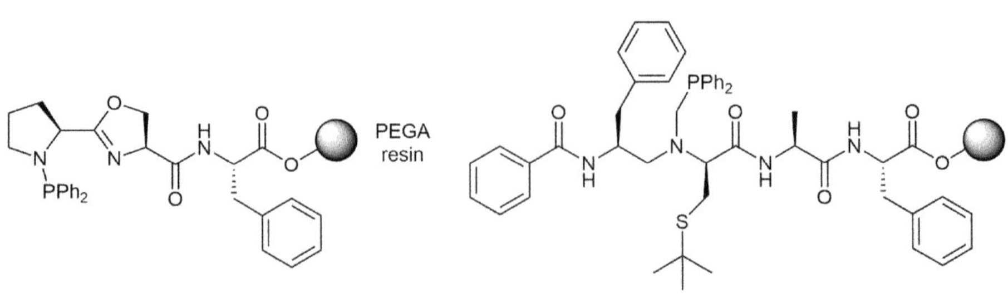

Figure 19.12 Resin-supported phosphine-oxazoline ligand and phosphine-thioether ligand.

hormone, consists of polyproline helix, turn, and α-helix segments, and forms an antiparallel homodimer in solution. It can be expected that the hydrophobic interface between the two subunits provides a chiral pocket for an asymmetric reaction. From the original sequence, C-terminal 5 residues were truncated, and the seventh Tyr was replaced with 3-pyridiylalanine to add a coordinating ability (Fig. 19.13). To prevent the formation of the aspartimide during solid-phase synthesis, the variant with Glu at 10 position was also prepared. In the presence of these ligands, copper-catalyzed Diels–Alder reaction and Michael addition were enhanced in water, and the products were obtained with good enantioselectivity.

Herrmann and coworkers developed cyclic peptide ligands for copper-catalyzed asymmetric Diels–Alder reaction and Friedel–Crafts reaction (Fig. 19.14) [58]. The cyclic structure was constructed by the formation of a disulfide bond between the N- and C-terminal Cys residues. Amino acids between the two Cys residues were randomly screened, and sequences with moderate enantioselectivity were

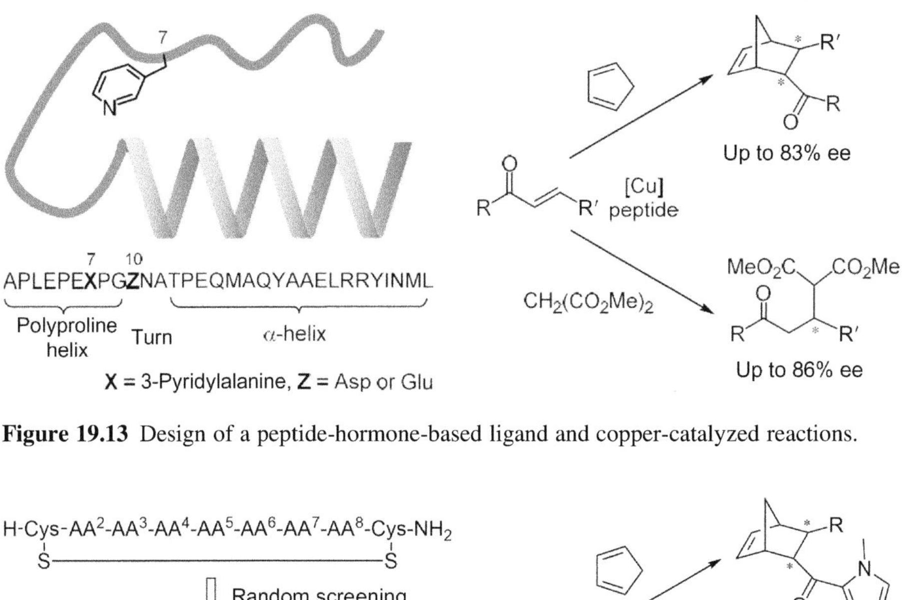

Figure 19.13 Design of a peptide-hormone-based ligand and copper-catalyzed reactions.

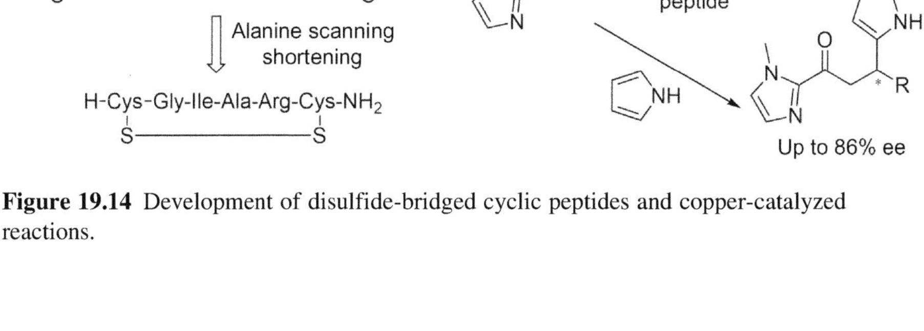

Figure 19.14 Development of disulfide-bridged cyclic peptides and copper-catalyzed reactions.

found. By conducting alanine scanning, important residues were identified. Then, the peptide was shortened to afford an optimum sequence for the highly enantioselective reactions. The ligands used in this study only contain natural amino acids; thus, significant synthetic efforts such as the preparation of non-natural building blocks and the postmodification of peptides are not necessary.

Ball and coworker used an α-helical peptide with two aspartates as the ligand for a rhodium-catalyzed carbenoid reaction (Fig. 19.15) [59]. The Asp residues were introduced at i and $i + 4$ positions for binding a dirhodium catalyst. When two peptide molecules were used for the formation of a dirhodium complex, parallel and antiparallel isomers were generated. These isomers could be separated by HPLC, and one isomer showed higher enantioselectivity than the other. Aspartate-peptide-bound dirhodium complexes were also applied to the selective functionalizations of a helical peptide [60] and a protein [61].

They performed the screenings of peptide libraries constructed on resin-beads to identify enantioselective ligands for dirhodium-catalyzed cyclopropanation [62]. Compared to ligand **L1** developed based on their previous study, ligand **L2** found by the library screening showed the opposite enantioselectivity (Fig. 19.16). They also applied the screening for the exploration of a tridentate peptide, and obtained ligand **L3** with a His residue [63]. In this case, a mono-peptide complex was

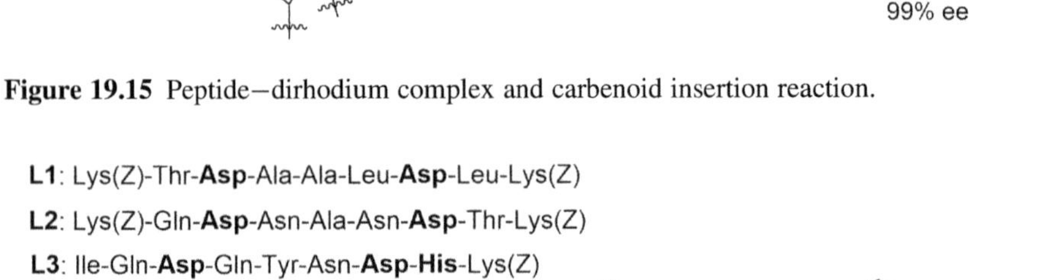

Figure 19.15 Peptide–dirhodium complex and carbenoid insertion reaction.

L1: Lys(Z)-Thr-**Asp**-Ala-Ala-Leu-**Asp**-Leu-Lys(Z)

L2: Lys(Z)-Gln-**Asp**-Asn-Ala-Asn-**Asp**-Thr-Lys(Z)

L3: Ile-Gln-**Asp**-Gln-Tyr-Asn-**Asp**-**His**-Lys(Z)

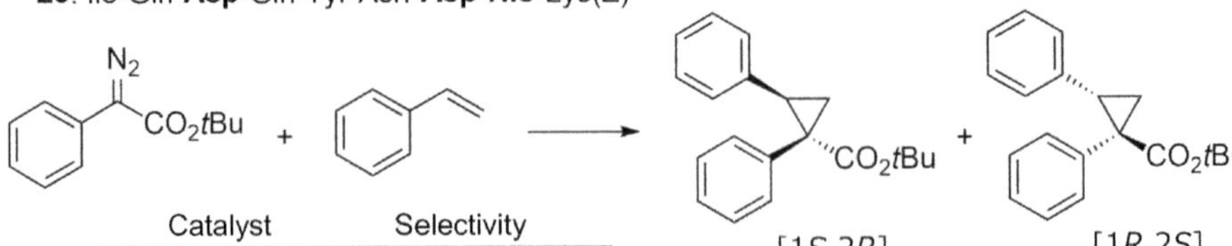

Catalyst	Selectivity
Rh_2(**L1**)$_2$	93% ee [1*S*,2*R*]
Rh_2(**L2**)$_2$	92% ee [1*R*,2*S*]
Rh_2(**L3**)$(OAc)_2$	92% ee [1*R*,2*S*]

Figure 19.16 Stereoselectivity of cyclopropanation by peptide–dirhodium complexes.

effective for the selective reaction. It is indicated that the coordination of His to rhodium induces the conformational change from the original α-helical structure of the peptide. As for the enantioselective cyclopropanation, Shi and coworkers reported dipeptide-promoted Simmons–Smith reaction of olefins [64]. Although a stoichiometric amount of the peptide was used in their first report, they later succeeded in reducing the peptide to a catalytic amount [65,66].

Heme is an important cofactor for oxidative enzymes, and its reactivity is precisely controlled by peptide chains around the iron core. In 1969, peptide-bound hemes were prepared and their catalytic activities were examined for oxidation of cytochrome(II) c and guaiacol [67]. After a few decades, a heme protein analogue with a porphyrin core and two helical peptide chains was synthesized by Pavone and coworkers (Fig. 19.17) [68]. The helical peptide contains a His residue at 5 position, and it can bind to the metal ion taken into the porphyrin center. The iron-complex of this molecule was used for oxidation of styrene [69]. The reaction gave styrene oxide and benzaldehyde, but no enantioselectivity was observed for the generation of styrene oxide. Other groups investigated the catalytic activity of heme with His-containing helical peptides [70] and heme with amyloid β peptides [71] for oxidation of guaiacol and cytochrome(II) c, respectively. Feringa and coworkers synthesized a nonheme iron complex consisting of a tetrapyridyl ligand and helical peptides, and used it for oxidation of tetramethoxybenzene [72].

Other examples of using peptides as ligands are briefly described. Gilbertson and coworkers synthesized the peptide ligands with an *N*-heterocyclic carbene (NHC) side chain [73,74]. Albrecht and coworker used the rhodium complex of an NHC-containing peptide for hydrosilylation of a ketone, although no asymmetric induction was observed [75]. Alper and coworker performed palladium-catalyzed asymmetric carbonylation of an allylic alcohol in the presence of a high-molecular-weight poly(Leu) ligand [76]. Nonaka and coworker used poly(Val)-coated platinum anodes for asymmetric oxidation of sulfides in electrochemical reactions [77,78]. This group also reported electrochemical asymmetric reduction with poly(Val)-coated graphite [79]. Baiker and coworkers modified the surface of alumina-supported platinum with peptides, and used it for enantioselective hydrogenation of ketones [80].

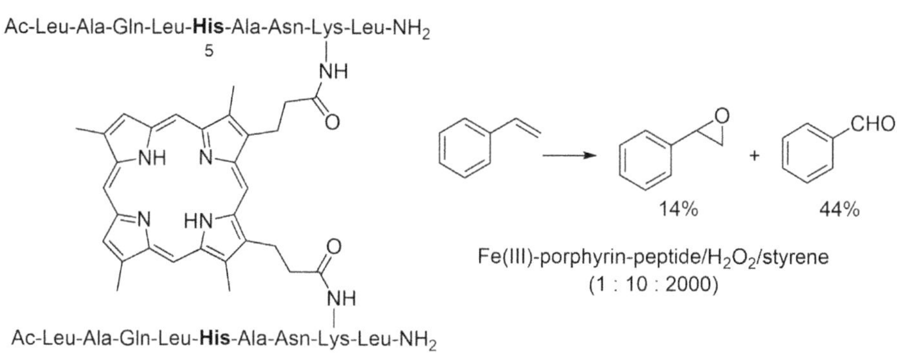

Figure 19.17 Peptide-bound porphyrin and oxidation by an iron-complex.

19.4 Peptides for organocatalysts

In cases that the functional groups of peptides are directly involved in the transformations of organic compounds, the peptides can be regarded as organocatalysts. In the field of peptide catalysis, the use of peptides as asymmetric organocatalysts is most actively investigated [81,82]. Compared to small-molecule catalysts, peptides can provide large reaction sites for recognizing the three-dimensional structures of substrates, which enables novel stereo- and regioselective transformations. In most cases, the peptides with specific secondary structures, such as α-helix and β-turn, are used for the construction of selective reaction sites.

19.4.1 Helical poly(amino acid)s for asymmetric reactions

α-Helix is a representative secondary structure observed in some poly(amino acid)s, which can be readily prepared by polymerization of *N*-carboxyanhydride monomers [83]. In 1975, Inoue and coworkers reported asymmetric thiol addition to an α,β-unsaturated ketone in the presence of α-helical poly[Glu(OBzl)] (Fig. 19.18A) [5,84]. The degree of polymerization largely affected the selectivity, which was ascribed to the conformational changes of peptides. In 1980, Juliá and coworkers reported highly enantioselective epoxidation of chalcone with poly(amino acid)s such as poly(Ala) (Fig. 19.18B) [85]. The reaction is called Juliá–Colonna epoxidation, and has been intensively studied to expand a substrate scope and improve reaction efficiency [86,87]. Commonly, poly(Leu) is used as a stable α-helical peptide catalyst. Mechanistic investigations indicate that the hydrogen bonds between reactants and a few amide protons at the N-terminus of the helical peptide are the key for promoting the enantioselective reaction [88–90]. To facilitate the preparation and reuse of the catalyst, various types of supported poly(Leu) have been employed [91–95]. In addition to Juliá–Colonna epoxidation, poly(Leu) was

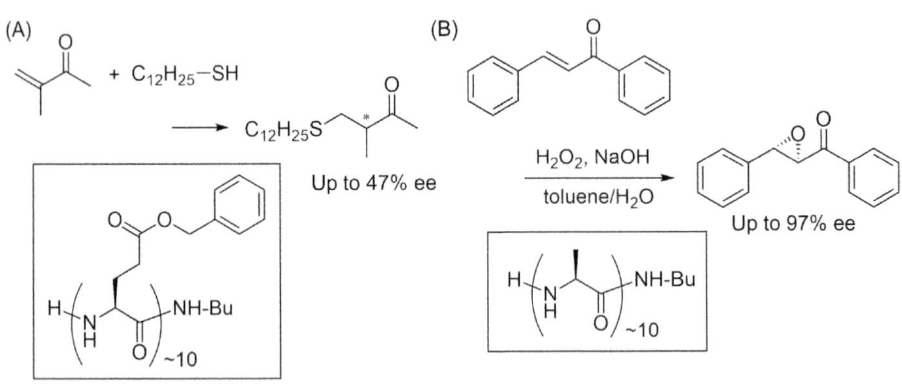

Figure 19.18 (A) α-Helical-peptide-catalyzed thiol addition. (B) Juliá–Colonna epoxidation.

applied to other asymmetric reactions such as an aldol reaction [96] and cyanosilylation of aldehydes [97].

19.4.2 Peptides for aldol reactions

Since List and coworkers reported a proline-catalyzed intermolecular aldol reaction in 2000 [6], the development of N-terminal prolyl peptides has become a trend in the field of peptide catalysis. The secondary amino group of a prolyl residue activates carbonyl compounds through the formation of enamine or iminium ion intermediates, and promotes a wide range of reactions (Fig. 19.19).

The asymmetric aldol reaction between an aldehyde and a ketone is the representative reaction proceeding through the enamine mechanism. A number of peptides have been reported as efficient and enantioselective catalysts for the aldol reaction (Fig. 19.20A) [10,11,98–113]. While most of the catalysts are N-terminal prolyl peptides, some peptides have a different residue at the N-terminus [100,107]. Wennemers and coworkers developed the strategy for screening a resin-supported peptide library in a high-throughput manner [114], and applied it to discover capable peptide catalysts for the asymmetric aldol reaction [115]. Two kinds of peptides were identified as highly reactive catalysts (Fig. 19.20B). Interestingly, although two peptides have a common L-Pro residue at the N-terminus, they gave opposite enantiomers as a major aldol product. Computational modelings indicate that both peptides adopt turn-like conformations, and the mirror image of the one peptide is close to the other. Reymond and coworkers developed a different screening method, and applied it to dendrimer-type peptide libraries [116]. Da and coworkers originally developed a tetrapeptide for the asymmetric aldol reaction [107]. Later, they used the same catalyst for kinetic resolution of racemic aldol products by dehydration [117].

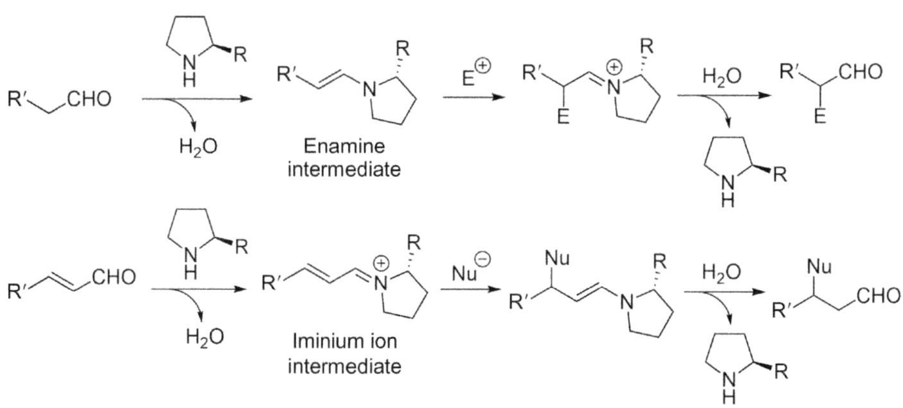

Figure 19.19 Prolyl-peptide-catalyzed reactions by enamine and iminium mechanisms.

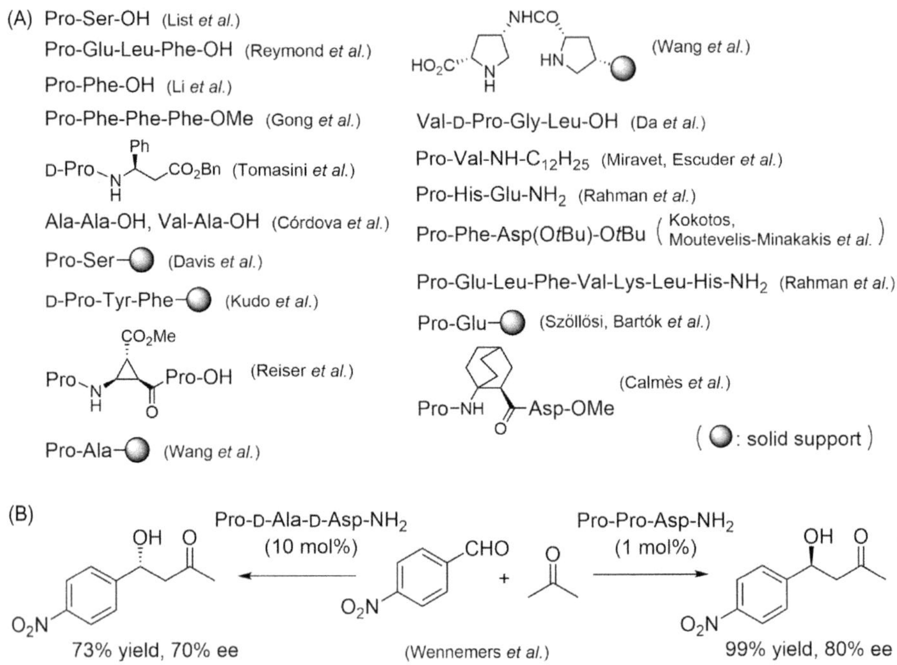

Figure 19.20 (A) Peptide catalysts for aldol reactions. (B) Aldol reaction by the peptides with complementary enantioselectivity.

19.4.3 Peptides for α-functionalizations

The enamine intermediate formed between an N-terminal prolyl peptide and an aldehyde reacts with an electrophile to give an α-functionalized product. Tanaka, Kudo, and coworkers used a resin-supported tetrapeptide for enantioselective α-amination of aldehydes (Fig.19.21A) [118]. The supported catalyst could be recovered by filtration and reused multiple times. They also performed α-oxyamination of aldehydes with a finely designed peptide catalyst (Fig. 19.21B) [119]. The peptide contains β-turn motif D-Pro-Aib and α-helical poly(Leu). The combination of these secondary structures was essential for high enantioselectivity. It is suggested that the presence of the helical part stabilizes the turn structure necessary for the selective reaction [120]. The peptide-catalyzed α-oxyamination was extended to sequential oxidation [121] and enzyme-cocatalyzed reactions [122,123].

19.4.4 Peptides for michael reactions via enamine mechanism

The enamine activation by an N-terminal prolyl peptide can be applied to Michael reactions. Wennemers and coworkers reported that tripeptides, D-Pro-Pro-Xaa, were effective for highly diastereo- and enantioselective Michael reactions (Fig. 19.22). They used D-Pro-Pro-Asp and D-Pro-Pro-Glu sequences for addition of aldehydes to

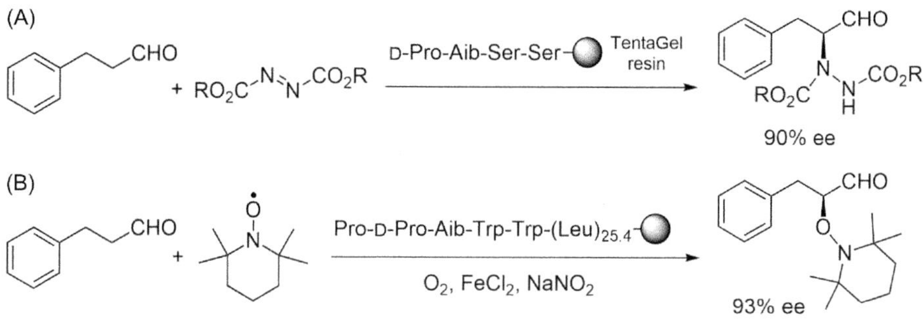

Figure 19.21 (A) Tetrapeptide-catalyzed α-amination. (B) Turn-helix-peptide-catalyzed α-oxyamination.

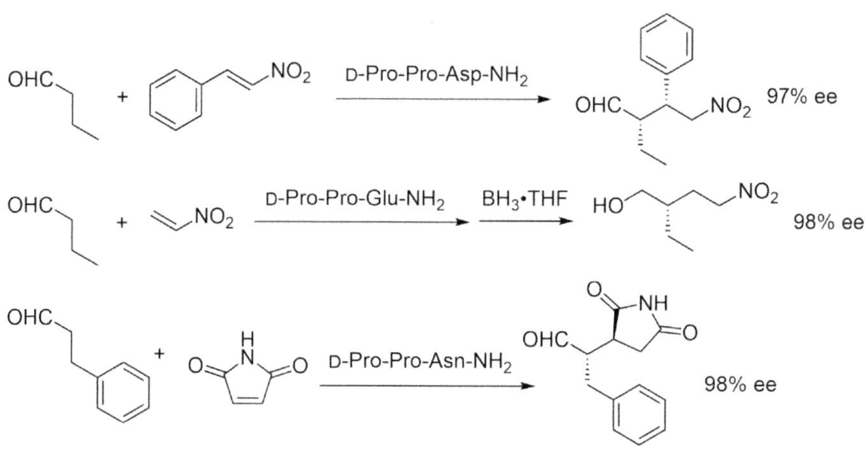

Figure 19.22 Michael additions by D-Pro-Pro-Xaa catalysts.

β-monosubstituted and nonsubstituted nitroolefins [124,125]. The catalytic efficiency of the peptide was improved by performing the reaction under dry conditions, in which the amount of the catalyst could be reduced to 0.1 mol% [126]. They extended the substrate scope to α,β- and β,β-disubstituted nitroolefins by modifying peptide sequences [127,128]. For Michael addition of aldehydes to maleimide, D-Pro-Pro-Asn sequence was used [129]. They attached the D-Pro-Pro-Glu sequence to TentaGel resin, and used it as a solid-supported catalyst with excellent reusability [130]. By changing the resin to crosslinked polystyrene, the catalyst was successfully applied to a continuous flow system [131]. Introducing an amphiphilic part at the C-terminus of the tripeptide caused the formation of an emulsion in aqueous media, which was effective for promoting the stereoselective reaction [132]. Other groups also reported peptide-catalyzed Michael addition of ketones or aldehydes to nitroolefins. Córdova and coworkers used Ala-Ala and Ala-D-Ala for diastereo- and enantioselective reactions [133]. Tsogoeva and coworkers performed an on-water reaction with Pro-Phe [134]. Rivera, Paixão, and coworkers constructed a series of peptide–peptoid hybrids by four-component Ugi reactions, and evaluated their catalytic abilities [135]. Cass, Paixão, and coworkers applied a silica-supported peptide catalyst for a continuous flow system [136]. Deschamp, Lecouvey, and coworkers synthesized a prolyl peptide with a phosphonic acid group, and used it as a recyclable bifunctional catalyst [137].

19.4.5 Peptides for michael reactions via iminium mechanism

In the case that an α,β-unsaturated aldehyde is activated by an N-terminal prolyl peptide through the formation of an iminium ion intermediate, it functions as a Michael acceptor and reacts with a variety of nucleophiles. Kudo and coworkers developed resin-supported turn-helix-peptides (Fig. 19.23A), and used them for transfer hydrogenation [138], Friedel–Crafts-type alkylation [139], epoxidation [140], addition of boronic acids [141], cyclopropanation [142], and addition of thiols [143]. Addition of nitromethane to β,β-disubstituted unsaturated aldehydes gave the products with an all-carbon quaternary stereocenter, which could be derivatized to β,β-disubstituted γ-amino acids [144]. They also applied these types of peptide catalysts to novel selective reactions such as regioselective asymmetric reduction of α,β,γ,δ-unsaturated aldehydes (Fig. 19.23B) [145], kinetic resolution of planar-chiral metallocenes [146], and the induction of planar chirality by desymmetrization of a prochiral ferrocenyl compound (Fig. 19.23C) [147]. These reactions are difficult to perform in a selective way by conventional small-molecule catalysts.

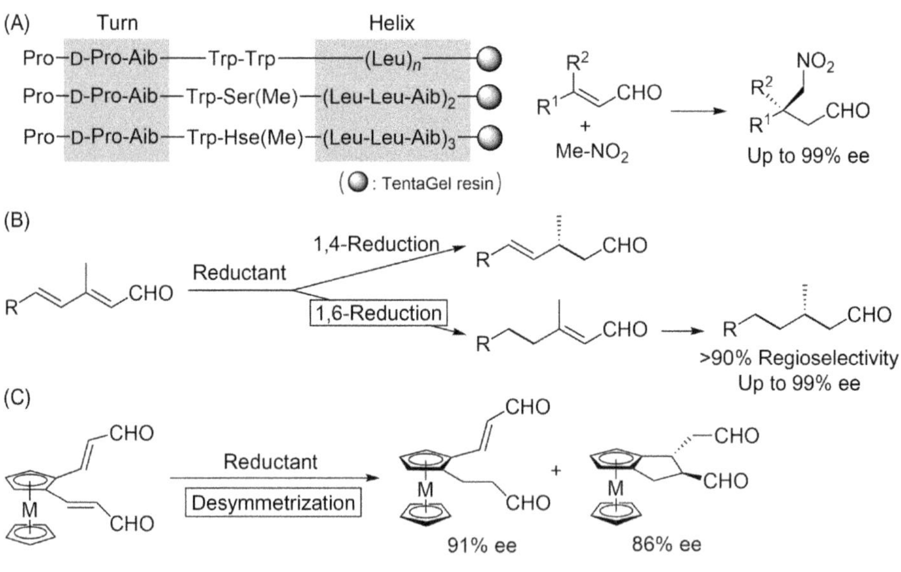

Figure 19.23 (A) Resin-supported peptides with turn and helix structures, and Michael addition of nitromethane. (B) Regioselective asymmetric reduction. (C) Induction of planar-chirality by desymmetrization.

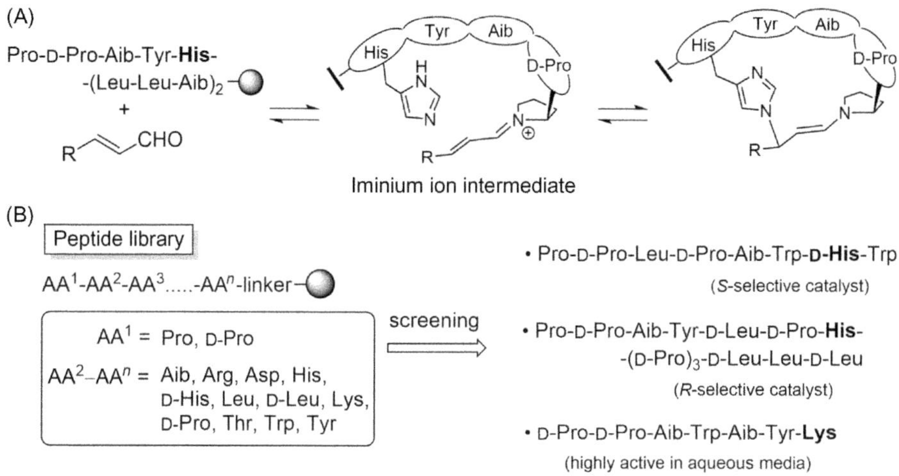

Figure 19.24 (A) His-containing bifunctional peptide and schematic illustration for binding a substrate. (B) Peptide sequences developed by library screenings.

To facilitate the optimization of peptide sequences, Kudo and coworkers developed a high-throughput method for screening a resin-supported peptide library [148]. In this method, a dye-labeled product is introduced to resin beads, and the catalytic activities of peptides can be evaluated by color. The screening of a turn-helix-peptide library afforded a His-containing peptide (Fig. 19.24A) as an efficient catalyst for asymmetric Michael addition of malonates. It is proposed that the peptide has the ability to bind a substrate through the formation of the iminium ion intermediate followed by reversible addition of the His residue. They applied the screening method to completely randomized peptide libraries for exploring new structural frameworks (Fig. 19.24B). The screening in organic solvent gave two types of His-containing peptides, which showed opposite enantioselectivity despite a common N-terminal Pro-D-Pro sequence [149]. A Lys-containing peptide obtained by the screening under aqueous conditions catalyzed an aqueous reaction with much higher efficiency than that of a previously developed peptide [150].

19.4.6 Helical peptides for primary-amine-catalyzed reactions

Although N-terminal prolyl peptides catalyze various types of reactions, sterically hindered substrates are not applicable due to the steric repulsion with the secondary amino group of Pro. The substrate scope is expected to be enlarged by introducing other amino acids at the N-terminus. Kudo and coworkers developed a resin-supported peptide with a 3_{10}-helical (Leu-Leu-Aib)$_3$ sequence for primary-amine-catalyzed reactions (Fig. 19.25A) [151]. With this peptide, α,β-unsaturated ketones and an α-substituted aldehyde could be used as the substrates for iminium- and enamine-catalyzed reactions. The helicity of the peptide is important for both the reaction efficiency and the enantioselectivity. They applied this type of peptide to

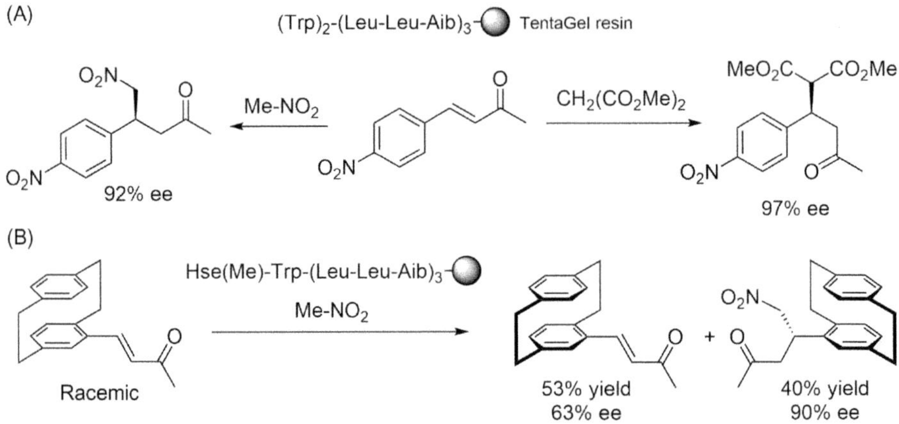

Figure 19.25 (A) Primary-amine-catalyzed asymmetric reactions with a resin-supported helical peptide. (B) Kinetic resolution of a planar-chiral cyclophane. *Aib*, 2-aminoisobutyric acid; *Hse(Me)*, homoserine methyl ether.

kinetic resolution of a planar-chiral cyclophane derivative by Michael addition of nitromethane (Fig. 19.25B) [152]. Planar chirality derives from the whole structure of a molecule; therefore, a large reaction space is necessary to discriminate enantiomers. While a small-molecule catalyst only showed poor selectivity, the helical peptide promoted the reaction in a highly enantioselective manner. They also used a related peptide for asymmetric epoxidation of α,β-unsaturated ketones [153]. Because of the difference in the reaction mechanisms, the major enantiomer of an epoxy-product was opposite to that obtained in Juliá–Colonna epoxidation [86,87]. Tanaka and coworkers incorporated α,α-disubstituted amino acids into peptides to increase α-helicity, and employed them for both Juliá–Colonna epoxidation [154] and iminium-catalyzed reactions [155]. There are a few examples that the amino groups in the side chain of helical peptides are used for the reactions proceeding through the formation of iminium ion intermediates. Decarboxylation of oxaloacetate [156] and the retro-aldol reaction of a pyruvate-benzaldehyde-adduct [157] were reported, although they were nonstereoselective.

19.4.7 Aspartyl peptides for oxidations

Miller and coworkers developed a novel oxidation system using the side chain of Asp as a catalyst [158,159]. In the catalytic cycle (Fig. 19.26A), the carboxylic acid of Asp is converted to a peracid by a carbodiimide and an oxidant such as hydrogen peroxide. Then, the peracid oxidizes a substrate and returns to the carboxylic acid. A tripeptide promoted enantioselective epoxidation of olefins with a carbamate functionality (Fig. 19.26B). The hydrogen bond between the amide group of the peptide and the carbamate part of the substrate is important for the selectivity.

They applied this oxidation system to site-selective epoxidation of polyenes. Remarkably, the peptide for 6,7-selective epoxidation of farnesol was identified

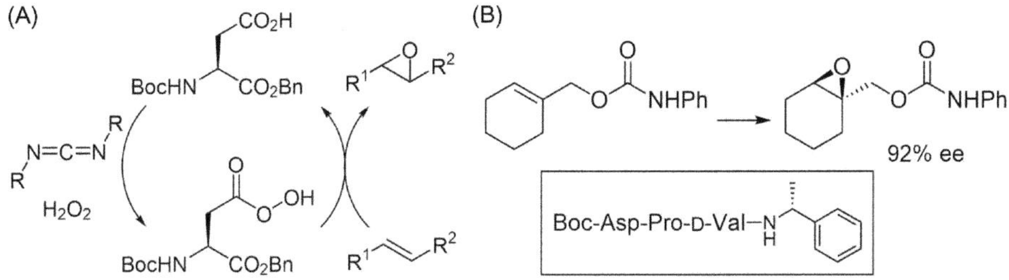

Figure 19.26 (A) Catalytic cycle of Asp-catalyzed oxidation. (B) Peptide-catalyzed asymmetric epoxidation.

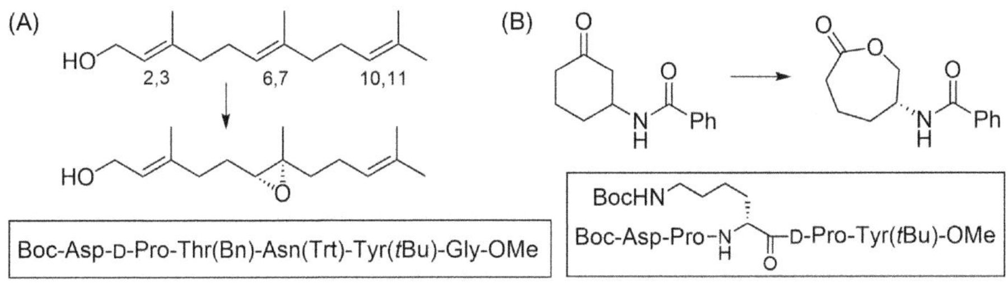

Figure 19.27 (A) Site-selective epoxidation of farnesol. (B) Regioselective Baeyer–Villiger oxidation.

through library screenings (Fig. 19.27A) [160,161]. The 6,7-position of farnesol is remote from the terminal hydroxy group that can be a directing group for 2,3-epoxidation [162]; thus, discriminating the 6,7-position for the selective conversion is challenging. It is proposed that the hydrogen bond network between the substrate, the protected side chain of Tyr, and the amide protons of the peptide chain directs the 6,7-position toward the terminal Asp functionality. The Asp-catalyzed system was also applied to enantioselective oxidation of indoles [163] and Baeyer–Villiger oxidation of ketones (Fig. 19.27B) [164]. In the latter reaction, the inherent regioselectivity for the insertion of an oxygen atom was changed by the peptide catalyst. As a related catalytic system, the peptide with a trifluoromethyl ketone group was used for epoxidation of olefins [165].

19.4.8 Peptides with imidazolyl groups

An N-alkylimidazolyl group accelerates acylation of alcohols through the formation of an acyl imidazolium intermediate with an acid anhydride. The peptides containing a His derivative have been used for selective acylation. In 1998, Miller and coworkers reported kinetic resolution of racemic alcohols by acylation with an imidazole-containing tripeptide [166]. The selectivity of the resolution was improved by the use of tetrapeptides with a His(3-Me) residue at the N-terminus (Fig. 19.28) [167]. The key for the high enantioselectivity is the formation of a

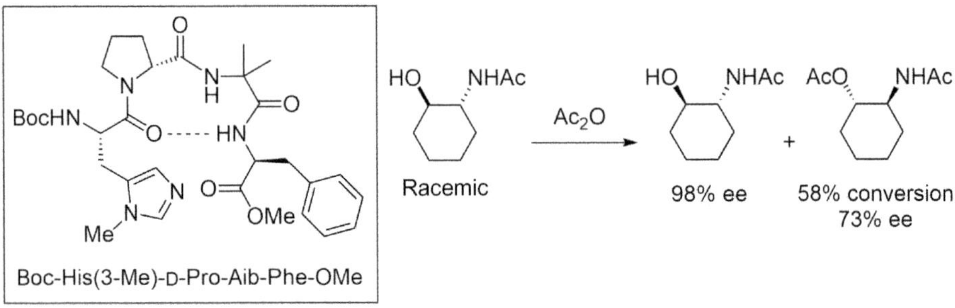

Figure 19.28 Peptide with a methylimidazolyl group and kinetic resolution of a racemic alcohol.

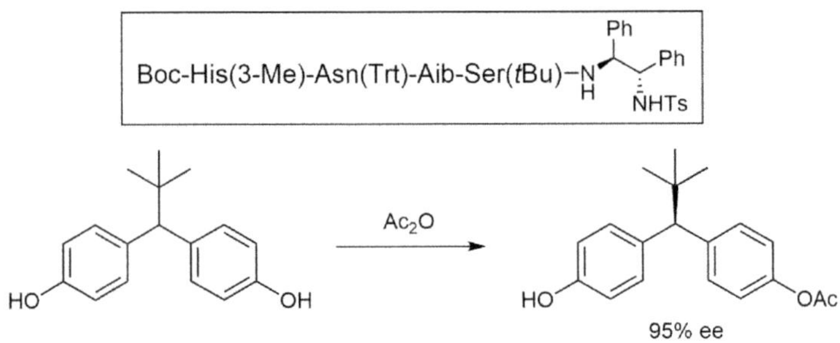

Figure 19.29 Desymmetrization of a bis(phenol).

type II′ β-turn structure, which is stabilized by a D-Pro-Aib sequence. Such turn-type peptides have been used for various selective transformations by Miller's group [168]. They developed a fluorescence-based screening method to find reactive acylation catalysts from a resin-supported peptide library [169]. With the aid of this method, they obtained an octapeptide effective for kinetic resolution of a range of alcohols [170]. A similar fluorescence-based assay was used to identify a pentapeptide for the resolution of tertiary alcohols [171]. Toniolo and coworkers [172] and Qu and coworker [173], respectively, modified the turn part of Miller's peptide, and examined the effects on kinetic resolution of alcohols.

Miller and coworkers performed a desymmetrization reaction of a prochiral glycerol derivative by peptide-catalyzed acylation [174]. To induce chirality and obtain an optically active product, one of the two hydroxy groups of the substrate was selectively converted to an ester with a pentapeptide. Later, they achieved desymmetrization of a more challenging bis(phenol) compound (Fig. 19.29) [175]. Because the prochiral stereocenter of this substrate is distant from hydroxy groups, selective acylation by discriminating the enantiotopic site is difficult. After five rounds of library screenings and the optimization of a C-terminal substituent, a highly selective peptide catalyst was obtained. As for the related desymmetrization reaction of diarylmethanes, they also performed copper-catalyzed Ullman cross-coupling with a peptide ligand having a guanidyl group [176].

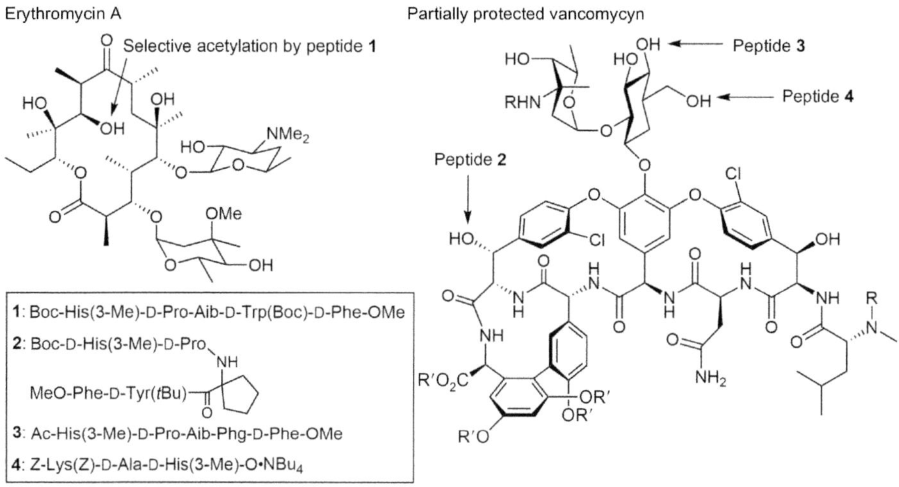

Figure 19.30 Positions of peptide-catalyzed mono-acylation. *Phg*, 2-phenylglycine.

His(3-Me)-containing peptides were employed for site-selective acylation of complex molecules with multiple hydroxy groups. Specific positions of glucosamine and glucose derivatives [177], erythromycin A [178], and partially protected vancomycin [179] were selectively acylated by the peptides that were respectively optimized for discriminating reaction sites (Fig. 19.30). The peptide catalysts changed the inherent reactivities of the hydroxy groups in these molecules, and promoted the selective reactions in a catalyst-controlled manner. As for the site-selective acylation, Kawabata and coworkers developed a peptide-mimetic catalyst with a pyridyl group for monoacylation of glucose derivatives [180,181]. They applied related catalysts for novel selective acylations such as length- [182], regio- [183], and geometry-selective [184] ones. Kirsch and coworker constructed a peptide library, in which the amino acid with a 4-aminopyridyl moiety was incorporated in the peptide sequences [185]. Compounds with multiple hydroxy groups were subjected to acylation, and site-selective peptide catalysts were chosen from the library.

The N-alkylimidazolyl group can be applied to other reactions proceeding through the formation of an imidazolium intermediate. Miller and coworkers performed desymmetrization of *meso*-1,3-diols by phosphorylation [186] and sulfonylation [187] with His(3-Me)-containing peptide catalysts (Fig. 19.31). The sense of enantioselectivity in phosphorylation could be reversed by changing peptide sequences, and both enantiomers of the monophosphate product were obtained [188]. Site-selective phosphorylation of a complex molecule was also realized; each of three distinct hydroxy groups of a teicoplanin derivative selectively underwent the reaction by three kinds of peptide catalysts [189].

The strategy for the site-selective functionalization of hydroxy groups was extended to deoxygenation of polyols. After performing site-selective thiocarbonylation of sugar derivatives with His(3-Me)-containing peptides, the oxygen atom was

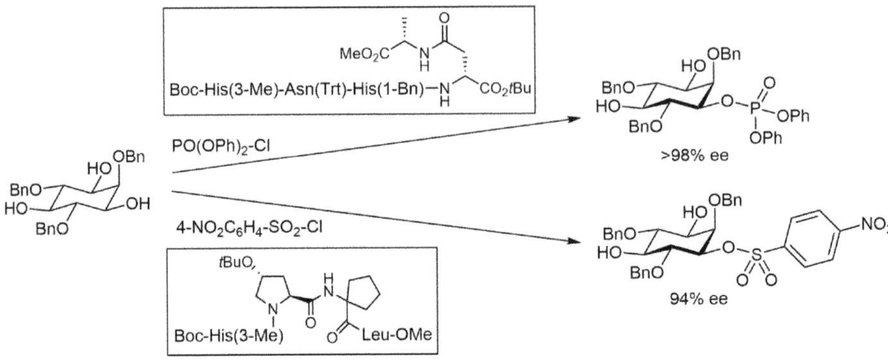

Figure 19.31 Desymmetrization by phosphorylation and sulfonylation.

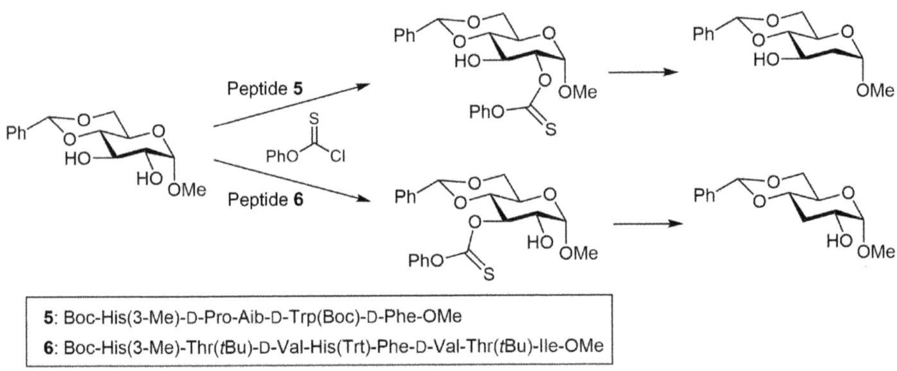

Figure 19.32 Site-selective thiocarbonylation and deoxygenation.

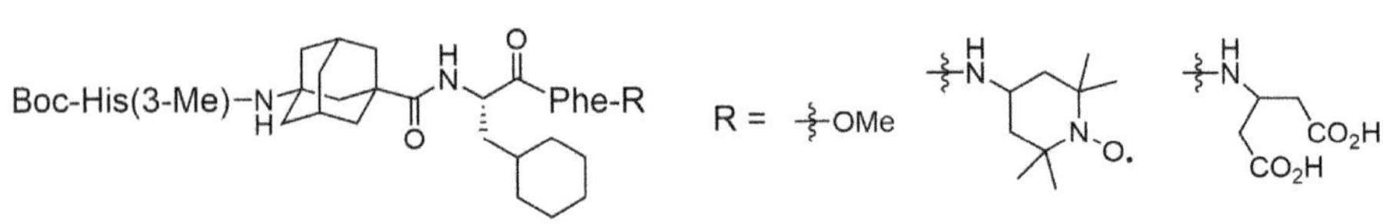

Figure 19.33 His(3-Me)-peptide with an adamantane framework and bifunctional peptide catalysts.

removed by the Barton–McCombie reaction (Fig. 19.32) [190]. This two-step procedure was employed for site-selective deoxygenation of protected vancomycin [191]. To improve a substrate scope of site-selective deoxygenation, a new protocol including phosphitylation of a hydroxy group with a tetrazole-containing peptide and radical deoxygenation was reported [192].

Schreiner and coworkers developed a tetrapeptide containing His(3-Me) and an adamantane amino acid for kinetic resolution of diols (Fig. 19.33) [193,194]. The catalyst was applied to Steglich esterification using a carboxylic acid and a carbodiimide as acylation reagents [195]. A related catalyst was used for enantioselective Dakin–West reaction [196]. They also synthesized the bifunctional peptide catalyst with a tetramethylpiperidine *N*-oxyl moiety [197] and one with carboxylic acid functionalities [198]. These catalysts were used for one-pot sequential reactions including an oxidation step.

In addition to the modification of alcohols, His-containing peptides have been employed for various types of reactions. In 1979, Inoue and coworkers reported hydrocyanation of an aldehyde catalyzed by a cyclic dipeptide, *cyclo*[Ala-His] [199]. After examining reaction conditions [200] and peptide structures, a highly enantioselective reaction was achieved by *cyclo*[Phe-His] [201]. Miller and coworkers used the peptides with alkylated His groups for asymmetric conjugate addition of azide ion [202,203], proline-cocatalyzed asymmetric Morita–Baylis–Hillman reaction [204,205], dynamic kinetic resolution of sulfinyl chloride by esterification [206], asymmetric conjugate addition of α-nitroketones [207], and kinetic resolution of formamides and thioformamides by N-acylation [208].

19.4.9 Peptide with other functional groups

Cyclic dipeptide *cyclo*[Phe-His] reported by Inoue and coworkers was not effective for Strecker reaction using an imine for hydrocyanation. Lipton and coworkers developed a cyclic dipeptide with a guanidyl group to increase the basicity of the catalyst, and successfully applied it to asymmetric Strecker reaction [209].

Miller and coworkers found that addition of an allenoate to an *N*-acylimine was catalyzed by an amino acid derivative with a pyridyl group, whereas an *N*-methylimidazolyl group did not promote the reaction [210]. By incorporating a pyridine-containing residue into a turn peptide, an enantioselective reaction was achieved (Fig. 19.34). A further tuning of the peptide structure extended the substrate scope [211].

Atropisomerism can be seen in biaryl compounds with *ortho*-substituents. When the rotational barrier of the bond between two aromatic rings is low, racemization of atropisomers easily takes place. Millar and coworkers performed bromination of a biaryl substrate to convert it into a stable atropisomeric product (Fig. 19.35) [212]. A tripeptide containing a dimethylamino group promoted the selective reaction for one atropisomer of the substrate. Rapid racemization of the other isomer

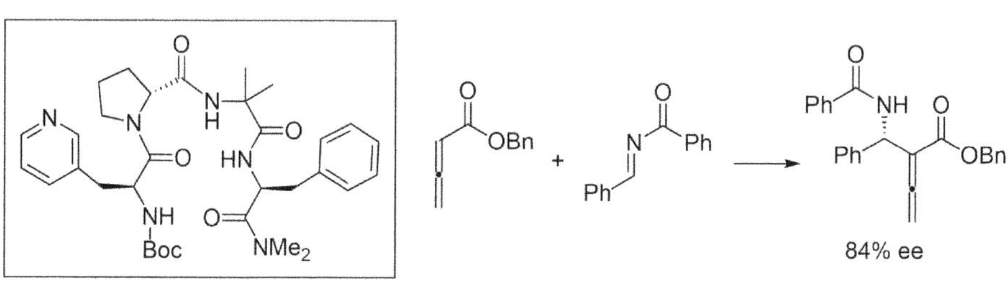

Figure 19.34 Peptide with a pyridyl group and addition of an allenoate to an imine.

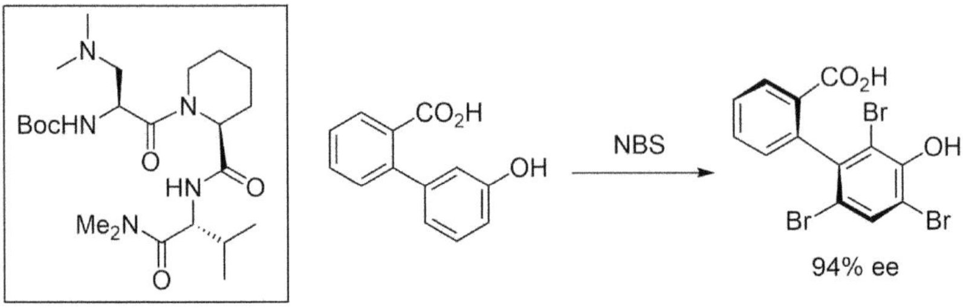

Figure 19.35 Peptide with a dimethylamino group and dynamic kinetic resolution of an atropisomeric compound.

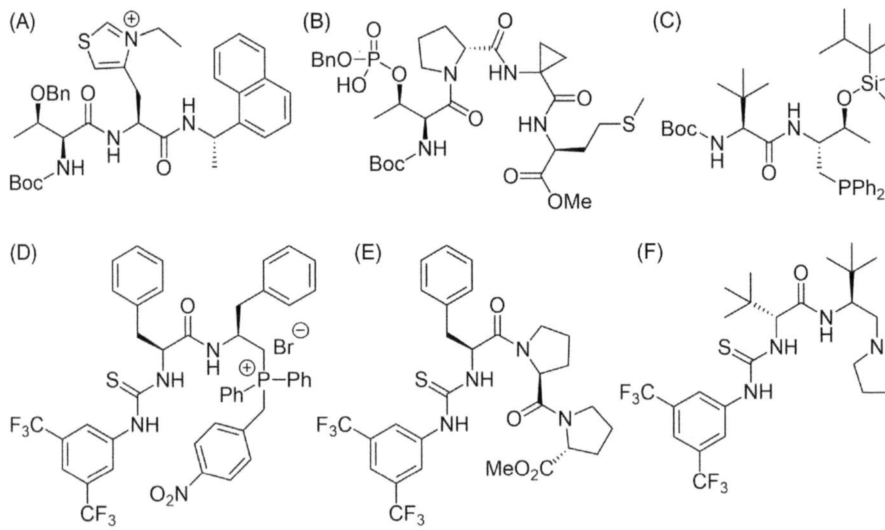

Figure 19.36 Peptides with various functionalities; (A) thiazolium ion, (B) phosphoric acid, (C) phosphine, (D) phosphonium ion, (E) thiourea, and (F) thiourea-amine.

enabled dynamic kinetic resolution, in which both isomers converged into the same enantiomer. Other atropisomeric compounds such as benzamides [213] and a quinazolinone [214] were employed for peptide-catalyzed enantioselective bromination. Site-selective bromination of teicoplanin, a natural product with biaryl substructures, was attained by tripeptides [215]. The peptide with a dimethylamino group was also used for dynamic kinetic resolution of azlactones by methanolysis [216].

Thiazolium salts are typical catalysts for the Stetter reaction. Miller and coworkers introduced a thiazolium moiety into the side chain of a peptide (Fig. 19.36A), and used it for intramolecular asymmetric Stetter reaction [217]. This type of peptide catalyst was also used for enantioselective coupling reactions of aldehydes and imines [218].

In addition to basic and nucleophilic catalysts, Miller and coworker developed a peptide for Brønsted acid catalysis. They attached phosphothreonine to the

N-terminus of a turn peptide (Fig. 19.36B), and employed it for asymmetric transfer hydrogenation [219]. The peptide promoted enantioselective reduction of quinolines with amine-derived functionalities.

Chiral phosphines derived from amino acids are versatile organocatalysts for asymmetric transformations [220,221]. Lu and coworkers developed a dipeptide-based phosphine catalyst (Fig. 19.36C), and used it for enantioselective [3 + 2] cycloaddition of allenoates and imines [222]. Zhao and coworkers used a dipeptide-derived compound with a quaternary phosphonium moiety (Fig. 19.36D) as a chiral phase-transfer catalyst [223]. The catalyst also has a thiourea functionality, which is considered to be a hydrogen bond donor for interacting with a substrate.

Chimni and coworkers introduced a thiourea moiety at the N-terminus of a Phe-Pro-D-Pro sequence (Fig. 19.36E) [224]. The peptide was used for the asymmetric ring-opening reaction of *meso*-stilbene oxide. Lee, Jiang, and coworkers synthesized a catalyst with thiourea and pyrrolidine functionalities based on a *tert*-leucine dimer (Fig. 19.36F) [225]. The bifunctional catalyst promoted a sequential Michael-Mannich-type reaction to give a product with multiple quaternary carbon stereocenters.

The application of the thiol group of a peptide for an organocatalytic reaction is limited, although it has been shown that a simple Cys derivative is effective for asymmetric Rauhut–Currier reaction [226,227]. Ghadiri and coworkers designed a Cys-containing α-helical peptide for the catalytic generation of a diketopiperazine (Fig. 19.37) [228]. The peptide forms a supramolecular assembly, and the reaction takes place at the interface of the peptide chains. The thiol groups of Cys bind amino acid thioesters by transthioesterification, and the resulting adducts undergo dimerization followed by cyclization to give the product.

The peptides for organocatalytic reactions are summarized in Table 19.1.

19.5 Conclusions

Peptides have been applied to various types of reactions as capable catalysts. Their structures and functionalities can be easily diversified by changing amino acid components, which ensures the high potential of peptide catalysis. Peptides can be rationally designed by using secondary structures, such as α-helix and β-turn, to construct appropriate reaction environments around catalytic centers. The structural frameworks of natural peptides are also available. By modifying the sequences of wild-type peptides, catalytic functions can be customized. Another powerful method for the development of peptide catalysts is the screening of peptide libraries. Combinatorial approaches facilitate both the discovery of new functional peptides and the optimization of sequences. In addition to conventional enantioselective reactions, fine tuned peptides promoted kinetic resolution of planar- and axial-chiral compounds, regioselective reactions, and site-selective conversions of complex natural products. Including such applications of peptides for novel transformations, the use of peptides as efficient and selective catalysts will be continuously explored in the future.

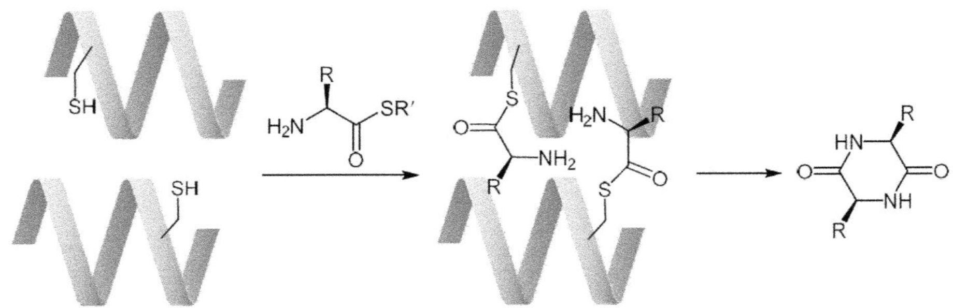

Figure 19.37 Helical-peptide-catalyzed formation of a diketopiperazine.

Table 19.1 Application of peptides for organocatalytic reactions

Peptide	Application	References
Poly[Glu(OBzl)]	Michael addition of thiol	[5,84]
Poly(Ala)	Juliá–Colonna epoxidation	[85]
Poly(Leu)	Juliá–Colonna epoxidation	[86,87]
	Aldol reaction, cyanosilylation	[96,97]
Pro-Ser-OH	Aldol reaction	[10]
Pro-Glu-Leu-Phe-OH	Aldol reaction	[11]
Pro-Phe-OH	Aldol reaction	[98]
Pro-Phe-Phe-Phe-OMe	Aldol reaction	[99]
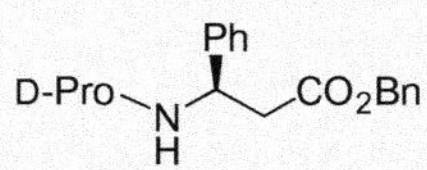	Aldol reaction	[100]
Ala-Ala-OH, Val-Ala-OH	Aldol reaction	[101]
Pro-Ser-resin	Aldol reaction	[102]
D-Pro-Tyr-Phe-resin	Aldol reaction	[103]
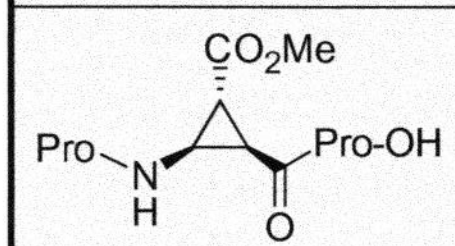	Aldol reaction	[104]
Pro-Ala-resin	Aldol reaction	[105]
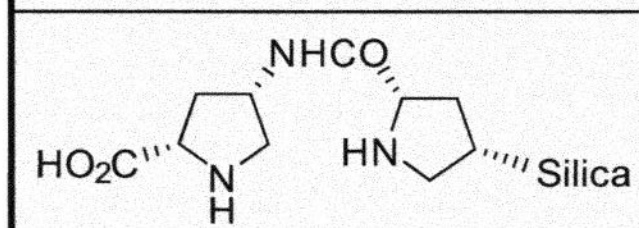	Aldol reaction	[106]
Val-D-Pro-Gly-Leu-OH	Aldol reaction, dehydration	[107,117]

(*Continued*)

Table 19.1 **(Continued)**

Peptide	Application	References
Pro-Val-NH-$C_{12}H_{25}$	Aldol reaction	[108]
Pro-His-Glu-NH_2	Aldol reaction	[109]
Pro-Phe-Asp(O*t*Bu)-O*t*Bu	Aldol reaction	[110]
Pro-Glu-Leu-Phe-Val-Lys-Leu-His-NH_2	Aldol reaction	[111]
Pro-Glu-resin	Aldol reaction	[112]
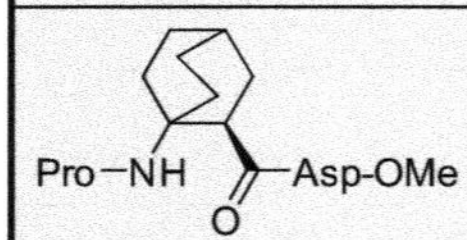	Aldol reaction	[113]
Pro-D-Ala-D-Asp-NH_2, Pro-Pro-Asp-NH_2	Aldol reaction	[115]
Dendrimer-type prolyl peptide	Aldol reaction	[116]
D-Pro-Aib-Ser-Ser-resin	α-Amination	[118]
Pro-D-Pro-Aib-Trp-Trp-(Leu)$_n$-resin	α-Oxyamination	[119]
	Sequential oxidation	[121]
	Enzyme-cocatalyzed reaction	[122,123]
D-Pro-Pro-Asp-NH_2	Michael addition to nitroolefin	[124]
D-Pro-Pro-Glu-NH_2	Michael addition to nitroolefin	[125,126]
D-Pro-Pro-Glu-resin		[130,131]
D-Pro-Pro-Glu-NH-$C_{12}H_{25}$		[132]
Pro-Pro-D-Gln-OH, Pro-Pro-Asn-OH	Michael addition to nitroolefin	[127]

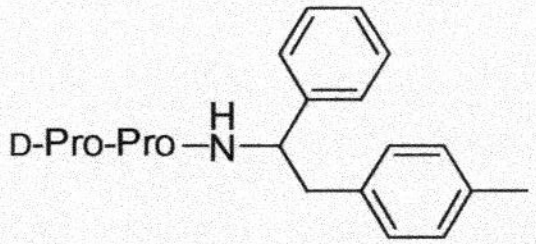	Michael addition to nitroolefin	[128]
D-Pro-Pro-Asn-NH_2	Michael addition to maleimide	[129]
Ala-Ala-OH, Ala-D-Ala-OH	Michael addition to nitroolefin	[133]
Pro-Phe-OH	Michael addition to nitroolefin	[134]
Prolyl peptide-peptoid hybrid	Michael addition to nitroolefin	[135]
Silica-supported prolyl peptide	Michael addition to nitroolefin	[136]
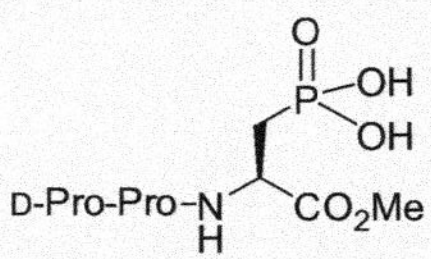	Michael addition to nitroolefin	[137]
Pro-D-Pro-Aib-Trp-Trp-$(Leu)_n$-resin	Transfer hydrogenation	[138]
Pro-D-Pro-Aib-Trp-Trp-$(Leu\text{-}Leu\text{-}Aib)_2$-resin	Friedel–Crafts alkylation	[120,139]
	Michael addition of thiol	[143]
	Michael addition of nitromethane	[144]
D-Pro-Ach-$[Ala(1\text{-}Pyn)]_3$-$(Leu)_n$-resin	Epoxidation	[140]
Pro-D-Pro-Aib-$(Ala)_5$-resin	Michael addition of boronic acid	[141]
Pro-D-Pro-Aib-Trp-Ser(Me)-$(Leu\text{-}Leu\text{-}Aib)_2$-resin	Cyclopropanation	[142]
Pro-D-Pro-Ach-$[Trp(5\text{-}OMe)]_2$-$(Leu)_6$-resin	Transfer hydrogenation	[145]
Pro-D-Pro-Aib-Trp-Hse(Me)-$(Leu\text{-}Leu\text{-}Aib)_3$-resin	Transfer hydrogenation	[146,147]
Pro-D-Pro-Aib-Tyr-His-$(Leu\text{-}Leu\text{-}Aib)_2$-resin	Michael addition of malonate	[148]

(*Continued*)

Table 19.1 (Continued)

Peptide	Application	References
Pro-D-Pro-Leu-D-Pro-Aib-Trp-D-His-Trp-resin	Michael addition of malonate	[149]
Pro-D-Pro-Aib-Tyr-D-Leu-D-Pro-His-(D-Pro)$_3$-D-Leu-Leu-D-Leu-resin		
D-Pro-D-Pro-Aib-Ala(1-Pyn)-Aib-Tyr-Lys-resin	Michael addition of malonate	[150]
Trp-Trp-(Leu-Leu-Aib)$_3$-resin	Michael addition of nitromethane	[151]
	Michael addition of malonate	
	α-Amination	
Hse(Me)-Trp-(Leu-Leu-Aib)$_3$-resin	Michael addition of nitromethane	[152]
Ala(1-Pyn)-Pro-(Leu-Leu-Aib)$_2$-resin	Epoxidation	[153]
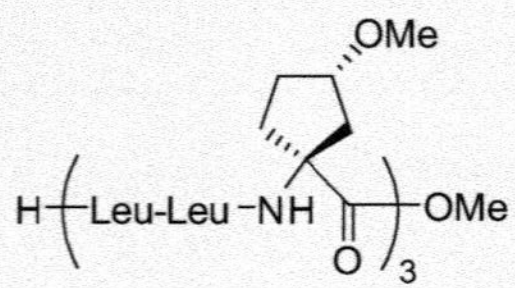	Juliá–Colonna epoxidation	[154]
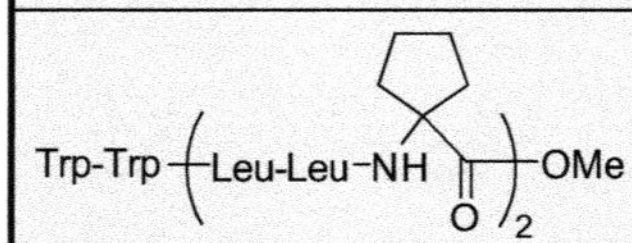	Michael addition of nitromethane	[155]
	Michael addition of malonate	
Leu-Ala-Lys-Leu-Leu-Lys-Ala-Leu-Ala-Lys-Leu-Leu-Lys-Lys-NH$_2$	Decarboxylation	[156]
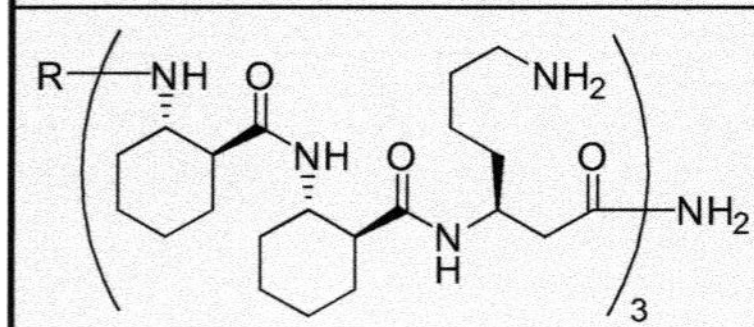	Retro-aldol reaction	[157]

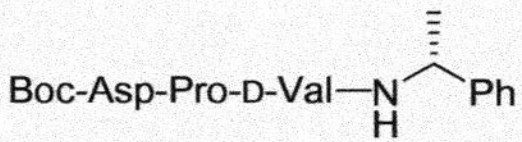	Epoxidation	[158,159]
Boc-Asp-D-Pro-Thr(Bn)-Asn(Trt)-Tyr(*t*Bu)-Gly-OMe	Epoxidation	[160,161]
Boc-Asp-Pro-D-Val-Leu-Val-OMe	Oxidation of indole	[163]
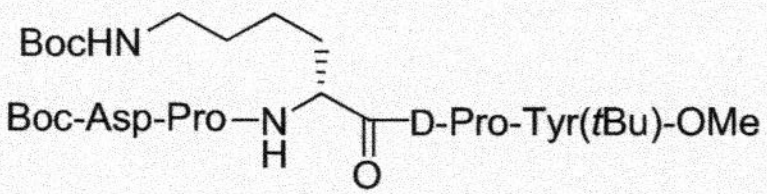	Baeyer–Villiger oxidation	[164]
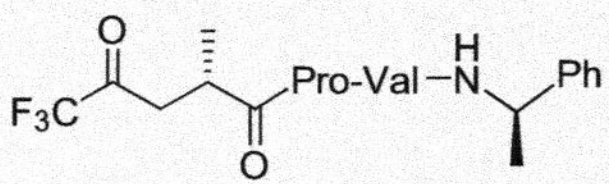	Epoxidation	[165]
Boc-His(3-Me)-D-Pro-Aib-Phe-OMe	Acylation	[167]
Boc-His(3-Me)-Thr(*t*Bu)-D-Val-His(Trt)-D-Phe-D-Val-Thr(*t*Bu)-Ile-OMe	Acylation	[170]
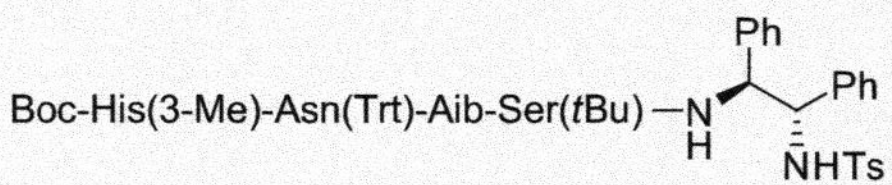	Acylation	[175]
Boc-His(3-Me)-D-Pro-Aib-D-Trp(Boc)-D-Phe-OMe	Acylation	[178]
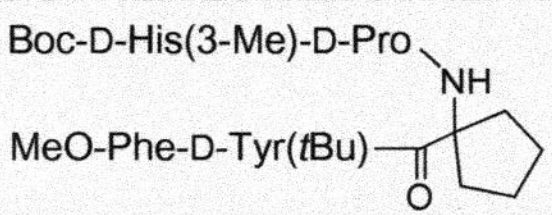 Ac-His(3-Me)-D-Pro-Aib-Phg-D-Phe-OMe Z-Lys(Z)-D-Ala-D-His(3-Me)-O•NBu_4	Acylation	[179]

(*Continued*)

Table 19.1 **(Continued)**

Peptide	Application	References
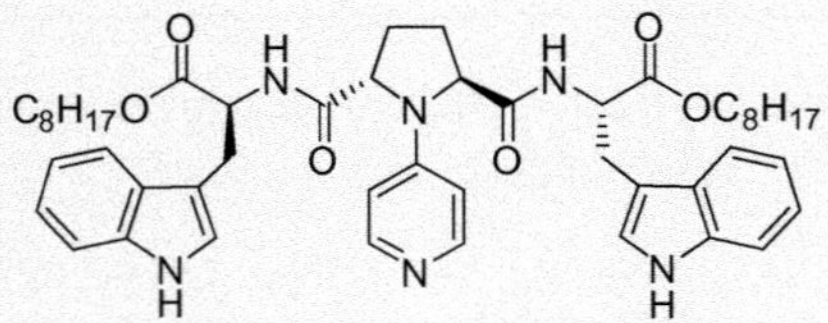	Acylation	[180–184]
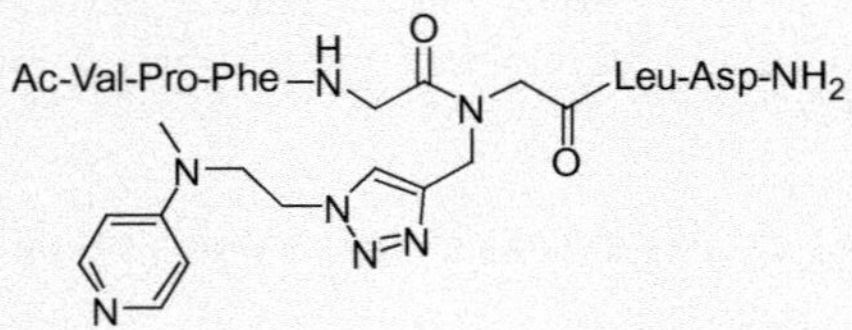 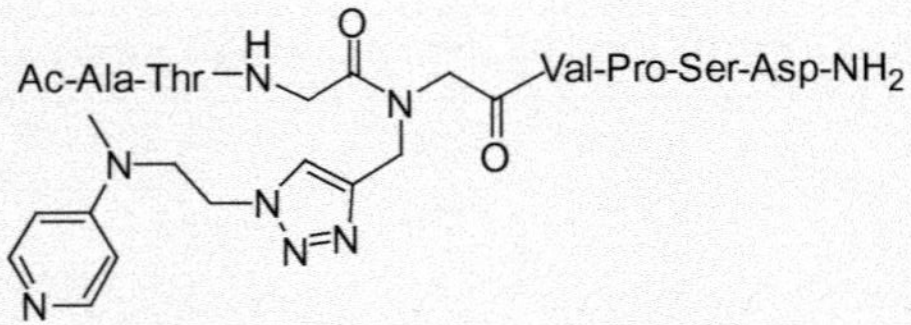	Acylation	[185]
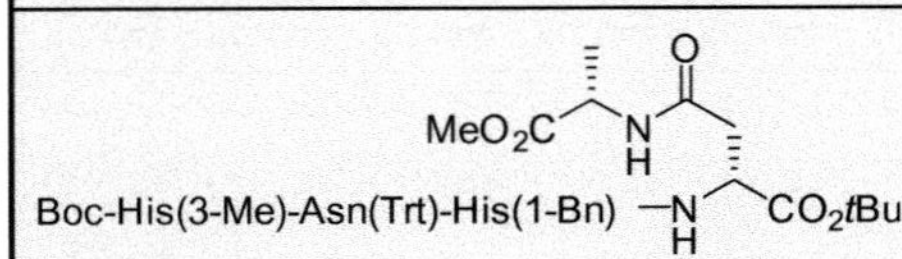	Phosphorylation	[186]
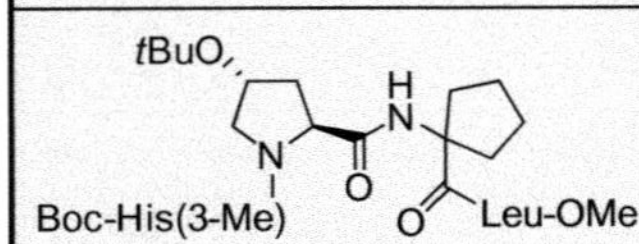	Sulfonylation	[187,188]
Boc-His(3-Me)-D-Pro-Aib-D-Trp(Boc)-D-Phe-OMe	Thiocarbonylation	[190,191]
Boc-His(3-Me)-Thr(*t*Bu)-D-Val-His(Trt)-Phe-D-Val-Thr(*t*Bu)-Ile-OMe		

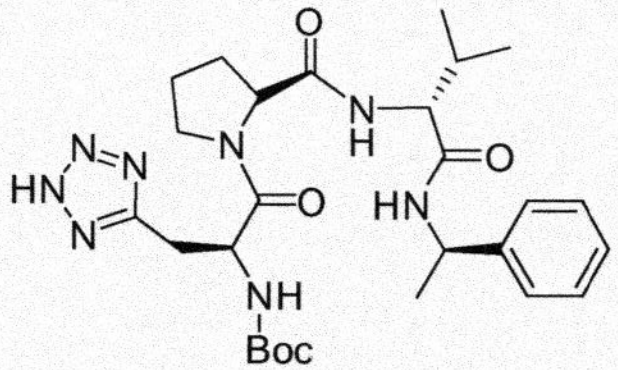	Phosphitylation	[192]
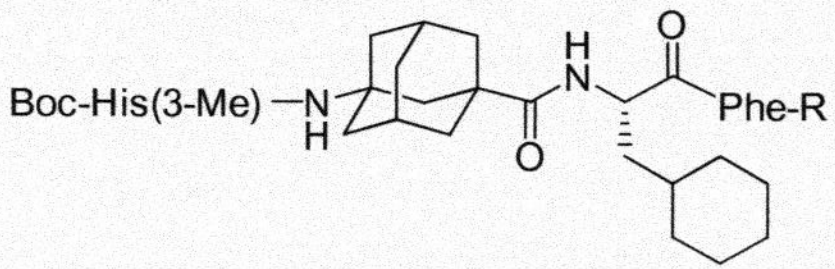	Acylation	[193,194]
	Steglich esterification	[195]
	Sequential reaction	[197,198]
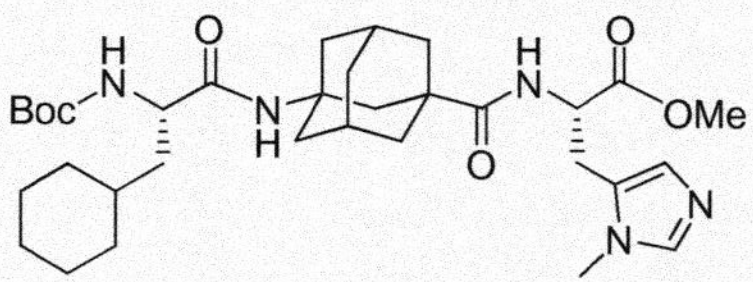	Dakin–West reaction	[196]
Miller's His-containing peptides	Michael addition of azide ion	[202,203]
	Morita–Baylis–Hillman reaction	[204,205]
	Esterification	[206]
	Michael addition of nitroketone	[207]
	N-acylation	[208]
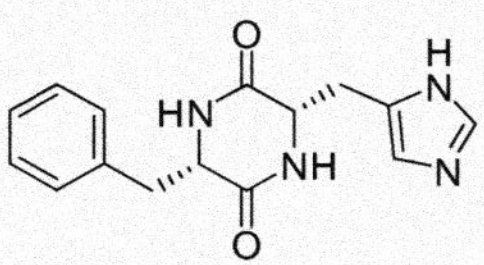	Hydrocyanation	[201]

(Continued)

Table 19.1 **(Continued)**

Peptide	Application	References
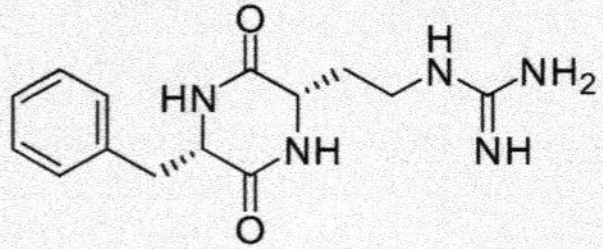	Strecker reaction	[209]
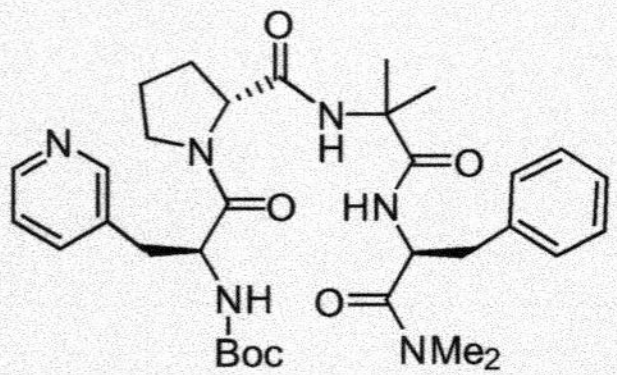	Addition of allenoate to imine	[210,211]
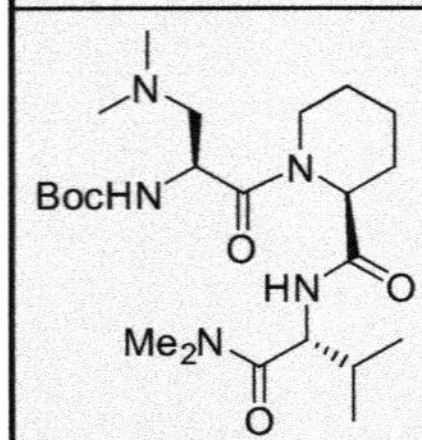	Bromination	[212–215]
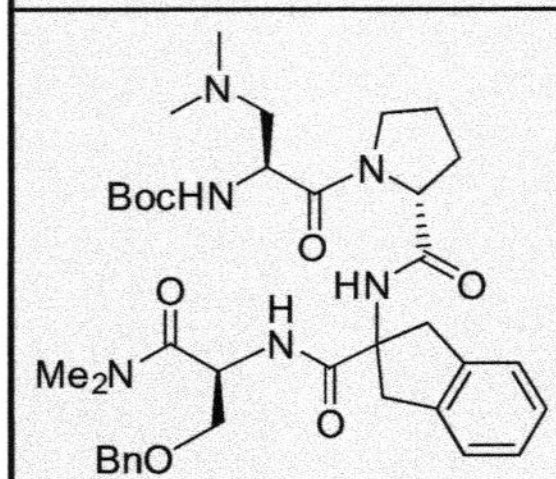	Methanolysis of azlactone	[216]

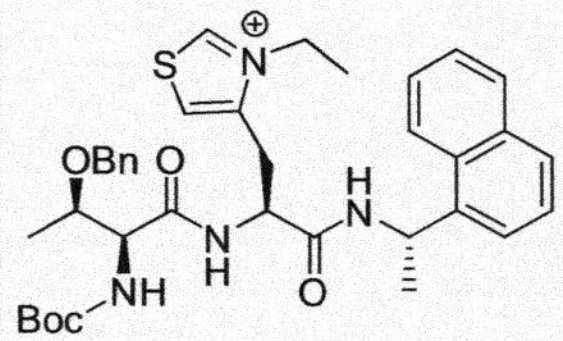	Stetter reaction	[217]
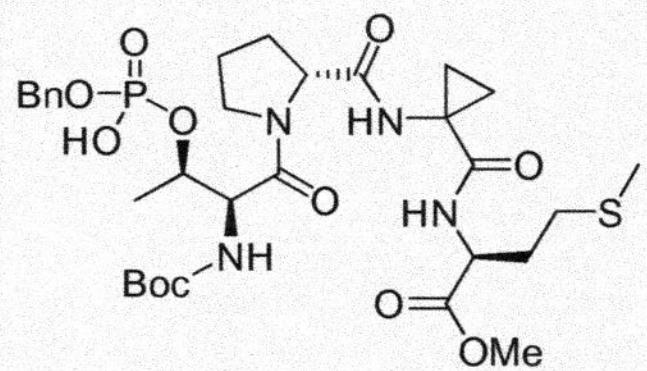	Transfer hydrogenation	[219]
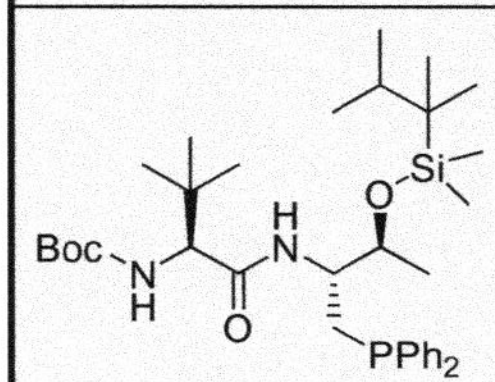	Cycloaddition of allenoate and imine	[222]
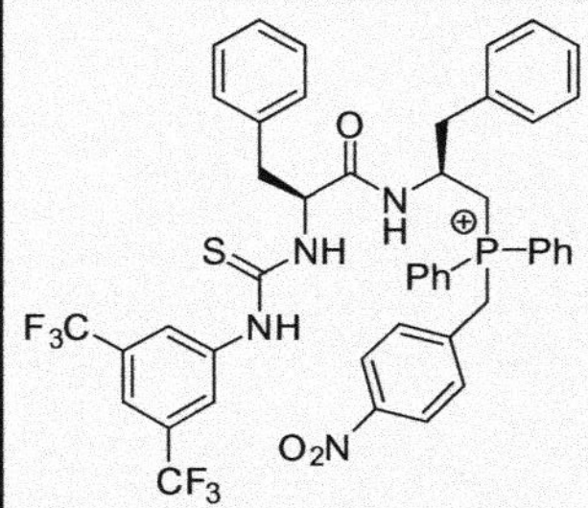	Michael addition of malonate	[223]

(*Continued*)

Table 19.1 **(Continued)**

Peptide	Application	References
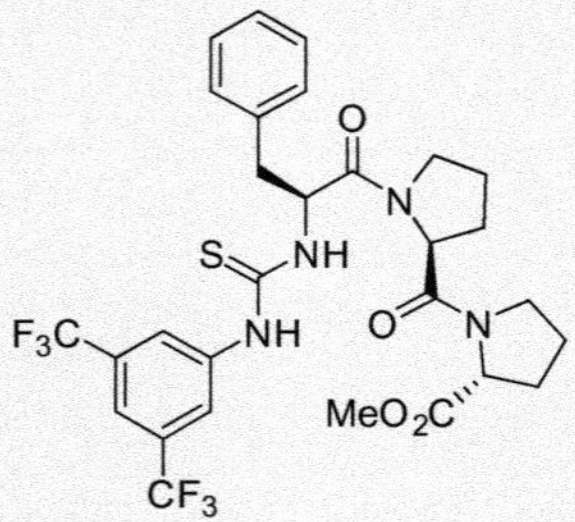	Ring-opening of epoxide	[224]
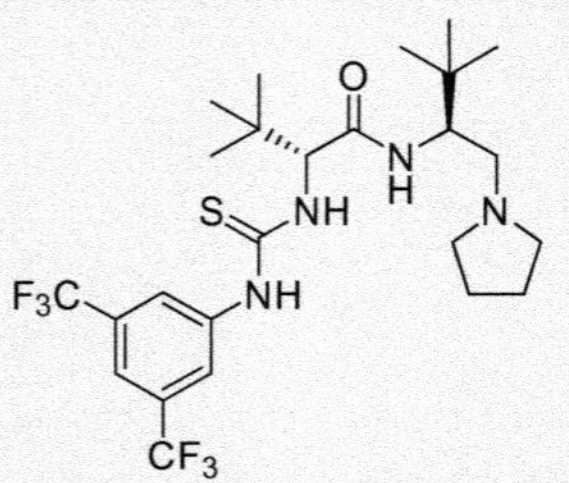	Michael–Mannich reaction	[225]
Cys-containing α-helical peptide	Formation of diketopiperazine	[228]

References

[1] B.M. Nestl, S.C. Hammer, B.A. Nebel, B. Hauer, New generation of biocatalysts for organic synthesis, Angew. Chem. Int. Ed. Engl. 53 (12) (2014) 3070–3095. Available from: http://dx.doi.org/10.1002/anie.201302195.

[2] B.W. Matthews, P.B. Sigler, R. Henderson, D.M. Blow, Three-dimensional structure of tosyl-α-chymotrypsin, Nature 214 (5089) (1967) 652–656. Available from: http://dx.doi.org/10.1038/214652a0.

[3] H. Nozaki, H. Takaya, S. Moriuti, R. Noyori, Homogeneous catalysis in the decomposition of diazo compounds by copper chelates: asymmetric carbenoid reactions, Tetrahedron 24 (9) (1968) 3655–3669, doi:10.1016/s0040-4020(01)91998-2.

[4] W.S. Knowles, M.J. Sabacky, Catalytic asymmetric hydrogenation employing a soluble, optically active, rhodium complex, Chem. Commun. (London) (22) (1968) 1445–1446. Available from: http://dx.doi.org/10.1039/c19680001445.

[5] H. Fukushima, S. Ohashi, S. Inoue, Asymmetric synthesis catalyzed by poly(5-benzyl L-glutamate), Die Makromol. Chem. 176 (9) (1975) 2751–2753. Available from: http://dx.doi.org/10.1002/macp.1975.021760923.

[6] B. List, R.A. Lerner, C.F. Barbas III, Proline-catalyzed direct asymmetric aldol reactions, J. Am. Chem. Soc. 122 (10) (2000) 2395–2396. Available from: http://dx.doi.org/10.1021/ja994280y.

[7] K.A. Ahrendt, C.J. Borths, D.W.C. MacMillan, New strategies for organic catalysis: the first highly enantioselective organocatalytic Diels–Alder reaction, J. Am. Chem. Soc. 122 (17) (2000) 4243–4244. Available from: http://dx.doi.org/10.1021/ja000092s.

[8] S. Mukherjee, J.W. Yang, S. Hoffmann, B. List, Asymmetric enamine catalysis, Chem. Rev. 107 (12) (2007) 5471–5569. Available from: http://dx.doi.org/10.1021/cr0684016.

[9] A. Erkkilä, I. Majander, P.M. Pihko, Iminium catalysis, Chem. Rev. 107 (12) (2007) 5416–5470. Available from: http://dx.doi.org/10.1021/cr068388p.

[10] B. List, H.J. Martin, Mining sequence space for asymmetric aminocatalysis: *N*-terminal prolyl-peptides efficiently catalyze enantioselective Aldol and Michael reactions, Synlett (12) (2003) 1901–1902. Available from: http://dx.doi.org/10.1055/s-2003-41490.

[11] J. Kofoed, J. Nielsen, J.-L. Reymond, Discovery of new peptide-based catalysts for the direct asymmetric Aldol reaction, Bioorg. Med. Chem. Lett. 13 (15) (2003) 2445–2447, doi:10.1016/s0960-894x(03)00498-0.

[12] K. Ohkubo, K. Sugahara, K. Yoshinaga, R. Ueoka, Enantioselective deacylation of long chain *p*-nitrophenyl *N*-acylphenyl-alanates by *N*-(*N*-dodecanoyl-L-histidyl)-L-leucine and a cationic chiral surfactant, J. Chem. Soc., Chem. Commun. (13) (1980) 637–639. Available from: http://dx.doi.org/10.1039/c39800000637.

[13] K. Ohkubo, N. Matsumoto, H. Ohta, High stereoselectivity in the deacylation of *p*-nitrophenyl *N*-acylphenylalanates by bilayer vesicular systems which include dipeptide-type nucleophiles, J. Chem. Soc., Chem. Commun. (13) (1982) 738–740. Available from: http://dx.doi.org/10.1039/c39820000738.

[14] R. Ueoka, Y. Matsumoto, Y. Ihara, Dramatically enhanced enantioselective hydrolysis of amino acid esters with tripeptide nucleophile by controlling the reaction field at room temperature, Chem. Lett. 13 (10) (1984) 1807–1810. Available from: http://dx.doi.org/10.1246/cl.1984.1807.

[15] K. Ohkubo, K. Urabe, S. Usui, T. Sagawa, The role of the membrane-assisted hydrophobic interaction between di-, tri-, or tetrapeptide catalysts and amino acid esters for the

enhancement of stereoselective hydrolysis reactions, Macromol. Rapid Commun. 17 (2) (1996) 109–116. Available from: http://dx.doi.org/10.1002/marc.1996.030170206.
[16] M.C. Cleij, W. Drenth, R.J.M. Nolte, Mechanism of enantioselective ester cleavage by histidine-containing dipeptides at a micellar interface, J. Org. Chem. 56 (12) (1991) 3883–3891. Available from: http://dx.doi.org/10.1021/jo00012a019.
[17] K. Hahn, W. Klis, J. Stewart, Design and synthesis of a peptide having chymotrypsin-like esterase activity, Science 248 (4962) (1990) 1544–1547. Available from: http://dx.doi.org/10.1126/science.2360048.
[18] K.S. Broo, L. Brive, P. Ahlberg, L. Baltzer, Catalysis of hydrolysis and transesterification reactions of *p*-nitrophenyl esters by a designed helix–loop–helix dimer, J. Am. Chem. Soc. 119 (47) (1997) 11362–11372. Available from: http://dx.doi.org/10.1021/ja970854s.
[19] S. Olofsson, G. Johansson, L. Baltzer, Design, synthesis and solution structure of a helix–loop–helix dimer—a template for the rational design of catalytically active polypeptides, J. Chem. Soc., Perkin Trans. 2 (11) (1995) 2047–2056. Available from: http://dx.doi.org/10.1039/p29950002047.
[20] K.S. Broo, H. Nilsson, J. Nilsson, A. Flodberg, L. Baltzer, Cooperative nucleophilic and general-acid catalysis by the $HisH^+$–His pair and arginine transition state binding in catalysis of ester hydrolysis reactions by designed helix–loop–helix motifs, J. Am. Chem. Soc. 120 (17) (1998) 4063–4068. Available from: http://dx.doi.org/10.1021/ja9737580.
[21] L. Baltzer, A.-C. Lundh, K. Broo, S. Olofsson, P. Ahlberg, Polypeptides with supersecondary structures as templates in rational catalyst design. Catalysis of self functionalization by designed helix–loop–helix motifs, J. Chem. Soc., Perkin Trans. 2 (8) (1996) 1671–1676. Available from: http://dx.doi.org/10.1039/p29960001671.
[22] K.S. Broo, H. Nilsson, J. Nilsson, L. Baltzer, Substrate recognition and saturation kinetics in de novo designed histidine-based four-helix bundle catalysts, J. Am. Chem. Soc. 120 (40) (1998) 10287–10295. Available from: http://dx.doi.org/10.1021/ja980682e.
[23] P. Rossi, P. Tecilla, L. Baltzer, P. Scrimin, De novo metallonucleases based on helix–loop–helix motifs, Chemistry 10 (17) (2004) 4163–4170. Available from: http://dx.doi.org/10.1002/chem.200400160.
[24] J. Razkin, H. Nilsson, L. Baltzer, Catalysis of the cleavage of uridine 3′-2,2,2-trichloroethylphosphate by a designed helix–loop–helix motif peptide, J. Am. Chem. Soc. 129 (47) (2007) 14752–14758. Available from: http://dx.doi.org/10.1021/ja075478i.
[25] J. Razkin, J. Lindgren, H. Nilsson, L. Baltzer, Enhanced complexity and catalytic efficiency in the hydrolysis of phosphate diesters by rationally designed helix-loop-helix motifs, Chembiochem 9 (12) (2008) 1975–1984. Available from: http://dx.doi.org/10.1002/cbic.200800057.
[26] B. Barbier, A. Brack, Basic polypeptides accelerate the hydrolysis of ribonucleic acids, J. Am. Chem. Soc. 110 (20) (1988) 6880–6882. Available from: http://dx.doi.org/10.1021/ja00228a045.
[27] P. Malon, J.-M. Bonmatin, A. Brack, Synthesis and characterization of cyclic peptides with hydrolytic activity, Tetrahedron Lett. 32 (39) (1991) 5337–5340, doi:10.1016/s0040-4039(00)92379-7.
[28] B. Barbier, A. Brack, Conformation-controlled hydrolysis of polyribonucleotides by sequential basic polypeptides, J. Am. Chem. Soc. 114 (9) (1992) 3511–3515. Available from: http://dx.doi.org/10.1021/ja00035a053.
[29] P.S.P. Wang, J.B. Nguyen, A. Schepartz, Design and high-resolution structure of a β^3-peptide bundle catalyst, J. Am. Chem. Soc. 136 (19) (2014) 6810–6813. Available from: http://dx.doi.org/10.1021/ja5013849.

[30] H. Tsutsumi, K. Hamasaki, H. Mihara, A. Ueno, Rate enhancement and enantioselectivity in ester hydrolysis catalysed by cyclodextrin–peptide hybrids, J. Chem. Soc., Perkin Trans. 2 (9) (2000) 1813–1818. Available from: http://dx.doi.org/10.1039/b004031m.

[31] S. Bezer, M. Matsumoto, M.W. Lodewyk, S.J. Lee, D.J. Tantillo, M.R. Gagné, et al., Identification and optimization of short helical peptides with novel reactive functionality as catalysts for acyl transfer by reactive tagging, Org. Biomol. Chem. 12 (9) (2014) 1488–1494. Available from: http://dx.doi.org/10.1039/c3ob41421c.

[32] M. Matsumoto, S.J. Lee, M.L. Waters, M.R. Gagné, A catalyst selection protocol that identifies biomimetic motifs from β-hairpin libraries, J. Am. Chem. Soc. 136 (45) (2014) 15817–15820. Available from: http://dx.doi.org/10.1021/ja503012g.

[33] C.M. Rufo, Y.S. Moroz, O.V. Moroz, J. Stöhr, T.A. Smith, X. Hu, et al., Short peptides self-assemble to produce catalytic amyloids, Nat. Chem. 6 (4) (2014) 303–309. Available from: http://dx.doi.org/10.1038/nchem.1894.

[34] C. Zhang, X. Xue, Q. Luo, Y. Li, K. Yang, X. Zhuang, et al., Self-assembled peptide nanofibers designed as biological enzymes for catalyzing ester hydrolysis, ACS Nano 8 (11) (2014) 11715–11723. Available from: http://dx.doi.org/10.1021/nn5051344.

[35] A. Nomura, Y. Sugiura, Hydrolytic reaction by zinc finger mutant peptides: successful redesign of structural zinc sites into catalytic zinc sites, Inorg. Chem. 43 (5) (2004) 1708–1713. Available from: http://dx.doi.org/10.1021/ic034931y.

[36] A. Esposito, E. Delort, D. Lagnoux, F. Djojo, J.-L. Reymond, Catalytic peptide dendrimers, Angew. Chem. Int. Ed. Engl. 42 (12) (2003) 1381–1383. Available from: http://dx.doi.org/10.1002/anie.200390354.

[37] E. Delort, T. Darbre, J.-L. Reymond, A strong positive dendritic effect in a peptide dendrimer-catalyzed ester hydrolysis reaction, J. Am. Chem. Soc. 126 (48) (2004) 15642–15643. Available from: http://dx.doi.org/10.1021/ja044652p.

[38] E. Delort, N.-Q. Nguyen-Trung, T. Darbre, J.-L. Reymond, Synthesis and activity of histidine-containing catalytic peptide dendrimers, J. Org. Chem. 71 (12) (2006) 4468–4480. Available from: http://dx.doi.org/10.1021/jo060273y.

[39] A. Clouet, T. Darbre, J.-L. Reymond, A combinatorial approach to catalytic peptide dendrimers, Angew. Chem. Int. Ed. Engl. 43 (35) (2004) 4612–4615. Available from: http://dx.doi.org/10.1002/anie.200460177.

[40] N. Maillard, R. Biswas, T. Darbre, J.-L. Reymond, Combinatorial discovery of peptide dendrimer enzyme models hydrolyzing isobutyryl fluorescein, ACS Comb. Sci. 13 (3) (2011) 310–320. Available from: http://dx.doi.org/10.1021/co200006z.

[41] C. Schmuck, U. Michels, J. Dudaczek, Hydrolytic activity of histidine-containing octapeptides in water identified by quantitative screening of a combinatorial library, Org. Biomol. Chem. 7 (21) (2009) 4362–4368. Available from: http://dx.doi.org/10.1039/b904415a.

[42] K. Akagawa, K. Kudo, Solvolysis of formylphenyl esters by a bifunctional peptide catalyst, Chem. Lett. 45 (3) (2016) 300–302. Available from: http://dx.doi.org/10.1246/cl.151154.

[43] Q.-L. Zhou (Ed.), Privileged Chiral Ligands and Catalysts, Wiley-VCH Verlag GmbH, Weinheim, Germany, 2011.

[44] B.M. Cole, K.D. Shimizu, C.A. Krueger, J.P.A. Harrity, M.L. Snapper, A.H. Hoveyda, Discovery of chiral catalysts through ligand diversity: Ti-catalyzed enantioselective addition of TMSCN to *meso* epoxides, Angew. Chem. Int. Ed. Engl. 35 (15) (1996) 1668–1671. Available from: http://dx.doi.org/10.1002/anie.199616681.

[45] N.S. Josephsohn, K.W. Kuntz, M.L. Snapper, A.H. Hoveyda, Mechanism of enantioselective Ti-catalyzed Strecker reaction: peptide-based metal complexes as bifunctional

catalysts, J. Am. Chem. Soc. 123 (47) (2001) 11594–11599. Available from: http://dx.doi.org/10.1021/ja011875w.

[46] H. Deng, M.P. Isler, M.L. Snapper, A.H. Hoveyda, Aluminum-catalyzed asymmetric addition of TMSCN to aromatic and aliphatic ketones promoted by an easily accessible and recyclable peptide ligand, Angew. Chem. Int. Ed. Engl. 41 (6) (2002) 1009–1012, doi:10.1002/1521-3773(20020315)416 <1009AID-ANIE1009> 3.0.CO;2-F.

[47] C.A. Luchaco-Cullis, H. Mizutani, K.E. Murphy, A.H. Hoveyda, Modular pyridinyl peptide ligands in asymmetric catalysis: enantioselective synthesis of quaternary carbon atoms through copper-catalyzed allylic substitutions, Angew. Chem. Int. Ed. Engl. 40 (8) (2001) 1456–1460, doi:10.1002/1521-3773(20010417)408 <1456AID-ANIE1456> 3.0.CO;2-T.

[48] S.J. Degrado, H. Mizutani, A.H. Hoveyda, Modular peptide-based phosphine ligands in asymmetric catalysis: efficient and enantioselective Cu-catalyzed conjugate additions to five-, six-, and seven-membered cyclic enones, J. Am. Chem. Soc. 123 (4) (2001) 755–756. Available from: http://dx.doi.org/10.1021/ja003698p.

[49] T. Soeta, K. Selim, M. Kuriyama, K. Tomioka, Peptidic amidomonophosphane ligand for copper-catalyzed asymmetric conjugate addition of diorganozincs to cycloalkenones, Adv. Synth. Catal. 349 (4–5) (2007) 629–635. Available from: http://dx.doi.org/10.1002/adsc.200600562.

[50] S.R. Gilbertson, R.V. Pawlick, Synthesis of thiophosphoryl derivatives of proline: building blocks for phosphanyl-substituted peptides with β-turns, Angew. Chem. Int. Ed. Engl. 35 (8) (1996) 902–904. Available from: http://dx.doi.org/10.1002/anie.199609021.

[51] S.R. Gilbertson, S.E. Collibee, A. Agarkov, Asymmetric catalysis with libraries of palladium β-turn phosphine complexes, J. Am. Chem. Soc. 122 (27) (2000) 6522–6523. Available from: http://dx.doi.org/10.1021/ja992306f.

[52] A. Agarkov, S.J. Greenfield, T. Ohishi, S.E. Collibee, S.R. Gilbertson, Catalysis with phosphine-containing amino acids in various "turn" motifs, J. Org. Chem. 69 (23) (2004) 8077–8085. Available from: http://dx.doi.org/10.1021/jo049103g.

[53] G. Guisado-Barrios, B.K. Muñoz, P.C.J. Kamer, B. Lastdrager, G. van der Marel, M. Overhand, et al., Cyclic decapeptide gramicidin S derivatives containing phosphines: novel ligands for asymmetric catalysis, Dalton Trans. 42 (6) (2013) 1973–1978. Available from: http://dx.doi.org/10.1039/c2dt31782f.

[54] A.C. Laungani, J.M. Slattery, I. Krossing, B. Breit, Supramolecular bidentate ligands by metal-directed in situ formation of antiparallel β-sheet structures and application in asymmetric catalysis, Chemistry 14 (15) (2008) 4488–4502. Available from: http://dx.doi.org/10.1002/chem.200800359.

[55] J.M. Benito, C.A. Christensen, M. Meldal, Versatile solid-phase synthesis of peptide-derived 2-oxazolines. Application in the synthesis of ligands for asymmetric catalysis, Org. Lett. 7 (4) (2005) 581–584. Available from: http://dx.doi.org/10.1021/ol047675h.

[56] C.A. Christensen, M. Meldal, Solid-phase synthesis of a peptide-based P,S-ligand system designed for generation of combinatorial catalyst libraries, J. Comb. Chem. 9 (1) (2007) 79–85. Available from: http://dx.doi.org/10.1021/cc0600627.

[57] D. Coquière, J. Bos, J. Beld, G. Roelfes, Enantioselective artificial metalloenzymes based on a bovine pancreatic polypeptide scaffold, Angew. Chem. Int. Ed. Engl. 48 (28) (2009) 5159–5162. Available from: http://dx.doi.org/10.1002/anie.200901134.

[58] L. Zheng, A. Marcozzi, J.Y. Gerasimov, A. Herrmann, Conformationally constrained cyclic peptides: powerful scaffolds for asymmetric catalysis, Angew. Chem. Int. Ed. Engl. 53 (29) (2014) 7599–7603. Available from: http://dx.doi.org/10.1002/anie.201403829.

[59] R. Sambasivan, Z.T. Ball, Metallopeptides for asymmetric dirhodium catalysis, J. Am. Chem. Soc. 132 (27) (2010) 9289–9291. Available from: http://dx.doi.org/10.1021/ja103747h.
[60] B.V. Popp, Z.T. Ball, Structure-selective modification of aromatic side chains with dirhodium metallopeptide catalysts, J. Am. Chem. Soc. 132 (19) (2010) 6660–6662. Available from: http://dx.doi.org/10.1021/ja101456c.
[61] F. Vohidov, J.M. Coughlin, Z.T. Ball, Rhodium(II) metallopeptide catalyst design enables fine control in selective functionalization of natural SH3 domains, Angew. Chem. Int. Ed. Engl. 54 (15) (2015) 4587–4591. Available from: http://dx.doi.org/10.1002/anie.201411745.
[62] R. Sambasivan, Z.T. Ball, Screening rhodium metallopeptide libraries "on bead": asymmetric cyclopropanation and a solution to the enantiomer problem, Angew. Chem. Int. Ed. Engl. 51 (34) (2012) 8568–8572. Available from: http://dx.doi.org/10.1002/anie.201202512.
[63] R. Sambasivan, W. Zheng, S.J. Burya, B.V. Popp, C. Turro, C. Clementi, et al., A tripodal peptide ligand for asymmetric Rh(II) catalysis highlights unique features of on-bead catalyst development, Chem. Sci. 5 (4) (2014) 1401–1407. Available from: http://dx.doi.org/10.1039/c3sc53354a.
[64] J. Long, Y. Yuan, Y. Shi, Asymmetric Simmons–Smith cyclopropanation of unfunctionalized olefins, J. Am. Chem. Soc. 125 (45) (2003) 13632–13633. Available from: http://dx.doi.org/10.1021/ja030488e.
[65] J. Long, H. Du, K. Li, Y. Shi, Catalytic asymmetric Simmons–Smith cyclopropanation of unfunctionalized olefins, Tetrahedron Lett. 46 (16) (2005) 2737–2740. Available from: http://dx.doi.org/10.1016/j.tetlet.2005.02.161.
[66] H. Du, J. Long, Y. Shi, Catalytic asymmetric Simmons–Smith cyclopropanation of silyl enol ethers. Efficient synthesis of optically active cyclopropanol derivatives, Org. Lett. 8 (13) (2006) 2827–2829. Available from: http://dx.doi.org/10.1021/ol0609659.
[67] Y. Baba, H. Mizushima, H. Watanabe, Catalytic properties of cytochrome c heme peptides, Chem. Pharm. Bull. (Tokyo) 17 (1) (1969) 82–88. Available from: http://dx.doi.org/10.1248/cpb.17.82.
[68] F. Nastri, A. Lombardi, G. Morelli, O. Maglio, G. D'Auria, C. Pedone, et al., Hemoprotein models based on a covalent helix-heme-helix sandwich: 1. Design, synthesis, and characterization, Chemistry 3 (3) (2006) 340–349. Available from: http://dx.doi.org/10.1002/chem.19970030305.
[69] F. Nastri, A. Lombardi, G. Morelli, C. Pedone, V. Pavone, G. Chottard, et al., Hemoprotein models based on a covalent helix-heme-helix sandwich. 3. Coordination properties, reactivity and catalytic application of Fe(III)- and Fe(II)-mimochrome I, J. Biol. Inorg. Chem. 3 (6) (1998) 671–681. Available from: http://dx.doi.org/10.1007/s007750050281.
[70] S. Sakamoto, A. Ueno, H. Mihara, Design and synthesis of haem-binding peptides. Relationship between haem-binding properties and catalytic activities, J. Chem. Soc., Perkin Trans. 2 (11) (1998) 2395–2404. Available from: http://dx.doi.org/10.1039/a805669b.
[71] M. Seal, C. Ghosh, O. Basu, S.G. Dey, Cytochrome c peroxidase activity of heme bound amyloid β peptides, J. Biol. Inorg. Chem. 21 (5–6) (2016) 683–690. Available from: http://dx.doi.org/10.1007/s00775-016-1367-6.
[72] C.T. Choma, E.P. Schudde, R.M. Kellogg, G.T. Robillard, B.L. Feringa, A functional mimic of natural peroxidases: synthesis and catalytic activity of a non-heme iron/peptide

hydroperoxide complex, J. Chem. Soc., Perkin Trans. 1 (4) (1998) 769–774. Available from: http://dx.doi.org/10.1039/a706416k.

[73] G. Xu, S.R. Gilbertson, Development of building blocks for the synthesis of N-heterocyclic carbene ligands, Org. Lett. 7 (21) (2005) 4605–4608. Available from: http://dx.doi.org/10.1021/ol0516521.

[74] Y. Zhao, S.R. Gilbertson, Synthesis of proline-based N-heterocyclic carbene ligands, Org. Lett. 16 (4) (2014) 1033–1035. Available from: http://dx.doi.org/10.1021/ol403465d.

[75] A. Monney, M. Albrecht, A chelating tetrapeptide rhodium complex comprised of a histidylidene residue: biochemical tailoring of an NHC-Rh hydrosilylation catalyst, Chem. Commun. 48 (89) (2012) 10960–10962. Available from: http://dx.doi.org/10.1039/c2cc35491h.

[76] H. Alper, N. Hamel, Poly-L-leucine as an added chiral ligand for the palladium catalysed carbonylation of allylic alcohols, J. Chem. Soc., Chem. Commun. (2) (1990) 135–136. Available from: http://dx.doi.org/10.1039/c39900000135.

[77] T. Komori, T. Nonaka, Electroorganic reactions on organic electrodes. 3. Electrochemical asymmetric oxidation of phenyl cyclohexyl sulfide on poly(L-valine)-coated platinum electrodes, J. Am. Chem. Soc. 105 (17) (1983) 5690–5691. Available from: http://dx.doi.org/10.1021/ja00355a029.

[78] T. Komori, T. Nonaka, Electroorganic reactions on organic electrodes. 6. Electrochemical asymmetric oxidation of unsymmetric sulfides to the corresponding chiral sulfoxides on poly(amino acid)-coated electrodes, J. Am. Chem. Soc. 106 (9) (1984) 2656–2659. Available from: http://dx.doi.org/10.1021/ja00321a028.

[79] S. Abe, T. Nonaka, T. Fuchigami, Electroorganic reactions on organic electrodes. 1. Asymmetric reduction of prochiral activated olefins on a poly-L-valine-coated graphite, J. Am. Chem. Soc. 105 (11) (1983) 3630–3632. Available from: http://dx.doi.org/10.1021/ja00349a046.

[80] C. Mondelli, A. Vargas, G. Santarossa, A. Baiker, Fundamental aspects of the chiral modification of platinum with peptides: asymmetric induction in hydrogenation of activated ketones, J. Phys. Chem. C 113 (34) (2009) 15246–15259. Available from: http://dx.doi.org/10.1021/jp902855a.

[81] E.A.C. Davie, S.M. Mennen, Y. Xu, S.J. Miller, Asymmetric catalysis mediated by synthetic peptides, Chem. Rev. 107 (12) (2007) 5759–5812. Available from: http://dx.doi.org/10.1021/cr068377w.

[82] H. Wennemers, Asymmetric catalysis with peptides, Chem. Commun. 47 (44) (2011) 12036–12041. Available from: http://dx.doi.org/10.1039/c1cc15237h.

[83] H.R. Kricheldorf, Polypeptides and 100 years of chemistry of α-amino acid *N*-carboxyanhydrides, Angew. Chem. Int. Ed. Engl. 45 (35) (2006) 5752–5784. Available from: http://dx.doi.org/10.1002/anie.200600693.

[84] K. Ueyanagi, S. Inoue, Asymmetric addition of 1-dodecanethiol to isopropenyl methyl ketone catalyzed by poly(α-amino acid)s: determination of the enantiomeric purity of the product, Die Makromol. Chem. 178 (1) (1977) 235–239. Available from: http://dx.doi.org/10.1002/macp.1977.021780123.

[85] S. Juliá, J. Masana, J.C. Vega, "Synthetic enzymes". Highly stereoselective epoxidation of chalcone in a triphasic toluene-water-poly[(S)-alanine] system, Angew. Chem. Int. Ed. Engl. 19 (11) (1980) 929–931. Available from: http://dx.doi.org/10.1002/anie.198009291.

[86] M.J. Porter, S.M. Roberts, J. Skidmore, Polyamino acids as catalysts in asymmetric synthesis, Bioorg. Med. Chem. 7 (10) (1999) 2145–2156, doi:10.1016/s0968-0896(99)00144-3.

[87] D.R. Kelly, S.M. Roberts, Oligopeptides as catalysts for asymmetric epoxidation, Biopolymers 84 (1) (2006) 74–89. Available from: http://dx.doi.org/10.1002/bip.20373.

[88] D.R. Kelly, S.M. Roberts, The mechanism of polyleucine catalysed asymmetric epoxidation, Chem. Commun. (18) (2004) 2018–2020. Available from: http://dx.doi.org/10.1039/b404390c.

[89] S. Colonna, D. Perdicchia, E. Di Mauro, Enantioselective reactions catalyzed by synthetic enzymes. A model for chemical evolution, Tetrahedron: Asymmetry 20 (15) (2009) 1709–1714. Available from: http://dx.doi.org/10.1016/j.tetasy.2009.07.007.

[90] A. Weyer, D. Díaz, A. Nierth, N.E. Schlörer, A. Berkessel, The L-Leu hexamer, a short and highly enantioselective peptide catalyst for the Juliá-Colonna epoxidation: stabilization of a helical conformation in DMSO, ChemCatChem 4 (3) (2011) 337–340. Available from: http://dx.doi.org/10.1002/cctc.201100243.

[91] S. Itsuno, M. Sakakura, K. Ito, Polymer-supported poly(amino acids) as new asymmetric epoxidation catalyst of α,β-unsaturated ketones, J. Org. Chem. 55 (24) (1990) 6047–6049. Available from: http://dx.doi.org/10.1021/jo00311a024.

[92] T. Geller, S.M. Roberts, A new procedure for the Juliá–Colonna stereoselective epoxidation reaction under non-aqueous conditions: the development of a catalyst comprising polyamino acid on silica (PaaSiCat), J. Chem. Soc., Perkin Trans. 1 (11) (1999) 1397–1398. Available from: http://dx.doi.org/10.1039/a903219c.

[93] A. Berkessel, N. Gasch, K. Glaubitz, C. Koch, Highly enantioselective enone epoxidation catalyzed by short solid phase-bound peptides: dominant role of peptide helicity, Org. Lett. 3 (24) (2001) 3839–3842. Available from: http://dx.doi.org/10.1021/ol0166451.

[94] H. Yi, G. Zou, Q. Li, Q. Chen, J. Tang, M. He, Asymmetric epoxidation of α,β-unsaturated ketones catalyzed by silica-grafted poly-(L)-leucine catalysts, Tetrahedron Lett. 46 (34) (2005) 5665–5668. Available from: http://dx.doi.org/10.1016/j.tetlet.2005.06.096.

[95] W. Qiu, L. He, Q. Chen, W. Luo, Z. Yu, F. Yang, et al., Imidazolium-modified poly (L-leucine) catalyst: an efficient and recoverable catalyst for Juliá–Colonna reactions, Tetrahedron Lett. 50 (37) (2016) 5225–5227. Available from: http://dx.doi.org/10.1016/j.tetlet.2009.06.127.

[96] G. Carrea, G. Ottolina, A. Lazcano, V. Pironti, S. Colonna, Enantioselective aldol reaction catalysed by polyleucines, Tetrahedron: Asymmetry 18 (11) (2007) 1265–1268. Available from: http://dx.doi.org/10.1016/j.tetasy.2007.05.024.

[97] K. Akagawa, K. Kudo, Asymmetric induction by helical poly(amino acid)s in cyanosilylation of aldehydes, Tetrahedron Lett. 53 (45) (2012) 5981–5983. Available from: http://dx.doi.org/10.1016/j.tetlet.2012.07.034.

[98] L.-X. Shi, Q. Sun, Z.-M. Ge, Y.-Q. Zhu, T.-M. Cheng, R.-T. Li, Dipeptide-catalyzed direct asymmetric Aldol reaction, Synlett (12) (2004) 2215–2217. Available from: http://dx.doi.org/10.1055/s-2004-831324.

[99] Z. Tang, Z.-H. Yang, L.-F. Cun, L.-Z. Gong, A.-Q. Mi, Y.-Z. Jiang, Small peptides catalyze highly enantioselective direct Aldol reactions of aldehydes with hydroxyacetone: unprecedented regiocontrol in aqueous media, Org. Lett. 6 (13) (2004) 2285–2287. Available from: http://dx.doi.org/10.1021/ol049141m.

[100] G. Luppi, P.G. Cozzi, M. Monari, B. Kaptein, Q.B. Broxterman, C. Tomasini, Dipeptide-catalyzed asymmetric Aldol condensation of acetone with (N-alkylated) isatins, J. Org. Chem. 70 (18) (2005) 7418–7421. Available from: http://dx.doi.org/10.1021/jo050257l.

[101] W. Zou, I. Ibrahem, P. Dziedzic, H. Sundén, A. Córdova, Small peptides as modular catalysts for the direct asymmetric aldol reaction: ancient peptides with aldolase enzyme activity, Chem. Commun. (39) (2005) 4946–4948. Available from: http://dx.doi.org/10.1039/b509920j.

[102] M.R.M. Andreae, A.P. Davis, Heterogeneous catalysis of the asymmetric aldol reaction by solid-supported proline-terminated peptides, Tetrahedron: Asymmetry 16 (14) (2005) 2487–2492. Available from: http://dx.doi.org/10.1016/j.tetasy.2005.06.031.

[103] K. Akagawa, S. Sakamoto, K. Kudo, Direct asymmetric aldol reaction in aqueous media using polymer-supported peptide, Tetrahedron Lett. 46 (47) (2005) 8185–8187. Available from: http://dx.doi.org/10.1016/j.tetlet.2005.09.113.

[104] V. D'Elia, H. Zwicknagl, O. Reiser, Short α/β-peptides as catalysts for intra- and intermolecular Aldol reactions, J. Org. Chem. 73 (8) (2008) 3262–3265. Available from: http://dx.doi.org/10.1021/jo800168h.

[105] L. Wang, J. Yan, Merrifield resin supported dipeptides: efficient and recyclable organocatalysts for asymmetric Aldol reactions under neat reaction conditions, Synthesis (13) (2008) 2065–2072. Available from: http://dx.doi.org/10.1055/s-2008-1067104.

[106] J. Yan, L. Wang, Asymmetric Aldol reactions catalyzed by efficient and recyclable silica-supported proline-based peptides, Chirality 21 (4) (2009) 413–420. Available from: http://dx.doi.org/10.1002/chir.20603.

[107] F.-C. Wu, C.-S. Da, Z.-X. Du, Q.-P. Guo, W.-P. Li, L. Yi, et al., *N*-primary-amine-terminal β-turn tetrapeptides as organocatalysts for highly enantioselective Aldol reaction, J. Org. Chem. 74 (13) (2009) 4812–4818. Available from: http://dx.doi.org/10.1021/jo9005766.

[108] F. Rodríguez-Llansola, J.F. Miravet, B. Escuder, A supramolecular hydrogel as a reusable heterogeneous catalyst for the direct aldol reaction, Chem. Commun. (47) (2009) 7303–7305. Available from: http://dx.doi.org/10.1039/b916250j.

[109] S. Bayat, B.A. Tejo, A.B. Salleh, E. Abdmalek, Y.M. Normi, M.B.A. Rahman, Various polar tripeptides as asymmetric organocatalyst in direct Aldol reactions in aqueous media, Chirality 25 (11) (2013) 726–734. Available from: http://dx.doi.org/10.1002/chir.22205.

[110] A. Psarra, C.G. Kokotos, P. Moutevelis-Minakakis, *tert*-Butyl esters of tripeptides based on Pro-Phe as organocatalysts for the asymmetric aldol reaction in aqueous or organic medium, Tetrahedron 70 (3) (2014) 608–615. Available from: http://dx.doi.org/10.1016/j.tet.2013.12.007.

[111] S. Bayat, B.A. Tejo, E. Abdulmalek, A.B. Salleh, Y.M. Normi, M.B.A. Rahman, Rational design of mimetic peptides based on aldo-ketoreductase enzyme as asymmetric organocatalysts in aldol reactions, RSC Adv. 4 (73) (2014) 38859–38868. Available from: http://dx.doi.org/10.1039/c4ra04866k.

[112] A. Gurka, I. Bucsi, L. Kovács, G. Szőllősi, M. Bartók, Reversal of the enantioselectivity in aldol addition over immobilized di- and tripeptides: studies under continuous flow conditions, RSC Adv. 4 (106) (2014) 61611–61618. Available from: http://dx.doi.org/10.1039/c4ra07188c.

[113] P. Milbeo, K. Maurent, L. Moulat, A. Lebrun, C. Didierjean, E. Aubert, et al., *N*-Pyrrolidine-based α/β-peptides incorporating ABOC, a constrained bicyclic β-amino acid, for asymmetric aldol reaction catalysis, Tetrahedron 72 (13) (2016) 1706–1715. Available from: http://dx.doi.org/10.1016/j.tet.2016.02.027.

[114] P. Krattiger, C. McCarthy, A. Pfaltz, H. Wennemers, Catalyst–substrate coimmobilization: a strategy for catalysts discovery in split-and-mix libraries, Angew. Chem. Int. Ed. Engl. 42 (15) (2003) 1722–1724. Available from: http://dx.doi.org/10.1002/anie.200250422.

[115] P. Krattiger, R. Kovasy, J.D. Revell, S. Ivan, H. Wennemers, Increased structural complexity leads to higher activity: peptides as efficient and versatile catalysts for asymmetric Aldol reactions, Org. Lett. 7 (6) (2005) 1101–1103. Available from: http://dx.doi.org/10.1021/ol0500259.
[116] J. Kofoed, T. Darbre, J.-L. Reymond, Artificial aldolases from peptide dendrimer combinatorial libraries, Org. Biomol. Chem. 4 (17) (2006) 3268–3281. Available from: http://dx.doi.org/10.1039/b607342e.
[117] Z.-X. Du, L.-Y. Zhang, X.-Y. Fan, F.-C. Wu, C.-S. Da, Highly enantioselective biomimetic intramolecular dehydration: kinetic resolution of β-hydroxy ketones catalyzed by β-turn tetrapeptides, Tetrahedron Lett. 54 (22) (2013) 2828–2832. Available from: http://dx.doi.org/10.1016/j.tetlet.2013.03.087.
[118] T. Tanaka, K. Akagawa, M. Mitsuda, K. Kudo, Asymmetric α-amination of aldehydes by a recyclable resin-supported peptide catalyst, Adv. Synth. Catal. 355 (2–3) (2013) 294–296. Available from: http://dx.doi.org/10.1002/adsc.201200611.
[119] K. Akagawa, T. Fujiwara, S. Sakamoto, K. Kudo, Efficient asymmetric α-oxyamination of aldehydes by resin-supported peptide catalyst in aqueous media, Org. Lett. 12 (8) (2010) 1804–1807. Available from: http://dx.doi.org/10.1021/ol100415h.
[120] K. Akagawa, R. Suzuki, K. Kudo, Effect of the helical tether of a resin-supported peptide catalyst for Friedel–Crafts-type alkylation in water, Adv. Synth. Catal. 354 (7) (2012) 1280–1286. Available from: http://dx.doi.org/10.1002/adsc.201100950.
[121] K. Akagawa, T. Fujiwara, S. Sakamoto, K. Kudo, One-pot sequential alcohol oxidation and asymmetric α-oxyamination in aqueous media using recyclable resin-supported peptide catalyst, Chem. Commun. 46 (42) (2010) 8040–8042. Available from: http://dx.doi.org/10.1039/c0cc02301a.
[122] K. Akagawa, K. Kudo, Peptide/laccase cocatalyzed asymmetric α-oxyamination of aldehydes, Org. Lett. 13 (13) (2011) 3498–3501. Available from: http://dx.doi.org/10.1021/ol2012956.
[123] K. Akagawa, R. Umezawa, K. Kudo, Asymmetric one-pot sequential Friedel–Crafts-type alkylation and α-oxyamination catalyzed by a peptide and an enzyme, Beilstein J. Org. Chem. 8 (2012) 1333–1337. Available from: http://dx.doi.org/10.3762/bjoc.8.152.
[124] M. Wiesner, J.D. Revell, H. Wennemers, Tripeptides as efficient asymmetric catalysts for 1,4-addition reactions of aldehydes to nitroolefins: a rational approach, Angew. Chem. Int. Ed. Engl. 47 (10) (2008) 1871–1874. Available from: http://dx.doi.org/10.1002/anie.200704972.
[125] M. Wiesner, J.D. Revell, S. Tonazzi, H. Wennemers, Peptide catalyzed asymmetric conjugate addition reactions of aldehydes to nitroethylene: a convenient entry into γ^2-amino acids, J. Am. Chem. Soc. 130 (17) (2008) 5610–5611. Available from: http://dx.doi.org/10.1021/ja801027s.
[126] M. Wiesner, G. Upert, G. Angelici, H. Wennemers, Enamine catalysis with low catalyst loadings – high efficiency via kinetic studies, J. Am. Chem. Soc. 132 (1) (2010) 6–7. Available from: http://dx.doi.org/10.1021/ja9068112.
[127] J. Duschmalé, H. Wennemers, Adapting to substrate challenges: peptides as catalysts for conjugate sddition reactions of aldehydes to α,β-disubstituted nitroolefins, Chemistry 18 (4) (2011) 1111–1120. Available from: http://dx.doi.org/10.1002/chem.201102484.
[128] R. Kastl, H. Wennemers, Peptide-catalyzed stereoselective conjugate addition reactions generating all-carbon quaternary stereogenic centers, Angew. Chem. Int. Ed. Engl. 52 (28) (2013) 7228–7232. Available from: http://dx.doi.org/10.1002/anie.201301583.

[129] C.E. Grünenfelder, J.K. Kisunzu, H. Wennemers, Peptide-catalyzed stereoselective conjugate addition reactions of aldehydes to maleimide, Angew. Chem. Int. Ed. Engl. 55 (30) (2016) 8571–8574. Available from: http://dx.doi.org/10.1002/anie.201602230.

[130] Y. Arakawa, M. Wiesner, H. Wennemers, Efficient recovery and reuse of an immobilized peptidic organocatalyst, Adv. Synth. Catal. 353 (8) (2011) 1201–1206. Available from: http://dx.doi.org/10.1002/adsc.201100118.

[131] Y. Arakawa, H. Wennemers, Enamine catalysis in flow with an immobilized peptidic catalyst, ChemSusChem 6 (2) (2012) 242–245. Available from: http://dx.doi.org/10.1002/cssc.201200740.

[132] J. Duschmalé, S. Kohrt, H. Wennemers, Peptide catalysis in aqueous emulsions, Chem. Commun. 50 (60) (2014) 8109–8112. Available from: http://dx.doi.org/10.1039/c4cc01759e.

[133] Y. Xu, W. Zou, H. Sundén, I. Ibrahem, A. Córdova, Small peptide-catalyzed enantioselective addition of ketones to nitroolefins, Adv. Synth. Catal. 348 (4–5) (2006) 418–424. Available from: http://dx.doi.org/10.1002/adsc.200505373.

[134] M. Freund, S. Schenker, S.B. Tsogoeva, Enantioselective nitro-Michael reactions catalyzed by short peptides on water, Org. Biomol. Chem. 7 (20) (2009) 4279–4284. Available from: http://dx.doi.org/10.1039/b910249c.

[135] A.F. de la Torre, D.G. Rivera, M.A.B. Ferreira, A.G. Corrêa, M.W. Paixão, Multicomponent combinatorial development and conformational analysis of prolyl peptide–peptoid hybrid catalysts: application in the direct asymmetric Michael addition, J. Org. Chem. 78 (20) (2013) 10221–10232. Available from: http://dx.doi.org/10.1021/jo401609z.

[136] G.S. Scatena, A.F. de la Torre, Q.B. Cass, D.G. Rivera, M.W. Paixão, Multicomponent approach to silica-grafted peptide catalysts: a 3D continuous-flow organocatalytic system with on-line monitoring of conversion and stereoselectivity, ChemCatChem 6 (11) (2014) 3208–3214. Available from: http://dx.doi.org/10.1002/cctc.201402501.

[137] M. Cortes-Clerget, O. Gager, M. Monteil, J.-L. Pirat, E. Migianu-Griffoni, J. Deschamp, et al., Novel easily recyclable bifunctional phosphonic acid carrying tripeptides for the stereoselective Michael addition of aldehydes with nitroalkenes, Adv. Synth. Catal. 358 (1) (2015) 34–40. Available from: http://dx.doi.org/10.1002/adsc.201500794.

[138] K. Akagawa, H. Akabane, S. Sakamoto, K. Kudo, Organocatalytic asymmetric transfer hydrogenation in aqueous media using resin-supported peptide having a polyleucine tether, Org. Lett. 10 (10) (2008) 2035–2037. Available from: http://dx.doi.org/10.1021/ol800031p.

[139] K. Akagawa, T. Yamashita, S. Sakamoto, K. Kudo, Friedel–Crafts-type alkylation in aqueous media using resin-supported peptide catalyst having polyleucine, Tetrahedron Lett. 50 (40) (2009) 5602–5604. Available from: http://dx.doi.org/10.1016/j.tetlet.2009.07.071.

[140] K. Akagawa, K. Kudo, Asymmetric epoxidation of α,β-unsaturated aldehydes in aqueous media catalyzed by resin-supported peptide- containing unnatural amino acids, Adv. Synth. Catal. 353 (6) (2011) 843–847. Available from: http://dx.doi.org/10.1002/adsc.201000805.

[141] K. Akagawa, M. Sugiyama, K. Kudo, Asymmetric Michael addition of boronic acids to a γ-hydroxy-α,β-unsaturated aldehyde catalyzed by resin-supported peptide, Org. Biomol. Chem. 10 (25) (2012) 4839–4843. Available from: http://dx.doi.org/10.1039/c2ob25431j.

[142] K. Akagawa, S. Takigawa, I.S. Nagamine, R. Umezawa, K. Kudo, Peptide-catalyzed diastereo- and enantioselective cyclopropanation of aromatic α,β-unsaturated aldehydes, Org. Lett. 15 (19) (2013) 4964–4967. Available from: http://dx.doi.org/10.1021/ol402227y.
[143] K. Akagawa, N. Nishi, J. Sen, K. Kudo, Peptide-catalyzed consecutive 1,6- and 1,4-additions of thiols to α,β,γ,δ-unsaturated aldehydes, Org. Biomol. Chem. 12 (22) (2014) 3581–3585. Available from: http://dx.doi.org/10.1039/c4ob00565a.
[144] K. Akagawa, K. Kudo, Construction of an all-carbon quaternary stereocenter by the peptide-catalyzed asymmetric Michael addition of nitromethane to β-disubstituted α,β-unsaturated aldehydes, Angew. Chem. Int. Ed. Engl. 51 (51) (2012) 12786–12789. Available from: http://dx.doi.org/10.1002/anie.201206916.
[145] K. Akagawa, J. Sen, K. Kudo, Peptide-catalyzed regio- and enantioselective reduction of α,β,γ,δ-unsaturated aldehydes, Angew. Chem. Int. Ed. Engl. 52 (44) (2013) 11585–11588. Available from: http://dx.doi.org/10.1002/anie.201305004.
[146] M. Akiyama, K. Akagawa, H. Seino, K. Kudo, Peptide-catalyzed kinetic resolution of planar-chiral metallocenes, Chem. Commun. 50 (58) (2014) 7893–7896. Available from: http://dx.doi.org/10.1039/c4cc03266g.
[147] K. Akagawa, M. Akiyama, K. Kudo, Peptide-catalyzed desymmetrization of an achiral ferrocenyl compound to induce planar chirality, Eur. J. Org. Chem. (18) (2015) 3894–3898. Available from: http://dx.doi.org/10.1002/ejoc.201500428.
[148] K. Akagawa, N. Sakai, K. Kudo, Histidine-containing peptide catalysts developed by a facile library screening method, Angew. Chem. Int. Ed. Engl. 54 (6) (2014) 1822–1826. Available from: http://dx.doi.org/10.1002/anie.201410268.
[149] K. Akagawa, J. Satou, K. Kudo, Exploration of structural frameworks for reactive and enantioselective peptide catalysts by library screenings, J. Org. Chem. 81 (19) (2016) 9396–9401. Available from: http://dx.doi.org/10.1021/acs.joc.6b01591.
[150] K. Akagawa, Y. Iwasaki, K. Kudo, Library screening in aqueous media to develop a highly active peptide catalyst for enantioselective Michael addition of a malonate, Eur. J. Org. Chem. (26) (2016) 4460–4464. Available from: http://dx.doi.org/10.1002/ejoc.201600828.
[151] K. Akagawa, R. Suzuki, K. Kudo, Development of a peptide-based primary aminocatalyst with a helical structure, Asian J. Org. Chem. 3 (4) (2014) 514–522. Available from: http://dx.doi.org/10.1002/ajoc.201400028.
[152] K. Akagawa, N. Nishi, I. Yoshikawa, K. Kudo, Kinetic resolution of a planar-chiral [2.2]paracyclophane derivative by helical-peptide-catalyzed Michael addition of nitromethane, Eur. J. Org. Chem. (23) (2015) 5055–5059. Available from: http://dx.doi.org/10.1002/ejoc.201500594.
[153] K. Akagawa, T. Hirata, K. Kudo, Asymmetric epoxidation of enones by peptide-based catalyst: a strategy inverting Juliá–Colonna stereoselectivity, Synlett 27 (08) (2015) 1217–1222. Available from: http://dx.doi.org/10.1055/s-0035-1560597.
[154] M. Nagano, M. Doi, M. Kurihara, H. Suemune, M. Tanaka, Stabilized α-helix-catalyzed enantioselective epoxidation of α,β-unsaturated ketones, Org. Lett. 12 (15) (2010) 3564–3566. Available from: http://dx.doi.org/10.1021/ol101435w.
[155] A. Ueda, T. Umeno, M. Doi, K. Akagawa, K. Kudo, M. Tanaka, Helical-peptide-catalyzed nantioselective Michael addition reactions and their mechanistic insights, J. Org. Chem. 81 (15) (2016) 6343–6356. Available from: http://dx.doi.org/10.1021/acs.joc.6b00982.
[156] K. Johnsson, R.K. Allemann, H. Widmer, S.A. Benner, Synthesis, structure and activity of artificial, rationally designed catalytic polypeptides, Nature 365 (6446) (1993) 530–532. Available from: http://dx.doi.org/10.1038/365530a0.

[157] M.M. Müller, M.A. Windsor, W.C. Pomerantz, S.H. Gellman, D. Hilvert, A rationally designed aldolase foldamer, Angew. Chem. Int. Ed. Engl. 48 (5) (2009) 922–925. Available from: http://dx.doi.org/10.1002/anie.200804996.

[158] G. Peris, C.E. Jakobsche, S.J. Miller, Aspartate-catalyzed asymmetric epoxidation reactions, J. Am. Chem. Soc. 129 (28) (2007) 8710–8711. Available from: http://dx.doi.org/10.1021/ja073055a.

[159] C.E. Jakobsche, G. Peris, S.J. Miller, Functional analysis of an aspartate-based epoxidation catalyst with amide-to-alkene peptidomimetic catalyst analogues, Angew. Chem. Int. Ed. Engl. 47 (35) (2008) 6707–6711. Available from: http://dx.doi.org/10.1002/anie.200802223.

[160] P.A. Lichtor, S.J. Miller, Combinatorial evolution of site- and enantioselective catalysts for polyene epoxidation, Nat. Chem. 4 (12) (2012) 990–995. Available from: http://dx.doi.org/10.1038/nchem.1469.

[161] P.A. Lichtor, S.J. Miller, Experimental lineage and functional analysis of a remotely directed peptide epoxidation catalyst, J. Am. Chem. Soc. 136 (14) (2014) 5301–5308. Available from: http://dx.doi.org/10.1021/ja410567a.

[162] N.C. Abascal, P.A. Lichtor, M.W. Giuliano, S.J. Miller, Function-oriented investigations of a peptide-based catalyst that mediates enantioselective allylic alcohol epoxidation, Chem. Sci. 5 (11) (2014) 4504–4511. Available from: http://dx.doi.org/10.1039/c4sc01440e.

[163] F. Kolundzic, M.N. Noshi, M. Tjandra, M. Movassaghi, S.J. Miller, Chemoselective and enantioselective oxidation of indoles employing aspartyl peptide catalysts, J. Am. Chem. Soc. 133 (23) (2011) 9104–9111. Available from: http://dx.doi.org/10.1021/ja202706g.

[164] D.K. Romney, S.M. Colvin, S.J. Miller, Catalyst control over regio- and enantioselectivity in Baeyer–Villiger oxidations of functionalized ketones, J. Am. Chem. Soc. 136 (40) (2014) 14019–14022. Available from: http://dx.doi.org/10.1021/ja508757g.

[165] D.K. Romney, S.J. Miller, A peptide-embedded trifluoromethyl ketone catalyst for enantioselective epoxidation, Org. Lett. 14 (4) (2012) 1138–1141. Available from: http://dx.doi.org/10.1021/ol3000712.

[166] S.J. Miller, G.T. Copeland, N. Papaioannou, T.E. Horstmann, E.M. Ruel, Kinetic resolution of alcohols catalyzed by tripeptides containing the *N*-alkylimidazole substructure, J. Am. Chem. Soc. 120 (7) (1998) 1629–1630. Available from: http://dx.doi.org/10.1021/ja973892k.

[167] G.T. Copeland, E.R. Jarvo, S.J. Miller, Minimal acylase-like peptides. Conformational control of absolute stereospecificity, J. Org. Chem. 63 (20) (1998) 6784–6785. Available from: http://dx.doi.org/10.1021/jo981642w.

[168] J.T. Blank, S.J. Miller, Studies of folded peptide-based catalysts for asymmetric organic synthesis, Biopolymers 84 (1) (2006) 38–47. Available from: http://dx.doi.org/10.1002/bip.20392.

[169] G.T. Copeland, S.J. Miller, A chemosensor-based approach to catalyst discovery in solution and on solid support, J. Am. Chem. Soc. 121 (17) (1999) 4306–4307. Available from: http://dx.doi.org/10.1021/ja984139+.

[170] G.T. Copeland, S.J. Miller, Selection of enantioselective acyl transfer catalysts from a pooled peptide library through a fluorescence-based activity assay: an approach to kinetic resolution of secondary alcohols of broad structural scope, J. Am. Chem. Soc. 123 (27) (2001) 6496–6502. Available from: http://dx.doi.org/10.1021/ja0108584.

[171] E.R. Jarvo, C.A. Evans, G.T. Copeland, S.J. Miller, Fluorescence-based screening of asymmetric acylation catalysts through parallel enantiomer analysis. Identification of

a catalyst for tertiary alcohol resolution, J. Org. Chem. 66 (16) (2001) 5522–5527. Available from: http://dx.doi.org/10.1021/jo015803z.

[172] F. Formaggio, A. Barazza, A. Bertocco, C. Toniolo, Q.B. Broxterman, B. Kaptein, et al., Role of secondary structure in the asymmetric acylation reaction catalyzed by peptides based on chiral C^{α}-tetrasubstituted α-amino acids, J. Org. Chem. 69 (11) (2004) 3849–3856. Available from: http://dx.doi.org/10.1021/jo040107v.

[173] P. Chen, J. Qu, Backbone modification of β-hairpin-forming tetrapeptides in asymmetric acyl transfer reactions, J. Org. Chem. 76 (9) (2011) 2994–3004. Available from: http://dx.doi.org/10.1021/jo200403g.

[174] C.A. Lewis, B.R. Sculimbrene, Y. Xu, S.J. Miller, Desymmetrization of glycerol derivatives with peptide-based acylation catalysts, Org. Lett. 7 (14) (2005) 3021–3023. Available from: http://dx.doi.org/10.1021/ol051061a.

[175] C.A. Lewis, J.L. Gustafson, A. Chiu, J. Balsells, D. Pollard, J. Murry, et al., A case of remote asymmetric induction in the peptide-catalyzed desymmetrization of a bis(phenol), J. Am. Chem. Soc. 130 (48) (2008) 16358–16365. Available from: http://dx.doi.org/10.1021/ja807120z.

[176] B. Kim, A.J. Chinn, D.R. Fandrick, C.H. Senanayake, R.A. Singer, S.J. Miller, Distal stereocontrol using guanidinylated peptides as multifunctional ligands: desymmetrization of diarylmethanes via Ullman cross-coupling, J. Am. Chem. Soc. 138 (25) (2016) 7939–7945. Available from: http://dx.doi.org/10.1021/jacs.6b03444.

[177] K.S. Griswold, S.J. Miller, A peptide-based catalyst approach to regioselective functionalization of carbohydrates, Tetrahedron 59 (45) (2003) 8869–8875. Available from: http://dx.doi.org/10.1016/j.tet.2003.05.002.

[178] C.A. Lewis, S.J. Miller, Site-selective derivatization and remodeling of erythromycin A by using simple peptide-based chiral catalysts, Angew. Chem. Int. Ed. Engl. 45 (34) (2006) 5616–5619. Available from: http://dx.doi.org/10.1002/anie.200601490.

[179] S. Yoganathan, S.J. Miller, Structure diversification of vancomycin through peptide-catalyzed, site-selective lipidation: a catalysis-based approach to combat glycopeptide-resistant pathogens, J. Med. Chem. 58 (5) (2015) 2367–2377. Available from: http://dx.doi.org/10.1021/jm501872s.

[180] T. Kawabata, W. Muramatsu, T. Nishio, T. Shibata, H. Schedel, A catalytic one-step process for the chemo- and regioselective acylation of monosaccharides, J. Am. Chem. Soc. 129 (42) (2007) 12890–12895. Available from: http://dx.doi.org/10.1021/ja074882e.

[181] H. Takeuchi, K. Mishiro, Y. Ueda, Y. Fujimori, T. Furuta, T. Kawabata, Total synthesis of ellagitannins through regioselective sequential functionalization of unprotected glucose, Angew. Chem. Int. Ed. Engl. 54 (21) (2015) 6177–6180. Available from: http://dx.doi.org/10.1002/anie.201500700.

[182] K. Yoshida, T. Furuta, T. Kawabata, Organocatalytic chemoselective monoacylation of 1,*n*-linear diols, Angew. Chem. Int. Ed. Engl. 50 (21) (2011) 4888–4892. Available from: http://dx.doi.org/10.1002/anie.201100700.

[183] K. Yoshida, T. Shigeta, T. Furuta, T. Kawabata, Catalyst-controlled reversal of chemoselectivity in acylation of 2-aminopentane-1,5-diol derivatives, Chem. Commun. 48 (55) (2012) 6981–6983. Available from: http://dx.doi.org/10.1039/c2cc32525j.

[184] M. Yamanaka, U. Yoshida, M. Sato, T. Shigeta, K. Yoshida, T. Furuta, et al., Origin of high *E*-selectivity in 4-pyrrolidinopyridine-catalyzed tetrasubstituted α,α′-alkenediol: a computational and experimental study, J. Org. Chem. 80 (6) (2015) 3075–3082. Available from: http://dx.doi.org/10.1021/jo5029453.

[185] F. Huber, S.F. Kirsch, Site-selective acylations with tailor-made catalysts, Chemistry 22 (17) (2016) 5914–5918. Available from: http://dx.doi.org/10.1002/chem.201600790.

[186] B.R. Sculimbrene, S.J. Miller, Discovery of a catalytic asymmetric phosphorylation through selection of a minimal kinase mimic: a concise total synthesis of D-*myo*-inositol-1-phosphate, J. Am. Chem. Soc. 123 (41) (2001) 10125–10126. Available from: http://dx.doi.org/10.1021/ja016779+.

[187] K.W. Fiori, A.L.A. Puchlopek, S.J. Miller, Enantioselective sulfonylation reactions mediated by a tetrapeptide catalyst, Nat. Chem. 1 (8) (2009) 630–634. Available from: http://dx.doi.org/10.1038/nchem.410.

[188] B.R. Sculimbrene, A.J. Morgan, S.J. Miller, Enantiodivergence in small-molecule catalysis of asymmetric phosphorylation: concise total syntheses of the enantiomeric D-*myo*-inositol-1-phosphate and D-*myo*-inositol-3-phosphate, J. Am. Chem. Soc. 124 (39) (2002) 11653–11656. Available from: http://dx.doi.org/10.1021/ja027402m.

[189] S. Han, S.J. Miller, Asymmetric catalysis at a distance: catalytic, site-selective phosphorylation of teicoplanin, J. Am. Chem. Soc. 135 (33) (2013) 12414–12421. Available from: http://dx.doi.org/10.1021/ja406067v.

[190] M. Sánchez-Roselló, A.L.A. Puchlopek, A.J. Morgan, S.J. Miller, Site-selective catalysis of phenyl thionoformate transfer as a tool for regioselective deoxygenation of polyols, J. Org. Chem. 73 (5) (2008) 1774–1782. Available from: http://dx.doi.org/10.1021/jo702334z.

[191] B.S. Fowler, K.M. Laemmerhold, S.J. Miller, Catalytic site-selective thiocarbonylations and deoxygenations of vancomycin reveal hydroxyl-dependent conformational effects, J. Am. Chem. Soc. 134 (23) (2012) 9755–9761. Available from: http://dx.doi.org/10.1021/ja302692j.

[192] P.A. Jordan, S.J. Miller, An approach to the site-selective deoxygenation of hydroxy groups based on catalytic phosphoramidite transfer, Angew. Chem. Int. Ed. Engl. 51 (12) (2012) 2907–2911. Available from: http://dx.doi.org/10.1002/anie.201109033.

[193] C.E. Müller, L. Wanka, K. Jewell, P.R. Schreiner, Enantioselective kinetic resolution of *trans*-cycloalkane-1,2-diols, Angew. Chem. Int. Ed. Engl. 47 (33) (2008) 6180–6183. Available from: http://dx.doi.org/10.1002/anie.200800641.

[194] C.E. Müller, D. Zell, R. Hrdina, R.C. Wende, L. Wanka, S.M.M. Schuler, et al., Lipophilic oligopeptides for chemo- and enantioselective acyl transfer reactions onto alcohols, J. Org. Chem. 78 (17) (2013) 8465–8484. Available from: http://dx.doi.org/10.1021/jo401195c.

[195] R. Hrdina, C.E. Müller, P.R. Schreiner, Kinetic resolution of *trans*-cycloalkane-1,2-diols *via* Steglich esterification, Chem. Commun. 46 (15) (2010) 2689–2690. Available from: http://dx.doi.org/10.1039/c001075h.

[196] R.C. Wende, A. Seitz, D. Niedek, S.M.M. Schuler, C. Hofmann, J. Becker, et al., The enantioselective Dakin-West reaction, Angew. Chem. Int. Ed. Engl. 55 (8) (2016) 2719–2723. Available from: http://dx.doi.org/10.1002/anie.201509863.

[197] C.E. Müller, R. Hrdina, R.C. Wende, P.R. Schreiner, A multicatalyst system for the one-pot desymmetrization/oxidation of *meso*-1,2-alkane diols, Chemistry 17 (23) (2011) 6309–6314. Available from: http://dx.doi.org/10.1002/chem.201100498.

[198] R. Hrdina, C.E. Müller, R.C. Wende, L. Wanka, P.R. Schreiner, Enantiomerically enriched *trans*-diols from alkenes in one pot: a multicatalyst approach, Chem. Commun. 48 (19) (2012) 2498–2500. Available from: http://dx.doi.org/10.1039/c2cc17435a.

[199] J. Oku, N. Ito, S. Inoue, Asymmetric cyanohydrin synthesis catalyzed by synthetic dipeptides, 1, Die Makromol. Chem. 180 (4) (1979) 1089–1091. Available from: http://dx.doi.org/10.1002/macp.1979.021800431.

[200] J. Oku, S. Inoue, Asymmetric cyanohydrin synthesis catalysed by a synthetic cyclic dipeptide, J. Chem. Soc., Chem. Commun. (5) (1981) 229–230. Available from: http://dx.doi.org/10.1039/c39810000229.

[201] K. Tanaka, A. Mori, S. Inoue, The cyclic dipeptide *cyclo*[(*S*)-phenylalanyl-(*S*)-histidyl] as a catalyst for asymmetric addition of hydrogen cyanide to aldehydes, J. Org. Chem. 55 (1) (1990) 181–185. Available from: http://dx.doi.org/10.1021/jo00288a030.

[202] T.E. Horstmann, D.J. Guerin, S.J. Miller, Asymmetric conjugate addition of azide to α,β-unsaturated carbonyl compounds catalyzed by simple peptides, Angew. Chem. Int. Ed. Engl. 39 (20) (2000) 3635–3638, doi:10.1002/1521-3773(20001016)39:20<3635::aid-anie3635>3.0.co;2-y.

[203] D.J. Guerin, S.J. Miller, Asymmetric azidation–cycloaddition with open-chain peptide-based catalysts. A sequential enantioselective route to triazoles, J. Am. Chem. Soc. 124 (10) (2002) 2134–2136. Available from: http://dx.doi.org/10.1021/ja0177814.

[204] J.E. Imbriglio, M.M. Vasbinder, S.J. Miller, Dual catalyst control in the amino acid-peptide-catalyzed enantioselective Baylis–Hillman reaction, Org. Lett. 5 (20) (2003) 3741–3743. Available from: http://dx.doi.org/10.1021/ol035466b.

[205] M.M. Vasbinder, J.E. Imbriglio, S.J. Miller, Amino acid-peptide-catalyzed enantioselective Morita–Baylis–Hillman reactions, Tetrahedron 62 (49) (2006) 11450–11459. Available from: http://dx.doi.org/10.1016/j.tet.2006.05.005.

[206] J.W. Evans, M.B. Fierman, S.J. Miller, J.A. Ellman, Catalytic enantioselective synthesis of sulfinate esters through the dynamic resolution of *tert*-butanesulfinyl chloride, J. Am. Chem. Soc. 126 (26) (2004) 8134–8135. Available from: http://dx.doi.org/10.1021/ja047845l.

[207] B.R. Linton, M.H. Reutershan, C.M. Aderman, E.A. Richardson, K.R. Brownell, C.W. Ashley, et al., Asymmetric Michael addition of α-nitro-ketones using catalytic peptides, Tetrahedron Lett. 48 (11) (2007) 1993–1997. Available from: http://dx.doi.org/10.1016/j.tetlet.2007.01.073.

[208] B.S. Fowler, P.J. Mikochik, S.J. Miller, Peptide-catalyzed kinetic resolution of formamides and thioformamides as an entry to nonracemic amines, J. Am. Chem. Soc. 132 (9) (2010) 2870–2871. Available from: http://dx.doi.org/10.1021/ja9107897.

[209] M.S. Iyer, K.M. Gigstad, N.D. Namdev, M. Lipton, Asymmetric catalysis of the Strecker amino acid synthesis by a cyclic dipeptide, J. Am. Chem. Soc. 118 (20) (1996) 4910–4911. Available from: http://dx.doi.org/10.1021/ja952686e.

[210] B.J. Cowen, L.B. Saunders, S.J. Miller, Pyridylalanine (pal)-peptide catalyzed enantioselective allenoate additions to *N*-acyl imines, J. Am. Chem. Soc. 131 (17) (2009) 6105–6107. Available from: http://dx.doi.org/10.1021/ja901279m.

[211] C.T. Mbofana, S.J. Miller, Diastereo- and enantioselective addition of anilide-functionalized allenoates to *N*-acylimines catalyzed by a pyridylalanine-based peptide, J. Am. Chem. Soc. 136 (8) (2014) 3285–3292. Available from: http://dx.doi.org/10.1021/ja412996f.

[212] J.L. Gustafson, D. Lim, S.J. Miller, Dynamic kinetic resolution of biaryl atropisomers via peptide-catalyzed asymmetric bromination, Science 328 (5983) (2010) 1251–1255. Available from: http://dx.doi.org/10.1126/science.1188403.

[213] K.T. Barrett, S.J. Miller, Enantioselective synthesis of atropisomeric benzamides through peptide-catalyzed bromination, J. Am. Chem. Soc. 135 (8) (2013) 2963–2966. Available from: http://dx.doi.org/10.1021/ja400082x.

[214] A.J. Metrano, N.C. Abascal, B.Q. Mercado, E.K. Paulson, S.J. Miller, Structural studies of β-turn-containing peptide catalysts for atroposelective quinazolinone bromination, Chem. Commun. 52 (26) (2016) 4816–4819. Available from: http://dx.doi.org/10.1039/c6cc01428c.
[215] T.P. Pathak, S.J. Miller, Chemical tailoring of teicoplanin with site-selective reactions, J. Am. Chem. Soc. 135 (22) (2013) 8415–8422. Available from: http://dx.doi.org/10.1021/ja4038998.
[216] A.J. Metrano, S.J. Miller, Peptide-catalyzed conversion of racemic oxazol-5(4*H*)-ones into enantiomerically enriched α-amino acid derivatives, J. Org. Chem. 79 (4) (2014) 1542–1554. Available from: http://dx.doi.org/10.1021/jo402828f.
[217] S.M. Mennen, J.T. Blank, M.B. Tran-Dubé, J.E. Imbriglio, S.J. Miller, A peptide-catalyzed asymmetric Stetter reaction, Chem. Commun. (2) (2005) 195–197. Available from: http://dx.doi.org/10.1039/b414574g.
[218] S.M. Mennen, J.D. Gipson, Y.R. Kim, S.J. Miller, Thiazolylalanine-derived catalysts for enantioselective intermolecular aldehyde–imine cross-couplings, J. Am. Chem. Soc. 127 (6) (2005) 1654–1655. Available from: http://dx.doi.org/10.1021/ja042650z.
[219] C.R. Shugrue, S.J. Miller, Phosphothreonine as a catalytic residue in peptide-mediated asymmetric transfer hydrogenations of 8-aminoquinolines, Angew. Chem. Int. Ed. Engl. 54 (38) (2015) 11173–11176. Available from: http://dx.doi.org/10.1002/anie.201505898.
[220] L.-W. Xu, Powerful amino acid derived bifunctional phosphine catalysts bearing a hydrogen bond donor in asymmetric synthesis, ChemCatChem 5 (10) (2013) 2775–2784. Available from: http://dx.doi.org/10.1002/cctc.201300198.
[221] B.J. Cowen, S.J. Miller, Enantioselective [3 + 2]-cycloadditions catalyzed by a protected, multifunctional phosphine-containing α-amino acid, J. Am. Chem. Soc. 129 (36) (2007) 10988–10989. Available from: http://dx.doi.org/10.1021/ja0734243.
[222] X. Han, F. Zhong, Y. Wang, Y. Lu, Versatile enantioselective [3 + 2] cyclization between imines and allenoates catalyzed by dipeptide-based phosphines, Angew. Chem. Int. Ed. Engl. 51 (3) (2011) 767–770. Available from: http://dx.doi.org/10.1002/anie.201106672.
[223] D. Cao, J. Zhang, H. Wang, G. Zhao, Dipeptide-derived multifunctional quaternary phosphonium salt catalyzed asymmetric cyclizations via a tandem Michael addition/S_N2 sequence, Chemistry 21 (28) (2015) 9998–10002. Available from: http://dx.doi.org/10.1002/chem.201501806.
[224] S.S. Chimni, V. Kumar, N. Bala, Design of peptidyl thiourea derivatives as organocatalysts for the asymmetric ring opening of *meso*-stilbene oxides, Asian J. Org. Chem. 3 (6) (2014) 700–705. Available from: http://dx.doi.org/10.1002/ajoc.201402040.
[225] B. Zhu, S. Qiu, J. Li, M.L. Coote, R. Lee, Z. Jiang, Asymmetric [4 + 2] annulation of 5*H*-thiazol-4-ones with a chiral dipeptide-based Brønsted base catalyst, Chem. Sci. 7 (9) (2016) 6060–6067. Available from: http://dx.doi.org/10.1039/c6sc02039a.
[226] C.E. Aroyan, S.J. Miller, Enantioselective Rauhut–Currier reactions promoted by protected cysteine, J. Am. Chem. Soc. 129 (2) (2007) 256–257. Available from: http://dx.doi.org/10.1021/ja067139f.
[227] C.E. Aroyan, A. Dermenci, S.J. Miller, Development of a cysteine-catalyzed enantioselective Rauhut–Currier reaction, J. Org. Chem. 75 (17) (2010) 5784–5796. Available from: http://dx.doi.org/10.1021/jo101018t.
[228] Z.-Z. Huang, L.J. Leman, M.R. Ghadiri, Biomimetic catalysis of diketopiperazine and dipeptide syntheses, Angew. Chem. Int. Ed. Engl. 47 (9) (2008) 1758–1761. Available from: http://dx.doi.org/10.1002/anie.200704266.

Peptide-based biosensors

20

Qian Zhang, Yanli Lu, Shuang Li, Jiajia Wu and Qingjun Liu
Biosensor National Special Laboratory, Key Laboratory for Biomedical Engineering of Education Ministry, Department of Biomedical Engineering, Zhejiang University, Hangzhou, China

20.1 Introduction

Biosensors are analytical devices incorporating biological sensing elements with the processing power of microelectronics [1,2]. The main components of biosensors are biological recognition elements and physical/chemical transducers. Due to these biorecognition components, biosensors have significant advantages such as high sensitivity, high selectivity, and the ability for high-throughput processing [3,4]. Therefore, they have important applications in various fields such as clinical diagnosis and environmental monitoring. The construction of biosensors mainly relies on the incorporation of bioactive elements, which can be divided primarily into tissues, cells, and proteins [5,6]. These components can maintain high biosensitivity, however, biosensors fabricated with them still suffer from tough acquirement of biological components, short survival time, high cost, and other difficulties, which brought significant obstacles to the practical applications.

In recent years, with the development of biosynthetic technologies, including specific peptide screening and artificial peptide synthesis, the field of biosensor development has seen significant advances. Through standard Fmoc and *t*-Boc solid-phase synthesis, a wide variety of peptide chains with high specificity and versatility can be produced on a large scale and can be custom-made to meet specific requirements. Compared to other biological materials, peptides are stable and active for prolonged periods of time, and can be produced at low cost. Besides, peptides with high specificity can be facilely prepared, modified through mature and diverse methods, and used for repaid detections. Meanwhile, with the development of modern science and technology, peptides can be designed to have specific three-dimensional structures like proteins, thereby having new and improved biorecognition capabilities. As a result, increasing numbers of researchers have set out to develop peptide-based biosensors that can be commercialized.

Generally, the binding activities between peptides and target analytes (such as small molecules, nanoparticles, heavy metals, DNA, and proteins) will induce changes that cannot be easily detected. Thus, transducers or molecular beacons are usually conjugated with peptides to convert these changes into detectable signals, thus facilitating subsequent analysis and applications. Such conjugation will lead to the construction of peptide-based biosensors that combine the properties of peptides and sensors. Depending on the conjugation method, peptide-based biosensors can

Peptide Applications in Biomedicine, Biotechnology and Bioengineering.
DOI: http://dx.doi.org/10.1016/B978-0-08-100736-5.00024-7

be classified into electrochemical biosensors, optical biosensors, and other kinds of biosensors. In the following sections, a brief introduction to the development of each of these types of peptide-based biosensor is provided.

20.2 Peptide-based electrochemical biosensors

As one of the main sensing detection tools in modern analytical science, electrochemical detection technology has been proved to be of great value in the detection of important analytes such as proteins, metabolites, nucleic acids, and metal ions. It is an excellent quantitative detection method, with advantages that include low cost, high reliability, simple operation, and low limit of detection (LOD). Moreover, it provides direct information of the collected electrical signals, while its detection systems can be easily miniaturized and integrated. Hence, sensors based on electrochemical detection have been widely used in many fields including environmental monitoring, health care, and food regulation. Researchers have increasingly focused on the combination of peptides and electrochemical sensors in recent years. Depending on the sensing technology, peptide-based electrochemical biosensors can be divided into sensors operating on differential pulse voltammetry (DPV) mode, square wave voltammetry (SWV) mode, electrochemical impedance spectroscopy (EIS) mode, chronoamperometry mode, cyclic voltammetry (CV) mode, anodic stripping voltammetry (ASV), and so on. Some typical peptide-based electrochemical biosensing systems are described in Table 20.1.

20.2.1 Peptide-based DPV biosensors

As can be seen in Table 20.1, DPV is the most popular technique used in peptide-based electrochemical biosensors. In DPV, the base potential is incremented from the initial potential toward the final potential, with a potential pulse applied on it. Current is sampled before the potential pulse and at the end of the pulse. The difference between the two current samples is recorded as the response. This method can reduce the influence of background interferences such as the redox current induced by impurities. Therefore it gains a higher detection sensitivity and lower detection limit than other electrochemical approaches, which makes it a more sensitive tool in quantitative detection compared with chromatographic methods.

Based on these properties, DPV has advantages such as simultaneous detection of multiple substances, low demand for electrolytes, and less analytes needed. As a result, it has been widely used in the field of analysis, and is mainly used for quantitative determination of a variety of substances, the study of adsorption phenomena, the study of the mechanism of complex electrode reactions, and the detection of trace analytes. For example, Zhao et al. proposed a peptide-based DPV biosensor for the detection of prostate-specific antigen (PSA), which is correlated with prostate cancer [7]. As Fig. 20.1 shows, ferrocene-functionalized helix peptides with a specific molecular recognition site for PSA were assembled on the gold electrode

Table 20.1 **Typical peptide-based optical biosensors**

Sensing technique	Analytes	Peptides	References
DPV	Prostate-specific antigen (PSA)	CHSSLKQK	[7]
	Matrix metalloproteinase-2 (MMP-2)	Specific peptide (PLGVR)	[8,9]
	Protein kinase A	Kemptide or CGGALRRASLG	[10,11]
	Renin	Peptide aptamer (KLASSPLS)	[12]
	The breast cancer marker BRCA1	Zwitterionic peptide (EKEKEKE)	[13]
	Apoptotic cells	FNFRLKAGAKIRFGRGC and AFGNRGRAAKNFHARGC	[14]
SWV	Human papillomavirus DNA type 16	Peptide nucleic acid (AQ-CATACACCTCCAGC-LysNH_2)	[15,16]
	BIR3 domain of X-linked inhibitor of apoptosis protein (XIAP-BIR3)	AVPFAQKG	[17]
	Caspase-3	Biotinylated DEVD-peptide (biotin-GDGDEVDGC)	[18]
EIS	*Escherichia coli* O157:H7	Antimicrobial peptide	[19]
	Human noroviral capsid proteins (rP2) and human norovirus	Noro-1 peptide (QHKMHKPHKNTKGGGGSC)	[20]
	Thrombin	Coiled-coil peptide (EKKLAQLEWENQALEKELAQGG-*LVPRGS*-GGAQLKKKLQANKKELAQLKWK)	[21]
Chronoamperometry	H_2O_2	Peptide nanofiber (AEAKAEAKYWYAFAEAKAEAK)	[22]
	Phenolic compounds (catechol, phenol, and bisphenol A)	Silk peptide (a partly hydrolysis product of silk protein)	[23]
	Ochratoxin A	NFO4 synthetic peptide (VYMNRKYYKCCK)	[24]
CV	MMP-9	MB-peptide (methylene blue-GPLGMWSRC)	[25]
	Plasmin	KTFKGGGGGC	[26]
ASV	Caspase 3	Biotinylated DEVD-peptide (biotin-GDGDEVDGC)	[27]

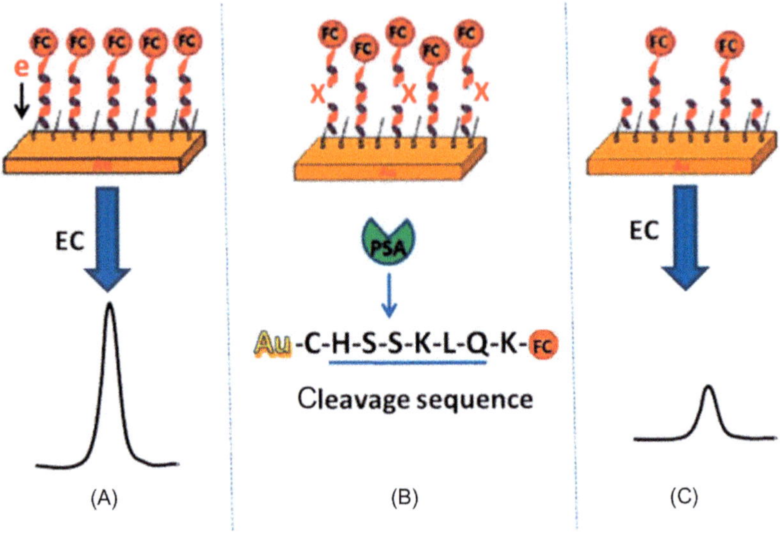

Figure 20.1 Sensing mechanism of the biosensor proposed by N. Zhao et al. (A) Ferrocene-functionalized helix peptides were assembled on a gold electrode surface and the DPV response showed a large current; (B) PSA can recognize the specific sequence on the peptides and cleave at the site; (C) after cleavage of PSA, the cleaved ferrocene and peptide fragments were removed from the electrode surface, resulting in a smaller current than before.

and could be cleaved when there was PSA in the environment. This led to a decrease in the DPV response of the biosensor. With the help of DPV and peptides, the biosensor could linearly respond to the analytes at a concentration range of 0.5–40 ng/mL with a competitive detection limit of 0.2 ng/mL.

Due to the small sizes of peptides, changes induced by interactions between peptides and anayltes are usually small. DPV seems to be an ideal electrochemical approach to apply in peptide-based biosensors, as it can eliminate background noises, leading to a large signal-to-noise rate (SNR), and can be applied in the detection of substances with low electrical activity.

20.2.2 Peptide-based SWV biosensors

SWV is also a common detection method which is used for the construction of peptide-based electrochemical biosensors. It takes a rapid scanning stepped voltage as the baseline potential, on each step of which a small symmetric square wave is superimposed. At the end of the forward and reverse steps of the potential, currents are sampled and their differences are plotted against the voltage to obtain an SWV curve. SWV is more effective than DPV in suppressing background currents. Thus, with advantages such as high sensitivity and rapid detection, the method is widely used in quantitative analyses of substances and kinetic studies, and it can detect organic and inorganic compounds with redox properties.

Wang et al. reported a reagentless SWV-based peptide sensor that detected the BIR3 domain of X-linked inhibitor of apoptosis protein (XIAP-BIR3) [17]. As Fig. 20.2 shows, the bioreceptor was based on a conducting copolymer film electrosynthesized from juglone (JUG) and a juglone–peptide conjugate (JP) that was designed for this experiment. The peptide–protein interactions generated an important increase in steric hindrance at the interface and a current decrease (signal off) of the redox reaction from quinone which was embedded in the polymer backbone. The biosensor allowed a specific and sensitive detection of XIAP-BIR3 with a detection limit of 1 nM. The peptide–protein complex could then be dissociated by adding the free precursor peptide (AVPFAQKG) into the solution, causing a shift-back on the signal, i.e., an increase in the current intensity (signal-on). This "off–on" detection sequence was used in this work as a double verification of the specificity and this approach can be employed as a general way to increase the reliability of results.

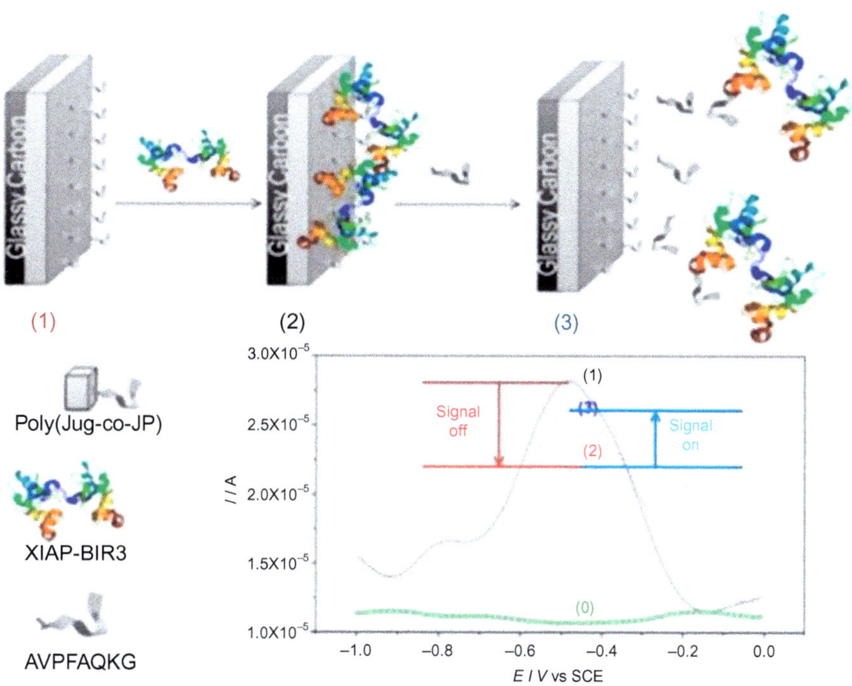

Figure 20.2 Strategy for label-free electrochemical detection of XIAP-BIR3 based on surface hindrance of the poly(JUG-co-JP)-modified electrode and their SWV response curves: (0) bare electrode (open circles); (1) poly(JUG-co-JP)-modified electrode; (2) after complexation with 10 nM XIAP-BIR3 (dash, signal-off); (3) after addition of 10 nM peptide into the assay solution (dash dot, signal-on).

20.2.3 Peptide-based EIS biosensors

EIS has been one of the most common electrochemical detection approaches in recent years. When the electrochemical electrode system is disturbed by an AC signal of a sine voltage (typically <10 mV) or current, a corresponding current or voltage response signal is generated from which the impedance or conductance of the electrode can be obtained. Then, the impedance spectrum can also be calculated from the disturbance of sine signals at a series of frequencies, known as EIS. Due to the small-amplitude disturbance, great impacts on the system are avoided, and an approximate linear relationship is shown between the disturbance and the system responses, which makes the data processing of the measurements straightforward. Responses of EIS can be used to deduce the equivalent circuit of the electrode, by which the kinetic process and excitation of the system can be analyzed. Kinetic parameters, such as capacitances of double electric layers, reaction resistances of the charge transfer process, and parameters of the diffusion mass-transfer process, can also be estimated by relevant element values in the equivalent circuit.

EIS can detect impedance spectra with a wide frequency range to study the electrode system and can provide more information of the kinetics and electrode interface structure than other conventional electrochemical methods. Therefore, EIS is suitable for the rapid detection of electrode processes (mass transfer process, diffusion process, etc.). For example, Kongsuphol et al. fabricated an ultrasensitive, disposable, electrochemical thrombin biosensor with a comb-structured gold microelectrode array (CSGMA), which was functionalized with a self-assembled monolayer of thiol terminated coiled-coil peptide (CCP) [21]. Containing a thrombin-specific cleavage site (LVPRGS), the CCP provided sites for thrombin capture and detection, as shown in Fig. 20.3. The EIS technique was chosen to detect thrombin at different concentrations. Results revealed the sensor can respond to thrombin with a LOD of 28 fM and was able to detect the catalytic activity of thrombin within 30 min. CCP/CSGMA electrodes were found to be selective against other antibodies such as the p-53 antibody (clone DO1) and the hemagglutinin (HA) antibody. Thus, the sensor provided high specificity toward thrombin detection and mechanistic details of the binding and cleavage process.

Although EIS is a rapid detection tool in electrochemical sensors, it shows lower sensitivity and larger background noises than DPV and SWV. Thus, it needs more effort to be applied in peptide-based electrochemical sensors.

20.2.4 Peptide-based chronoamperometry biosensors

As the simplest method in electrochemical detection, chronoamperometry is a potential-controlled analytical method that measures the functional relationship between current responses and time by applying a single-potential step or a double-potential step to the working electrode. It is often used in research on electron transfer kinetics. With general usage of electrodes of fixed area, chronoamperometry is suitable to study the electrode process coupling chemical reactions and the reaction mechanism of organic electrochemistry.

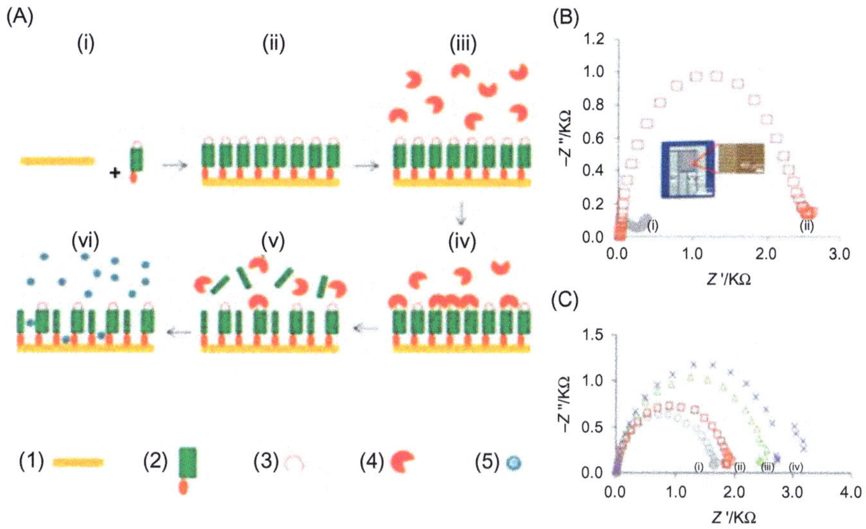

Figure 20.3 (A) Scheme of the sensor fabrication and thrombin detection mechanism. Symbols represent (1) gold electrode, (2) thiol-terminated CCP, (3) thrombin-specific cleavage site, (4) thrombin, and (5) redox molecule. (B) EIS verification of (i) CSGMA electrode, (ii) CCP/CSGMA electrode. Inset shows the optical image of a CSGMA electrode chip. (C) Nyquist plot of (i) CCP/CSGMA, (ii) 10 pg/mL p-53 antibody (clone DO1), (iii) 100 fg/mL thrombin, and (iv) 10 pg/mL DO1 + 100 fg/mL thrombin.

Li et al. designed a high-performance electrochemical hydrogen peroxide (H_2O_2) biosensor based on motif-designed peptides. These peptides form nanofibers (PNFs) and recognize graphene quantum dots (GQDs) and graphene oxide (GO) nanosheets specifically [22]. By designing the peptide sequence, ternary GQD-PNF-GO nanohybrids were synthesized and applied to modify the working electrodes for the detection of H_2O_2 with the chronoamperometry method (Fig. 20.4). The biosensor exhibited high sensitivity and selectivity, a low detection limit, and wide linear range for sensing H_2O_2, demonstrating the usefulness of peptides as biosensors.

The chronoamperometry approach is simple and easy-to-perform, however the method provides lower sensitivity and less information for biosensing detection compared with other electrochemical tools. As a result, it acts more like a quick test to investigate the performance of peptides in biosensors.

20.2.5 Other peptide-based electrochemical biosensors

Besides the methods introduced above, there have also been other electrochemical tools applied in peptide-based biosensors. For example, Lee et al. constructed a biosensor for detection of matrix metalloproteinase-9 (MMP-9) with CV [25]. As shown in Fig. 20.5, the biosensor was integrated with a methylene blue (MB)-labeled peptide, which was specific to MMP-9. Peptide cleavage by the protease caused the release of electroactive MB from the gold electrodes and

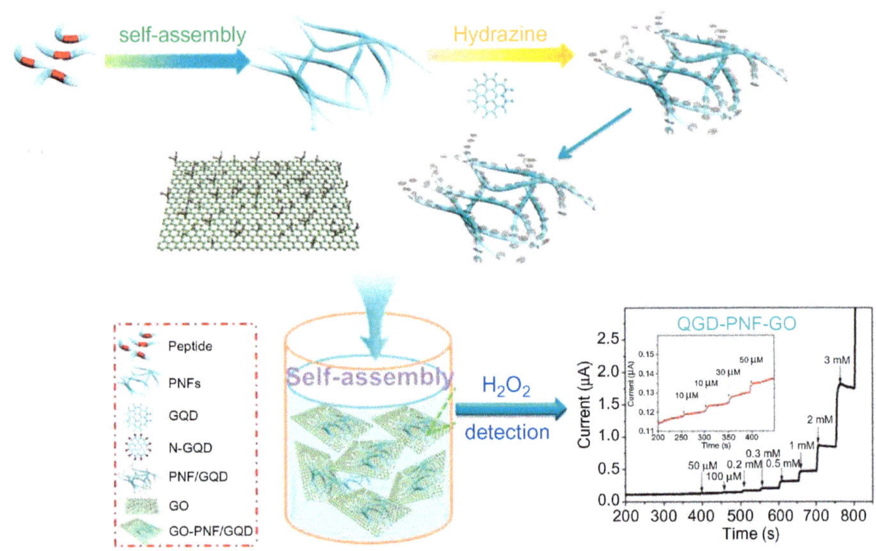

Figure 20.4 Schematic representation for the synthesis of PNFs, binary GQD-PNF nanohybrids, and ternary GQD-PNF-GO nanohybrids, and the chronoamperometry responses to different doses of H_2O_2 detected with GQD-PNF-GO.

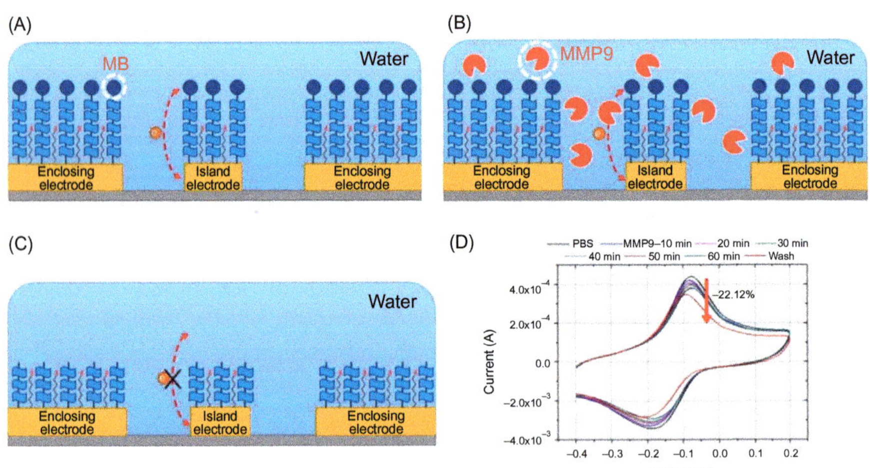

Figure 20.5 Sensing mechanism of the biosensor proposed by Lee et al. and its CV responses to MMP-9. (A) Tunneling of electrons in the presence of MB-peptide on an Au electrode. (B) Peptide cleavage reaction of the MB-peptide with MMP-9. (C) Decrease in the current of electron tunneling after the peptide cleavage reaction. (D) CV results after incubating an MB-peptide immobilized electrode challenged with 1 nM MMP-9.

resulted in a decrease in the electrical tunneling current. As CV can determine changes to voltage and current during the electrolysis process through repeated scanning with a triangular potential at different rates, this method was chosen to carry out quantitative and qualitative analysis of thrombin. With the help of a self-designed gold electrode with self-gating effect, the electrochemical biosensor exhibited a linear correlation between the concentration of MMP-9 and the current reduction in a range from 1 pM to 1 nM, and the LOD was determined to be at 7 pM.

In another study by Zhang et al., ASW was introduced to a biosensing platform for the determination of caspase 3 activity and inhibition using a DEVD-peptide and quantum dots (QDs) [27]. As shown in Fig. 20.6, a gold electrode was immobilized by a biotinylated peptide which contained the specific cleavage site (DEVD) of caspase 3 and then blocked by 6-mercapto-1-hexanol (MCH). Subsequently, the electrode was immersed into apoptotic cell lysates containing active caspase 3, which could specifically recognize and cleave the peptide. This procedure led to the loss of the biotin label on the electrode surface. The ratio of the uncleaved biotin-DEVD on the electrode surface was related to the activity of caspase 3 and evaluated by a signaling probe composed of CdTe QDs, carbon nanotubes (CNTs), and

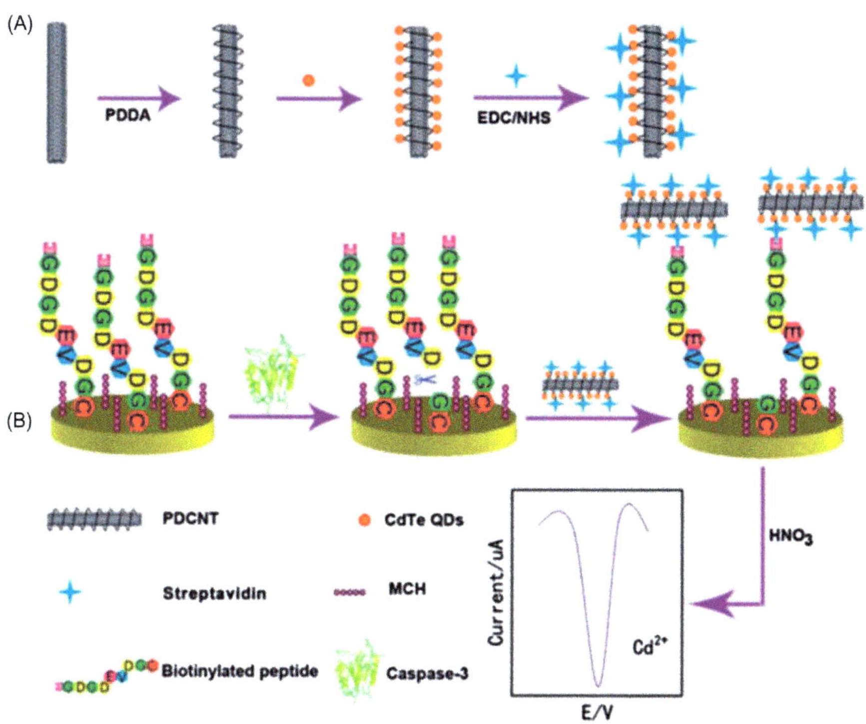

Figure 20.6 (A) Preparation process of CNT-QD-SA bioconjugates. (B) Schematic illustration of the ASV strategy for sensing caspase 3 activity.

streptavidin (SA) that can be captured by the uncleaved biotin-DEVD. Finally, the cadmic component obtained by subsequent dissolution of the electrode in HNO_3 can be quantified by ASV. The ASV method is suitable for detection of metal elements in dilute solutions. This electrochemical strategy allowed for the study of drug action in activating caspase 3 within a short time and at a low concentration, presenting a significant tool for efficient screening of potential caspase 3 inhibitors and anticancer drugs.

Although the sensors introduced above can also be used effectively in detection, electrochemical techniques in general are not widely applied due to difficulties in controlling the experimental conditions. However, taking all the electrochemical tools together, the peptide-based electrochemical sensors have broad potential in various fields because of their advantages such as high sensitivity, accurate quantitation, and simple operation. The significant progress in the field of portable electrochemical detection systems suggests that peptide-based electrochemical biosensors will likely see rapid development.

20.3 Peptide-based optical biosensors

Compared to electrochemical techniques, optical tools seem to be more popular in the development of peptide-based biosensors. With the help of molecules with significant optical characteristics, peptide-based optical biosensors are easier to construct than electrochemical ones. According to their different sensing technology, peptide-based optical biosensors can be mainly divided into fluorescence biosensors, colorimetric biosensors, surface plasmon resonance (SPR) biosensors, and surface-enhanced Raman scattering (SERS) biosensors. Some typical peptide-based optical biosensing systems are described in Table 20.2.

20.3.1 Peptide-based fluorescence biosensors

Biosensors utilizing fluorescence signals represent the most common approach for the development of peptide based biosensors. This is due to the fact that the technology for the detection of fluorescence emitted by peptides tagged with a fluorescent probe is well established and optimized. There are several modes of this technology including fluorescence "turn-on" techniques, fluorescence resonance energy transfer (FRET), fluorescence quenching, fluorescence enhancement, and fluorescence lifetime imaging microscopy (FLIM).

Fluorescence "turn-on" indicates a process like turning on a light. The general preparation of such a biosensor is to modify each end of a peptide with a fluorescent probe and a fluorescent quencher. When target analytes are present in the solution, they will interact with the peptides, which results in separation of the probe and the quencher, leading to the "turning-on" of fluorescence. For example, Wang et al. reported a novel intracellular protease fluorescence "turn-on" sensor based on the nanoconjugate of GO and peptide substrates [30]. Apoptosis-related caspase-3 activation was chosen as the case study. GO is a carrier for delivering peptide

Table 20.2 **Typical peptide-based optical biosensors**

Sensing technique	Signal strategy	Analytes	Peptides	References
Fluorescence	Fluorescence turn-on	Cyclin A	H_2N-AKRRLI-Dap(4-DMAP)-E	[28]
		Broad-spectrum cancer-related protein LAPTM4B	IHGHHIISVG (referred to as AP2H)	[29]
		Caspase-3	DEVD	[30]
		Human carbonic anhydrase II (HCAII)	KE2b-D(15)C(25)-5, a specific designed polypeptide	[31–33]
		Protein tyrosine kinases (PTKs)		[34]
		Protein kinase activity		[35]
	Fluorescence quenching	Matrix metalloproteinase-2 (MMP-2) and MMP-9	A triple-helical peptide with the type V collagen sequence GPPG ~ VVGEKGEQ	[36]
		Alkaline phosphatase (ALP)	Fmoc-K(FITC)FFYp	[37]
		Protease activity	CCAAAA	[38]
		Human immunodeficiency virus type 1 (HIV-1) matrix protein p17	CEKIRLRC	[39]
	Fluorescence enhancing	Protein kinase activity	Phosphorylated peptide	[40,41]
		Src and ab1 PTKs		[42]

(Continued)

Table 20.2 (Continued)

Sensing technique	Signal strategy	Analytes	Peptides	References
	FRET	Lipopolysaccharide (LPS, also known as bacterial endotoxin)	Tamra-QVPAQLLVGALRVLAYSRLK-εK(αFluo)-NH_2 or Tamra-QVAAKLLVKALRKLAYKRLK-εK (αFluo)-NH_2	[43]
		Hg	GGTLAVPGMTCAACPITVKKGGW-$CONH_2$	[44,45]
		Protein kinase Cα (PKCα)	FKKQGSFAKKK-NH_2	[46]
	FLIM	Abl kinase	GGEAIYAAPC_{Cy5}GGRKKRRQRRRPQ	[47]
Colorimetric	Color changes/ density	Matrix metalloproteinase-7 (MMP-7)	HHHHHHRPLALWRSC	[48]
		HCAII		[49]
		Escherichia coli O157:H7 (*E. coli* O157:H7)	NH2-Ahx-KVSRRRRRGGDKVDRRRRRGGD-Ahx-Cys	[50]
		Zn^{2+} and chymotrypsin	CCPGCAR-NH_2, CAYRA-NH_2, etc.	[51]
		Thrombin	CLVPRGSC	[52]
SPR	Refractive angle	His-tagged proteins	3-Mercaptopropionyl-LHDLHD	[53]
		Staphylococcal enterotoxin B (SEB)	LLADTTHHRPWTLLADTTHHRPWT	[54]
		Leukocyte function-associated antigen-1 (LFA-1)-derived peptides	$CD11a_{237-261}$ (ITDGEATDSGNIDAAKDIIRYIIGI) and $CD11a_{456-465}$ (APLFYGEQRG)	[55]
SERS	Raman intensity and wavelength shift	Phosphorylation catalyzed by Abl kinase	EAIYAAPVAKKKKGGGGC	[56]
		Heparin, trypsin, and thrombin	RKGSGRRLVKC, RCFRGGDD, and RCFLVPRGSDD	[57–59]

cargoes inside living cells. Therefore, the peptide is cleaved from the GO−peptide conjugate after being transported into cells by intracellular proteases. This results in a significant increase of the fluorescence emission due to the release of the fluorophore from GO.

Based on intracellular delivery of GO−peptide conjugate and caspase-3-mediated cleavage of substrate peptide, a sensitive, simple, and robust intracellular protease sensor for high-contrast imaging of apoptotic signaling in live cells was developed (Fig. 20.7A). Assays were performed using mixtures of GO−peptide conjugate and caspase-3 of varying concentrations to demonstrate the quantitative nature of the protease sensor (Fig. 20.7B). The fluorescence peaks dynamically increased with increasing caspase-3 concentration in the range from 7.25 to 362 ng/mL. A plot of fluorescence at the maximum emission wavelength of 520 nm versus caspase-3 concentration revealed a linear response characteristic of the protease sensor, with a readily achieved detection limit of 7.25 ng/mL.

The fluorescence quenching mode is a process which involves two submodes: self-quenching and probe-pair quenching. These two submodes can be simply distinguished based on the quenching mechanism. For self-quenching, Oh et al. proposed a peptide beacon sensor for detection of human immunodeficiency virus type 1 (HIV-1) matrix protein p17 with a self-quenching fluorescence at 475 nm (Fig. 20.8A) [39]. The excimer-based PB architecture is composed of two *N*-(1-pyrenyl)maleimides conjugated to a recognition peptide. Intramolecular collisions and, perhaps, hydrophobic affinity occurring in the absence of target (left), allow for excimer formation and thus longer wavelength excimer emission. Upon target binding (right), the pyrene moieties are segregated, thereby eliminating the longer wavelength excimer emission. Complex formation between the PB (3 nM) and its target (here an antibody directed against the HIV protein p17) results in a twofold decrease in excimer emission (Fig. 20.8B). In the case of probe-quencher quenching mode, Gu et al. reported a novel fluorescent "turn-off" biosensing platform for protease activity assay using peptide-templated gold nanoclusters (AuNCs) [38]. As shown in Fig. 20.8C, the AuNCs synthesized using the peptides as templates formed a core−shell structure through interactions between the amino acids

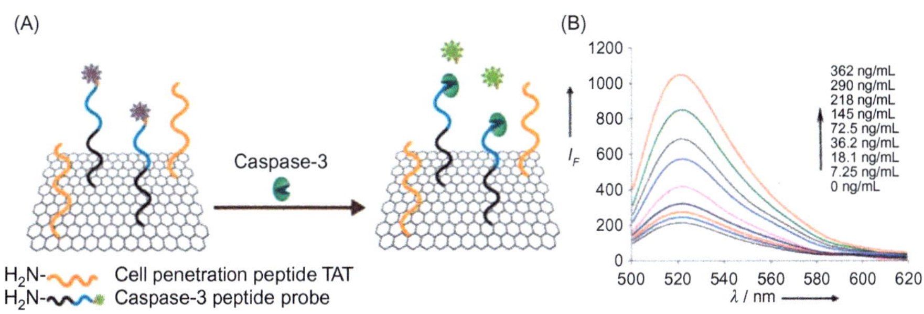

Figure 20.7 (A) Caspase-3 detection using GO−peptide conjugate. (B) Fluorescence spectral responses of the conjugate of GO with caspase-3-specific substrate peptide to caspase-3 of varying concentrations.

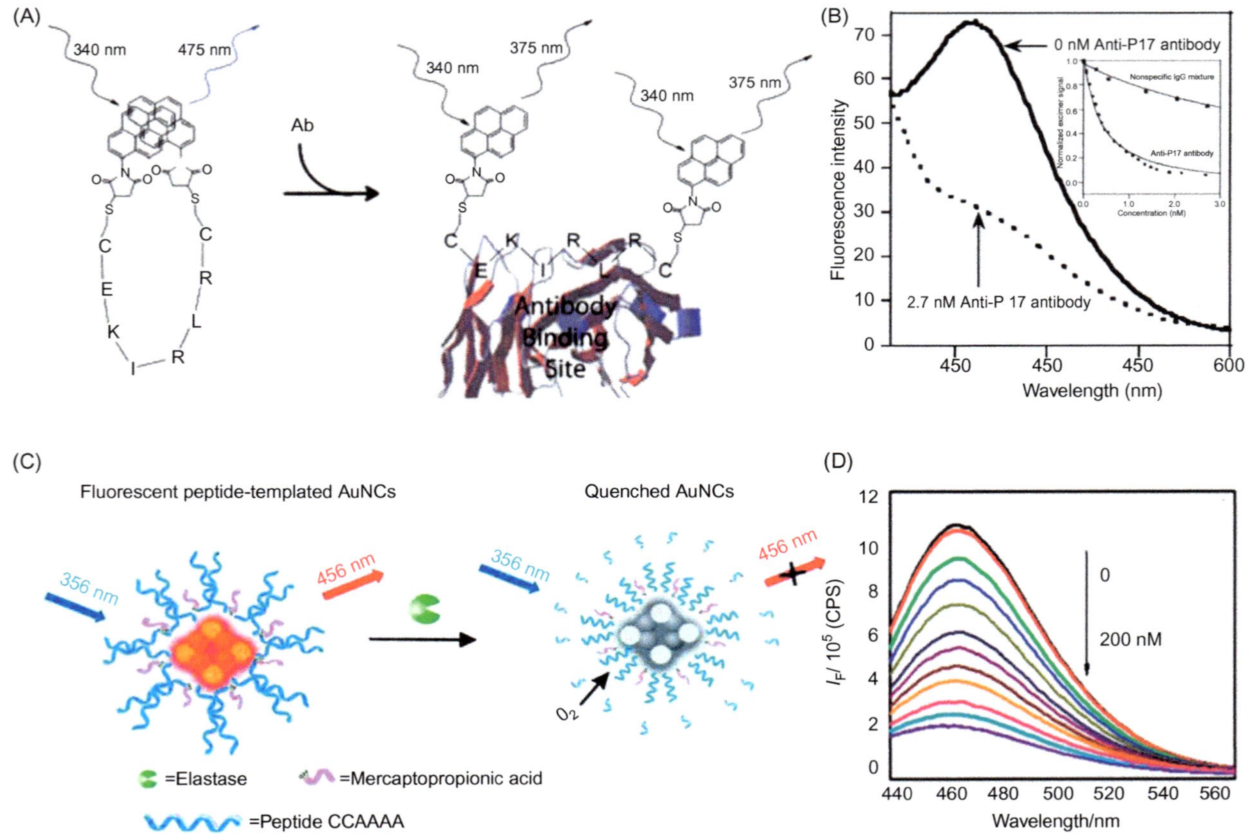

Figure 20.8 (A) The excimer-based peptide beacon sensor composition and recognition mechanism of antibody. (B) The fluorescent response of the biosensor to 2.7 nM Anti-P7 antibody. (C) Schematic illustration of the sensing mechanism for proteases using peptide-protected AuNCs. (D) Fluorescence responses of peptide–AuNCs to different concentrations of elastase (range from 0 to 200 nM).

and the AuNC core. The peptides acted as a compact coating layer, protecting the AuNCs from O_2-mediated fluorescence quenching. Cleavage of the substrate peptides by elastase removed the protective peptide coating, which allows O_2 to diffuse and interact with the AuNC cores and quench their fluorescence in a concentration-dependent mode (Fig. 20.8D).

Fluorescence enhancement, or sensitized fluorescence, is different from fluorescence "turn-on" as its initial fluorescence intensity is not zero. Shults and Imperiali introduced a versatile fluorescence-enhanced peptide probe of protein kinase activity (Fig. 20.9) [40]. The probe can be modified to target a desired kinase by changing the kinase recognition motif in the peptide sequence. The reporter motif contains the Sox amino acid, which generates a fluorescence signal when bound to Mg^{2+} when it is present in the reaction mixture. The phosphorylated peptide exhibits a much greater affinity for Mg^{2+} than its unphosphorylated analogue, resulting in greater fluorescence intensity. Product formation during phosphorylation by the kinase was easily followed by the increase in fluorescence intensity over time. These probes exhibited a three- to fivefold increase in fluorescence intensity upon phosphorylation, the magnitude of which depended on the substrate. Peptides containing the reporter functionality were phosphorylated on serine by protein kinase C and cAMP-dependent protein kinase and were shown to be good substrates for these enzymes.

Generally speaking, FRET is an energy transfer phenomenon taking place between two fluorescent molecules. Nonradioactive energy transfer will appear when the emission spectrum of the fluorescent donor molecule overlaps with the absorption spectrum of the fluorescent receptor molecule and the distance between the two moieties is within 10 nm. The FRET phenomenon results in significantly decreased fluorescence intensity of the donor while the emission intensity of the receptor is greatly enhanced. In the field of life sciences, FRET is a powerful tool for detection of the nanoscale distance and distance changes between large biomolecules in vivo and can be used to detect the direct interaction of two protein molecules inside the cell.

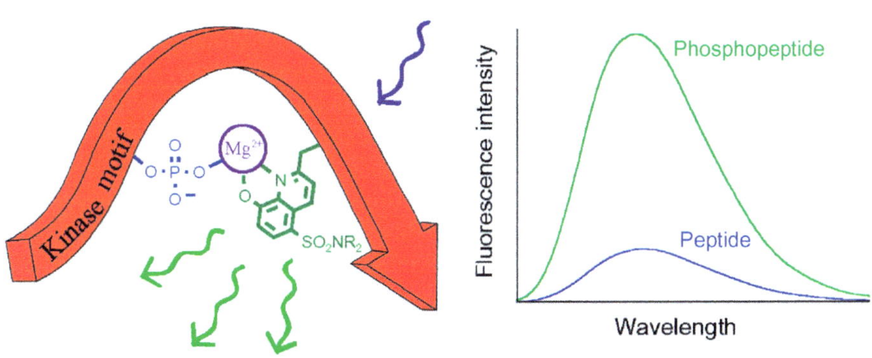

Figure 20.9 Schematic of the sensing mechanism of the fluorescence-enhanced peptide probe and its fluorescence response curve.

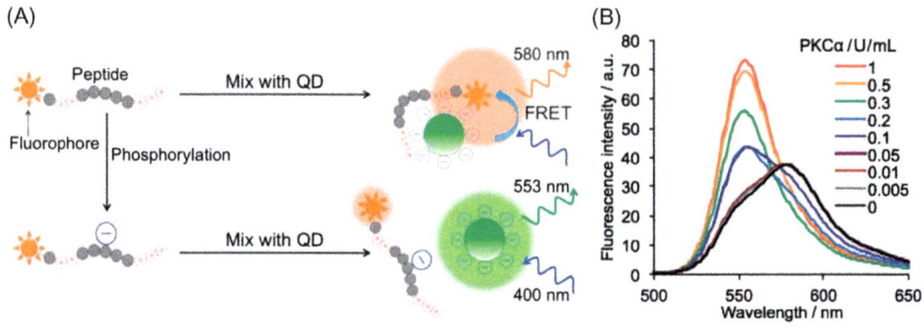

Figure 20.10 (A) Schematic illustration of the FRET-based detection principle. QDs and fluorophore-modified peptides electrostatically interact to induce FRET. (B) PKCα concentration-dependent emission spectra of peptide S–QD mixtures after the phosphorylation reaction with ATP.

Based on FRET, Shiosaki et al. developed a novel antibody-free assay using QDs without any surface modification [46]. As shown in Fig. 20.10A, a cationic substrate peptide modified with a fluorophore interacts electrostatically with anionic QDs causing FRET from the QDs to the fluorophore. In contrast, a phosphorylated peptide with reduced cationic charge due to the anionic phosphate group cannot induce efficient FRET. This simple assay was applied for the evaluation of protein kinase Cα (PKCα) inhibitors. The phosphorylation reaction of peptide was carried out at various PKCα concentrations for 60 min at 30°C. The resulting solution was mixed with a solution containing QDs and it was allowed to interact for 15 min before measuring the emission spectra. Fig. 20.10B shows the emission spectra of the solutions. The QD fluorescence intensity increased with increasing PKCα concentration.

The fluorescence lifetime of the fluorescent probes can also be used for monitoring biological processes. This phenomenon is called FLIM. Fluorescence lifetime refers to the average time that a molecule can stay in an excited state upon excitation with light pulses before it returns to the ground state. Fluorescence lifetime depends on the microenvironment of the fluorescent molecules. Some researchers believe that FLIM can be more reliable than monitoring fluorescence intensity. Damayanti et al. reported a cell-penetrating peptide biosensor for dynamic monitoring of phosphorylation by Abl kinase [47]. FLIM, which is not confounded by photobleaching or cellular autofluorescence, was employed to detect phosphorylation-dependent fluorophore lifetime shifts (1–2 ns) in living cells (Fig. 20.11). The authors of this work studied the dependence of the fluorophore lifetime shift on phosphorylation by Abl kinase, mapped the fluorophore intensity and lifetime components to quantify subcellular phosphorylation, and monitored kinase inhibition in real time. This approach, which can be used for the study of other kinases, provides a new method for interrogating real-time, subcellular signaling activities in cell populations that are not amenable to expression of genetically engineered biosensor proteins.

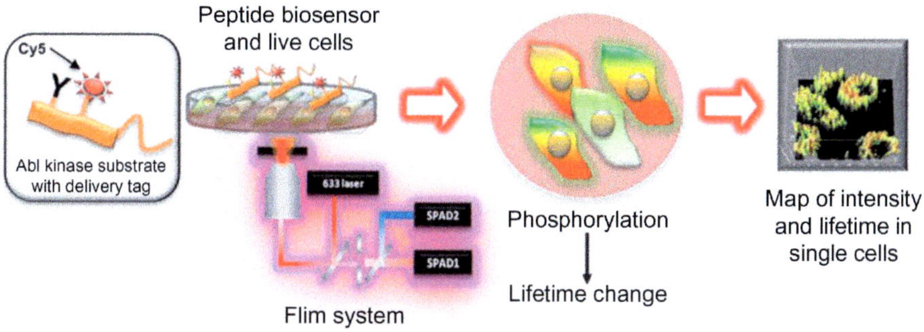

Figure 20.11 FLIM to detect phosphorylation-dependent fluorophore lifetime shifts for biosensors in living cells. A fluorophore-labeled peptide substrate is delivered to living cells by a cell-penetrating peptide. Phosphorylation of the substrate results in increased fluorophore lifetime by FLIM.

It can be seen from the above that there are various kinds of peptide-based fluorescence biosensors, which can be used for different occasions and requirements. However, the quantitative detection of fluorescence is not so convenient, thus demanding the cooperation of portable detection equipment with these sensors to expand its follow-up applications.

20.3.2 Peptide-based colorimetric biosensors

Colorimetry is an approach that determines the content of the target analytes by measuring or comparing colored substance solutions. It is based on Lambert's law and depends on the generation or cooperation of colored compounds to complete the colorimetric detection. Aili et al. described a novel strategy for the colorimetric sensing of proteins, based on peptide-functionalized gold nanoparticles [49]. Recognition is accomplished using a polypeptide sensor scaffold designed to specifically bind to the model analyte, human carbonic anhydrase II (HCAII). The extent of particle aggregation, induced by the Zn^{2+}-triggered dimerization and folding of a second polypeptide which is also present on the surface of the gold nanoparticles, gives a readily detectable colorimetric shift that is dependent on the concentration of the target protein. In the absence of HCAII, particle aggregation results in a significant red-shift of the plasmon peak, whereas analyte binding prevented the formation of dense aggregates, significantly reducing the magnitude of the red-shift. The versatility of the technique is demonstrated using a second model system based on the recognition of a peptide sequence from the tobacco mosaic virus coat protein (TMVP) by a recombinant antibody fragment (Fab57P). Concentrations down to 10 and 25 nM are detected for HCAII and Fab57P, respectively. This strategy is proposed as a generic platform for robust and specific protein analysis that can be further developed to monitor a wide range of target proteins (Fig. 20.12).

Peptide-based colorimetric biosensors are mainly based on the combination of peptides and chromogenic nanomaterials, for example, gold or silver nanoparticles.

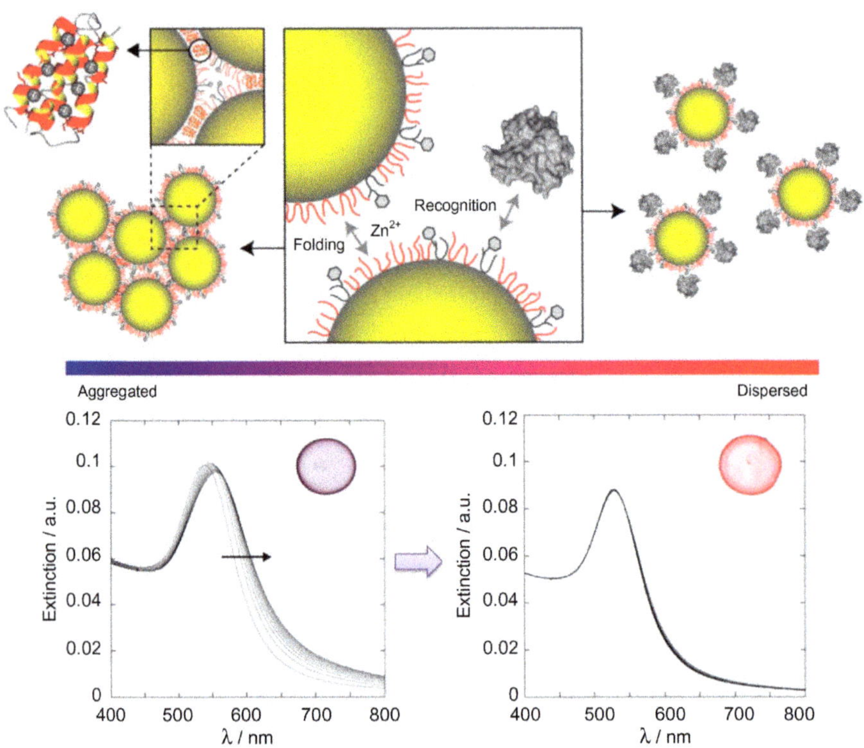

Figure 20.12 Gold nanoparticles were modified with a peptide designed to allow folding-induced particle aggregation triggered by Zn^{2+} and a polypeptide-based synthetic receptor for binding of protein analytes. In the absence of the analyte, addition of Zn^{2+} triggers dimerization and folding, resulting in particle aggregation and a rapid color shift of the suspension from red to purple. Analyte binding precludes aggregation and the dispersion remains red. Analyte binding can thus easily be determined even by the naked eye. The two spectra were recorded with 2-min intervals for 20 min after exposing the particles to 10 mM Zn^{2+} (pH 7), in the absence (left) and in the presence (right) of HCAII. The concentration of HCAII was 70 nM after dilution in the Zn^{2+}-containing buffer. The insets show the corresponding suspensions after being dried on nitrocellulose membranes.

Compared with the fluorescence method, it has many advantages such as ease of the method, simple detection procedure, low cost, and ease of miniaturization. However, sensitivity is still its demerit, and more ingenious sensor designs are needed to improve its detection accuracy.

20.3.3 Peptide-based SPR biosensors

SPR is an opto-physical phenomenon with high sensitivity and high specificity. When a beam of P-polarized light is incident on the prism side within a certain angular range, a surface plasma wave can be generated at the interface between the prism and the metal film (Au or Ag). When the propagation constant of the incident light matches that of the surface plasma wave, free electrons in the metal film will resonate, namely SPR. In biosensing detection, biological recognition molecules will be immobilized on the surface of the sensor chip, and can interact with target molecules. This can cause refractive index changes to the metal film surface, which ultimately leads to changes in the SPR angle. Through detection of the SPR angle changes, information on the analytes, such as dose, affinity, kinetic constants, and specificity can be obtained.

Dudak et al. developed a label-free biosensor based on SPR to evaluate the potential of peptide ligands as biorecognition molecules for the detection of bacterial toxins [54]. A 24-mer peptide, previously identified from a phage display library and modified for better binding was used for the detection of staphylococcal enterotoxin B (SEB). The peptide was immobilized covalently onto a gold-coated surface modified with self-assembly monolayer. The LOD for SEB was found to be 20 μg/mL. The selectivity of the peptide-based sensor for SEB was verified by injecting bovine serum albumin (BSA) and the sensor response for the BSA was negligible. Further, to probe the effectiveness of the sensor in complex food environments, the binding phenomenon was tested using milk samples spiked with SEB and the results showed that the system can be used in complex food matrices. The results of this study demonstrate the potential of peptides as recognition agents for detecting bacterial toxins using biosensors (Fig. 20.13).

Due to the sensitivity of the method, SPR can be used for the construction of peptide-based biosensors. Refraction angle changes appear to be more rapid and sensitive than wavelength shifts of optical spectra, which is an important factor for precise quantification. However, the SPR methodology is more complex.

20.3.4 Peptide-based SERS biosensors

SERS refers to the approach in which the sample is adsorbed on the surface of colloidal metal particles (e.g., gold, silver, or copper) or on the rough surface of sheet metal and then the sample is measured by conventional Raman spectroscopy. Due to the electromagnetic field enhancement of the surface or near surface of the

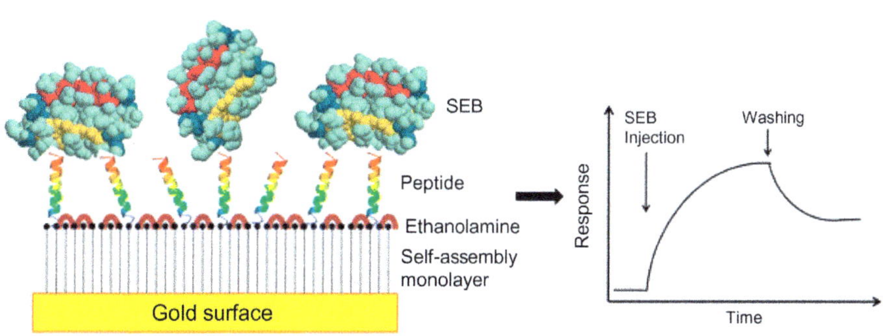

Figure 20.13 Schematic illustration of the peptide-based SPR sensor system for detection of SEB.

sample in the excitation region, SERS can increase the detection of Raman intensity by 10^3-10^6 times and can provide structural information that is not easily accessible through conventional biophysical methods of analysis. SERS has been widely used in conformational and structural studies of biomolecules.

Hu's group has developed a series of peptide-based SERS biosensors with 4-mercaptobenzoic acid (4-MBA)-labeled gold nanoparticles (AuNPs). The method employed aggregation of 4-MBA-modified AuNPs by peptides after treatment with target protease. SERS signals of 4-MBA are sharply decreased or increased because of the increased or decreased electrostatic stability of AuNPs. Through this strategy, facile "turn-off" or "turn-on" SERS biosensors for proteases can be established with sensitivity, selectivity, and simplicity. Compared with other methods, this newly proposed method has improved sensitivity and can be easily adapted for the detection of many proteases. The method demonstrated the capability for application in complex matrix samples. The results also illustrated the potential and superiority of SERS biosensors as a general approach for proteases based on enzyme activity [57–59] (Fig. 20.14).

20.4 Other peptide-based biosensors

The two techniques described above are the most important techniques used in peptide-based biosensors. However, other technologies have also been developed such as electrochemical luminescence (ECL), microcantilever-based biosensors, quartz crystal microbalance (QCM), and field-effect transistors (FET). Some typical such peptide-based biosensing systems are described in Table 20.3.

ECL is a specific luminescence reaction initiated by electrochemical phenomena on the electrode surface involving both electrochemistry and chemiluminescence. Advantages of ECL include the ease of operation and imaging, fast analysis, high throughput, and ease of automating the process. ECL has been widely used in the fields of biomedicine, food safety, environment monitoring, and clinical diagnostics. Wu et al. proposed a simple and novel "signal-on" ECL biosensor for the detection of PSA. The system is composed of multiwalled carbon nanotubes (MWCNTs), polyamidoamine (PAMAM) dendrimers, and Au nanoparticle (NP) film on a glassy carbon electrode (GCE) to improve the electron transfer, provide abundant amine group for the immobilization of biomolecules, and amplify the ECL signal. Au nanorod (Au NR)-labeled peptide is modified on the electrode surface to serve as an ECL-RET acceptor due to the excellent overlap between the ECL emission spectrum of Ru(bpy)32+ and the absorption spectrum of Au NRs, leading to a significant reduction of the ECL signal. Upon cleavage of the peptide by PSA, both Au NRs and peptides are released from the electrode surface, resulting in the high recovery efficiency of the ECL signal. The proposed approach exhibits a wide linear range from 0.1 pg/mL to 10 ng/mL with a detection limit of 0.03 pg/mL. The results revealed that the recoveries were between 95% and 108%, indicating good accuracy of the method

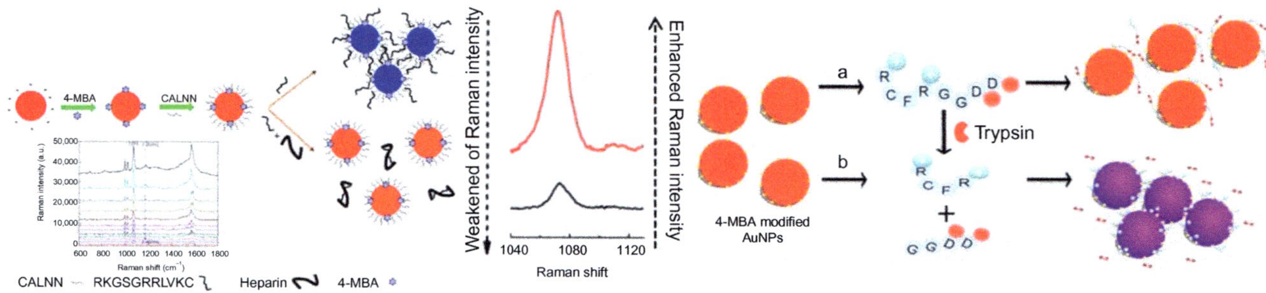

Figure 20.14 Schematic diagram of SERS-based enzymatic assay for heparin and trypsin.

Table 20.3 Other typical peptide-based biosensors

Sensing technique	Analytes	Peptides	References
Electrochemical luminescence	Cyclin A_2 (CA2)	RRLIFGGGGG & TGQSWPESLIRKTGYTLESLKPCLMGGGGG	[60]
	PSA	CHSSKLQK	[61,62]
	MMP-2	CGPLGVRGK	[63]
	Casein kinase II (CK2) and cAMP-dependent protein kinase (PKA)	CRRRADDSDDDDD and CLRRASLG	[64]
QCM	Rheumatoid arthritis (RA)	A 14-mer cyclic filaggrin peptide derived from the amino acid sequence deduced from cDNA of human filaggrin	[65]
	Hepatitis B virus	5′-Bio-OO-TCCTTTTT-OOO-TTTTTCCT-Lys	[66]
	Recombinant human interferon-β	NH_2-IAASLQESKQVVAH-OH and NH_2-IAASLEESKEVVAH-OH	[67]
Microcantilever	Vancomycin	CGGGG-L-Lys-D-Ala-D-Ala	[68]
	Breast cancer cells	Peptide **18-4** (WxEAAYQrFL)	[69]
FET	Streptavidin	IMVTESSDYSSY	[70]
	Trinitrotoluene (TNT)	HSSYWYAFNNKTGGGGWFVI	[71]

for PSA detection. In addition, the biosensor exhibited specificity for the detection of PSA [62] (Fig. 20.15).

In QCM the resonation frequency of the quartz crystal will change depending on the size and mass of the molecules deposited on the surface. QCM can detect very small mass changes on the surface of the sensor in real time and is commonly used for detection of unlabeled molecules and the study of biomolecular interactions. Drouvalakis et al. demonstrated a label-free peptide-coated carbon nanotube-based immunosensor for the direct assay of human serum [66]. A rheumatoid arthritis (RA)-specific (cyclic citrulline-containing) peptide was immobilized onto functionalized single-walled carbon nanotubes deposited on a QCM sensing crystal. Serum from RA patients was used and antibody binding was detected by QCM sensing. Specific antibody binding was determined by comparing the assay of two groups (normal and diseased sera) and the native unmodified peptide. The sensitivity of the nanotube-based sensor (detection in femtomolar range) was higher than that of the established enzyme-linked immunosorbent assay (ELISA) and recently described microarray assay systems, detecting 34.4% and 37.5% more RA patients with anticitrullinated peptide antibodies than those found by ELISA and microarray, respectively. The performance of their label-free biosensor enables its application in the direct assay of sera in research and diagnostics (Fig. 20.16).

In the case of the microcantilever technology, the resonance frequency of the microcantilever shifts in response to mass loading from molecular interaction as in the case of any other biosensors. Through vibration changes, the microcantilever platform provides information about receptor–ligand, antibody–antigen, or enzyme–substrate interactions. Its dynamic range is large, with a resonant frequency ranging from tens of hertz to several megahertz and it can perform analyses with high resolution. In the study of Bai et al. a reversible detection method for vancomycin was developed utilizing the cantilever array sensor functionalized by a designed peptide consisting of a cysteine (Cys-), a space linker (-GGGG-), and a molecular recognition ligand (-L-Lys-D-Ala-D-Ala) [68]. It was found that the peptide space linker was necessary for the response of the cantilever array sensor. The sensing cantilevers in the array were functionalized with the peptide, while the

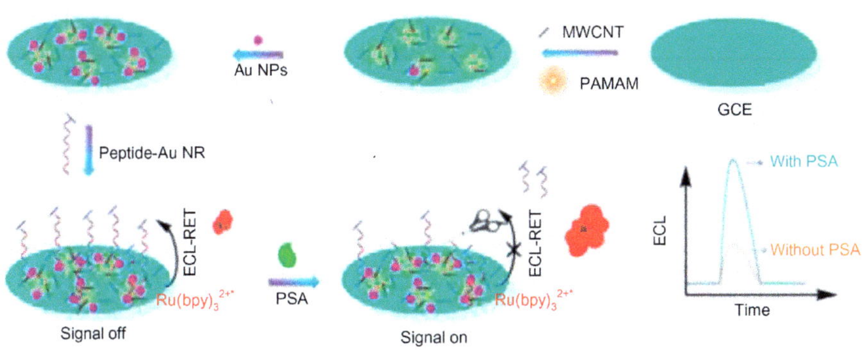

Figure 20.15 A simple "signal-on" ECL biosensor for detection of PSA.

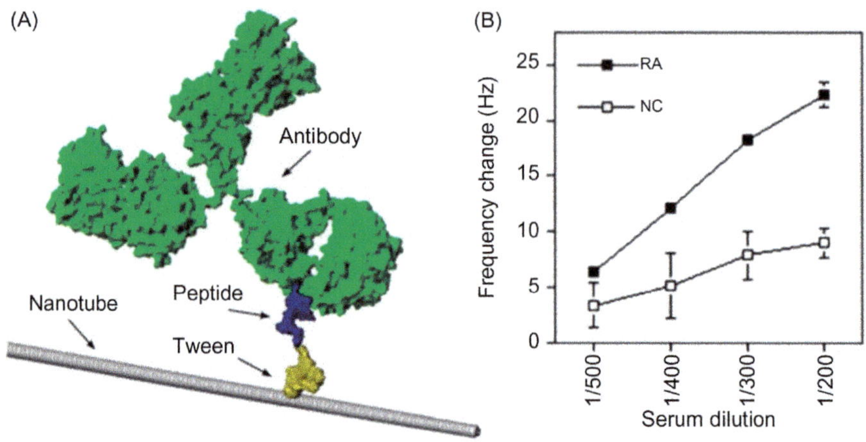

Figure 20.16 (A) A visual representation of an antibody bound to a peptide–Tween–nanotube complex that comprises the nanotube biosensor. (B) A sample representation of a dilution curve of an RA and normal control (NC) serum in the nanotube assay measuring antibody binding to citrullinated peptide-coated nanotubes as a change in frequency (Hz).

reference cantilevers were modified by 6-mercapto-1-hexanol (MCH) to eliminate the influence of environmental disturbances. The binding between vancomycin and the peptide induced a change of surface stresses in the sensing cantilevers, resulting in a differential deflection between the sensing and reference cantilevers. The reciprocal of the differential deflection is linear with the reciprocal of vancomycin concentration within the range of 2–100 μM at a detection limit of 0.2 μM. The reversible detection was performed by regenerating the sensing cantilevers with running buffer solution. Other antibiotics such as doxycycline, streptomycin, and kanamycin have a negligible effect on the response of the sensor. The sensor can also be utilized for the reversible detection of vancomycin in serum, which clearly shows the potential of the sensor for vancomycin detection in real biological samples (Fig. 20.17).

FET sensors are based on the field effect principle and can be fabricated with semiconductor technology, which has advantages such as high density and dynamical detection. The technology is also used in combination with nanomaterials for higher sensitivity. Kuang et al. described the rational design of a peptide recognition element (PRE) that is capable of noncovalently attaching to SWCNTs on the FET sensor as well as binding to trinitrotoluene (TNT) [71]. The PRE contains two domains, a TNT binding domain derived from the binding pocket of the honeybee odor binding protein ASP1, and an SWCNT binding domain previously identified from the phage peptide display library. The PRE structure in the presence of SWCNT was investigated by performing classical all-atom molecular dynamics simulations, circular dichroism spectroscopy, and atomic force microscopy. Both computational and experimental analyses demonstrate that the peptide retains two

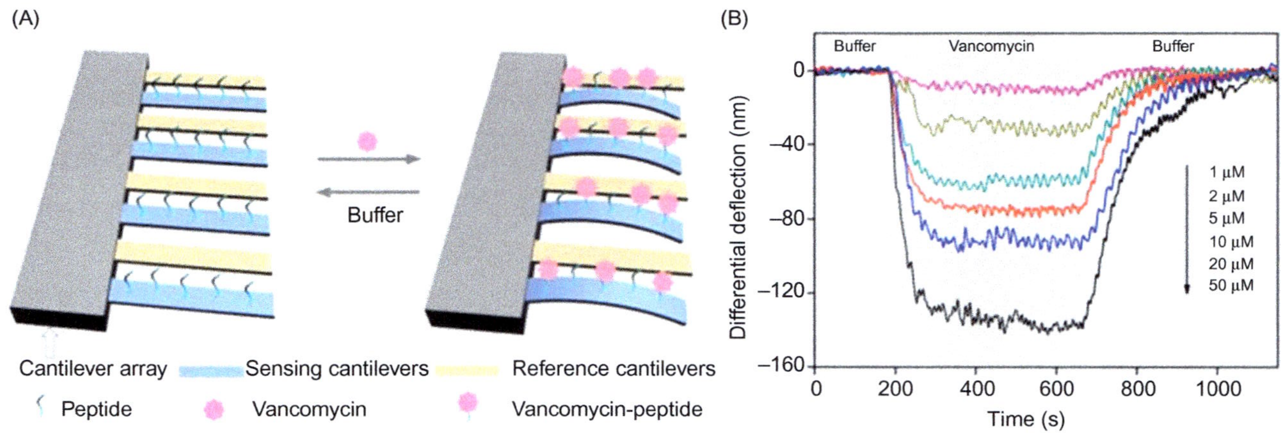

Figure 20.17 (A) Schematic illustration of reversible vancomycin detection. Sensing cantilevers were functionalized with peptides while reference cantilevers were modified with MCH. Binding of vancomycin to the peptide induced a change in cantilever surface stress and produced a differential cantilever deflection between the sensing and reference cantilevers. (B) Differential deflection as a function of time for different concentrations of vancomycin: 1, 2, 5, 10, 20, and 50 μM (downward).

functional domains for SWCNT and TNT binding. The binding motif of the peptide to SWCNT and to TNT was revealed from interaction energy calculations by molecular dynamics simulations. The potential application of the peptide for the detection of TNT is theoretically predicted and experimentally validated using an SWCNT-FET sensor functionalized with a designer PRE. Results from this study demonstrate the creation of chemosensors using designed PRE as selective surface coatings for targeted analytes (Fig. 20.18).

All these biosensors were developed on the basis of traditional sensing technology with various forms and prominent features. Although most of these biosensors exhibited very high sensitivity, their construction and operation procedures are complicated. Other methodologies, such as the peptide-based ECL biosensors are in rapid development, and show broad potential in a variety of fields.

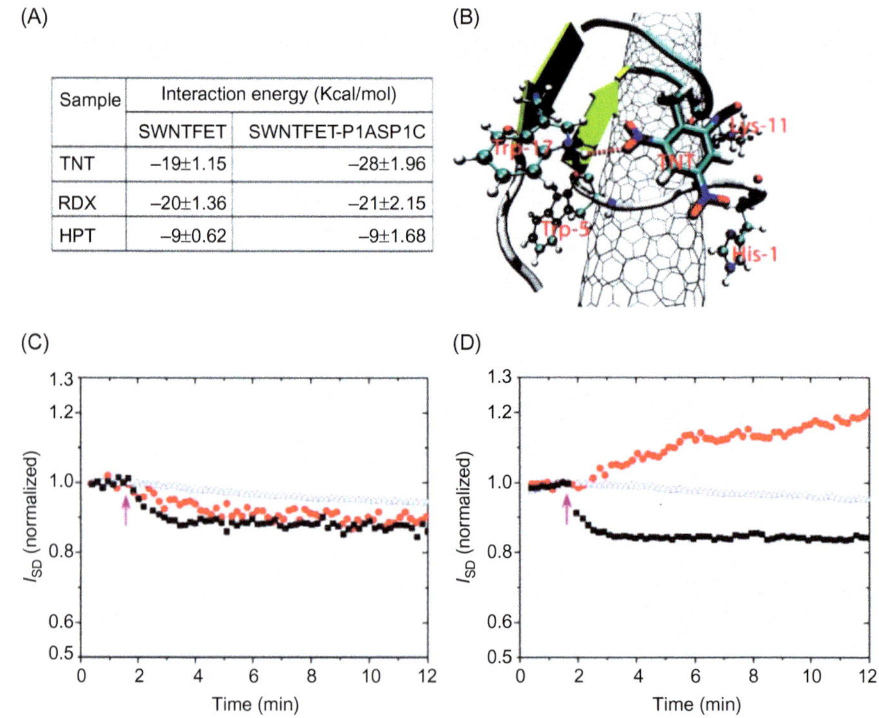

(A)

Sample	Interaction energy (Kcal/mol)	
	SWNTFET	SWNTFET-P1ASP1C
TNT	−19±1.15	−28±1.96
RDX	−20±1.36	−21±2.15
HPT	−9±0.62	−9±1.68

Figure 20.18 (A) Interaction energy of TNT, RDX, and HPT with bare SWCNT or P1ASP1C-decorated SWCNT. (B) Computational modeling predicts that TNT binds to P1ASP1C−SWCNT hybrid through a H bond with Trp17 and $\pi-\pi$ interaction with the SWCNT surface. Response of (C) bare SWCNT-FET and (D) P1ASP1C-coated SWCNT-FET to TNT (red circles), RDX (blue triangles), and HPT (black squares). Arrow indicates when the vapor was introduced into the device. Note that the chemical agents were exposed under ambient conditions.

20.5 Construction of peptide-based biosensors

Based on the diverse advantages of peptide-based sensors, they will continue to be the focus of development of biosensors in the next few years. Therefore, the construction approaches of peptide-based biosensors will be important and helpful to researchers. This section introduces some simple methods to fabricate peptide-based sensors with applications ranging from environment monitoring to disease diagnosis using examples from our work. The aim of this section is to offer technical support to those who want to work in this direction.

20.5.1 Environment monitoring

Environment monitoring has always been one of the main applications of biosensors because they can detect analytes in complex environments with high sensitivity, fast responses, and specificity. In the case of explosive detection, such as 2,4,6-trinitrotoluene (TNT), our group has constructed a novel nano-scale optical biosensing platform for TNT detection with TNT-sensitive peptides and derivatives of graphene oxide (GO) [72].

GO is a 2D nanomaterial with many oxygen-containing groups on single graphene sheets. These polar oxygen-functional groups give GO unique optoelectronic properties and make it easy to modify using nanoparticles, nucleic acids, and proteins [73]. Moreover, it is also an electronically hybrid material in which a large fraction of carbon atoms that are sp^3 hybridized bind to oxygen in the form of epoxy and hydroxyl groups, whereas the other carbon atoms that are sp^2 hybridized bind to neighboring carbon atoms or to oxygen in the form of carboxyl and carbonyl groups [74]. The strongly heterogeneous atomic and electronic characteristics of the material give GO unique optoelectronic and chemical properties. Researchers have confirmed that tuning the ratio of the sp^2 and sp^3 fractions can tune the bandgap of GO and therefore fine-tune the electrical, optical, and/or chemical properties of GO, such as enhancement of the conductivity and shift of the absorption peak [75]. In our design, GO was used as a transducer based on its tunable optical property.

According to the report of Jaworski et al., peptides containing the WHWQRPLMPVSI sequence can specifically interact with TNT molecules through partial charge−charge interactions or hydrogen bonding [76]. We synthesized TNT-specific peptides by solid-phase peptide synthesis (SPSS) and conjugated them with GO through a dehydration condensation reaction. Using *N*-hydroxysuccinimide (NHS) and *N*-ethyl-*N*′-[3-(dimethylamino)propyl]carbodiimide (EDC) the amino groups of the peptides can form stable covalent bonds with the carboxyl groups on the GO sheets, as shown in Fig. 20.19A. Through the ultraviolet (UV) absorbance spectra (Fig. 20.19B), it can be observed that the absorption peak of peptide-functionalized GO had a red shift in the wavelength compared to GO, which was due to changes in the molecular structure of GO and confirmed the tunable properties of the material. The peptide-modified GO combined the sensitivity of peptides

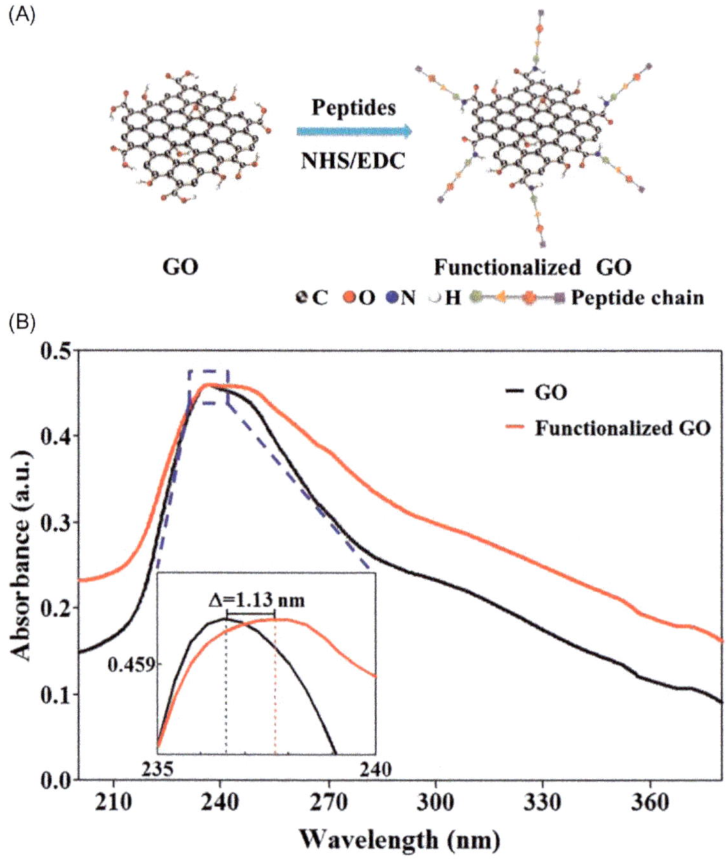

Figure 20.19 Schematic of synthesis (A) and UV absorption spectra (B) of biofunctionalized GO. The insert show that the absorption peak of biofunctionalized GO has a red shift in the wavelength.

and the optical properties of GO and it was used as an optical biosensor for TNT detection.

In the presence of TNT molecules, the peptide chains that are tethered to the GO sheets specifically interact and capture TNT molecules, as described in Fig. 20.20A. Donor–acceptor and π–π interactions between peptide chains and TNT molecules result in changes in the optical properties of the peptide–GO complex which resulted in biosensor response upon binding TNT (Fig. 20.20B). The results showed that the addition of TNT can change the absorption magnitude of the biosensor, showing a linear relationship against log doses of TNT ranging from 4.40×10^{-9} to 4.40×10^{-4} mM with a detection limit of 4.40×10^{-12} mM in 3δ/slope. Therefore, through modification of GO with designed peptides, an easy-to-construct peptide-based optical biosensor can be fabricated with a relatively high-sensitivity response to molecules such as TNT.

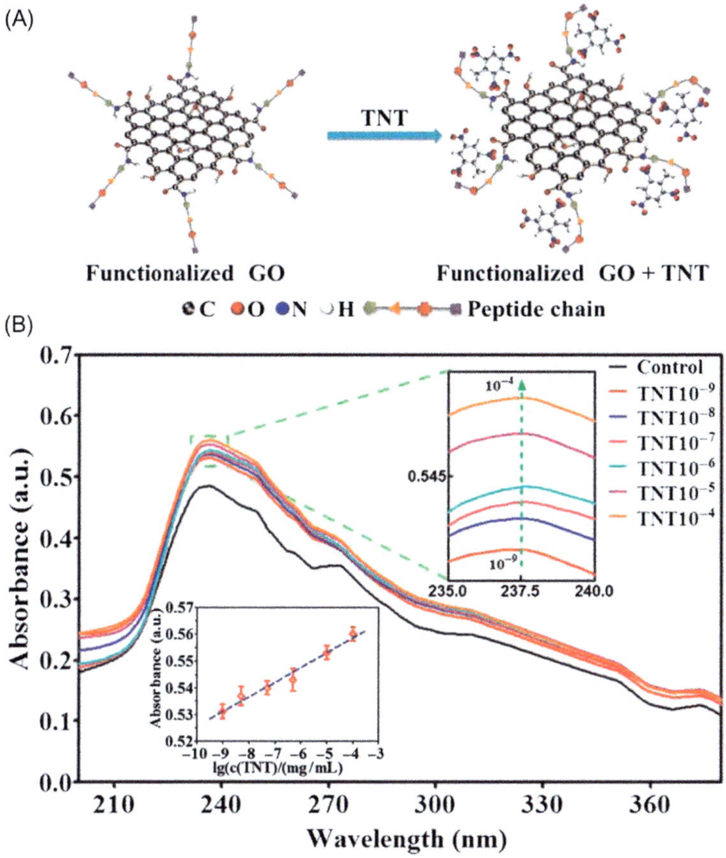

Figure 20.20 Schematic illustration (A) and UV absorption spectra (B) of TNT detection. The insert is the linear plot of statistical results of the absorption peak values ($n = 4$).

The peptide-based optical biosensor for TNT detection was based on the unique characteristics of GO and of the peptides. This biosensor was characterized by high sensitivity, versatility, and easy construction and modification. With the development of nanomaterials such as GO, such biosensors can exhibit great potential in high-throughput detection, not only in environmental monitoring but also in other fields such as food safety and health care.

20.5.2 Disease diagnosis

Disease diagnosis is one of the most important applications of biosensors. The detection of various biomarkers with different biosensing technologies represents one of the major challenges in the field of biosensor technology. Using the detection of thrombin detection as an example, we present here the development of a peptide-based optical biosensor for colorimetric detection.

Thrombin is a serine protease that plays a key role in the blood-clotting process. Modern medical research shows that thrombin is associated with diseases such as arteriosclerosis, liver fibrosis, and dementia [77]. Numerous efforts have been made towards thrombin detection and, among them, peptide-based biosensors stand out for their biosensitivity, high selectivity, and fast response. Thrombin can selectively cleave the arginine–glycine (RG) bond at Arg residues, which allows for the development of peptide-based biosensor construction.

As suggested previously, GO can be used in combination with peptides for the development of biosensor devices [78]. The GO surface can also be modified compounded with nanoparticles to obtain unique physical and chemical properties such as ultrasensitive features and excellent catalytic properties [79]. Herein, we proposed a peptide-based optical biosensor for thrombin detection [52]. Different from the biosensor constructed above, this sensor was fabricated by linking peptides containing specific cleavage sites of thrombin with composites of GO and gold nanoparticles (GNPs). The composition of GO and GNPs can not only change the responses of the sensor from the UV region to the visible region but also it can further improve the sensor sensitivity and allow for colorimetric detection.

Fig. 20.21A shows the synthesis and thrombin detection process of the biosensor. The nanocomposites of GO and GNPs (GO/GNPs) were prepared by the reduction of chloroauric acid with trisodium citrate on the surface of graphene sheets. Such a process resulted in nucleation and growth of gold particles attaching to the oxygen groups of the graphene material without internal bridges due to electrostatic interactions [80]. GO/GNPs can exhibit two absorption peaks at both the UV and visible regions. With a blue shift compared to the absorption peak of GO and a broader absorption band compared to that of GNPs, the nanocomposites were able to combine the optical properties of both materials (Fig. 20.21B,C). Then, specifically designed octapeptides (CLVPRGSC) were synthesized by SPSS with the recognition site of thrombin and cysteine groups at both ends of the peptides. Thus, the peptides can self-assemble on the nanocomposites covalently bonded to the GNPs through the sulfydryl groups. Self-assembly could also induce the aggregation of the composites when two gold nanoparticles on different GO/GNPs were connected by one peptide chain (Fig. 20.21A). This may be seen from the spectral comparison of GO/GNPs in Fig. 20.21C, which is based on the localized plasmon surface resonance (LSPR) characteristics of GNPs. Subsequently, when thrombin molecules were added, the peptides could be cleaved, resulting in separation of the aggregated GO/GNP complexes. Thus, composites clustered together by peptides would redisperse in the solution, and the spectrum of the solution could be expected to change and resemble the spectrum of the individual GO/GNPs (Fig. 20.21C). As a result, a peptide-based biosensor for thrombin detection was constructed successfully based on GO and GNPs.

When the biosensor was used for the spectroscopic detection of thrombin, the proteolytic activity of thrombin which cleaves the octapeptides at Arg-Gly sites results in redispersion of the nanocomposite complexes which aggregated due to the presence of the octapeptides (Fig. 20.21C). An increase of the thrombin

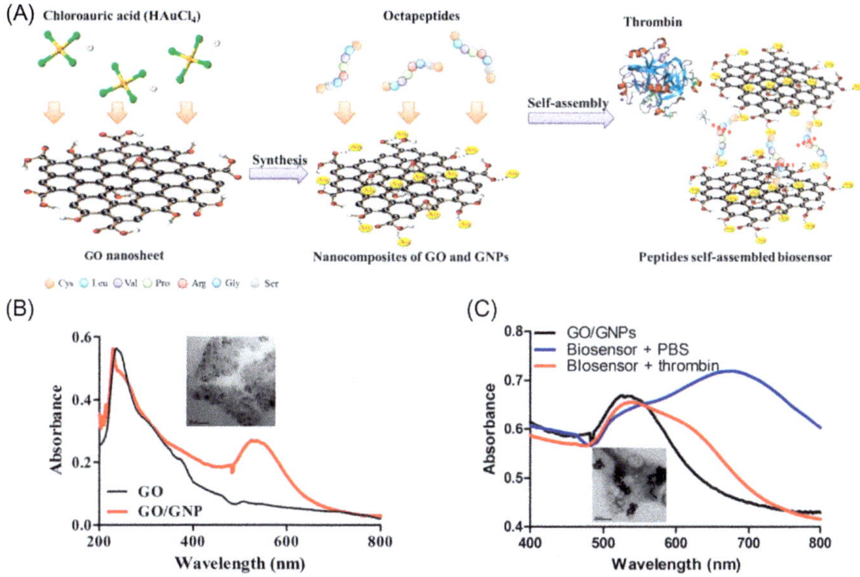

Figure 20.21 (A) Synthetic process of composites of graphene and oxide, self-assembly process of octapeptides on the nanocomposites, and detection mechanism of the biosensor for thrombin. (B) Optical spectroscopy for characterization of the nanocomposites and graphene oxide. The TEM image of the composites is also shown. (C) Optical spectroscopy for characterization of the octapeptide self-assembled biosensor before and after addition of thrombin at 27 μM. The TEM image shows the peptide self-assembled nanocomposites before addition of thrombin.

concentration in the solution of the nanocomposite complexes resulted in an increase of the absorption peak at ~528 nm, while the absorbance decreased at wavelengths larger than 600 nm (Fig. 20.22A). With a 528–680 nm absorbance ratio as the sensitive detection parameter, the biosensor had good linearity, stability, and specificity against logarithms of thrombin concentrations (Fig. 20.22B), showing good performance in detection of such proteases.

As the biosensor fabricated here was composed of GO, GNPs, and thrombin-specific peptides, all three elements made a contribution to the performance of the biosensor. With various oxygen-containing groups on the surface of GO, the latter acted as the building block for GNPs and peptides and enabled the stability of the whole system. GNPs provided the nanocomposite system with an absorption peak in the visible region and aggregation-to-discoloration characteristics, which was the detection basis of the biosensor. Peptides played the role of biorecognition to thrombin. These three factors combined into one system gave rise to a sensor for thrombin detection with high sensitivity. Such a biosensor could also be designed for sensitive detection of other proteases and even other biomolecules such as important biomarkers of diseases.

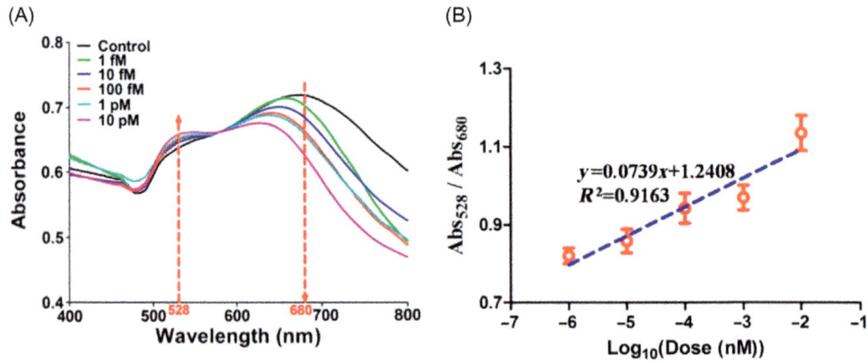

Figure 20.22 Dose-dependent measurements for thrombin detection with the biosensor system that is based on the aggregation of nanocompostie complexes consisting of GO, GNPs, and octapeptides. (A) Absorption spectra for detection of thrombin at different concentrations (1 fM−10 pM). (B) The statistical results of the 528−680 nm absorbance ratios (Abs528/Abs680) compared with those of the control ($n = 3$).

20.6 Conclusion

In this chapter, a number of peptide-based biosensors have been discussed including electrochemical, optical, etc. We aimed at providing the reader with an introduction to the peptide-based biosensor field. We believe that peptide-based biosensors will be further developed to advance sensitivity.

References

[1] A.P. Turner, Biosensors: sense and sensibility, Chem. Soc. Rev. 42 (2013) 3184−3196.
[2] A.P. Turner, Biosensors—sense and sensitivity, Science 290 (2000) 1315−1317.
[3] H. Razavi, S. Janfaza, Medical nanobiosensors: a tutorial review, Nanomed. J. 2 (2015) 74−87.
[4] H.L. Schmidt, W. Schuhmann, F.W. Scheller, F. Schubert, Specific features of biosensors, in: Sensors Set: A Comprehensive Survey, 2008, pp. 717−817.
[5] J.P. Chambers, B.P. Arulanandam, L.L. Matta, A. Weis, J.J. Valdes, Biosensor recognition elements, DTIC Document, 2008.
[6] A. Touhami, Biosensors and nanobiosensors: design and application, Nanomedicine (2014) 374−400.
[7] N. Zhao, Y. He, X. Mao, Y. Sun, X. Zhang, C.-Z. Li, et al., Electrochemical assay of active prostate-specific antigen (PSA) using ferrocene-functionalized peptide probes, Electrochem. Commun. 12 (2010) 471−474.
[8] S. Xie, J. Zhang, Y. Yuan, Y. Chai, R. Yuan, An electrochemical peptide cleavage-based biosensor for prostate specific antigen detection via host-guest interaction between ferrocene and β-cyclodextrin, Chem. Commun. (Camb.) 51 (2015) 3387−3390.

[9] W. Xu, P. Jing, H. Yi, S. Xue, R. Yuan, Bimetallic Pt/Pd encapsulated mesoporous-hollow CeO_2 nanospheres for signal amplification toward electrochemical peptide-based biosensing for matrix metalloproteinase 2, Sens. Actuators B Chem. 230 (2016) 345–352.
[10] H. Yin, M. Wang, B. Li, Z. Yang, Y. Zhou, S. Ai, A sensitive electrochemical biosensor for detection of protein kinase A activity and inhibitors based on Phos-tag and enzymatic signal amplification, Biosens. Bioelectron. 63 (2015) 26–32.
[11] Y. Zhou, H. Yin, X. Li, Z. Li, S. Ai, H. Lin, Electrochemical biosensor for protein kinase A activity assay based on gold nanoparticles-carbon nanospheres, phos-tag-biotin and β-galactosidase, Biosens. Bioelectron. 86 (2016) 508–515.
[12] M. Biyani, K. Kawai, K. Kitamura, M. Chikae, M. Biyani, H. Ushijima, et al., PEP-on-DEP: a competitive peptide-based disposable electrochemical aptasensor for renin diagnostics, Biosens. Bioelectron. 84 (2016) 120–125.
[13] G. Wang, R. Han, X. Su, Y. Li, G. Xu, X. Luo, Zwitterionic peptide anchored to conducting polymer PEDOT for the development of antifouling and ultrasensitive electrochemical DNA sensor, Biosens. Bioelectron. 92 (2017) 396–401.
[14] P. Miao, J. Yin, L. Ning, X. Li, Peptide-based electrochemical approach for apoptosis evaluation, Biosens. Bioelectron. 62 (2014) 97–101.
[15] S. Jampasa, W. Wonsawat, N. Rodthongkum, W. Siangproh, P. Yanatatsaneejit, T. Vilaivan, et al., Electrochemical detection of human papillomavirus DNA type 16 using a pyrrolidinyl peptide nucleic acid probe immobilized on screen-printed carbon electrodes, Biosens. Bioelectron. 54 (2014) 428–434.
[16] P. Teengam, W. Siangproh, A. Tuantranont, C. Henry, T. Vilaivan, O. Chailapakul, Electrochemical paper-based peptide nucleic acid biosensor for detecting human papillomavirus, Anal. Chim. Acta. 952 (2017) 32.
[17] X. Wang, B. Piro, S. Reisberg, G. Anquetin, H. De Rocquigny, P. Jiang, et al., Direct, reagentless electrochemical detection of the BIR3 domain of X-linked inhibitor of apoptosis protein using a peptide-based conducting polymer sensor, Biosens. Bioelectron. 61 (2014) 57–62.
[18] B. Khalilzadeh, H.N. Charoudeh, N. Shadjou, R. Mohammad-Rezaei, Y. Omidi, K. Velaei, et al., Ultrasensitive caspase-3 activity detection using an electrochemical biosensor engineered by gold nanoparticle functionalized MCM-41: its application during stem cell differentiation, Sens. Actuators B Chem. 231 (2016) 561–575.
[19] Y. Li, R. Afrasiabi, F. Fathi, N. Wang, C. Xiang, R. Love, et al., Impedance based detection of pathogenic *E. coli* O157: H7 using a ferrocene-antimicrobial peptide modified biosensor, Biosens. Bioelectron. 58 (2014) 193–199.
[20] H. Hwang, M. Ryu, C. Park, J. Ahn, H. Park, C. Choi, et al., High sensitive and selective electrochemical biosensor: label-free detection of human norovirus using affinity peptide as molecular binder, Biosens. Bioelectron. 87 (2017) 164–170.
[21] P. Kongsuphol, S.K. Arya, C.C. Wong, L.J. Polla, M.K. Park, Coiled-coil peptide based sensor for ultra-sensitive thrombin detection, Biosens. Bioelectron. 55 (2014) 26–31.
[22] Y. Li, W. Zhang, L. Zhang, J. Li, Z. Su, G. Wei, Sequence-designed peptide nanofibers bridged conjugation of graphene quantum dots with graphene oxide for high performance electrochemical hydrogen peroxide biosensor, Adv. Mater. Interfaces 4 (2017) 1600895.
[23] Y. Qu, M. Ma, Z. Wang, G. Zhan, B. Li, X. Wang, et al., Sensitive amperometric biosensor for phenolic compounds based on graphene–silk peptide/tyrosinase composite nanointerface, Biosens. Bioelectron. 44 (2013) 85–88.

[24] S.A. Tria, D. Lopez-Ferber, C. Gonzalez, I. Bazin, A. Guiseppi-Elie, Microfabricated biosensor for the simultaneous amperometric and luminescence detection and monitoring of Ochratoxin A, Biosens. Bioelectron. 79 (2016) 835–842.
[25] J. Lee, J.Y. Yun, W.C. Lee, S. Choi, J. Lim, H. Jeong, et al., A reference electrode-free electrochemical biosensor for detecting MMP-9 using a concentric electrode device, Sens. Actuators B Chem. 240 (2017) 735–741.
[26] G. Castillo, K. Pribransky, G. Mező, L. Kocsis, A. Csámpai, K. Németh, et al., Electrochemical and photometric detection of plasmin by specific peptide substrate, Electroanalysis 27 (2015) 789–798.
[27] J. Zhang, T. Zheng, F. Cheng, J. Zhu, Electrochemical sensing for caspase 3 activity and inhibition using quantum dot functionalized carbon nanotube labels, Chem. Commun. (Camb.) 47 (2011) 1178–1180.
[28] E. Pazos, M. Pérez, H. Gutiérrez-de-Terán, M. Orzáez, T. Guevara, J.L. Mascareñas, et al., Rational design of a cyclin A fluorescent peptide sensor, Org. Biomol. Chem. 9 (2011) 7629–7632.
[29] Y. Huang, F. Hu, R. Zhao, G. Zhang, H. Yang, D. Zhang, Tetraphenylethylene conjugated with a specific peptide as a fluorescence turn-on bioprobe for the highly specific detection and tracing of tumor markers in live cancer cells, Chemistry 20 (2014) 158–164.
[30] H. Wang, Q. Zhang, X. Chu, T. Chen, J. Ge, R. Yu, Graphene oxide-peptide conjugate as an intracellular protease sensor for caspase-3 activation imaging in live cells, Angew. Chem. Int. Ed. Engl. 50 (2011) 7065–7069.
[31] K. Enander, G.T. Dolphin, L. Baltzer, Designed, functionalized helix–loop–helix motifs that bind human carbonic anhydrase II: a new class of synthetic receptor molecules, J. Am. Chem. Soc. 126 (2004) 4464–4465.
[32] L.K. Andersson, G.T. Dolphin, L. Baltzer, Multifunctional folded polypeptides from peptide synthesis and site-selective self-functionalization: practical scaffolds in aqueous solution, Chembiochem 3 (2002) 741–751.
[33] K. Enander, G.T. Dolphin, B. Liedberg, I. Lundström, L. Baltzer, A versatile polypeptide platform for integrated recognition and reporting: affinity arrays for protein–ligand interaction analysis, Chemistry 10 (2004) 2375–2385.
[34] Q. Wang, Z. Dai, S.M. Cahill, M. Blumenstein, D.S. Lawrence, Light-regulated sampling of protein tyrosine kinase activity, J. Am. Chem. Soc. 128 (2006) 14016–14017.
[35] V. Sharma, R.S. Agnes, D.S. Lawrence, Deep quench: an expanded dynamic range for protein kinase sensors, J. Am. Chem. Soc. 129 (2007) 2742–2743.
[36] W.J. Akers, B. Xu, H. Lee, G.P. Sudlow, G.B. Fields, S. Achilefu, et al., Detection of MMP-2 and MMP-9 activity in vivo with a triple-helical peptide optical probe, Bioconjug. Chem. 23 (2012) 656–663.
[37] L. Dong, Q. Miao, Z. Hai, Y. Yuan, G. Liang, Enzymatic hydrogelation-induced fluorescence turn-off for sensing alkaline phosphatase in vitro and in living cells, Anal. Chem. 87 (2015) 6475–6478.
[38] Y. Gu, Q. Wen, Y. Kuang, L. Tang, J. Jiang, Peptide-templated gold nanoclusters as a novel label-free biosensor for the detection of protease activity, RSC Adv. 4 (2014) 13753–13756.
[39] K.J. Oh, K.J. Cash, K.W. Plaxco, Excimer-based peptide beacons: a convenient experimental approach for monitoring polypeptide–protein and polypeptide–oligonucleotide interactions, J. Am. Chem. Soc. 128 (2006) 14018–14019.
[40] M.D. Shults, B. Imperiali, Versatile fluorescence probes of protein kinase activity, J. Am. Chem. Soc. 125 (2003) 14248–14249.

[41] M.D. Shults, D. Carrico-Moniz, B. Imperiali, Optimal Sox-based fluorescent chemosensor design for serine/threonine protein kinases, Anal. Biochem. 352 (2006) 198–207.
[42] M.S. Tremblay, M. Lee, D. Sames, A luminescent sensor for tyrosine phosphorylation, Org. Lett. 10 (2008) 5–8.
[43] S. Voss, R. Fischer, G. Jung, K.-H. Wiesmüller, R. Brock, A fluorescence-based synthetic LPS sensor, J. Am. Chem. Soc. 129 (2007) 554–561.
[44] B.R. White, H.M. Liljestrand, J.A. Holcombe, A "turn-on" FRET peptide sensor based on the mercury binding protein MerP, Analyst 133 (2008) 65–70.
[45] B.P. Joshi, J. Park, W.I. Lee, K.-H. Lee, Ratiometric and turn-on monitoring for heavy and transition metal ions in aqueous solution with a fluorescent peptide sensor, Talanta 78 (2009) 903–909.
[46] S. Shiosaki, T. Nobori, T. Mori, R. Toita, Y. Nakamura, C.W. Kim, et al., A protein kinase assay based on FRET between quantum dots and fluorescently-labeled peptides, Chem. Commun. 49 (2013) 5592–5594.
[47] N.P. Damayanti, L.L. Parker, J.M. Irudayaraj, Fluorescence lifetime imaging of biosensor peptide phosphorylation in single live cells, Angew. Chem. Int. Ed. Engl. 125 (2013) 4023–4026.
[48] W. Cheng, Y. Chen, F. Yan, L. Ding, S. Ding, H. Ju, et al., Ultrasensitive scanometric strategy for detection of matrix metalloproteinases using a histidine tagged peptide–Au nanoparticle probe, Chem. Commun. 47 (2011) 2877–2879.
[49] D. Aili, R. Selegård, L. Baltzer, K. Enander, B. Liedberg, Colorimetric protein sensing by controlled assembly of gold nanoparticles functionalized with synthetic receptors, Small 5 (2009) 2445–2452.
[50] G.A. Suaifan, S. Alhogail, M. Zourob, Paper-based magnetic nanoparticle-peptide probe for rapid and quantitative colorimetric detection of *Escherichia coli* O157: H7, Biosens. Bioelectron. 92 (2017) 702–708.
[51] Y. Li, W. Li, K.-Y. He, P. Li, Y. Huang, Z. Nie, et al., A biomimetic colorimetric logic gate system based on multi-functional peptide-mediated gold nanoparticle assembly, Nanoscale 8 (2016) 8591–8599.
[52] Q. Zhang, D. Zhang, G. Xu, Y. Xu, Y. Lu, S. Li, et al., Spectroscopic detection of thrombin with peptides self-assembled on gold nanoparticles hybridized graphene oxide, Sens. Actuators B Chem. 242 (2017) 443–449.
[53] O.R. Bolduc, P. Lambert-Lanteigne, D.Y. Colin, S.S. Zhao, C. Proulx, D. Boeglin, et al., Modified peptide monolayer binding His-tagged biomolecules for small ligand screening with SPR biosensors, Analyst 136 (2011) 3142–3148.
[54] F.C. Dudak, İ.H. Boyaci, Peptide-based surface plasmon resonance biosensor for detection of staphylococcal enterotoxin B, Food Anal. Methods 7 (2014) 506–511.
[55] S.-H. Wu, D. Núnez, S.-Y. Hu, M.P. Domingo, Y.-C. Chen, P.-K. Wei, et al., The effect of acidic pH on the inhibitory efficacy of peptides against the interaction ICAM-1/LFA-1 studied by surface plasmon resonance (SPR), Biosens. Bioelectron. 56 (2014) 159–166.
[56] M. Joseph, Kinase phosphorylation monitoring with i-motif DNA cross-linked SERS probes, Chem. Commun. 52 (2016) 410–413.
[57] Z. Wu, Y. Liu, Y. Liu, H. Xiao, A. Shen, X. Zhou, et al., A simple and universal "turn-on" detection platform for proteases based on surface enhanced Raman scattering (SERS), Biosens. Bioelectron. 65 (2015) 375–381.
[58] G. Qu, G. Zhang, Z. Wu, A. Shen, J. Wang, J. Hu, A "turn-off" SERS assay of heparin with high selectivity based on heparin–peptide complex and Raman labelled gold nanoparticles, Biosens. Bioelectron. 60 (2014) 124–129.

[59] Z. Wu, Y. Liu, X. Zhou, A. Shen, J. Hu, A "turn-off" SERS-based detection platform for ultrasensitive detection of thrombin based on enzymatic assays, Biosens. Bioelectron. 44 (2013) 10–15.

[60] H. Wang, Y. Yuan, Y. Chai, R. Yuan, Sandwiched electrochemiluminescent peptide biosensor for the detection of prognostic indicator in early-stage cancer based on hollow, magnetic, and self-enhanced nanosheets, Small 11 (2015) 3703–3709.

[61] H. Qi, M. Li, M. Dong, S. Ruan, Q. Gao, C. Zhang, Electrogenerated chemiluminescence peptide-based biosensor for the determination of prostate-specific antigen based on target-induced cleavage of peptide, Anal. Chem. 86 (2014) 1372–1379.

[62] M.-S. Wu, R.-N. Chen, Y. Xiao, Z.-X. Lv, Novel "signal-on" electrochemiluminescence biosensor for the detection of PSA based on resonance energy transfer, Talanta 161 (2016) 271–277.

[63] Q. Dang, H. Gao, Z. Li, H. Qi, Q. Gao, C. Zhang, Simple and sensitive electrogenerated chemiluminescence peptide-based biosensor for detection of matrix metalloproteinase 2 released from living cells, Anal. Bioanal. Chem. 408 (2016) 7067–7075.

[64] M. Dong, X. Liu, Q. Dang, H. Qi, Y. Huang, Q. Gao, et al., Sensitive and versatile electrogenerated chemiluminescence biosensing platform for protein kinase based on Ru(bpy)3(2+) functionalized gold nanoparticles mediated signal transduction, Anal. Chim. Acta 906 (2016) 72–79.

[65] K.A. Drouvalakis, S. Bangsaruntip, W. Hueber, L.G. Kozar, P.J. Utz, H. Dai, Peptide-coated nanotube-based biosensor for the detection of disease-specific autoantibodies in human serum, Biosens. Bioelectron. 23 (2008) 1413–1421.

[66] C. Yao, T. Zhu, J. Tang, R. Wu, Q. Chen, M. Chen, et al., Hybridization assay of hepatitis B virus by QCM peptide nucleic acid biosensor, Biosens. Bioelectron. 23 (2008) 879–885.

[67] J. Luo, Q. Zhang, Y. Huang, G. Liu, R. Zhao, Quartz crystal microbalance biosensor for recombinant human interferon-β detection based on antisense peptide approach, Anal. Chim. Acta 590 (2007) 91–97.

[68] X. Bai, B. Lu, X. Chen, B. Zhang, J. Tang, Reversible detection of vancomycin using peptide-functionalized cantilever array sensor, Biosens. Bioelectron. 62 (2014) 145–150.

[69] H. Etayash, K. Jiang, S. Azmi, T. Thundat, K. Kaur, Real-time detection of breast cancer cells using peptide-functionalized microcantilever arrays, Sci. Rep. 5 (2015) 13967.

[70] D. Khatayevich, T. Page, C. Gresswell, Y. Hayamizu, W. Grady, M. Sarikaya, Selective detection of target proteins by peptide-enabled graphene biosensor, Small 10 (2014) 1505–1513.

[71] Z. Kuang, S.N. Kim, W.J. Crookes-Goodson, B.L. Farmer, R.R. Naik, Biomimetic chemosensor: designing peptide recognition elements for surface functionalization of carbon nanotube field effect transistors, ACS Nano 4 (2009) 452–458.

[72] Q. Zhang, D. Zhang, Y. Lu, Y. Yao, S. Li, Q. Liu, Graphene oxide-based optical biosensor functionalized with peptides for explosive detection, Biosens. Bioelectron. 68 (2015) 494–499.

[73] Y. Tao, Y. Lin, J. Ren, X. Qu, Self-assembled, functionalized graphene and DNA as a universal platform for colorimetric assays, Biomaterials 34 (2013) 4810–4817.

[74] G. Eda, M. Chhowalla, Chemically derived graphene oxide: towards large-area thin-film electronics and optoelectronics, Adv. Mater. 22 (2010) 2392–2415.

[75] K.P. Loh, Q. Bao, G. Eda, M. Chhowalla, Graphene oxide as a chemically tunable platform for optical applications, Nat. Chem. 2 (2010) 1015–1024.

[76] J.W. Jaworski, Molecular Engineering of Selective Recognition Elements as Coatings for Sensor Platforms, University of California, Berkeley, CA, 2009.
[77] J.A. Huntington, Thrombin plasticity, Biochim. Biophys. Acta 1824 (2012) 246–252.
[78] D.R. Dreyer, S. Park, C.W. Bielawski, R.S. Ruoff, The chemistry of graphene oxide, Chem. Soc. Rev. 39 (2010) 228–240.
[79] Y. Tang, H. Hu, M.G. Zhang, J. Song, L. Nie, S. Wang, et al., An aptamer-targeting photoresponsive drug delivery system using "off–on" graphene oxide wrapped mesoporous silica nanoparticles, Nanoscale 7 (2015) 6304–6310.
[80] G. Goncalves, P.A. Marques, C.M. Granadeiro, H.I. Nogueira, M. Singh, J. Gracio, Surface modification of graphene nanosheets with gold nanoparticles: the role of oxygen moieties at graphene surface on gold nucleation and growth, Chem. Mater. 21 (2009) 4796–4802.

Peptides from plants and their applications

21

Maria Hayes[1] *and Stephen Bleakley*[1,2]
[1]Teagasc Food Research Centre, Dublin, Ireland, [2]Dublin Institute of Technology, Dublin, Ireland

21.1 Introduction

A number of plant protein sources are associated with bioactive peptides and mild but significant health benefits such as lowering of blood pressure levels and inhibition of enzymes within the renin–angiotensin aldosterone system (RAAS) including renin and angiotensin-I-converting enzyme (ACE-I). Other bioactivities associated with plant-derived bioactive peptides include cholesterol-lowering effects [1,2]; antiinflammatory activity [3]; anticancer [4]; and immunomodulatory activity [5], as well as prevention of and protection against oxidative damage through free radical scavenging activities [6]; antimicrobial activity and others as reviewed recently by Cicero et al. [7]. These bioactivities are inert within the parent protein and must be released by proteolytic enzyme activities in food-processing strategies including hydrolysis and fermentation. Once released, the peptides can exert their "hormone-like" beneficial bioactivities.

However, within the plant a number of secreted peptides are also present where they act as regulators of signaling events and cell-to-cell communication [8]. These peptides are classified as cysteine-rich peptides and posttranslationally modified peptides. These peptides are thought to have potential for use as antimicrobial agents in the pharmaceutical and agricultural industries [9].

This chapter will focus primarily on the exogenous plant-derived peptides that can be generated from food protein sources by processing methods using proteolytic enzymes and will look at the actions and uses of these peptides for human health in functional foods, pharmaceuticals, and medicines.

21.2 Heart health beneficial peptides derived from plant proteins

Cardiovascular disease (CVD) is a global health problem and is thought to be responsible for almost half of all deaths in Europe alone. Primary prevention is a preferred public health strategy and functional foods are a key part of this strategy for preventative healthcare. Functional foods or nutraceuticals may be defined as foods or food components that impart a health benefit to the consumer that goes

Peptide Applications in Biomedicine, Biotechnology and Bioengineering.
DOI: http://dx.doi.org/10.1016/B978-0-08-100736-5.00025-9

above and beyond basic human nutrition. Indeed, the World Health Organization (WHO) stresses the importance of a healthy diet and nutrition as a method for the control of CVD in the current and future decades [10].

The renin–angiotensin–aldosterone system (RAAS) is central to the control of human blood pressure and salt:water balance. Within this system of the body two enzymes, renin and angiotensin-I-converting enzyme (ACE-I) play important physiological roles in control of cardiovascular functions. Firstly, renin converts angiotensinogen to angiotensin I. Angiotensin I is a vasodilatory decapeptide. Subsequently, ACE-I converts this decapeptide to angiotensin II, an octapeptide with vasoconstricting activity. Renin is the rate-limiting enzyme in this cascade and ACE-I results in the formation of angiotensin II, which results in hypotension and increased blood pressure and salt retention. In addition, ACE-I acts on the vasodilator bradykinin, which causes vascular endothelial nitric oxide (NO) release and increased blood pressure (Fig. 21.1).

A number of ACE-I- and renin-inhibiting bioactive peptides have been identified from plant sources to date. These include renin- and ACE-I-inhibitory peptides generated from sources including potatoes [11], sweet potatoes [12], yams [13], rapeseed [14], marine plants including the red seaweed *Palmaria palmata* [15], chick pea [16], lentils [17], soy [18], and several cereals [19]. These peptides, their generation, bioactivities, and the concentration of these peptides that provide 50% inhibition of either renin or ACE-I (IC_{50} value) can be found within the database BIOPEP. A detailed list of plant-derived bioactive peptides with heart-beneficial bioactivities is presented in Table 21.1.

Most of the ACE-I- and renin-inhibitory peptides derived from plant proteins are generated using hydrolysis. Indeed, several of these peptides were first generated using commercially available enzymes including Alcalase, pepsin, chymotrypsin, or others [10]. Hydrolysis is performed using either a batch or continuous approach with a suitable enzyme. It is important that optimal hydrolysis conditions are used during the process so that desirable functional and bioactive properties are attained and temperature, pH, the degree of hydrolysis (DH) (which can be measured using several different approaches including the pH Stat method or the OPA method), and duration (time) must be carefully monitored. Following hydrolysis, a bioactive peptide containing permeate can be obtained using molecular weight cut-off (MWCO) filtration with 3- or 10-kDa membranes. In most cases, the ACE-I- or renin-inhibitory bioactivities increase when peptides are concentrated using MWCO filtration. Other filtration methods including nano-, ion-exchange-, and column chromatography can be used to enhance bioactivity and purify the peptides. Indeed, if the peptide-containing product is to be sold in Europe or the United States it is desirable that the molecular weight profile of the bioactive peptide containing hydrolysate is determined and that the bioactive peptide sequences and amino acid content are determined. This can be achieved by using mass spectrometry and reverse-phase HPLC methods.

The structure–activity relationship and mechanism of action of ACE-I- and renin-inhibitory peptides should also be determined. ACE-I-inhibitory peptides are usually short and consist largely of proline amino acid residues. Indeed, the active site of

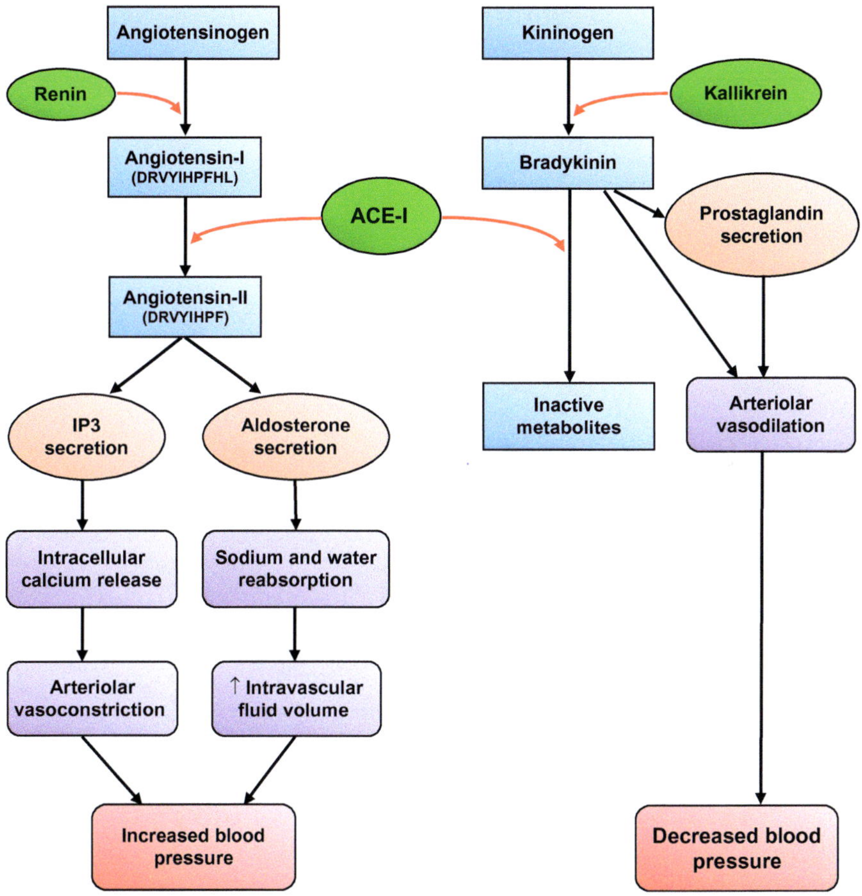

Figure 21.1 The renin–angiotensin–aldosterone system (RAAS). The RAAS and kallikrein–kinin system. Angiotensinogen is converted to the decapeptide angiotensin-I by renin. ACE-I cleaves the C-terminal dipeptide His-Leu of angiotensin-I to form angiotensin-II. Binding of angiotensin-II to its receptor (AT1) stimulates the secretion of inositol 1,4,5-triphosphate (IP3) and aldosterone, which induce arteriolar vasoconstriction and increase intravascular fluid volume, respectively, resulting in increased blood pressure. Within the kallikrein–kinin system, kallikrein converts kininogen to bradykinin, which induces arteriolar vasodilation by prostaglandin secretion and binding of bradykinin with its receptor, resulting in decreased blood pressure. However, the hypotensive effect of bradykinin is largely dependent on the rate of degradation by ACE-I, which hydrolyzes bradykinin to form inactive metabolites [78].

ACE-I cannot accommodate large peptides. The C-terminal end of the peptide is important in competitive binding to the ACE-I active site and often consists of hydrophobic, aromatic, and branched-side chains. Amino acids found often at the C-terminal end of ACE-I-inhibitory peptides include tyrosine (Tyr), phenylalanine (Phe), tryptophan (Trp), lysine (Lys), isoleucine (Iso), and proline (Pro). Renin-inhibitory peptides

Table 21.1 ACE-I and antihypertensive peptides derived from plant proteins

Peptide sequence and source	ACE-I inhibition	Renin inhibition	Antihypertensive	References
VWIS rapeseed protein	Yes	Not tested	Yes	[20]
VW rapeseed protein	Yes	Not tested	Yes	[20]
IY rapeseed protein	Yes	Not tested	Yes	[20]
RIY rapeseed protein	Yes	Not tested	Yes	[20]
QGR from flaxseed	Yes	Not tested	Yes	[21]
RW from flaxseed	Yes	Not tested	Yes	[21]
SVR from flaxseed	Yes	Not tested	Yes	[21]
GQMRQPI from flaxseed	Yes	Not tested	Yes	[21]
QQQG from flaxseed	Yes	Not tested	Yes	[21]
ASVRT from flaxseed	Yes	Not tested	Yes	[21]
KDYRL from mung bean protein	Yes	Not tested	Yes	[22]
VTPALR from mung bean protein	Yes	Not tested	Yes	[22]
KLPAGTLR from mung bean protein	Yes	Not tested	Yes	[22]
DLP from soy protein	Yes	Not tested	Yes	[18]
DG from soy protein	Yes	Not tested	Yes	[18]
AW from soybean fermentation	Yes	Not tested	Yes	[23]

(Continued)

Table 21.1 **(Continued)**

Peptide sequence and source	ACE-I inhibition	Renin inhibition	Antihypertensive	References
GW from soybean fermentation	Yes	Not tested	Yes	[23]
AY and SY from soybean fermentation	Yes	Not tested	Yes	[23]
VPP, IPP, LQP from rye sourdough	Yes	Not tested	Unknown	[24]
LQP, IQP, LRP form wheat bran	Yes	Not tested	Yes	[25]
TQVY from rice	Yes	Not tested	Yes	[26]
IRLIIVLMPILMA from the red seaweed *Palmaria palmata* (dulse)	No	Yes	Yes	[15]
Ala-Ile-Tyr-Lys from *Undaria pinnatifida*	Yes	Unknown	Unknown	[27]

usually are also proline rich. Once the structure of the peptide is identified, it is important that the bioactivity is confirmed in vivo to ensure that there is an actual antihypertensive effect. The usual model used to establish this uses spontaneously hypertensive rats (SHRs). To date, few studies have been carried out on plant-derived peptides in animal models or indeed in human clinical trials. Despite this, a number of plant-derived bioactive peptides have been discovered to date and their bioactivities in terms of ACE-I- and renin-inhibitory activities are comparable to inhibitory IC_{50} values obtained for peptides derived from other sources including dairy and marine proteins.

21.3 Plant-derived antioxidant peptides

Reactive oxygen species (ROS) include the superoxide anion, hydroxyl radical, and hydrogen peroxide which are all produced as a result of the metabolism of oxygen

[28]. Reactive nitrogen species (RNS) are derived from nitric oxide and, of these, peroxynitrite is considered to have the most oxidative effect [29]. A build-up of ROS/RNS in the body can cause oxidative/nitrosative damage to cellular components and is thought to be one of the major causes of many chronic diseases, including cancer, myocardial infarction, diabetes, Parkinson's, renal failure, and several inflammatory conditions such as cardiovascular disease, certain cancers, bowel diseases, and osteoporosis [29–32]. Antioxidants sourced from our diets often contain radical chain reaction inhibitors, metal chelators, oxidative enzyme inhibitors, and antioxidant enzyme cofactors that can prevent damage caused by oxidative stress [33]. Synthetic antioxidants have an association with toxic and carcinogenic effects, and therefore alternatives from naturally sourced antioxidants are being sought [32]. Bioactive compounds of interest produced by plants are often stress-induced [34].

Oxidative stress is thought to be responsible for a number of cardiovascular, neurodegenerative, and inflammatory diseases including cancer. Oxidative stress is caused by reactive oxygen species (ROS) that can cause physical damage to DNA, lipids, and proteins of the body [35]. A number of antioxidant peptides are listed in BIOPEP and interest in plant-derived antioxidant peptides is increasing. Similar to ACE-I- and renin-inhibitory peptides antioxidant peptides can be generated through protein hydrolysis, digestion, and fermentation. In addition, a number of amino acids have known antioxidant activities including histidine (His), tryptophan (Trp), tyrosine (Tyr), and lysine (Lys). These amino acids can chelate metallic ions and Cys thanks to thiolic groups. Udenigwe and colleagues underline that amino acid and peptide antioxidant activities strongly rely on environmental conditions. For example, the antioxidant activity of wheat peptides was lost when heated and malondialdehyde was found to oxidize peptides and led to loss of antioxidant activity [35]. Cereal peptides have also been identified with antioxidant activity. Lunasin was shown to scavenge both peroxyl and superoxide radicals and was found to protect Caco-2 cell viability and could have use as a protective agent at the intestinal level [36].

21.4 Antimicrobial and anticancer plant-derived peptides

Cancer is a leading cause of death worldwide and is characterized by the rapid growth of abnormal cells that extend beyond their usual limits and invade adjoining parts of the body or spread to other organs, a process known as metastasis. Cancer treatment requires careful selection of one or more therapeutic modalities, such as surgery, radiotherapy, or chemotherapy. Chemotherapeutic drugs used in cancer treatment are highly toxic and multidrug resistance can develop. Antimicrobial peptides (AMPs) represent a novel alternative for cancer treatment.

AMPs are bioactive peptides, usually amphipathic in nature that are produced by a wide variety of organisms as part of the first line of defense in eukaryotes or for use in competition for nutrients and resources in prokaryotes. There are over 2400 known AMPs reported in the Antimicrobial Peptide Database (URL: http://aps.unmc.edu/AP/main.php) and over 300 AMPs of plant origin have been reported to date [37]. Groups of plant-derived AMPs include the thionins, plant defensins, cyclotides, and small cationic peptides.

21.4.1 Thionins

These are cysteine-rich amino acids containing peptides usually around 5 kDa in size that display both toxic and antimicrobial activities. There are four different types of thionins: type 1 is basic, contains 45 amino acids and 8 cysteine amino acids within its sequence that can form four disulfide bonds; type 2 thionins contain 46–47 amino acids and can form four disulfide bonds also; type 3 thionins contain 45–46 amino acids and can have three or four disulfide bonds; and type 4 thionins are neutral and contain three disulfide bonds. The activities of thionins include plant protection, seed maturation, dormancy, germination, packaging of storage proteins, and act as secondary messengers during signal transduction [37]. However, in the treatment of cancer, pyrularia thionin from mistletoe (*Pyrularia pubera*) displayed anticancer activity against cervical cancer cells (HeLa) and mouse melanoma cells (B16) with an IC_{50} of 50 μg/mL (half maximal inhibitory concentration) [38].

21.4.2 Plant defensins

These are antimicrobial peptides that resemble the defensins produced by fungi, vertebrates, and invertebrates that are classified into two groups depending on the structure of the precursor protein from which they are derived. They inhibit growth of fungi and are less toxic to mammalian and plant cells. A defensin sesquin from *Vigna sesquipedalis* was the first ever reported anticancer defensin peptide. Other defensin peptides have been isolated from lima beans and are known as lunatusin but this defensin peptide is thought to be toxic to normal tissue and other cell types. Other defensin peptides are listed in Table 21.2.

21.4.3 Cyclotides

These are macrocyclic peptides around 30 amino acids in length that are often isolated from the Rubiaceae and Violaceae plant families and possess a cysteine knot within their structure. This knot imparts excellent chemical and biological stability to cyclotides, which is significant in terms of drug development and design for the pharmaceutical sector. Cyclotides are thought to be important in host defense and up to 1 g/kg of leaf material cyclotides can be expressed. They have been examined for use in the treatment of HIV, anthelmintic, insecticidal, and other pharmacological applications as outlined in Table 21.2.

Protein- and antibody-based therapeutics dominate the pharmaceutical market for treatment of immune and inflammatory diseases. More than 30% of the licensed pharmaceutical products are biologics that target autoimmune and inflammatory diseases and cancer. However, large portions of these come with immune-related safety

Table 21.2 **Anticancer peptides derived from plant sources**

Peptide name	Classed as	Bioactivity	Cytotoxic	Anticancer	References
Pyrularia from *Pyrularia pubera* (buffalo nut)	Thionin	B16, HeLa, rat hepatocytes, lymphocytes	Yes	Yes	[38]
Viscotoxin B2 from *Viscum coloratum*	Thionin	Rat sarcoma cells	Unknown	Yes	[39]
Viscotoxins 1-PS A1, A2, A3, B from *Viscum album* (mistletoe)	Thionin	Human lymphocytes	Yes	Unknown	[40]
Viscotoxin C1 from *Coloratum ohwi* (variant of mistletoe)	Thionin	Rat sarcoma cells	Unknown	Yes	[41]
Thi2.1 from *Arabidopsis thaliana* (mouse ear cress—Brassicaceae)	Thionin	HeLa, A549, MCF-7, bovine mammary epithelial cells	Yes	Yes	[42]
β-Purothionin from *Tricum aestivum* (common wheat)	Thionin	p388	Unknown	Yes	[43]
Phoratoxins C, D, E, F from *Phoradendron tomentosum* (Christmas mistletoe)	Thionin	10 cancerous cell lines	Unknown	Yes	[44]
Sesquin from *Vigna sesquipedalis* (yardlong bean)	Plant defensin	MCF-7 and M1	Unknown	Yes	[45]
Limenin from *Phaseolus limensis* (lima bean)	Plant defensin	L1210 and M1	Unknown	Yes	[45]
Mitogenic defensin from *Phaseolus vulgaris* (common bean)	Plant defensin	MCF-7, murine splenocytes	Yes	Yes	[45]
Lunatusin from *Phaseolus lunatus* (lima bean variety)	Plant defensin	MCF-7 rabbit reticulocyte	Yes	Yes	[45]

Coccinin from *Phaseolus coccineus* cv. "Major" (scarlet runner bean)	Plant defensin	HL60 and L1210	Unknown	Yes	[46]
Cycloviolacin O2 from *Viola odorata* (sweet violet)	Cyclotide	U-937 and HeLa	Yes	Yes	[47]
Viphi A, Viphi F, Viphi G from *Viola philippica*	Cyclotide	MM96L, HeLa, BgC-823, HFF-1	Yes	Yes	[48]
MCoTI-I from *Momordica cochinchinensis*	Cyclotide	LNCaP and HCT116	Unknown	Yes	[49]
Cliotides T1-T4 from *Clitoria ternatea* (blue pea)	Cyclotide	HeLa and human erythrocytes	Yes	Yes	[50]
Kalata B1 and Kalata B2 from *Oldenlandia affinis* (blue diamond flower)	Cyclotide	U-937, GTB, HT-29, and Ht116	Yes	Yes	[51]
HB7 from *Hedyotis biflora* (Kinwoj)	Cyclotide	Capan2 and PANC1	Unknown	Yes	[52]
Cn-AMP1 from *Cocos nucifera* (coconut water)	Small cationic peptide	Caco-2, RAW264.7, MCF-7, HCT-116 cells	No	Yes	[53]
Cr-ACP from *Cycas revoluta* (Japanese plant)	Small cationic peptide	Human epidermoid cancer (Hep2)	Unknown	Yes	[54]
Cherimolacyclopeptide C from *Annona cherimola* (Cherimoya—Inca plant) seeds	Small cationic peptides	KB cells in vitro	Unknown	Yes	[55]

warnings [56]. There are more than 60 peptide-based approved drugs on the market and approximately 140 in clinical trials. As there are more than 100 autoimmune diseases immunomodulation by peptides and peptidomimetics is an area that has very high potential [56]. Peptidomimetics are compounds whose essential elements or pharmacophore mimic a natural peptide or protein but which retain the ability to interact with the biological target and produce the same bioactive effect [57].

21.5 Plant-derived peptides and inhibition of enzymes important in the development of type II diabetes

Plant-derived peptides may also play a role in the control of blood sugar levels in patients with type II diabetes mellitus by inhibiting the enzyme dipeptidyl peptidase-IV (DPP-IV; EC 3.4.14.5). The number of adults with diabetes worldwide is predicted to reach 439 million by 2030 and diabetes is predicted to become a global epidemic [58]. Glucagon-like peptide 1 (GLP-1) and glucose-dependent insulinotropic peptide (GIP) belong to a family of gut peptides called incretins, which are secreted from K cells of the upper small intestine and L cells of the lower intestine, respectively [59]. Incretins play an important role in decreasing blood glucose levels after eating by promoting the release of insulin from pancreatic β-cells (see Fig. 21.3. DPP-IV is a serine exopeptidase that cleaves dipeptides from the *N*-terminus of regulatory peptides with the sequence Xaa-Pro- or, to a lesser extent, Xaa-Ala- (where Xaa represents any amino acid) [60]. GLP-1 and GIP are included in these affected peptides, resulting in the inactive metabolites GLP-1 (9–36) amide and GIP (3–42), and leading to a hyperglycemic state [61] (Fig. 21.2).

There are already several pharmaceutical drugs to inhibit DPP-IV. Indeed, 11 classes of glucose-lowering agents are currently available for the management of diabetes. Of these, dipeptidyl peptidase-IV (DPP-IV; EC 3.4.14.5) inhibitors are among the newest medications that have been introduced for control of diabetes type II and medication available that is peptidic in nature includes Alogliptin (Nesina), Saxagliptin (Onglyza), Sitagliptin (Januvia), and Vildagliptin (Galvus), although these are often associated with side effects such as nausea, anxiety, depression, and seizures [62]. Peptides that inhibit DPP-IV have therefore become an attractive alternative for the treatment of diabetes [63].

Medical plants play an important role in the management of type II diabetes mellitus (T2DM) by delaying the development of diabetic complications and correcting metabolic abnormalities and several peptidic DPP-IV-inhibitory drugs have been isolated from plants to date [64]. The most commonly studied hypoglycemic plants are *Opuntia streptacantha*, *Trigonella foenum-graecum*, *Momordica charantia*, *Ficus bengalensis*, *Polygala senega*, and *Gymnema sylvestre* [64]. Gausch and colleagues recently identified 18 potential DPP-IV inhibitors in 18 different plant extracts (12 of these plants have known antidiabetic properties, whereas, for the remaining 6, antidiabetic activity has been reported for other plant species from the same genus). Moreover, none of the 18 molecules exhibits chemical similarity with a large group of known DPP-IV inhibitors.

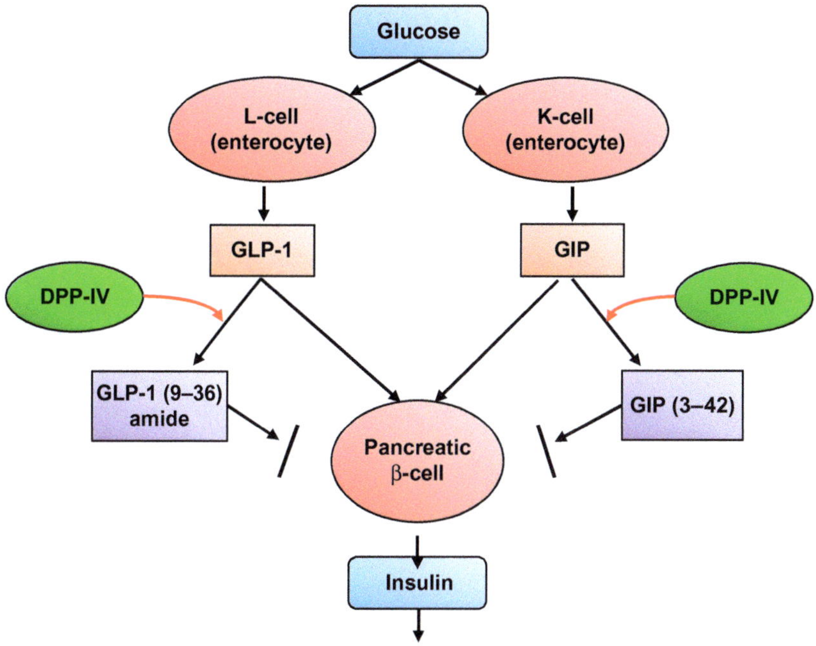

Figure 21.2 Schematic for the inactivation of GLP-1 and GIP by DPP-IV. GLP-1 and GIP are released postprandially from L- and K-cells in the intestine, respectively. GLP-1 and GIP promote the release of insulin from pancreatic β-cells in order to maintain glucose homeostasis. However, GLP-1 and GIP can be rapidly degraded by DPP-IV to form GLP-1 (9−36) amide and GIP (3−42), which are unable to stimulate insulin release from pancreatic β-cells.

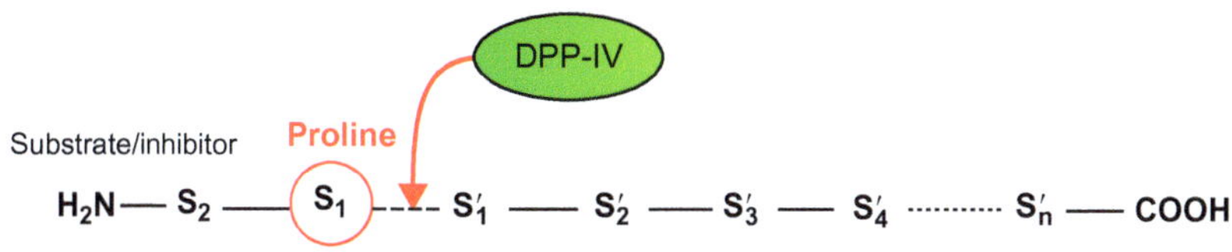

Figure 21.3 Substrate/inhibitor cleavage sites of DPP-IV. DPP-IV is an exopeptidase that cleaves the scissile bond on the C-terminal side of a proline residue, represented as S_1, when it is located at the second residue from the *N*-terminus.

Several plant-derived peptides can be seen in the GLP-1 agonist field, which represents an established drug class with several products, including Byetta (exenatide), Bydureon (exenatide), Victoza (liraglutide), Lyxumia (lixisenatide), and most recently Tanzeum (albiglutide) commercially available (Fossgerau and Hoffmann, 2015).

Proline residues induce conformational changes and kinks in the secondary structure of proteins meaning that they require specific proteases to be hydrolyzed. DPP-IV is one such proline-specific endopeptidase that cleaves dipeptides from the *N*-terminus (Fig. 21.3) [65].

21.6 In silico strategies in plant protein evaluation as peptide precursors

Iwaniak and Dziuba [66] presented a modern in silico approach for the evaluation of proteins as a source of ACE-I inhibitors previously. They found that using the criteria for protein evaluation outlined in their work it was possible to identify plant proteins that could serve as potentially good sources of peptides with ACE-I-inhibitory bioactivities. Other studies on protein sources such as those performed by Lafarga et al. [67] outlined a strategy for the predictive release of bioactive peptides from a protein source and methods to identify the most suitable enzyme for use in the generation of bioactive peptides from a given protein source (once the amino acid sequence of the protein is available in PubMed); the potency of the released peptides; the potential allergenicity and even the toxicity of these peptides as well as known bioactivities associated with the identified peptide sequence. This strategy is outlined in Fig. 21.4.

21.7 Cereal-derived bioactive peptides

Cereals are rich in bioactive peptides and a number of these are documented in the database BIOPEP (http://www.uwm.edu.pl/biochemia). Bioactivities associated with cereal proteins include antioxidant, antiinflammatory, cholesterol-lowering, satiety, antidiabetic, and others, as reviewed recently by Cavazos and Gonzalez de Mejia [19]. They have a wide range of physiological effects that have been previously demonstrated in animal models. However, prolamins from some cereals including wheat, barley, and rye give rise to biologically active antinutritional peptides following proteolysis that are able to adversely affect in vivo the intestinal mucosa of celiac patients, whereas prolamins from other cereals such as maize and rice do not. This should be taken into account when developing cereal protein hydrolysate products for potential health applications. Bioactivities associated with oat, barley, and wheat protein-derived peptides include opioid activity while rice protein contains the sequence RGD and lacks celiac toxicity which could be beneficial in health and nutrition. Several groups have already documented the use and potential of cereal-derived bioactive peptides [10,19,35,66,68]. From an agronomical viewpoint, issues including cultivar use, environmental conditions, and agricultural practices can affect the bioactive peptide content found within cereal proteins and the content could be improved through improved breeding practices and optimization of growing and processing technologies. In terms of determining a physiological effect in humans, knowledge concerning the bioavailability of peptides and research on randomized clinical trials is necessary in order to understand the full potential of these proteins for health benefits and applications.

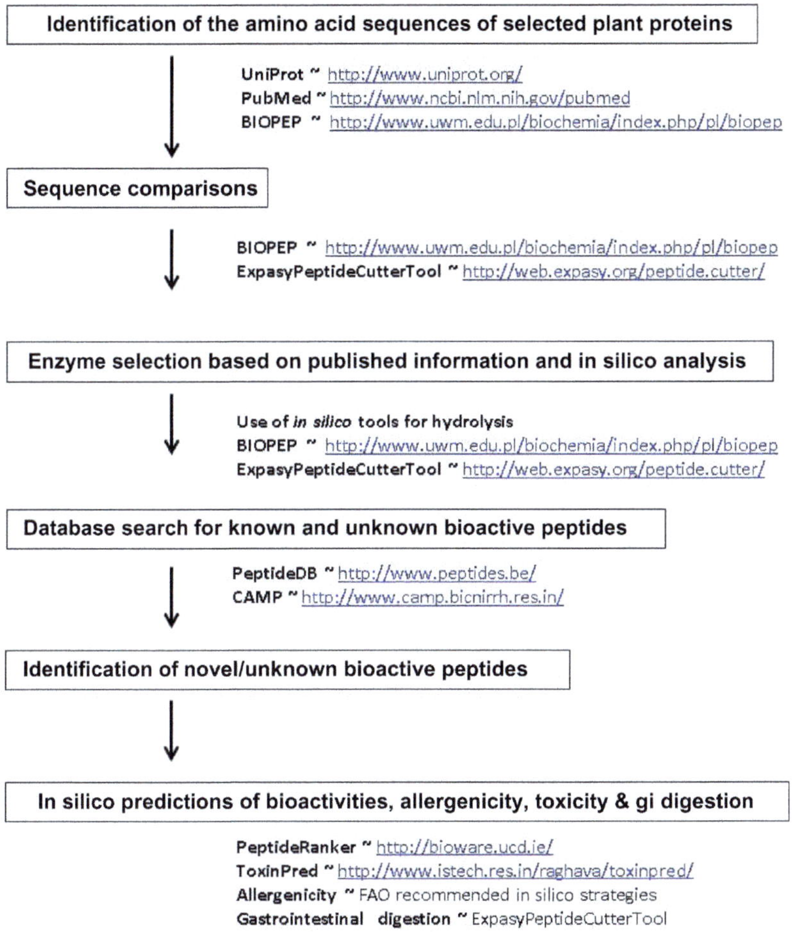

Figure 21.4 In silico strategies useful for prediction of bioactive peptides from plant protein sources.

21.8 Soy- and lupine-derived cholesterol-lowering peptides

Dyslipidemia is an important risk factor in the development of heart disease. Several soy- and lupine-derived bioactive peptides have been isolated and provide clinical evidence for inducing a reduction in cholesterolemia [1,2]. In particular, the soy protein B-conglycinin demonstrated a cholesterol-lowering effect and was found to lower both LDL-cholesterol and HDL-cholesterol. The bioactive peptide lunasin derived from soy and characterized by a RGD sequence followed by eight aspartate residues is thought to be responsible for this observed cholesterol-lowering effect when tested in animal and cell line models [69].

Cholesterol lowering is attributed to upregulation of LDL receptors, regulation of the sterol regulatory element-binding protein-2 pathway, inhibition of hydroxymethylglutaril-CoA reductase enzyme, and increased fecal excretion of bile salts [70]. Peptides derived from cow pea protein were also found to inhibit the synthesis of cholesterol and its solubilization into micelles [71] and a peptide derived from the plant *Mucana pruriens* (a tropical legume native to Africa) was found to bind with micelle formation and prevents the absorption of exogenous cholesterol [72].

21.9 Soy-derived bioactive peptides in wound healing and tissue engineering

Recent research has focused on the development of acellular and cellular bioactive wound matrices for improving the wound-healing process [73,74]. These matrices can enhance wound healing in the absence of exogenous cells and have advantages over traditional wound matrices as they don't have long culture times and short shelf-lives. There are several bioactive wound matrices in clinical use currently but few can induce regeneration of skin tissue and they are expensive and complicated to use. Recently, Lelkes et al. [75] filed a patent which details a method of using soy protein isolates for drug delivery and wound healing. Their soy isolate was produced using reverse-phase high-pressure liquid chromatography and the composition of the isolate consists of water-soluble soy protein isolate known as WSsoy. The researchers found that WSsoy forms gel-like matrices when suspended with certain concentration ranges in aqueous environments and that this suspension promotes natural healing of wounds and had a low-risk profile.

21.10 Challenges associated with the development of plant-derived bioactive peptides

Plant-derived bioactive peptides show great promise for use in the treatment of a number of diseases ranging from cancer treatment and immunotherapy to maintenance of heart health. However, there are challenges. For example, lack of scalable purification and enrichment methods means that production of bioactive peptides, however effective and safe for use, is often expensive. It is thought that scale-up accounts for 70% of the capital and operating costs for peptide production [76]. This is why chemical synthesis of peptides identified from plant sources is normally employed at commercial scales. In addition to this, determination of the mechanism of action of plant-derived peptides is often unknown but required for peptides to be used as drug and functional food ingredients. Moreover, it is widely assumed that bioactive peptides are nontoxic but as they are small in size, they may react with other food components in a food matrix and could become toxic [77]. The chemical characteristics of peptides can also be a challenge for food and drug formulation. Most peptidic drugs are not currently consumed but injected, as survival in the gastrointestinal tract is not possible due to breakdown by proteolytic enzymes. However, intranasal absorption of peptidic drugs is an area that is developing fast and it is likely that with intelligent assembly of known modules with unique properties and functions, the construction of multifunctional peptidic medicines with targeted delivery and improved efficacy is possible. Furthermore, the use of small molecules or antibodies as methods for novel peptide therapy such as in oncology is an exciting area where safety and efficacy of peptides has been improved. Indeed, currently there are more than 20 peptide conjugates in clinical trials.

21.11 Conclusion

Plant-derived peptides display a myriad of activities ranging from antimicrobial, anticancer, anticholesterol, and heart health beneficial (ACE-I and renin inhibition) and studies in humans have established the effects of identified plant peptides on cancer, cell proliferation, and their use as preventative control agents for diseases such as diabetes and high blood pressure. However, up-scale technologies are required to ensure cost of production of these bioactives remains feasible and studies concerning toxicity, bioavailability, and absorption are necessary to ensure their efficacy in vivo. Peptide drug development and commercialization of peptide hydrolysate products in recent times such as WSsoy and DuPont Danisco Soy Protein in recent times highlights the importance of natural product discovery in contributing to developing drugs and functional foods for improved health of the global populace and expansion of available food ingredients.

References

[1] D.N. Butteiger, A.A. Hibberd, N.J. McGraw, N. Napawan, J.M. Hall-Porter, E.S. Krul, Soy protein compared with milk protein in a western diet increases gut microbial diversity and reduces serum lipids in golden Syrian hamsters, J. Nutr. 146 (2016) 697–705.

[2] C. Lammi, C. Zanoni, G.M. Scigliuolo, A. D'Amato, A. Arnoldi, Lupin peptides lower low-density lipoprotein (LDL) cholesterol through an up-regulation of the LDL receptor/sterol regulatory element binding protein (SREBP2) Pathway at HepG2 cell line, J. Agric. Food Chem. 62 (2014) 7151–7159.

[3] K. Aihara, H. Ishii, M. Yoshida, Casein-derived tripeptide, Val-Pro-Pro (VPP), modulates monocyte adhesion to vascular endothelium, J. Atheroscler. Thromb. 16 (2009) 594–603.

[4] A. Tyagi, A. Tuknait, P. Anand, S. Gupta, M. Sharma, D. Mathur, et al., CancerPPD: a database of anticancer peptides and proteins, Nucleic Acids Res. 43 (2015) D837–D843.

[5] Y.K. Wang, H.L. He, G.F. Wang, H. Wu, B.C. Zhou, X.L. Chen, et al., Oyster (*Crassostrea gigas*) hydrolysates produced on a plant scale have antitumor activity and immunostimulating effects in BALB/c mice, Mar. Drugs 8 (2010) 255–268.

[6] M.J. Garcia-Nebot, I. Recio, B. Hernandez-Ledesma, Antioxidant activity and protective effects of peptide lunasin against oxidative stress in intestinal Caco-2 cells, Food Chem. Toxicol. 65 (2014) 155–161.

[7] A.F. Cicero, A. Colletti, M. Rosticci, M. Cagnati, R. Urso, M. Giovannini, et al., Effect of lactotripeptides (isoleucine-proline-proline/valine-proline-proline) on blood pressure and arterial stiffness changes in subjects with suboptimal blood pressure control and metabolic syndrome: a double-blind, randomized, crossover clinical trial, Metab. Syndr. Relat. Disord. 14 (2016) 161–166.

[8] E. Murphy, S. Smith, I. De Smet, Small signalling peptides in *Arabidopsis* development: how cells communicate over a short distance, Plant Cell 24 (8) (2012) 3198–3217.

[9] J. Weidmann, D.J. Craik, Discovery, structure, function and applications of cyclotides: circular proteins from plants, J. Exp. Bot. 67 (16) (2016) 4801–4812.

[10] A. Pihlanto, S. Mäkinen, Antihypertensive properties of plant protein derived peptides, in: B. Hernandez-Ledesma, C. Hsieh (Eds.), Bioactive Food Peptides in Health and Disease, INTECH Publishers, Rijeka, Croatia, 2013, pp. 145–182in: B. Hernandez-Ledesma, C. Hsieh (Eds.), Bioactive Food Peptides in Health and Disease, INTECH Publishers, Rijeka, Croatia, 2013, pp. 145–182. Available from: http://dx.doi.org/10.5772/54565.

[11] S. Mäkinen, T. Johansson, G. Vegarud, J.-M. Pihlava, A. Pihlanto, Angiotensin-I-converting enzyme inhibitory and antioxidant properties of rapeseed hydrolysates, J. Funct. Foods 4 (2002) 575–583.

[12] K. Ishiguro, Y. Sameshima, T. Kume, K. Ikeda, J. Matsumoto, M. Yoshimoto, Hypotensive effect of a sweet potato protein digest in spontaneously hypertensive rats and purification of angiotensin I-converting enzyme inhibitory peptides, Food Chem. 131 (2012) 774–779.

[13] Y. Liu, Y. Lin, D. Liu, C. Han, C. Chen, M. Fan, et al., Effects of different types of yam (*Dioscorea alata*) products on the blood pressure of spontaneously hypertensive rats, Biosci. Biotech. Biochem. 73 (2009) 1371–1376.

[14] C. Megias, J. Pedroche, M.D.M. Yust, M. Alaiz, J. Giron-Calle, F. Millan, et al., Affinity purification of angiotensin converting enzyme inhibitory peptides using immobilized ACE, J. Agric. Food Chem. 54 (2006) 7120–7124.

[15] C. Fitzgerald, R.E. Aluko, M. Hossain, D.K. Rai, M. Hayes, Potential of a renin inhibitory peptide from the red seaweed *Palmaria palmata* as a functional food ingredient following confirmation and characterisation of a hypotensive effect in spontaneously hypertensive rats (SHRs), J. Agric. Food Chem. 62 (33) (2014) 8352–8356.

[16] M.M. Yust, J. Pedroche, J. Giron-Calle, M. Alaiz, M. Francisco, J. Vioque, Production of ace inhibitory peptides by digestion of chickpea legumin with alcalase, Food Chem. 81 (2003) 363–369.

[17] C. Barbana, J.I. Boye, Angiotensin I-converting enzyme inhibitory activity of chickpea and pea protein hydrolysates, Food Res. Int. 43 (2010) 1642–1649.

[18] J. Wu, X. Ding, Characterization of inhibition and stability of soy-protein-derived angiotensin I-converting enzyme inhibitory peptides, Food Res. Int. 35 (2002) 367–375.

[19] A. Cavazos, E. Gonzalez de Mejia, Identification of bioactive peptides from cereal storage proteins and their potential role in prevention of chronic diseases, Comp. Rev. Food Sci. Food Saf. 12 (2013) 364–380.

[20] E. Marczak, H. Usui, H. Fujita, Y. Yang, M. Yokoo, A. Lipkowski, et al., New antihypertensive peptides isolated from rapeseed, Peptides 24 (2003) 791–798.

[21] C.C. Udenigwe, R.E. Aluko, Multifunctional cationic peptide fractions from flaxseed protein hydrolysates, Plant Foods Hum. Nutr. 67 (2012) 1–9.

[22] G.H. Li, J.Z. Wan, G.W. Le, Y.H. Shi, Novel angiotensin-I-converting enzyme inhibitory peptides isolated from alcalase hydrolysate of mung bean protein, J. Pept. Sci. 12 (8) (2006) 509–514.

[23] T. Nakahara, A. Sano, H. Yamaguchi, K. Sugimoto, H. Chikata, E. Kinoshita, et al., Antihypertensive effect of peptide-enriched soy sauce-like seasoning and identification of its angiotensin I-converting enzyme inhibitory substances, J. Agric. Food Chem. 58 (2010) 821–827.

[24] I.W.Y. Cheung, S. Nakayama, M.N.K. Hsu, A.G.P. Samaranayaka, E.C.Y. Li-Chan, Angiotensin-I converting enzyme inhibitory activity of hydrolysates from oat (*Avena sativa*) proteins by *in silico* and *in vitro* analyses, J. Agric. Food Chem. 57 (2009) 9234–9242.

[25] Y. Nogata, T. Nagamine, K. Sekiya, Antihypertensive effect of angiotensin I converting enzyme inhibitory peptides derived from wheat bran in spontaneously hypertensive rats, J. Jpn. Soc. Food Sci. Technol. 58 (2011) 67–70.

[26] T. Matsui, C.H. Li, T. Tanaka, T. Maki, Y. Osajima, K. Matsumoto, Depressor effect of wheat germ hydrolysate and its novel angiotensin I-converting enzyme inhibitory peptide, Ile-Val-Tyr, and the metabolism in rat and human plasma, Biol. Pharm. Bull. 23 (2000) 427–431.

[27] K. Suetsuna, T. Nakano, Identification of an antihypertensive peptide from peptidic digest of wakame (*Undaria pinnatifida*), J. Nutr. Biochem. 11 (9) (2000) 450–454.

[28] S.-J. Heo, E.-J. Park, K.-W. Lee, Y.-J. Jeon, Antioxidant activities of enzymatic extracts from brown seaweeds, Bioresour. Technol. 96 (2005) 1613–1623.

[29] L.M. Walker, J.L. York, S.Z. Imam, S.F. Ali, K.L. Muldrew, P.R. Mayeux, Oxidative stress and reactive nitrogen species generation during renal ischemia, Toxicol. Sci. 17 (2001) 300–312.

[30] P. Libby, Inflammatory mechanisms: the molecular basis of inflammation and disease, Nutr. Rev. 65 (2007) S140–S146.

[31] A.R. Kim, T.-S. Shin, M.-S. Lee, J.-Y. Park, K.-E. Park, N.-Y. Yoon, et al., Isolation and identification of phlorotannins from *Ecklonia stolonifera* with antioxidant and anti-inflammatory properties, J. Agric. Food Chem. 57 (2009) 3483–3489.

[32] M. Yangthong, N. Hutadilok-Towatana, W. Phromkunthong, Antioxidant activities of four edible seaweeds from the southern coast of Thailand, Plant Foods Hum. Nutr. 64 (2009) 218–223.

[33] D. Huang, B. Ou, R.L. Prior, The chemistry behind anti-oxidant capacity assays, J. Agric. Food Chem. 53 (2005) 1841–1856.

[34] M.S. Tierney, A.K. Croft, M. Hayes, A review of antihypertensive and antioxidant activities in macroalgae, Botanica Marina 53 (2010) 387–408.

[35] M. Malaguti, G. Dinelli, E. Leoncini, V. Bregola, S. Bosi, A.F.G. Cicero, et al., Bioactive peptides in cereals and legumes: agronomical, biochemical and clinical aspects, Int. J. Mol. Sci. 15 (2014) 21120–21135.

[36] H. Yin, X. Pan, Z. Song, S. Wang, L. Yang, G. Sun, Protective effect of wheat peptides against indomethacin-induced oxidative stress in IEC-6 cells, Nutrients 6 (2014) 564–574.

[37] J.J. Guzmán-Rodríguez, A. Ochoa-Zarzosa, R. López-Gómez, López, J.E. Meza, Plant antimicrobial peptides as potential anticancer agents, Biomed. Res. Int. 2015 (2015) 735087. Available from: http://dx.doi.org/10.1155/2015/735087.

[38] J. Evans, Y.D. Wang, K.P. Shaw, L.P. Vernon, Cellular responses to pyrularia thionin are mediated by Ca2 + influx and phospholipase A2 activation and are inhibited by thionin tyrosine iodination, Proc. Natl. Acad. Sci. U.S.A. 86 (15) (1989) 5849–5853.

[39] J.L. Kong, X.B. Du, C.X. Fan, et al., Purification and primary structure determination of a novel polypeptide isolated from mistletoe *Viscum coloratum*, Chin. Chem. Lett. 15 (11) (2004) 1311–1314.

[40] A. Bussing, G.M. Stein, M. Wagner, Accidental cell death and generation of reactive oxygen intermediates in human lymphocytes induced by thionins from *Viscum album* L, Eur. J. Biochem. 262 (1) (1999) 79–87.

[41] S. Romagnoli, F. Fogolari, M. Catalano, NMR solution structure of viscotoxin C1 from *Viscum album* species *Coloratum ohwi*: toward a structure-function analysis of viscotoxins, Biochemistry 42 (43) (2003) 12503–12510.

[42] H. Loeza-Angeles, E. Sagrero-Cisneros, L. Lara-Zárate, E. Villagómez-Gómez, J.E. López-Meza, A. Ochoa-Zarzosa, Thionin Thi2.1 from *Arabidopsis thaliana* expressed in endothelial cells shows antibacterial, antifungal and cytotoxic activity, Biotechnol. Lett. 30 (10) (2008) 1713–1719.

[43] P. Hughes, E. Dennis, M. Whitecross, D. Llewellyn, P. Gage, The cytotoxic plant protein, β-purothionin, forms ion channels in lipid membranes, J. Biol. Chem. 275 (2) (2000) 823–827.

[44] S. Johansson, J. Gullbo, P. Lindholmet, Small, novel proteins from the mistletoe *Phoradendron tomentosum* exhibit highly selective cytotoxicity to human breast cancer cells, Cell. Mol. Life Sci. 60 (1) (2003) 165–175.

[45] J.H. Wong, T.B. Ng, Limenin, a defensin-like peptide with multiple exploitable activities from shelf beans, J. Pept. Sci. 12 (5) (2006) 341–346.

[46] P.H.K. Ngai, T. Ng, A napin-like polypeptide from dwarf Chinese white cabbage seeds with translation-inhibitory, trypsin-inhibitory, and antibacterial activities, Peptides 25 (2) (2004) 171–176.

[47] E. Svangard, R. Burman, S. Gunasekera, H. Lovborg, J. Gullbo, U. Goransson, Mechanism of action of cytotoxic cyclotides: cycloviolacin O2 disrupts lipid membranes, J. Nat. Prod. 70 (4) (2007) 643–647.

[48] W. He, L.Y. Chan, G. Zeng, N.L. Daly, D.J. Craik, N. Tan, Isolation and characterization of cytotoxic cyclotides from *Viola philippica*, Peptides 32 (8) (2011) 1719–1723.

[49] Y. Ji, S. Majumder, M. Millard, *In vivo* activation of the p53 tumor suppressor pathway by an engineered cyclotide, J. Am. Chem. Soc. 135 (31) (2013) 11623–11633.

[50] G.K.T. Nguyen, S. Zhang, N.T.K. Nguyen, Discovery and characterization of novel cyclotides originated from chimeric precursors consisting of albumin-1 chain a and cyclotide domains in the fabaceae family, J. Biol. Chem. 286 (27) (2011) 24275–24287.

[51] R. Burman, A.A. Stromstedt, M. Malmsten, U. Goransson, Cyclotide-membrane interactions: defining factors of membrane binding, depletion and disruption, Biochim. Biophys. Acta 1808 (11) (2011) 2665–2673.

[52] X. Ding, D. Bai, J. Qian, Novel cyclotides from *Hedyotis biflora* inhibit proliferation and migration of pancreatic cancer cell *in vitro* and *in vivo*, Med. Chem. Res. 23 (3) (2014) 1406–1413.

[53] O.N. Silva, W.F. Porto, L. Migliolo, Cn-AMP1: a new promiscuous peptide with potential for microbial infection treatment, Biopolymers 98 (4) (2012) 322–331.

[54] S.M. Mandal, L. Migliolo, S. Das, M. Mandal, O.L. Franco, T.K. Hazra, Identification and characterization of a bactericidal and proapoptotic peptide from *Cycas revoluta* seeds with DNA binding properties, J. Cell. Biochem. 113 (1) (2012) 184–193.

[55] A. Wélé, Y. Zhang, I. Ndoye, J.P. Brouard, J.L. Pousset, B. Bodo, A cytotoxic cyclic heptapeptide from the seeds of *Annona cherimola*, J. Nat. Prod. 67 (9) (2004) 1577–1579.

[56] A. Gokhale, S. Kanthala, J. Latendresse, V. Taneja, S. Satyanarayanajois, Immunosuppression by co-stimulatory molecules: inhibition of CD2-CD48/CD58 interaction by peptides from CD2 to suppress progression of collagen-induced arthritis in mice, Chem. Biol. Drug Des. 82 (2013) 106–118.

[57] J. Vagner, H. Qu, V.J. Hruby, Peptidomimetics, a synthetic tool of drug discovery, Curr. Opin. Chem. Biol. 12 (3) (2008) 292–296.

[58] J.E. Shaw, R.A. Sicree, P.Z. Zimmet, Global estimates of the prevalence of diabetes for 2010 and 2030, Diabetes Res. Clin. Pract. 87 (1) (2010) 4–14.

[59] Y. Seino, H. Kuwata, D. Yabe, Incretin based drugs for type 2 diabetes: focus on East Asian perspectives, J. Diabetes Investig. 7 (2016) 102–109.

[60] R. Mentlein, Dipeptidyl-peptidase IV (CD26) – role in the inactivation of regulatory peptides, Regul. Pept. 30 (1) (1999) 9–24.

[61] M. Terasaki, M. Hiromura, Y. Mori, K. Kohashi, M. Nagashima, H. Kushima, Amelioration of hyperglycemia with a sodium-glucose cotransporter 2 inhibitor prevents macrophage-driven atherosclerosis through macrophage foam cell formation suppression in type 1 and type 2 diabetic mice, PLoS One 10 (2015) e0143396. Available from: http://dx.doi.org/10.1371/journal.pone.0143396.

[62] D.J. Drucker, M.A. Narck, The incretin system: glucagon-like peptide-1 receptor agonists and dipeptidyl peptidase IV inhibitors in type 2 diabetes, Lancet 368 (2006) 1696–1705.

[63] K. Fosgerau, T. Hoffmann, Peptide therapeutics: current status and future directions, Drug Discov. Today 20 (1) (2015) 122–128.

[64] L. Guasch, E. Sala, M.J. Ojeda, C. Valls, C. Bladé, M. Mulero, Identification of novel human dipeptidyl peptidase-IV inhibitors of natural origin (part II): in silico prediction in antidiabetic extracts, PLoS One 7 (9) (2012) e44972. Available from: http://dx.doi.org/10.1371/journal.pone.0044972.

[65] J. Lawandi, S. Gerber-Lemaire, L. Juillerat-Jeanneret, N. Moitessier, Inhibitors of prolyl oligopeptidase for the therapy of human disease: defining diseases and inhibitors, J. Med. Chem. 53 (9) (2010) 3423–3438.

[66] A. Iwaniak, B. Dziuba, Motifs with potential physiological activity in food proteins – BIOPEP database, Acta Sci. Pol. Technol. Aliment. 8 (3) (2009) 59–85.

[67] T. Lafarga, P.M. O'Connor, M. Hayes, *In silico* methods to identify meat derived prolyl endopeptidase inhibitors, Food Chem. 175 (2015) 337–343.

[68] N. Gangopadhyay, K. Wynne, P. O'Connor, E. Gallagher, N.P. Brunton, D.K. Rai, et al., *In silico* and *in vitro* analysis of the angiotensin-I-converting enzyme inhibitory activity of hydrolysates generated from crude Barley (*Hordeum vulgare*) protein concentrates., Food Chem. 203 (2016) 367–374.

[69] V.K. Lule, S. Garg, S.D. Pophaly, Hitesh, S.K. Tomar, Potential health benefits of lunasin: a multifaceted soy-derived bioactive peptide, J. Food Sci. 80 (2015) R485–R494.

[70] T.G. Marsh, R.K. Straub, F. Villalobos, M.Y. Hong, Soy protein supports cardiovascular health by downregulating hydroxyme thylglutaryl-coenzyme A reductase and sterol regulatory element-binding protein-2 and increasing antioxidant enzyme activity in rats with dextran sodium sulfate-induced mild systemic inflammation, Nutr. Res. 31 (2011) 922–928.

[71] M.R. Marques, R.A. Soares Freitas, C.A.C. Correa, E.S. Siquemoto, G.G. Fontamar, Peptides from cowpea present antioxidant activity, inhibit cholesterol synthesis and its solubilisation into micelles, Food Chem. 168 (2015) 288–293.

[72] F. Herrera-Chalé, J.C. Ruiz-Ruiz, D. Betancur-Ancona, M.R. Sequra-Campos, Potential therapeutic applications of *Mucana pruriens* peptide fractions purified by high performance liquid chromatography as angiotensin-converting enzyme inhibitors, antioxidants, anti-thrombotic and hypocholesterolemic agents, J. Med. Food 19 (2) (2016) 187–195.

[73] T.N. Demidova-Rice, M.R. Hamblin, I.M. Herman, Acute and impaired wound healing: pathophysiology and current methods for drug delivery, Part I: normal and chronic wounds: biology, causes and approaches to care, Adv. Skin Wound Care 25 (7) (2012) 304–314.

[74] N. Mayet, Y.E. Choonara, P. Kumar, L.K. Tomar, C. Tyagi, L.C. Du Toit, et al., A comprehensive review of advanced biopolymeric wound healing systems, J. Pharm. Sci. 103 (2014) 2211–2230.

[75] P.I. Lelkes, Soy-derived bioactive peptides for use in compositional methods for wound healing, tissue engineering and regenerative medicine, W-2017015488A1, 2017.

[76] H. Korhonen, Milk-derived bioactive peptides: from science to applications, J. Funct. Foods 1 (2009) 177–187.

[77] R. Brady, B. Woonton, M.L. Gee, A.J. O'Connor, Hierarchical mesoporous silica materials for separation of functional food ingredients – a review, Innov. Food Sci. Emerg. Technol. 9 (2) (2008) 243–248.

[78] S. Bleakley, M. Hayes, Algal proteins: extraction, application, and challenges concerning production, Foods 6 (5) (2017) 1–34.

Index

Note: Page numbers followed by "*f*" and "*t*" refer to figures and tables.

G

H

I

J

K

L